A BRIEF TABLE OF LAPLACE TRA

(See Chapter 5 for more details.)

$f(t)$	$F(s)$	$f(t)$	$F(s)$
1. $h(t)$	$\dfrac{1}{s}$	**13.** $f(t-\alpha)h(t-\alpha)$	$e^{-\alpha s}F(s)$
2. t^n	$\dfrac{n!}{s^{n+1}}$	**14.** $h(t-\alpha)$	$\dfrac{e^{-\alpha s}}{s}$
3. $e^{\alpha t}$	$\dfrac{1}{s-\alpha}$	**15.** $f'(t)$	$sF(s)-f(0)$
4. $\sin \omega t$	$\dfrac{\omega}{s^2+\omega^2}$	**16.** $\displaystyle\int_0^t f(u)\,du$	$\dfrac{F(s)}{s}$
5. $\cos \omega t$	$\dfrac{s}{s^2+\omega^2}$	**17.** $\sin \omega t - \omega t \cos \omega t$	$\dfrac{2\omega^3}{(s^2+\omega^2)^2}$
6. $\sinh \alpha t$	$\dfrac{\alpha}{s^2-\alpha^2}$	**18.** $t\sin \omega t$	$\dfrac{2\omega s}{(s^2+\omega^2)^2}$
7. $\cosh \alpha t$	$\dfrac{s}{s^2-\alpha^2}$	**19.** $tf(t)$	$-F'(s)$
8. $e^{\alpha t}f(t)$	$F(s-\alpha)$	**20.** $\dfrac{1}{t}f(t)$	$\displaystyle\int_s^\infty F(u)\,du$
9. $e^{\alpha t}h(t)$	$\dfrac{1}{s-\alpha}$	**21.** $f(\alpha t)$	$\dfrac{1}{\alpha}F\left(\dfrac{s}{\alpha}\right)$
10. $e^{\alpha t}t^n$	$\dfrac{n!}{(s-\alpha)^{n+1}}$	**22.** $(f*g)(t)$	$F(s)G(s)$
11. $e^{\alpha t}\sin \omega t$	$\dfrac{\omega}{(s-\alpha)^2+\omega^2}$	**23.** $f(t+T)=f(t)$	$\dfrac{\displaystyle\int_0^T e^{-st}f(t)\,dt}{1-e^{-sT}}$
12. $e^{\alpha t}\cos \omega t$	$\dfrac{(s-\alpha)}{(s-\alpha)^2+\omega^2}$	**24.** $\delta(t-t_0)$	e^{-st_0}

ELEMENTARY
DIFFERENTIAL
EQUATIONS

ELEMENTARY DIFFERENTIAL EQUATIONS

Second Edition

WERNER KOHLER
Virginia Tech

LEE JOHNSON
Virginia Tech

PEARSON

Addison Wesley

Boston   San Francisco   New York
London   Toronto   Sydney   Tokyo   Singapore   Madrid
Mexico City   Munich   Paris   Cape Town   Hong Kong   Montreal

Publisher: Greg Tobin
Senior Acquisitions Editor: William Hoffman
Executive Project Manager: Christine O'Brien
Editorial Assistant: Emily Portwood
Managing Editor: Karen Wernholm
Senior Production Supervisor: Peggy McMahon
Senior Designer: Barbara T. Atkinson
Digital Assets Manager: Marianne Groth
Media Producer: Michelle Murray
Marketing Manager: Phyllis Hubbard
Marketing Coordinator: Celena Carr
Senior Author Support/Technology Specialist: Joe Vetere
Senior Prepress Supervisor: Caroline Fell
Senior Manufacturing Buyer: Evelyn Beaton
Cover and Text Design: Barbara T. Atkinson
Production Coordination: Lifland et al., Bookmakers
Composition and Illustration: Techsetters, Inc.

Many of the designations used by manufacturers and sellers to distinguish their products are claimed as trademarks. Where those designations appear in this book, and Addison-Wesley was aware of a trademark claim, the designations have been printed in initial caps or all caps.

Library of Congress Cataloging-in-Publication Data

Kohler, W. E. (Werner E.), 1939–
    Elementary differential equations / Werner Kohler and Lee Johnson.—2nd ed.
        p. cm.
    Includes index.
    ISBN 0-321-29044-5 (alk. paper)
    1. Differential equations—Textbooks. I. Johnson, Lee W. II. Title.

QA371.K5854 2006
515'.35—dc22

2005045848

2 3 4 5 6 7 8 9 10—VHP—09 08 07 06

Book ISBN 0-321-29044-5
Book/CD package ISBN 0-321-39849-1

We especially want to acknowledge our families for making it fun.

Thanks to Abbie, Larry, Tom, Liz, Paul, Maggie, Cathy, Connie, Luke, Cano, Kiera, Mike, and baby Maya.

Thanks to Rochelle, Eric, Mark, Ali, Hannah, Quinn, and Casey.

Contents

Preface

This book is designed for the undergraduate differential equations course taken by students majoring in science and engineering. A year of calculus is the prerequisite.

The main goal of the text is to help students integrate the underlying theory, solution procedures, and computational aspects of differential equations as seamlessly as possible. Since we want the text to be easy to read and understand, we discuss the theory as simply as possible and emphasize how to use it. When developing models, we try to guide the reader carefully through the physical principles underlying the mathematical model.

We also emphasize the importance of common sense, intuition, and "back of the envelope" checks. When solving problems, we remind the student to ask "Does my answer make sense?" Where appropriate, examples and exercises ask the student to anticipate and subsequently interpret the physical content of their solution. (For example, "Should an equilibrium solution exist for this application? If so, why? What should its value be?") We believe that developing this mind-set is particularly important in resisting the temptation to accept almost any computer-generated output as correct.

New Features

As in the first edition, we have made a determined effort to write a text that is easy to understand. In response to the suggestions of first edition users and reviewers, this second edition offers even more support for both students and instructors.

- We have followed the advice of our reviewers to provide a more concise presentation. First order differential equations (linear and nonlinear) are now discussed in a single chapter, Chapter 2. Similarly, second order and higher order linear equations are discussed in one chapter, Chapter 3. The bulk of Chapter 3 develops the theory for the second order equations; the last three sections extend these ideas to higher order linear equations.

- The introductory discussion for linear systems (Chapter 4) has been streamlined to reach the computational aspects of the theory as quickly as possible.

- We have included a review of core material in the form of a set of review exercises at the end of Chapters 2, 3, and 4. These exercises, consisting of initial value problems for first order equations, higher order linear equations, and first order linear systems, respectively, require the student to select as well as apply the appropriate solution technique developed in the chapter.

- We have added a number of new exercises, ranging from routine drill exercises to those with applications to a variety of different disciplines. Answers to the odd-numbered exercises are again given at the back of the text.

- A brief look at boundary value problems appears as a project at the end of Chapter 3. This brief introductory overview of linear two-point boundary value problems highlights how these problems differ from their initial-value counterparts.

- We have added projects. There are now short projects at the end of each chapter. Some of these are challenging applications. Others are intended to expand the student's mathematical horizons, showing how the material in the chapter can be generalized. In certain applications, such as food processing, the project exposes the student to the mathematics aspects of current research.

A Multilevel Development of Certain Topics

Numerical Methods. The basic ideas underlying numerical methods and their use in applications are presented early for both scalar problems and systems. In Chapters 2 and 4, after Euler's method is developed, the route to more accurate algorithms is briefly outlined. The Runge-Kutta algorithm is then offered as an example of such an improved algorithm; accompanying exercises allow the student to apply the algorithm and experience its increased accuracy at an introductory level. Chapter 7 subsequently builds upon this introduction, developing a comprehensive treatment of one-step methods.

Phase Plane. An introduction to the phase plane is provided in Chapter 4 as the different solutions of the homogeneous constant coefficient linear system are developed. These ideas are then revisited and extended in the Chapter 6 discussion of autonomous nonlinear systems.

Supplements

The *Student's Solutions Manual* (0-321-29045-3) contains detailed solutions to the odd-numbered problems.

The *Instructor's Solutions Manual* (0-321-28937-4) contains detailed solutions to most problems.

The *Online Technology Resource Manual* includes suggestions for how to use a computer algebra system with the text. Specific instructions are given for MATLAB and Mathematica. It is available at http://www.aw-bc.com/kohler/.

Acknowledgments

Many people helped and encouraged us in this effort. Besides the support provided by our families, we are especially thankful for the editorial and developmental assistance of William Hoffman, our editor at Pearson Addison-Wesley.

We are very grateful to our reviewers, who made many insightful suggestions that improved the text:

John Baxley	*Wake Forest University*
William Beckner	*University of Texas at Austin*
Jerry L. Bona	*University of Illinois, Chicago*
Thomas Branson	*University of Iowa*
Dennis Brewer	*University of Arkansas*
Peter Brinkmann	*University of Illinois, Urbana-Champaign*
Almut Burchard	*University of Virginia*
Donatella Danielli	*Purdue University*
Charles Friedman	*University of Texas at Austin*
Moses Glasner	*Pennsylvania State University—University Park*
Weimin Han	*University of Iowa*
Harumi Hattori	*West Virginia University*
Jason Huffman	*Georgia College & State University*
Gregor Kovacic	*Rensselaer Polytechnic Institute*
Yang Kuang	*Arizona State University*
Zhiqin Lu	*University of California—Irvine*
Peter Mucha	*Georgia Institute of Technology*
John Neuberger	*Northern Arizona University*
Lei Ni	*University of California—San Diego*
Joan Remski	*University of Michigan—Dearborn*
Sinai Robins	*Temple University*
Robert Snider	*Virginia Polytechnic Institute & State University*
A. Shadi Tahvildar-Zadeh	*Rutgers University*
Jason Whitt	*Rhodes College*
Tamas Wiandt	*Rochester Institute of Technology*
Jennifer Zhao	*University of Michigan—Dearborn*
Fangyang Zheng	*Ohio State University*
Kehe Zhu	*State University of New York at Albany*

Special thanks are due to Peter Mucha. His ongoing interest and constructive feedback helped us greatly during the revision process. Our good friend and colleague George Flick continued to provide us with examples of applications from the life sciences; we greatly appreciate his encouragement and assistance. Tiri Chinyoka and Ermira Cami helped us at Virginia Tech with proofreading and problem-checking. We also thank Jeremy Bourdon and our Virginia Tech colleague Terri Bourdon for revising the solutions manuals.

We are very grateful for the professional expertise provided by the personnel at Addison-Wesley. Christine O'Brien was our project manager, and she made certain that we maintained our schedule. We are grateful to Peggy McMahon, our Production Supervisor, to Barbara Atkinson for the design of the text, to Jeanne Yost for her careful copyediting, to Rena Lam at Techsetters, Inc. for superb typesetting, and to Sally Lifland at Lifland et al., Bookmakers for her careful oversight and coordination of the revision process.

ELEMENTARY
DIFFERENTIAL
EQUATIONS

1

Introduction to Differential Equations

1.1 Introduction

Scientists and engineers develop mathematical models for physical processes as an aid to understanding and predicting the behavior of the processes. In this book we discuss mathematical models that help us understand, among other things, decay of radioactive substances, electrical networks, population dynamics, dispersion of pollutants, and trajectories of moving objects. Modeling a physical process often leads to equations that involve not only the physical quantity of interest but also some of its derivatives. Such equations are referred to as **differential equations**.

In Section 1.2, we give some simple examples that show how mathematical models are derived. We also begin our study of differential equations by introducing the corresponding terminology and by presenting some concrete examples of differential equations. Section 1.3 introduces the idea of a direction field for a differential equation. The concept of direction fields allows us to visualize, in geometric terms, the graphs of solutions of differential equations.

1.2 Examples of Differential Equations

When we apply Newton's second law of motion, $ma = f$, to an object moving in a straight line, we obtain a differential equation of the form

$$my''(t) = f(t, y(t), y'(t)). \tag{1}$$

1

In equation (1), $y(t)$ represents the position, at time t, of the object. As expressed in equation (1), the product of mass m and acceleration $y''(t)$ is equal to the sum of the applied forces. The applied forces [the right-hand side of equation (1)] often depend on time t, position $y(t)$, and velocity $y'(t)$.

E X A M P L E

1

One of the simplest examples of linear motion is an object falling under the influence of gravity. Let $y(t)$ represent the height of the object above the surface of the earth, and let g denote the constant acceleration due to gravity (32 ft/sec² or 9.8 m/s²). See Figure 1.1.

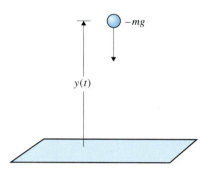

FIGURE 1.1

The only force acting on the falling body is its weight. The body's position, $y(t)$, is governed by the differential equation $y'' = -g$.

Since the only force acting on the body is assumed to be its weight, $W = mg$, equation (1) reduces to $my''(t) = -mg$, or

$$y''(t) = -g. \tag{2}$$

The negative sign appears on the right-hand side of the equation because the acceleration due to gravity is positive downward, while we assumed y to be positive in the upward direction. (Again, see Figure 1.1.)

Equation (2) is solved easily by taking successive antiderivatives. The first antiderivative gives the object's velocity,

$$y'(t) = -gt + C_1.$$

Another antidifferentiation gives the object's position,

$$y(t) = -\tfrac{1}{2}gt^2 + C_1 t + C_2.$$

Here, C_1 and C_2 represent arbitrary constants of integration. ❖

Notice in Example 1 that the solution involves two undetermined constants. This means that, by itself, differential equation (2) does not completely specify the solution $y(t)$. This makes sense physically since, to completely determine the motion, we also need some information about the initial state of the object. The arbitrary constants of integration that arise are often specified by prescribing velocity and position at some initial time, say $t = 0$. For example, if the object's initial velocity is $y'(0) = v_0$ and its initial position is $y(0) = y_0$, then we obtain a complete description of velocity and position:

$$y'(t) = -gt + v_0, \qquad y(t) = -\tfrac{1}{2}gt^2 + v_0 t + y_0.$$

Unless an application suggests otherwise, we normally use t to represent the independent variable and y to represent the dependent variable. Thus, in a typical differential equation, we are searching for a solution $y(t)$.

As is common in a mathematics text, we use a variety of notations to denote derivatives. For instance, we may use d^2y/dt^2 instead of $y''(t)$ or d^4y/dt^4 instead of $y^{(4)}(t)$. In addition, we often suppress the independent variable t and simply write y and y' instead of $y(t)$ and $y'(t)$. An example using this notation is the differential equation

$$y'' + \frac{1}{t}y' + t^3 y = 5.$$

EXAMPLE 2

Scientists have observed that radioactive materials have an instantaneous rate of decay (that is, a rate of decrease) that is proportional to the amount of material present. If $Q(t)$ represents the amount of material present at time t, then dQ/dt is proportional to $Q(t)$; that is,

$$\frac{dQ}{dt} = -kQ, \qquad k > 0. \tag{3}$$

The negative sign in equation (3) arises because Q is both positive and decreasing; that is, $Q(t) > 0$ and $Q'(t) < 0$.

Unlike equation (2), differential equation (3) cannot be solved by integrating the right-hand side, $-kQ(t)$, because $Q(t)$ is not known. Instead, equation (3) requires that we somehow find a function $Q(t)$ whose derivative, $Q'(t)$, is a constant multiple of $Q(t)$.

Recall that the exponential function has a derivative that is a constant multiple of itself. For example, if $y = Ce^{-kt}$, then $y' = -kCe^{-kt} = -ky$. Therefore, we see that a solution of equation (3) is

$$Q(t) = Ce^{-kt}, \tag{4}$$

where C can be any constant. ❖

As in Example 1, the differential equation by itself does not completely specify the solution. But setting $t = 0$ in (4) leads to $Q(0) = C$. Therefore, the quantity $Q(t)$ given in equation (4) is completely determined once the amount of material initially present is specified.

The Form of an *n*th Order Differential Equation

We now state the formal definition of a differential equation and point to some issues that need to be addressed. An equation of the form

$$y^{(n)} = f(t, y, y', \ldots, y^{(n-1)}) \tag{5}$$

is called an ***n*th order ordinary differential equation**.

In equation (5), t is the **independent variable**, while y is the **dependent variable**. A **solution** of the differential equation (5) is any function $y(t)$ that satisfies the equation on our t-interval of interest. For instance, Example 2 showed that $Q(t) = Ce^{-kt}$ is a solution of $Q' = -kQ$ for any value of the constant C; the t-interval of interest for Example 2 is typically the interval $0 \le t < \infty$.

The **order** of a differential equation is the order of the highest derivative that appears in the equation. For example, $y'' = -g$ is a second order differential equation. Similarly, $Q' = -kQ$ is a first order differential equation.

The form of the nth order ordinary differential equation (5) is not the most general one. In particular, an nth order ordinary differential equation is any equation of the form

$$G(t, y, y', y'', \ldots, y^{(n)}) = 0.$$

For example, the following equation is classified as a second order ordinary differential equation:

$$t^2 \sin y'' + y \ln y'' = 1.$$

Notice that it is not possible to rewrite this equation in the explicit form

$$y'' = f(t, y, y').$$

In our study, however, we usually consider only equations of the form

$$y^{(n)} = f(t, y, y', y'', \ldots, y^{(n-1)}),$$

where the nth derivative is given explicitly in terms of t, y, and lower order derivatives of y.

Differential equation (5) is called **ordinary** because the equation involves just a single independent variable, t. This is in contrast to other equations called **partial differential equations**, which involve two or more independent variables. An example of a partial differential equation is the one-dimensional wave equation

$$\frac{\partial^2 u(x,t)}{\partial x^2} - \frac{\partial^2 u(x,t)}{\partial t^2} = 0.$$

Here, the dependent variable u is a function of two independent variables, the spatial coordinate x and time t.

Initial Value Problems

What we have seen about differential equations thus far raises some important questions that we will address throughout this book. One such question is "What constitutes a properly formulated problem?" Examples 1 and 2 illustrate that auxiliary **initial conditions** are required if the differential equation is to have a unique solution. The differential equation, together with the proper number of initial conditions, constitutes what is known as an **initial value problem**.

For instance, an initial value problem associated with the falling object in Example 1 consists of the differential equation together with initial conditions specifying the object's initial position and velocity:

$$\frac{d^2y}{dt^2} = -g, \qquad y(0) = y_0, \qquad y'(0) = v_0.$$

Similarly, an initial value problem associated with the radioactive decay process in Example 2 consists of the differential equation together with a specification of the initial amount of the substance:

$$\frac{dQ}{dt} = -kQ, \qquad Q(0) = Q_0.$$

These examples suggest that the number of initial conditions we need to specify must be equal to the order of the differential equation. When we address the question of properly formulating problems, it will be apparent that this is the case. Once we understand how to properly formulate the problem to be solved, the obvious next question is "How do we go about solving this problem?" Answering the two questions

1. How do we properly formulate the problem?
2. How do we solve the problem?

is central to the study of differential equations.

Solving Initial Value Problems

As Chapters 2, 3, and 4 show, certain special types of differential equations have formulas for the general solution. The **general solution** is an expression containing arbitrary constants (or parameters) that can be adjusted so as to give every solution of the equation. Finding the general solution is often the first step in solving an initial value problem. The next example illustrates this idea.

E X A M P L E

3

Consider the initial value problem

$$y' + 3y = 6t + 5, \qquad y(0) = 3. \tag{6}$$

(a) Show, for any constant C, that

$$y = Ce^{-3t} + 2t + 1 \tag{7}$$

is a solution of the differential equation $y' + 3y = 6t + 5$.

(b) Use expression (7) to solve the initial value problem (6).

Solution:

(a) Inserting expression (7) into the differential equation $y' + 3y = 6t + 5$, we find

$$y' + 3y = (Ce^{-3t} + 2t + 1)' + 3(Ce^{-3t} + 2t + 1)$$
$$= (-3Ce^{-3t} + 2) + (3Ce^{-3t} + 6t + 3)$$
$$= 6t + 5.$$

Therefore, for any value C, $y = Ce^{-3t} + 2t + 1$ is a solution of $y' + 3y = 6t + 5$.

(b) Imposing the constraint $y(0) = 3$ on $y(t) = Ce^{-3t} + 2t + 1$ leads to $y(0) = C + 1 = 3$. Therefore, $C = 2$, and a solution of the initial value problem is

$$y = 2e^{-3t} + 2t + 1. \; ❖$$

We will show later that $y = Ce^{-3t} + 2t + 1$ is the general solution of the differential equation in Example 3. A geometric interpretation is given in Figure 1.2, which shows graphs of the general solution for representative values of C. The solution whose graph passes through the point $(t, y) = (0, 3)$ is the one that solves the initial value problem posed in Example 3.

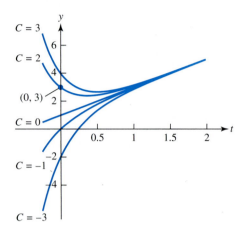

FIGURE 1.2

For any constant C, $y = Ce^{-3t} + 2t + 1$ is a solution of $y' + 3y = 6t + 5$. Solution curves are displayed for several values of C. For $C = 2$, the curve passes through the point $(t, y) = (0, 3)$; this is the solution of the initial value problem posed in Example 3.

EXERCISES

Exercises 1–4:

What is the order of the differential equation?

1. $y'' + 3ty^3 = 1$

2. $t^4 y' + y \sin t = 6$

3. $(y')^3 + t^5 \sin y = y^4$

4. $(y''')^4 - \dfrac{t^2}{(y')^4 + 4} = 0$

Exercises 5–8:

For what value(s) of the constant k, if any, is $y(t)$ a solution of the given differential equation?

5. $y' + 2y = 0$, $y(t) = e^{kt}$

6. $y'' - y = 0$, $y(t) = e^{kt}$

7. $y' + (\sin 2t)y = 0$, $y(t) = e^{k \cos 2t}$

8. $y' + y = 0$, $y(t) = ke^{-t}$

9. (a) Show that $y(t) = Ce^{t^2}$ is a solution of $y' - 2ty = 0$ for any value of the constant C.

   (b) Determine the value of C needed for this solution to satisfy the initial condition $y(1) = 2$.

10. Solve the differential equation $y''' = 2$ by computing successive antiderivatives. What is the order of this differential equation? How many arbitrary constants arise in the antidifferentiation solution process?

11. (a) Show that $y(t) = C_1 \sin 2t + C_2 \cos 2t$ is a solution of the differential equation $y'' + 4y = 0$, where C_1 and C_2 are arbitrary constants.

   (b) Find values of the constants C_1 and C_2 so that the solution satisfies the initial conditions $y(\pi/4) = 3, y'(\pi/4) = -2$.

12. Suppose $y(t) = 2e^{-4t}$ is the solution of the initial value problem $y' + ky = 0$, $y(0) = y_0$. What are the constants k and y_0?

13. Consider $t > 0$. For what value(s) of the constant c, if any, is $y(t) = c/t$ a solution of the differential equation $y' + y^2 = 0$?

14. Let $y(t) = -e^{-t} + \sin t$ be a solution of the initial value problem $y' + y = g(t)$, $y(0) = y_0$. What must the function $g(t)$ and the constant y_0 be?

15. Consider $t > 0$. For what value(s) of the constant r, if any, is $y(t) = t^r$ a solution of the differential equation $t^2 y'' - 2ty' + 2y = 0$?

16. Show that $y(t) = C_1 e^{2t} + C_2 e^{-2t}$ is a solution of the differential equation $y'' - 4y = 0$, where C_1 and C_2 are arbitrary constants.

Exercises 17–18:

Use the result of Exercise 16 to solve the initial value problem.

17. $y'' - 4y = 0$, $\quad y(0) = 2$, $\ y'(0) = 0$ **18.** $y'' - 4y = 0$, $\quad y(0) = 1$, $\ y'(0) = 2$

Exercises 19–20:

Use the result of Exercise 16 to find a function $y(t)$ that satisfies the given conditions.

19. $y'' - 4y = 0$, $\quad y(0) = 3$, $\ \lim_{t \to \infty} y(t) = 0$

20. $y'' - 4y = 0$, $\quad y(0) = 10$, $\ \lim_{t \to -\infty} y(t) = 0$

Exercises 21–22:

The graph shows the solution of the given initial value problem. In each case, m is an integer. In Exercise 21, determine m, y_0, and $y(t)$. In Exercise 22, determine m, t_0, and $y(t)$.

21. $y'(t) = m + 1$, $\quad y(1) = y_0$ **22.** $y'(t) = mt$, $\quad y(t_0) = -1$

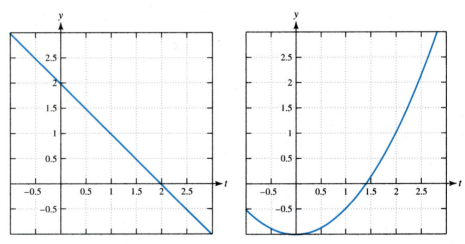

Figure for Exercise 21   Figure for Exercise 22

23. At time $t = 0$, an object having mass m is released from rest at a height y_0 above the ground. Let g represent the (constant) gravitational acceleration. Derive an expression for the impact time (the time at which the object strikes the ground). What is the velocity with which the object strikes the ground? (Express your answers in terms of the initial height y_0 and the gravitational acceleration g.)

24. A car, initially at rest, begins moving at time $t = 0$ with a constant acceleration down a straight track. If the car achieves a speed of 60 mph (88 ft/sec) at time $t = 8$ sec, what is the car's acceleration? How far down the track will the car have traveled when its speed reaches 60 mph?

1.3   Direction Fields

Before beginning a systematic study of differential equations, we consider a geometric entity called a direction field, which will assist in understanding the first order differential equation

$$y' = f(t, y).$$

A **direction field** is a way of predicting the *qualitative* behavior of solutions of a differential equation. A good way to understand the idea of a direction field is to recall the "iron filings" experiment that is often done in science classes to illustrate magnetism. In this experiment, iron filings (minute filaments of iron) are sprinkled on a sheet of cardboard, beneath which two magnets of opposite polarity are positioned. When the cardboard sheet is gently tapped, the iron filings align themselves so that their axes are tangent to the magnetic field lines. At a given point on the sheet, the orientation of an iron filing indicates the direction of the magnetic field line. The totality of oriented iron filings gives a good picture of the flow of magnetic field lines connecting the two magnetic poles. Figure 1.3 illustrates this experiment.

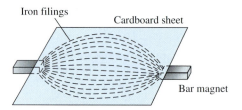

FIGURE 1.3

The orientation of iron filings gives a good picture of the flow of magnetic field lines connecting two magnetic poles.

The Direction Field for a Differential Equation

What is the connection between the iron filings experiment illustrated in Figure 1.3 and a qualitative understanding of differential equations? From calculus we know that if we graph a differentiable function $y(t)$, the slope of the curve at the point $(t, y(t))$ is $y'(t)$. If $y(t)$ is a solution of a differential equation $y' = f(t, y)$, then we can calculate this slope by simply evaluating the right-hand side $f(t, y)$ at the point $(t, y(t))$.

For example, suppose $y(t)$ is a solution of the equation

$$y' = 1 + 2ty \tag{1}$$

and suppose the graph of $y(t)$ passes through the point $(t, y) = (2, y(2)) = (2, -1)$. For differential equation (1), the right-hand side is $f(t, y) = 1 + 2ty$. Thus, we find

$$y'(2) = f(2, y(2)) = f(2, -1) = 1 + 2(2)(-1) = -3.$$

Even though we have not solved $y' = 1 + 2ty$, the preceding calculation has taught us something about the specific solution $y(t)$ passing through $(t, y) = (2, -1)$: it is decreasing (with slope equal to -3) when it passes through the point $(t, y) = (2, -1)$.

To exploit this idea, suppose we systematically evaluate the right-hand side $f(t, y)$ at a large number of points (t, y) throughout a region of interest in the ty-plane. At each point, we evaluate the function $f(t, y)$ to determine the slope of the solution curve passing through that point. We then sketch a tiny line segment at that point, oriented with the given slope $f(t, y)$. The resulting picture, called a direction field, is similar to that illustrated in Figure 1.3. Using such a direction field, we can create a good qualitative picture of the flow of solution curves throughout the region of interest.

E X A M P L E

1

(a) Sketch a direction field for $y' = 1 + 2ty$ in the square $-2 \leq t \leq 2$, $-2 \leq y \leq 2$.

(b) Using the direction field, sketch your guess for the solution curve passing through the point $P = (-2, 2)$. Also, using the direction field, sketch your guess for the solution curve passing through the point $Q = (0, -1)$.

Solution: The direction field for $y' = 1 + 2ty$ shown in Figure 1.4(a) was computer generated. There are a number of computer programs available for drawing direction fields. Figure 1.4(b) shows our guesses for the solutions of the initial value problems in part (b).

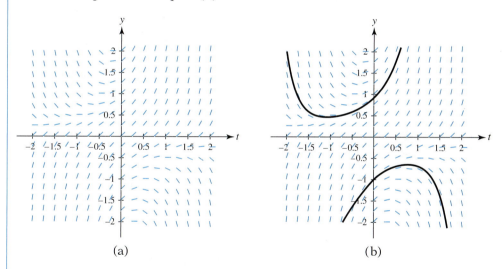

(a)   (b)

FIGURE 1.4

(a) The direction field for $y' = 1 + 2ty$. (b) Using the direction field, we have drawn our guess for the solution of $y' = 1 + 2ty, y(-2) = 2$ and for the solution of $y' = 1 + 2ty, y(0) = -1$.   ❖

Isoclines

The "method of isoclines" is helpful when you need to draw a direction field by hand. An **isocline** of the differential equation $y' = f(t, y)$ is a curve of the form

$$f(t, y) = c,$$

where c is a constant. For example, consider the differential equation

$$y' = y - t^2.$$

In this case, curves of the form $y - t^2 = c$ are isoclines of the differential equation. (These curves, $y = t^2 + c$, are parabolas opening upward. Each has its vertex on the y-axis.) Isoclines are useful because, at every point on an isocline, the associated direction field filaments have the same slope, namely $f(t, y) = c$. (In fact, the word "isocline" means "equal inclination" or "equal slope.")

To carry out the method of isoclines, we first sketch, for various values of c, the corresponding curves $f(t, y) = c$. Then, at representative points on these curves, we sketch direction field filaments having slope $f(t, y) = c$.

E X A M P L E

2

(a) Use the method of isoclines to sketch the direction field for $y' = y - t$. Restrict your direction field to the square defined by $-2 \leq t \leq 2$, $-2 \leq y \leq 2$.

(b) Using the direction field, sketch your guess for the solution curve passing through the point $(-1, \frac{1}{2})$. Also, sketch your guess for the solution curve passing through the point $(-1, -\frac{1}{2})$.

Solution: For the equation $y' = y - t$, lines of the form $y = t + c$ are isoclines. In Figure 1.5(a) we have drawn the isoclines $y = t + 3, y = t + 2, \ldots,$ $y = t - 2$. At selected points along an isocline of the form $y = t + c$, we have drawn direction field filaments, each having slope c.

Figure 1.5(b) shows our guesses for the solutions of the initial value problems in part (b). In addition, note that the line $y = t + 1$ appears to be a solution curve.

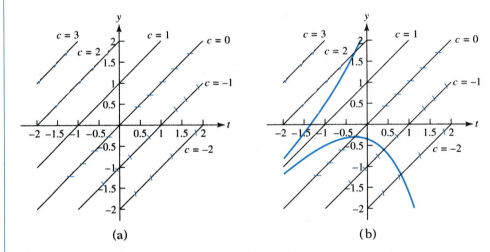

(a)   (b)

FIGURE 1.5

(a) The method of isoclines was used to sketch the direction field for $y' = y - t$. (b) Using the direction field, we have sketched our guesses for the solutions of the initial value problems in part (b) of Example 2.

Direction Fields for Autonomous Equations

The method of isoclines is particularly well suited for differential equations that have the special form

$$y' = f(y). \tag{2}$$

For equations of this form, the isoclines are *horizontal lines*. That is, if b is any number in the domain of $f(y)$, then the horizontal line $y = b$ is an isocline of the equation $y' = f(y)$. In particular, y' has the same value $f(b)$ all along the horizontal line $y = b$.

Differential equations of the form (2), where the right-hand side does not depend explicitly on t, are called **autonomous differential equations**. An example of an autonomous differential equation is

$$y' = y^2 - 3y.$$

By contrast, the differential equation $y' = y + 2t$ is not autonomous. Autonomous differential equations are quite important in applications, and we study them in Chapter 2.

As noted with respect to the autonomous equation $y' = f(y)$, all the slopes of direction field filaments along the horizontal line $y = b$ are equal. This fact is illustrated in Figure 1.6, which shows the direction field for the differential equation

$$y' = y(2 - y).$$

For instance, the filaments along the line $y = 1$ all have slope equal to 1. Similarly, the filaments along the line $y = 2$ all have slope equal to 0. In fact, looking at Figure 1.6, the horizontal lines $y = 0$ and $y = 2$ appear to be solution curves for the differential equation $y' = y(2 - y)$. This is indeed the case, as we show in the next subsection.

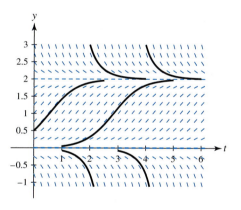

FIGURE 1.6

The direction field for the autonomous equation $y' = y(2 - y)$, together with portions of the graphs of some typical solutions. For an autonomous equation, the slopes are constant along horizontal lines.

Equilibrium Solutions

Consider the autonomous differential equation $y' = y(2 - y)$ whose direction field is shown in Figure 1.6. The horizontal lines $y = 0$ and $y = 2$ appear to be solution curves for this differential equation. In fact, by substituting either of the constant functions $y(t) = 0$ or $y(t) = 2$ into the differential equation, we see that it is a solution of $y' = y(2 - y)$.

In general, consider the autonomous differential equation

$$y' = f(y).$$

If the real number β is a root of the equation $f(y) = 0$, then the constant function $y(t) = \beta$ is a solution of $y' = f(y)$. Conversely, if the constant function $y(t) = \beta$ is a solution of $y' = f(y)$, then β must be a root of $f(y) = 0$. Constant solutions of an autonomous differential equation are known as **equilibrium solutions**.

REMARK: It is possible for differential equations that are not autonomous to have constant solutions. For example, $y(t) = 0$ is a solution of $y' = ty + \sin y$ and $y(t) = 1$ is a solution of $y' = (y - 1)t^2$. We will refer to any constant solution of a differential equation (autonomous or not) as an equilibrium solution.

E X A M P L E

3

Find the equilibrium solutions (if any) of

$$y' = y^2 - 4y + 3.$$

Solution: The right-hand side of the differential equation is

$$f(y) = y^2 - 4y + 3 = (y - 1)(y - 3).$$

Therefore, the equilibrium solutions are the constant functions $y(t) = 1$ and $y(t) = 3$. ❖

EXERCISES

Exercises 1–6:

(a) State whether or not the equation is autonomous.

(b) Identify all equilibrium solutions (if any).

(c) Sketch the direction field for the differential equation in the rectangular portion of the ty-plane defined by $-2 \leq t \leq 2$, $-2 \leq y \leq 2$.

1. $y' = -y + 1$ **2.** $y' = t - 1$ **3.** $y' = \sin y$

4. $y' = y^2 - y$ **5.** $y' = -1$ **6.** $y' = -ty$

Exercises 7–9:

(a) Determine and sketch the isoclines $f(t, y) = c$ for $c = -1, 0$, and 1.

(b) On each of the isoclines drawn in part (a), add representative direction field filaments.

7. $y' = -y + 1$ **8.** $y' = -y + t$ **9.** $y' = y^2 - t^2$

Exercises 10–13:

Find an autonomous differential equation that possesses the specified properties. [Note: There are many possible solutions for each exercise.]

10. Equilibrium solutions at $y = 0$ and $y = 2$; $y' > 0$ for $0 < y < 2$; $y' < 0$ for $-\infty < y < 0$ and $2 < y < \infty$.

11. An equilibrium solution at $y = 1$; $y' < 0$ for $-\infty < y < 1$ and $1 < y < \infty$.

12. A differential equation with no equilibrium solutions and $y' > 0$ for all y.

13. Equilibrium solutions at $y = n/2, n = 0, \pm 1, \pm 2, \ldots$.

Exercises 14–19:

Consider the six direction field plots shown. Associate a direction field with each of the following differential equations.

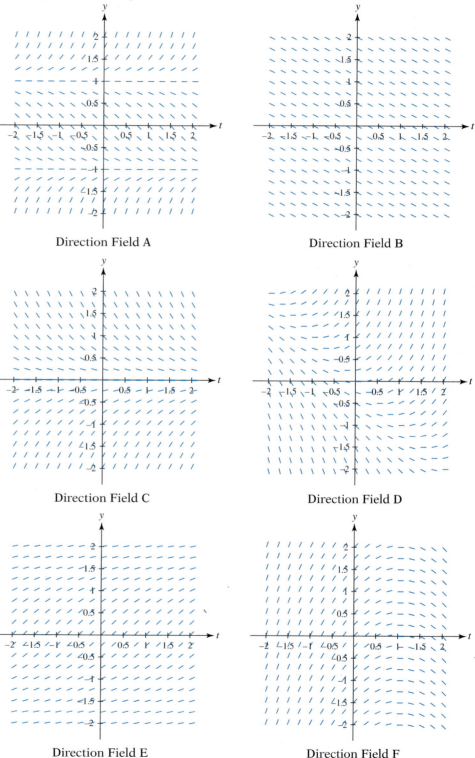

Direction Field A

Direction Field B

Direction Field C

Direction Field D

Direction Field E

Direction Field F

Figure for Exercises 14–21

14. $y' = -y$ **15.** $y' = -t + 1$ **16.** $y' = y^2 - 1$

17. $y' = -\dfrac{1}{2}$ **18.** $y' = y + t$ **19.** $y' = \dfrac{1}{1 + y^2}$

20. For each of the six direction fields shown, assume we are interested in the solution that satisfies the initial condition $y(0) = 0$. Use the graphical information contained in the plots to roughly estimate $y(1)$.

21. Repeat Exercise 20 with $y(0) = 0$ as before, but this time estimate $y(-1)$.

First Order Differential Equations

2.1   Introduction

First order differential equations arise in modeling a wide variety of physical phenomena. In this chapter we study the differential equations that model applications such as population dynamics, radioactive decay, belt friction, and mixing and cooling.

Chapter 2 has two main parts. The first part, consisting of Sections 2.1–2.4, focuses on first order *linear* differential equations and their applications. The second part, consisting of Sections 2.5–2.9, treats first order *nonlinear* equations. The final section, Section 2.10, introduces numerical techniques, such as Euler's method and Runge-Kutta methods, that can be used to approximate the solution of a first order differential equation.

First Order Linear Differential Equations

A differential equation of the form

$$y' + p(t)y = g(t) \tag{1}$$

is called a **first order linear differential equation**. In equation (1), $p(t)$ and $g(t)$ are functions defined on some t-interval of interest, $a < t < b$.

If the function $g(t)$ on the right-hand side of (1) is the zero function, then equation (1) is called **homogeneous**. If $g(t)$ is not the zero function, then equation (1) is **nonhomogeneous**.

A first order equation that can be put into the form of equation (1) by algebraic manipulations is also called a first order linear differential equation. For example, the following are first order linear differential equations:

$$\text{(a)} \ \ e^{-t}y' + 3ty = \sin t \qquad \text{(b)} \ \ \frac{1}{y}y' + t^2 = \frac{\ln t}{y}.$$

A first order differential equation that cannot be put into the form of equation (1) is called **nonlinear**. As we will see, it is possible to find an explicit representation for the solution of a first order *linear* equation. By contrast, most first order *nonlinear* equations cannot be solved explicitly. In Sections 2.6 and 2.7, we will discuss solution techniques for certain special types of first order nonlinear differential equations.

E X A M P L E

1

Is the differential equation linear or nonlinear? If the equation is linear, decide whether it is homogeneous or nonhomogeneous.

$$\text{(a)} \ \ y' = ty^2 \qquad \text{(b)} \ \ y' = t^2 y \qquad \text{(c)} \ \ (\cos t)y' + e^t y = \sin t \qquad \text{(d)} \ \ \frac{y'}{y} + t^3 = \sin t$$

Solution:

(a) This equation is nonlinear because of the presence of the y^2 term.

(b) This equation is linear and homogeneous; it can be put in the form $y' - t^2 y = 0$.

(c) This equation can be put into the form of equation (1),

$$y' + \frac{e^t}{\cos t}y = \tan t.$$

Therefore, the equation is linear and nonhomogeneous.

(d) This equation can be rewritten as

$$y' + (t^3 - \sin t)y = 0.$$

Therefore, the equation is linear and homogeneous. ❖

Existence and Uniqueness for First Order Linear Initial Value Problems

Before looking at how to solve a first order linear equation, we want to address the following question: "What constitutes a properly formulated problem?" This question is answered by the following theorem, which we state now and prove in the Exercises at the end of Section 2.2.

Theorem 2.1

Let $p(t)$ and $g(t)$ be continuous functions on the interval (a, b), and let t_0 be in (a, b). Then the initial value problem

$$y' + p(t)y = g(t), \qquad y(t_0) = y_0$$

has a unique solution on the entire interval (a, b).

Notice that the theorem states three conclusions. A solution exists, it is unique, and this unique solution exists on the entire interval (a, b). We will see that determining intervals of existence is considerably more complicated for nonlinear differential equations.

The importance of Theorem 2.1 lies in the fact that it defines the framework within which we can construct solutions. In particular, suppose we are given a linear differential equation $y' + p(t)y = g(t)$ with coefficient functions $p(t)$ and $g(t)$ that are continuous on (a, b). If we impose an initial condition of the form $y(t_0) = y_0$, where $a < t_0 < b$, the theorem tells us there is one and only one solution. Therefore, if we are able to construct a solution by using some technique we have discovered, the theorem guarantees that it is the only solution—there is no other solution we might have overlooked, one obtainable perhaps by a technique other than the one we are using.

EXAMPLE

2

Consider the initial value problem

$$y' + \frac{1}{t(t+2)}y = \frac{1}{t-5}, \qquad y(3) = 1.$$

What is the largest interval (a, b) on which Theorem 2.1 guarantees the existence of a unique solution?

Solution: The coefficient function $p(t) = t^{-1}(t+2)^{-1}$ has discontinuities at $t = 0$ and $t = -2$ but is continuous everywhere else. Similarly, $q(t) = (t-5)^{-1}$ has a discontinuity at $t = 5$ but is continuous for all other values t. Therefore, Theorem 2.1 guarantees that a unique solution exists on each of the following t-intervals:

$$(-\infty, -2), \qquad (-2, 0), \qquad (0, 5), \qquad (5, \infty).$$

Since the initial condition is imposed at $t = 3$, we are guaranteed that a unique solution exists on the interval $0 < t < 5$. (The solution might actually exist over a larger interval, but we cannot ascertain this without actually solving the initial value problem.) ❖

EXERCISES

Exercises 1–10:

Classify each of the following first order differential equations as linear or nonlinear. If the equation is linear, decide whether it is homogeneous or nonhomogeneous.

1. $y' - \sin t = t^2 y$ **2.** $y' - \sin t = ty^2$ **3.** $\dfrac{y'}{y} - y \cos t = t$

4. $y' \sin y = (t^2 + 1)y$ **5.** $y' \sin t = \dfrac{t^2 + 1}{y}$ **6.** $2ty + e^t y' = \dfrac{y}{t^2 + 4}$

7. $yy' = t^3 + y \sin 3t$ **8.** $2ty + e^y y' = \dfrac{y}{t^2 + 4}$ **9.** $\dfrac{ty'}{(t^4 + 2)y} = \cos t + \dfrac{e^{3t}}{y}$

10. $\dfrac{y'}{(t^2 + 1)y} = \cos t$

Exercises 11–14:

Consider the following first order linear differential equations. For each of the initial conditions, determine the largest interval $a < t < b$ on which Theorem 2.1 guarantees the existence of a unique solution.

11. $y' + \dfrac{t}{t^2 + 1} y = \sin t$

    (a) $y(-2) = 1$ (b) $y(0) = \pi$ (c) $y(\pi) = 0$

12. $y' + \dfrac{t}{t^2 - 4} y = 0$

    (a) $y(6) = 2$ (b) $y(1) = -1$ (c) $y(0) = 1$ (d) $y(-6) = 2$

13. $y' + \dfrac{t}{t^2 - 4} y = \dfrac{e^t}{t - 3}$

    (a) $y(5) = 2$ (b) $y(-\tfrac{3}{2}) = 1$ (c) $y(0) = 0$

    (d) $y(-5) = 4$ (e) $y(\tfrac{3}{2}) = 3$

14. $y' + (t - 1)y = \dfrac{\ln \left| t + t^{-1} \right|}{t - 2}$

    (a) $y(3) = 0$ (b) $y\left(\tfrac{1}{2}\right) = -1$ (c) $y(-\tfrac{1}{2}) = 1$ (d) $y(-3) = 2$

15. If $y(t) = 3e^{t^2}$ is known to be the solution of the initial value problem

$$y' + p(t)y = 0, \qquad y(0) = y_0,$$

what must the function $p(t)$ and the constant y_0 be?

16. (a) For what value of the constant C and exponent r is $y = Ct^r$ the solution of the initial value problem

$$2ty' - 6y = 0, \qquad y(-2) = 8?$$

    (b) Determine the largest interval of the form (a, b) on which Theorem 2.1 guarantees the existence of a unique solution.

    (c) What is the actual interval of existence for the solution found in part (a)?

17. If $p(t)$ is any function continuous on an interval of the form $a < t < b$ and if t_0 is any point lying within this interval, what is the unique solution of the initial value problem

$$y' + p(t)y = 0, \qquad y(t_0) = 0$$

on this interval? [Hint: If, by inspection, you can identify one solution of the given initial value problem, then Theorem 2.1 tells you that it must be the only solution.]

2.2   First Order Linear Differential Equations

In this section, we solve the first order linear homogeneous differential equation

$$y' + p(t)y = 0, \tag{1}$$

and then we build on this result to solve the nonhomogeneous equation $y' + p(t)y = g(t)$.

Solving the Linear Homogeneous Equation

Consider the homogeneous first order linear equation $y' + p(t)y = 0$, which we rewrite as

$$y' = -p(t)y. \tag{2}$$

We assume that $p(t)$ is continuous on the t-interval of interest.

To solve equation (2), we need to find a function $y(t)$ whose derivative is equal to $-p(t)$ times $y(t)$. Recall from calculus that

$$\frac{d}{dt}e^{-P(t)} = -P'(t)e^{-P(t)}.$$

The function $y = e^{-P(t)}$ has the property that

$$y' = -P'(t)y.$$

Therefore, if we choose a function $P(t)$ such that $P'(t) = p(t)$, then

$$y = e^{-P(t)} \tag{3}$$

is a solution of $y' = -p(t)y$.

If $P'(t) = p(t)$, then $P(t)$ is an **antiderivative** of $p(t)$ and is usually denoted by the integral notation, $P(t) = \int p(t)\,dt$. So a solution of $y' = -p(t)y$ can be expressed as

$$y = e^{-\int p(t)\,dt}.$$

E X A M P L E

1

Find a solution of the differential equation

$$y' + 2ty = 0.$$

Solution: For this linear equation, $p(t) = 2t$. For $P(t)$ we can choose any convenient antiderivative of $p(t)$. If we select

$$P(t) = t^2,$$

then, using (3), we obtain the solution

$$y = e^{-t^2}.$$

As a check, let $y = e^{-t^2}$. Then $y' = -2te^{-t^2} = -2ty$. Thus, we have verified that $y = e^{-t^2}$ is a solution of $y' + 2ty = 0$. Figure 2.1 shows the direction field for this differential equation, as well as a graph of the solution.

(continued)

(continued)

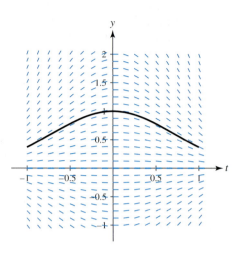

FIGURE 2.1

The direction field for the differential equation in Example 1 and the graph of a solution, $y = e^{-t^2}$.   ❖

The General Solution

Equation (3) represents *one* solution of $y' + p(t)y = 0$. But, in order to solve initial value problems, we need to develop a method for finding *all* the solutions. Observe that if we multiply solution (3) by any constant C, then the resulting function,

$$y = Ce^{-P(t)}, \tag{4}$$

is also a solution. In fact (see Exercises 47–48), Theorem 2.1 can be used to show that every solution of $y' + p(t)y = 0$ has the form (4) for some constant C. We call (4) the **general solution** of $y' + p(t)y = 0$.

EXAMPLE
2

Find the general solution of

$$y' + (\cos t)y = 0.$$

Solution: A convenient antiderivative for $p(t) = \cos t$ is $P(t) = \sin t$. Thus, the general solution is

$$y = Ce^{-\sin t}. \quad ❖$$

REMARK: Let $P(t)$ be an antiderivative of $p(t)$. From calculus we know that any other antiderivative of $p(t)$ has the form $P(t) + K$, where K is some constant. For instance, in Example 2, we chose $P(t) = \sin t$ as an antiderivative of $p(t) = \cos t$. We could just as well have chosen $P(t) = 2 + \sin t$ as the antiderivative. In that case, the general solution would have had the form

$$y = Ce^{-(2+\sin t)} = Ce^{-2}e^{-\sin t} = C_1 e^{-\sin t}.$$

In this expression, C is an arbitrary constant. Since $C_1 = Ce^{-2}$, we can regard C_1 as an arbitrary constant as well. Thus, no matter which antiderivative we choose, the general solution is still the product of an arbitrary constant and the function $e^{-\sin t}$.

Using the General Solution to Solve Initial Value Problems

An initial value problem for a homogeneous first order linear equation can be solved by first forming the general solution

$$y = Ce^{-P(t)}$$

and then choosing the constant C so as to satisfy the initial condition.

E X A M P L E

3

Solve the initial value problem

$$ty' + 2y = 0, \qquad y(1) = 5.$$

Solution: Notice that the differential equation is not in the standard form for a first order linear equation. In order to use equation (4) to represent the general solution, we need to rewrite the differential equation as

$$y' + \frac{2}{t}y = 0, \qquad y(1) = 5.$$

As rewritten, $p(t) = 2/t$. A convenient antiderivative is

$$P(t) = \int \frac{2}{t}\,dt = 2\ln|t| = \ln|t|^2 = \ln t^2.$$

Having an antiderivative $P(t)$, we obtain the general solution

$$y = Ce^{-P(t)} = Ce^{-\ln t^2} = Ct^{-2}.$$

The initial condition $y(1) = 5$ requires that $C = 5$. Therefore, the unique solution of the initial value problem is

$$y = \frac{5}{t^2}. \quad ❖$$

Example 3 illustrates a point about Theorem 2.1. The differential equation has a coefficient function, $p(t) = 2/t$, that is not defined and certainly not continuous at $t = 0$. Therefore, Theorem 2.1 cannot be used to *guarantee* that solutions exist across any interval containing $t = 0$. In fact, for this initial value problem, the solution, $y(t) = 5/t^2$, is not defined at $t = 0$.

However, if we change the initial condition in Example 3 to $y(1) = 0$, we find that the solution is the zero function, $y(t) = 0$ (see Figure 2.2). Thus, even though this particular initial value problem does not satisfy the conditions of Theorem 2.1 on $(-\infty, \infty)$, it does in fact have a solution that is defined for all t. It is important to realize that the failure of Theorem 2.1 to apply to an initial value problem does not imply that the solution must necessarily "behave badly." The logical distinction is important. Theorem 2.1 asserts that if the hypotheses are satisfied, "good things will happen." It does *not* assert that when the hypotheses are not satisfied, "bad things must happen."

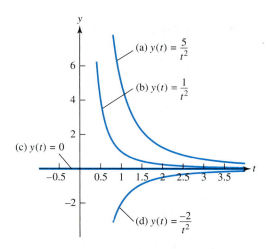

FIGURE 2.2

Some solutions of the problem $ty' + 2y = 0$, $y(1) = y_0$, posed in Example 3. When y_0 is nonzero, the solution is defined only for $t > 0$. But if $y_0 = 0$, the solution is the zero function and is defined for all t.

First Order Linear Nonhomogeneous Equations

We now solve the *nonhomogeneous* linear equation

$$y' + p(t)y = g(t).$$

We assume that $p(t)$ and $g(t)$ are continuous on the t-interval of interest.

Integrating Factors

As preparation for solving the nonhomogeneous equation, we reconsider the homogeneous equation from a slightly different point of view. In particular, the homogeneous equation has the form

$$y' + p(t)y = 0. \tag{5}$$

Let $P(t)$ be some antiderivative of $p(t)$, and define a new function $\mu(t)$ by

$$\mu(t) = e^{P(t)}. \tag{6}$$

The function $\mu(t) = e^{P(t)}$ is called an **integrating factor**. We will shortly see the reason for this name.

Note from equation (6) that

$$\mu'(t) = P'(t)e^{P(t)} = p(t)\mu(t). \tag{7a}$$

We multiply equation (5) by the integrating factor $\mu(t)$ to obtain a new equation,

$$\mu(t)y' + \mu(t)p(t)y = 0.$$

From (7a), $\mu'(t) = p(t)\mu(t)$, and therefore

$$\mu(t)y' + \mu'(t)y = 0. \tag{7b}$$

The left-hand side of equation (7b) is the derivative of a product and can be rewritten as

$$\frac{d}{dt}(\mu(t)y(t)) = 0. \tag{8}$$

If the derivative of a function is identically zero, then the function must be constant. Therefore, equation (8) implies

$$\mu(t)y(t) = C,$$

where C is a constant. Since $\mu(t)y(t) = C$ and since $\mu(t) = e^{P(t)}$ is nonzero, we can solve for $y(t)$:

$$y(t) = \frac{1}{\mu(t)}C$$
$$= Ce^{-P(t)}.$$

The derivation of equation (8) explains why the function $\mu(t) = e^{P(t)}$ is called an "integrating factor." That is, we multiply equation (5) by $\mu(t)$ to obtain the new equation (8), which can be integrated. Also note that this derivation leads to the same general solution for equation (5) that we found in equation (4), $y = Ce^{-P(t)}$. The purpose of the derivation is to introduce the concept of an integrating factor.

Using an Integrating Factor to Solve the Nonhomogeneous Equation

Now consider the nonhomogeneous equation

$$y' + p(t)y = g(t). \tag{9}$$

If we multiply equation (9) by the integrating factor $\mu(t) = e^{P(t)}$, we obtain $\mu(t)y' + \mu(t)p(t)y = \mu(t)g(t)$. Since $\mu'(t) = \mu(t)p(t)$, we have

$$\mu(t)y' + \mu'(t)y = \mu(t)g(t),$$

or

$$\frac{d}{dt}(\mu(t)y(t)) = \mu(t)g(t).$$

Integrating both sides gives

$$\mu(t)y(t) = \int \mu(t)g(t)\,dt + C,$$

where C is a constant and where $\int \mu(t)g(t)\,dt$ represents some particular antiderivative of $\mu(t)g(t)$. Solving for $y(t)$, we are led to the *general solution* of the nonhomogeneous equation (9):

$$y = e^{-P(t)}\int e^{P(t)}g(t)\,dt + Ce^{-P(t)}. \tag{10}$$

REMARKS:

1. Don't be confused by the notation. In particular, the terms $e^{-P(t)}$ and $e^{P(t)}$ in (10) do not cancel; $e^{P(t)}$ is part of the function $e^{P(t)}g(t)$ whose antiderivative must be determined. Once this antiderivative has been calculated, it is multiplied by the term $e^{-P(t)}$.

2. Notice that the general solution given by (10) is the sum of two terms, $e^{-P(t)}\int e^{P(t)}g(t)\,dt$ and $Ce^{-P(t)}$. The first term is some *particular* solution of the nonhomogeneous equation, while the second term represents the

general solution of the homogeneous equation. We'll see this same solution structure again when we study higher order linear equations and systems of linear equations.

3. Observe that the general solution contains only one arbitrary constant, C. This constant is determined by imposing an initial condition.

4. Although expression (10) is the general solution of the nonhomogeneous equation, you should not try to memorize it. Instead, remember the steps leading to (10).

E X A M P L E

4

Find the general solution and then solve the initial value problem

$$y' + 2ty = 4t, \qquad y(0) = 5.$$

Solution: For this differential equation, $p(t) = 2t$. An antiderivative is $P(t) = t^2$, and so an integrating factor is

$$\mu(t) = e^{t^2}.$$

Multiplying the differential equation by $\mu(t)$, we obtain

$$e^{t^2}y' + 2te^{t^2}y = 4te^{t^2} \quad \text{or} \quad (e^{t^2}y)' = 4te^{t^2}.$$

Therefore,

$$e^{t^2}y = 2e^{t^2} + C.$$

Solving for y, we obtain the general solution

$$y = 2 + Ce^{-t^2}.$$

Imposing the initial condition $y(0) = 5$, we find

$$y = 2 + 3e^{-t^2}.$$

The solution is graphed in Figure 2.3.

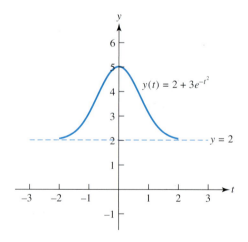

FIGURE 2.3

The solution of the problem posed in Example 4 is $y = 2 + 3e^{-t^2}$.   ❖

Example 4 illustrates the second remark following equation (10). The general solution we found (namely $y = 2 + Ce^{-t^2}$) is the sum of some particular solution of the nonhomogeneous equation (namely the constant function $y = 2$) and the general solution of the homogeneous equation (namely $y = Ce^{-t^2}$). Note that the initial condition was imposed on the general solution as the last step. This will always be the case.

Discontinuous Coefficient Functions

In some applications, physical conditions undergo abrupt changes. For example, a hot metal object might be plunged suddenly into a cooling bath, or we might throw a switch and abruptly change the source voltage in an electrical network.

Such applications are often modeled by an initial value problem

$$y' + p(t)y = g(t), \qquad y(a) = y_0, \qquad a \le t \le b,$$

where one or both of the functions $p(t)$ and $g(t)$ have a jump discontinuity at some point, say $t = c$, where $a < c < b$. In such cases, even though $y'(t)$ is not continuous at $t = c$, we expect on physical grounds that the solution $y(t)$ is continuous at $t = c$. For these problems we first solve the initial value problem on the interval $a \le t < c$; the solution $y(t)$ will have a one-sided limit,

$$\lim_{t \to c^-} y(t) = y(c^-).$$

To complete the solution, we use the limiting value $y(c^-)$ as the initial condition on the subinterval $[c, b]$ and then solve a second initial value problem on $[c, b]$.

E X A M P L E

5

Solve the following initial value problem on the interval $0 \le t \le 2$:

$$y' - y = g(t), \qquad y(0) = 0, \qquad \text{where } g(t) = \begin{cases} 1, & 0 \le t < 1 \\ -2, & 1 \le t \le 2. \end{cases}$$

Solution: The graph of $g(t)$ is shown in Figure 2.4(a); it has a jump discontinuity at $t = 1$. On the interval $[0, 1)$, the differential equation reduces to $y' - y = 1$. The general solution is

$$y(t) = Ce^t - 1.$$

Imposing the initial condition, we obtain $y(t) = e^t - 1, 0 \le t < 1$. As t approaches 1 from the left, $y(t)$ approaches the value $e - 1$. Therefore, to complete the solution process, we solve a second initial value problem,

$$y' - y = -2, \qquad y(1) = e - 1, \qquad 1 \le t \le 2.$$

The solution of this initial value problem is

$$y(t) = \left(1 - \frac{3}{e}\right)e^t + 2, \qquad 1 \le t \le 2.$$

Combining the individual solutions of these two initial value problems, we obtain the solution for the entire interval $0 \le t \le 2$:

$$y(t) = \begin{cases} e^t - 1, & 0 \le t < 1 \\ \left(1 - \dfrac{3}{e}\right)e^t + 2, & 1 \le t \le 2. \end{cases}$$

(continued)

(continued)

The graph of $y(t)$ is shown in Figure 2.4(b). Note that $y(t)$ is continuous on the entire t-interval of interest. However, $y(t)$ is not differentiable at $t = 1$.

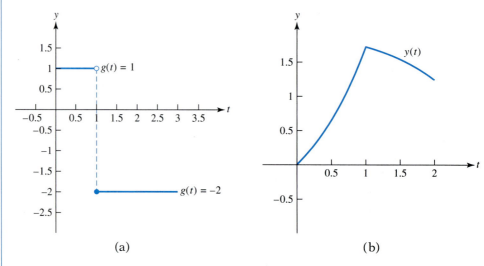

(a)   (b)

FIGURE 2.4

(a) The coefficient function $g(t)$ of the differential equation $y' - y = g(t)$ in Example 5 has a jump discontinuity at $t = 1$. (b) The solution of $y' - y = g(t)$, $y(0) = 0$ is continuous on the interval $0 \le t \le 2$, but is not differentiable at $t = 1$. ❖

EXERCISES

Exercises 1–10:

For each initial value problem,

(a) Find the general solution of the differential equation.

(b) Impose the initial condition to obtain the solution of the initial value problem.

1. $y' + 3y = 0$, $y(0) = -3$ **2.** $2y' - y = 0$, $y(-1) = 2$

3. $2ty - y' = 0$, $y(1) = 3$ **4.** $ty' - 4y = 0$, $y(1) = 1$

5. $y' - 3y = 6$, $y(0) = 1$ **6.** $y' - 2y = e^{3t}$, $y(0) = 3$

7. $2y' + 3y = e^{t}$, $y(0) = 0$ **8.** $y' + y = 1 + 2e^{-t}\cos 2t$, $y(\pi/2) = 0$

9. $2y' + (\cos t)y = -3\cos t$, $y(0) = -4$ **10.** $y' + 2y = e^{-t} + t + 1$, $y(-1) = e$

Exercises 11–24:

Find the general solution.

11. $ty' + 4y = 0$ **12.** $y' + (1 + \sin t)y = 0$ **13.** $y' - 2(\cos 2t)y = 0$

14. $(t^2 + 1)y' + 2ty = 0$ **15.** $\dfrac{y'}{(t^2 + 1)y} = 3$ **16.** $y + e^{t}y' = 0$

17. $y' + 2y = 1$ **18.** $y' + 2y = e^{-t}$ **19.** $y' + 2y = e^{-2t}$

20. $y' + 2ty = t$ **21.** $ty' + 2y = t^2$, $t > 0$ **22.** $(t^2 + 4)y' + 2ty = t^2(t^2 + 4)$

23. $y' + y = t$ **24.** $y' + 2y = \cos 3t$

25. Consider the three direction fields shown. Match each of the direction field plots with one of the following differential equations:

   (a) $y' + y = 0$ (b) $y' + t^2 y = 0$ (c) $y' - y = 0$

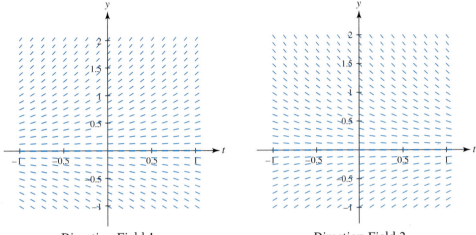

Direction Field 1   Direction Field 2

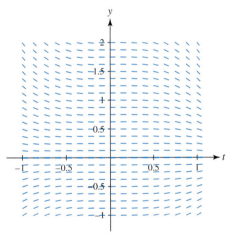

Direction Field 3

Figure for Exercise 25

Exercises 26–27:

The graph of the solution of the given initial value problem is known to pass through the (t, y) points listed. Determine the constants α and y_0.

26. $y' + \alpha y = 0$, $y(0) = y_0$.   Solution graph passes through the points $(1, 4)$ and $(3, 1)$.

27. $ty' - \alpha y = 0$, $y(1) = y_0$.   Solution graph passes through the points $(2, 1)$ and $(4, 4)$.

28. Following are four graphs of $y(t)$ versus t, $0 \le t \le 10$, corresponding to solutions of the four differential equations (a)–(d). Match the graphs to the differential equations. For each match, identify the initial condition, $y(0)$.

   (a) $2y' + y = 0$ (b) $y' + (\cos 2t)y = 0$

   (c) $10y' - (1 - \cos 2t)y = 0$ (d) $10y' - y = 0$

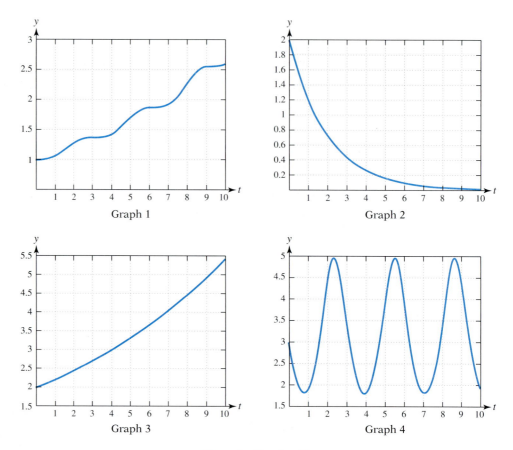

Graph 1

Graph 2

Graph 3

Graph 4

Figure for Exercise 28

29. Antioxidants   Active oxygen and free radicals are believed to be exacerbating factors in causing cell injury and aging in living tissue.[1] These molecules also accelerate the deterioration of foods. Researchers are therefore interested in understanding the protective role of natural antioxidants. In the study of one such antioxidant (Hsian-tsao leaf gum), the antioxidation activity of the substance has been found to depend on concentration in the following way:

$$\frac{dA(c)}{dc} = k[A^* - A(c)], \qquad A(0) = 0.$$

In this equation, the dependent variable A is a quantitative measure of antioxidant activity at concentration c. The constant A^* represents a limiting or equilibrium value of this activity, and k is a positive rate constant.

(a) Let $B(c) = A(c) - A^*$ and reformulate the given initial value problem in terms of this new dependent variable, B.

(b) Solve the new initial value problem for $B(c)$ and then determine the quantity of interest, $A(c)$. Does the activity $A(c)$ ever exceed the value A^*?

(c) Determine the concentration at which 95% of the limiting antioxidation activity is achieved. (Your answer is a function of the rate constant k.)

30. The solution of the initial value problem $ty' + 4y = \alpha t^2$, $y(1) = -\frac{1}{3}$ is known to exist on $-\infty < t < \infty$. What is the constant α?

---

[1]Lih-Shiuh Lai, Su-Tze Chou, and Wen-Wan Chao, "Studies on the Antioxidative Activities of Hsian-tsao (*Mesona procumbens Hemsl*) Leaf Gum," *J. Agric. Food Chem.*, Vol. 49, 2001, pp. 963–968.

Exercises 31–33:

In each exercise, the general solution of the differential equation $y' + p(t)y = g(t)$ is given, where C is an arbitrary constant. Determine the functions $p(t)$ and $g(t)$.

31. $y(t) = Ce^{-2t} + t + 1$ **32.** $y(t) = Ce^{t^2} + 2$ **33.** $y(t) = Ct^{-1} + 1, \quad t > 0$

Exercises 34–35:

In each exercise, the unique solution of the initial value problem $y' + y = g(t), y(0) = y_0$ is given. Determine the constant y_0 and the function $g(t)$.

34. $y(t) = e^{-t} + t - 1$ **35.** $y(t) = -2e^{-t} + e^t + \sin t$

Exercises 36–37:

In each exercise, discuss the behavior of the solution $y(t)$ as t becomes large. Does $\lim_{t \to \infty} y(t)$ exist? If so, what is the limit?

36. $y' + y + y\cos t = 1 + \cos t, \quad y(0) = 3$

37. $\dfrac{y' - e^{-t} + 2}{y} = -2, \quad y(0) = -2$

38. The solution of the initial value problem $y' + y = e^{-t}, y(0) = y_0$ has a maximum value of $e^{-1} = 0.367\ldots$, attained at $t = 1$. What is the initial condition y_0?

39. Let $y(t)$ be a nonconstant solution of the differential equation $y' + \lambda y = 1$, where λ is a real number. For what values of λ is $\lim_{t \to \infty} y(t)$ finite? What is the limit in this case?

Exercises 40–43:

As in Example 5, find a solution to the initial value problem that is continuous on the given interval $[a, b]$.

40. $y' + \dfrac{1}{t}y = g(t), \quad y(1) = 1; \qquad g(t) = \begin{cases} 3t, & 1 \le t \le 2 \\ 0, & 2 < t \le 3; \end{cases} \qquad [a, b] = [1, 3]$

41. $y' + (\sin t)y = g(t), \quad y(0) = 3; \quad g(t) = \begin{cases} \sin t, & 0 \le t \le \pi \\ -\sin t, & \pi < t \le 2\pi; \end{cases} \qquad [a, b] = [0, 2\pi]$

42. $y' + p(t)y = 2, \quad y(0) = 1; \qquad p(t) = \begin{cases} 0, & 0 \le t \le 1 \\ \dfrac{1}{t}, & 1 < t \le 2; \end{cases} \qquad [a, b] = [0, 2]$

43. $y' + p(t)y = 0, \quad y(0) = 3; \qquad p(t) = \begin{cases} 2t - 1, & 0 \le t \le 1 \\ 0, & 1 < t \le 3 \\ -\dfrac{1}{t}, & 3 < t \le 4; \end{cases} \qquad [a, b] = [0, 4]$

Exercises 44–45:

In each exercise, you are asked to express the solution in terms of a "special function" [the function Si(t) in Exercise 44 and erf(t) in Exercise 45]. Such **special functions** are sufficiently important in applications to warrant giving them names and studying their properties. (A book such as *Handbook of Mathematical Functions* by Abramowitz and Stegun[2] gives the definitions for many important special functions, lists their properties, and has tables of their values. Scientific software such as MATLAB, Mathematica, Maple, and Derive has subroutines for evaluating special functions.)

---

[2] Milton Abramowitz and Irene Stegun, *Handbook of Mathematical Functions* (New York: Dover Publications, 1965).

44. Solve $y' - \dfrac{1}{t}y = \sin t$, $y(1) = 3$. Express your answer in terms of the sine integral, Si(t), where Si$(t) = \displaystyle\int_0^t \dfrac{\sin s}{s}\, ds$.   [Note that Si$(t) = $ Si$(1) + \displaystyle\int_1^t \dfrac{\sin s}{s}\, ds$.]

45. Solve $y' - 2ty = 1$, $y(0) = 2$. Express your answer in terms of the error function, erf(t), where erf$(t) = \dfrac{2}{\sqrt{\pi}} \displaystyle\int_0^t e^{-s^2}\, ds$.

46. Superposition   First order linear differential equations possess important superposition properties. Show the following:

(a) If $y_1(t)$ and $y_2(t)$ are any two solutions of the homogeneous equation $y' + p(t)y = 0$ and if c_1 and c_2 are any two constants, then the sum $c_1 y_1(t) + c_2 y_2(t)$ is also a solution of the homogeneous equation.

(b) If $y_1(t)$ is a solution of the homogeneous equation $y' + p(t)y = 0$ and $y_2(t)$ is a solution of the nonhomogeneous equation $y' + p(t)y = g(t)$ and c is any constant, then the sum $cy_1(t) + y_2(t)$ is also a solution of the nonhomogeneous equation.

(c) If $y_1(t)$ and $y_2(t)$ are any two solutions of the nonhomogeneous equation $y' + p(t)y = g(t)$, then the sum $y_1(t) + y_2(t)$ is *not* a solution of the nonhomogeneous equation.

Exercises 47–48:

Outline of a Proof of Theorem 2.1   The discussion of integrating factors in this section provides a basis for establishing the existence-uniqueness result stated in Theorem 2.1. In particular, consider the initial value problem $y' + p(t)y = g(t), y(t_0) = y_0$, where $p(t)$ and $g(t)$ are continuous on the interval (a, b) and where t_0 is in the interval (a, b). Let $P(t)$ denote the specific antiderivative of $p(t)$ that vanishes at t_0,

$$P(t) = \int_{t_0}^t p(s)\, ds. \tag{11}$$

Since p is continuous on (a, b), it follows from calculus that $P(t)$ is defined and differentiable for all t in (a, b). As an instance of equation (10), define $y(t)$ by

$$y(t) = y_0 e^{-P(t)} + e^{-P(t)} \int_{t_0}^t e^{P(s)} g(s)\, ds. \tag{12}$$

Since g is continuous on (a, b) and $P(t)$ is differentiable on (a, b), it follows from calculus that $G(t) = \int_{t_0}^t e^{P(s)} g(s)\, ds$ is defined and differentiable for all t in (a, b) and that $dG/dt = e^{P(t)} g(t)$.

47. Use the facts above to show that $y(t)$ defined in equation (12) is a solution of the initial value problem $y' + p(t)y = g(t), y(t_0) = y_0$. This explicit construction establishes that at least one solution of the initial value problem exists on the entire interval (a, b).

48. To establish the uniqueness part of Theorem 2.1, assume $y_1(t)$ and $y_2(t)$ are two solutions of the initial value problem $y' + p(t)y = g(t), y(t_0) = y_0$. Define the difference function $w(t) = y_1(t) - y_2(t)$.

(a) Show that $w(t)$ is a solution of the homogeneous linear differential equation $w' + p(t)w = 0$.

(b) Multiply the differential equation $w' + p(t)w = 0$ by the integrating factor $e^{P(t)}$, where $P(t)$ is defined in equation (11), and deduce that $e^{P(t)} w(t) = C$, where C is a constant.

(c) Evaluate the constant C in part (b) and show that $w(t) = 0$ on (a, b). Therefore, $y_1(t) = y_2(t)$ on (a, b), establishing that the solution of the initial value problem is unique.

2.3   Introduction to Mathematical Models

Differential equations often serve as mathematical models that are used to describe and make predictions about physical systems. This section and the next one focus on models based on first order linear differential equations, while Sections 2.8 and 2.9 consider models involving first order nonlinear differential equations. Later chapters present models involving higher order differential equations and systems of differential equations.

We saw a simple example of a mathematical model in Chapter 1:

$$\frac{d^2y}{dt^2} = -g, \qquad y(0) = y_0, \qquad y'(0) = v_0.$$

This initial value problem is a mathematical model, derived from Newton's[3] second law of motion, for an object falling under the influence of gravity. The starting position of the object is $y(0) = y_0$, and its initial velocity is $y'(0) = v_0$. The solution predicts how the object's position, $y(t)$, and its velocity, $y'(t)$, vary with time as the object falls.

In this section, we study two important problems modeled by first order linear differential equations—mixing problems and cooling problems. These problems arise when we model phenomena such as the mixing of solutes and solvents in flow systems, the spread and removal of pollutants in air and water, and the cooking and sterilization of foods.

Modeling

We can divide the art of mathematical modeling into three phases:

1. *Formulation*   After observing the physical system, we need to identify the appropriate independent and dependent variables. Then we need to develop a mathematical description of how these variables interact. Often, a differential equation (along with appropriate initial conditions) will serve as a mathematical description of the system.

2. *Solution*   Once we have formulated the modeling problem, we need to solve it. This involves recognizing the mathematical structure of the problem and bringing the appropriate analytical and/or numerical techniques to bear.

3. *Validation and Interpretation*   Once we have solved the problem, the solution needs to be examined carefully. Does it make sense? Is it consistent with our physical intuition about what should be expected? The solution needs to be scrutinized for its physical content: What does it say about the physical phenomenon being modeled?

---

[3]Sir Isaac Newton (1643–1727) profoundly influenced the development of mathematics and science. Newton, along with Gottfried Leibniz, is generally credited with laying the foundations of differential and integral calculus. His work *De Methodis Serierum et Fluxionum* was completed in 1671 but was not published until 1736. *Optiks*, published in 1704, summarizes Newton's research in the theory of light and color. His greatest work, *Philosophiae naturalis principia mathematica* (or simply *Principia*), was published in 1687. This work summarizes his research in physics and celestial mechanics. It contains his laws of motion and the law of universal gravitation. The *Principia* is arguably the greatest scientific work ever published.

Mixing Problems

Consider the fluid mixing problem shown schematically in Figure 2.5. A tank initially contains a volume of fluid, within which is dissolved a certain amount of solute. For definiteness, consider the liquid to be water, having the units of gallons. Consider the solute to be salt, having the units of pounds. Time is measured in minutes.

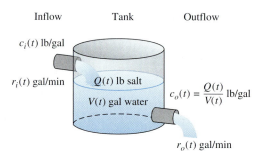

Inflow   Tank   Outflow

$c_i(t)$ lb/gal

$r_i(t)$ gal/min $Q(t)$ lb salt

$V(t)$ gal water $c_o(t) = \dfrac{Q(t)}{V(t)}$ lb/gal

$r_o(t)$ gal/min

FIGURE 2.5

> A salt solution enters the tank at a certain inflow rate and the well-stirred solution leaves the tank at a certain outflow rate. How much salt is in the tank at a given time t?

At some starting time, say $t = 0$, a salt solution enters the tank at a certain inflow rate and the well-stirred solution flows out of the tank at some outflow rate. The phrase **well stirred** means that the concentration of salt is uniform within the tank; the concentration depends only on time and not on spatial location within the tank. In other words, any salt entering the tank is instantaneously dispersed throughout the tank (through either mechanical mixing or diffusion). This is a reasonable approximation if the salt dissolves and disperses into solution very quickly relative to the speed at which the solution enters or leaves the tank.

Our objective in this mixing problem is to determine the amount of salt in the tank, as a function of time.

Modeling the Mixing Problem

To model the dynamics of the mixing process shown in Figure 2.5, we invoke a "conservation of salt" law:

$$\boxed{\text{Rate of change of salt in the tank}} = \boxed{\text{Rate at which salt enters the tank}} - \boxed{\text{Rate at which salt leaves the tank.}} \qquad \textbf{(1)}$$

We need to translate the words of equation (1) into mathematics. To that end, let

$Q(t)$ = amount of salt (pounds) in the tank at time t (minutes),

$V(t)$ = volume of water (gallons) in the tank at time t,

$c_i(t)$ = inflow salt concentration (pounds/gallon) at time t,

$c_o(t)$ = outflow salt concentration (pounds/gallon) at time t,

$r_i(t)$ = inflow rate (gallons/minute) at time t,

$r_o(t)$ = outflow rate (gallons/minute) at time t.

Using these definitions, we can convert each term of equation (1) into a mathematical statement and obtain a mathematical model for the mixing process.

The rate at which salt enters the tank is given by the product of the inflow rate and the inflow salt concentration. That is,

$$\text{Rate at which salt enters the tank} = r_i(t)c_i(t).$$

$$\left(\text{Dimensionally,} \quad \frac{\text{pounds}}{\text{minute}} = \frac{\text{gallons}}{\text{minute}} \cdot \frac{\text{pounds}}{\text{gallon}}.\right)$$

Similarly, the rate at which salt leaves the tank is the product of the outflow rate and the outflow salt concentration. But, while the inflow salt concentration is a known function, $c_i(t)$, the outflow salt concentration is *not* a known function of t. In particular, outflow salt concentration, $c_o(t)$, is determined by volume $V(t)$ and by how much salt is in the tank at time t:

$$c_o(t) = \frac{Q(t)}{V(t)}.$$

Thus,

$$\text{Rate at which salt leaves the tank} = r_o(t)c_o(t) = r_o(t)\frac{Q(t)}{V(t)}.$$

Combining these two calculations, we have a mathematical model for the mixing process:

$$\frac{dQ}{dt} = r_i(t)c_i(t) - r_o(t)\frac{Q}{V(t)}, \qquad Q(0) = Q_0. \tag{2}$$

In equation (2), $Q(0) = Q_0$ gives the amount of salt in the tank at the starting time, $t = 0$.

In equation (2), the volume of water in the tank, $V(t)$, is related to the flow rates by the differential equation

$$\frac{dV}{dt} = r_i(t) - r_o(t).$$

Solving this equation by antidifferentiation, we obtain an expression for $V(t)$,

$$V(t) = V(0) + \int_0^t [r_i(s) - r_o(s)]\, ds.$$

Having $V(t)$, we can solve the first order linear equation (2) for $Q(t)$. [In many cases, the inflow rate and the outflow rate are equal. In such cases, $V(t) = V(0)$ is constant.]

EXAMPLE

1

A tank initially contains 1000 gal of water in which is dissolved 20 lb of salt. A valve is opened and water containing 0.2 lb of salt per gallon flows into the tank at a rate of 5 gal/min. The mixture in the tank is well stirred and drains from the tank at a rate of 5 gal/min.

(a) Find $Q(t)$, the amount of salt in the tank after t minutes.

(b) Find the limiting value: $\lim_{t\to\infty} Q(t)$. Why should you expect such a limit to exist?

(continued)

(continued)

(c) Let the limit in part (b) be designated as Q_L. How long will it be until $Q(t)$ is within 1% of Q_L?

Solution: Since the inflow rate and the outflow rate are the same, the volume of water in the tank remains constant at 1000 gal. Equation (2), together with the condition that there was 20 lb of salt in the tank at time $t = 0$, leads to the following initial value problem:

$$\frac{dQ}{dt} = (5)(0.2) - 5\,\frac{Q}{1000}, \qquad Q(0) = 20.$$

(a) Solving the nonhomogeneous differential equation using the techniques of Section 2.2, we obtain the general solution

$$Q(t) = 200 + Ce^{-t/200}.$$

Imposing the initial condition, we find

$$Q(t) = 200 - 180e^{-t/200} \text{ lb.}$$

(b) From part (a) we see that $Q(t) \to 200$ as $t \to \infty$. The discussion following this example comments on the physical significance of the limit, why the limit should exist, and why the limit should be $Q_L = 200$ lb.

(c) Given $Q_L = 200$, we need to determine when $Q(t)$ is within 1% of 200. Now, from the solution graphed in Figure 2.6, $Q(t)$ is an increasing function. Thus, $Q(t)$ is within 1% of Q_L when $Q(t) \geq 198$. Solving the equation $Q(t) = 198$, we find

$$-t = 200 \ln\left(\tfrac{2}{180}\right),$$

or $t = 899.96\ldots$ min. Therefore, after about 900 min (15 hr), there will be at least 198 lb of salt in the tank.

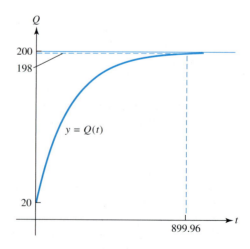

FIGURE 2.6

For the mixing problem in Example 1, there is $Q(t)$ lb of salt in the tank after t min, where $Q(t) = 200 - 180e^{-t/200}$.

The general solution of the differential equation in Example 1 is

$$Q(t) = 200 + Ce^{-t/200}.$$

Thus, $Q(t) \to 200$ as $t \to \infty$, regardless of the value of C. Note that the constant function $Q(t) = 200$ is an equilibrium solution of the differential equation. This is the amount of salt needed to make the concentration of salt in the tank equal to the inflow concentration.

This limiting behavior also can be seen in more general situations. For instance, suppose the inflow and outflow rates are constant and equal, say at r gal/min. Let the inflow concentration, c_i, be constant as well. Let V_0 denote the volume of water in the tank, and let $Q(0) = Q_0$ denote the amount of salt in the tank at time $t = 0$. Under these conditions, the mixing model becomes

$$\frac{dQ}{dt} = rc_i - r\frac{Q}{V_0}, \qquad Q(0) = Q_0. \tag{3a}$$

The solution of this initial value problem is

$$Q(t) = c_i V_0 + (Q_0 - c_i V_0)e^{-(r/V_0)t}. \tag{3b}$$

Note that the solution $Q(t)$ tends to the same limiting value, $c_i V_0$, regardless of the flow rate r. In fact, the constant function $Q(t) = c_i V_0$ is an equilibrium solution of differential equation (3a). It is the amount of salt needed to make the concentration of the solution in the tank equal to the inflow concentration. Think about flushing out a tank with a salt solution. No matter how much salt is initially in the tank, it is flushed out as time increases and the concentration of salt in the tank approaches the inflow concentration of c_i lb/gal. Hence, equation (3b) is consistent with our physical intuition. Increasing or decreasing the flow rate r affects how rapidly the limiting value is approached but does not affect the limiting value itself.

Although we've talked only about tanks and salt, problems of this sort arise in a variety of circumstances, such as environmental applications where the tank is actually a body of water (such as a lake) and the solute is some pollutant entering and leaving via connecting streams.

Cooling Problems

Imagine a bowl of hot soup placed on a kitchen table and left there to cool. Suppose you wanted to develop a mathematical model to predict how the temperature of the soup changes as time progresses. How would you proceed?

At any instant of time, we would expect the temperature at all points within the soup itself to be approximately the same. (This is the thermal equivalent of "well stirred.") Therefore, we assume that the temperature of the soup is described by a function of time alone. We make the same assumption about the kitchen surroundings, and thus its temperature can be described by a second function of time (quite possibly a constant function). Moreover, the kitchen surroundings are sufficiently large that kitchen temperature is basically unchanged by introducing the bowl of hot soup.

With this example as a guide, we'll now set up the general framework for **Newton's law of cooling**. Instead of soup and a kitchen, we speak of an **object** and its **surroundings**.

Let $\Theta(t)$ and $S(t)$ denote the temperatures of the object and its surroundings, respectively. The basic assumption underlying Newton's law of cooling is that the rate of change of the object's temperature is proportional to the difference in temperatures between the object and its surroundings. Expressed mathematically, Newton's law of cooling is

$$\Theta'(t) = k[S(t) - \Theta(t)]. \tag{4}$$

In the model given by equation (4), we assume that the temperature of the surroundings, $S(t)$, is known for all time of interest and is unaffected by the presence of the object.

In equation (4) we also assume that the constant of proportionality, k, is positive. Does that make sense to you? Suppose that, at some instant in time, the temperature of the surroundings is less than the temperature of the object. Should the object's temperature be increasing or decreasing at that instant?

E X A M P L E

2

A metal object is heated to 200°C and then placed in a large room to cool. The temperature of the room is held constant at 20°C. After 10 min, the object's temperature is 100°C. How long will it take the object to cool to 25°C?

Solution: Using equation (4), we can model the object's temperature, $\Theta(t)$, by

$$\Theta'(t) = k[20 - \Theta(t)], \qquad \Theta(0) = 200.$$

Here, t is measured in minutes and temperature in degrees Celsius.

The general solution of $\Theta'(t) = k[20 - \Theta(t)]$ is

$$\Theta(t) = 20 + Ce^{-kt}.$$

Imposing the initial condition, we get $\Theta(t) = 20 + 180e^{-kt}$. Knowing $\Theta(10) = 100$, we determine the rate constant k, finding $k = 0.08109$ (min)$^{-1}$. Thus, $\Theta(t)$ is given (approximately) by

$$\Theta(t) = 20 + 180e^{-0.0811t}.$$

Solving $\Theta(t) = 25$, we find that the metal object cools to 25°C after 44.186... min. ❖

Consider a cooling problem (such as the one in Example 2) where the surrounding temperature $S(t)$ is constant. In particular, let $S(t) = S_0$ for all t of interest. The initial value problem modeling constant-temperature surroundings is

$$\Theta'(t) = k[S_0 - \Theta(t)], \qquad \Theta(0) = \Theta_0. \tag{5}$$

From a mathematical point of view, equations (3a) and (5) are the same. We need only identify k with r/V_0 and S_0 with $V_0 c_i$. Equation (5) has equilibrium solution $\Theta(t) = S_0$. Therefore, if the object has the same initial temperature as the surroundings (that is, if $\Theta_0 = S_0$), then its temperature will remain constant. If the object's initial temperature is not equal to that of the surroundings, we expect the object's temperature to approach S_0 as $t \to \infty$. These common-sense checks are satisfied by the solution of (5):

$$\Theta(t) = S_0 + (\Theta_0 - S_0)e^{-kt}.$$

You can also gain insight into this behavior by examining the direction field for the differential equation, shown in Figure 2.7. Observe from the direction field that $\Theta(t) = S_0$ is an equilibrium solution. This analysis together with Figure 2.7 clearly suggests that the temperature of the body tends toward the temperature of the surroundings as time evolves.

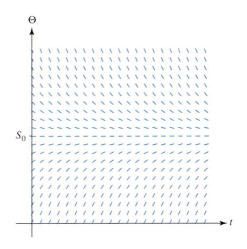

FIGURE 2.7

The direction field for the differential equation $\Theta'(t) = k[S_0 - \Theta(t)]$ that models a cooling problem. The constant function $\Theta(t) = S_0$ is an equilibrium solution. The direction field shows that the temperature of the body, $\Theta(t)$, tends toward the temperature of the surroundings, S_0.

EXERCISES

1. A tank originally contains 100 gal of fresh water. At time $t = 0$, a solution containing 0.2 lb of salt per gallon begins to flow into the tank at a rate of 3 gal/min and the well-stirred mixture flows out of the tank at the same rate.

   (a) How much salt is in the tank after 10 min?

   (b) Does the amount of salt approach a limiting value as time increases? If so, what is this limiting value and what is the limiting concentration?

2. A tank initially holds 500 gal of a brine solution having a concentration of 0.1 lb of salt per gallon. At some instant, fresh water begins to enter the tank at a rate of 10 gal/min and the well-stirred mixture leaves at the same rate. How long will it take before the concentration of salt is reduced to 0.01 lb/gal?

3. An auditorium is 100 m in length, 70 m in width, and 20 m in height. It is ventilated by a system that feeds in fresh air and draws out air at the same rate. Assume that airborne impurities form a well-stirred mixture. The ventilation system is required to reduce air pollutants present at any instant to 1% of their original concentration in 30 min. What inflow (and outflow) rate is required? What fraction of the total auditorium air volume must be vented per minute?

4. A tank originally contains 5 lb of salt dissolved in 200 gal of water. Starting at time $t = 0$, a salt solution containing 0.10 lb of salt per gallon is to be pumped into the tank at a constant rate and the well-stirred mixture is to flow out of the tank at the same rate.

(a) The pumping is to be done so that the tank contains 15 lb of salt after 20 min of pumping. At what rate must the pumping occur in order to achieve this objective?

(b) Suppose the objective is to have 25 lb of salt in the tank after 20 min. Is it possible to achieve this objective? Explain.

5. A 5000-gal aquarium is maintained with a pumping system that passes 100 gal of water per minute through the tank. To treat a certain fish malady, a soluble antibiotic is introduced into the inflow system. Assume that the inflow concentration of medicine is $10te^{-t/50}$ mg/gal, where t is measured in minutes. The well-stirred mixture flows out of the aquarium at the same rate.

(a) Solve for the amount of medicine in the tank as a function of time.

(b) What is the maximum concentration of medicine achieved by this dosing and when does it occur?

(c) For the antibiotic to be effective, its concentration must exceed 100 mg/gal for a minimum of 60 min. Was the dosing effective?

6. A tank initially contains 400 gal of fresh water. At time $t = 0$, a brine solution with a concentration of 0.1 lb of salt per gallon enters the tank at a rate of 1 gal/min and the well-stirred mixture flows out at a rate of 2 gal/min.

(a) How long does it take for the tank to become empty? (This calculation determines the time interval on which our model is valid.)

(b) How much salt is present when the tank contains 100 gal of brine?

(c) What is the maximum amount of salt present in the tank during the time interval found in part (a)? When is this maximum achieved?

7. A tank, having a capacity of 700 gal, initially contains 10 lb of salt dissolved in 100 gal of water. At time $t = 0$, a solution containing 0.5 lb of salt per gallon flows into the tank at a rate of 3 gal/min and the well-stirred mixture flows out of the tank at a rate of 2 gal/min.

(a) How much time will elapse before the tank is filled to capacity?

(b) What is the salt concentration in the tank when it contains 400 gal of solution?

(c) What is the salt concentration at the instant the tank is filled to capacity?

Exercises 8–10:

A tank, containing 1000 gal of liquid, has a brine solution entering at a constant rate of 2 gal/min. The well-stirred solution leaves the tank at the same rate. The concentration within the tank is monitored and is found to be the function of time specified. In each exercise, determine

(a) the amount of salt initially present within the tank.

(b) the inflow concentration $c_i(t)$, where $c_i(t)$ denotes the concentration of salt in the brine solution flowing into the tank.

8. $c(t) = \dfrac{e^{-t/500}}{50}$ lb/gal   9. $c(t) = \dfrac{1}{20}(1 - e^{-t/500})$ lb/gal   10. $c(t) = \dfrac{te^{-t/500}}{500}$ lb/gal

11. A 500-gal aquarium is cleansed by the recirculating filter system schematically shown in the figure. Water containing impurities is pumped out at a rate of 15 gal/min, filtered, and returned to the aquarium at the same rate. Assume that passing through the filter reduces the concentration of impurities by a fractional amount α, as shown in the figure. In other words, if the impurity concentration upon entering the filter is $c(t)$, the exit concentration is $\alpha c(t)$, where $0 < \alpha < 1$.

(a) Apply the basic conservation principle (rate of change = rate in − rate out) to obtain a differential equation for the amount of impurities present in the aquarium at time t. Assume that filtering occurs instantaneously. If the outflow concentration at any time is $c(t)$, assume that the inflow concentration at that same instant is $\alpha c(t)$.

(b) What value of filtering constant α will reduce impurity levels to 1% of their original values in a period of 3 hr?

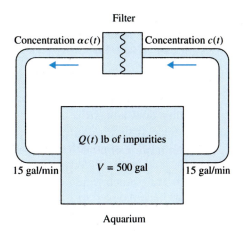

Figure for Exercise 11

12. Consider the mixing process shown in the figure. A mixing chamber initially contains 2 gal of a clear fluid. Clear fluid flows into the chamber at a rate of 10 gal/min. A dye solution having a concentration of 4 oz/gal is injected into the mixing chamber at a rate of r gal/min. When the mixing process is started, the well-stirred mixture is pumped from the chamber at a rate of $10 + r$ gal/min.

(a) Develop a mathematical model for the mixing process.

(b) The objective is to obtain a dye concentration in the outflow mixture of 1 oz/gal. What injection rate r is required to achieve this equilibrium solution? Would this equilibrium value of r be different if the fluid in the chamber at time $t = 0$ contained some dye?

(c) Assume the mixing chamber contains 2 gal of clear fluid at time $t = 0$. How long will it take for the outflow concentration to rise to within 1% of the desired concentration?

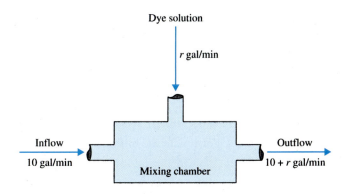

Figure for Exercise 12

13. Series Connections of Tanks   Consider the sketch shown below, where two ponds are connected and fed by a single stream flowing through them. Pond A holds 500,000 gal of water, while Pond B holds 200,000 gal of water. The fresh water stream flows through these ponds at a rate of 1000 gal/hr. Assume that at some time, say $t = 0$, 1000 lb of a toxin is spilled into Pond A and disperses rapidly enough that a well-stirred assumption is reasonable.

(a) Let $Q_A(t)$ and $Q_B(t)$ denote the amounts of toxin in Ponds A and B, respectively, at time t. Apply the "conservation of salt" principle to each pond and formulate initial value problems governing how the amount of toxin in each pond varies with time.

(b) Solve the two initial value problems for $Q_A(t)$ and $Q_B(t)$. (Because of the way the two ponds are connected by the feeder stream, the problem for Pond A can be solved independently of that for Pond B and the solution, in turn, used to specify the problem for Pond B.)

(c) What is the maximum amount of toxin present in Pond B and at what time after the spill is this maximum value reached?

(d) How much time must elapse before the concentration of toxin in both ponds has been reduced to 1 lb per million gallons?

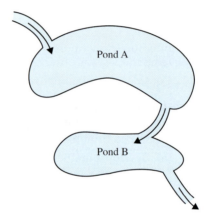

Pond A

Pond B

Figure for Exercise 13

14. Oscillating Flow Rate   A tank initially contains 10 lb of solvent in 200 gal of water. At time $t = 0$, a pulsating or oscillating flow begins. To model this flow, we assume that the input and output flow rates are both equal to $3 + \sin t$ gal/min. Thus, the flow rate oscillates between a maximum of 4 gal/min and a minimum of 2 gal/min; it repeats its pattern every $2\pi \approx 6.28$ min. Assume that the inflow concentration remains constant at 0.5 lb of solvent per gallon.

(a) Does the amount of solution in the tank, V, remain constant or not? Explain.

(b) Let $Q(t)$ denote the amount of solvent (in pounds) in the tank at time t (in minutes). Explain, on the basis of physical reasoning, whether you expect the amount of solvent in the tank to approach an equilibrium value or not. In other words, do you expect $\lim_{t \to \infty} Q(t)$ to exist and, if so, what is this limit?

(c) Formulate the initial value problem to be solved.

(d) Solve the initial value problem. Determine $\lim_{t \to \infty} Q(t)$ if it exists.

15. Oscillating Inflow Concentration   A tank initially contains 10 lb of salt dissolved in 200 gal of water. Assume that a salt solution flows into the tank at a rate of 3 gal/min and the well-stirred mixture flows out at the same rate. Assume that the inflow concentration oscillates in time, however, and is given by $c_i(t) = 0.2(1 + \sin t)$ lb of salt per gallon. Thus, as time evolves, the concentration oscillates back and forth between 0 and 0.4 lb of salt per gallon.

(a) Make a conjecture, on the basis of physical reasoning, as to whether or not you expect the amount of salt in the tank to reach a constant equilibrium value as time increases. In other words, will $\lim_{t \to \infty} Q(t)$ exist?

(b) Formulate the corresponding initial value problem.

(c) Solve the initial value problem.

(d) Plot $Q(t)$ versus t. How does the amount of salt in the tank vary as time becomes increasingly large? Is this behavior consistent with your intuition?

Assume Newton's law of cooling applies in Exercises 16–23.

16. A chef removed an apple pie from the oven and allowed it to cool at room temperature (72°F). The pie had a temperature of 350°F when removed from the oven; 10 min later, the pie had cooled to 290°F. How long will it take for the pie to cool to 120°F?

17. The temperature of an object is raised from 70°F to 150°F in 10 min when placed within a 300°F oven. What oven temperature will raise the object's temperature from 70°F to 150°F in 5 min?

18. An object, initially at 150°F, was placed in a constant-temperature bath. After 2 min, the temperature of the object had dropped to 100°F; after 4 min, the object's temperature was observed to be 90°F. What is the temperature of the bath?

Exercises 19–21:

A metal casting is placed in an environment maintained at a constant temperature, S_0. Assume the temperature of the casting varies according to Newton's law of cooling. A thermal probe attached to the casting records the temperature $\theta(t)$ listed. Use this information to determine

(a) the initial temperature of the casting.

(b) the temperature of the surroundings.

19. $\theta(t) = 70 + 270e^{-t}$ °F   **20.** $\theta(t) = 390e^{-t/2}$ °F   **21.** $\theta(t) = 80 - 40e^{-2t}$ °F

22. Food, initially at a temperature of 40°F, was placed in an oven preheated to 350°F. After 10 min in the oven, the food had warmed to 120°F. After 20 min, the food was removed from the oven and allowed to cool at room temperature (72°F). If the ideal serving temperature of the food is 110°F, when should the food be served?

23. A student performs the following experiment using two identical cups of water. One cup is removed from a refrigerator at 34°F and allowed to warm in its surroundings to room temperature (72°F). A second cup is simultaneously taken from room temperature surroundings and placed in the refrigerator to cool. The time at which each cup of water reached a temperature of 53°F is recorded. Are the two recorded times the same or not? Explain.

2.4   Population Dynamics and Radioactive Decay

In this section, we study simple population models based on first order linear equations. We also examine models for radioactive decay and applications such as radiocarbon dating.

From Discrete to Continuous Models

Many physical systems are inherently discrete in nature. For example, a population is composed of an integer number of individuals and the size of a population changes in time by integer jumps. However, we sometimes can use differential equations to model such discrete systems. Differential equation models provide useful approximations when the population is large and individual births and deaths occur frequently during our time interval of interest.

Population Models

Let $P(t)$ represent the population of a species at time t. We assume the population lives in some well-defined environment that we call a *colony*. For this introductory model we will be quite naïve, making no distinction among population members as to age, gender, health, or location within the colony.

Assume that the population can change in time through births, deaths, and migration in and out of the colony. Also assume that the population is sufficiently large to warrant describing its evolution in time by a differential equation based on the following "conservation of population" law:

$$\begin{array}{ccccc} \text{Rate of change} & = & \text{Rate of} & - & \text{Rate of} \\ \text{of population} & & \text{population increase} & & \text{population decrease.} \end{array} \qquad (1)$$

For our model, population can increase through either births or migration into the colony. Similarly, population decreases through either deaths or migration out of the colony.

To translate the principle in equation (1) into mathematics, we'll need to introduce some notation. Let r_b and r_d be positive constants representing the birth and death rates per unit population. In other words, $r_b P(t)$ represents the *rate* of population increase through births at time t. Similarly, $r_d P(t)$ represents, at time t, the *rate* of population decrease through deaths.

Let $M(t)$ denote the migration rate at time t. Note that $M(t)$ can be positive or negative, depending on whether or not the rate of immigration into the colony exceeds the exodus rate. Combining this notation with the conservation of population principle in equation (1), we obtain the differential equation

$$\frac{dP}{dt} = r_b P - r_d P + M(t),$$

or

$$\frac{dP}{dt} = (r_b - r_d)P + M(t). \qquad (2)$$

If there is no migration, then equation (2) takes the form

$$\frac{dP}{dt} = kP. \qquad (3)$$

When k is positive in equation (3), we often refer to it as the **growth rate**; if k is negative, we refer to it as the **decay rate**. As Example 1 illustrates, we can sometimes use data about the population to estimate this rate k.

**EXAMPLE
1**

For a population of about 100,000 bacteria in a petri dish, we decide to model population growth by the differential equation

$$\frac{dP}{dt} = kP.$$

Suppose, 2 days later, that the population has grown to about 150,000 bacteria. Find the growth rate k and estimate the bacteria population after 7 days.

Solution: The general solution of $P' = kP$ is

$$P(t) = Ce^{kt},$$

where, for this problem, t is measured in days. Imposing the initial condition, $P(0) = 100,000$, we obtain

$$P(t) = 100,000e^{kt}.$$

Knowing $P(2) = 150,000$, we find

$$150,000 = 100,000e^{2k}, \quad \text{or} \quad 1.5 = e^{2k}.$$

Therefore,

$$k = \tfrac{1}{2} \ln{(1.5)} \text{ days}^{-1}.$$

Having k, we arrive at an expression for the bacteria population:

$$P(t) = 100,000e^{(\ln(1.5)/2)t} = 100,000(1.5)^{t/2}.$$

At the end of 7 days, there will be about 413,000 bacteria. [The formula for $P(t)$ gives $P(7) = 413,351$.] ❖

**EXAMPLE
2**

An aquaculture firm raises catfish in ponds. At the beginning of the year, the ponds contain approximately 500,000 catfish. The net growth rate coefficient, $r_b - r_d$, is estimated to be about 6.1 per 1000 per week. The firm wants to harvest at a constant rate of R fish per week, but it also wants to increase the population to about 600,000 fish by the end of the year. Find the appropriate harvest rate, R.

Solution: Using the model in equation (2), we have

$$\frac{dP}{dt} = (r_b - r_d)P + M(t).$$

For our problem, t is measured in weeks. The growth rate coefficient, $r_b - r_d$, is 6.1 per 1000 per week. Therefore,

$$r_b - r_d = \frac{6.1}{1000} = 0.0061 \text{ week}^{-1}.$$

For this problem, the migration rate is the same as the harvest rate. Since the harvest rate is constant, we set $M(t) = -R$, where R is a positive constant. Therefore, we arrive at the following model for the catfish population:

$$\frac{dP}{dt} = 0.0061P - R, \qquad P(0) = 500,000. \tag{4}$$

(continued)

(continued)

Our objective is to choose R so that $P(52) = 600{,}000$. [In the absence of any harvesting (that is, with $R = 0$), the population $P(t)$ would be

$$P(t) = 500{,}000e^{0.0061t}.$$

So, with no harvesting, the population at the end of the year would be about $P(52) = 500{,}000e^{0.3172}$, or about 685,000 fish.]

To determine a reasonable harvest rate R, we first find the general solution for nonhomogeneous differential equation (4):

$$P(t) = Ce^{0.0061t} + \frac{R}{0.0061}.$$

Imposing the initial condition $P(0) = 500{,}000$, we obtain

$$P(t) = \left(500{,}000 - \frac{R}{0.0061}\right)e^{0.0061t} + \frac{R}{0.0061},$$

or

$$P(t) = \frac{R}{0.0061}\left(1 - e^{0.0061t}\right) + 500{,}000e^{0.0061t}.$$

We want to choose R so that $P(52) = 600{,}000$. Thus,

$$600{,}000 = \frac{R}{0.0061}\left(1 - e^{0.3172}\right) + 500{,}000e^{0.3172}.$$

Solving for R, we arrive at a harvest rate of $R = 1416$ per week. Thus, the firm can harvest at a rate of about 1400 fish per week and still see its fish population grow to about 600,000 by year's end. ❖

Radioactive Decay

The process of radioactive decay is, in many respects, like the behavior of a large population in which there are deaths but no births. At the atomic level, individual atoms of a radioactive element spontaneously undergo change, transforming themselves into new material.

At the macroscopic level (the level of continuous modeling), we'll let $Q(t)$ represent the amount of radioactive material present at time t. It has been observed empirically that the rate of decrease of radioactive material is proportional to the amount present. That is, the mathematical model is

$$\frac{dQ}{dt} = -kQ, \qquad k > 0. \tag{5}$$

We can obtain differential equation (5) by invoking the same basic conservation law, equation (1), that was used to derive the population model $P'(t) = (r_b - r_d)P(t) + M(t)$. Here, the birth rate r_b is zero since no radioactive material is being created. The death rate constant r_d has been replaced by k. Likewise, we are tacitly assuming that no material is being added or taken away, and therefore the migration rate $M(t)$ is also zero.

EXAMPLE 3

Initially, 50 mg of a radioactive substance is present. Five days later, the quantity has decreased to 43 mg. How much will remain after 30 days?

Solution: The general solution of equation (5) is

$$Q(t) = Ce^{-kt},$$

where t is measured in days. Imposing the initial condition, we obtain

$$Q(t) = 50e^{-kt}.$$

As in Example 1, we use the fact that $Q(5) = 43$ mg to determine the decay rate k:

$$k = -\tfrac{1}{5} \ln \tfrac{43}{50} = 0.03016\ldots \text{ days}^{-1}.$$

After 30 days, therefore, we expect to have

$$Q(30) = 50e^{-k30} = 20.228\ldots \text{ mg.} \ \clubsuit$$

The **half-life** of a radioactive substance is the length of time it takes a given amount of the substance to be reduced to one half of its original amount. Thus, the half-life τ is defined by the equation

$$Q(t + \tau) = \tfrac{1}{2}Q(t).$$

Since $Q(t) = Ce^{-kt}$, this equation reduces to $e^{-k\tau} = 0.5$ and hence

$$\tau = \frac{\ln 2}{k}.$$

For example, the substance in Example 3 has a half-life of about

$$\frac{\ln 2}{0.0302} = 22.95 \text{ days.}$$

If we had 300 mg of the substance at some given time, we would have about 150 mg of the substance 22.95 days later and 75 mg of the substance after 45.9 days.

EXERCISES

Assume the populations in Exercises 1–4 evolve according to the differential equation $P' = kP$.

1. A colony of bacteria initially has 10,000,000 members. After 5 days, the population increases to 11,000,000. Estimate the population after 30 days.

2. How many days will it take the colony in Exercise 1 to double in size?

3. A colony of bacteria is observed to increase in size by 30% over a 2-week period. How long will the colony take to triple its initial size?

4. A colony of bacteria initially has 100,000 members. After 6 days, the population has decreased to 80,000. At that time, 50,000 new organisms are added to replenish its size. How many bacteria will be in the colony after an additional 6 days?

5. Initially, 100 g of a radioactive material is present. After 3 days, only 75 g remains. How much additional time will it take for radioactive decay to reduce the amount present to 30 g?

6. Radioactive decay reduces an initial amount of material by 20% over a period of 90 days. What is the half-life of this material?

7. A radioactive material has a half-life of 2 weeks. After 5 weeks, 20 g of the material is seen to remain. How much material was initially present?

8. After 30 days of radioactive decay, 100 mg of a radioactive substance was observed to remain. After 120 days, only 30 mg of this substance was left.

   (a) How much of the substance was initially present?

   (b) What is the half-life of this radioactive substance?

   (c) How long will it take before only 1% of the original amount remains?

9. Initially, 100 g of material A and 50 g of material B were present. Material A is known to have a half-life of 30 days, while material B has a half-life of 90 days. At some later time it was observed that equal amounts of the two radioactive materials were present. When was this observation made?

10. The evolution of a population with constant migration rate M is described by the initial value problem

$$\frac{dP}{dt} = kP + M, \qquad P(0) = P_0.$$

   (a) Solve this initial value problem; assume k is constant.

   (b) Examine the solution $P(t)$ and determine the relation between the constants k and M that will result in $P(t)$ remaining constant in time and equal to P_0. Explain, on physical grounds, why the two constants k and M must have opposite signs to achieve this constant equilibrium solution for $P(t)$.

11. Assume that the population of fish in an aquaculture farm can be modeled by the differential equation $dP/dt = kP + M(t)$, where k is a positive constant. The manager wants to operate the farm in such a way that the fish population remains constant from year to year. The following two harvesting strategies are under consideration.

   Strategy I: Harvest the fish at a constant and continuous rate so that the population itself remains constant in time. Therefore, $P(t)$ would be a constant and $M(t)$ would be a negative constant; call it $-M$. (Refer to Exercise 10.)

   Strategy II: Let the fish population evolve without harvesting throughout the year, and then harvest the excess population at year's end to return the population to its value at the year's beginning.

   (a) Determine the number of fish harvested annually with each of the two strategies. Express your answer in terms of the population at year's beginning; call it P_0. (Assume that the units of k are year^{-1}.)

   (b) Suppose, as in Example 2, that $P_0 = 500,000$ fish and $k = 0.0061 \times 52 = 0.3172$ year^{-1}. Assume further that Strategy I, with its steady harvesting and return, provides the farm with a net profit of \$0.75/fish while Strategy II provides a profit of only \$0.60/fish. Which harvesting strategy will ultimately prove more profitable to the farm?

12. Assume that two colonies each have P_0 members at time $t = 0$ and that each evolves with a constant relative birth rate $k = r_b - r_d$. For colony 1, assume that individuals migrate into the colony at a rate of M individuals per unit time. Assume that this immigration occurs for $0 \le t \le 1$ and ceases thereafter. For colony 2, assume that a similar migration pattern occurs but is delayed by one unit of time; that is, individuals migrate at a rate of M individuals per unit time, $1 \le t \le 2$. Suppose we are interested in comparing the evolution of these two populations over the time

interval $0 \leq t \leq 2$. The initial value problems governing the two populations are

$$\frac{dP_1}{dt} = kP_1 + M_1(t), \qquad P_1(0) = P_0, \qquad M_1(t) = \begin{cases} 1, & 0 \leq t \leq 1 \\ 0, & 1 < t \leq 2; \end{cases}$$

$$\frac{dP_2}{dt} = kP_2 + M_2(t), \qquad P_2(0) = P_0, \qquad M_2(t) = \begin{cases} 0, & 0 \leq t < 1 \\ 1, & 1 \leq t \leq 2. \end{cases}$$

(a) Solve both problems to determine P_1 and P_2 at time $t = 2$.

(b) Show that $P_1(2) - P_2(2) = (M/k)\left(e^k - 1\right)^2$. If $k > 0$, which population is larger at time $t = 2$? What happens if $k < 0$?

(c) Suppose that there is a fixed number of individuals that can be introduced into a population at any time through migration and that the objective is to maximize the population at some fixed future time. Do the calculations performed in this problem suggest a strategy (based on the relative birth rate) for accomplishing this?

13. Radiocarbon Dating Carbon-14 is a radioactive isotope of carbon produced in the upper atmosphere by radiation from the sun. Plants absorb carbon dioxide from the air, and living organisms, in turn, eat the plants. The ratio of normal carbon (carbon-12) to carbon-14 in the air and in living things at any given time is nearly constant. When a living creature dies, however, the carbon-14 begins to decrease as a result of radioactive decay. By comparing the amounts of carbon-14 and carbon-12 present, the amount of carbon-14 that has decayed can therefore be ascertained.

Let $Q(t)$ denote the amount of carbon-14 present at time t after death. If we assume its behavior is modeled by the differential equation $Q'(t) = -kQ(t)$, then $Q(t) = Q(0)e^{-kt}$. Knowing the half-life of carbon-14, we can determine the constant k. Given a specimen to be dated, we can measure its radioactive content and deduce $Q(t)$. Knowing the amount of carbon-12 present enables us to determine $Q(0)$. Therefore, we can use the solution of the differential equation $Q(t) = Q(0)e^{-kt}$ to deduce the age, t, of the radioactive sample.

(a) The half-life of carbon-14 is nominally 5730 years. Suppose remains have been found in which it is estimated that 30% of the original amount of carbon-14 is present. Estimate the age of the remains.

(b) The half-life of carbon-14 is not known precisely. Let us assume that its half-life is 5730 ± 30 years. Determine how this half-life uncertainty affects the age estimate you computed in (a); that is, what is the corresponding uncertainty in the age of the remains?

(c) It is claimed that radiocarbon dating cannot be used to date objects older than about 60,000 years. To appreciate this practical limitation, compute the ratio $Q(60,000)/Q(0)$, assuming a half-life of 5730 years.

14. Suppose that 50 mg of a radioactive substance, having a half-life of 3 years, is initially present. More of this material is to be added at a constant rate so that 100 mg of the substance is present at the end of 2 years. At what constant rate must this radioactive material be added?

15. Iodine-131, a fission product created in nuclear reactors and nuclear weapons explosions, has a half-life of 8 days. If 30 micrograms of iodine-131 is detected in a tissue site 3 days after ingestion of the radioactive substance, how much was originally present?

16. U-238, the dominant isotope of natural uranium, has a half-life of roughly 4 billion years. Determine how long it takes for a sample to be reduced in amount by 1% through radioactive decay.

2.5   First Order Nonlinear Differential Equations

Thus far we have studied first order linear differential equations, equations of the form

$$y' + p(t)y = g(t).$$

We now consider first order *non*linear differential equations. The term "nonlinear differential equation" encompasses all differential equations that are not linear. In particular, a **first order nonlinear differential equation** has the form

$$y' = f(t, y),$$

where $f(t, y) \neq -p(t)y + g(t)$.

   Three examples of first order nonlinear differential equations are

$$\text{(a) } y' = t^2 + y^2 \qquad \text{(b) } y' = t + \cos y \qquad \text{(c) } y' = \frac{t}{y}.$$

Nonlinear differential equations arise in many models of physical phenomena, such as population dynamics influenced by environmental constraints and one-dimensional motion in the presence of air resistance. We'll consider such applications in Sections 2.8 and 2.9.

   Because the set of nonlinear differential equations is so diverse, the type of theoretical statement that can be made about the behavior of their solutions is less comprehensive than that made in Theorem 2.1 for linear equations. In addition, unlike the situation for linear equations, we cannot derive a general solution procedure that applies to the entire class of nonlinear equations. We therefore concentrate on certain subclasses of nonlinear differential equations for which solution procedures do exist.

Existence and Uniqueness

We begin our study of nonlinear equations by considering questions of existence and uniqueness for initial value problems. In particular, given the initial value problem

$$y' = f(t, y), \qquad y(t_0) = y_0, \tag{1}$$

what conditions on the function $f(t, y)$ guarantee that problem (1) has a unique solution? On what t-interval does this unique solution exist? The answers to these questions provide a framework within which we can work.

   For example, if we use some special technique to find a solution of problem (1), then it is essential to know whether the solution we found is the *only* solution. In fact, if the initial value problem does not have a unique solution, then it probably is not a good mathematical model for the physical phenomenon under consideration.

   Existence and uniqueness are also important considerations if we need to use numerical methods to approximate a solution. For example, if a numerical solution "blows up," we want to know whether this behavior arises from inaccuracies in the numerical method or correctly depicts the behavior of the solution.

We now state a theorem that guarantees the existence of a unique solution to an initial value problem. The proof of this theorem is usually studied in a more advanced course in differential equations; we do not give a proof here.

Theorem 2.2

Let R be the open rectangle defined by $a < t < b$, $\alpha < y < \beta$. Let $f(t, y)$ be a function of two variables defined on R, where $f(t, y)$ and the partial derivative $\partial f / \partial y$ are continuous on R. Suppose (t_0, y_0) is a point in R. Then there is an open t-interval (c, d), contained in (a, b) and containing t_0, in which there exists a unique solution of the initial value problem

$$y' = f(t, y), \qquad y(t_0) = y_0.$$

A typical open rectangle R, initial point (t_0, y_0), and interval (c, d) are shown in Figure 2.8. (The rectangle R is called an **open rectangle** because it does not contain the four line segments forming its boundary.)

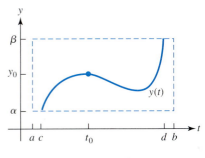

FIGURE 2.8

The open rectangle R, defined by $a < t < b$, $\alpha < y < \beta$, contains the initial point (t_0, y_0). If the hypotheses of Theorem 2.2 hold on R, we are guaranteed a unique solution to the initial value problem on some open interval (c, d).

Although presented in the context of nonlinear differential equations, Theorem 2.2 makes no distinction between linear and nonlinear differential equations. It applies to linear first order equations where

$$f(t, y) = -p(t)y + g(t),$$

as well as to nonlinear first order equations.

Two important observations can be made about Theorem 2.2.

1. The hypotheses of Theorem 2.2 are a natural generalization of those made in Theorem 2.1 for linear differential equations. That is, if $f(t, y) = -p(t)y + g(t)$, then $\partial f / \partial y = -p(t)$. Therefore, requiring $f(t, y)$ and $\partial f / \partial y$ to be continuous on the rectangle R means that any linear differential equation satisfying the hypotheses of Theorem 2.2 also satisfies the hypotheses of Theorem 2.1. Conversely, a linear differential equation satisfying the hypotheses of Theorem 2.1 also satisfies the hypotheses of Theorem 2.2.

2. The conclusions of Theorem 2.2, however, differ substantially from those of Theorem 2.1. Since we have broadened our perspective to encompass nonlinear differential equations, the corresponding conclusions of Theorem 2.2 are weaker than those of Theorem 2.1. Theorem 2.1 guarantees existence and uniqueness on the entire (a, b) interval. Theorem 2.2 guarantees existence and uniqueness only on *some* subinterval (c, d) of (a, b) containing t_0; it does *not* guarantee existence and uniqueness on the entire (a, b) interval. Moreover, Theorem 2.2 gives no insight into how large (c, d) is or how we might go about estimating it.

Although Theorem 2.2 leaves many questions unanswered, it does provide us with the framework we need to study solution techniques for certain classes of nonlinear differential equations. Examining these special cases will give us valuable insight into the behavior of solutions of nonlinear equations.

Autonomous Differential Equations

First order autonomous equations have the form $y' = f(y)$. The right-hand side of the differential equation does not explicitly depend on the independent variable t. Solution curves for an autonomous differential equation have the important geometric property that they can be translated parallel to the t-axis.

As an example, consider the autonomous equation

$$y' = y(2 - y).$$

The direction field for this equation, along with portions of some solution curves, is shown in Figure 2.9. As observed in Section 1.3, the slopes of the direction field filaments for an autonomous equation remain constant along horizontal lines. For instance (see Figure 2.9), at every point along the line $y = 1$, the direction field filaments have slope equal to 1. Similarly, at every point along the line $y = 3$, the direction field filaments have slope equal to -3.

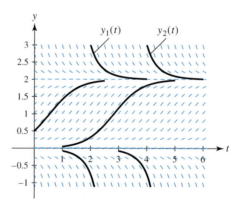

FIGURE 2.9

The direction field for the autonomous equation $y' = y(2 - y)$, together with portions of some typical solutions. Notice that the graph of $y_1(t)$, when translated to the right, looks as though it coincides with the graph of $y_2(t)$.

Besides illustrating that horizontal lines are isoclines for the autonomous equation, Figure 2.9 illustrates an important property of solutions to autonomous differential equations. That is, it looks as though the graph of $y_1(t)$, when translated about 2 units to the right, will fall exactly on the graph of $y_2(t)$. This is indeed the case, and we show in Theorem 2.3 that the solution $y_2(t)$ is related to the solution $y_1(t)$ by

$$y_2(t) = y_1(t - c),$$

where c is a constant.

Theorem 2.3

Let the initial value problem

$$y' = f(y), \qquad y(0) = y_0$$

satisfy the conditions of Theorem 2.2, and let $y_1(t)$ be the unique solution, where the interval of existence for $y_1(t)$ is $a < t < b$, with $a < 0 < b$.
  Consider the initial value problem

$$y' = f(y), \qquad y(t_0) = y_0. \tag{2}$$

Then the function $y_2(t)$ defined by $y_2(t) = y_1(t - t_0)$ is the unique solution of initial value problem (2) and has an interval of existence

$$t_0 + a < t < t_0 + b.$$

● **PROOF:**   Since $y_1(t)$ is defined for $a < t < b$, we know that $y_2(t) = y_1(t - t_0)$ is defined for $a < t - t_0 < b$ and hence for $t_0 + a < t < t_0 + b$. We next observe that $y_2(t)$ satisfies the initial condition of (2), since $y_2(t_0) = y_1(t_0 - t_0) = y_1(0) = y_0$. Therefore, to complete the proof of Theorem 2.3, we need to show that $y_2(t)$ is a solution of the differential equation $y' = f(y)$.

Using the definition of $y_2(t)$, the chain rule, and the fact that $y_1(t)$ solves the differential equation $y' = f(y)$, we have

$$y_2'(t) = \frac{d}{dt} y_1(t - t_0) = y_1'(t - t_0) \frac{d}{dt} (t - t_0) = y_1'(t - t_0) = f(y_1(t - t_0)) = f(y_2(t)).$$

Therefore, the function $y_2(t) = y_1(t - t_0)$ is a solution of the initial value problem. ●

The important conclusion to be reached from Theorem 2.3 is that the solution of the autonomous initial value problem $y' = f(y), y(t_0) = y_0$ depends on the independent variable t and the initial condition time t_0 as a function of the combination $t - t_0$. What matters is time t measured relative to the initial time t_0. As a simple example, recall that the solution of the linear autonomous equation $y' = ky, y(t_0) = y_0$ is

$$y(t) = y_0 e^{k(t - t_0)}.$$

Bernoulli Equations

We conclude this section by studying a class of nonlinear differential equations known as Bernoulli equations. By making an appropriate change of dependent variable, these nonlinear equations can be transformed into first order linear equations and solved using the techniques described in Section 2.2.

Bernoulli differential equations[4] are first order differential equations having the special structure

$$\frac{dy}{dt} + p(t)y = q(t)y^n,$$

where n is an integer. We do not consider $n = 0$ and $n = 1$ since, in those cases, the Bernoulli equation is a first order linear equation.

A simple example of a Bernoulli equation is

$$\frac{dy}{dt} + e^{2t}y = y^3 \sin t.$$

Bernoulli equations arise in applications such as population models and models of one-dimensional motion influenced by drag forces.

Consider a Bernoulli equation

$$\frac{dy}{dt} + p(t)y = q(t)y^n, \tag{3}$$

where n is a given integer ($n \neq 0$ and $n \neq 1$). We look for a change of dependent variable of the form $v(t) = y(t)^m$, where m is a constant to be determined. Using the chain rule, we have

$$\frac{dv}{dt} = my^{m-1}\frac{dy}{dt}$$

and therefore

$$\frac{dy}{dt} = m^{-1}y^{1-m}\frac{dv}{dt} = m^{-1}v^{(1-m)/m}\frac{dv}{dt}.$$

Equation (3) transforms into the following differential equation for $v(t)$:

$$\frac{dv}{dt} + mp(t)v = mq(t)v^{(m+n-1)/m}. \tag{4}$$

At first glance, it may seem that our change of variables has accomplished little. The structure of equation (4) seems similar to what we started with. However, we are free to choose the constant m. In particular, if we select $m = 1 - n$, then equation (4) reduces to the first order linear equation

$$\frac{dv}{dt} + (1-n)p(t)v = (1-n)q(t). \tag{5}$$

We can solve this equation for $v(t)$ and then obtain the desired solution, $y(t) = v(t)^{1/(1-n)}$.

---

[4]Jacob Bernoulli (1654–1705) is one of eight members of the extended Bernoulli family remembered for their contributions to mathematics and science. While at the University of Basel, Jacob made important contributions to such areas as infinite series, probability theory, geometry, and differential equations. In 1696 he solved the differential equation that now bears his name. Jacob always had a particular fascination for the logarithmic spiral and requested that this curve be carved on his tombstone.

E X A M P L E

1

Solve the initial value problem

$$y' + y = ty^3, \qquad y(0) = 2.$$

Solution: The differential equation is a Bernoulli equation with $n = 3$.
We make the change of dependent variable $v = y^{1-n}$ or, since $n = 3, v = y^{-2}$.
The initial value problem for $v(t)$ then becomes [recall equation (5)]

$$v' - 2v = -2t, \qquad v(0) = y(0)^{-2} = \tfrac{1}{4}.$$

The general solution is

$$v = Ce^{2t} + \left(t + \tfrac{1}{2}\right).$$

Imposing the initial condition, we have

$$v = -\tfrac{1}{4}e^{2t} + \left(t + \tfrac{1}{2}\right).$$

Finally, since $y = v^{-1/2}$, we arrive at the desired solution

$$y = \left[-\tfrac{1}{4}e^{2t} + \left(t + \tfrac{1}{2}\right)\right]^{-1/2}. ❖$$

EXERCISES

Exercises 1–8:

For the given initial value problem,

(a) Rewrite the differential equation, if necessary, to obtain the form

$$y' = f(t, y), \qquad y(t_0) = y_0.$$

Identify the function $f(t, y)$.

(b) Compute $\partial f/\partial y$. Determine where in the ty-plane both $f(t, y)$ and $\partial f/\partial y$ are continuous.

(c) Determine the largest open rectangle in the ty-plane that contains the point (t_0, y_0) and in which the hypotheses of Theorem 2.2 are satisfied.

1. $3y' + 2t \cos y = 1, \quad y(\pi/2) = -1$

2. $3ty' + 2 \cos y = 1, \quad y(\pi/2) = -1$

3. $2t + (1 + y^2)y' = 0, \quad y(1) = 1$

4. $2t + (1 + y^3)y' = 0, \quad y(1) = 1$

5. $y' + ty^{1/3} = \tan t, \quad y(-1) = 1$

6. $(y^2 - 9)y' + e^{-y} = t^2, \quad y(2) = 2$

7. $(\cos y)y' = 2 + \tan t, \quad y(0) = 0$

8. $(\cos 2t)y' = 2 + \tan y, \quad y(\pi) = 0$

9. Consider the initial value problem $t^2 y' - y^2 = 0, y(1) = 1$.

(a) Determine the largest open rectangle in the ty-plane, containing the point $(t_0, y_0) = (1, 1)$, in which the hypotheses of Theorem 2.2 are satisfied.

(b) A solution of the initial value problem is $y(t) = t$. This solution exists on $-\infty < t < \infty$. Does this fact contradict Theorem 2.2? Explain your answer.

10. The solution of the initial value problem $y' = f(y), y(0) = 8$ is known to be $y(t) = (4 + t)^{3/2}$. Let $\bar{y}(t)$ represent the solution of the initial value problem $y' = f(y), y(t_0) = 8$. Suppose we know that $\bar{y}(0) = 1$. What is t_0?

11. The solution of the initial value problem $y' = f(y), y(0) = 2$ is known to be $y(t) = 2/\sqrt{1 - t}$. Let $\bar{y}(t)$ represent the solution of the initial value problem $y' = f(y)$, $y(1) = 2$. What is the value of $\bar{y}(0)$?

12. The graph shows the solution of $y' = -1/(2y)$, $y(0) = 1$. Use the graph to answer questions (a) and (b).

   (a) If $z_1(t)$ is the solution of $z_1' = -1/(2z_1)$, $z_1(-2) = 1$, what is $z_1(-5)$?

   (b) If $z_2(t)$ is the solution of $z_2' = -1/(2z_2)$, $z_2(2) = 1$, what is $z_2(3)$?

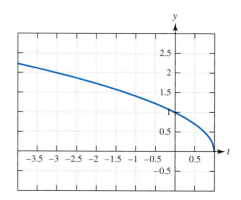

Figure for Exercise 12

Exercises 13–19:

(a) Solve the initial value problem by

   (i)   transforming the given Bernoulli differential equation and initial condition into a first order linear differential equation with its corresponding initial condition,

   (ii)   solving the new initial value problem,

   (iii) transforming back to the dependent variable of interest.

(b) Determine the interval of existence.

13. $y' = y(2 - y)$, $y(0) = 1$ **14.** $y' = 2ty(1 - y)$, $y(0) = -1$

15. $y' = -y + e^t y^2$, $y(-1) = -1$ **16.** $y' = y + y^{-1}$, $y(0) = -1$

17. $ty' + y = t^3 y^{-2}$, $y(1) = 1$ **18.** $y' - y = ty^{1/3}$, $y(0) = -8$

19. $y' = -(y + 1) + t(y + 1)^{-2}$, $y(0) = 1$ [Hint: Let $z = y + 1$.]

20. The initial value problem $y' + y = q(t)y^2$, $y(0) = y_0$ is known to have solution

$$y(t) = \frac{3}{(1 - 3t)e^t}$$

on the interval $-\infty < t < \frac{1}{3}$. Determine the coefficient function $q(t)$ and the initial value y_0.

2.6   Separable First Order Equations

In Section 2.2, we obtained an explicit representation for the solution of a first order linear differential equation; recall equation (10) in Section 2.2. By contrast, there is no all-encompassing technique that leads to an explicit representation for the solution of a first order nonlinear differential equation.

    For certain types of nonlinear equations, however, techniques have been discovered that give us some information about the solution. We have already seen one type, the Bernoulli equations, in the previous section. In this section we study another type, called separable differential equations.

Separable Equations

The term **separable differential equation** is used to describe any first order differential equation that can be put into the form

$$n(y) \frac{dy}{dt} + m(t) = 0.$$

For example, the differential equations

$$\text{(a)} \ \frac{dy}{dt} + t^2 \sin y = 0 \quad \text{and} \quad \text{(b)} \ y' + e^{t+y} = e^y \sin t$$

are separable since they can be rewritten (respectively) as

$$\text{(a')} \ \csc y \ \frac{dy}{dt} + t^2 = 0 \quad \text{and} \quad \text{(b')} \ e^{-y}y' + [e^t - \sin t] = 0.$$

A simple example of a nonseparable differential equation is

$$y' = 2ty^2 + 1.$$

The structure of a separable differential equation,

$$n(y) \frac{dy}{dt} + m(t) = 0,$$

gives the equation its name. The first term is the product of dy/dt and a term $n(y)$ that involves only the dependent variable y. The second term, $m(t)$, involves only the independent variable t. In this sense, the variables "separate."

Solving a Separable Differential Equation

We can get some information about the solution of a separable equation by "reversing the chain rule." We illustrate this technique in Example 1 and then describe the general procedure.

E X A M P L E
1

Solve the initial value problem

$$y' = 2ty^2, \qquad y(0) = 1.$$

Solution: First, notice that $f(t, y) = 2ty^2$ is continuous on the entire ty-plane, as is the partial derivative $\partial f/\partial y = 4ty$. Therefore, the conditions of Theorem 2.2 are satisfied on the open rectangle R defined by $-\infty < t < \infty$, $-\infty < y < \infty$. Theorem 2.2 guarantees the existence of a unique solution of the initial value problem, but it provides no insight into the interval of existence of the solution.

The differential equation is separable. It can be rewritten as

$$\frac{y'}{y^2} - 2t = 0.$$

(continued)

(continued)

To emphasize the fact that y is the dependent variable, we express the equation as

$$\frac{y'(t)}{y(t)^2} - 2t = 0.$$

Taking antiderivatives, we find

$$\int \frac{y'(t)}{y(t)^2}\, dt - \int 2t\, dt = C.$$

Evaluating the integrals on the left-hand side yields

$$\frac{-1}{y(t)} - t^2 = C.$$

Solving for $y(t)$, we obtain a family of solutions

$$y(t) = \frac{-1}{t^2 + C}.$$

Imposing the initial condition, $y(0) = 1$, yields the unique solution of the initial value problem:

$$y(t) = \frac{1}{1 - t^2}.$$

Having determined the solution, we are now able to see that the interval of existence is $-1 < t < 1$. ❖

The solution process of Example 1 can be viewed as reversing the chain rule. To explain, we return to the general separable differential equation,

$$n(y)\frac{dy}{dt} + m(t) = 0. \tag{1}$$

Let y be a differentiable function of t, and let $N(y)$ be any antiderivative of $n(y)$. By the chain rule,

$$\frac{d}{dt}N(y) = n(y)\frac{dy}{dt}. \tag{2a}$$

Similarly, let $M(t)$ be any antiderivative of $m(t)$,

$$\frac{d}{dt}M(t) = m(t). \tag{2b}$$

Combining (2a) and (2b), we can rewrite the left-hand side of equation (1) as

$$n(y)\frac{dy}{dt} + m(t) = \frac{d}{dt}N(y) + \frac{d}{dt}M(t) = \frac{d}{dt}[N(y) + M(t)].$$

Therefore, equation (1) reduces to

$$\frac{d}{dt}[N(y) + M(t)] = 0.$$

Since the term $N(y) + M(t)$ is a function of t whose derivative vanishes identically, we have

$$N(y) + M(t) = C,$$ **(3)**

where C is an arbitrary constant.

Equation (3) provides us with information about the solution $y(t)$. It is not an explicit expression for the solution; rather, it is an equation that the solution must satisfy. An equation in y and t, such as (3), is called an **implicit solution**. Sometimes (as in Example 1 and in Example 2 below) we can "unravel" the implicit solution and solve for $y(t)$ as an explicit function of the independent variable t. In other cases, we must be content with the implicit solution.

Whether or not we can unravel the implicit solution given by equation (3), we can always determine the constant C by imposing the initial condition $y(t_0) = y_0$, finding

$$C = N(y_0) + M(t_0).$$

E X A M P L E

2

Solve the initial value problem

$$\frac{dy}{dt} = -\frac{t}{y}, \qquad y(0) = -2.$$

Solution: Separating the variables, we obtain

$$y\frac{dy}{dt} + t = 0.$$

Integrating, we find an implicit solution

$$\frac{y^2}{2} + \frac{t^2}{2} = C.$$

Imposing the initial condition, we find $C = 2$. Thus, an implicit solution of the initial value problem is given by

$$y^2 + t^2 = 4.$$

Suppose we want an explicit solution. Solving the equation above, we find

$$y = \pm\sqrt{4 - t^2}.$$

Which root should we take? To satisfy the initial condition, $y(0) = -2$, we must take the negative root. Thus, the solution is

$$y = -\sqrt{4 - t^2}.$$ **(4)**

This choice of roots is also obvious geometrically, since the graph of $y^2 + t^2 = 4$ is a circle of radius 2 in the ty-plane, as shown in Figure 2.10. The solution of the initial value problem has the lower semicircle as its graph. The function given by equation (4) is defined and continuous on $[-2, 2]$. It is differentiable and satisfies the differential equation on the open interval $(-2, 2)$.

(continued)

(continued)

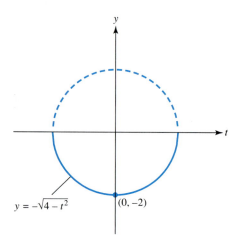

$y = -\sqrt{4 - t^2}$ (0, −2)

FIGURE 2.10

The implicit solution of the initial value problem in Example 2 is $y^2 + t^2 = 4$; its graph is a circle of radius 2. The explicit solution of the initial value problem is $y = -\sqrt{4 - t^2}$; its graph is the lower semicircle. ❖

E X A M P L E

3

Solve the initial value problem

$$(2 + \sin y)y' + t = 0, \qquad y(2) = 0.$$

Solution: Computing the antiderivatives yields

$$2y - \cos y + \frac{t^2}{2} = C.$$

Imposing the initial condition, we obtain the implicit solution

$$2y - \cos y + \frac{t^2}{2} = 1. \tag{5}$$

Although we cannot unravel this equation and determine an explicit solution, we can plot the graph of equation (5); see Figure 2.11. Observe that if the cosine term in (5) were absent, the graph would be that of a concave-down parabola. Loosely speaking, therefore, the cosine term creates the ripples displayed by the graph in Figure 2.11. ❖

Differences between Linear and Nonlinear Differential Equations

We can use Examples 1–3 to make several points about Theorem 2.2 and to illustrate some of the differences between nonlinear and linear differential equations.

1. *The Interval of Existence May Not Be Obvious*   If the coefficient functions for a *linear* differential equation are continuous on an interval (a, b), where (a, b) contains the initial point t_0, then a unique solution

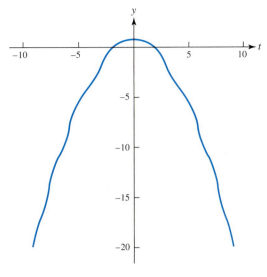

FIGURE 2.11

The initial value problem posed in Example 3 has an implicitly defined solution given by equation (5). The graph of equation (5) is shown above.

of the initial value problem $y' + p(t)y = g(t), y(t_0) = y_0$ exists and is defined on all of (a, b).

By way of contrast, consider the initial value problem in Example 1. The function $f(t, y) = 2ty^2$ is continuous everywhere and is about as nice a nonlinear function as we can expect. However, the solution of the initial value problem has vertical asymptotes at $t = -1$ and at $t = 1$. Note that we cannot predict that the interval of existence is $(-1, 1)$ by simply looking at the equation $y' = 2ty^2$. In fact, if the initial condition is changed from $y(0) = 1$ to $y(0) = -1$, then the interval of existence changes to $(-\infty, \infty)$. (See Exercise 4.)

Example 2 provides another illustration. The interval of existence in Example 2 is $-2 < t < 2$. Suppose we leave the initial condition alone and simply change the sign on the right-hand side of the differential equation. This change produces a new initial value problem

$$\frac{dy}{dt} = \frac{t}{y}, \qquad y(0) = -2.$$

In this case, the solution is defined for all t. (See Exercise 12.)

In each of these examples, a harmless-looking change in the differential equation or in the initial condition leads to a pronounced change in the nature of the solution and in the interval of existence.

2. *There May Not Be a Single Formula That Gives All Solutions*   Note that the family of solutions found in Example 1,

$$y(t) = \frac{-1}{t^2 + C}, \tag{6}$$

does not include the zero function. The zero function is, however, a solution of $y' = 2ty^2$. In particular, given the initial value problem $y' = 2ty^2$,

$y(0) = 0$, there is no choice for C in equation (6) that yields the unique solution, $y(t) = 0$, that is guaranteed by Theorem 2.2.

3. *We May Have to Be Content with Implicitly Defined Solutions*   In Examples 1 and 2 we were able to find an explicit formula for the solution, $y(t)$, of the initial value problem. However, as Example 3 illustrates, it may not be possible to obtain an explicit formula for the solution. By contrast [see equation (10) in Section 2.2], there is an explicit formula for the solution of any first order *linear* differential equation.

EXERCISES

Exercises 1–17:

(a) Obtain an implicit solution and, if possible, an explicit solution of the initial value problem.

(b) If you can find an explicit solution of the problem, determine the t-interval of existence.

1. $y \dfrac{dy}{dt} - \sin t = 0, \quad y(\pi/2) = -2$

2. $\dfrac{dy}{dt} = \dfrac{1}{y^2}, \quad y(1) = 2$

3. $y' + \dfrac{1}{y+1} = 0, \quad y(1) = 0$

4. $y' - 2ty^2 = 0, \quad y(0) = -1$

5. $y' - ty^3 = 0, \quad y(0) = 2$

6. $\dfrac{dy}{dt} + e^y t = e^y \sin t, \quad y(0) = 0$

7. $\dfrac{dy}{dt} = 1 + y^2, \quad y(\pi/4) = -1$

8. $t^2 y' + \sec y = 0, \quad y(-1) = 0$

9. $\dfrac{dy}{dt} = t - ty^2, \quad y(0) = \dfrac{1}{2}$

10. $3y^2 \dfrac{dy}{dt} + 2t = 1, \quad y(-1) = -1$

11. $\dfrac{dy}{dt} = e^{t-y}, \quad y(0) = 1$

12. $\dfrac{dy}{dt} = \dfrac{t}{y}, \quad y(0) = -2$

13. $e^t y' + (\cos y)^2 = 0, \quad y(0) = \pi/4$

14. $(2y - \sin y)y' + t = \sin t, \quad y(0) = 0$

15. $e^y y' + \dfrac{t}{y+1} = \dfrac{2}{y+1}, \quad y(1) = 2$

16. $(\ln y)y' + t = 1, \quad y(3) = e$

17. $e^y y' = 1 + e^y, \quad y(2) = 0$

18. For what values of the constants α, y_0 and integer n is the function $y(t) = (4 + t)^{-1/2}$ a solution of the initial value problem

$$y' + \alpha y^n = 0, \qquad y(0) = y_0?$$

19. For what values of the constants α, y_0 and integer n is the function $y(t) = 6/(5 + t^4)$ a solution of the initial value problem

$$y' + \alpha t^n y^2 = 0, \qquad y(1) = y_0?$$

20. State an initial value problem, with initial condition imposed at $t_0 = 2$, having implicit solution $y^3 + t^2 + \sin y = 4$.

21. State an initial value problem, with initial condition imposed at $t_0 = 0$, having implicit solution $ye^y + t^2 = \sin t$.

22. Consider the initial value problem

$$y' = 2y^2, \qquad y(0) = y_0.$$

For what value(s) y_0 will the solution have a vertical asymptote at $t = 4$ and a t-interval of existence $-\infty < t < 4$?

23. (a) A first order autonomous differential equation has the form $y' = f(y)$. Show that such an equation is separable.

(b) Solve $y' = y(2 - y), \quad y(2) = 1$.

Exercises 24–26:

A differential equation of the form

$$y' = p_1(t) + p_2(t)y + p_3(t)y^2$$

is known as a **Riccati equation**.[5] Equations of this form arise when we model one-dimensional motion with air resistance; see Section 2.9. In general, this equation is not separable. In certain cases, however (such as in Exercises 24–26), the equation does assume a separable form.

Solve the given initial value problem and determine the t-interval of existence.

24. $y' = 2 + 2y + y^2, \quad y(0) = 0$ **25.** $y' = t(5 + 4y + y^2), \quad y(0) = -3$

26. $y' = (y^2 + 2y + 1)\sin t, \quad y(0) = 0$

27. Let $Q(t)$ represent the amount of a certain reactant present at time t. Suppose that the rate of decrease of $Q(t)$ is proportional to $Q^3(t)$. That is, $Q' = -kQ^3$, where k is a positive constant of proportionality. How long will it take for the reactant to be reduced to one half of its original amount? Recall that, in problems of radioactive decay where the differential equation has the form $Q' = -kQ$, the half-life was independent of the amount of material initially present. What happens in this case? Does half-life depend on $Q(0)$, the amount initially present?

28. The rate of decrease of a reactant is proportional to the square of the amount present. During a particular reaction, 40% of the initial amount of this chemical remained after 10 sec. How long will it take before only 25% of the initial amount remains?

29. Consider the differential equation $y' = |y|$.

(a) Is this differential equation linear or nonlinear? Is the differential equation separable?

(b) A student solves the two initial value problems $y' = |y|, y(0) = 1$ and $y' = y$, $y(0) = 1$ and then graphs the two solution curves on the interval $-1 \le t \le 1$. Sketch what she observes.

(c) She next solves both problems with initial condition $y(0) = -1$. Sketch what she observes in this case.

30. Consider the following autonomous first order differential equations:

$$y' = -y^2, \qquad y' = y^3, \qquad y' = y(4 - y).$$

Match each of these equations with one of the solution graphs shown. Note that each solution satisfies the initial condition $y(0) = 1$. Can you match them without solving the differential equations?

---

[5] Jacopo Riccati (1676–1754) worked on many differential equations, including the one that now bears his name. His work in hydraulics proved useful to his native city of Venice.

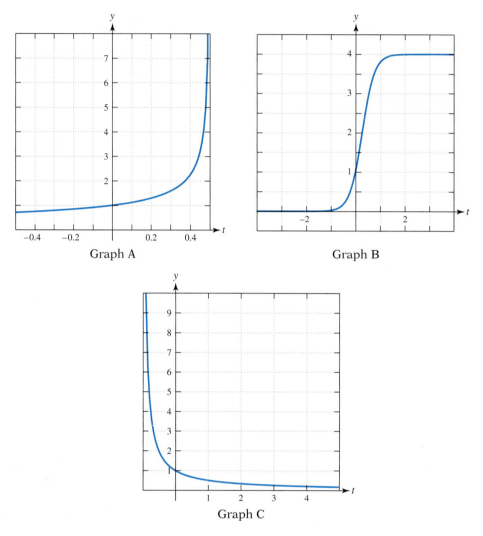

Graph A

Graph B

Graph C

Figure for Exercise 30

31. Let $S(t)$ represent the amount of a chemical reactant present at time t, $t \geq 0$. Assume that $S(t)$ can be determined by solving the initial value problem

$$S' = -\frac{\alpha S}{K + S}, \qquad S(0) = S_0,$$

where α, K, and S_0 are positive constants. Obtain an implicit solution of the initial value problem. (The differential equation, often referred to as the Michaelis-Menten equation, arises in the study of biochemical reactions.)

32. Change of Dependent Variable   Sometimes a change of variable can be used to convert a differential equation $y' = f(t, y)$ into a separable equation.

(a) Consider a differential equation of the form $y' = f(\alpha t + \beta y + \gamma)$, where α, β, and γ are constants. Use the change of variable $z = \alpha t + \beta y + \gamma$ to rewrite the differential equation as a separable equation of the form $z' = g(z)$. List the function $g(z)$.

(b) A differential equation that can be written in the form $y' = f(y/t)$ is called an equidimensional differential equation. Use the change of variable $z = y/t$ to rewrite the equation as a separable equation of the form $tz' = g(z)$. List the function $g(z)$.

Exercises 33–38:

Use the ideas of Exercise 32 to solve the given initial value problem. Obtain an explicit solution if possible.

33. $y' = \dfrac{y-t}{y+t}, \quad y(2) = 2$

34. $y' = \dfrac{y+t}{y+t+1}, \quad y(-1) = 0$

35. $y' = (t+y)^2 - 1, \quad y(1) = 2$

36. $y' = \dfrac{1}{2t+3y+1}, \quad y(1) = 0$

37. $y' = 2t + y + \dfrac{1}{2t+y}, \quad y(1) = 1$

38. $t^2 y' = y^2 - ty, \quad y(-2) = 2$

39. Consider the initial value problem

$$y' = \sqrt{1-y^2}, \qquad y(0) = 0.$$

(a) Show that $y = \sin t$ is an explicit solution on the t-interval $-\pi/2 \le t \le \pi/2$.

(b) Show that $y = \sin t$ is *not* a solution on either of the intervals $-3\pi/2 < t < -\pi/2$ or $\pi/2 < t < 3\pi/2$.

(c) What are the equilibrium solutions of $y' = \sqrt{1-y^2}$? Suppose a solution $y(t)$ reaches an equilibrium value at $t = t^*$. What happens to the graph of $y(t)$ for $t > t^*$?

(d) Show that the solution of the initial value problem is given by

$$y = \begin{cases} -1, & -\infty < t < -\pi/2 \\ \sin t, & -\pi/2 \le t \le \pi/2 \\ 1, & \pi/2 < t < \infty. \end{cases}$$

2.7   Exact Differential Equations

The class of differential equations referred to as exact includes separable equations as a special case. As with separable equations, the solution procedure for this new class consists of reversing the chain rule. This time, however, we use a chain rule that involves a function of two variables.

The Extended Chain Rule

Suppose $H(t, y)$ is a function of two independent variables t and y, where $H(t, y)$ has continuous partial derivatives with respect to t and y. If the second independent variable y is replaced with a differentiable function of t, call it $y(t)$, we obtain a composition $H(t, y(t))$ which is now a function of t only. What is dH/dt?

The appropriate chain rule is

$$\frac{d}{dt} H(t, y(t)) = \frac{\partial H(t, y(t))}{\partial t} + \frac{\partial H(t, y(t))}{\partial y} \frac{dy(t)}{dt}. \tag{1}$$

To understand this equation, note that the partial derivatives on the right-hand side refer to H viewed as a function of the two independent variables t and y. Once these partial derivatives are computed, the variable y is replaced by the function $y(t)$.

Formula (1) is an extension of the chain rule for functions of a single variable. If the function H has the form $H(y)$ so that it is only a function of the single variable y, then the first term on the right-hand side of (1) vanishes and the formula reverts to the usual chain rule for the composite function $H(y(t))$.

Solving Exact Differential Equations

The basic idea underlying the solving of exact differential equations is to reverse the extended chain rule when possible. To that end, consider a differential equation of the form

$$M(t, y) + N(t, y) \frac{dy}{dt} = 0. \tag{2}$$

Notice the similarity between the form of differential equation (2) and that of the chain rule (1). Suppose there exists some function, call it $H(t, y)$, that satisfies the following two conditions:

$$\frac{\partial H}{\partial t} = M(t, y) \quad \text{and} \quad \frac{\partial H}{\partial y} = N(t, y). \tag{3}$$

Because of (3), we can rewrite differential equation (2) as

$$\frac{\partial H}{\partial t} + \frac{\partial H}{\partial y} \frac{dy}{dt} = 0. \tag{4}$$

By the chain rule (1), equation (4) is the same as

$$\frac{d}{dt} H(t, y) = 0.$$

Therefore, we obtain an implicitly defined solution of equation (4),

$$H(t, y) = C. \tag{5}$$

If there is a function $H(t, y)$ satisfying the conditions in (3), then differential equation (2) is called an **exact differential equation**. If we can identify the function $H(t, y)$, then an implicitly defined solution is given by (5).

Recognizing an Exact Differential Equation

Once we know that a given differential equation is exact and once we identify a function $H(t, y)$ satisfying the conditions in (3), then we can write down an implicit solution, $H(t, y) = C$, for the differential equation. Two basic questions therefore need to be answered.

1. Given a differential equation of the form $M(t, y) + N(t, y)y' = 0$, how do we know whether or not it is exact? That is, how do we determine whether there is a function $H(t, y)$ satisfying the conditions

$$\frac{\partial H}{\partial t} = M(t, y) \quad \text{and} \quad \frac{\partial H}{\partial y} = N(t, y)?$$

2. Suppose we are somehow assured that such a function exists. How do we go about finding $H(t, y)$?

The answer to the first question is given in Theorem 2.4, which is stated without proof. To answer the second question, we will use a process of "anti-partial-differentiation."

Theorem 2.4

Consider the differential equation $M(t, y) + N(t, y)y' = 0$. Let the functions $M, N, \partial M/\partial y$, and $\partial N/\partial t$ be continuous in an open rectangle R of the ty-plane. Then the differential equation is exact in R if and only if

$$\frac{\partial M}{\partial y} = \frac{\partial N}{\partial t} \tag{6}$$

for all points (t, y) in R.

Theorem 2.4 provides an easy test for whether or not a given differential equation is exact. The theorem does not, however, tell how to construct the implicitly defined solution $H(t, y) = C$.

E X A M P L E

1

Which of the following differential equations is (are) exact?

(a) $y + t + ty' = 0$ (b) $y + \sin t + (y\cos t)y' = 0$ (c) $\sin y + (2y + t\cos y)y' = 0$

Solution:

(a) Using the notation of Theorem 2.4, we have

$$M(t, y) = y + t \quad \text{and} \quad N(t, y) = t.$$

Calculating the partial derivatives, we find

$$\frac{\partial M}{\partial y} = 1 \quad \text{and} \quad \frac{\partial N}{\partial t} = 1.$$

Therefore, by Theorem 2.4, the differential equation is exact.

(b) For this equation,

$$M(t, y) = y + \sin t \quad \text{and} \quad N(t, y) = y\cos t.$$

Calculating the partial derivatives yields

$$\frac{\partial M}{\partial y} = 1 \quad \text{and} \quad \frac{\partial N}{\partial t} = -y\sin t.$$

Since the partial derivatives are not equal, the differential equation is not exact.

(c) Calculating the partial derivatives, we have

$$\frac{\partial M}{\partial y} = \cos y \quad \text{and} \quad \frac{\partial N}{\partial t} = \cos y.$$

Since the partial derivatives are equal, the differential equation is exact. ❖

REMARK: Recall that a separable differential equation is one that can be written in the form $n(y)\, dy/dt + m(t) = 0$. Notice that

$$\frac{\partial m}{\partial y} = 0 \quad \text{and} \quad \frac{\partial n}{\partial t} = 0.$$

Thus, by Theorem 2.4, any separable differential equation is also exact.

Anti-Partial-Differentiation

We can use Theorem 2.4 to determine whether the differential equation $M(t, y) + N(t, y)y' = 0$ is exact. If it is exact, then we know there must be a function $H(t, y)$ such that

$$\frac{\partial H}{\partial t} = M(t, y) \quad \text{and} \quad \frac{\partial H}{\partial y} = N(t, y).$$

Once we determine $H(t, y)$, we have an implicitly defined solution,

$$H(t, y) = C.$$

A process of **anti-partial-differentiation** can be used to construct H.

As an illustration, recall from Example 1 that the following differential equation is exact:

$$\sin y + (2y + t \cos y)y' = 0. \tag{7}$$

Thus, there is a function $H(t, y)$ such that

$$\frac{\partial H}{\partial t} = \sin y \quad \text{and} \quad \frac{\partial H}{\partial y} = 2y + t \cos y. \tag{8}$$

Choose one of these equalities, say $\partial H/\partial y = 2y + t \cos y$, and compute an "anti-partial-derivative." Antidifferentiating $2y + t \cos y$ with respect to y, we obtain

$$H(t, y) = y^2 + t \sin y + g(t), \tag{9}$$

where $g(t)$ is an arbitrary function of t. [Note: The "constant of integration" in equation (9) is an arbitrary function of t since t is treated as a constant when the partial derivative with respect to y is computed.]

We now determine $g(t)$ so that the representation (9) for H satisfies the first equality in (8). Taking the partial derivative of H with respect to t, we find

$$\frac{\partial H}{\partial t} = \frac{\partial}{\partial t}\left[y^2 + t \sin y + g(t)\right] = \sin y + \frac{dg}{dt}.$$

Comparing the preceding result with the first condition of equation (8), it follows that we need

$$\frac{dg}{dt} = 0$$

or $g(t) = C_1$, where C_1 is an arbitrary constant. Thus, from equation (9), we know

$$H(t, y) = y^2 + t \sin y + C_1. \tag{10}$$

We can drop the arbitrary constant C_1, since we will eventually set $H(t, y)$ equal to an arbitrary constant in the implicit solution. Therefore, $H(t, y) = y^2 + t \sin y$.

In this illustration, we started with the second equation in (8), $\partial H/\partial y = 2y + t \cos y$. We could just as well have started with the first equation, $\partial H/\partial t = \sin y$. If we had done so, we would have arrived at the same function H; see Exercise 21.

E X A M P L E

2

Consider the initial value problem

$$1 + y^2 + 2(t + 1)y \frac{dy}{dt} = 0, \qquad y(0) = 1.$$

Verify that the differential equation is exact and solve the initial value problem.

Solution: To verify that the differential equation is exact, we appeal to Theorem 2.4, using $M(t,y) = 1 + y^2$ and $N(t,y) = 2(t+1)y$. The functions M, N, $\partial M/\partial y$, and $\partial N/\partial t$ are continuous in the entire ty-plane. Since

$$\frac{\partial M}{\partial y} = 2y \quad \text{and} \quad \frac{\partial N}{\partial t} = 2y,$$

the differential equation is exact.

We now find $H(t,y)$. Since the equation is exact,

$$\frac{\partial H}{\partial t} = 1 + y^2.$$

Antidifferentiating with respect to t yields

$$H(t,y) = t(1+y^2) + g(y). \tag{11}$$

To determine $g(y)$, we differentiate (11) with respect to y, finding

$$\frac{\partial H}{\partial y} = 2ty + \frac{dg}{dy}.$$

Since the differential equation is exact,

$$2ty + \frac{dg}{dy} = 2(t+1)y.$$

Therefore,

$$g(y) = y^2 + C_1.$$

Without loss of generality, we let $C_1 = 0$ to obtain

$$H(t,y) = t(1+y^2) + y^2 = (t+1)y^2 + t.$$

Thus, we have the following implicitly defined solution of the differential equation:

$$(t+1)y^2 + t = C.$$

Imposing the initial condition, we obtain an implicit solution of the initial value problem:

$$(t+1)y^2 + t = 1.$$

We can solve for y:

$$y^2 = \frac{1-t}{1+t}, \quad \text{or} \quad y = \pm\sqrt{\frac{1-t}{1+t}}.$$

Choosing the positive sign so as to satisfy the initial condition $y(0) = 1$, we arrive at an explicit solution of the initial value problem:

$$y = \sqrt{\frac{1-t}{1+t}}. \quad \diamond$$

The explicit solution $y = \sqrt{(1-t)/(1+t)}$ has $-1 < t < 1$ as an interval of existence. This should not be surprising in light of the existence-uniqueness statement in Theorem 2.2. The differential equation in Example 2 has the form $y' = f(t, y)$, where

$$f(t, y) = -\frac{1 + y^2}{2(t + 1)y}.$$

Since the initial condition point is $(t_0, y_0) = (0, 1)$, the hypotheses of Theorem 2.2 are satisfied in the infinite rectangle defined by $-1 < t < \infty$, $0 < y < \infty$. Along the lines $t = -1$ and $y = 0$, both f and $\partial f / \partial y$ are undefined. As Figure 2.12 shows, the solution curve has a vertical asymptote at the boundary line $t = -1$ and approaches the boundary line $y = 0$ as $t \to 1$ from the left.

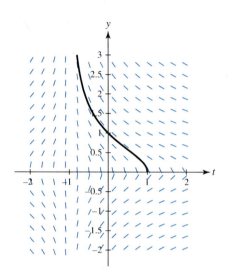

FIGURE 2.12

A portion of the direction field for the differential equation in Example 2. The solution $y = \sqrt{(1 - t)/(1 + t)}$ is shown as a solid curve.

EXERCISES

Exercises 1–10:

Show that the given nonlinear differential equation is exact. (Some algebraic manipulation may be required. Also, recall the remark that follows Example 1.) Find an implicit solution of the initial value problem and (where possible) an explicit solution.

1. $(2y - t)y' - y + 2t = 0$, $y(1) = 0$
2. $(t + y^3)y' + y + t^3 = 0$, $y(0) = -2$
3. $y' = (3t^2 + 1)(y^2 + 1)$, $y(0) = 1$
4. $(y^3 + \cos t)y' = 2 + y \sin t$, $y(0) = -1$
5. $(e^{t+y} + 2y)y' + (e^{t+y} + 3t^2) = 0$, $y(0) = 0$
6. $(y^3 - t^3)y' = 3t^2 y + 1$, $y(-2) = -1$
7. $(e^{2y} + t^2 y)y' + ty^2 + \cos t = 0$, $y(\pi/2) = 0$
8. $y' = -\dfrac{y \cos(ty) + 1}{t \cos(ty) + 2ye^{y^2}}$, $y(\pi) = 0$
9. $\left(2ty + \dfrac{1}{y}\right)y' + y^2 = 1$, $y(1) = 1$
10. $(2y \ln t - t \sin y)y' + t^{-1}y^2 + \cos y = 0$, $y(2) = 0$

Exercises 11–16:

Determine the general form of the function $M(t, y)$ or $N(t, y)$ that will make the given differential equation exact.

11. $(2t + y)y' + M(t, y) = 0$

12. $(t^2 + y^2 \sin t)y' + M(t, y) = 0$

13. $(te^y + t + 2y)y' + M(t, y) = 0$

14. $N(t, y)y' + 2t + y = 0$

15. $N(t, y)y' + t^2 + y^2 \sin t = 0$

16. $N(t, y)y' + e^{y^2} + 2ty = 1$

Exercises 17–20:

The given equation is an implicit solution of $N(t, y)y' + M(t, y) = 0$, satisfying the given initial condition. Assuming the equation $N(t, y)y' + M(t, y) = 0$ is exact, determine the functions $M(t, y)$ and $N(t, y)$, as well as the possible value(s) of y_0.

17. $t^3 y + e^t + y^2 = 5, \quad y(0) = y_0$

18. $2ty + \cos(ty) + y^2 = 2, \quad y(0) = y_0$

19. $\ln(2t + y) + t^2 + e^{yt} = 1, \quad y(0) = y_0$

20. $y^3 + 4ty + t^4 + 1 = 0, \quad y(0) = y_0$

21. The differential equation $(2y + t \cos y)y' + \sin y = 0$ is exact, and thus there exists a function $H(t, y)$ such that $\partial H / \partial t = \sin y$ and $\partial H / \partial y = 2y + t \cos y$. Antidifferentiating the second equation with respect to y, we ultimately arrived at $H(t, y) = y^2 + t \sin y + C_1$; see equation (10). Show that the same result would be obtained if we began the solution process by antidifferentiating the first equation, $\partial H / \partial t = \sin y$, with respect to t.

22. Making a Differential Equation Exact   Suppose the differential equation $N(t, y)y' + M(t, y) = 0$ is not exact; that is, $N_t(t, y) \neq M_y(t, y)$. Is it possible to multiply the equation by a function, call it $\mu(t, y)$, so that the resulting equation is exact?

(a) Show that if $\mu(t, y)N(t, y)y' + \mu(t, y)M(t, y) = 0$ is exact, the function μ must be a solution of the partial differential equation

$$N(t, y)\mu_t - M(t, y)\mu_y = [M_y(t, y) - N_t(t, y)]\mu.$$

Parts (b) and (c) of this exercise discuss special cases where the function μ can be chosen to be a function of a single variable. In these special cases, the partial differential equation in part (a) reduces to a first order linear ordinary differential equation and can be solved using the techniques of Section 2.2.

(b) Suppose the quotient $[M_y(t, y) - N_t(t, y)]/N(t, y)$ is just a function of t, call it $p(t)$. Let $P(t)$ be an antiderivative of $p(t)$. Show that μ can be chosen as $\mu(t) = e^{P(t)}$.

(c) Suppose the quotient $[N_t(t, y) - M_y(t, y)]/M(t, y)$ is just a function of y, call it $q(y)$. Let $Q(y)$ be an antiderivative of $q(y)$. Show that μ can be chosen as $\mu(y) = e^{Q(y)}$.

Exercises 23–28:

In each exercise,

(a) Show that the given differential equation is not exact.

(b) Multiply the equation by the function μ, if it is provided, and show that the resulting differential equation is exact. If the function μ is not given, use the ideas of Exercise 22 to determine μ.

(c) Solve the given problem, obtaining an explicit solution if possible.

23. $4tyy' + y^2 - t = 0, \quad y(1) = 1, \quad \mu(t, y) = t^{-1/2}$

24. $(t^2 y^2 + 1)y' + ty^3 = 0, \quad y(0) = 1, \quad \mu(t, y) = y^{-1}$

25. $(2t + y)y' + y = 0, \quad y(2) = -3$

26. $ty^2 y' + 2y^3 = 1, \quad y(1) = -1$

27. $tyy' + y^2 + e^t = 0, \quad y(1) = -2$

28. $(3ty + 2)y' + y^2 = 0, \quad y(-1) = -1$

2.8   The Logistic Population Model

The art of mathematical modeling involves a trade-off between realism and complexity. A model should incorporate enough reality to make it useful as a predictive tool. However, the model must also be tractable mathematically; if it's too complex to analyze and if we cannot deduce any useful information from it, then the model is worthless.

A key assumption of the population model described in Section 2.4, often referred to as the Malthusian[6] model, is that relative birth rate is independent of population. (Recall that the relative birth rate, $r_b - r_d$, is the difference between birth and death rates per unit population.) The assumption that $r_b - r_d$ is independent of population leads to uninhibited exponential growth or decay of solutions. Such behavior is often a poor approximation of reality.

For a colony possessing limited resources, a more realistic model is one that accounts for the impact of population on the relative birth rate. When population size is small, resources are relatively plentiful and the population should thrive and grow. When the population becomes larger, however, we expect that resources become scarcer, the population becomes stressed, and the relative birth rate begins to decline (eventually becoming negative). In this section we consider a model that attempts to account for this effect. This model leads to a nonlinear differential equation.

The Verhulst, or Logistic, Model

The **Verhulst population model**[7] assumes that the population $P(t)$ evolves according to the differential equation

$$\frac{dP}{dt} = r\left(1 - \frac{P}{P_e}\right)P. \tag{1}$$

In equation (1), r and P_e are positive constants. Equation (1) is also known as the **logistic equation**. Comparing equation (1) with the Malthus equation

$$\frac{dP}{dt} = (r_b - r_d)P, \tag{2}$$

we see that the constant relative birth rate $r_b - r_d$ of equation (2) has been replaced by the *population-dependent* relative birth rate

$$r\left(1 - \frac{P(t)}{P_e}\right),$$

where r is a positive constant. If $P(t) > P_e$, then $dP/dt < 0$ and the population is decreasing. Conversely, if $P(t) < P_e$, then $dP/dt > 0$ and the population is increasing.

The qualitative behavior of solutions of the logistic equation can also be deduced from the direction field. Figure 2.13, for example, shows the direction field for the logistic equation for the parameters $r = 3$ and $P_e = 1$.

---

[6]Thomas Malthus (1766–1834) was an English political economist whose *Essay on the Principle of Population* had an important influence on Charles Darwin's thinking. Malthus believed that human population growth, if left unchecked, would inevitably lead to poverty and famine.
[7]Pierre Verhulst (1804–1849) was a Belgian scientist remembered chiefly for his research on population growth. He deduced and studied the nonlinear differential equation named after him.

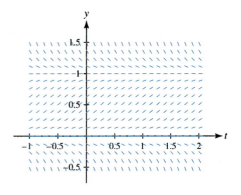

FIGURE 2.13

The direction field for the logistic equation $P' = 3(1 - P)P$. The equilibrium solutions are $P = 0$ and $P = 1$.

The logistic equation has two equilibrium solutions, the trivial solution $P(t) = 0$ and the nontrivial equilibrium solution $P(t) = P_e$. If $P(t_0) = P_e$, then the population remains equal to this value as time evolves. The constant population, P_e, is called an **equilibrium population**. In terms of its structure, the logistic equation is a nonlinear separable differential equation. It is also a Bernoulli equation. We solve it below as a separable equation. Its treatment as a Bernoulli equation is left to the exercises.

Solution of the Logistic Equation

To solve equation (1), we separate variables, obtaining

$$\frac{1}{\left(1 - \dfrac{P}{P_e}\right)P} \frac{dP}{dt} = r.$$

An implicit solution is therefore

$$\int \frac{dP}{\left(1 - \dfrac{P}{P_e}\right)P} = rt + K, \tag{3}$$

where K is an arbitrary constant. Recalling the method of partial fractions from calculus (also see Section 5.3), we obtain

$$\int \frac{dP}{\left(1 - \dfrac{P}{P_e}\right)P} = \int \left(\frac{1}{P} - \frac{1}{P - P_e}\right) dP = \ln\left|\frac{P}{P - P_e}\right|.$$

Therefore,

$$\ln\left|\frac{P}{P - P_e}\right| = rt + K.$$

Hence,

$$\left|\frac{P(t)}{P(t) - P_e}\right| = Ce^{rt}, \tag{4}$$

where $C = e^K$ is an arbitrary positive constant. We can remove the absolute value signs by arguing as follows. The exponential function e^{rt} is never zero. Therefore, the quotient $P(t)/[P(t) - P_e]$ is never zero and, being a continuous function of t, never changes sign; it is either positive or negative for all t. Therefore, we can remove the absolute value signs if we allow the constant C to be either positive or negative.

Assume that the initial population $P(0) = P_0$ is positive and $P_0 \neq P_e$. Then

$$C = \frac{P_0}{P_0 - P_e}.$$

After some algebraic manipulation, we obtain the explicit solution

$$P(t) = \frac{P_0 P_e}{P_0 - (P_0 - P_e)e^{-rt}}. \tag{5}$$

The derivation leading to equation (5) tacitly assumes that $P(t)$ never takes on the value 0 or P_e. The final expression (5), however, is valid for any value of P_0 (including the equilibrium values 0 and P_e).

What behavior is predicted by the solution (5)? Since $r > 0$, we see that

$$\lim_{t \to \infty} P(t) = P_e$$

for all positive values of P_0. Therefore, any given nonzero population will tend to the equilibrium population value, P_e, as time increases. Figure 2.13 illustrates this behavior, showing the direction field for the special case $P' = 3(1 - P)P$.

We can ask additional questions. What happens if we allow temporal variations in the relative birth rate by allowing $r = r(t)$ in equation (1)? What happens if migration is introduced into the logistic model? When migration is introduced, the population model has the form

$$\frac{dP}{dt} = r\left(1 - \frac{P}{P_*}\right)P + M. \tag{6}$$

If r, P_*, and M are constants, the equation is separable. Moreover, the question of equilibrium populations is relevant. What happens to the equilibria as the migration level M changes? Some of these questions are addressed in the exercises.

Example 1 illustrates how sketching a diagram similar to a direction field enables us to predict the behavior of solutions of equation (6) without actually solving the equation. The basis for such predictions is found in Exercises 14–15. These exercises set $r = 1, P_* = 1$ in equation (6), considering the equation $P' = -(P - P_1)(P - P_2)$ with equilibrium populations $P_1 \geq P_2 > 0$. These exercises show the following:

1. If $P(0) > P_2$, then $P(t)$ exists for $0 \leq t < \infty$ and $P(t) \to P_1$ as $t \to \infty$.
2. If $0 \leq P(0) < P_2$, then $\lim_{t \to t^*} P(t) = -\infty$ for some finite $t^* > 0$.

EXAMPLE
1

Let $P(t)$ denote the population of a colony, where P is measured in units of 100,000 individuals and time t is in years. Assume that population is modeled by the logistic model with constant out-migration,

$$\frac{dP}{dt} = (1 - P)P - \frac{2}{9}, \qquad P(0) = 2.$$

(a) Determine all the equilibrium populations (that is, the nonnegative equilibrium solutions). Sketch a diagram indicating those regions in the first quadrant of the tP-plane where the population is increasing $[P'(t) > 0]$ and those regions where the population is decreasing $[P'(t) < 0]$. This diagram gives a rough indication of how the solutions behave.

(b) Without solving the initial value problem, use the results of Exercise 14 (as summarized above) to determine $\lim_{t\to\infty} P(t)$.

Solution:

(a) The equilibrium solutions are the constant solutions $P(t) = \frac{2}{3}$ and $P(t) = \frac{1}{3}$. Since these equilibrium solutions are nonnegative, they correspond to equilibrium populations. The diagram sketched in Figure 2.14 indicates the regions in the first quadrant where $P'(t) > 0$ and where $P'(t) < 0$.

(b) From Figure 2.14, it follows that the solution of the initial value problem $P' = (1 - P)P - \frac{2}{9}, P(0) = 2$ is decreasing at $t = 0$, since $P'(0) = -\frac{20}{9}$. This solution curve cannot cross the line $P = \frac{2}{3}$, since $P = \frac{2}{3}$ is also a solution of $P' = (1 - P)P - \frac{2}{9}$. By Exercise 14, either solutions of equation (6) are unbounded or they tend to an equilibrium value as $t \to \infty$. The diagram in Figure 2.14 indicates that

$$\lim_{t\to\infty} P(t) = \tfrac{2}{3}.$$

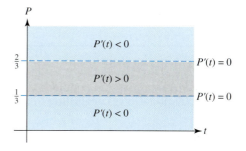

FIGURE 2.14

Consider the population model $P' = (1 - P)P - \frac{2}{9}$. The equilibrium populations are $P = \frac{2}{3}$ and $P = \frac{1}{3}$; see Example 1. The diagram shows the regions where population is increasing $[P'(t) > 0]$ and where it is decreasing $[P'(t) < 0]$. The solution satisfying the initial condition $P(0) = 2$ is decreasing and bounded below by the line $P = \frac{2}{3}$. Therefore, by Exercise 14, $P(t) \to \frac{2}{3}$ as $t \to \infty$.

(continued)

(continued)

> [Similarly, if $P(0) = P_0$ where $\frac{1}{3} < P_0 < \frac{2}{3}$, we again obtain $\lim_{t\to\infty} P(t) = \frac{2}{3}$. However, if $P(0) = P_0$ where $P_0 < \frac{1}{3}$, then the population becomes extinct when the graph of $P(t)$ intersects the t-axis. At that point, the population model ceases to be meaningful.] ❖

The Spread of an Infectious Disease

The logistic differential equation also arises in modeling the spread of an infectious disease. Suppose we have a constant population of N individuals and at time t the number of infected members is $P(t)$. The corresponding number of uninfected individuals is then $N - P(t)$. A reasonable assumption is that the rate of spread of the disease at time t is proportional to the product of noninfected and infected individuals. This assumption leads to the differential equation

$$\frac{dP}{dt} = k(N - P)P,$$

where k is the constant of proportionality. If the equation is rewritten as

$$\frac{dP}{dt} = kN\left(1 - \frac{P}{N}\right)P, \tag{7}$$

we can see that it is similar to the logistic equation (1). The corresponding initial value problem is completed by specifying the number of infected individuals at some initial time, $P(t_0) = P_0$.

Note that the differential equation (7) has two equilibrium solutions, $P = 0$ and $P = N$. This certainly makes sense. If no one is infected or if everyone is infected, the disease will not spread. We consider aspects of this infectious disease model in the exercises.

EXERCISES

Exercises 1–3:

Let $P(t)$ represent the population of a colony, in millions of individuals. Assume that the population evolves according to the equation

$$\frac{dP}{dt} = 0.1\left(1 - \frac{P}{3}\right)P,$$

with time t measured in years. Use the explicit solution given in equation (5) to answer the questions.

1. Suppose the colony starts with 100,000 individuals. How long will it take the population to reach 90% of its equilibrium value?

2. Suppose the colony is initially overpopulated, starting with 5,000,000 individuals. How long will it take for the population to decrease to 110% of its equilibrium value?

3. Suppose, after 3 years of existence, the colony population is found to be 2,000,000 individuals. What was the initial population?

Exercises 4–10:

Constant Migration   Consider a population modeled by the initial value problem

$$\frac{dP}{dt} = (1 - P)P + M, \qquad P(0) = P_0, \tag{8}$$

where the migration rate M is constant. [The model (8) is derived from equation (6) by setting the constants r and P_* to unity. We did this so that we can focus on the effect M has on the solutions.]

For the given values of M and $P(0)$,

(a) Determine all the equilibrium populations (the nonnegative equilibrium solutions) of differential equation (8). As in Example 1, sketch a diagram showing those regions in the first quadrant of the tP-plane where the population is increasing [$P'(t) > 0$] and those regions where the population is decreasing [$P'(t) < 0$].

(b) Describe the qualitative behavior of the solution as time increases. Use the information obtained in (a) as well as the insights provided by the figures in Exercises 11–13 (these figures provide specific but representative examples of the possibilities).

4. $M = -\frac{3}{16}, \quad P(0) = \frac{3}{2}$ **5.** $M = -\frac{3}{16}, \quad P(0) = \frac{1}{2}$ **6.** $M = -\frac{1}{4}, \quad P(0) = \frac{1}{4}$

7. $M = -\frac{1}{4}, \quad P(0) = 1$ **8.** $M = -\frac{1}{4}, \quad P(0) = \frac{1}{2}$ **9.** $M = 2, \quad P(0) = 0$

10. $M = 2, \quad P(0) = 4$

Exercises 11–13:

The direction fields shown correspond to the differential equation

$$\frac{dP}{dt} = \left(1 - \frac{P}{P_*}\right)P + M.$$

Determine the constants P_* and M.

11. The equilibrium solutions are   **12.** The equilibrium solution is
$P(t) = 2$ and $P(t) = 1$. $P(t) = 2$.

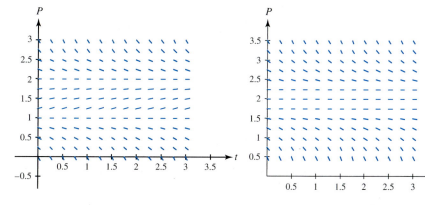

13. The equilibrium solutions are
$P(t) = 2$ and $P(t) = -1$.

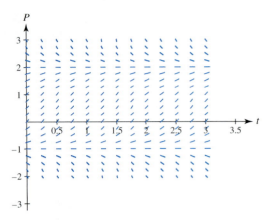

14. Consider the initial value problem (8). Let P_1 and P_2 denote the roots of $P^2 - P - M = 0$, and assume that $P_1 > P_2 > 0$. [Since P_1 and P_2 are positive constant solutions of $P' = (1 - P)P + M$, they correspond to equilibrium populations.]

(a) What does the assumption $P_1 > P_2 > 0$ imply about the sign of M? Is migration occurring at a constant rate into or out of the colony?

(b) Solve initial value problem (8), assuming that $P(0) \neq P_1$ and $P(0) \neq P_2$. Note that the differential equation is separable. Your solution will have the form

$$\frac{|P(t) - P_1|}{|P(t) - P_2|} = Ke^{-\lambda t},$$

where K and λ are positive constants. Unravel this implicit solution. [Hint: The graph of $P(t)$ cannot cross the lines $P = P_1$ and $P = P_2$. Therefore, the terms $P(t) - P_1$ and $P(t) - P_2$ have the same signs as $P_0 - P_1$ and $P_0 - P_2$, respectively. In order to unravel the solution, consider the separate cases $P_0 > P_1$, $P_1 > P_0 > P_2$, and $P_2 > P_0$ and remove the absolute values.]

(c) Use the explicit solutions found in part (b) to show that if $P_0 > P_1$, then $P(t) \to P_1$ as $t \to \infty$. Similarly, show that if $P_1 > P_0 > P_2$, then $P(t) \to P_1$ as $t \to \infty$. Finally, show that if $P_2 > P_0$, then $\lim_{t \to t^*} P(t) = -\infty$ at some finite time t^*. Since populations are nonnegative, what actually happens to the population in this last case?

15. Repeat the analysis of Exercise 14 for the case of a positive repeated root of $P^2 - P - M = 0$ (that is, for the case of $P_1 = P_2 > 0$).

(a) What does the assumption $P_1 = P_2 > 0$ imply about the sign of M? Is migration occurring at a constant rate into or out of the colony?

(b) Solve initial value problem (8) for the case in which $P(0) \neq P_1$. Argue that $P_1 = P_2$ can happen only if $P_1 = P_2 = \frac{1}{2}$.

(c) By taking limits of the explicit solution found in part (b), show that $P(t) \to P_1$ as $t \to \infty$ if $P_0 \geq P_1$. What happens if $P_0 < P_1$?

Variable Birth Rates   In certain situations, a population's relative birth rate may vary with time. For example, environmental conditions that change over a period of time may affect the birth rate. When food is stored, variations in temperature may affect the growth of harmful bacteria. It is of interest, therefore, to understand the behavior of the logistic equation when the relative birth rate $r(t)$ is variable.

16. Consider the initial value problem

$$\frac{dP}{dt} = r(t)\left(1 - \frac{P}{P_e}\right)P, \qquad P(0) = P_0.$$

Observe that the differential equation is separable. Let $R(t) = \int_0^t r(s)\,ds$. Solve the initial value problem. Note that your solution will involve the function $R(t)$.

17. Solve the initial value problem in Exercise 16 for the particular case of $r(t) = 1 + \sin 2\pi t$, $P_e = 1$, and $P_0 = \frac{1}{4}$. Here, time is measured in years and population in millions of individuals. (The varying relative birth rate might reflect the impact of seasonal changes on the population's ability to reproduce over the course of a year.) How does the population behave as time increases? In particular, does $\lim_{t \to \infty} P(t)$ exist, and if so, what is this limit?

18. Let $P(t)$ represent the number of individuals who, at time t, are infected with a certain disease. Let N denote the total number of individuals in the population. Assume that the spread of the disease can be modeled by the initial value problem

$$\frac{dP}{dt} = k(N - P)P, \qquad P(0) = P_0,$$

where k is a constant. Obtain an explicit solution of this initial value problem.

19. Consider the special case of the infectious disease model in Exercise 18, where $N = 2,000,000$ and $P_0 = 100,000$. Suppose that after 1 year, the number of infected individuals had increased to 200,000. How many members of the population will be infected after 5 years?

20. Consider a chemical reaction of the form $A + B \to C$, in which the rates of change of the two chemical reactants, A and B, are described by the two differential equations

$$A' = -kAB, \qquad B' = -kAB,$$

where k is a positive constant. Assume that 5 moles of reactant A and 2 moles of reactant B are present at the beginning of the reaction.

(a) Show that the difference $A(t) - B(t)$ remains constant in time. What is the value of this constant?

(b) Use the observation made in (a) to derive an initial value problem for reactant A.

(c) It was observed, after the reaction had progressed for 1 sec, that 4 moles of reactant A remained. How much of reactants A and B will be left after 4 sec of reaction time?

21. Solve the initial value problem

$$\frac{dP}{dt} = r\left(1 - \frac{P}{P_e}\right)P, \qquad P(0) = P_0$$

by viewing the differential equation as a Bernoulli equation.

2.9   Applications to Mechanics

In Chapter 1, we considered a falling object acted upon only by gravity. For that case, with $y(t)$ used to represent the vertical position of the body at time t and $v(t)$ to represent its velocity, Newton's second law of motion, $ma = F$, reduces to

$$m\frac{dv}{dt} = -mg.$$

Here, m is the mass of the object and g the acceleration due to gravity, nominally 32 ft/sec^2 or 9.8 m/s^2. The minus sign on the right-hand side is present because we measure $y(t)$ and $v(t)$ as positive upward while the force of gravity acts downward. For the simple model above, we solved for $v(t)$ and $y(t)$ by computing successive antiderivatives.

A more realistic model of one-dimensional motion is one that incorporates the effects of air resistance. As an object moves through air, a retarding aerodynamic force is created by the combination of pressure and frictional forces. The air close to the object exerts a normal pressure force upon it. Likewise, friction creates a tangential force that opposes the motion of air past the object. The combination of these effects creates a drag force that acts to reduce the speed of the moving object.

The drag force depends on the velocity $v(t)$ of the object and acts on it in such a way as to reduce its speed, $|v(t)|$. We consider two idealized models of drag force that are consistent with these ideas.

Case 1: Drag Force Proportional to Velocity

Assume that velocity $v(t)$ is positive in the upward direction and that the drag force is proportional to velocity. If k is the positive constant of proportionality, Newton's second law of motion leads to

$$m\frac{dv}{dt} = -mg - kv. \tag{1}$$

Does the model of drag that we have postulated act as we want it to? If the object is moving upward [that is, if $v(t) > 0$], then the drag force $-kv(t)$ is negative and thus acts downward. Conversely, if $v(t) < 0$, then the object is moving downward. In this case, the drag force $-kv(t) = k|v(t)|$ is a positive (upward) force, as it should be. Therefore, whether the object is moving upward or downward, drag acts to slow the object; this drag model is consistent with our ideas of how drag should act. See Figure 2.15.

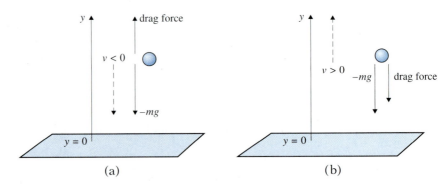

FIGURE 2.15

Assume a drag force of the form $kv(t), k > 0$; see equation (1). (a) When the object is moving downward, drag acts upward and tends to slow the object. (b) When the object is moving upward, drag acts downward and likewise tends to slow the object.

This drag model leads to equation (1), a first order linear constant coefficient equation; we can solve it using the ideas of Section 2.2. Before we do, however, let's consider the question of equilibrium solutions to see if the model makes sense in that regard. The only constant solution of equation (1) is $v(t) = -mg/k$. At the velocity $v(t) = -mg/k$, the drag force and gravitational force acting on the object are equal and opposite. This equilibrium velocity is often referred to as the **terminal velocity** of the object.

The initial value problem corresponding to equation (1) is

$$m\frac{dv}{dt} = -mg - kv, \qquad v(0) = v_0. \tag{2}$$

Solving initial value problem (2), we obtain

$$v(t) = -\frac{mg}{k} + \left(v_0 + \frac{mg}{k}\right)e^{-(k/m)t}.$$

From this explicit solution it is clear that, for any initial velocity v_0, $v(t)$ tends to the terminal velocity

$$\lim_{t\to\infty} v(t) = -\frac{mg}{k}.$$

The concept of terminal velocity is a mathematical abstraction since, in any application, the time interval of interest is finite. However, whether the object is initially moving up or down, it eventually begins to fall and its velocity approaches terminal velocity as the time of falling increases. The velocity that the object actually has the instant before it strikes the ground is called the **impact velocity**. Once impact occurs, the mathematical model (2) is no longer applicable.

Case 2: Drag Force Proportional to the Square of Velocity

A drag force having magnitude $\kappa v^2(t)$ (where the constant of proportionality κ is assumed to be positive) has been found in many cases to be a reasonably good approximation to reality over a range of velocities. However, since it involves an even power of velocity, incorporating this model of drag into the equations of motion requires more care.

Drag must act to reduce the speed of the moving object. Therefore, if the object is moving upward, the drag force acts downward and should be $-\kappa v^2(t)$. If the object is moving downward, the drag force acts upward and should be $\kappa v^2(t)$. In other words, when we use this model for drag, Newton's second law leads to

$$m\frac{dv}{dt} = -mg - \kappa v^2, \qquad v(t) \geq 0,$$

$$m\frac{dv}{dt} = -mg + \kappa v^2, \qquad v(t) \leq 0.$$

Here again we can ask about equilibrium solutions and see if the answer makes sense physically. Note that no equilibrium solution exists if $v(t) > 0$ since $-mg - \kappa v^2$ is never zero. When $v(t) < 0$, however, there is an equilibrium solution:

$$v(t) = -\sqrt{\frac{mg}{\kappa}}.$$

This equilibrium solution is again a terminal velocity corresponding to downward motion; at this velocity, drag and gravity exert equal and opposite forces.

Each of the preceding equations is a first order separable equation. If the problem involves a falling (rising) object, then velocity is always nonpositive (nonnegative) and a single equation is valid for the entire problem (that is, over the entire t-interval of interest). If, however, the problem involves both upward and downward motion, then both equations will ultimately be needed. The first equation must be used to model the upward dynamics, the behavior of the projectile from launch until the time t_m at which it reaches its highest point [that is, when $v(t_m) = 0$]. After time t_m, the projectile begins to fall and the second differential equation [with initial condition $v(t_m) = 0$] is needed to model the descending dynamics.

E X A M P L E

1

Assume a projectile having mass m is launched vertically upward from ground level at time $t = 0$ with initial velocity $v(0) = v_0$. Further assume that the drag force experienced by the projectile is proportional to the square of its velocity, with drag coefficient κ. Determine the maximum height reached by the projectile and the time at which this maximum height is reached.

Solution: The projectile motion being considered involves only upward motion. Therefore,

$$mv' = -mg - \kappa v^2, \qquad v(0) = v_0.$$

Separating variables yields

$$\frac{v'}{1 + \dfrac{\kappa}{mg}v^2} = -g.$$

Integrating, we find

$$\int \frac{dv}{1 + \dfrac{\kappa}{mg}v^2} = \sqrt{\frac{mg}{\kappa}}\,\tan^{-1}\left(\sqrt{\frac{\kappa}{mg}}\,v\right) = -gt + C.$$

Imposing the initial condition gives

$$C = \sqrt{\frac{mg}{\kappa}}\,\tan^{-1}\left(\sqrt{\frac{\kappa}{mg}}\,v_0\right).$$

Finally, after some algebra, we obtain the explicit solution

$$v(t) = \sqrt{\frac{mg}{\kappa}}\,\tan\left[\tan^{-1}\left(\sqrt{\frac{\kappa}{mg}}\,v_0\right) - \sqrt{\frac{\kappa g}{m}}\,t\right]. \tag{3}$$

As a check on (3), note that $v(0)$ does reduce to the given initial velocity v_0. As a further check, one can show (using L'Hôpital's rule[8]) that for every fixed time t,

$$\lim_{\kappa \to 0} v(t) = v_0 - gt,$$

which is the velocity in the absence of drag.

Let t_m denote the time when the maximum height is reached; that is, $v(t_m) = 0$. From equation (3), we see that the maximum height is attained at the first positive value t where the argument of the tangent function is zero. Thus,

$$\tan^{-1}\left(\sqrt{\frac{\kappa}{mg}}\,v_0\right) - \sqrt{\frac{\kappa g}{m}}\,t_m = 0,$$

---

[8]Guillaume de L'Hôpital (1661–1704) wrote the first textbook on differential calculus, *Analyse des infiniment petits pour l'intelligence des lignes courbes*, which appeared in 1692. The book contains the rule that bears his name.

or

$$t_m = \sqrt{\frac{m}{\kappa g}} \tan^{-1}\left(\sqrt{\frac{\kappa}{mg}}\, v_0\right).$$ **(4)**

We next want to determine the maximum height, $y(t_m)$, reached by the projectile. To find $y(t_m)$, we need an expression for position, $y(t)$. To determine position, $y(t)$, we integrate velocity:

$$y(t) - y(0) = \int_0^t v(s)\, ds.$$

For our problem, $y(0) = 0$. Therefore,

$$y(t_m) = \int_0^{t_m} v(t)\, dt,$$ **(5)**

where $v(t)$ is given by equation (3). To carry out the integration in equation (5), we use the fact that $\int \tan u\, du = -\ln|\cos u| + C$, finding

$$y(t_m) = \frac{m}{2\kappa} \ln\left(1 + \frac{\kappa v_0^2}{mg}\right).$$

Here again, as a check, one can show using L'Hôpital's rule that

$$\lim_{\kappa \to 0} y(t_m) = \frac{v_0^2}{2g}$$ **(6)**

is the maximum height reached in the absence of drag. Figure 2.16 shows how the size of the drag constant affects the quantities t_m and $y(t_m)$ for the case of a 2-lb object launched upward from ground level with an initial velocity of 60 mph ($v_0 = 88$ ft/sec).

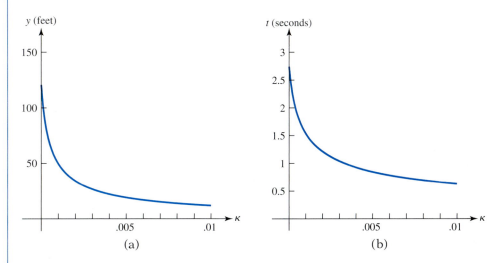

FIGURE 2.16

A 2-lb object is launched upward from ground level with an initial velocity of 88 ft/sec. The graph in (a) shows the maximum altitude as a function of the drag constant κ; see equation (5). The graph in (b) shows the time when the maximum altitude is reached as a function of the drag constant κ; see equation (4).   ❖

The computations in Example 1 are relatively complicated. The task of finding maximum projectile height can be simplified by transforming the problem to one in which height rather than time is the independent variable. This transformation is discussed in the next subsection.

One-Dimensional Dynamics with Distance as the Independent Variable

In many applications, velocity is always nonnegative or always nonpositive over the entire time interval of interest. For the present discussion, assume that motion occurs in a straight line, along the x-axis. Then position $x(t)$ is a monotonic function of time. If velocity is nonnegative, the position of the object is an increasing function of time; if velocity is nonpositive, then position is a decreasing function of time. In these cases, we can simplify the differential equation describing the object's motion if we use position x as the independent variable rather than time t.

Suppose (for definiteness) that position $x(t)$ is an increasing function of time t [a similar analysis is valid if $x(t)$ is decreasing]. Then an inverse function exists and we can express time as a function of position. We can ultimately view velocity as a function of position and use the chain rule to relate dv/dt to dv/dx:

$$\frac{dv}{dt} = \frac{dv}{dx}\frac{dx}{dt} = v\frac{dv}{dx}. \tag{7}$$

This change of independent variable is useful when the net force acting on the body is a function of velocity or position or both but does not explicitly depend on time. In such a case, when $F = F(x, v)$, the equation $m\dfrac{dv}{dt} = F$ transforms into

$$mv\frac{dv}{dx} = F(x, v). \tag{8}$$

We adopt the customary notation and write $v = v(x)$ when referring to velocity as a function of position and $v = v(t)$ when referring to velocity as a function of time.

Equation (8) is typically supplemented by an initial condition that prescribes velocity at some initial position $v(x_0) = v_0$. Equation (8) and the initial condition define an initial value problem on the underlying x-interval of interest.

To illustrate the simplifications that can be realized by adopting distance as the independent variable, we reconsider the problem solved in Example 1, that of determining the maximum height reached by a projectile when subjected to a drag force proportional to the square of the velocity.

EXAMPLE 2

Consider the problem treated in Example 1. The initial value problem describing the projectile's upward motion is

$$m\frac{dv}{dt} = -mg - \kappa v^2, \qquad v(0) = v_0,$$

where $v = dy/dt$. Determine the maximum height reached by the projectile.

Solution: We view velocity as a function of position y. Since $y = 0$ when $t = 0$, the initial value problem satisfied by $v(y)$ becomes

$$mv\frac{dv}{dy} = -mg - \kappa v^2, \qquad v(0) = v_0.$$

Therefore,

$$\frac{m}{\kappa} \int \frac{v}{v^2 + (mg/\kappa)}\, dv = \frac{m}{2\kappa} \ln\,[v^2 + (mg/\kappa)] = -y + C.$$

Imposing the initial condition yields

$$\frac{m}{2\kappa} \ln\,[v^2 + (mg/\kappa)] = -y + \frac{m}{2\kappa} \ln\,[v_0^2 + (mg/\kappa)]. \tag{9}$$

If we would like to have velocity expressed as a function of height y, we can "unravel" the implicit solution (9). However, to determine y_m, the maximum height reached by the projectile, we need only note that the velocity is zero at the maximum height. Setting $v = 0$ in (9) leads to

$$\frac{m}{2\kappa} \ln\,[mg/\kappa] = -y_m + \frac{m}{2\kappa} \ln\,[v_0^2 + (mg/\kappa)], \quad \text{or} \quad y_m = \frac{m}{2\kappa} \ln\left[1 + \frac{\kappa v_0^2}{mg}\right].$$

This is the same expression we found in Example 1. ❖

Impact Velocity

The next example, concerning an object falling through the atmosphere, shows that using position as the independent variable may convert a problem we cannot solve into one that we can solve.

The force of Earth's gravitational attraction diminishes as a body moves higher above the surface. For objects near the surface, the usual assumption of constant gravity is fairly reasonable. However, for a body falling from a great height (such as a satellite reentering the atmosphere), a constant gravity assumption is not accurate and may lead to erroneous predictions of the reentry trajectory.

While most of the realistic gravity models used in aerospace applications are quite complicated, they are all based on Newton's law of gravitation, which states that the force of mutual attraction between two bodies is

$$F = \frac{GmM}{r^2}. \tag{10}$$

In equation (10), $G = 6.673 \times 10^{-11}$ m^3/(kg · s^2) is the universal gravitational constant, m and M are the masses of the two bodies, and r is the distance between the centers of mass of the two bodies. [If the bodies are moving relative to each other, then r varies with time and therefore $r = r(t)$.]

As a first approximation to the force of gravitational attraction for an object falling to the surface of Earth, we assume Earth is a sphere of homogeneous material. Under this assumption, we can use equation (10) to model gravity. The mass of Earth is $M_e = 5.976 \times 10^{24}$ kg, and its radius is $R_e = 6371$ km or 6.371×10^6 m.

E X A M P L E

3

Consider an object having mass $m = 100$ kg which is released from rest at an altitude of $h = 200$ km above the surface of Earth. Neglecting drag and considering only the force of gravitational attraction, calculate the impact velocity of the object at Earth's surface.

Solution: In this problem we do not assume that the force of gravitational attraction is constant. Rather, we take into account the variation of this force with the separation distance between the two bodies. See Figure 2.17.

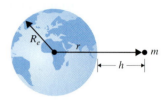

FIGURE 2.17

An object of mass m is released at an altitude of 200 km above the surface of Earth. As the object falls, its distance from the center of Earth is r. Earth has radius R_e, and therefore $r = R_e + h$ defines the object's altitude, h, above Earth. The quantity r is positive in the direction of increasing altitude.

As shown in Figure 2.17, separation distance r is measured positively from Earth's center to the 100-kg body. If $v = dr/dt$, then the application of Newton's second law of motion leads to the differential equation

$$m \frac{dv}{dt} = -\frac{GmM_e}{r^2}. \tag{11}$$

If time is retained as the independent variable, we obtain a differential equation for the dependent variable $r(t)$ by using the fact that $v = dr/dt$. The resulting differential equation, however, is second order and nonlinear:

$$m \frac{d^2r}{dt^2} = -\frac{GmM_e}{r^2}.$$

Note that the separation distance $r(t)$ is a decreasing function of time. If we transform differential equation (11) into one in which distance r is the independent variable and use the fact that $v = 0$ when $r = R_e + h$, we obtain the following initial value problem:

$$mv \frac{dv}{dr} = -\frac{GmM_e}{r^2}, \qquad R_e < r < R_e + h, \tag{12}$$
$$v(R_e + h) = 0.$$

The differential equation in (12), while nonlinear, is a first order *separable* equation for the quantity of interest, v. Solving, we obtain

$$\frac{v^2}{2} = \frac{GM_e}{r} + C.$$

Imposing the initial condition, we find the implicit solution

$$\frac{v^2}{2} = GM_e \left(\frac{1}{r} - \frac{1}{R_e + h} \right).$$

Since separation distance is decreasing with time, velocity is negative. Therefore, an explicit solution of the problem is

$$v = -\sqrt{2GM_e \left(\frac{1}{r} - \frac{1}{R_e + h} \right)}.$$

The impact velocity is found by evaluating velocity at $r = R_e$:

$$v_{\text{impact}} = -\sqrt{2GM_e \left(\frac{1}{R_e} - \frac{1}{R_e + h} \right)}.$$

Using the values given, we obtain $v_{\text{impact}} = -1952$ m/s (a speed of about 4350 mph). [Note: The impact velocity does not depend on the mass of the object, but only on its distance from Earth when it is released.] ❖

In the Exercises, we ask you to consider the same problem in the presence of a drag force that is proportional to the square of the velocity. In particular, the governing differential equation becomes

$$m \frac{dv}{dt} = -\frac{GmM_e}{r^2} + \kappa v^2. \tag{13}$$

After the independent variable is changed from time to distance, the resulting differential equation can be recast as a Bernoulli equation and then solved; see Exercise 16.

EXERCISES

1. An object of mass m is dropped from a high altitude. How long will it take the object to achieve a velocity equal to one half of its terminal velocity if the drag force is assumed to be proportional to the velocity?

2. A drag chute must be designed to reduce the speed of a 3000-lb dragster from 220 mph to 50 mph in 4 sec. Assume that the drag force is proportional to the velocity.
   (a) What value of the drag coefficient k is needed to accomplish this?
   (b) How far will the dragster travel in the 4-sec interval?

3. Repeat Exercise 2 for the case in which the drag force is proportional to the square of the velocity. Determine both the drag coefficient κ and the distance traveled.

4. A projectile of mass m is launched vertically upward from ground level at time $t = 0$ with initial velocity v_0 and is acted upon by gravity and air resistance. Assume the drag force is proportional to velocity, with drag coefficient k. Derive an expression for the time, t_m, when the projectile achieves its maximum height.

5. Derive an expression for the maximum height, $y_m = y(t_m)$, achieved in Exercise 4.

6. An object of mass m is dropped from a high altitude and is subjected to a drag force proportional to the square of its velocity. How far must the object fall before its velocity reaches one half its terminal velocity?

7. An object is dropped from altitude y_0. Determine the impact velocity if air resistance is neglected—that is, if we assume no drag force.

Exercises 8–11:

An object undergoes one-dimensional motion along the x-axis subject to the given decelerating forces. At time $t = 0$, the object's position is $x = 0$ and its velocity is $v = v_0$. In each case, the decelerating force is a function of the object's position $x(t)$ or its velocity $v(t)$ or both. Transform the problem into one having distance x as the independent variable. Determine the position x_f at which the object comes to rest. (If the object does not come to rest, $x_f = \infty$.)

8. $m\dfrac{dv}{dt} = -kx^2v$ **9.** $m\dfrac{dv}{dt} = -kxv^2$ **10.** $m\dfrac{dv}{dt} = -ke^{-x}$ **11.** $m\dfrac{dv}{dt} = -\dfrac{kv}{1+x}$

12. A boat having mass m is pushed away from a dock with an initial velocity v_0. The water exerts on the boat a drag force that is proportional to the square of its velocity. Determine the velocity of the boat when it is a distance d from the dock.

13. An object is dropped from altitude y_0.

   (a) Determine the impact velocity if the drag force is proportional to the square of velocity, with drag coefficient κ.

   (b) If the terminal velocity is known to be -120 mph and the impact velocity was -90 mph, what was the initial altitude y_0? (Recall that we take velocity to be negative when the object is moving downward.)

14. An object is dropped from altitude y_0.

   (a) Assume that the drag force is proportional to velocity, with drag coefficient k. Obtain an implicit solution relating velocity and altitude.

   (b) If the terminal velocity is known to be -120 mph and the impact velocity was -90 mph, what was the initial altitude y_0?

15. We need to design a ballistics chamber to decelerate test projectiles fired into it. Assume the resistive force encountered by the projectile is proportional to the square of its velocity and neglect gravity. As the figure indicates, the chamber is to be constructed so that the coefficient κ associated with this resistive force is not constant but is, in fact, a linearly increasing function of distance into the chamber. Let $\kappa(x) = \kappa_0 x$, where κ_0 is a constant; the resistive force then has the form $\kappa(x)v^2 = \kappa_0 x v^2$. If we use time t as the independent variable, Newton's law of motion leads us to the differential equation

$$m\frac{dv}{dt} + \kappa_0 x v^2 = 0 \quad \text{with} \quad v = \frac{dx}{dt}. \tag{14}$$

   (a) Adopt distance x into the chamber as the new independent variable and rewrite differential equation (14) as a first order equation in terms of the new independent variable.

   (b) Determine the value κ_0 needed if the chamber is to reduce projectile velocity to 1% of its incoming value within d units of distance.

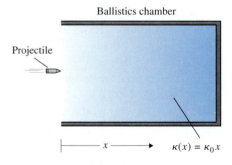

Ballistics chamber

Projectile

x

$\kappa(x) = \kappa_0 x$

Figure for Exercise 15

16. The motion of a body of mass m, gravitationally attracted to Earth in the presence of a resisting drag force proportional to the square of its velocity, is given by

$$m\frac{dv}{dt} = -\frac{GmM_e}{r^2} + \kappa v^2$$

[recall equation (13)]. In this equation, r is the radial distance of the body from the center of Earth, G is the universal gravitational constant, M_e is the mass of Earth, and $v = dr/dt$. Note that the drag force is positive, since it acts in the positive r direction.

(a) Assume that the body is released from rest at an altitude h above the surface of Earth. Recast the differential equation so that distance r is the independent variable. State an appropriate initial condition for the new problem.

(b) Show that the impact velocity can be expressed as

$$v_{\text{impact}} = -\left[2GM_e \int_0^h \frac{e^{-2(\kappa/m)s}}{(R_e + s)^2}\, ds\right]^{1/2},$$

where R_e represents the radius of Earth. (The minus sign reflects the fact that $v = dr/dt < 0$.)

17. On August 24, 1894, Pop Shriver of the Chicago White Stockings caught a baseball dropped (by Clark Griffith) from the top of the Washington Monument. The Washington Monument is 555 ft tall and a baseball weighs $5\frac{1}{8}$ oz.

(a) If we ignore air resistance and assume the baseball was acted upon only by gravity, how fast would the baseball have been traveling when it was 7 ft above the ground?

(b) Suppose we now include air resistance in our model, assuming that the drag force is proportional to velocity with a drag coefficient $k = 0.0018$ lb-sec/ft. How fast is the baseball traveling in this case when it is 7 ft above the ground?

18. A 180-lb skydiver drops from a hot-air balloon. After 10 sec of free fall, a parachute is opened. The parachute immediately introduces a drag force proportional to velocity. After an additional 4 sec, the parachutist reaches the ground. Assume that air resistance is negligible during free fall and that the parachute is designed so that a 200-lb person will reach a terminal velocity of -10 mph.

(a) What is the speed of the skydiver immediately before the parachute is opened?

(b) What is the parachutist's impact velocity?

(c) At what altitude was the parachute opened?

(d) What is the balloon's altitude?

19. When modeling the action of drag chutes and parachutes, we have assumed that the chute opens instantaneously. Real devices take a short amount of time to fully open and deploy.

   In this exercise, we try to assess the importance of this distinction. Consider again the assumptions of Exercise 2. A 3000-lb dragster is moving on a straight track at a speed of 220 mph when, at time $t = 0$, the drag chute is opened. If we assume that the drag force is proportional to velocity and that the chute opens instantaneously, the differential equation to solve is $mv' = -kv$.

   If we assume a short deployment time to open the chute, a reasonable differential equation might be $mv' = -k(\tanh t)v$. Since $\tanh(0) = 0$ and $\tanh(1) \approx 0.76$, it will take about 1 sec for the chute to become 76% deployed in this model.

   Assume $k = 25$ lb-sec/ft. Solve the two differential equations and determine in each case how long it takes the vehicle to slow to 50 mph. Which time do you anticipate will be larger? (Explain.) Is the idealization of instantaneous chute deployment realistic?

20. An object of mass m is dropped from a high platform at time $t = 0$. Assume the drag force is proportional to the square of the velocity, with drag coefficient κ. As in Example 1, derive an expression for the velocity $v(t)$.

21. Assume the action of a parachute can be modeled as a drag force proportional to the square of the velocity. Determine the drag coefficient κ of the parachute needed so that a 200-lb person using the parachute will have a terminal velocity of -10 mph.

Pendulum Motion: Conservation of Energy   In Exercises 22 and 23, our goal is to describe the rotational motion of the pendulum shown in the figure. We neglect the weight of the rod when modeling the motion of this pendulum.

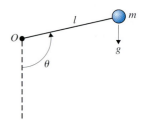

Figure for Exercises 22 and 23

Applying the rotational version of Newton's laws to the pendulum leads to the second order differential equation

$$ml^2\theta'' = -mgl\sin\theta. \tag{15}$$

In equation (15), the right-hand side is negative because it acts to cause clockwise rotation—that is, rotation in the negative θ direction.

22. Suppose that at some initial time the pendulum is located at angle θ_0 with an angular velocity $d\theta/dt = \omega_0$ radians/sec.

(a) Equation (15) is a second order differential equation. Rewrite it as a first order separable equation by adopting angle θ as the independent variable, using the fact that

$$\theta'' = \frac{d}{dt}\left(\frac{d\theta}{dt}\right) = \frac{d\omega}{dt} = \frac{d\omega}{d\theta}\frac{d\theta}{dt} = \omega\frac{d\omega}{d\theta}.$$

Complete the specification of the initial value problem by specifying an appropriate initial condition.

(b) Obtain the implicit solution

$$ml^2\frac{\omega^2}{2} - mgl\cos\theta = ml^2\frac{\omega_0^2}{2} - mgl\cos\theta_0. \tag{16}$$

The pendulum is a conservative system; that is, energy is neither created nor destroyed. Equation (16) is a statement of conservation of energy. At a position defined by the angle θ, the quantity $ml^2\omega^2/2$ is the kinetic energy of the pendulum while the term $-mgl\cos\theta$ is the potential energy, referenced to the horizontal position $\theta = \pi/2$. The constant right-hand side is the total initial energy.

(c) Determine the angular velocity at the instant the pendulum reaches the vertically downward position, $\theta = 0$. Express your answer in terms of the constants ω_0 and θ_0.

23. A pendulum, 2 ft in length and initially in the downward position, is launched with an initial angular velocity ω_0. If it achieves a maximum angular displacement of 135 degrees, what is ω_0?

2.10   Euler's Method

Up to this point, we have focused on *analytical* techniques for solving the initial value problem

$$y' = f(t, y), \qquad y(t_0) = y_0. \tag{1}$$

In Section 2.2, we saw that there is an explicit representation, in terms of integrals, for the solution of a linear differential equation. However, we also saw that it might be difficult to work with this representation if the integrand does not have an elementary antiderivative.

Sections 2.5–2.7 discussed analytical techniques for solving certain special types of nonlinear differential equations (Bernoulli equations, separable equations, exact equations, etc.). These techniques, however, often lead to implicit solutions, and it may be difficult to "unravel" the implicit solution in order to obtain an explicit solution. In addition, there are many nonlinear differential equations that do not belong to the special categories for which analytical techniques have been developed.

Therefore, it's clear that the analytical methods we've discussed, while important and useful, are not totally adequate. We need tools that enable us to obtain quantitative information about the solution of (1) in the general case. Numerical methods are one such tool. We introduce numerical techniques by considering Euler's method,[9] perhaps the simplest and most intuitive numerical method.

Besides serving as an introduction to numerical techniques, Euler's method also provides us with a means (albeit a relatively crude one) to analyze initial value problems for which analytical methods are not applicable. Numerical methods are discussed in greater detail in Chapter 7.

We begin by asking the most basic question: "What is a numerical solution of initial value problem (1)?" As we will see, a numerical method for (1) typically generates values

$$y_1, \quad y_2, \quad \ldots, \quad y_n$$

that approximate corresponding solution values

$$y(t_1), \quad y(t_2), \quad \ldots, \quad y(t_n)$$

---

[9]Leonhard Euler (1707–1783) was one of the most gifted individuals in the history of mathematics and science; his 866 publications (books and articles) make him arguably the most prolific as well. He established the foundations of mathematical analysis, the theory of special functions, and analytical mechanics. Complex analysis, number theory, differential equations, differential geometry, lunar theory, elasticity, acoustics, fluid mechanics, and the wave theory of light are some of the other areas to which Euler made important contributions. A backlog of his work continued to be published for nearly 50 years after his death.

Euler's achievements become even more remarkable when one appreciates the circumstances surrounding his work. He was involved in the world about him. In Russia, he worked on state projects involving cartography, science education, magnetism, fire engines, machines, and shipbuilding. Later, in Berlin, he served as an advisor to the government on state lotteries, insurance, annuities, and pensions. He fathered thirteen children, although only five survived infancy. Euler claimed that he made some of his greatest discoveries while holding an infant in his arms with other children playing at his feet. In 1771 Euler became totally blind, but he was able to maintain a prodigious output of work until his death.

at designated abscissa values $t_0 < t_1 < t_2 < \cdots < t_n$. Figure 2.18 illustrates the output of a numerical method, where the solid curve indicates the actual (unknown) solution of initial value problem (1).

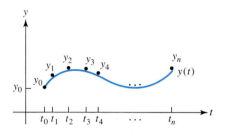

FIGURE 2.18

The solid curve indicates the actual (unknown) solution of initial value problem (1). For $i = 1, 2, \ldots, n$, the points (t_i, y_i) are approximations to corresponding points $(t_i, y(t_i))$ on the actual solution curve. We say that y_i is a numerical approximation to the unknown value $y(t_i)$.

In Figure 2.18, the values $y_1, y_2, y_3, \ldots$ are found sequentially—first y_1, then y_2, and so forth. Note that the starting value $y_0 = y(t_0)$ is known exactly, since it is specified in the initial value problem (1). The numerical problem therefore reduces to this question:

Given (t_0, y_0), how do we determine (t_1, y_1)? And in general, given an approximation (t_k, y_k), how do we determine the next approximation, (t_{k+1}, y_{k+1})?

The simplest answer, the simplest numerical algorithm, is **Euler's method** (also known as the **tangent line method**).

Derivation of Euler's Method

The geometric ideas underlying Euler's method can be understood in terms of direction fields. At the initial condition point (t_0, y_0), the differential equation specifies the slope of the solution curve at (t_0, y_0); it is equal to $f(t_0, y_0)$. Therefore, the line tangent to the solution curve $y(t)$ at the point (t_0, y_0) has equation

$$y = y_0 + f(t_0, y_0)(t - t_0). \qquad (2)$$

We follow this tangent line over a short time interval, to $t = t_1$, where $t_0 < t_1$. At $t = t_1$, we reach the point (t_1, y_1), where

$$y_1 = y_0 + f(t_0, y_0)(t_1 - t_0).$$

The value found above, y_1, is the Euler's method approximation to the solution value $y(t_1)$. (See Figure 2.19.)

While the new point (t_1, y_1) does not exactly coincide with the point $(t_1, y(t_1))$, it is generally close to that point (assuming t_1 is sufficiently close to t_0). Moreover, since $f(t, y)$ is continuous, the direction field filament at (t_1, y_1)

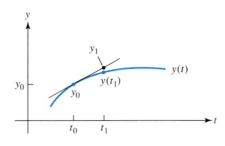

FIGURE 2.19

The line tangent to $y(t)$ at the initial point (t_0, y_0) has slope $f(t_0, y_0)$. Following the tangent line to time t_1, we arrive at the point (t_1, y_1) and have an approximation, y_1, to the solution value, $y(t_1)$.

has nearly the same slope as the filament at $(t_1, y(t_1))$. Hence, although we do not know the exact value of $y(t)$ when $t = t_1$, we are close to it and we do have a good idea of which direction the graph of $y(t)$ is heading. At $t = t_1$, the graph of $y(t)$ has slope $f(t_1, y(t_1))$, which is nearly equal to $f(t_1, y_1)$. So, in an attempt to follow the solution curve $y(t)$, we proceed from (t_1, y_1) along a line having slope $f(t_1, y_1)$:

$$y = y_1 + f(t_1, y_1)(t - t_1).$$

Following this line until $t = t_2$, we reach a point (t_2, y_2), where

$$y_2 = y_1 + f(t_1, y_1)(t_2 - t_1).$$

This process is repeated, leading to the algorithm

$$y_{k+1} = y_k + f(t_k, y_k)(t_{k+1} - t_k), \qquad k = 0, 1, 2, \ldots. \tag{3}$$

Iteration (3) is known as Euler's method and is illustrated in Figure 2.20. Euler's method amounts to tracing a polygonal path through the direction field.

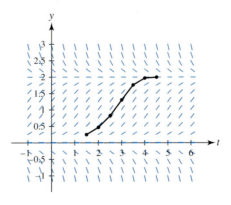

FIGURE 2.20

Starting on the solution curve at (t_0, y_0), Euler's method attempts to track the solution $y(t)$, tracing a polygonal path through the direction field. As the path proceeds, its direction is constantly corrected by sampling the direction field; see equation (3).

Apply Euler's method to the initial value problem

$$y' = t^2 + y, \qquad y(2) = 1.$$

Use $t_1 = 2.1$, $t_2 = 2.2$, and $t_3 = 2.3$. Generate approximations y_1 to $y(2.1)$, y_2 to $y(2.2)$, and y_3 to $y(2.3)$.

Solution: The actual (unknown) solution starts at $(t_0, y_0) = (2, 1)$ and has a starting slope of $f(t_0, y_0) = f(2, 1) = 5$. Following the line of slope 5 passing through $(2, 1)$, we obtain, by (3),

$$y_1 = y_0 + f(t_0, y_0)(t_1 - t_0) = 1 + 5(0.1) = 1.5.$$

Having $(t_1, y_1) = (2.1, 1.5)$, we take the next step to (t_2, y_2):

$$y_2 = y_1 + f(t_1, y_1)(t_2 - t_1) = 1.5 + 5.91(0.1) = 2.091.$$

Having $(t_2, y_2) = (2.2, 2.091)$, we take the next step to (t_3, y_3):

$$y_3 = y_2 + f(t_2, y_2)(t_3 - t_2) = 2.091 + 6.931(0.1) = 2.7841.$$

Note that the differential equation in this example is linear, and hence a formula for the solution can be found. The exact solution is

$$y(t) = 11e^{t-2} - (t^2 + 2t + 2).$$

Figure 2.21 compares the Euler's method approximations with the exact solution at the points $t = 2, 2.1, 2.2,$ and 2.3.

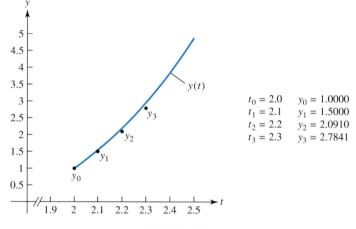

FIGURE 2.21

The values $y_1, y_2,$ and y_3 are the Euler's method approximations found in Example 1.

Implementing Euler's Method

The simplest way to organize an Euler's method calculation is to choose an appropriate **step size**, h, and then use h to define the equally spaced sample points:

$$t_1 = t_0 + h, \qquad t_2 = t_1 + h,$$

and, in general,

$$t_{k+1} = t_k + h, \qquad k = 0, 1, \ldots, n - 1.$$

In Example 1 we used a step size of $h = 0.1$ to define sample points $t_0 = 2.0$, $t_1 = 2.1$, $t_2 = 2.2$, and $t_3 = 2.3$.

For a constant step size h, the term $t_{k+1} - t_k$ in (3) is equal to h. Thus, Euler's method takes the form

$$y_{k+1} = y_k + hf(t_k, y_k), \qquad k = 0, 1, \dots, n-1. \tag{4}$$

We anticipate that Euler's method should become more accurate when we take smaller steps, sampling the direction field more often and using this information to correct the "Euler path" that is tracking the solution $y(t)$ (see Exercise 15). Using a small step size h, however, may lead to a significant amount of computation. Therefore, numerical methods are usually programmed and run on a computer or programmable calculator.

In this section, we have assumed a constant step size h in order to simplify the discussion. Many implementations of numerical methods, however, use variable-step algorithms rather than a fixed-step algorithm. Such variable-step algorithms use error estimates that monitor errors as the algorithm proceeds. When errors are increasing, the steplength is reduced; when errors are decreasing, the steplength is increased.

E X A M P L E

2

Apply Euler's method to the initial value problem

$$y' = y(2 - y), \qquad y(0) = 0.1. \tag{5}$$

Use $h = 0.2$ and approximate the solution on the interval $0 \le t \le 4$.

Solution: Using a fixed step size of $h = 0.2$, Euler's method samples the direction field at time values $t_1 = 0.2$, $t_2 = 0.4, \dots, t_{19} = 3.8$, and $t_{20} = 4.0$. For this test case, initial value problem (5) can be solved, since the differential equation is separable. Figure 2.22 shows a portion of the direction field for $y' = y(2 - y)$. The solid curve in Figure 2.22 is the graph of the actual solution of (5), and the dots show the Euler path through the direction field.

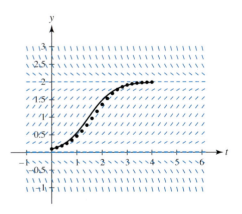

FIGURE 2.22

The direction field for $y' = y(2 - y)$. The curve denotes the solution of $y' = y(2 - y)$, $y(0) = 0.1$, the initial value problem posed in Example 2. The dots are the points generated by Euler's method, using a step size of $h = 0.2$.

❖

Runge-Kutta Methods

Euler's method is conceptually important, but it is a relatively crude numerical algorithm. The question then becomes "How do we systematically develop algorithms that are more accurate?"

A key to systematically achieving greater accuracy is to view Euler's method from another, slightly different perspective. Consider the initial value problem (1). Suppose we assume not only that the solution $y(t)$ is differentiable, but also that it has a Taylor series expansion that converges in the interval of interest, $a \leq t \leq b$. We then have, for t and $t + h$ in $[a, b]$,

$$y(t + h) = y(t) + y'(t)h + \frac{y''(t)}{2!}h^2 + \frac{y'''(t)}{3!}h^3 + \cdots. \tag{6}$$

If we set $t = t_k, t + h = t_{k+1}$ and use the fact that $y'(t) = f(t, y(t))$, we obtain

$$y(t_{k+1}) = y(t_k) + f(t_k, y(t_k))h + \frac{y''(t_k)}{2!}h^2 + \frac{y'''(t_k)}{3!}h^3 + \cdots. \tag{7}$$

As shown in (7), Euler's method can be viewed as truncating the Taylor series after the linear term. Equation (7) also shows that the error made in taking one step of Euler's method is the infinite sum that begins with the term $y''(t_k)h^2/2!$. In most applications, it can be shown (using Taylor's theorem) that this error is bounded by a constant multiple of h^2.

Equation (7) provides a blueprint for improving accuracy; an improved algorithm should somehow incorporate more terms of the Taylor series. For example, if the algorithm incorporated the term $y''(t_k)h^2/2!$, then it would be including concavity information in addition to the slope information given by the term $y'(t_k)h$. Moreover, such an algorithm would have an error bounded by a constant multiple of h^3 instead of a multiple of h^2. Since h is small, we would expect increased accuracy.

It is shown in Chapter 7 that the terms $y^{(n)}(t_k)h^n/n!$ can be obtained directly from the differential equation $y' = f(t, y)$. In principle, therefore, retaining additional terms in (7) in order to create a more accurate algorithm is straightforward. There are two practical difficulties, however. The computations necessary to determine

$$y^{(n)}(t_k) = \left. \frac{d^{(n-1)}}{dt^{(n-1)}} f(t, y(t)) \right|_{t=t_k} \tag{8}$$

very quickly become unwieldy as n increases. Moreover, the resulting algorithm is problem specific, since the calculations in (8) have to be redone every time the method is applied to a new differential equation.

The challenge facing the numerical analyst is to retain more terms of the Taylor series (7), but in a way that is both problem independent and computationally friendly. One operation that computers perform very easily is the evaluation of functions. Knowing this, numerical analysts have achieved the desired objectives by creating algorithms that use nested compositions of functions to approximate the higher derivatives $y^{(n)}(t_k), n = 2, 3, \ldots$. Several such algorithms are discussed in Chapter 7. One popular example is the fourth-order

Runge-Kutta[10] method listed in (9). For initial value problem (1), this method has the form

$$y_{k+1} = y_k + \frac{h}{6}[K_1 + 2K_2 + 2K_3 + K_4], \tag{9a}$$

where

$$
\begin{aligned}
K_1 &= f(t_k, y_k) \\
K_2 &= f(t_k + h/2, y_k + (h/2)K_1) \\
K_3 &= f(t_k + h/2, y_k + (h/2)K_2) \\
K_4 &= f(t_k + h, y_k + hK_3).
\end{aligned}
\tag{9b}
$$

Note that the terms K_j in (9b) are formed by sequentially evaluating compositions of the function f; for example,

$$K_3 = f(t_k + h/2, y_k + (h/2)f(t_k + h/2, y_k + (h/2)f(t_k, y_k))).$$

When the compositions in (9) are unraveled, it can be seen that the algorithm correctly replicates the Taylor series expansion (7) up to and including the term $y^{(4)}(t_k)h^4/4!$.

E X A M P L E

3

As a test case to illustrate how the Runge-Kutta philosophy can improve accuracy, consider the initial value problem

$$y' = 2ty + 1, \qquad y(0) = 2.$$

(a) Solve this initial value problem mathematically.

(b) Solve this initial value problem numerically on the interval $0 \le t \le 2$, using Euler's method and the Runge-Kutta method (9). Use a constant step size of $h = 0.05$.

(c) Tabulate the exact solution values and both sets of numerical approximations from $t = 0$ to $t = 2$ in steps of $\Delta t = 0.25$. Is the Runge-Kutta method more accurate than Euler's method for this test case?

Solution:

(a) This initial value problem was given in Exercise 45 in Section 2.2. The exact solution is

$$y(t) = e^{t^2}\left[2 + \frac{\sqrt{\pi}}{2}\mathrm{erf}(t)\right],$$

where $\mathrm{erf}(t)$ denotes the error function, $\mathrm{erf}(t) = \dfrac{2}{\sqrt{\pi}}\displaystyle\int_0^t e^{-s^2}\,ds$.

(continued)

---

[10]Carle David Tolmé Runge (1856–1927) was a German scientist whose initial interest in pure mathematics was eventually supplanted by an interest in spectroscopy and applied mathematics. During his career, he held positions at universities in Hanover and Gottingen. Runge was a particularly fit and active man; it is reported that he entertained grandchildren on his seventieth birthday by doing handstands.

Martin Wilhelm Kutta (1867–1944) held positions at Munich, Jena, Aachen, and Stuttgart. In addition to the Runge-Kutta method (1901), he is remembered for his work in the study of airfoils.

(continued)

(b) We coded the Runge-Kutta method (9) for this initial value problem, using MATLAB as a programming environment; a listing, together with a discussion of the practical aspects of coding a Runge-Kutta, is given in the next subsection. (We also provide brief tutorials for MATLAB and Mathematica in our Technical Resource Manual. You may view these tutorials at http://www.aw-bc.com/kohler/.)

(c) The results are shown in Table 2.1. As is shown in the table, the Runge-Kutta results for the constant step size $h = 0.05$ are considerably more accurate than those of Euler's method using $h = 0.05$.

TABLE 2.1

The Results of Example 3
Here, y^E denotes the Euler's method estimates of $y(t)$, y^{RK} denotes the Runge-Kutta method estimates, and y^T the true values.

t	y^E	y^{RK}	y^T
0.0000	2.0000	2.0000	2.0000
0.2500	2.3594	2.3897	2.3897
0.5000	3.0726	3.1603	3.1603
0.7500	4.3960	4.6162	4.6162
1.0000	6.9084	7.4666	7.4666
1.2500	11.9543	13.4434	13.4434
1.5000	22.8447	27.0987	27.0988
1.7500	48.3243	61.4573	61.4577
2.0000	113.2709	157.3539	157.3563

❖

Coding a Runge-Kutta Method

We conclude with a short discussion of the practical aspects of writing a program to implement a Runge-Kutta method. Figures 2.23 and 2.24 list the programs used to generate the numerical solution in Example 3. This particular code was written in MATLAB, but the principles are the same for any programming language.

We note first that no matter what numerical method we decide to use for the initial value problem

$$y' = f(t, y), \qquad y(t_0) = y_0,$$

we need to write a subprogram (or module) that evaluates $f(t, y)$. Such a module is listed in Figure 2.24 for the initial value problem of Example 3. Figure 2.23 lists a MATLAB program that executes 40 steps of the fourth-order Runge-Kutta method (9) for the initial value problem of Example 3.

The program listed in Figure 2.23 is not as general as it could be. Normally a Runge-Kutta code is written as a subprogram or module that we can call

```
%
%  Set the initial conditions for the
%    initial value problem of Example 3
%
t=0;
y=2;
h=0.05;
output=[t,y];
%
%  Execute the fourth-order Runge-Kutta method
%    on the interval [0,2]
%
for i=1:40
    ttemp=t;
    ytemp=y;
    k1=f(ttemp,ytemp);
    ttemp=t+h/2;
    ytemp=y+(h/2)*k1;
    k2=f(ttemp,ytemp);
    ttemp=t+h/2;
    ytemp=y+(h/2)*k2;
    k3=f(ttemp,ytemp);
    ttemp=t+h;
    ytemp=y+h*k3;
    k4=f(ttemp,ytemp);
    y=y+(h/6)*(k1+2*k2+2*k3+k4);
    t=t+h;
    output=[output;t,y];
end
```

FIGURE 2.23

A Runge-Kutta code for the initial value problem in Example 3.

```
function yp=f(t,y)
yp=2*t*y+1;
```

FIGURE 2.24

A function subprogram that evaluates $f(t, y)$ for the differential equation of Example 3.

whenever we have an initial value problem to solve numerically. Such subprograms allow the user to input a step size h, the number of steps to execute, and the initial conditions. They can be used over and over again and do not have to be modified whenever the initial value problem changes. We did not list a general module for Figure 2.23 because we wanted the basic steps of a Runge-Kutta program to be obvious.

Observe that the code listed in Figure 2.23 stays as close as possible to the notation and format of the fourth-order Runge-Kutta method (9). In general, it is a good idea to use variable names (such as K_1 and K_2) that match the names in the algorithm. Beyond the choice of variable names, the code in Figure 2.23 also mimics the steps of algorithm (9) as closely as possible. Adhering to such conventions makes programs much easier to read and debug.

EXERCISES

Exercises 1–6:

In each exercise,

(a) Write the Euler's method iteration $y_{k+1} = y_k + hf(t_k, y_k)$ for the given problem. Also, identify the values t_0 and y_0.

(b) Using step size $h = 0.1$, compute the approximations y_1, y_2, and y_3.

(c) Solve the given problem analytically.

(d) Using the results from (b) and (c), tabulate the errors $e_k = y(t_k) - y_k$ for $k = 1, 2, 3$.

1. $y' = 2t - 1$, $y(1) = 0$
2. $y' = -y$, $y(0) = 1$
3. $y' = -ty$, $y(0) = 1$
4. $y' = -y + t$, $y(0) = 0$
5. $y' = y^2$, $y(0) = 1$
6. $y' = y$, $y(-1) = -1$

Exercises 7–10:

Reducing the Step Size   These exercises examine graphically the effects of reducing step size on the accuracy of the numerical solution. A computer or programmable calculator is needed.

(a) Use Euler's method to obtain numerical solutions on the specified time interval for step sizes $h = 0.1$, $h = 0.05$, and $h = 0.025$.

(b) Solve the problem analytically and plot the exact solution and the three numerical solutions on a single graph. Does the error appear to be getting smaller as h is reduced?

7. $y' = 2y - 1$, $y(0) = 1$, $0 \leq t \leq 0.5$
8. $y' = y + e^{-t}$, $y(0) = 0$, $0 \leq t \leq 1$
9. $y' = y^{-1}$, $y(0) = 1$, $0 \leq t \leq 1$
10. $y' = -y^2$, $y(-1) = 2$, $-1 \leq t \leq 0$

11. Assume we are considering the direction field of an autonomous first order differential equation.

(a) Suppose we can qualitatively establish, by examining this direction field, that all solution curves $y(t)$ in a given region of the ty-plane have one of the following four types of behavior:

(i)   increasing, concave up
(ii)   increasing, concave down
(iii) decreasing, concave up
(iv) decreasing, concave down.

Suppose we implement an Euler's method approximation to one of the solution curves in the region, using some reasonable step size h. Consider each of the four cases. In each case, will the values y_k underestimate the exact values or overestimate the exact values or is it impossible to reach a definite conclusion?

(b) What do you think will happen if Euler's method is used to approximate an "S-shaped solution curve" similar to the logistic curve shown in Figure 2.22 on page 93. In that case, a solution curve changes from increasing and concave up to increasing and concave down. Are your answers to part (a) consistent with the behavior exhibited by the Euler approximation shown in the figure?

Exercises 12–15:

A programmable calculator or computer is needed for these exercises.

12. Use Euler's method with step size $h = 0.01$ to numerically solve the initial value problem

$$y' - ty = \sin 2\pi t, \qquad y(0) = 1, \qquad 0 \le t \le 1.$$

Graph the numerical solution. [Note: Although the differential equation in this problem is a first order linear equation and we can get an explicit representation for the exact solution, the representation involves antiderivatives that we cannot express in terms of known functions. From a quantitative point of view, the representation itself is of little use.]

13. Assume a tank having a capacity of 200 gal initially contains 90 gal of fresh water. At time $t = 0$, a salt solution begins flowing into the tank at a rate of 6 gal/min and the well-stirred mixture flows out at a rate of 1 gal/min. Assume that the inflow concentration is given by $c(t) = 2 - \cos \pi t$ oz/gal, where time t is in minutes. Formulate the appropriate initial value problem for $Q(t)$, the amount of salt (in ounces) in the tank at time t. Use Euler's method to approximately determine the amount of salt in the tank when the tank contains 100 gal of liquid. Use a step size of $h = 0.01$.

14. Let $P(t)$ denote the population of a certain colony, measured in millions of members. The colony is modeled by

$$P' = 0.1 \left(1 - \frac{P}{3}\right) P + M(t), \qquad P(0) = P_0,$$

where time t is measured in years. Assume that the colony experiences a migration influx that is initially strong but that soon tapers off. Specifically assume that $M(t) = e^{-t}$. Suppose the colony had 500,000 members initially. Use Euler's method to estimate its size after 2 years.

15. In Chapter 7, we will examine how the error in numerical algorithms, such as Euler's method, depends on step size h. In this exercise, we further examine the dependence of errors on step size by studying a particular example,

$$y' = y + 1, \qquad y(0) = 0.$$

(a) Use Euler's method to obtain approximate solutions to this initial value problem on the interval $0 \le t \le 1$, using step sizes $h_1 = 0.02$ and $h_2 = 0.01$. You will therefore obtain two sets of points,

$$(t_k^{(1)}, y_k^{(1)}), \qquad k = 0, \ldots, 50$$
$$(t_k^{(2)}, y_k^{(2)}), \qquad k = 0, \ldots, 100$$

where $t_k^{(1)} = 0.02k, k = 0, 1, \ldots, 50$ and $t_k^{(2)} = 0.01k, k = 0, 1, \ldots, 100$.

(b) Determine the exact solution, $y(t)$.

(c) Print a table of the errors at the common points, $t_k^{(1)}, k = 0, 1, \ldots, 50$:

$$e^{(1)}(t_k^{(1)}) = y(t_k^{(1)}) - y_k^{(1)} \quad \text{and} \quad e^{(2)}(t_k^{(1)}) = y(t_k^{(1)}) - y_{2k}^{(2)}.$$

(d) Note that the approximations $y_{2k}^{(2)}$ were found using a step size equal to one half of the step size used to obtain the approximations $y_k^{(1)}$; that is, $h_2 = h_1/2$. Compute the corresponding error ratios. In particular, compute

$$\left| \frac{e^{(2)}(t_k^{(1)})}{e^{(1)}(t_k^{(1)})} \right|, \qquad k = 1, \ldots, 50.$$

On the basis of these computations, conjecture how halving the step size affects the error of Euler's method.

16. This exercise treats the simple initial value problem $y' = \lambda y, y(0) = y_0$, where we can see the behavior of the numerical solution as the step size h approaches zero.

(a) Show that the solution of the initial value problem is $y = e^{\lambda t}y_0$.

(b) Apply Euler's method to the initial value problem, using a step size h. Show that y_n is related to the initial value by the formula $y_n = (1 + h\lambda)^n y_0$.

(c) Consider any fixed time t^*, where $t^* > 0$. Let $h = t^*/n$ so that $t^* = nh$. (The exact solution is $y(t) = y_0 e^{\lambda t}$.) Show that letting $h \to 0$ and $n \to \infty$ in such a way that t^* remains fixed leads to

$$\lim_{\substack{h \to 0 \\ t^* = nh}} y_n = y(t^*).$$

17. Consider the initial value problem $y' = 2t - 1, y(1) = 0$.

(a) Solve this initial value problem.

(b) Suppose Euler's method is used to solve this problem numerically on the interval $1 \le t \le 5$, using step size $h = 0.1$. Will the numerical solution values be the same as the exact solution values found in part (a)? That is, will $y_k = y(t_k), k = 1, 2, \ldots, 40$? Explain.

(c) What will be the answer to the question posed in part (b) if the Runge-Kutta method (9) is used instead of Euler's method?

Exercises 18–22:

In each exercise,

(a) Using step size $h = 01$, compute the first estimate y_1 using Euler's method and the Runge-Kutta method (9). Let these estimates be denoted by y_1^E and y_1^{RK}, respectively.

(b) Solve the problem analytically.

(c) Compute the errors $|y(t_1) - y_1|$ for the two estimates obtained in (a).

18. $y' = -y, \quad y(0) = 1$ **19.** $y' = -ty, \quad y(0) = 1$ **20.** $y' = -y + t, \quad y(0) = 0$

21. $y' = y^2, \quad y(0) = 1$ **22.** $y' = y, \quad y(-1) = -1$

Exercises 23–27:

A computer or programmable calculator is needed for these exercises. For the given initial value problem, use the Runge-Kutta method (9) with a step size of $h = 0.1$ to obtain a numerical solution on the specified interval.

23. $y' = -ty + 1, \quad y(0) = 0, \quad 0 \le t \le 2$ **24.** $y' = y^3, \quad y(1) = 0.5, \quad 1 \le t \le 2$

25. $y' = -y + t, \quad y(1) = 0, \quad 1 \le t \le 5$ **26.** $y' + 2ty = \sin t, \quad y(0) = 0, \quad 0 \le t \le 3$

27. $y' = y^2, \quad y(0) = 1, \quad 0 \le t \le 0.9$

CHAPTER 2 REVIEW EXERCISES

These review exercises provide you with an opportunity to test your understanding of the concepts and solution techniques developed in this chapter. The end-of-section exercises deal with the topics discussed in the section. These review exercises, however, require you to identify an appropriate solution technique before solving the problem.

Exercises 1–30:

If the differential equation is linear, determine the general solution. If the differential equation is nonlinear, obtain an implicit solution (and an explicit solution if possible). If an initial condition is given, solve the initial value problem.

1. $y' + 2y = 6$

2. $\dfrac{y'}{t^2 + 4} = 3y$

3. $y' - 3t^2 y^{-1} = 0$

4. $y^2 y' = 1 + y^3, \quad y(0) = 0$

5. $y' + 2ty = 2t$

6. $y' - \dfrac{y}{\sqrt{t}} = 0, \quad y(1) = 3e^2$

7. $(t + 3y^2)y' + y + 2t = 0$

8. $y' = \sin t, \quad y(0) = 4$

9. $\dfrac{y'}{t^2 + 1} = 3y, \quad y(0) = 4$

10. $y' - 2y = 0$

11. $\sqrt{t}\, y' - \sqrt{y}\, t = 0$

12. $y' - 4y = 6e^{2t}$

13. $(t \sin y)y' - \cos y = 0$

14. $y' = e^{3t + 2y}, \quad y(0) = 0$

15. $y' + p(t)y = 0, \quad y(0) = 1; \quad p(t) = \begin{cases} 2, & 0 \le t < 1 \\ 1, & 1 \le t \le 2 \end{cases}$

16. $y' - y = g(t), \quad y(0) = 0; \quad g(t) = \begin{cases} 3e^t, & 0 \le t < 1 \\ 0, & 1 \le t \le 2 \end{cases}$

17. $y' = y^3, \quad y(0) = 1$

18. $ty' - 2y = 0, \quad t > 0$

19. $y' + (\cos t)y = \cos t$

20. $y^2 y' = 2t(1 + y^3)$

21. $y' = t\sqrt{y - 1}, \quad y(1) = 5$

22. $(2y + 3t^3)y' + 9yt^2 = 0$

23. $2\sqrt{t}\, y' - y = 8, \quad t > 0$

24. $2yy' = e^{t - y^2}$

25. $y' + y = 12, \quad y(0) = 5$

26. $t + 5yy' = 0$

27. $2yy' + 3t^2 = 4, \quad y(0) = 1$

28. $t^2(1 + 9y^2)y' + 2ty(1 + 3y^2) = 0$

29. $y' + (\sin 2t)y = \sin 2t, \quad y(\pi/4) = 4$

30. $t^2 y' + \csc y = 0, \quad t > 0$

PROJECTS

Project 1: Flushing Out a Radioactive Spill

A lake holding 5,000,000 gallons of water is fed by a stream. Assume that fresh water flows into the lake at a rate of 1000 gal/min and that water flows out at the same rate. At a certain instant, an accidental spill introduces 5 lb of soluble radioactive pollutant into the lake. Assume that the radioactive substance has a half-life of 2 days and dissolves in the lake water to form a well-stirred mixture.

1. Let $Q(t)$ denote the amount of radioactive material present within the lake at time t, measured in minutes from the instant of the spill. Use the conservation principle [rate of change of $Q(t)$ equals rate in minus rate out] to derive a differential equation describing how $Q(t)$ changes with time. Note that the solute is removed by both outflow and radioactive decay. Add the appropriate initial condition to obtain the initial value problem of interest.

2. Solve the initial value problem formulated in part 1.

3. How long will it take for the concentration of the radioactive pollutant to be reduced to 0.01% of its original value?

Project 2: Processing Seafood

Many foods, such as crabmeat, are sterilized by cooking. Harvested crabs are laden with bacteria, and the crabmeat must be steamed to reduce the bacteria population to an acceptable level. The longer the crabmeat is steamed, the lower the final bacteria count. But steaming forces moisture out of the meat, reducing the amount of crabmeat for sale. Excessive cooking also destroys taste and texture. The processor is therefore faced with a tradeoff when choosing an appropriate steaming time.

The basis for a choice of steaming time is the concept of "shelf life." After the steaming treatment is completed, the product is placed in a sterile package and refrigerated. Under refrigeration, the bacterial content in the meat slowly increases and eventually reaches a size where the crabmeat is no longer suitable for consumption. The time span during which packaged crabmeat is suitable for sale is called the *shelf life* of the product. We study the following problem: How long must the crabmeat be steamed to achieve a desired shelf life?

The first step in modeling shelf life is to choose a model that describes the population dynamics of the bacteria. For simplicity, assume

$$\frac{dP}{dt} = k(T)P, \tag{1}$$

where $P(t)$ denotes the bacteria population at time t. In equation (1), $k(T)$ represents the difference between birth and death rates per unit population per unit time. In this model, k is not constant; it is a function of T, where T denotes the temperature of the crabmeat. [Note that $k(T)$ is ultimately a function of time, since the temperature T of the crabmeat varies with time in the steamer and in the refrigeration case.]

We need to choose a reasonable model for the bacteria growth rate, $k(T)$. We do so by reasoning as follows. At low temperatures (near freezing), the rate of growth of the bacteria population is slow; that's why we refrigerate foods. Mathematically, $k(T)$ is a relatively small positive quantity at those temperatures. As temperature increases, the bacterial growth rate, $k(T)$, first increases, with the most rapid rate of growth occurring near 90°F. Beyond this temperature, the growth rate begins to decrease. Beyond about 145°F, the death rate exceeds the birth rate and the bacteria population begins to decline. A simple model that captures this qualitative behavior is the quadratic function

$$k(T) = k_0 + k_1(T - 34)(140 - T), \tag{2}$$

where k_0 and k_1 are positive constants that are typically determined experimentally.

We also need a model that describes the thermal behavior of the crabmeat—how the crabmeat temperature T varies in response to the temperature of the surroundings. Assuming Newton's law of cooling, we have

$$\frac{dT}{dt} = \eta[S(t) - T]. \tag{3}$$

In equation (3), η is a positive constant and $S(t)$ is the temperature of the surroundings. Note that the surrounding temperature is not constant, since the crabmeat is initially in the steamer and then in the refrigeration case.

We now apply this model to a specific set of circumstances. Assume the following:

(i)   Initially the crabmeat is at room temperature (75°F) and contains about 10^7 bacteria per cubic centimeter.

(ii) The steam bath is maintained at a constant 250°F temperature.

(iii) When the crabmeat is placed in the steam bath, it is observed that its temperature rises from 75°F to 200°F in 5 min.

(iv) When the crabmeat is kept at a constant 34°F temperature, the bacterial count in the crabmeat doubles in 60 hr.

(v) The bacterial count in the crabmeat begins to decline once the temperature exceeds 145°F; that is [see equation (2)], $k(145) = 0$.

(vi) A bacterial count of 10^5 bacteria per cubic centimeter governs shelf life. Once this bacterial count is reached, the crabmeat can no longer be offered for sale.

Determine how long the crabmeat must be steamed to achieve a shelf life of 16 days. Assume that the crabmeat goes directly from the 250°F steam bath to the 34°F refrigeration case. Assume the 16-day shelf life requirement includes transit time to the point of sale; that is, assume that the measurement of shelf life begins the moment the crabmeat is removed from the steam bath.

Project 3: Belt Friction

The slippage of flexible belts, cables, and ropes over shafts or pulleys of circular cross-section is an important consideration in the design of belt drives. When the frictional contact between the belt and the shaft is about to be broken (that is, when slippage is imminent), a belt drive is acting under the most demanding of conditions. The belt tension (force) is not constant along the contact region. Rather, it increases along the contact region between belt and shaft in the direction of the impending slippage.

Consider the belt drive shown in Figure 2.25. Suppose we ask the following question: "How much greater can we make tension T_2 relative to the opposing tension T_1 before the belt slips over the pulley in the direction of T_2?" The answer obviously depends in part on friction—that is, on the roughness of the belt-shaft contact surface.

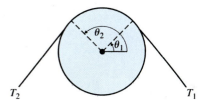

FIGURE 2.25

Consider the belt drive. How much greater can we make tension T_2 relative to the opposing tension T_1 before the belt slips over the pulley in the direction of T_2?

When slippage is imminent, the tension in the belt has been found to satisfy the differential equation

$$\frac{dT(\theta)}{d\theta} = \mu T(\theta),$$

where the angle θ (in radians) is measured in the direction of the impending slippage over the belt-pulley contact region. The parameter μ is an empirically determined constant known as the *coefficient of friction*. The larger the value of μ, the rougher the contact surface. In Figure 2.25, with slippage impending in the direction of T_2, the value of T_2 is determined relative to T_1 by solving the initial value problem

$$\frac{dT}{d\theta} = \mu T, \qquad T(\theta_1) = T_1, \qquad \theta_1 \le \theta \le \theta_2.$$

Consider the belt drive configurations shown in Figures 2.26 and 2.27. Assume that belt slippage is impending in the direction shown by the dashed arrow. Compute the belt tensions at the locations shown for the geometries and coefficients of friction given. Which configuration can support the greater load T_l before slipping?

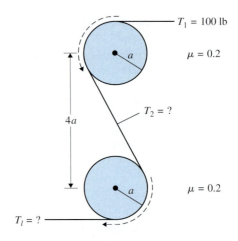

FIGURE 2.26

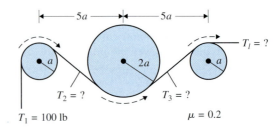

FIGURE 2.27

Project 4: The Baranyi Population Model

Milk and milk products depend on refrigeration for storage after pasteurization. Most microorganisms do not grow at refrigeration temperatures (0°C–7°C). One exception is *Listeria monocytogenes*. This anaerobic pathogen can multiply at refrigeration temperatures, and the microbe has been responsible for some recent outbreaks of listeriosis, caused by human consumption of contaminated milk products.

Food scientists are interested in developing predictive mathematical models that can accurately model the growth of harmful organisms. These models should be able to relate environmental conditions (such as temperature and pH) to the growth rate of a microbial population. In this regard, the modeler walks a fine line. On the one hand, there is an ongoing need to "build more reality" into the model. On the other hand, the model must be kept simple enough to be mathematically tractable and useful.

A population model currently being studied and used in food science research is the Baranyi population model.[11] It attempts to account for the way certain *critical substances* affect bacterial cell growth. The essence of the model is a pair of initial value problems,

$$\frac{dP(t)}{dt} = \mu \frac{q(t)}{1+q(t)} \left[1 - \left(\frac{P(t)}{P_e} \right)^m \right] P(t), \qquad P(0) = P_0,$$

$$\frac{dq(t)}{dt} = vq(t), \qquad q(0) = q_0.$$

(4)

[11]J. Baranyi, T. A. Roberts, and P. McClure, "A Non-autonomous Differential Equation to Model Bacterial Growth," *Food Microbiology*, Vol. 10, 1993, pp. 43–59.

In this model, $P(t)$ represents the population of bacteria at time t, while $q(t)$ represents the concentration of critical substance present. The positive constant v represents the growth rate of the critical substance. The parameter μ accounts for the effects of environmental conditions, such as temperature, on the growth rate of the bacteria. We shall, for simplicity, assume both μ and v to be constants. In (4), the initial values P_0 and q_0 represent the bacterial population and the amount of critical substance present at time $t = 0$, respectively. The integer exponent m is introduced into the relative birth rate to allow for greater modeling flexibility. (When $m = 1$, the differential equation reduces to the logistic equation.)

1. Solve the initial value problem for $q(t)$. For brevity, let

$$\alpha(t) = \frac{q(t)}{1 + q(t)}.$$

The differential equation for $P(t)$ now takes the form

$$\frac{dP(t)}{dt} = \mu\alpha(t)\left[1 - \left(\frac{P(t)}{P_e}\right)^m\right]P(t). \tag{5}$$

2. Show that

$$\lim_{t \to \infty} \alpha(t) = 1.$$

Therefore, in the model (5), the critical substance exerts only a certain limited effect on bacterial growth.

3. Make a "change-of-clock" change of independent variable by introducing the new independent variable τ, where

$$\frac{d\tau}{dt} = \mu\alpha(t), \qquad \tau(0) = 0.$$

Therefore,

$$\tau(t) = \mu\int_0^t \alpha(s)\,ds. \tag{6}$$

Using the chain rule, show that if we introduce the normalized dependent variable $p = P/P_e$ and view p to be a function of τ, then the initial value problem for p becomes

$$\frac{dp(\tau)}{d\tau} = [1 - p^m(\tau)]p(\tau), \qquad p(0) = \frac{P_0}{P_e}. \tag{7}$$

4. Solve initial value problem (7) for $p(\tau)$. Note that this differential equation is separable (as well as being a Bernoulli equation). For a general integer m, the equation is most easily solved as a Bernoulli equation. In particular, show that

$$p(\tau) = \left[1 + \left(\left[\frac{P_0}{P_e}\right]^{-m} - 1\right)e^{-m\tau}\right]^{-\frac{1}{m}}. \tag{8}$$

5. Noting that $P(t) = P_e p(\tau(t))$, determine $P(t)$.

C H A P T E R

3

Second and Higher Order Linear Differential Equations

3.1   Introduction

Most of this chapter discusses initial value problems of the form

$$y'' + p(t)y' + q(t)y = g(t), \qquad y(t_0) = y_0, \qquad y'(t_0) = y_0', \qquad a < t < b. \qquad \textbf{(1)}$$

In equation (1), $p(t)$, $q(t)$, and $g(t)$ are continuous functions on the interval $a < t < b$, and t_0 is some point in this t-interval of interest. The differential equation in problem (1) is called a **second order linear differential equation**. If $g(t)$ is the zero function, then the differential equation is a **homogeneous differential equation**; otherwise the equation is a **nonhomogeneous differential equation**. An initial value problem for a second order equation involves two supplementary or initial conditions, $y(t_0) = y_0$ and $y'(t_0) = y_0'$.

Second order differential equations owe much of their relevance and importance to Newton's laws of motion. Since acceleration is the second derivative of position, modeling one-dimensional dynamics often leads to a differential equation of the form

$$my'' = F(t, y, y').$$

When the applied force F is a linear function of position and velocity, we obtain a second order linear differential equation. The two supplementary conditions in (1) specify position and velocity at time t_0.

The concepts and techniques used to analyze the second order initial value problem (1) extend naturally to analogous problems involving higher order linear equations. We consider these extensions in Sections 3.11–3.13.

An Example: The Bobbing Motion of a Floating Object

We have all observed a cork, a block of wood, or some other object bobbing up and down in a liquid such as water. How do we mathematically model this bobbing motion?

In its rest or equilibrium state, a floating object is subjected to two equal and opposite forces—the weight of the object is counteracted by an upward buoyant force equal to the weight of the displaced liquid. (This is the law of buoyancy discovered by Archimedes.[1]) If we disturb this equilibrium state by pushing down or pulling up on the object and then releasing it, the object will begin to bob up and down. The physical principle governing the object's movement is Newton's second law of motion: $ma = F$, the product of the mass and acceleration of an object is equal to the sum of the forces acting on it.

For example, consider a cylindrical object having uniform mass density ρ, constant cross-sectional area A, and height L (see Figure 3.1). We assume the object is floating in a liquid having density ρ_l, where $\rho < \rho_l$. In its rest or equilibrium state, the object sinks into the liquid until the weight of the liquid

---

[1]Archimedes of Syracuse (287–212 B.C.) was a remarkable mathematician and scientist, contributing important results in geometry, mechanics, and hydrostatics. Archimedes developed an early form of integration that empowered his work. He was also an inventor, developing the compound pulley, a pump known as Archimedes' screw, and military machines used to defend his native Syracuse in Sicily from attack by the Romans. Archimedes was killed when Syracuse was captured by the Romans during the Second Punic War.

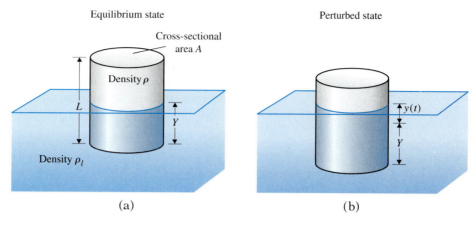

Equilibrium state   Perturbed state

(a)   (b)

FIGURE 3.1

(a) The floating object is in its equilibrium or rest state when the weight of the displaced liquid is equal to the weight of the object.
(b) The object is in a perturbed state when it is displaced from its equilibrium position. At any time t, the quantity $y(t)$ measures how far the object is from its equilibrium position.

displaced equals the weight of the object; we denote the depth to which the object sinks as Y. It can be shown (see Exercise 12) that

$$Y = \frac{\rho}{\rho_l}L. \tag{2}$$

Suppose now that the object is displaced from its equilibrium state, as illustrated in Figure 3.1(b). Let the depth to which the body is immersed in the liquid at time t be denoted by $Y + y(t)$. Thus, $y(t)$ represents the time-varying displacement of the object from its equilibrium state. For definiteness, we assume $y(t)$ to be positive in the downward direction. In the perturbed state, the net force acting upon the object is typically nonzero. In Exercise 12, you are asked to show that Newton's law, $ma = F$, leads to the equation

$$y''(t) + \omega^2 y(t) = 0, \qquad \omega^2 = \frac{\rho_l g}{\rho L}, \tag{3}$$

where g is the acceleration due to gravity. The perturbation depth, $y(t)$, is thus described by a second order linear homogeneous differential equation.

We need more than just the differential equation to uniquely characterize the motion of the bobbing object. Specifying the object's depth and velocity at some particular instant of time by specifying $y(t_0) = y_0$ and $y'(t_0) = y_0'$ would seem (on physical grounds) to uniquely characterize the motion. The discussion of existence and uniqueness issues given later shows that this physical intuition is, in fact, correct.

Consider the differential equation in (3). Assume for the present discussion that the initial value problem

$$y''(t) + \omega^2 y(t) = 0, \qquad y(0) = y_0, \qquad y'(0) = y_0' \tag{4}$$

has a unique solution on any time interval of interest. For simplicity, we've chosen the initial time to be $t_0 = 0$. Does the solution of initial value problem (4) describe a bobbing or oscillating motion that is consistent with our experience?

For some insight and a preview of what's to come, note that the functions $y(t) = \sin \omega t$ and $y(t) = \cos \omega t$ are each solutions of the differential equation in (4). Section 3.5 shows how to obtain these solutions. For now, the assertion can be verified by direct substitution. In fact, for any choice of constants C_1 and C_2, the function

$$y(t) = C_1 \sin \omega t + C_2 \cos \omega t \tag{5}$$

is a solution of $y'' + \omega^2 y = 0$.

You will see later that $y(t) = C_1 \sin \omega t + C_2 \cos \omega t$ is, in fact, the general solution of the differential equation; that is, any solution of the differential equation can be constructed by making an appropriate choice of the constants C_1 and C_2. For initial value problem (4), imposing the initial conditions upon the general solution (5) leads to the set of equations

$$y(0) = C_1 \sin(0) + C_2 \cos(0) = y_0$$
$$y'(0) = C_1 \omega \cos(0) - C_2 \omega \sin(0) = y_0'.$$

Solving this system of equations, we find $C_1 = y_0'/\omega$ and $C_2 = y_0$. The unique solution of initial value problem (4) is therefore

$$y(t) = \frac{y_0'}{\omega} \sin \omega t + y_0 \cos \omega t. \tag{6}$$

If either $y_0 = 0$ or $y_0' = 0$, it's obvious that the solution represents the type of sinusoidal oscillating behavior that is consistent with our experience. In general, as you will see later in this chapter, the solution (6) can always be written as a sinusoid. Figure 3.2, for example, shows the behavior of (6) for the special case $y_0 = y_0' = 1, \omega = 2$. Thus, the mathematical model (4) and its solution (6) do, in fact, predict an oscillatory behavior that is consistent with physical intuition about the motion of a bobbing body.

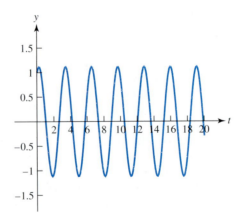

FIGURE 3.2

The graph of the solution of $y'' + 4y = 0, y(0) = 1, y'(0) = 1$. The solution is given by equation (6), using values $\omega = 2, y_0 = 1, y_0' = 1$.

Existence and Uniqueness

We begin to develop the necessary mathematical underpinnings by stating an existence-uniqueness theorem proved in advanced texts.

Theorem 3.1

Let $p(t)$, $q(t)$, and $g(t)$ be continuous functions on the interval (a, b), and let t_0 be in (a, b). Then the initial value problem

$$y'' + p(t)y' + q(t)y = g(t), \qquad y(t_0) = y_0, \qquad y'(t_0) = y_0'$$

has a unique solution defined on the entire interval (a, b).

Compare this theorem with Theorem 2.1, which states an analogous existence-uniqueness result for first order linear initial value problems. Both theorems assume that the coefficient functions and the nonhomogeneous term on the right-hand side are continuous on the interval of interest. Both theorems reach the same three conclusions:

1. The solution exists.
2. The solution is unique.
3. The solution exists on the entire interval (a, b).

Theorem 3.1 defines the framework within which we will work. It assures us that, given an initial value problem of the type described, there is one and only one solution. Our job is to find it. The similarity of Theorems 2.1 and 3.1 is no accident. You will see in Chapter 4 that these two theorems, as well as an analogous theorem stated for higher order linear initial value problems, can be viewed as special cases of a single, all-encompassing existence-uniqueness theorem for first order linear systems.

EXAMPLE 1

Determine the largest t-interval on which we can guarantee the existence of a solution of the initial value problem

$$ty'' + (\cos t)y' + t^2 y = t, \qquad y(-1) = -1, \qquad y'(-1) = 2.$$

Solution: Before we apply Theorem 3.1, we need to write the differential equation in standard form:

$$y'' + \frac{\cos t}{t}y' + ty = 1.$$

With the equation in this form, we can identify the coefficient functions in the hypotheses of Theorem 3.1. One of the coefficient functions is not continuous at $t = 0$, but there are no other points of discontinuity on the t-axis. Since the initial conditions are posed at the point $t_0 = -1$, it follows from Theorem 3.1 that the given initial value problem is guaranteed to have a unique solution on the interval $-\infty < t < 0$.

Figure 3.3 shows a numerical solution for this initial value problem on the interval $[-1, -0.002]$. As you can see, it appears that the solution is not defined at $t = 0$.

(continued)

(continued)

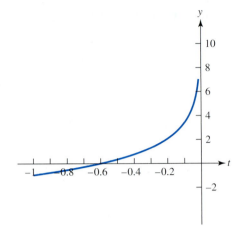

FIGURE 3.3

The graph of a numerical solution of the initial value problem in Example 1. The solution appears to have a vertical asymptote at $t = 0$.   ❖

EXERCISES

Exercises 1–4:

For each initial value problem, determine the largest t-interval on which Theorem 3.1 guarantees the existence of a unique solution.

1. $y'' + 3t^2 y' + 2y = \sin t$, $y(1) = 1$, $y'(1) = -1$

2. $y'' + y' + 3ty = \tan t$, $y(\pi) = 1$, $y'(\pi) = -1$

3. $e^t y'' + \dfrac{1}{t^2 - 1} y = \dfrac{4}{t}$, $y(-2) = 1$, $y'(-2) = 2$

4. $ty'' + \dfrac{\sin 2t}{t^2 - 9} y' + 2y = 0$, $y(1) = 0$, $y'(1) = 1$

5. Consider the initial value problem $t^2 y'' - ty' + y = 0, y(1) = 1, y'(1) = 1$.

   (a) What is the largest interval on which Theorem 3.1 guarantees the existence of a unique solution?

   (b) Show by direct substitution that the function $y(t) = t$ is the unique solution of this initial value problem. What is the interval on which this solution actually exists?

   (c) Does this example contradict the assertion of Theorem 3.1? Explain.

Exercises 6–7:

Let $y(t)$ denote the solution of the given initial value problem. Is it possible for the corresponding limit to hold? Explain your answer.

6. $y'' + \dfrac{1}{t^2 - 16} y = 0$, $y(0) = 1$, $y'(0) = 1$, $\displaystyle\lim_{t \to 3^-} y(t) = +\infty$

7. $y'' + 2y' + \dfrac{1}{t - 3} y = 0$, $y(1) = 1$, $y'(1) = 2$, $\displaystyle\lim_{t \to 0^+} y(t) = +\infty$

Exercises 8–10:

In each exercise, assume that $y(t) = C_1 \sin \omega t + C_2 \cos \omega t$ is the general solution of $y'' + \omega^2 y = 0$. Find the unique solution of the given initial value problem.

8. $y'' + y = 0, \quad y(\pi) = 0, \quad y'(\pi) = 2$ **9.** $y'' + 4y = 0, \quad y(0) = -2, \quad y'(0) = 0$

10. $y'' + 16y = 0, \quad y(\pi/4) = 1, \quad y'(\pi/4) = -4$

11. Concavity of the Solution Curve   In the discussion of direction fields in Section 1.3, you saw how the differential equation defines the slope of the solution curve at a point in the ty-plane. In particular, given the initial value problem $y' = f(t, y), y(t_0) = y_0$, the slope of the solution curve at initial condition point (t_0, y_0) is $y'(t_0) = f(t_0, y_0)$. In like manner, a second order equation provides direct information about the concavity of the solution curve. Given the initial value problem $y'' = f(t, y, y'), y(t_0) = y_0, y'(t_0) = y_0'$, it follows that the concavity of the solution curve at the initial condition point (t_0, y_0) is $y''(t_0) = f(t_0, y_0, y_0')$. (What is the slope of the solution curve at that point?)

    Consider the four graphs shown. Each graph displays a portion of the solution of one of the four initial value problems given. Match each graph with the appropriate initial value problem.

(a) $y'' + y = 2 - \sin t, \quad y(0) = 1, \quad y'(0) = -1$

(b) $y'' + y = -2t, \quad y(0) = 1, \quad y'(0) = -1$

(c) $y'' - y = t^2, \quad y(0) = 1, \quad y'(0) = 1$

(d) $y'' - y = -2\cos t, \quad y(0) = 1, \quad y'(0) = 1$

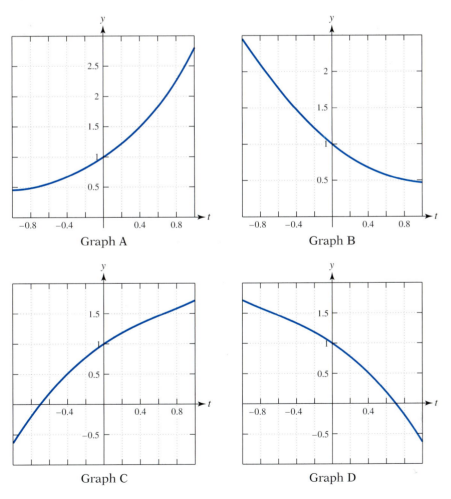

Figure for Exercise 11

12. The Bobbing Cylinder Model   Using Figure 3.1 for reference, carry out the following derivations.

(a) Derive expressions for the mass of the cylinder and the displaced liquid, in terms of the mass densities and cylinder geometry. Recall that weight W is given by $W = mg$. Apply the law of buoyancy to the equilibrium state shown in Figure 3.1(a) and establish equation (2), $Y = (\rho/\rho_l)L$.

(b) Apply Newton's law $ma = F$ to the cylinder shown in its perturbed state in Figure 3.1(b). Since y is positive downward, the net force F equals the cylinder weight minus the buoyant force. Show that

$$y'' + \frac{\rho_l g}{\rho L} y = 0.$$

[Hint: The equilibrium equality of part (a) can be used to simplify the differential equation obtained from $ma = F$.]

13. Since $\sin(\omega t + 2\pi) = \sin \omega t$ and $\cos(\omega t + 2\pi) = \cos \omega t$, the amount of time T it takes a bobbing object to go through one cycle of its motion is determined by the relation $\omega T = 2\pi$, or $T = 2\pi/\omega$. This time T is called the *period* of the motion (see Section 3.6). As the period decreases, the bobbing motion of the floating object becomes more rapid.

(a) Two identically shaped cylindrical drums, made of different material, are floating at rest as shown in part (a) of the figure.

(b) Two cylindrical drums, made of identical material, are floating at rest as shown in part (b).

For each case, when the drums are put into motion, is it possible to identify the drum that will bob up and down more rapidly? Explain.

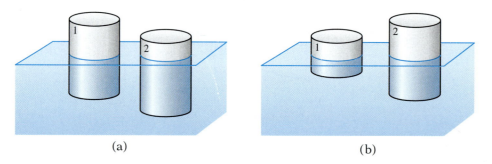

(a)   (b)

Figure for Exercise 13

14. A buoy having the shape of a right circular cylinder 3 ft in diameter and 5 ft in height is initially floating upright in water. When it was put into motion at time $t = 0$, the following 10-sec record of its displacement from equilibrium, measured in inches positive in the downward direction, was obtained.

(a) Determine the initial displacement y_0 and the period T of the motion (see Exercise 13).

(b) Determine the constant ω and the initial velocity y_0' of the buoy.

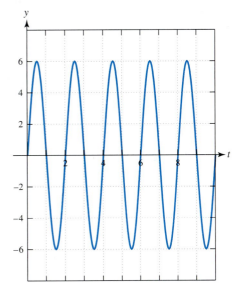

Figure for Exercise 14

3.2   The General Solution of Homogeneous Equations

Consider the second order linear homogeneous differential equation

$$y'' + p(t)y' + q(t)y = 0, \qquad a < t < b, \tag{1}$$

where $p(t)$ and $q(t)$ are are continuous on (a, b). We begin with this homogeneous equation because understanding its solution structure is basic to developing methods for solving linear differential equations, whether they are homogeneous or nonhomogeneous.

The general solution of equation (1) is often described as a "linear combination" of functions. In particular, let $f_1(t)$ and $f_2(t)$ be any two functions having a common domain, and let c_1 and c_2 be any two constants. A function of the form

$$f(t) = c_1 f_1(t) + c_2 f_2(t)$$

is called a **linear combination of the functions** f_1 and f_2. For example, the function $f(t) = 3 \sin t + 8 \cos t$ is a linear combination of the functions $\sin t$ and $\cos t$.

The Principle of Superposition

The first result we establish for the homogeneous equation (1) is a **superposition principle**. It shows that a linear combination of two solutions is also a solution. An analogous superposition principle is also valid for higher order linear homogeneous equations; see Section 3.11.

Theorem 3.2	Let $y_1(t)$ and $y_2(t)$ be any two solutions of

$$y'' + p(t)y' + q(t)y = 0$$

defined on the interval (a, b). Then, for any constants c_1 and c_2, the linear combination

$$y(t) = c_1 y_1(t) + c_2 y_2(t)$$

is also a solution on (a, b).

● **PROOF:**   The hypotheses state that $y_1(t)$ and $y_2(t)$ are both solutions of the homogeneous equation. Therefore,

$$y_1'' + p(t)y_1' + q(t)y_1 = 0 \quad \text{and} \quad y_2'' + p(t)y_2' + q(t)y_2 = 0.$$

Substituting $y(t) = c_1 y_1(t) + c_2 y_2(t)$ into the differential equation, we obtain

$$y'' + p(t)y' + q(t)y = (c_1 y_1 + c_2 y_2)'' + p(t)(c_1 y_1 + c_2 y_2)' + q(t)(c_1 y_1 + c_2 y_2). \quad \text{(2)}$$

Using basic properties from calculus, we can write the right-hand side of (2) as

$$c_1[y_1'' + p(t)y_1' + q(t)y_1] + c_2[y_2'' + p(t)y_2' + q(t)y_2] = c_1[0] + c_2[0] = 0.$$

Therefore, the linear combination $y(t) = c_1 y_1(t) + c_2 y_2(t)$ is also a solution. ●

It is important to understand that the superposition principle of Theorem 3.2 is valid for *homogeneous* linear equations. In general, a linear combination of solutions of a linear *nonhomogeneous* equation is not a solution of the nonhomogeneous equation. Similarly, a linear combination of solutions of a *nonlinear* differential equation is normally not a solution of the nonlinear equation.

Fundamental Sets of Solutions

Theorem 3.2 shows that we can form a linear combination of two solutions of equation (1) and create a new solution. We now turn this idea around and ask, "Is it possible to find two solutions, $y_1(t)$ and $y_2(t)$, such that every solution of

$$y'' + p(t)y' + q(t)y = 0, \qquad a < t < b \tag{3}$$

can be expressed as a linear combination of $y_1(t)$ and $y_2(t)$?" In other words, if we are given any solution, $y(t)$, of the homogeneous equation (3), can we determine constants c_1 and c_2 such that

$$y(t) = c_1 y_1(t) + c_2 y_2(t), \qquad a < t < b?$$

If there are two such solutions $y_1(t)$ and $y_2(t)$, we say that $\{y_1(t), y_2(t)\}$ is a **fundamental set of solutions** for equation (3). The term "fundamental set" is an appropriate one, since every solution of equation (3) can be constructed using the basic building blocks $y_1(t)$ and $y_2(t)$.

EXAMPLE

1

Consider the linear homogeneous differential equation

$$y'' + 4y = 0. \tag{4}$$

(a) Show, by direct substitution, that $y_1(t) = \cos 2t$ and $y_2(t) = \sin 2t$ are solutions of differential equation (4).

(b) Show, by direct substitution, that $y(t) = 3\cos[2t + (\pi/4)]$ is a solution of (4).

(c) It can be shown (see Example 2) that $y_1(t)$ and $y_2(t)$ form a fundamental set of solutions for equation (4). Find constants c_1 and c_2 such that $3\cos[2t + (\pi/4)] = c_1 \cos 2t + c_2 \sin 2t$.

Solution:

(a) Inserting $y_1(t) = \cos 2t$ into equation (4), we obtain

$$y_1'' + 4y_1 = (\cos 2t)'' + 4\cos 2t = -4\cos 2t + 4\cos 2t = 0.$$

A similar calculation shows that $y_2(t) = \sin 2t$ is also a solution of equation (4).

(b) We leave this part as an exercise.

(c) We want to find constants c_1 and c_2 such that

$$3\cos\left(2t + \frac{\pi}{4}\right) = c_1 \cos 2t + c_2 \sin 2t.$$

Rewriting the left-hand side of this equation using the trigonometric identity

$$\cos(\theta_1 + \theta_2) = \cos\theta_1 \cos\theta_2 - \sin\theta_1 \sin\theta_2$$

yields

$$3\cos\left(2t + \frac{\pi}{4}\right) = 3\cos 2t \cos\frac{\pi}{4} - 3\sin 2t \sin\frac{\pi}{4}$$

$$= \left(3\cos\frac{\pi}{4}\right)\cos 2t + \left(-3\sin\frac{\pi}{4}\right)\sin 2t.$$

The constants c_1 and c_2 are therefore

$$c_1 = 3\cos\frac{\pi}{4} = \frac{3\sqrt{2}}{2} \quad \text{and} \quad c_2 = -3\sin\frac{\pi}{4} = -\frac{3\sqrt{2}}{2}. \;\; ❖$$

Fundamental Sets and the General Solution

Assume that $\{y_1(t), y_2(t)\}$ is a fundamental set of solutions for the linear homogeneous equation

$$y'' + p(t)y' + q(t)y = 0, \qquad a < t < b, \tag{5}$$

where $p(t)$ and $q(t)$ are continuous on (a, b). Therefore, if $y(t)$ is any solution of (5), there are corresponding constants c_1 and c_2 such that

$$y(t) = c_1 y_1(t) + c_2 y_2(t), \qquad a < t < b. \tag{6}$$

Expression (6) is called the **general solution** of equation (5). Once we obtain the general solution, we can use it to solve an initial value problem such as

$$y'' + p(t)y' + q(t)y = 0, \qquad y(t_0) = y_0, \quad y'(t_0) = y_0'. \tag{7}$$

In particular, since every solution of the differential equation has the form (6), solving initial value problem (7) reduces to finding constants c_1 and c_2 such that

$$c_1 y_1(t_0) + c_2 y_2(t_0) = y_0$$
$$c_1 y_1'(t_0) + c_2 y_2'(t_0) = y_0'.$$

These equations can be written in matrix form as

$$\begin{bmatrix} y_1(t_0) & y_2(t_0) \\ y_1'(t_0) & y_2'(t_0) \end{bmatrix} \begin{bmatrix} c_1 \\ c_2 \end{bmatrix} = \begin{bmatrix} y_0 \\ y_0' \end{bmatrix}. \tag{8}$$

Theorem 3.1 guarantees that initial value problem (7) has a unique solution for any choice of the initial conditions. Therefore, equation (8) has a unique solution for any choice of y_0 and y_0', and this means that the coefficient matrix has a nonzero[2] determinant:

$$\begin{vmatrix} y_1(t_0) & y_2(t_0) \\ y_1'(t_0) & y_2'(t_0) \end{vmatrix} = y_1(t_0)y_2'(t_0) - y_1'(t_0)y_2(t_0) \neq 0. \tag{9}$$

The Wronskian

The determinant in (9) plays a key role in characterizing fundamental sets of solutions. As we saw above, if $\{y_1(t), y_2(t)\}$ is a fundamental set of solutions for equation (5), then the determinant in (9) is nonzero at every point t_0 in (a, b).

The converse is true as well. That is, if $\{y_1(t), y_2(t)\}$ is a set of solutions such that the determinant in (9) is nonzero at every point t in (a, b), then $\{y_1(t), y_2(t)\}$ is a fundamental set of solutions. To prove this, let $u(t)$ be any solution of equation (5), and let t_0 be a point in (a, b). Let $u(t_0) = \alpha, u'(t_0) = \beta$, and consider the initial value problem

$$y'' + p(t)y' + q(t)y = g(t), \qquad y(t_0) = \alpha, \qquad y'(t_0) = \beta. \tag{10}$$

Since we are assuming the determinant in (9) is nonzero, there are c_1 and c_2 such that

$$\begin{bmatrix} y_1(t_0) & y_2(t_0) \\ y_1'(t_0) & y_2'(t_0) \end{bmatrix} \begin{bmatrix} c_1 \\ c_2 \end{bmatrix} = \begin{bmatrix} \alpha \\ \beta \end{bmatrix}.$$

Let us define $\hat{y}(t) = c_1 y_1(t) + c_2 y_2(t)$. Therefore, $\hat{y}(t)$ is a solution of initial value problem (10). Since the solution of (10) is unique and since solutions $u(t)$ and $\hat{y}(t)$ both satisfy the initial conditions, it follows that $u(t) = \hat{y}(t) = c_1 y_1(t) + c_2 y_2(t)$. Since the solution $u(t)$ is a linear combination of $y_1(t)$ and $y_2(t)$, we have that $\{y_1(t), y_2(t)\}$ is a fundamental set of solutions for equation (5).

To summarize: Let $y_1(t)$ and $y_2(t)$ be solutions of (5). Then $\{y_1(t), y_2(t)\}$ is a fundamental set of solutions if and only if $W(t) \neq 0$ for all t in (a, b), where

---

[2]From linear algebra, if a (2×2) matrix equation $A\mathbf{x} = \mathbf{b}$ is consistent for every right-hand side $\mathbf{b}$, then A is invertible. Equivalently, the determinant of A is nonzero.

$$W(t) = \begin{vmatrix} y_1(t) & y_2(t) \\ y_1'(t) & y_2'(t) \end{vmatrix} = y_1(t)y_2'(t) - y_1'(t)y_2(t).$$ **(11)**

The determinant in (11), $W(t)$, is called the **Wronskian determinant** or simply the **Wronskian**[3] of $\{y_1(t), y_2(t)\}$.

In Section 3.11, we prove an important result known as Abel's theorem. Abel's theorem shows that if $y_1(t)$ and $y_2(t)$ are solutions of (5), then $W(t)$ either is identically zero on (a, b) or is never zero in (a, b). Therefore, if we want to decide whether or not solutions $y_1(t)$ and $y_2(t)$ form a fundamental set, all we need to do is check the value of $W(t)$ at some convenient test point, $t = t_0$. If $W(t_0) \neq 0$, then $\{y_1(t), y_2(t)\}$ is a fundamental set of solutions. If $W(t_0) = 0$, then $\{y_1(t), y_2(t)\}$ is not a fundamental set.

E X A M P L E

2

Example 1 showed that $y_1(t) = \cos 2t$ and $y_2(t) = \sin 2t$ are solutions of the homogeneous linear equation $y'' + 4y = 0$, $-\infty < t < \infty$. Show that $\{y_1(t), y_2(t)\}$ is a fundamental set of solutions.

Solution: We compute the Wronskian of the solution pair. If it is nonzero in the interval of interest, then $\{y_1(t), y_2(t)\}$ is a fundamental set of solutions. The Wronskian is

$$W(t) = \begin{vmatrix} \cos 2t & \sin 2t \\ -2\sin 2t & 2\cos 2t \end{vmatrix} = 2\cos^2 2t + 2\sin^2 2t = 2.$$

Since the Wronskian is nonzero on $-\infty < t < \infty$, we see that $\{y_1(t), y_2(t)\}$ is a fundamental set of solutions and the general solution of $y'' + 4y = 0$, $-\infty < t < \infty$ is

$$y(t) = c_1 \cos 2t + c_2 \sin 2t. \ ❖$$

E X A M P L E

3

Consider the initial value problem

$$y'' - \frac{1}{t}y' - \frac{3}{t^2}y = 0, \qquad y(1) = 4, \qquad y'(1) = 8, \qquad 0 < t < \infty.$$

(a) Verify that the functions $y_1(t) = t^3$ and $y_2(t) = t^{-1}$ form a fundamental set of solutions for the differential equation on the interval $0 < t < \infty$.

(b) Solve the initial value problem.

Solution:

(a) We leave it as an exercise to verify that $y_1(t)$ and $y_2(t)$ are solutions of the differential equation. To show that they form a fundamental set, we calculate the Wronskian:

$$W(t) = \begin{vmatrix} t^3 & t^{-1} \\ 3t^2 & -t^{-2} \end{vmatrix} = -4t.$$

(continued)

---

[3]Hoene Wronski (1778–1853) was born Josef Hoene but changed his name just after he married. The determinants we now know as Wronskians were given their name by Muir in 1882.

(continued)

Since the Wronskian is nonzero on $0 < t < \infty$, it follows that $\{y_1(t), y_2(t)\}$ is a fundamental set of solutions.

(b) By part (a), the general solution is $y(t) = c_1 t^3 + c_2 t^{-1}$. Imposing the initial conditions at $t = 1$, we find

$$c_1 + c_2 = 4$$
$$3c_1 - c_2 = 8.$$

Solving, we obtain $c_1 = 3$ and $c_2 = 1$. The solution of the initial value problem is

$$y(t) = 3t^3 + t^{-1}, \qquad 0 < t < \infty. \;\; ❖$$

EXERCISES

Exercises 1–15:

In these exercises, the t-interval of interest is $-\infty < t < \infty$ unless indicated otherwise.

(a) Verify that the given functions are solutions of the differential equation.

(b) Calculate the Wronskian. Do the two functions form a fundamental set of solutions?

(c) If the two functions form a fundamental set, determine the unique solution of the initial value problem.

1. $y'' - 4y = 0$; $y_1(t) = e^{2t}$, $y_2(t) = 2e^{-2t}$; $y(0) = 1$, $y'(0) = -2$

2. $y'' - y = 0$; $y_1(t) = 2e^t$, $y_2(t) = e^{-t+3}$; $y(-1) = 1$, $y'(-1) = 0$

3. $y'' + y = 0$; $y_1(t) = 0$, $y_2(t) = \sin t$; $y(\pi/2) = 1$, $y'(\pi/2) = 1$

4. $y'' + y = 0$; $y_1(t) = \cos t$, $y_2(t) = \sin t$; $y(\pi/2) = 1$, $y'(\pi/2) = 1$

5. $y'' - 4y' + 4y = 0$; $y_1(t) = e^{2t}$, $y_2(t) = te^{2t}$; $y(0) = 2$, $y'(0) = 0$

6. $2y'' - y' = 0$; $y_1(t) = 1$, $y_2(t) = e^{t/2}$; $y(2) = 0$, $y'(2) = 2$

7. $y'' - 3y' + 2y = 0$; $y_1(t) = 2e^t$, $y_2(t) = e^{2t}$; $y(-1) = 1$, $y'(-1) = 0$

8. $4y'' + y = 0$; $y_1(t) = \sin[(t/2) + (\pi/3)]$, $y_2(t) = \sin[(t/2) - (\pi/3)]$; $y(0) = 0$, $y'(0) = 1$

9. $ty'' + y' = 0$, $0 < t < \infty$; $y_1(t) = \ln t$, $y_2(t) = \ln 3t$; $y(3) = 0$, $y'(3) = 3$

10. $ty'' + y' = 0$, $0 < t < \infty$; $y_1(t) = \ln t$, $y_2(t) = \ln 3$; $y(1) = 0$, $y'(1) = 3$

11. $t^2 y'' - ty' - 3y = 0$, $-\infty < t < 0$; $y_1(t) = t^3$, $y_2(t) = -t^{-1}$; $y(-1) = 0$, $y'(-1) = -2$

12. $y'' + 2y' + y = 0$; $y_1(t) = e^{-t}$, $y_2(t) = 2e^{1-t}$; $y(0) = 1$, $y'(0) = 0$

13. $y'' = 0$; $y_1(t) = t + 1$, $y_2(t) = -t + 2$; $y(1) = 4$, $y'(1) = -1$

14. $y'' + \pi^2 y = 0$; $y_1(t) = \sin \pi t + \cos \pi t$, $y_2(t) = \sin \pi t - \cos \pi t$; $y\left(\frac{1}{2}\right) = 1$, $y'\left(\frac{1}{2}\right) = 0$

15. $4y'' + 4y' + y = 0$; $y_1(t) = e^{-t/2}$, $y_2(t) = te^{-t/2}$; $y(1) = 1$, $y'(1) = 0$

Exercises 16–18:

The given pair of functions $\{y_1, y_2\}$ forms a fundamental set of solutions of the differential equation.

(a) Show that the given function $\bar{y}(t)$ is also a solution of the differential equation.

(b) Determine coefficients c_1 and c_2 such that $\bar{y}(t) = c_1 y_1(t) + c_2 y_2(t)$.

16. $y'' + 4y = 0$; $y_1(t) = 2\cos 2t$, $y_2(t) = \sin 2t$; $\bar{y}(t) = \sin[2t + (\pi/4)]$

17. $t^2 y'' - t y' + y = 0$, $0 < t < \infty$; $y_1(t) = t$, $y_2(t) = t \ln t$; $\bar{y}(t) = 2t + t \ln 3t$

18. $4y'' - y = 0$; $y_1(t) = e^{-t/2}$, $y_2(t) = -2e^{t/2}$; $\bar{y}(t) = 2 \cosh(t/2)$

19. The functions $y_1(t) = e^{3t}$ and $y_2(t) = e^{-3t}$ are known to be solutions of $y'' + \alpha y' + \beta y = 0$, where α and β are constants. Determine α and β. [Hint: Obtain a system of two equations for the two unknown constants.]

20. The functions $y_1(t) = \sin(t + \alpha)$ and $y_2(t) = \sin(t - \alpha)$ are solutions of $y'' + y = 0$ on $-\infty < t < \infty$. For what values of the constant α, if any, is $\{y_1, y_2\}$ a fundamental set?

Exercises 21–22:

In each exercise, assume that y_1 and y_2 are solutions of $y'' + p(t)y' + q(t)y = 0$, where $p(t)$ and $q(t)$ are continuous on (a, b). Explain why $y_1(t)$ and $y_2(t)$ cannot form a fundamental set of solutions.

21. $y_1(t)$ and $y_2(t)$ have a common zero in (a, b); that is, $y_1(t_0) = 0$ and $y_2(t_0) = 0$ at some point t_0 in (a, b).

22. $y_1(t)$ and $y_2(t)$ achieve a local extremum at the same point t_0 in (a, b).

3.3   Constant Coefficient Homogeneous Equations

The discussion in Section 3.2 established the solution structure for second order linear homogeneous differential equations. We saw that to obtain the general solution, we need to find a fundamental set of solutions—that is, a pair of solutions whose Wronskian is nonzero on the t-interval of interest. This section, along with Sections 3.4 and 3.5, shows how to find a fundamental set of solutions for the important special case of a constant coefficient equation,

$$ay'' + by' + cy = 0. \tag{1}$$

In equation (1), a, b, and c are constants and we assume that $a \neq 0$. When discussing equation (1), we can assume the t-interval of interest is $-\infty < t < \infty$ or any subinterval of $(-\infty, \infty)$, since the coefficient functions are constant and hence continuous everywhere.

Finding Solutions of Second Order Constant Coefficient Equations

We look for solutions of the form $y(t) = e^{\lambda t}$, where λ is a constant to be determined. The motivation for assuming this form for a solution comes from observing how the function $e^{\lambda t}$ behaves under repeated differentiation:

$$\frac{d}{dt} e^{\lambda t} = \lambda e^{\lambda t} \quad \text{and} \quad \frac{d^2}{dt^2} e^{\lambda t} = \lambda^2 e^{\lambda t}.$$

Each differentiation of $e^{\lambda t}$ simply multiplies $e^{\lambda t}$ by a power of the constant, λ. Substituting $y(t) = e^{\lambda t}$ into differential equation (1) leads to

$$ay'' + by' + cy = a\lambda^2 e^{\lambda t} + b\lambda e^{\lambda t} + ce^{\lambda t} = e^{\lambda t}(a\lambda^2 + b\lambda + c) = 0. \tag{2}$$

Equation (2) must hold for all t in the interval of interest. Since the factor $e^{\lambda t}$ is never zero, equation (2) is valid only if λ is a root of the polynomial equation

$$a\lambda^2 + b\lambda + c = 0. \tag{3}$$

The quadratic equation (3) is called the **characteristic equation** for $ay'' + by' + cy = 0$, and the polynomial $P(\lambda) = a\lambda^2 + b\lambda + c$ is called the **characteristic polynomial**. The roots of the characteristic equation are exactly those values λ for which $y(t) = e^{\lambda t}$ is a solution of the differential equation $ay'' + by' + cy = 0$.

E X A M P L E

1

Consider the homogeneous linear differential equation

$$y'' + 8y' + 15y = 0.$$

(a) Find all values λ such that $y(t) = e^{\lambda t}$ is a solution of the differential equation.

(b) Do the functions found in part (a) form a fundamental set of solutions for the differential equation? If so, what is the general solution of the differential equation?

Solution:

(a) The characteristic polynomial for $y'' + 8y' + 15y = 0$ is

$$P(\lambda) = \lambda^2 + 8\lambda + 15 = (\lambda + 5)(\lambda + 3).$$

The roots of the characteristic equation are $\lambda_1 = -5$ and $\lambda_2 = -3$. Thus, the trial form $y(t) = e^{\lambda t}$ leads to two solutions of the differential equation:

$$y_1(t) = e^{-5t} \quad \text{and} \quad y_2(t) = e^{-3t}.$$

(b) To decide whether $\{y_1, y_2\}$ is a fundamental set of solutions, we form the Wronskian:

$$W(t) = \begin{vmatrix} e^{-5t} & e^{-3t} \\ -5e^{-5t} & -3e^{-3t} \end{vmatrix} = 2e^{-8t}.$$

Since $W(t) = 2e^{-8t}$ is never zero, $\{y_1, y_2\}$ is a fundamental set of solutions for $y'' + 8y' + 15y = 0$ on any t-interval. The general solution of $y'' + 8y' + 15y = 0$ is therefore

$$y(t) = c_1 e^{-5t} + c_2 e^{-3t}. \ \text{❖}$$

Roots of the Characteristic Equation

The function $y(t) = e^{\lambda_1 t}$ is a solution of $ay'' + by' + cy = 0$ provided λ_1 is a root of the characteristic equation

$$a\lambda^2 + b\lambda + c = 0. \tag{4}$$

As we know, the quadratic equation $a\lambda^2 + b\lambda + c = 0$ might have two distinct real roots, one real root, or two complex roots. Figure 3.4 shows the graph of $P(\lambda) = a\lambda^2 + b\lambda + c$ versus λ, illustrating each of these three cases.

We can obtain the roots of characteristic equation (4) from the quadratic formula,

$$\lambda_{1,2} = \frac{-b \pm \sqrt{b^2 - 4ac}}{2a}. \tag{5}$$

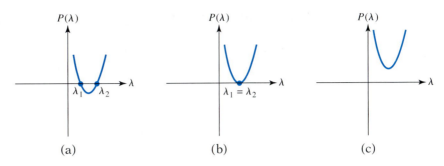

FIGURE 3.4

Three possibilities for the graph of $P(\lambda) = a\lambda^2 + b\lambda + c, a > 0$.
(a) The characteristic equation has two real distinct roots, λ_1 and λ_2.
(b) The characteristic equation has a single repeated real root, λ_1.
(c) The characteristic equation has two complex roots but no real roots.

As illustrated in Figure 3.4, there are three cases, depending on the value of the discriminant, $b^2 - 4ac$:

(a) Suppose $b^2 - 4ac > 0$. In this case, the two roots λ_1 and λ_2 are real and distinct. We obtain two solutions, $y_1(t) = e^{\lambda_1 t}$ and $y_2(t) = e^{\lambda_2 t}$. We show later in this section that these two solutions form a fundamental set of solutions for equation (1).

(b) Suppose $b^2 - 4ac = 0$. In this case, the two roots are equal,

$$\lambda_1 = \lambda_2 = \frac{-b}{2a}.$$

Our computation, based on the trial form $y(t) = e^{\lambda t}$, therefore yields only one solution, namely

$$y_1(t) = e^{-(b/2a)t}.$$

Since a fundamental set of solutions consists of two solutions having a nonvanishing Wronskian, we must find another solution having a different functional form. In Section 3.4, we will discuss this real "repeated root" case and show how to obtain the second function needed for a fundamental set.

(c) Suppose $b^2 - 4ac < 0$. In this case, the roots are complex-valued and we have

$$\lambda_{1,2} = -\frac{b}{2a} \pm i\frac{\sqrt{4ac - b^2}}{2a}.$$

Since a, b, and c are real constants, the roots $\lambda_{1,2}$ form a complex conjugate pair. For brevity, let

$$\alpha = -\frac{b}{2a}, \qquad \beta = \frac{\sqrt{4ac - b^2}}{2a}.$$

Then $\lambda_{1,2} = \alpha \pm i\beta$, where β is nonzero. Several questions arise. What mathematical interpretation do we give to expressions of the form $e^{(\alpha \pm i\beta)t}$? Once we make sense of such expressions mathematically, how do we obtain real-valued, physically meaningful solutions of equation (1)? These issues will be addressed in Section 3.5.

The General Solution When the Characteristic Equation Has Real Distinct Roots

We now consider case (a), where the discriminant $b^2 - 4ac$ is positive. In this case, the two roots λ_1 and λ_2 are real and distinct and $y_1(t) = e^{\lambda_1 t}$ and $y_2(t) = e^{\lambda_2 t}$ are two solutions of $ay'' + by' + cy = 0$. To determine whether $\{y_1, y_2\}$ forms a fundamental set of solutions, we calculate the Wronskian:

$$W(t) = \begin{vmatrix} e^{\lambda_1 t} & e^{\lambda_2 t} \\ \lambda_1 e^{\lambda_1 t} & \lambda_2 e^{\lambda_2 t} \end{vmatrix} = (\lambda_2 - \lambda_1)e^{(\lambda_1 + \lambda_2)t}.$$

The factor $(\lambda_2 - \lambda_1)$ is nonzero since the roots are distinct. In addition, the exponential function $e^{(\lambda_1 + \lambda_2)t}$ is nonzero for all t. This calculation establishes, once and for all, that the two exponential solutions obtained in the real, distinct root case form a fundamental set of solutions. There is no need to reestablish this fact for every particular example. The corresponding general solution of $ay'' + by' + cy = 0$ is

$$y(t) = c_1 e^{\lambda_1 t} + c_2 e^{\lambda_2 t}. \tag{6}$$

E X A M P L E

2

Solve the initial value problem

$$y'' + 4y' + 3y = 0, \qquad y(0) = 7, \qquad y'(0) = -17.$$

Solution: The characteristic equation is

$$\lambda^2 + 4\lambda + 3 = 0,$$

or

$$(\lambda + 1)(\lambda + 3) = 0.$$

Therefore, the general solution is

$$y(t) = c_1 e^{-t} + c_2 e^{-3t},$$

and its derivative is

$$y'(t) = -c_1 e^{-t} - 3c_2 e^{-3t}.$$

To satisfy the initial conditions, c_1 and c_2 must satisfy

$$c_1 + c_2 = 7$$
$$-c_1 - 3c_2 = -17.$$

We find $c_1 = 2$ and $c_2 = 5$. The unique solution of the initial value problem is

$$y(t) = 2e^{-t} + 5e^{-3t}. \; ❖$$

E X A M P L E

3

Solve the initial value problem

$$y'' + y' - 2y = 0, \qquad y(0) = y_0, \qquad y'(0) = y_0'.$$

For what values of the constants y_0 and y_0' can we guarantee that $\lim_{t \to \infty} y(t) = 0$?

Solution: The characteristic polynomial for $y'' + y' - 2y = 0$ is

$$\lambda^2 + \lambda - 2 = (\lambda + 2)(\lambda - 1).$$

Thus, the general solution is

$$y(t) = c_1 e^{-2t} + c_2 e^t.$$

Imposing the initial conditions, we have

$$c_1 + c_2 = y_0$$
$$-2c_1 + c_2 = y_0'.$$

The solution of this system of equations is $c_1 = (y_0 - y_0')/3, c_2 = (2y_0 + y_0')/3$. The solution of the initial value problem is therefore

$$y(t) = \left(\frac{y_0 - y_0'}{3}\right) e^{-2t} + \left(\frac{2y_0 + y_0'}{3}\right) e^t.$$

Since $\lim_{t \to \infty} e^{-2t} = 0$ and $\lim_{t \to \infty} e^t = +\infty$, the solution of the initial value problem will tend to zero as t increases if the coefficient of e^t in the solution is zero. Therefore, $\lim_{t \to \infty} y(t) = 0$ if $y_0' = -2y_0$. ❖

EXERCISES

Exercises 1–15:

(a) Find the general solution of the differential equation.

(b) Impose the initial conditions to obtain the unique solution of the initial value problem.

(c) Describe the behavior of the solution $y(t)$ as $t \to -\infty$ and as $t \to \infty$. Does $y(t)$ approach $-\infty$, $+\infty$, or a finite limit?

1. $y'' + y' - 2y = 0$, $y(0) = 3$, $y'(0) = -3$

2. $y'' - \frac{1}{4}y = 0$, $y(2) = 1$, $y'(2) = 0$

3. $y'' - 4y' + 3y = 0$, $y(0) = -1$, $y'(0) = 1$

4. $2y'' - 5y' + 2y = 0$, $y(0) = -1$, $y'(0) = -5$

5. $y'' - y = 0$, $y(0) = 1$, $y'(0) = -1$

6. $y'' + 2y' = 0$, $y(-1) = 0$, $y'(-1) = 2$

7. $y'' + 5y' + 6y = 0$, $y(0) = 1$, $y'(0) = -1$

8. $y'' - 5y' + 6y = 0$, $y(0) = 1$, $y'(0) = -1$

9. $y'' - 4y = 0$, $y(3) = 0$, $y'(3) = 0$

10. $8y'' - 6y' + y = 0$, $y(1) = 4$, $y'(1) = \frac{3}{2}$

11. $2y'' - 3y' = 0$, $y(-2) = 3$, $y'(-2) = 0$

12. $y'' - 6y' + 8y = 0$, $y(1) = 2$, $y'(1) = -8$

13. $y'' + 4y' + 2y = 0$, $y(0) = 0$, $y'(0) = 4$

14. $y'' - 4y' - y = 0$, $y(0) = 1$, $y'(0) = 2 + \sqrt{5}$

15. $2y'' - y = 0$, $y(0) = -2$, $y'(0) = \sqrt{2}$

16. Consider the initial value problem $y'' + \alpha y' + \beta y = 0, y(0) = 1, y'(0) = y'_0$, where α, β, and y'_0 are constants. It is known that one solution of the differential equation is $y_1(t) = e^{-3t}$ and that the solution of the initial value problem satisfies $\lim_{t \to \infty} y(t) = 2$. Determine the constants α, β, and y'_0.

17. Consider the initial value problem $y'' + \alpha y' + \beta y = 0, y(0) = 3, y'(0) = 5$. The differential equation has a fundamental set of solutions, $\{y_1(t), y_2(t)\}$. It is known that $y_1(t) = e^{-t}$ and that the Wronskian formed by the two members of the fundamental set is $W(t) = 4e^{2t}$.

(a) Determine the second member of the fundamental set, $y_2(t)$.

(b) Determine the constants α and β.

(c) Solve the initial value problem.

18. The three graphs display solutions of initial value problems on the interval $0 \le t \le 3$. Each solution satisfies the initial conditions $y(0) = 1, y'(0) = -1$. Match the differential equation with the graph of its solution.

(a) $y'' + 2y' = 0$ (b) $6y'' - 5y' + y = 0$ (c) $y'' - y = 0$

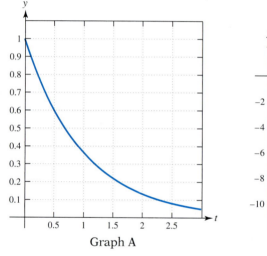

Graph A

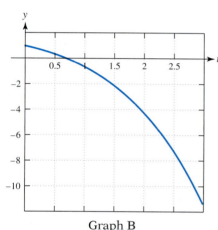

Graph B

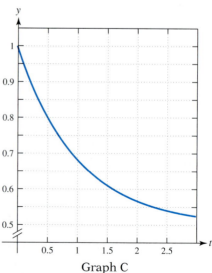

Graph C

Figure for Exercise 18

19. Obtain the general solution of $y''' - 5y'' + 6y' = 0$. [Hint: Make the change of dependent variable $u(t) = y'(t)$, determine $u(t)$, and then antidifferentiate to obtain $y(t)$.]

20. Rectilinear Motion with a Drag Force   In Chapter 2, we considered rectilinear motion in the presence of a drag force proportional to velocity. We solved the first order linear equation for velocity and antidifferentiated the solution to obtain distance as a function of time. We now consider directly the second order linear differential equation for the distance function.

A particle of mass m moves along the x-axis and is acted upon by a drag force proportional to its velocity. The drag constant is denoted by k. If $x(t)$ represents the particle position at time t, Newton's law of motion leads to the differential equation $mx''(t) = -kx'(t)$.

(a) Obtain the general solution of this second order linear differential equation.

(b) Solve the initial value problem if $x(0) = x_0$ and $x'(0) = v_0$.

(c) What is $\lim_{t \to \infty} x(t)$?

3.4   Real Repeated Roots; Reduction of Order

In Section 3.3, it was shown that the constant coefficient differential equation $ay'' + by' + cy = 0$ has the general solution

$$y(t) = c_1 e^{\lambda_1 t} + c_2 e^{\lambda_2 t}$$

whenever the characteristic equation $a\lambda^2 + b\lambda + c = 0$ has distinct real roots λ_1 and λ_2. In this section, we consider the case where the characteristic equation has a repeated real root (that is, when the discriminant $b^2 - 4ac = 0$). In this event, looking for solutions of the form $y(t) = e^{\lambda t}$ leads to only one solution, since the characteristic equation has just one distinct root. We must somehow find a second solution in order to form a fundamental set of solutions.

The Method of Reduction of Order

To obtain a second solution for $ay'' + by' + cy = 0$ in the repeated root case, we use a method called *reduction of order*. We apply the method first to the problem at hand,

$$ay'' + by' + cy = 0. \tag{1}$$

Then, at the end of this section, we'll discuss reduction of order as a technique for finding a second solution of the general homogeneous linear equation, $y'' + p(t)y' + q(t)y = 0$, given that we have somehow found one solution, $y_1(t)$, of the equation.

Assume, without loss of generality, that $a = 1$ in equation (1). Then, since $b^2 - 4c = 0$, we know that c is positive. We can represent c as $c = \alpha^2$ and choose $b = -2\alpha$. With these simplifications in notation, differential equation (1) becomes

$$y'' - 2\alpha y' + \alpha^2 y = 0. \tag{2}$$

The characteristic polynomial for equation (2) is

$$\lambda^2 - 2\alpha\lambda + \alpha^2 = (\lambda - \alpha)^2,$$

and therefore one solution is

$$y_1(t) = e^{\alpha t}.$$

To find a second solution, $y_2(t)$, we use the **method of reduction of order**. The basic idea underlying the method is to look for a second solution, $y_2(t)$, of the form

$$y_2(t) = y_1(t)u(t) = e^{\alpha t}u(t). \tag{3}$$

The function $u(t)$ in (3) must be chosen so that $y_2(t)$ is also a solution of equation (2).

At this point, there's no obvious reason to believe that this assumed form of the solution provides any simplification. We must substitute (3) into differential equation (2) and see what happens. Substituting, we obtain

$$y_2'' - 2\alpha y_2' + \alpha^2 y_2 = (e^{\alpha t}u)'' - 2\alpha(e^{\alpha t}u)' + \alpha^2(e^{\alpha t}u), \tag{4}$$

which simplifies to

$$y_2'' - 2\alpha y_2' + \alpha^2 y_2 = e^{\alpha t}u''. \tag{5}$$

Since the exponential function is nonzero everywhere, $y_2(t) = e^{\alpha t}u(t)$ is a solution of $y'' - 2\alpha y' + \alpha^2 y = 0$ if and only if $u'' = 0$. The equation $u'' = 0$ can be solved by antidifferentiation to obtain $u(t) = a_1 t + a_2$, where a_1 and a_2 are arbitrary constants.

Thus, the method of reduction of order has led us to a second solution,

$$y_2(t) = e^{\alpha t}(a_1 t + a_2) = a_1 t e^{\alpha t} + a_2 e^{\alpha t}.$$

Notice that the term $a_2 e^{\alpha t}$ is simply a constant multiple of $y_1(t)$. Since the general solution of the differential equation contains $y_1(t)$ multiplied by an arbitrary constant, we lose no generality by setting $a_2 = 0$. We can likewise take $a_1 = 1$ since $y_2(t)$ will also be multiplied by an arbitrary constant in the general solution. With these simplifications, the second solution is

$$y_2(t) = t e^{\alpha t}.$$

To verify that $\{y_1, y_2\} = \{e^{\alpha t}, te^{\alpha t}\}$ forms a fundamental set, we compute the Wronskian:

$$W(t) = \begin{vmatrix} e^{\alpha t} & te^{\alpha t} \\ \alpha e^{\alpha t} & (\alpha t + 1)e^{\alpha t} \end{vmatrix} = e^{2\alpha t}.$$

Since the Wronskian is nonzero, we have shown that the general solution is

$$y(t) = c_1 e^{\alpha t} + c_2 t e^{\alpha t}. \tag{6}$$

E X A M P L E

1

Solve the initial value problem

$$4y'' + 4y' + y = 0, \qquad y(2) = 1, \qquad y'(2) = 0.$$

Solution: Looking for solutions of the form $y(t) = e^{\lambda t}$ leads to the characteristic equation

$$4\lambda^2 + 4\lambda + 1 = (2\lambda + 1)^2 = 0.$$

Therefore, the characteristic equation has real repeated roots $\lambda_1 = \lambda_2 = -\frac{1}{2}$. By (6), the general solution is

$$y(t) = c_1 e^{-t/2} + c_2 t e^{-t/2}.$$

Imposing the initial conditions leads to

$$c_1 e^{-1} + c_2 2 e^{-1} = 1$$

$$-\frac{c_1}{2} e^{-1} \qquad = 0.$$

The solution is $c_1 = 0, c_2 = e/2$. The solution of the initial value problem is

$$y(t) = \frac{e}{2} t e^{-t/2} = \frac{t}{2} e^{1-t/2}. \quad \clubsuit$$

Method of Reduction of Order (General Case)

The method of reduction of order is not restricted to constant coefficient equations. It can be applied to the general second order linear homogeneous equation

$$y'' + p(t)y' + q(t)y = 0, \tag{7}$$

where $p(t)$ and $q(t)$ are continuous functions on the t-interval of interest. Suppose we know one solution of equation (7); call it $y_1(t)$. We again assume that there is another solution, $y_2(t)$, of the form

$$y_2(t) = y_1(t)u(t).$$

Substituting the assumed form into (7) leads (after some rearranging of terms) to

$$y_1 u'' + [2y_1' + p(t)y_1]u' + [y_1'' + p(t)y_1' + q(t)y_1]u = 0.$$

At first it seems as though this equation offers little improvement. Recall, however, that y_1 is not an arbitrary function; it is a solution of differential equation (7). Therefore, the factor multiplying u in the preceding equation vanishes, and we obtain a considerable simplification:

$$y_1 u'' + [2y_1' + p(t)y_1]u' = 0. \tag{8}$$

The structure of equation (8) is what gives the method its name. Although equation (8) is a second order linear differential equation for u, we can define a new dependent variable $v(t) = u'(t)$. Under this change of variables, equation (8) reduces to a first order linear differential equation for v,

$$y_1(t)v' + [2y_1'(t) + p(t)y_1(t)]v = 0. \tag{9}$$

Thus, the task of solving a second order linear differential equation has been replaced by that of solving a first order linear differential equation. Once we have $v = u'$, we obtain u (and ultimately y_2) by antidifferentiation.

E X A M P L E

2

Observe that $y_1(t) = t$ is a solution of the homogeneous linear differential equation

$$t^2 y'' - ty' + y = 0, \qquad 0 < t < \infty. \tag{10}$$

(a) Use reduction of order to obtain a second solution, $y_2(t)$. Does the pair $\{y_1, y_2\}$ form a fundamental set of solutions for this differential equation?

(continued)

(continued)

(b) If $\{y_1, y_2\}$ is a fundamental set of solutions, solve the initial value problem

$$t^2 y'' - t y' + y = 0, \qquad y(1) = 3, \qquad y'(1) = 8.$$

Solution: Note that the initial value problem, written in standard form, becomes

$$y'' - \frac{1}{t} y' + \frac{1}{t^2} y = 0, \qquad y(1) = 3, \qquad y'(1) = 8. \tag{11}$$

The coefficient functions are not continuous at $t = 0$. Our t-interval of interest, $0 < t < \infty$, is the largest interval containing $t_0 = 1$ on which we are guaranteed the existence of a unique solution of the initial value problem.

(a) Since one solution is known, we apply reduction of order. Assuming $y_2(t) = tu(t)$, we have

$$y_2' = u + t u' \quad \text{and} \quad y_2'' = 2u' + t u''.$$

Substituting these expressions into the differential equation, $t^2 y'' - t y' + y = 0$, we find

$$t^2(2u' + t u'') - t(u + t u') + tu = t^2(t u'' + u') = 0.$$

Therefore, $t u'' + u' = 0$. Setting $v = u'$ leads to the first order linear equation

$$t v' + v = 0. \tag{12}$$

The general solution of equation (12) is

$$v(t) = \frac{c}{t}.$$

Since $v(t) = u'(t)$, it follows that $u(t) = c \ln t + d$, and we obtain a second solution,

$$y_2(t) = tu(t) = t(c \ln t + d). \tag{13}$$

[Note that $\ln |t| = \ln t$ since $t > 0$.] Using the same rationale as before, we can take $c = 1, d = 0$ and let

$$y_2(t) = t \ln t.$$

The Wronskian of y_1 and y_2 is $W(t) = t$, which is nonzero on the interval $0 < t < \infty$. Therefore, the general solution is

$$y(t) = c_1 y_1(t) + c_2 y_2(t) = c_1 t + c_2 t \ln t, \qquad 0 < t < \infty. \tag{14}$$

(b) For $y(t) = c_1 t + c_2 t \ln t$, we have $y'(t) = c_1 + c_2(1 + \ln t)$. Imposing the initial conditions $y(1) = 3$ and $y'(1) = 8$, we obtain

$$c_1 \qquad\quad = 3$$
$$c_1 + c_2 = 8.$$

The solution of the initial value problem is

$$y(t) = 3t + 5t \ln t. \;\; ❖$$

EXERCISES

Exercises 1–9:

(a) Obtain the general solution of the differential equation.

(b) Impose the initial conditions to obtain the unique solution of the initial value problem.

(c) Describe the behavior of the solution as $t \to -\infty$ and $t \to \infty$. In each case, does $y(t)$ approach $-\infty$, $+\infty$, or a finite limit?

1. $y'' + 2y' + y = 0, \quad y(1) = 1, \quad y'(1) = 0$

2. $9y'' - 6y' + y = 0, \quad y(3) = -2, \quad y'(3) = -\frac{5}{3}$

3. $y'' + 6y' + 9y = 0, \quad y(0) = 2, \quad y'(0) = -2$

4. $25y'' + 20y' + 4y = 0, \quad y(5) = 4e^{-2}, \quad y'(5) = -\frac{3}{5}e^{-2}$

5. $4y'' - 4y' + y = 0, \quad y(1) = -4, \quad y'(1) = 0$

6. $y'' - 4y' + 4y = 0, \quad y(-1) = 2, \quad y'(-1) = 1$

7. $16y'' - 8y' + y = 0, \quad y(0) = -4, \quad y'(0) = 3$

8. $y'' + 2\sqrt{2}y' + 2y = 0, \quad y(0) = 1, \quad y'(0) = 0$

9. $y'' - 5y' + 6.25y = 0, \quad y(-2) = 0, \quad y'(-2) = 1$

10. Consider the simple differential equation $y'' = 0$.

(a) Obtain the general solution by successive antidifferentiation.

(b) View the equation $y'' = 0$ as a second order linear homogeneous equation with constant coefficients, where the characteristic equation has a repeated real root. Obtain the general solution using this viewpoint. Is it the same as the solution found in part (a)?

Exercises 11–12:

In each exercise, the graph shown is the solution of $y'' - 2\alpha y' + \alpha^2 y = 0$, $y(0) = y_0$, $y'(0) = y_0'$. Determine the constants α, y_0, and y_0' as well as the solution $y(t)$. In Exercise 11, the maximum point shown on the graph has coordinates $(2, 8e^{-1})$.

11.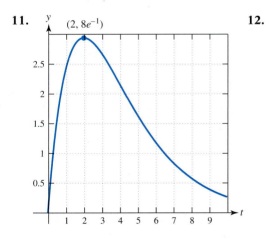

12.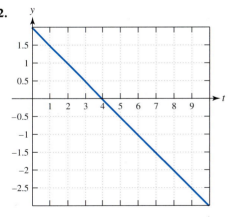

13. The graph of a solution $y(t)$ of the differential equation $4y'' + 4y' + y = 0$ passes through the points $(1, \; e^{-1/2})$ and $(2, 0)$. Determine $y(0)$ and $y'(0)$.

One solution, $y_1(t)$, of the differential equation is given.

(a) Use the method of reduction of order to obtain a second solution, $y_2(t)$.

(b) Compute the Wronskian formed by the solutions $y_1(t)$ and $y_2(t)$.

14. $ty'' - (2t + 1)y' + (t + 1)y = 0, \quad y_1(t) = e^t$

15. $t^2 y'' - ty' + y = 0, \quad y_1(t) = t$

16. $y'' - (2\cot t)y' + (1 + 2\cot^2 t)y = 0, \quad y_1(t) = \sin t$

17. $(t + 1)^2 y'' - 4(t + 1)y' + 6y = 0, \quad y_1(t) = (t + 1)^2$

18. $y'' + 4ty' + (2 + 4t^2)y = 0, \quad y_1(t) = e^{-t^2}$

19. $(t - 2)^2 y'' + (t - 2)y' - 4y = 0, \quad y_1(t) = (t - 2)^2$

20. $y'' - \left(2 + \dfrac{n-1}{t}\right)y' + \left(1 + \dfrac{n-1}{t}\right)y = 0$, where n is a positive integer, $y_1(t) = e^t$

3.5   Complex Roots

We now complete the discussion of finding the general solution for the differential equation

$$ay'' + by' + cy = 0,$$

by obtaining the general solution when the discriminant is negative.

Looking for solutions of the form $y(t) = e^{\lambda t}$ when the discriminant is negative leads to a characteristic equation $a\lambda^2 + b\lambda + c = 0$ having complex conjugate roots,

$$\lambda_{1,2} = -\frac{b}{2a} \pm i\frac{\sqrt{4ac - b^2}}{2a}.$$

Using $\alpha = -b/2a$ and $\beta = \sqrt{4ac - b^2}/2a$ for simplicity, we have for the roots

$$\lambda_{1,2} = \alpha \pm i\beta. \tag{1}$$

The Complex Exponential Function

The approach to solving $ay'' + by' + cy = 0$ has been based on looking for solutions of the form $y(t) = e^{\lambda t}$. When the roots (1) of the characteristic equation are complex, we are led to consider exponential functions with complex arguments:

$$y_1(t) = e^{(\alpha + i\beta)t} \quad \text{and} \quad y_2(t) = e^{(\alpha - i\beta)t}. \tag{2}$$

We need to clarify the mathematical meaning of these two expressions. How is the definition of the exponential function extended or broadened to accommodate complex as well as real arguments? Once such a generalization is understood mathematically, we then need to demonstrate for complex λ that the function $e^{\lambda t}$ is, in fact, a differentiable function of t satisfying the fundamental differentiation formula

$$\frac{d}{dt}e^{\lambda t} = \lambda e^{\lambda t}.$$

This differentiation formula was tacitly assumed in Section 3.3, and it is needed now if we want to show that $y = e^{(\alpha+i\beta)t}$ and $y = e^{(\alpha-i\beta)t}$ are, in fact, solutions of $ay'' + by' + cy = 0$.

A second issue that needs to be addressed is that of physical relevance. We will be able to show that the functions $\{e^{(\alpha+i\beta)t}, e^{(\alpha-i\beta)t}\}$ form a fundamental set of solutions. How do we use them to obtain real-valued solutions?

As a case in point, recall the buoyancy example discussed in Section 3.1. When we modeled the object's position, $y(t)$, we arrived at a differential equation of the form

$$y'' + \omega^2 y = 0.$$

The characteristic equation for this differential equation is $\lambda^2 + \omega^2 = 0$ and has roots $\lambda_1 = i\omega$ and $\lambda_2 = -i\omega$. How do we relate the two solutions

$$y_1(t) = e^{i\omega t} \quad \text{and} \quad y_2(t) = e^{-i\omega t}$$

to the real-valued general solution

$$y = A\sin\omega t + B\cos\omega t$$

used to describe the bobbing motion of the object?

The Definition of the Complex Exponential Function

In calculus, the exponential function $y = e^t$ is often introduced as the inverse of the natural logarithm function. For our present purposes, however, we want a representation of the exponential function that permits us to generalize from a real argument to a complex argument in a straightforward and natural way. In this regard, the power series representation of the function e^z is very convenient.

From calculus, the Maclaurin[4] series expansion for e^z is the infinite series

$$e^z = 1 + z + \frac{z^2}{2!} + \frac{z^3}{3!} + \cdots = \sum_{n=0}^{\infty} \frac{z^n}{n!}, \tag{3}$$

where z^0 and $0!$ are understood to be equal to 1. For a given value of z, the Maclaurin series (3) is interpreted as the limit of the sequence of partial sums,

$$e^z = \lim_{M\to\infty} \sum_{n=0}^{M} \frac{z^n}{n!}. \tag{4}$$

When limit (4) exists, we say that the series **converges**. It is shown in calculus that the Maclaurin series (3) converges to e^z for every real value of z. Although we do not do so here, it is possible to derive a number of familiar properties of the exponential function from the power series (3). For example, it can be shown that e^z is differentiable and that

$$e^{z_1+z_2} = e^{z_1}e^{z_2}.$$

---

[4]Colin Maclaurin (1698–1746) was a professor of mathematics at the University of Aberdeen and later at the University of Edinburgh. In a two-volume exposition of Newton's calculus, titled the *Treatise of Fluxions* (published in 1742), Maclaurin used the special form of Taylor series now bearing his name. Maclaurin also is credited with introducing the integral test for the convergence of an infinite series.

The importance of the power series representation (3) for e^z is that the representation is not limited to real values of z. It can be shown that power series (3) converges for all complex values of z. In this manner, the function $e^{\lambda t}$ is given meaning even for a complex value of λ. The power series representation can also be used to show that

$$e^{\lambda t} = e^{(\alpha+i\beta)t} = e^{\alpha t + i\beta t} = e^{\alpha t}e^{i\beta t}. \tag{5}$$

Euler's Formula

From equation (5),

$$e^{\lambda t} = e^{\alpha t + i\beta t} = e^{\alpha t}e^{i\beta t}.$$

This result simplifies our task of understanding the function $y = e^{\lambda t}$, since we already know the behavior of the factor $e^{\alpha t}$ when α is real. Thus, we focus on the other factor, $e^{i\beta t}$. Using $z = i\beta t$ in power series (3), we obtain

$$e^{i\beta t} = 1 + (i\beta t) + \frac{(i\beta t)^2}{2!} + \frac{(i\beta t)^3}{3!} + \frac{(i\beta t)^4}{4!} + \frac{(i\beta t)^5}{5!} + \cdots$$

$$= \left[1 - \frac{(\beta t)^2}{2!} + \frac{(\beta t)^4}{4!} - \cdots\right] + i\left[\beta t - \frac{(\beta t)^3}{3!} + \frac{(\beta t)^5}{5!} - \cdots\right] \tag{6}$$

$$= \sum_{n=0}^{\infty} \frac{(-1)^n(\beta t)^{2n}}{(2n)!} + i\sum_{n=0}^{\infty} \frac{(-1)^n(\beta t)^{2n+1}}{(2n+1)!}.$$

In (6), we used the fact that $i^2 = -1$ and regrouped the terms into real and imaginary parts. The two series on the right-hand side of (6) are Maclaurin series representations of familiar functions:

$$\cos\beta t = 1 - \frac{(\beta t)^2}{2!} + \frac{(\beta t)^4}{4!} - \cdots = \sum_{n=0}^{\infty} \frac{(-1)^n(\beta t)^{2n}}{(2n)!},$$

$$\sin\beta t = \beta t - \frac{(\beta t)^3}{3!} + \frac{(\beta t)^5}{5!} - \cdots = \sum_{n=0}^{\infty} \frac{(-1)^n(\beta t)^{2n+1}}{(2n+1)!}.$$

Using these results in equation (6), we obtain **Euler's formula**,

$$e^{i\beta t} = \cos\beta t + i\sin\beta t. \tag{7}$$

The symmetry properties of the sine and cosine functions [$\cos(-\theta) = \cos\theta$ and $\sin(-\theta) = -\sin\theta$], together with Euler's formula (7), lead to an analogous expression for $e^{-i\beta t}$,

$$e^{-i\beta t} = e^{i(-\beta t)} = \cos(-\beta t) + i\sin(-\beta t) = \cos\beta t - i\sin\beta t. \tag{8}$$

Equations (5), (7), and (8) can be used to express $e^{(\alpha+i\beta)t}$ and $e^{(\alpha-i\beta)t}$ in terms of familiar functions:

$$e^{(\alpha+i\beta)t} = e^{\alpha t}e^{i\beta t} = e^{\alpha t}(\cos\beta t + i\sin\beta t),$$
$$e^{(\alpha-i\beta)t} = e^{\alpha t}e^{-i\beta t} = e^{\alpha t}(\cos\beta t - i\sin\beta t). \tag{9}$$

We can use equation (9) to show that the Wronskian of $y_1(t) = e^{(\alpha+i\beta)t}$ and $y_2(t) = e^{(\alpha-i\beta)t}$ is $W(t) = -i2\beta e^{2\alpha t}$, which is nonzero for all t since $\beta \neq 0$. The two solutions, y_1 and y_2, therefore form a fundamental set of solutions.

From a mathematical point of view, we are done with the complex roots case since we have found a fundamental set. From a physical point of view, however, if the mathematical problem arises from a (real-valued) physical process and we seek real-valued, physically meaningful answers, the two solutions in (9) are not satisfactory. We want a fundamental set consisting of real-valued solutions.

The Real and Imaginary Parts of a Complex-Valued Solution Are Also Solutions

We now state and prove a theorem that shows how to obtain a fundamental set of real-valued solutions when the characteristic equation has complex roots.

Theorem 3.3

Let $y(t) = u(t) + iv(t)$ be a solution of the differential equation

$$y'' + p(t)y' + q(t)y = 0,$$

where $p(t)$ and $q(t)$ are real-valued and continuous on $a < t < b$ and where $u(t)$ and $v(t)$ are real-valued functions defined on (a, b). Then $u(t)$ and $v(t)$ are also solutions of the differential equation on this interval.

● **PROOF:**   Since $y(t)$ is known to be a solution, we have

$$(u + iv)'' + p(t)(u + iv)' + q(t)(u + iv) = 0.$$

Therefore,

$$u'' + iv'' + p(t)(u' + iv') + q(t)(u + iv) = 0.$$

Collecting real and imaginary parts, we have

$$[u'' + p(t)u' + q(t)u] + i[v'' + p(t)v' + q(t)v] = 0, \qquad a < t < b.$$

Since a complex quantity vanishes if and only if its real and imaginary parts both vanish, we know that

$$u'' + p(t)u' + q(t)u = 0, \qquad a < t < b$$

and

$$v'' + p(t)v' + q(t)v = 0, \qquad a < t < b.$$

Therefore, $u(t)$ and $v(t)$ are both solutions of $y'' + p(t)y' + q(t)y = 0$. ●

We now apply Theorem 3.3 to the equation $y'' + ay' + by = 0$ in the case where the characteristic equation has complex roots $\lambda_1 = \alpha + i\beta$ and $\lambda_2 = \alpha - i\beta$ and corresponding complex-valued solutions

$$y_1(t) = e^{\alpha t}(\cos \beta t + i \sin \beta t) \quad \text{and} \quad y_2(t) = e^{\alpha t}(\cos \beta t - i \sin \beta t).$$

Taking the real and imaginary parts of $y_1(t)$ [or, equivalently, of $y_2(t)$], we obtain a pair of real-valued solutions

$$u(t) = e^{\alpha t} \cos \beta t \quad \text{and} \quad v(t) = e^{\alpha t} \sin \beta t.$$

The Wronskian is

$$W(t) = \begin{vmatrix} e^{\alpha t} \cos \beta t & e^{\alpha t} \sin \beta t \\ e^{\alpha t}(\alpha \cos \beta t - \beta \sin \beta t) & e^{\alpha t}(\alpha \sin \beta t + \beta \cos \beta t) \end{vmatrix} = \beta e^{2\alpha t} \neq 0.$$

Therefore, the general solution of $ay'' + by' + cy = 0$ is

$$y(t) = Ae^{\alpha t} \cos \beta t + Be^{\alpha t} \sin \beta t = e^{\alpha t}(A \cos \beta t + B \sin \beta t). \qquad \textbf{(10)}$$

E X A M P L E 1

Find the general solution for the differential equation

$$y'' + 25y = 0.$$

Solution: The characteristic equation is $\lambda^2 + 25 = 0$. The roots are $\lambda_1 = 5i$ and $\lambda_2 = -5i$. Therefore, in equation (10), we have $\alpha = 0$ and $\beta = 5$. The general solution is

$$y(t) = A \cos 5t + B \sin 5t. \quad \diamondsuit$$

E X A M P L E 2

Solve the initial value problem

$$y'' + 2y' + 5y = 0, \qquad y(0) = 2, \qquad y'(0) = 2.$$

Solution: The characteristic equation is $\lambda^2 + 2\lambda + 5 = 0$. The roots are

$$\lambda_{1,2} = \frac{-2 \pm \sqrt{4 - 20}}{2} = -1 \pm 2i.$$

The general solution of the differential equation is

$$y(t) = e^{-t}(A \cos 2t + B \sin 2t).$$

In order to impose the initial conditions, we differentiate $y(t)$, obtaining

$$y'(t) = -e^{-t}(A \cos 2t + B \sin 2t) + e^{-t}(-2A \sin 2t + 2B \cos 2t).$$

The initial conditions at $t = 0$ lead to the equations

$$A \qquad = 2$$
$$-A + 2B = 2.$$

Solving this system, we find that the unique solution of the initial value problem is

$$y(t) = 2e^{-t}(\cos 2t + \sin 2t). \quad \diamondsuit$$

The graph of the solution found in Example 2, $y(t) = 2e^{-t}(\cos 2t + \sin 2t)$, is shown in Figure 3.5. Notice that the dashed curves, which are actually the graphs of $y = \pm 2\sqrt{2}\,e^{-t}$, represent an envelope that describes the (decreasing) size of the sinusoidal oscillations. We will discuss this envelope in the next subsection.

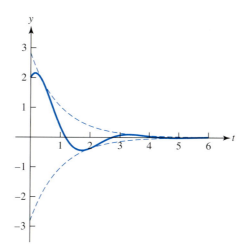

FIGURE 3.5

The graph of the solution found in Example 2, $y(t) = 2e^{-t}(\cos 2t + \sin 2t)$. The dashed curves, $y = 2\sqrt{2}\,e^{-t}$ and $y = -2\sqrt{2}\,e^{-t}$, constitute an envelope containing the graph of the solution.

Amplitude and Phase

Consider an initial value problem whose characteristic equation has complex roots, $\lambda_{1,2} = \alpha \pm i\beta$. The solution has the form

$$y(t) = e^{\alpha t}(A \cos \beta t + B \sin \beta t), \tag{11}$$

where A and B are constants determined from the given initial conditions. We now show that this solution can also be expressed as

$$y(t) = Re^{\alpha t} \cos(\beta t - \delta), \tag{12}$$

where R and δ are positive constants. In this form, the behavior of the solution as a function of time is more easily understood—it is the product of a sinusoidally oscillating function, $\cos(\beta t - \delta)$, and a term, $Re^{\alpha t}$, that is increasing with time when $\alpha > 0$, constant when $\alpha = 0$, and decreasing with time when $\alpha < 0$.

In (12), the term $Re^{\alpha t}$ is often referred to as the **amplitude** of the oscillations. The constant δ is referred to as the **phase**, the **phase angle**, or the **phase shift**. The term "phase shift" reflects the fact that we obtain the graph of $\cos(\beta t - \delta)$ by shifting the graph of $\cos \beta t$ to the right by an amount $t = \delta/\beta$. To see how equation (11) can be recast as (12), first recall the trigonometric identity

$$\cos(\theta_1 - \theta_2) = \cos \theta_1 \cos \theta_2 + \sin \theta_1 \sin \theta_2.$$

Using this trigonometric identity on the right-hand side of equation (12), we see that $R \cos(\beta t - \delta) = R \cos \beta t \cos \delta + R \sin \beta t \sin \delta$. Equating the corresponding expressions in (11) and (12), we obtain

$$R(\cos \beta t \cos \delta + \sin \beta t \sin \delta) = A \cos \beta t + B \sin \beta t.$$

Comparing like terms, we see R and δ must be chosen so that

$$R \cos \delta = A \quad \text{and} \quad R \sin \delta = B. \tag{13}$$

From (13), it follows that

$$R = \sqrt{A^2 + B^2},$$
$$\tan \delta = \frac{B}{A}, \quad A \neq 0. \tag{14}$$

We need to examine the signs of both $\cos \delta = A/R$ and $\sin \delta = B/R$ in order to determine the quadrant in which the angle δ lies, since there are two different choices for an angle δ that satisfies $\tan \delta = B/A$:

$$\delta = \tan^{-1}\left(\frac{B}{A}\right) \quad \text{or} \quad \delta = \tan^{-1}\left(\frac{B}{A}\right) + \pi. \tag{15}$$

EXAMPLE
3

Consider the solution of the initial value problem found in Example 2, $y(t) = 2e^{-t}(\cos 2t + \sin 2t)$. The solution is graphed in Figure 3.5. Rewrite $y(t)$ in the form

$$y(t) = Re^{\alpha t} \cos(\beta t - \delta).$$

Use this form to identify the main features of the graph of $y(t)$.

Solution: Compare the solution, $y(t) = 2e^{-t}(\cos 2t + \sin 2t)$, with expression (11). We see that $A = 2, B = 2, \alpha = -1$, and $\beta = 2$. Therefore,

$$R = 2\sqrt{2} \quad \text{and} \quad \tan \delta = 1.$$

Since $R \cos \delta = 2$ and $R \sin \delta = 2$, it follows that $\cos \delta$ and $\sin \delta$ are both positive. Therefore, δ is in the first quadrant and not in the third quadrant. Having identified the proper quadrant, we obtain $\delta = \tan^{-1}(1) = \pi/4$.

Knowing R, β, and δ, we can rewrite $y(t) = 2e^{-t}(\cos 2t + \sin 2t)$ as

$$y(t) = 2\sqrt{2}\, e^{-t} \cos\left(2t - \frac{\pi}{4}\right). \tag{16}$$

From expression (16), we can readily deduce the main features of the graph in Figure 3.5. As mentioned earlier, the envelope function $y(t) = 2\sqrt{2}\, e^{-t}$ governs the amplitude of the oscillations, and this fact is clear from equation (16). The phase angle, $\pi/4$, determines the shift of the cosine function. For instance, in $t > 0, y(t)$ is zero at $t = 3\pi/8, 7\pi/8, 11\pi/8, \ldots$. ❖

EXAMPLE
4

Solve the initial value problem

$$y'' + y = 0, \quad y(0) = -1, \quad y'(0) = -\sqrt{3}$$

and put the solution in the form $Re^{\alpha t} \cos(\beta t - \delta)$.

Solution: The characteristic equation is $\lambda^2 + 1 = 0$ and has roots $\lambda_{1,2} = \pm i$. The general solution is

$$y(t) = A \cos t + B \sin t.$$

Imposing the initial conditions, we obtain the solution

$$y(t) = -\cos t - \sqrt{3} \sin t.$$

Using $R\cos\delta = -1$ and $R\sin\delta = -\sqrt{3}$, we have

$$R = 2 \quad \text{and} \quad \tan\delta = \sqrt{3}.$$

Since $\cos\delta$ and $\sin\delta$ are both negative, the phase angle δ must lie in the third quadrant,

$$\delta = \tan^{-1}\sqrt{3} + \pi = \frac{4\pi}{3}.$$

Hence [see (12)],

$$y(t) = 2\cos\left(t - \frac{4\pi}{3}\right).$$

The graph of $y(t)$ is shown in Figure 3.6. It is the graph of $2\cos t$ shifted to the right by $4\pi/3 \approx 4.19$ radians.

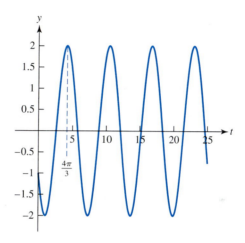

FIGURE 3.6

The graph of the solution found in Example 4, $y(t) = 2\cos[t - (4\pi/3)]$. ❖

EXERCISES

The identity $e^{z_1+z_2} = e^{z_1}e^{z_2}$, from which we obtain $(e^z)^n = e^{nz}$, is useful in some of the exercises.

1. Write each of the complex numbers in the form $\alpha + i\beta$, where α and β are real numbers.

   (a) $2e^{i\pi/3}$

   (b) $-2\sqrt{2}\,e^{-i\pi/4}$

   (c) $(2-i)e^{i3\pi/2}$

   (d) $\dfrac{1}{2\sqrt{2}}\,e^{i7\pi/6}$

   (e) $\left(\sqrt{2}\,e^{i\pi/6}\right)^4$

2. Write each of the functions in the form $Ae^{\alpha t}\cos\beta t + iBe^{\alpha t}\sin\beta t$, where α, β, A, and B are real numbers.

   (a) $2e^{i\sqrt{2}t}$

   (b) $\dfrac{2}{\pi}e^{-(2+3i)t}$

   (c) $-\dfrac{1}{2}e^{2t+i(t+\pi)}$

   (d) $\left(\sqrt{3}\,e^{(1+i)t}\right)^3$

   (e) $\left(-\dfrac{1}{\sqrt{2}}\,e^{i\pi t}\right)^3$

Exercises 3–12:

For the given differential equation,

(a) Determine the roots of the characteristic equation.

(b) Obtain the general solution as a linear combination of real-valued solutions.

(c) Impose the initial conditions and solve the initial value problem.

3. $y'' + 4y = 0$, $y(\pi/4) = -2$, $y'(\pi/4) = 1$

4. $y'' + 2y' + 2y = 0$, $y(0) = 3$, $y'(0) = -1$

5. $9y'' + y = 0$, $y(\pi/2) = 4$, $y'(\pi/2) = 0$

6. $2y'' - 2y' + y = 0$, $y(-\pi) = 1$, $y'(-\pi) = -1$

7. $y'' + y' + y = 0$, $y(0) = -2$, $y'(0) = -2$

8. $y'' + 4y' + 5y = 0$, $y(\pi/2) = 1/2$, $y'(\pi/2) = -2$

9. $9y'' + 6y' + 2y = 0$, $y(3\pi) = 0$, $y'(3\pi) = 1/3$

10. $y'' + 4\pi^2 y = 0$, $y(1) = 2$, $y'(1) = 1$

11. $y'' - 2\sqrt{2}\,y' + 3y = 0$, $y(0) = -1/2$, $y'(0) = \sqrt{2}$

12. $9y'' + \pi^2 y = 0$, $y(3) = 2$, $y'(3) = -\pi$

Exercises 13–21:

The function $y(t)$ is a solution of the initial value problem $y'' + ay' + by = 0$, $y(t_0) = y_0$, $y'(t_0) = y_0'$, where the point t_0 is specified. Determine the constants a, b, y_0, and y_0'.

13. $y(t) = \sin t - \sqrt{2}\cos t$, $t_0 = \pi/4$

14. $y(t) = 2\sin 2t + \cos 2t$, $t_0 = \pi/4$

15. $y(t) = e^{-2t}\cos t - e^{-2t}\sin t$, $t_0 = 0$

16. $y(t) = e^{t-\pi/6}\cos 2t - e^{t-\pi/6}\sin 2t$, $t_0 = \pi/6$

17. $y(t) = \sqrt{3}\cos \pi t - \sin \pi t$, $t_0 = 1/2$

18. $y(t) = \sqrt{2}\cos(2t - \pi/4)$, $t_0 = 0$

19. $y(t) = 2e^t\cos(\pi t - \pi)$, $t_0 = 1$

20. $y(t) = e^{-t}\cos(\pi t - \pi)$, $t_0 = 0$

21. $y(t) = 3e^{-2t}\cos(t - \pi/2)$, $t_0 = 0$

Exercises 22–26:

Rewrite the function $y(t)$ in the form $y(t) = Re^{\alpha t}\cos(\beta t - \delta)$, where $0 \le \delta < 2\pi$. Use this representation to sketch a graph of the given function, on a domain sufficiently large to display its main features.

22. $y(t) = \sin t + \cos t$

23. $y(t) = \cos \pi t - \sin \pi t$

24. $y(t) = e^t\cos t + \sqrt{3}e^t\sin t$

25. $y(t) = -e^{-t}\cos t + \sqrt{3}e^{-t}\sin t$

26. $y(t) = e^{-2t}\cos 2t - e^{-2t}\sin 2t$

Exercises 27–29:

In each exercise, the figure shows the graph of the solution of an initial value problem $y'' + ay' + by = 0, y(0) = y_0, y'(0) = y_0'$. Use the information given to express the solution in the form $y(t) = R\cos(\beta t - \delta)$, where $0 \le \delta < 2\pi$. Determine the constants a, b, y_0, and y_0'.

27. The graph has a maximum value at $(0, 2)$ and a t-intercept at $(1, 0)$.

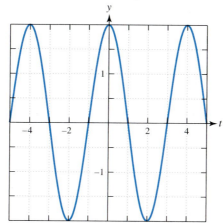

28. The graph has a maximum value at $(\pi/12, 1)$ and a t-intercept at $(5\pi/12, 0)$.

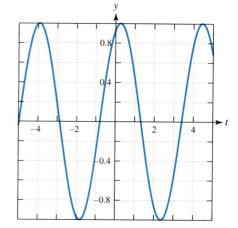

29. The graph has a maximum value at $(5\pi/12, 1/2)$ and a t-intercept at $(\pi/6, 0)$.

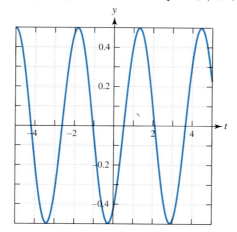

30. Consider the differential equation $y'' + ay' + by = 0$, where a and b are positive real constants. Show that $\lim_{t \to \infty} y(t) = 0$ for every solution of this equation.

31. Consider the differential equation $y'' + ay' + 9y = 0$, where a is a real constant. Suppose we know that the Wronskian of a fundamental set of solutions for this equation is constant. What is the general solution for this equation?

32. Buoyancy Problems with Drag Force   We discussed modeling the bobbing motion of floating cylindrical objects in Section 3.1. Bobbing motion will not persist indefinitely, however. One reason is the drag resistance a floating object experiences as it moves up and down in the liquid. If we assume a drag force proportional to velocity, an application of Newton's second law of motion leads to the differential equation $y'' + \mu y' + \omega^2 y = 0$, where $y(t)$ is the downward displacement of the object from its static equilibrium position, μ is a positive constant describing the drag force, and ω^2 is a positive constant determined by the mass densities of liquid and object and the vertical extent of the cylindrical object. (See Figure 3.1.)

(a) Obtain the general solution of this differential equation, assuming that $\mu^2 < 4\omega^2$.

(b) Assume that a cylindrical floating object is initially displaced downward a distance y_0 and released from rest [so the initial conditions are $y(0) = y_0, y'(0) = 0$].

Obtain, in terms of μ, ω, and y_0, the solution $y(t)$ of the initial value problem. Show that $\lim_{t \to \infty} y(t) = 0$.

33. This section has focused on the differential equation $ay'' + by' + cy = 0$, where a, b, and c are real constants. The fact that the roots of the characteristic equation occur in conjugate pairs when they are complex is due to the fact that the coefficients a, b, and c are real numbers. To see that this may not be true if the coefficients are allowed to be complex, determine the roots of the characteristic equation for the differential equation

$$y'' + 4iy' + 5y = 0.$$

Find two complex-valued solutions of this equation.

3.6   Unforced Mechanical Vibrations

In this section, we model the motion of a simple mechanical system—that of a mass suspended from the end of a hanging spring and subjected to some initial disturbance. The resulting up-and-down motion of the mass will be similar to the bobbing motion of a floating cylindrical object.

Hooke's Law

A spring hangs vertically from a ceiling. We assume that the weight of the spring is negligibly small. The natural or unstretched length of the spring is denoted by l, as in Figure 3.7. Suppose we now apply a vertical force to the end of the spring. If the force is directed downward, the spring will stretch. As it stretches, the spring develops an upward restoring force that resists this stretching or elongation. Conversely, if the applied force is directed upward, the spring compresses or shortens in length. In this case, the spring develops a counteracting downward restoring force that tends to resist compression.

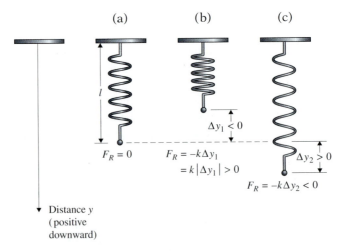

FIGURE 3.7

(a) A spring with natural length l. (b) The restoring force, $F_R = -k\Delta y_1$, is positive (directed downward) when the spring is compressed. (c) The restoring force, $F_R = -k\Delta y_2$, is negative (directed upward) when the spring is stretched.

We need an equation that relates the restoring force developed by the spring to the amount of elongation or compression that has occurred. The relation we use is **Hooke's law,**[5] which assumes the restoring force is proportional to the amount of stretching or compression that the spring has undergone. We assume that the displacement Δy is positive when the spring is stretched and negative when it is compressed. Whether the spring is stretched or compressed, its length is given by the quantity $l + \Delta y$.

Hooke's law states that the restoring force is

$$F_R = -k\Delta y, \tag{1}$$

where Δy is the displacement and k is a positive constant of proportionality, called the **spring constant**. The negative sign in equation (1) arises because the restoring force acts to counteract the displacement of the spring end.

The spring constant k in equation (1) has the dimensions of force per unit length and represents a measure of spring stiffness. A stiffer spring has a larger value of k, and the same restoring force arises from a smaller displacement.

Hooke's law is a useful description of reality when the displacement magnitude, $|\Delta y|$, is reasonably small. It cannot remain valid for arbitrarily large $|\Delta y|$ since one cannot stretch or compress a spring indefinitely. We assume in all our modeling and computations that displacement magnitudes are small enough to permit the use of Hooke's law.

A Mathematical Model for the Spring-Mass System

An object having mass m is attached to the end of the unstretched spring, as in Figure 3.8. The weight of the object is

$$W = mg,$$

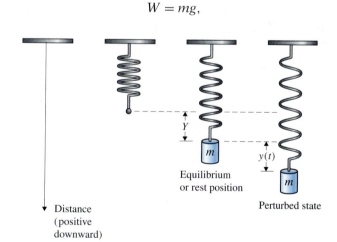

Equilibrium or rest position

Perturbed state

Distance (positive downward)

FIGURE 3.8

As the mass moves, the quantity $y(t)$ measures its displacement from the equilibrium position.

---

[5]Robert Hooke (1635–1703) served as professor of geometry at Gresham College, London, for 30 years. He worked on problems in elasticity, optics, and simple harmonic motion. Hooke invented the conical pendulum and was the first to build a Gregorian reflecting telescope. He was a competent architect and helped Christopher Wren rebuild London after the Great Fire of 1666.

where g is the acceleration due to gravity. The spring stretches until it achieves a new rest or equilibrium configuration. Let Y represent the distance the spring stretches to achieve this new equilibrium position. The displacement, Y, is determined by Hooke's law—the spring stretches until the restoring force exactly counteracts the object's weight:

$$W + F_R = mg - kY = 0. \tag{2}$$

It follows that $Y = mg/k$.

When the spring-mass system in Figure 3.8 is perturbed from its equilibrium position, we have from Newton's second law of motion

$$m\frac{d^2}{dt^2}(Y + y) = W + F_R = mg - k(Y + y).$$

Using (2) and noting that Y is a constant, we can reduce this equation to

$$my'' + ky = 0. \tag{3}$$

If we define $\omega^2 = k/m$, we obtain the differential equation $y'' + \omega^2 y = 0$, which characterizes the bobbing motion of a floating object.

Suppose we now assume a damping mechanism, a dashpot, is attached and suppresses the vibrating motion of the spring-mass system. It is shown schematically in Figure 3.9. We assume the damping force is proportional to velocity,

$$F_D = -\gamma \frac{dy(t)}{dt}. \tag{4}$$

In (4), γ is a positive constant of proportionality, referred to as the **damping coefficient**. The negative sign is present because the damping force acts to oppose the motion. A similar model of velocity damping was assumed in the linear model of projectile motion with air resistance discussed in Section 2.9.

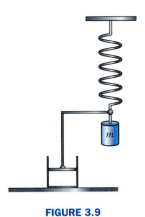

FIGURE 3.9

A spring-mass-dashpot system.

The differential equation describing the motion of the mass, again obtained from Newton's second law of motion, is

$$m\frac{d^2y}{dt^2} = W + F_R + F_D = mg - k(Y + y) - \gamma\frac{dy}{dt}. \tag{5}$$

Since $mg - kY = 0$, equation (5) simplifies to

$$m\frac{d^2y}{dt^2} + \gamma\frac{dy}{dt} + ky = 0. \tag{6}$$

Equation (6) is a second order linear constant coefficient differential equation. It is homogeneous because we are considering only the **unforced** or **free vibration** of the system. That is, there are no external forces applied to the system.

In Section 3.10, we will consider **forced vibrations**, where equation (6) is modified to include a time-varying applied force, $F(t)$. Inclusion of an applied force leads to a nonhomogeneous equation of the form

$$m\frac{d^2y}{dt^2} + \gamma\frac{dy}{dt} + ky = F(t). \tag{7}$$

Behavior of the Model

In this subsection, we discuss the solutions of equation (6), $my'' + \gamma y' + ky = 0$. These solutions describe how the mass-spring-dashpot system behaves—predicting the position, $y(t)$, and the velocity, $y'(t)$, of the moving mass at any time t. The characteristic equation,

$$m\lambda^2 + \gamma\lambda + k = 0,$$

has roots

$$\lambda_{1,2} = \frac{-\gamma \pm \sqrt{\gamma^2 - 4mk}}{2m}. \tag{8}$$

The corresponding mass-spring-dashpot system exhibits different behavior, depending on the roots of the characteristic equation. The roots, in turn, are determined by the relative values of the mass, spring constant, and damping coefficient.

Case 1   If $\gamma^2 > 4km$ (if damping is relatively strong), the characteristic equation has two negative real roots

$$\lambda_{1,2} = \frac{-\gamma \pm \sqrt{\gamma^2 - 4mk}}{2m}.$$

(Both roots are negative since $\sqrt{\gamma^2 - 4mk}$ is less than γ.) As shown in Section 3.3, the general solution is given by

$$y(t) = c_1 e^{\lambda_1 t} + c_2 e^{\lambda_2 t}. \tag{9a}$$

Therefore, the strong damping suppresses any vibratory motion of the mass. The general solution is a linear combination of two decreasing exponential functions. This case is referred to as the **overdamped case**.

Case 2   If $\gamma^2 = 4km$, then the roots are real and repeated. As we saw in Section 3.4, the general solution is given by

$$y(t) = c_1 e^{\lambda_1 t} + c_2 t e^{\lambda_1 t}, \tag{9b}$$

where $\lambda_1 = -\gamma/2m$. In this case, known as the **critically damped case**, damping is also sufficiently strong to suppress oscillatory vibrations of the mass.

Case 3 If $\gamma^2 < 4km$, then the roots are complex conjugates,

$$\lambda_{1,2} = \frac{-\gamma}{2m} \pm i\frac{\sqrt{4mk - \gamma^2}}{2m} = \alpha \pm i\beta.$$

As we saw in Section 3.5, the general solution is given by

$$y(t) = e^{\alpha t}(c_1 \cos \beta t + c_2 \sin \beta t). \tag{9c}$$

In this case, known as the **underdamped case**, damping is too weak to totally suppress the vibrations of the mass. Note that the underdamped case also includes the case where there is no damping whatsoever—that is, the case where $\gamma = 0$. Later, we refer to this as the **undamped case**. Finally, recall from Section 3.5 that solution (9c) can be restated in amplitude-phase form as

$$y(t) = Re^{\alpha t} \cos(\beta t - \delta), \tag{10}$$

where $R = \sqrt{c_1^2 + c_2^2}$, $R\cos\delta = c_1$, and $R\sin\delta = c_2$. If damping is present (that is, if $\gamma > 0$), then $\alpha = -\gamma/2m < 0$ and the motion of the mass described by equation (10) consists of damped vibrations (oscillations that decrease in amplitude as time progresses). If there is no damping, then $\alpha = 0$ and the oscillations do not decrease in magnitude.

Representative examples of the motion that occurs in these three cases are shown in Figure 3.10, parts (a)–(c). When damping is present, it follows from equations (9a)–(9c) that $\lim_{t\to\infty} y(t) = 0$ for any choices of c_1 and c_2. This is to be expected, since energy is dissipated and any initial disturbance will diminish in strength as time increases.

Vibrations and Periodic Functions

As a special case, assume that damping is absent; that is, $\gamma = 0$. In this case, $\alpha = 0$ and solution (10) reduces to

$$y(t) = R\cos\left(\sqrt{\frac{k}{m}}t - \delta\right). \tag{11}$$

If we set $\omega = \sqrt{k/m}$, equation (11) becomes

$$y(t) = R\cos(\omega t - \delta). \tag{12}$$

In this case, the amplitude of the vibrations remains constant and equal to R. The function $y(t)$ in (12) is an example of a *periodic function*.

In general, let $f(t)$ be defined on $-\infty < t < \infty$ or $a \leq t < \infty$ for some a. The function $f(t)$ is called a **periodic function** if there exists a positive constant T, called the **period**, such that

$$f(t + T) = f(t) \tag{13}$$

for all values of t in the domain. The smallest value of the constant T satisfying (13) is called the **fundamental period of the function**.

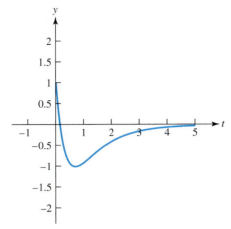

Equation: $y'' + 4y' + 3y = 0$
$\quad\quad\quad\quad\quad y(0) = 1, \quad y'(0) = -9$
Solution: $y(t) = 4e^{-3t} - 3e^{-t}$

(a) Overdamped motion

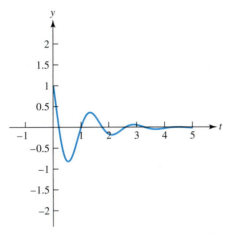

Equation: $y'' + 2y' + 17y = 0$
$\quad\quad\quad\quad\quad y(0) = 1, \quad y'(0) = -5$
Solution: $y(t) = e^{-t}(\cos 4t - \sin 4t)$
$\quad\quad\quad\quad = \sqrt{2}\,e^{-t}\cos(4t + \pi/4)$

(b) Underdamped motion

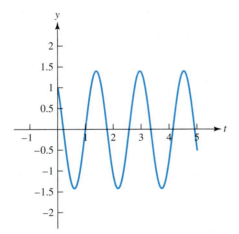

Equation: $y'' + 16y = 0$
$\quad\quad\quad\quad\quad y(0) = 1, \quad y'(0) = -4$
Solution: $y(t) = \cos 4t - \sin 4t$
$\quad\quad\quad\quad = \sqrt{2}\,\cos(4t + \pi/4)$

(c) Undamped motion

FIGURE 3.10

Examples of (a) overdamped motion, (b) underdamped motion with
nonzero damping, and (c) undamped motion.

The basic qualitative feature of a periodic function is that its graph repeats itself. If we know what the graph looks like on any time segment of duration T, we can obtain the graph on the entire domain simply by replicating this segment. Figure 3.11 provides an illustration.

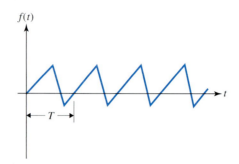

FIGURE 3.11

The graph of a periodic function repeats itself over any time period of duration T, where T is the fundamental period of the function.

Consider again the solution $y(t) = R\cos(\omega t - \delta)$ in equation (12). Since the cosine function repeats itself whenever its argument changes by 2π, $y(t)$ is a periodic function. To find the period for $y(t)$, we set $y(t + T) = y(t)$:

$$
\begin{aligned}
y(t + T) &= R\cos[\omega(t + T) - \delta] \\
&= R\cos[\omega t + \omega T - \delta] \\
&= y(t), \qquad \text{if } \omega T = 2n\pi, \qquad n = 1, 2, \ldots.
\end{aligned}
$$

Therefore, the function $y(t)$ has fundamental period $T = 2\pi/\omega$. In terms of the spring-mass system, $T = 2\pi/\omega$ is referred to as the **fundamental period of the motion** or simply the **period**. The period represents the time required for the mass to execute one cycle of its oscillatory motion. The motion itself is often referred to as **periodic motion**. The reciprocal of the period, $f = 1/T$, is called the **frequency** of the oscillations. The frequency represents the number of cycles of the periodic motion executed per unit time. For example, if $T = 0.01$ sec, the system completes 100 cycles of its motion per second. In current terminology, one cycle per second is referred to as one Hertz.[6] Therefore, we would say that the system oscillations have a frequency of 100 Hertz (100 Hz). From the relations $T = 2\pi/\omega$ and $f = 1/T$, it follows that $\omega = 2\pi f$. The constant ω is called the **angular frequency** or the **radian frequency**. It represents the change, in radians, that $\cos(\omega t - \delta)$ undergoes in one period.

It's worthwhile to check that the model predictions are consistent with our everyday experience. In the absence of damping, angular frequency is $\omega = \sqrt{k/m}$ and frequency is $f = (1/2\pi)\sqrt{k/m}$. Frequency therefore increases as either k increases or m decreases. Thus, when a given mass is attached in turn to two springs of differing stiffness, the model predicts it will vibrate more rapidly when suspended from the stiffer spring. Likewise, if two bodies of dif-

---

[6]Heinrich Hertz (1857–1894) was a German physicist who confirmed Maxwell's theory of electromagnetism by producing and studying radio waves. He demonstrated that these waves travel at the velocity of light and can be reflected, refracted, and polarized. The unit of frequency was named in his honor.

fering mass (and therefore weight) are suspended from the same spring, the smaller mass will vibrate more rapidly than the larger mass.

Now consider what happens when damping is added. The solution $y(t)$ in equation (10) has the form

$$y(t) = Re^{\alpha t}\cos(\beta t - \delta),\tag{14}$$

where

$$\alpha = \frac{-\gamma}{2m}\quad\text{and}\quad\beta = \frac{\sqrt{4km - \gamma^2}}{2m}.$$

Equation (14) predicts that when two different masses are attached to a spring-dashpot system having the same damping coefficient γ and spring constant k, the solution envelope of the larger mass (the heavier body) will decrease more slowly with time because the associated value α is smaller.

Note also that the introduction of damping changes the cosine term in equation (12) from $\cos[(\sqrt{k/m})t - \delta]$, when damping is absent, to

$$\cos\left(\frac{\sqrt{4km - \gamma^2}}{2m}t - \delta\right),$$

when damping is present. Since

$$\sqrt{\frac{k}{m}} > \frac{\sqrt{4km - \gamma^2}}{2m},$$

the introduction of damping causes the vibrations to "slow down" while simultaneously being reduced in amplitude.

Are these model predictions consistent with your everyday experience? What experiments might test these predictions, both qualitatively and quantitatively?

We conclude this section with two examples illustrating the motion of a spring-mass-dashpot system.

E X A M P L E

1

A block weighing 8 lb is attached to the end of a spring, causing the spring to stretch 6 in. beyond its natural length. The block is then pulled down 3 in. and released. Determine the motion of the block, assuming there are no damping forces or external applied forces.

Solution: The motion of the block is governed by equation (3),

$$my'' + ky = 0,$$

along with the initial conditions of the problem. Assuming the gravitational constant to be $g = 32$ ft /sec^2 and noting that the weight is given by $W = mg$, we find the block has mass

$$m = \frac{1}{4}\frac{\text{lb-sec}^2}{\text{ft}}.$$

[Note that 1 lb-sec^2/ft $= 1$ slug.] The spring constant k can be determined by the fact that an 8-lb force (the weight of the block) causes the spring to stretch

(continued)

(continued)

6 in.; Hooke's law implies $k = \frac{8}{6}$ lb/in. = 16 lb/ft. Since 3 in. $= \frac{1}{4}$ ft, the initial value problem governing the motion is

$$\tfrac{1}{4}y'' + 16y = 0, \qquad y(0) = \tfrac{1}{4}, \qquad y'(0) = 0. \tag{15}$$

The units of $y(t)$ are feet and of $y'(t)$ are feet per second. The solution of the initial value problem is $y(t) = \frac{1}{4}\cos 8t$. A graph of the block's position, $y(t)$, is shown in Figure 3.12(a). ❖

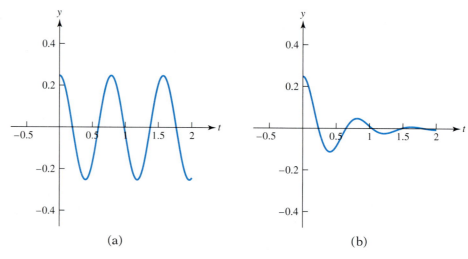

(a)   (b)

FIGURE 3.12

(a) The position, $y(t)$, of the mass in Example 1 (no damping).
(b) The position, $y(t)$, of the mass in Example 2 (includes damping).

E X A M P L E

2

Consider the spring-mass system in Example 1. Assume that damping is present and that the damping coefficient is given by $\gamma = 1$ lb-sec/ft. Determine the motion of the block.

Solution: To account for the assumed damping force, the equation of Example 1 will be modified to include a damping term $\gamma y'$, where $\gamma = 1$:

$$\tfrac{1}{4}y'' + y' + 16y = 0.$$

Therefore, the initial value problem governing the motion of the block is

$$y'' + 4y' + 64y = 0, \qquad y(0) = \tfrac{1}{4}, \qquad y'(0) = 0.$$

The general solution is

$$y(t) = e^{-2t}\left[c_1 \cos(2\sqrt{15}\,t) + c_2 \sin(2\sqrt{15}\,t)\right].$$

Imposing the initial conditions $y(0) = \frac{1}{4}$ and $y'(0) = 0$, we obtain

$$y(t) = \frac{1}{4}e^{-2t}\left[\cos(2\sqrt{15}\,t) + \frac{1}{\sqrt{15}}\sin(2\sqrt{15}\,t)\right].$$

The graph of $y(t)$ is shown in Figure 3.12(b). The block still oscillates about its equilibrium position, but the envelope of the oscillations decreases with time. ❖

EXERCISES

1. The given function $f(t)$ is periodic with fundamental period T; therefore, $f(t + T) = f(t)$. Use the information given to sketch the graph of $f(t)$ over the time interval $0 \leq t \leq 4T$.

   (a) $f(t) = t(2 - t)$, $0 \leq t < 2$, $T = 2$

   (b) $f(t) = \begin{cases} 1, & 0 \leq t \leq \frac{3}{2} \\ 0, & \frac{3}{2} < t < 2 \end{cases}$, $T = 2$

   (c) $f(t) = \begin{cases} t, & 0 \leq t \leq 2 \\ 4 - t, & 2 < t < 4 \end{cases}$, $T = 4$

   (d) $f(t) = -1 + t$, $0 \leq t < 2$, $T = 2$

   (e) $f(t) = 2e^{-t}$, $0 \leq t < 1$, $T = 1$

   (f) $f(t) = \sin t$, $0 \leq t < \pi$, $T = \pi$

   (g) $f(t) = \begin{cases} 2 \sin \pi t, & 0 \leq t \leq 1 \\ 0, & 1 < t < 2 \end{cases}$, $T = 2$

Exercises 2–8:

These exercises deal with undamped vibrations of a spring-mass system,

$$my'' + ky = 0, \qquad y(0) = y_0, \qquad y'(0) = y_0'. \tag{16}$$

Use a value of 9.8 m/s^2 or 32 ft/sec^2 for the acceleration due to gravity.

2. A 10-kg mass, when attached to the end of a spring hanging vertically, stretches the spring 30 mm. Assume the mass is then pulled down another 70 mm and released (with no initial velocity).

   (a) Determine the spring constant k.

   (b) State the initial value problem (giving numerical values for all constants) for $y(t)$, where $y(t)$ denotes the displacement (in meters) of the mass from its equilibrium rest position. Assume that y is measured positive in the downward direction.

   (c) Solve the initial value problem formulated in part (b).

3. A 3-kg mass is attached to a spring having spring constant $k = 300$ N/m. At time $t = 0$, the mass is pulled down 10 cm and released with a downward velocity of 100 cm/s.

   (a) Determine the resulting displacement, $y(t)$.

   (b) Solve the equation $y'(t) = 0, t > 0$, to find the time when the maximum downward displacement of the mass from its equilibrium position is first achieved.

   (c) What is the maximum downward displacement?

4. A 20-kg mass was initially at rest, attached to the end of a vertically hanging spring. When given an initial downward velocity of 2 m/s from its equilibrium rest position, the mass was observed to attain a maximum displacement of 0.2 m from its equilibrium position. What is the value of the spring constant?

5. A 9-lb weight, suspended from a spring having spring constant $k = 32$ lb/ft, is perturbed from its equilibrium state with a certain upward initial velocity. The amplitude of the resulting vibrations is observed to be 4 in.

   (a) What is the initial velocity?

   (b) What are the period and frequency of the vibrations?

6. (a) Derive an expression for the amplitude of the undamped vibrations modeled by equation (16). [Hint: From equations (10)–(12), the general solution of $my'' + ky = 0$ is $y = c_1 \cos \beta t + c_2 \sin \beta t$. The amplitude R is given by $R = \sqrt{c_1^2 + c_2^2}$. Use the initial conditions in equation (16) to determine c_1 and c_2. Your expression for R will involve y_0, y_0', and $\beta = \sqrt{k/m}$.]

(b) Two experiments are performed. A mass is given an initial downward displacement y_0 and then released with a downward initial velocity y_0'. Next, the mass is given the same downward displacement y_0, but this time released with an upward initial velocity $-y_0'$. Which experiment (if any) would you expect to yield the larger amplitude? Using the result of part (a), compare the amplitudes of the resulting vibrations.

7. A 4-kg mass was attached to a spring and set in motion. A record of the displacements was made and found to be described by $y(t) = 25\cos(2t - \pi/6)$, with displacement measured in centimeters and time in seconds. Determine the initial displacement y_0, initial velocity y_0', spring constant k, and period T of the vibrations.

8. The graph shows the displacement from equilibrium of a mass-spring system as a function of time after the vertically hanging system was set in motion at $t = 0$. Assume that the units of time and displacement are seconds and centimeters, respectively.

(a) What is the period T of the periodic motion?

(b) What is the frequency f (in Hertz)? What is the angular frequency ω (in rad/sec)?

(c) Determine the amplitude R and the phase angle δ (in radians), and express the displacement in the form $y(t) = R\cos(\omega t - \delta)$, with y in meters.

(d) With what initial displacement $y(0)$ and initial velocity $y'(0)$ was the system set into motion?

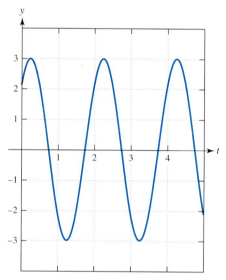

Figure for Exercise 8

The first t-intercept is $\left(\frac{3}{4}, 0\right)$, and the first minimum has coordinates $\left(\frac{5}{4}, -3\right)$.

9. A spring-mass-dashpot system consists of a 10-kg mass attached to a spring with spring constant $k = 100$ N/m; the dashpot has damping constant 7 kg/s. At time $t = 0$, the system is set into motion by pulling the mass down 0.5 m from its equilibrium rest position while simultaneously giving it an initial downward velocity of 1 m/s.

(a) State the initial value problem to be solved for $y(t)$, the displacement from equilibrium (in meters) measured positive in the downward direction. Give numerical values to all constants involved.

(b) Solve the initial value problem. What is $\lim_{t\to\infty} y(t)$? Explain why your answer for this limit makes sense from a physical perspective.

(c) Plot your solution on a time interval long enough to determine how long it takes for the magnitude of the vibrations to be reduced to 0.1 m. In other words, estimate the smallest time, τ, for which $|y(t)| \le 0.1m, \tau \le t < \infty$.

10. A spring and dashpot system is to be designed for a 32-lb weight so that the overall system is critically damped.

(a) How must the damping constant γ and spring constant k be related?

(b) Assume the system is to be designed so that the mass, when given an initial velocity of 4 ft/sec from its rest position, will have a maximum displacement of 6 in. What values of damping constant γ and spring constant k are required?

11. A 4-kg mass is attached to a spring having spring constant $k = 100$ N/m. The system is set in motion and measurements are taken. A dashpot is then attached and the experiment repeated. It is observed that the time interval between successive zero crossings is 20% larger for the damped vibration displacement than for the undamped vibration displacement. What is the damping constant γ?

12. A spring-mass-dashpot system is released from rest with an initial displacement given by $y(0) = y_0$. Consider the following question: "What happens to the displacement $y(t)$ if we keep the values of mass m and spring constant k fixed but increase the damping constant γ?" In particular, select an arbitrary but fixed time $t > 0$, think of the solution $y(t)$ as being a function of the damping constant γ, and determine $\lim_{\gamma\to\infty} y(t)$. Do you have any intuitive insight as to what the answer should be?

Develop the answer with the following steps.

(a) Show that the roots of the characteristic polynomial are

$$\lambda_1 = -\frac{\gamma}{2m} - \frac{\sqrt{\gamma^2 - 4mk}}{2m}, \qquad \lambda_2 = -\frac{\gamma}{2m} + \frac{\sqrt{\gamma^2 - 4mk}}{2m}.$$

Solve the initial value problem, assuming the system to be overdamped. Express the solution in terms of the two roots λ_1 and λ_2. (Since we are interested in the system's behavior for large values of γ, the overdamped assumption is appropriate.)

(b) Show that $\lim_{\gamma\to\infty} \lambda_1 = -\infty$ and $\lim_{\gamma\to\infty} \lambda_2 = 0$.

(c) Use the results of parts (a) and (b) to determine $\lim_{\gamma\to\infty} y(t)$ (with k, m, and $t > 0$ fixed). What is the physical meaning of your answer? Does it agree with your intuition? Does it make physical sense in retrospect?

13. In this problem, we explore computationally the question posed in Exercise 12. Consider the initial value problem

$$y'' + \gamma y' + y = 0, \qquad y(0) = 1, \qquad y'(0) = 0,$$

where, for simplicity, we have given the mass, spring constant, and initial displacement all a numerical value of unity.

(a) Determine γ_{crit}, the damping constant value that makes the given spring-mass-dashpot system critically damped.

(b) Use computational software to plot the solution of the initial value problem for $\gamma = \gamma_{\text{crit}}$, $2\gamma_{\text{crit}}$, and $20\gamma_{\text{crit}}$ over a common time interval sufficiently large to display the main features of each solution. What trend do you observe in the behavior of the solutions as γ increases? Is it consistent with the conclusions reached in Exercise 12?

3.7   The General Solution of a Linear Nonhomogeneous Equation

We consider the linear second order nonhomogeneous differential equation

$$y'' + p(t)y' + q(t)y = g(t), \qquad a < t < b \tag{1}$$

and ask, "What is the general solution of this equation?"

The General Solution

We begin by posing a second question: "To what extent can two solutions of equation (1) differ from one another?" Once we can answer this question, we will know how every solution of equation (1) is related to a single particular solution that we may somehow have found.

Assume we have two solutions of nonhomogeneous equation (1); call them $u(t)$ and $v(t)$. Since both are solutions,

$$u'' + p(t)u' + q(t)u = g(t) \quad \text{and} \quad v'' + p(t)v' + q(t)v = g(t), \qquad a < t < b. \tag{2}$$

Subtracting, we obtain

$$[u'' - v''] + p(t)[u' - v'] + q(t)[u - v] = g(t) - g(t) = 0, \qquad a < t < b.$$

Therefore, the difference function, $w(t) = u(t) - v(t)$, is a solution of the associated linear *homogeneous* equation:

$$y'' + p(t)y' + q(t)y = 0. \tag{3}$$

Now, let $y_P(t)$ be a particular solution of equation (1) that we somehow have found. Let $y(t)$ be any solution whatsoever of equation (1). As we saw above, the difference function $y(t) - y_P(t)$ is a solution of equation (3). Let $y_1(t)$ and $y_2(t)$ form a fundamental set of solutions for the homogeneous equation (3). Since $\{y_1, y_2\}$ is a fundamental set of solutions, there are constants c_1 and c_2 such that

$$y(t) - y_P(t) = c_1 y_1(t) + c_2 y_2(t).$$

Equivalently,

$$y(t) = [c_1 y_1(t) + c_2 y_2(t)] + y_P(t). \tag{4}$$

Since $y(t)$ was any solution whatsoever of equation (1), it follows that the general solution of (1) is given by (4). We can express result (4) in the following schematic form:

The general solution of the nonhomogeneous equation	=	The general solution of the homogeneous equation	+	A particular solution of the nonhomogeneous equation.

Note that the right-hand side of equation (4) contains two arbitrary constants c_1 and c_2 that we can select to satisfy given initial conditions, $y(t_0) = y_0$ and $y'(t_0) = y_0'$; we illustrate this point in Example 1.

We call the general solution of the homogeneous equation the **complementary solution** and denote it y_C. The solution of the nonhomogeneous equation that we somehow have found is called a **particular solution** and is denoted by y_P. If we use y to represent the general solution of nonhomogeneous equation (1), then the preceding schematic statement has the structural form

$$y(t) = y_C(t) + y_P(t). \tag{5}$$

In Sections 3.8 and 3.9, we will discuss methods for finding a particular solution, y_P. For now, we illustrate equation (5) with an example.

E X A M P L E

1

(a) Verify that $y_P(t) = 3t - 4$ is a solution of $y'' - y' - 2y = 5 - 6t$.

(b) Use the result of (a) together with equation (5) to solve the initial value problem

$$y'' - y' - 2y = 5 - 6t, \qquad y(0) = 3, \qquad y'(0) = 11.$$

Solution:

(a) Inserting y_P into the differential equation, we obtain

$$y_P'' - y_P' - 2y_P = (0) - (3) - 2(3t - 4) = 5 - 6t.$$

Therefore, $y_P = 3t - 4$ is a particular solution of the nonhomogeneous differential equation.

(b) The complementary solution of $y'' - y' - 2y = 0$ is $y_C(t) = c_1 e^{-t} + c_2 e^{2t}$. Therefore, by equation (5), the general solution of $y'' - y' - 2y = 5 - 6t$ is $y(t) = y_C(t) + y_P(t)$, or

$$y(t) = c_1 e^{-t} + c_2 e^{2t} + 3t - 4.$$

Imposing the initial conditions, we obtain the system of equations

$$c_1 + c_2 - 4 = 3$$
$$-c_1 + 2c_2 + 3 = 11.$$

The solution of this system is $c_1 = 2$ and $c_2 = 5$. Thus, the solution of the initial value problem is

$$y(t) = 2e^{-t} + 5e^{2t} + 3t - 4. \; ❖$$

The Superposition of Particular Solutions

As we noted in Section 3.2, the principle of superposition does not apply to *nonhomogeneous* linear equations. If $u_1(t)$ and $u_2(t)$ are two solutions of the nonhomogeneous equation

$$y'' + p(t)y' + q(t)y = g(t), \tag{6}$$

where $g(t)$ is nonzero, the sum $w(t) = u_1(t) + u_2(t)$ is not a solution of equation (6). In fact, when $w = u_1 + u_2$ is inserted into the left-hand side of (6), we obtain

$$w'' + p(t)w' + q(t)w = g(t) + g(t) = 2g(t).$$

Therefore, $w = u_1 + u_2$ is not a solution of equation (6).

There is, however, a different form of superposition that applies to nonhomogeneous equations. We state this result formally as a theorem since we will find it useful in Sections 3.8 and 3.9, which deal with the practical aspects of finding a particular solution, y_P, for equation (6). The proof is left as Exercise 13.

Theorem 3.4

Let $u(t)$ be a solution of $y'' + p(t)y' + q(t)y = g_1(t), a < t < b$. Let $v(t)$ be a solution of $y'' + p(t)y' + q(t)y = g_2(t), a < t < b$. Let a_1 and a_2 be any constants. Then the function $y_P(t) = a_1 u(t) + a_2 v(t)$ is a particular solution of

$$y'' + p(t)y' + q(t)y = a_1 g_1(t) + a_2 g_2(t).$$

Theorem 3.4 is a simplifying principle. For example, suppose we need to find a particular solution of

$$y'' + p(t)y' + q(t)y = e^{2t} + \cos t. \tag{7}$$

It often is simpler to separately find a particular solution u_1 that solves

$$y'' + p(t)y' + q(t)y = e^{2t}$$

and then find a particular solution u_2 that solves

$$y'' + p(t)y' + q(t)y = \cos t.$$

The desired particular solution of (7) is $y_P(t) = u_1(t) + u_2(t)$.

EXAMPLE 2

We ask you to show in Exercise 3 that $u(t) = 2e^{4t}$ is a particular solution of $y'' - y' - 2y = 20e^{4t}$. Recall from Example 1 that $v(t) = 3t - 4$ is a particular solution of $y'' - y' - 2y = 5 - 6t$. Find the general solution of

$$y'' - y' - 2y = -5e^{4t} + 20 - 24t.$$

Solution: Applying Theorem 3.4, we know that a particular solution of the equation

$$y'' - y' - 2y = -5e^{4t} + 20 - 24t$$

is

$$y_P(t) = -\tfrac{1}{4}u(t) + 4v(t) = -\tfrac{1}{2}e^{4t} + 12t - 16.$$

Therefore, the general solution is

$$y(t) = y_C(t) + y_P(t) = c_1 e^{-t} + c_2 e^{2t} - \tfrac{1}{2}e^{4t} + 12t - 16. ❖$$

EXERCISES

Exercises 1–12:

(a) Verify that the given function, $y_P(t)$, is a particular solution of the differential equation.

(b) Determine the complementary solution, $y_C(t)$.

(c) Form the general solution and impose the initial conditions to obtain the unique solution of the initial value problem.

1. $y'' - 2y' - 3y = -9t - 3$, $y(0) = 1$, $y'(0) = 3$, $y_p(t) = 3t - 1$

2. $y'' - 2y' - 3y = e^{2t}$, $y(0) = 1$, $y'(0) = 0$, $y_p(t) = -e^{2t}/3$

3. $y'' - y' - 2y = 20e^{4t}$, $y(0) = 0$, $y'(0) = 1$, $y_p(t) = 2e^{4t}$

4. $y'' - y' - 2y = 10$, $y(-1) = 0$, $y'(-1) = 1$, $y_p(t) = -5$

5. $y'' + y' = 2t$, $y(1) = 1$, $y'(1) = -2$, $y_p(t) = t^2 - 2t$

6. $y'' + y' = 2e^{-t}$, $y(0) = 2$, $y'(0) = 2$, $y_p(t) = -2te^{-t}$

7. $y'' + y = 2t - 3\cos 2t$, $y(0) = 0$, $y'(0) = 0$, $y_p(t) = 2t + \cos 2t$

8. $y'' + 4y = 10e^{t-\pi}$, $y(\pi) = 2$, $y'(\pi) = 0$, $y_p(t) = 2e^{t-\pi}$

9. $y'' - 2y' + 2y = 10t^2$, $y(0) = 0$, $y'(0) = 0$, $y_p(t) = 5(t + 1)^2$

10. $y'' - 2y' + 2y = 5\sin t$, $y(\pi/2) = 1$, $y'(\pi/2) = 0$, $y_p(t) = 2\cos t + \sin t$

11. $y'' - 2y' + y = e^t$, $y(0) = -2$, $y'(0) = 2$, $y_p(t) = \frac{1}{2}t^2 e^t$

12. $y'' - 2y' + y = t^2 + 4 + 2\sin t$, $y(0) = 1$, $y'(0) = 3$, $y_p(t) = t^2 + 4t + 10 + \cos t$

13. Assume that $u(t)$ and $v(t)$ are, respectively, solutions of the differential equations

$$u'' + p(t)u' + q(t)u = g_1(t) \quad \text{and} \quad v'' + p(t)v' + q(t)v = g_2(t),$$

where $p(t), q(t), g_1(t),$ and $g_2(t)$ are continuous on the t-interval of interest. Let a_1 and a_2 be any two constants. Show that the function $y_p(t) = a_1 u(t) + a_2 v(t)$ is a particular solution of the differential equation

$$y'' + p(t)y' + q(t)y = a_1 g_1(t) + a_2 g_2(t)$$

on the same t-interval.

Exercises 14–16:

The functions $u_1(t), u_2(t),$ and $u_3(t)$ are solutions of the differential equations

$$u_1'' + p(t)u_1' + q(t)u_1 = 2e^t + 1, \qquad u_2'' + p(t)u_2' + q(t)u_2 = 4,$$
$$u_3'' + p(t)u_3' + q(t)u_3 = 3t.$$

Use the functions $u_1(t), u_2(t),$ and $u_3(t)$ to construct a particular solution of the given differential equation.

14. $y'' + p(t)y' + q(t)y = e^t$ | **15.** $y'' + p(t)y' + q(t)y = t + 2$

16. $y'' + p(t)y' + q(t)y = e^t + t + 1$

Exercises 17–21:

The function $y_p(t)$ is a particular solution of the given differential equation. Determine the function $g(t)$.

17. $y'' + y' - y = g(t)$, $y_p(t) = e^{2t} - t^2$

18. $y'' - 2y' = g(t)$, $y_p(t) = 3t + \sqrt{t}$, $t > 0$

19. $ty'' + e^t y' + 2y = g(t)$, $y_p(t) = 3t$, $t > 0$

20. $y'' + y = g(t)$, $y_p(t) = \ln(1 + t)$, $t > -1$

21. $y'' + (\sin t)y' + ty = g(t)$, $y_p(t) = t + 1$

Exercises 22–26:

The general solution of the nonhomogeneous differential equation $y'' + \alpha y' + \beta y = g(t)$ is given, where c_1 and c_2 are arbitrary constants. Determine the constants α and β and the function $g(t)$.

22. $y(t) = c_1 e^t + c_2 e^{2t} + 2e^{-2t}$ | **23.** $y(t) = c_1 + c_2 e^{-t} + t^2$

24. $y(t) = c_1 e^t + c_2 t e^t + t^2 e^t$ **25.** $y(t) = c_1 e^t \cos t + c_2 e^t \sin t + e^t + \sin t$

26. $y(t) = c_1 \sin 2t + c_2 \cos 2t - 1 + \sin t$

3.8   The Method of Undetermined Coefficients

In Section 3.7, we discussed the structure of the general solution for the non-homogeneous equation

$$y'' + p(t)y' + q(t)y = g(t), \qquad a < t < b.$$

We saw that the general solution has the form

$$y(t) = y_C(t) + y_P(t), \tag{1}$$

where y_C is the general solution of the homogeneous equation $y'' + p(t)y' + q(t)y = 0$ and y_P is a particular solution of the nonhomogeneous equation $y'' + p(t)y' + q(t)y = g(t)$.

In this section, we describe a technique that often can be used to find a particular solution, y_P. The technique is known as the **method of undetermined coefficients**.

We first illustrate the method through a series of examples. Later, we summarize the method in tabular form. In Section 3.9, we describe a different technique for finding a particular solution, the method of variation of parameters.

Before we discuss these two methods for obtaining a particular solution, it's worth stating the procedure that should be followed to obtain the general solution (1):

1. The first step is to find the complementary solution, y_C. As you will see, knowledge of the complementary solution is a prerequisite for using either the method of undetermined coefficients or the method of variation of parameters.

2. Next, use undetermined coefficients or variation of parameters (or anything else that works) to find a particular solution, y_P.

3. Finally, obtain the general solution by forming $y_C + y_P$.

If you are solving an initial value problem, the initial conditions are imposed as a last step, step 4.

Introduction to the Method of Undetermined Coefficients

Consider the nonhomogeneous differential equation

$$ay'' + by' + cy = g(t), \tag{2}$$

where a, b, and c are constants. You will see that we can *guess* the form of a particular solution for certain types of functions $g(t)$. Example 1 introduces some of the main ideas.

EXAMPLE

1

Find the general solution of the nonhomogeneous equation

$$y'' - y' - 2y = 8e^{3t}.$$

Solution: We always begin by finding the complementary solution, y_C. The characteristic polynomial for $y'' - y' - 2y = 0$ is $\lambda^2 - \lambda - 2 = (\lambda + 1)(\lambda - 2)$. Therefore, the complementary solution is

$$y_C(t) = c_1 e^{-t} + c_2 e^{2t}.$$

We now look for a particular solution, a function $y_P(t)$ such that

$$y_P'' - y_P' - 2y_P = 8e^{3t}. \tag{3}$$

Since all derivatives of $y = e^{3t}$ are again multiples of e^{3t}, it seems reasonable that a particular solution might have the form

$$y_P = Ae^{3t},$$

where A is a coefficient to be determined. Inserting $y_P = Ae^{3t}$ into equation (3), we obtain

$$9Ae^{3t} - 3Ae^{3t} - 2(Ae^{3t}) = 8e^{3t}.$$

Collecting terms on the left-hand side yields $4Ae^{3t} = 8e^{3t}$. Therefore, $A = 2$, and $y_P(t) = 2e^{3t}$ is a particular solution of the nonhomogeneous equation.

Having the complementary solution y_C and a particular solution y_P, we form the general solution of the nonhomogeneous equation

$$y(t) = y_C(t) + y_P(t) = c_1 e^{-t} + c_2 e^{2t} + 2e^{3t}. \ \text{❖}$$

Example 1 suggests a reasonable approach to finding a particular solution, y_P. If the right-hand side of nonhomogeneous equation (2) is of a certain special type, then it might be possible to guess an appropriate form for y_P. The method of undetermined coefficients amounts to a recipe for choosing the *form* of y_P. This recipe involves unknown (or undetermined) coefficients that must be evaluated by inserting the form, y_P, into the differential equation. (The role that the complementary solution plays in this process will be clarified shortly.)

Trial Forms for the Particular Solution

The method of undetermined coefficients can be applied to differential equations of the form

$$ay'' + by' + cy = g(t),$$

where a, b, and c are constants and where the nonhomogeneous term $g(t)$ is one of several possible types. It's important to understand what types of functions $g(t)$ are suitable and why. To gain insight, we start with some examples. The first few examples will treat, for various right sides $g(t)$, the differential equation

$$y'' - y' - 2y = g(t). \tag{4}$$

Equation (4) with $g(t) = 8e^{3t}$ was discussed in Example 1; the complementary solution of (4) is

$$y_C(t) = c_1 e^{-t} + c_2 e^{2t}.$$

E X A M P L E

2

Find the general solution of

$$y'' - y' - 2y = 4t^2.$$

Solution: We already know the complementary solution from Example 1. Therefore, we can form the general solution after finding a particular solution.

Following the approach taken in Example 1, we are tempted to look for a solution of the form

$$y_P(t) = At^2,$$

where A is an unknown (undetermined) coefficient. The guess $y_P(t) = At^2$, however, does not work because forming the first and second derivatives of t^2 generates multiples of some new functions, t and 1. In particular, this guess leads to a contradiction when the trial form $y_P(t) = At^2$ is inserted into the differential equation.

Instead, we assume a particular solution of the form

$$y_P(t) = At^2 + Bt + C,$$

where the constants A, B, and C must be chosen so that

$$y_P'' - y_P' - 2y_P = 4t^2.$$

Substituting y_P, we obtain the condition

$$(2A) - (2At + B) - 2(At^2 + Bt + C) = 4t^2$$

or, after collecting terms,

$$-2At^2 - (2A + 2B)t + (2A - B - 2C) = 4t^2. \tag{5}$$

This equality must hold for all t in the interval of interest. Therefore, the coefficients of t^2, t, and 1 on the left-hand side of this equation must equal their counterparts on the right-hand side. We obtain the following three equations for the three unknown coefficients A, B, and C.

$$
\begin{aligned}
-2A &\qquad\qquad = 4 \\
-2A - 2B &\qquad\quad = 0 \\
2A - \;\; B - 2C &= 0.
\end{aligned}
\tag{6}
$$

The solution of this system is $A = -2$, $B = 2$, $C = -3$. Therefore, a particular solution is given by

$$y_P(t) = -2t^2 + 2t - 3.$$

The desired general solution is $y(t) = y_C(t) + y_P(t)$, or

$$y(t) = c_1 e^{-t} + c_2 e^{2t} - 2t^2 + 2t - 3. \; ❖$$

REMARKS ABOUT EXAMPLE 2:

1. In retrospect, it is clear why our first guess, $y_p(t) = At^2$, failed. Substitution into the nonhomogeneous equation leads [see system (6)] to the contradictory constraints $-2A = 4, -2A = 0, 2A = 0$. The emergence of such contradictions is a clear indicator that the assumed form for the particular solution is incorrect.

2. The assumed form $y_p(t) = At^2 + Bt + C$ of the particular solution is appropriate because of another fact that we did not mention—none of the functions $1, t$, or t^2 is a part of the complementary solution, y_C. If one of these functions *had been* part of the complementary solution, then (see Examples 4 and 5) our guess would have failed.

E X A M P L E
3

Find the general solution of
$$y'' - y' - 2y = -20 \sin 2t.$$

Solution: The complementary solution is
$$y_C(t) = c_1 e^{-t} + c_2 e^{2t}.$$

In choosing a guess for the particular solution, we observe that differentiation of the right-hand side,
$$g(t) = -20 \sin 2t,$$

produces a multiple of $\cos 2t$ but that continued differentiation of the set of functions $\{\sin 2t, \cos 2t\}$ simply produces multiples of the functions in the set. In addition, neither of the functions $\sin 2t$ or $\cos 2t$ appears as part of the complementary solution. Therefore, we choose the following trial form for a particular solution:
$$y_p(t) = A \sin 2t + B \cos 2t.$$

Substituting the trial form $y_p(t) = A \sin 2t + B \cos 2t$, we obtain
$$y_P'' - y_P' - 2y_P = -20 \sin 2t,$$

or
$$(-4A \sin 2t - 4B \cos 2t) - (2A \cos 2t - 2B \sin 2t) - 2(A \sin 2t + B \cos 2t) = -20 \sin 2t.$$

Collecting like terms reduces this equation to
$$(-6A + 2B) \sin 2t - (2A + 6B) \cos 2t = -20 \sin 2t.$$

Since this equation must hold for all t in the interval of interest, it follows that
$$-6A + 2B = -20$$
$$-2A - 6B = \quad 0.$$

The solution of this system is $A = 3$ and $B = -1$, leading to a particular solution
$$y_p(t) = 3 \sin 2t - \cos 2t$$

and the general solution
$$y(t) = c_1 e^{-t} + c_2 e^{2t} + 3 \sin 2t - \cos 2t. ❖$$

Our next two examples illustrate how the trial form for y_P must be modified if portions of $g(t)$ or derivatives of $g(t)$ are present in the complementary solution.

EXAMPLE 4

Find the general solution of

$$y'' - y' - 2y = 4e^{-t}.$$

Solution: In this case, we observe that the function e^{-t} is a solution of the homogeneous equation. To illustrate that the trial form $y_P(t) = Ae^{-t}$ is not correct, we substitute it into the differential equation, obtaining

$$Ae^{-t} - (-Ae^{-t}) - 2Ae^{-t} = 4e^{-t},$$

or

$$0Ae^{-t} = 4e^{-t}.$$

Since the condition $0A = 4$ cannot hold for any value A, the assumed form of the trial solution is not correct.

We obtain the correct form for a particular solution of $y'' - y' - 2y = 4e^{-t}$ if we multiply e^{-t} by t—that is, if we assume a trial solution of the form

$$y_P(t) = Ate^{-t}. \tag{7}$$

At first glance, it may seem surprising that this form is correct. Nevertheless, when we substitute $y_P(t) = Ate^{-t}$, we obtain

$$(Ate^{-t} - 2Ae^{-t}) - (-Ate^{-t} + Ae^{-t}) - 2Ate^{-t} = 4e^{-t}, \tag{8}$$

or

$$-3Ae^{-t} = 4e^{-t}.$$

This equation is satisfied by choosing $A = -\frac{4}{3}$, leading to a particular solution

$$y_P(t) = -\frac{4}{3}te^{-t}$$

and the general solution

$$y(t) = y_C(t) + y_P(t) = c_1e^{-t} + c_2e^{2t} - \frac{4}{3}te^{-t}. \; ❖$$

REMARK: In retrospect, it should be clear why the te^{-t} terms vanish on the left-hand side of equation (8). After $y_P(t) = Ate^{-t}$ is substituted into the differential equation, the terms that survive as te^{-t} terms are precisely those in which differentiation under the product rule has acted on the e^{-t} factor and not on the t factor. Such terms ultimately vanish because e^{-t} is a solution of the homogeneous equation.

EXAMPLE 5

Find the general solution of

$$y'' + 2y' + y = 2e^{-t}.$$

Solution: For this problem, the homogeneous equation has characteristic equation $\lambda^2 + 2\lambda + 1 = (\lambda + 1)^2 = 0$. We obtain two real repeated roots, $\lambda_1 = \lambda_2 = -1$, and the complementary solution is

$$y_C(t) = c_1e^{-t} + c_2te^{-t}.$$

It's clear that a guess of the form

$$y_P(t) = Ae^{-t}$$

will not be the appropriate form for a particular solution, because e^{-t} is a solution of the homogeneous equation. A guess of the form $y_P(t) = Ate^{-t}$ will fail for the same reason. It's perhaps not surprising that a guess of the form

$$y_P(t) = At^2e^{-t}$$

does work. Substituting this form of the trial solution leads to

$$(At^2e^{-t} - 4Ate^{-t} + 2Ae^{-t}) + 2(-At^2e^{-t} + 2Ate^{-t}) + At^2e^{-t} = 2e^{-t}.$$

Simplifying, we obtain

$$2Ae^{-t} = 2e^{-t},$$

so $2A = 2$ and hence $y_P(t) = t^2e^{-t}$. The general solution is

$$y(t) = c_1e^{-t} + c_2te^{-t} + t^2e^{-t}. \; \diamondsuit$$

A Table Summarizing the Method of Undetermined Coefficients

We summarize the method of undetermined coefficients in Table 3.1. The method applies to the nonhomogeneous linear differential equation

$$ay'' + by' + cy = g(t),$$

where a, b, and c are constants and $g(t)$ has one of the forms listed on the left-hand side of the table. The corresponding form to assume for the particular solution is listed on the right-hand side of the table. The forms listed in Table 3.1 will work; that is, they will always yield a particular solution. In the Exercises, we ask you to solve problems using Table 3.1. The role of the factor t^r in the

TABLE 3.1

The right-hand column gives the proper form to assume for a particular solution of $ay'' + by' + cy = g(t)$. In the right-hand column, choose r to be the smallest nonnegative integer such that no term in the assumed form is a solution of the homogeneous equation $ay'' + by' + cy = 0$. The value of r will be 0, 1, or 2.

Form of $g(t)$	Form to Assume for a Particular Solution $y_P(t)$
$a_nt^n + \cdots + a_1t + a_0$	$t^r[A_nt^n + \cdots + A_1t + A_0]$
$[a_nt^n + \cdots + a_1t + a_0]e^{\alpha t}$	$t^r[A_nt^n + \cdots + A_1t + A_0]e^{\alpha t}$
$[a_nt^n + \cdots + a_1t + a_0]\sin\beta t$ or $[a_nt^n + \cdots + a_1t + a_0]\cos\beta t$	$t^r[(A_nt^n + \cdots + A_1t + A_0)\sin\beta t + (B_nt^n + \cdots + B_1t + B_0)\cos\beta t]$
$e^{\alpha t}\sin\beta t$ or $e^{\alpha t}\cos\beta t$	$t^r[Ae^{\alpha t}\sin\beta t + Be^{\alpha t}\cos\beta t]$
$e^{\alpha t}[a_nt^n + \cdots + a_0]\sin\beta t$ or $e^{\alpha t}[a_nt^n + \cdots + a_0]\cos\beta t$	$t^r[(A_nt^n + \cdots + A_0)e^{\alpha t}\sin\beta t + (B_nt^n + \cdots + B_0)e^{\alpha t}\cos\beta t]$

right-hand column of Table 3.1 is to ensure that no term in the assumed form for y_p is present in the complementary solution. You need to choose the proper value for r; the procedure for doing so is described in the table.

E X A M P L E

6

Using Table 3.1, choose an appropriate form for a particular solution of

(a) $y'' + 4y = t^2 e^{3t}$ $\qquad\qquad$ (b) $y'' + 4y = te^{2t} \cos t$

(c) $y'' + 4y = 2t^2 + 5 \sin 2t + e^{3t}$ $\qquad$ (d) $y'' + 4y = t^2 \cos 2t$

Solution: We note first that the complementary solution for each of parts (a)–(d) is

$$y_C(t) = c_1 \sin 2t + c_2 \cos 2t.$$

(a) For $g(t) = t^2 e^{3t}$, Table 3.1 specifies $y_p(t) = t^r[A_2 t^2 + A_1 t + A_0]e^{3t}$. If $r = 0$, no term in the assumed form for y_p is present in the complementary solution. So the appropriate form for a trial particular solution is

$$y_p(t) = [A_2 t^2 + A_1 t + A_0]e^{3t}.$$

(b) For $g(t) = te^{2t} \cos t$, the specified form is

$$y_p(t) = t^r[(A_1 t + A_0)e^{2t} \sin t + (B_1 t + B_0)e^{2t} \cos t].$$

If $r = 0$, no term in the assumed form for y_p is present in the complementary solution. So the appropriate form for a trial particular solution is

$$y_p(t) = (A_1 t + A_0)e^{2t} \sin t + (B_1 t + B_0)e^{2t} \cos t.$$

(c) Note that the nonhomogeneous term $g(t) = 2t^2 + 5 \sin 2t + e^{3t}$ does not match any of the forms listed in Table 3.1. However, we can use the superposition principle described by Theorem 3.4. Suppose $u(t)$ is a particular solution of

$$y'' + 4y = 2t^2,$$

$v(t)$ is a particular solution of

$$y'' + 4y = 5 \sin 2t,$$

and $w(t)$ is a particular solution of

$$y'' + 4y = e^{3t}.$$

By Theorem 3.4, $y_p(t) = u(t) + v(t) + w(t)$ is a particular solution of

$$y'' + 4y = 2t^2 + 5 \sin 2t + e^{3t}. \tag{9}$$

To determine the individual particular solutions $u(t)$, $v(t)$, and $w(t)$, we turn to Table 3.1 to find suitable trial forms. In particular, an appropriate trial form for $y'' + 4y = 2t^2$ is $u(t) = A_2 t^2 + A_1 t + A_0$. A suitable trial form for $y'' + 4y = 5 \sin 2t$ is the function $v(t) = B_0 t \cos 2t + C_0 t \sin 2t$ and a suitable trial form for $y'' + 4y = e^{3t}$ is $w(t) = D_0 e^{3t}$. (In the first and last cases, $r = 0$.

In the second case, $r = 1$.) Therefore,

$$y_P(t) = A_2 t^2 + A_1 t + A_0 + B_0 t \cos 2t + C_0 t \sin 2t + D_0 e^{3t}.$$

(d) For $g(t) = t^2 \cos 2t$, Table 3.1 prescribes the form

$$y_P(t) = t^r [(A_2 t^2 + A_1 t + A_0) \sin 2t + (B_2 t^2 + B_1 t + B_0) \cos 2t].$$

If we set $r = 0$, the assumed form for $y_P(t)$ will contain two terms, $A_0 \sin 2t$ and $B_0 \cos 2t$, that are solutions of the homogeneous equation. Therefore, r cannot be zero. With $r = 1$, we see that no term in the assumed form is a solution of the homogeneous equation. Therefore, the appropriate form is

$$y_P(t) = t[(A_2 t^2 + A_1 t + A_0) \sin 2t + (B_2 t^2 + B_1 t + B_0) \cos 2t]$$
$$= (A_2 t^3 + A_1 t^2 + A_0 t) \sin 2t + (B_2 t^3 + B_1 t^2 + B_0 t) \cos 2t. \quad ❖$$

Although we presented Table 3.1 in the context of discussing nonhomogeneous constant coefficient *second order* linear differential equations, the method of undetermined coefficients is not restricted to second order equations; the ideas extend naturally to nonhomogeneous constant coefficient linear equations of order higher than two.

EXERCISES

Exercises 1–15:

For the given differential equation,

(a) Determine the complementary solution.

(b) Use the method of undetermined coefficients to find a particular solution.

(c) Form the general solution.

1. $y'' - 4y = 4t^2$

2. $y'' - 4y = \sin 2t$

3. $y'' + y = 8e^t$

4. $y'' + y = e^t \sin t$

5. $y'' - 4y' + 4y = e^{2t}$

6. $y'' - 4y' + 4y = 8 + \sin 2t$

7. $y'' + 2y' + 2y = t^3$

8. $2y'' - 5y' + 2y = te^t$

9. $y'' + 2y' + 2y = \cos t + e^{-t}$

10. $y'' + y' = 6t^2$

11. $2y'' - 5y' + 2y = -6e^{t/2}$

12. $y'' + y' = \cos t$

13. $9y'' - 6y' + y = 9te^{t/3}$

14. $y'' + 4y' + 5y = 5t + e^{-t}$

15. $y'' + 4y' + 5y = 2e^{-2t} + \cos t$

Exercises 16–22:

For the given differential equation,

(a) Determine the complementary solution.

(b) List the form of particular solution prescribed by the method of undetermined coefficients; you need not evaluate the constants in the assumed form. [Hint: In Exercises 20 and 22, rewrite the hyperbolic functions in terms of exponential functions. In Exercise 21, use trigonometric identities.]

16. $y'' - 2y' - 3y = 2e^{-t} \cos t + t^2 + te^{3t}$

17. $y'' + 9y = t^2 \cos 3t + 4 \sin t$

18. $y'' - y' = t^2 (2 + e^t)$

19. $y'' - 2y' + 2y = e^{-t} \sin 2t + 2t + te^{-t} \sin t$

20. $y'' - y = \cosh t + \sinh 2t$

21. $y'' + 4y = \sin t \cos t + \cos^2 2t$

22. $y'' + 4y = 2 \sinh t \cosh t + \cosh^2 t$

Exercises 23–27:

Consider the differential equation $y'' + \alpha y' + \beta y = g(t)$. In each exercise, the complementary solution, $y_C(t)$, and nonhomogeneous term, $g(t)$, are given. Determine α and β and then find the general solution of the differential equation.

23. $y_C(t) = c_1 e^{-t} + c_2 e^{2t}$, $g(t) = 4t$

24. $y_C(t) = c_1 + c_2 e^{-t}$, $g(t) = t$

25. $y_C(t) = c_1 e^{-2t} + c_2 t e^{-2t}$, $g(t) = 5\sin t$

26. $y_C(t) = c_1 \cos t + c_2 \sin t$, $g(t) = t + \sin 2t$

27. $y_C(t) = c_1 e^{-t} \cos 2t + c_2 e^{-t} \sin 2t$, $g(t) = 8e^{-t}$

Exercises 28–30:

Consider the differential equation $y'' + \alpha y' + \beta y = g(t)$. In each exercise, the nonhomogeneous term, $g(t)$, and the form of the particular solution prescribed by the method of undetermined coefficients are given. Determine the constants α and β.

28. $g(t) = t + e^{3t}$, $y_P(t) = A_1 t^2 + A_0 t + B_0 t e^{3t}$

29. $g(t) = 3e^{2t} - e^{-2t} + t$, $y_P(t) = A_0 t e^{2t} + B_0 t e^{-2t} + C_1 t + C_0$

30. $g(t) = -e^t + \sin 2t + e^t \sin 2t$,

$y_P(t) = A_0 e^t + B_0 t \cos 2t + C_0 t \sin 2t + D_0 e^t \cos 2t + E_0 e^t \sin 2t$

31. Consider the initial value problem $y'' + 4y = e^{-t}, y(0) = y_0, y'(0) = y_0'$. Suppose we know that $y(t) \to 0$ as $t \to \infty$. Determine the initial conditions y_0 and y_0' as well as the solution $y(t)$.

32. Consider the initial value problem $y'' - 4y = e^{-t}, y(0) = 1, y'(0) = y_0'$. Suppose we know that $y(t) \to 0$ as $t \to \infty$. Determine the initial condition y_0' as well as the solution $y(t)$.

33. Consider the initial value problem $y'' - y' + 2y = 3, y(0) = y_0, y'(0) = y_0'$. Suppose we know that $|y(t)| \leq 2$ for all $t \geq 0$. Determine the initial conditions y_0 and y_0' as well as the solution $y(t)$.

Exercises 34–36:

Consider the initial value problem $y'' + \omega^2 y = g(t), y(0) = 0, y'(0) = 0$, where ω is a real nonnegative constant. For the given function $g(t)$, determine the values of ω, if any, for which the solution satisfies the constraint $|y(t)| \leq 2, 0 \leq t < \infty$.

34. $g(t) = 1$ **35.** $g(t) = \cos 2\omega t$ **36.** $g(t) = \sin \omega t$

37. Each of the five graphs on the next page is the solution of one of the five differential equations listed. Each solution satisfies the initial conditions $y(0) = 1, y'(0) = 0$. Match each graph with one of the differential equations.

(a) $y'' + 3y' + 2y = \sin t$ (b) $y'' + y = \sin t$ (c) $y'' + y = \sin \dfrac{3t}{2}$

(d) $y'' + 3y' + 2y = e^{t/10}$ (e) $y'' + y' + y = \sin \dfrac{t}{3}$

Complex-Valued Solutions   Although we have emphasized the need to obtain real-valued, physically relevant solutions to problems of interest, it is sometimes computationally convenient to consider differential equations with complex-valued nonhomogeneous terms. The corresponding particular solutions will then likewise be complex-valued functions. Exercises 38 and 39 illustrate some aspects of this type of calculation. Exercises 40–44 provide some additional examples.

38. Consider the differential equation $y'' + p(t)y' + q(t)y = g_1(t) + ig_2(t)$, where $p(t)$, $q(t), g_1(t)$, and $g_2(t)$ are all real-valued functions continuous on some t-interval of interest. Assume that $y_P(t)$ is a particular solution of this equation. Generally, $y_P(t)$

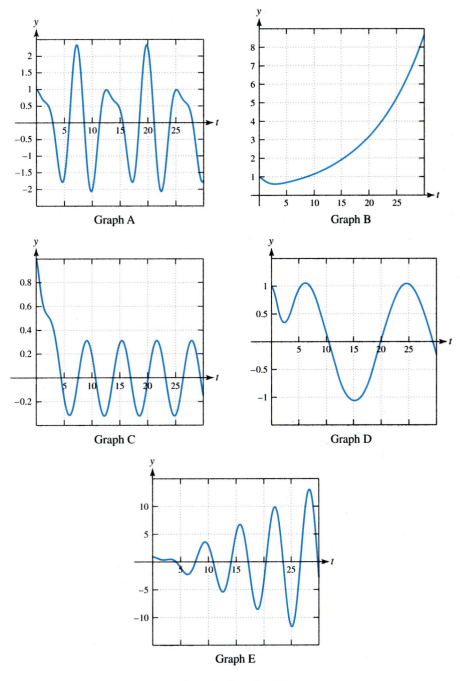

Graph A

Graph B

Graph C

Graph D

Graph E

Figure for Exercise 37

will be a complex-valued function. Let $y_p(t) = u(t) + iv(t)$, where $u(t)$ and $v(t)$ are real-valued functions. Show that

$$u'' + p(t)u' + q(t)u = g_1(t) \quad \text{and} \quad v'' + p(t)v' + q(t)v = g_2(t).$$

That is, show that the real and imaginary parts of the complex-valued particular solution, $y_p(t)$, are themselves particular solutions corresponding to the real and imaginary parts of the complex-valued nonhomogeneous term, $g(t)$.

39. Consider the nonhomogeneous differential equation $y'' - y = e^{i2t}$. The complementary solution is $y_C = c_1 e^t + c_2 e^{-t}$. Recall from Euler's formula that $e^{i2t} = \cos 2t + i \sin 2t$. Therefore, the right-hand side is a (complex-valued) linear combination of functions for which the method of undetermined coefficients is applicable.

(a) Assume a particular solution of the form $y_p = A e^{i2t}$, where A is an undetermined (generally complex) coefficient. Substitute this trial form into the differential equation and determine the constant A.

(b) With the constant A as determined in part (a), write $y_p(t) = A e^{i2t}$ in the form $y_p(t) = u(t) + iv(t)$, where $u(t)$ and $v(t)$ are real-valued functions.

(c) Show that $u(t)$ and $v(t)$ are themselves particular solutions of the following differential equations:

$$u'' - u = \text{Re}[e^{i2t}] = \cos 2t \quad \text{and} \quad v'' - v = \text{Im}[e^{i2t}] = \sin 2t.$$

Therefore, the single computation with the complex-valued nonhomogeneous term yields particular solutions of the differential equation for the two real-valued nonhomogeneous terms forming its real and imaginary parts.

Exercises 40–44:

For each exercise,

(a) Use the indicated trial form for $y_p(t)$ to obtain a (complex-valued) particular solution for the given differential equation with complex-valued nonhomogeneous term $g(t)$.

(b) Write $y_p(t)$ as $y_p(t) = u(t) + iv(t)$, where $u(t)$ and $v(t)$ are real-valued functions. Show that $u(t)$ and $v(t)$ are particular solutions of the given differential equation with nonhomogeneous terms $\text{Re}[g(t)]$ and $\text{Im}[g(t)]$, respectively.

40. $y'' + 2y' + y = e^{it}$, $y_p(t) = A e^{it}$ **41.** $y'' + 4y = e^{it}$, $y_p(t) = A e^{it}$

42. $y'' + 4y = e^{i2t}$, $y_p(t) = A t e^{i2t}$ **43.** $y'' + y' = e^{-i2t}$, $y_p(t) = A e^{-i2t}$

44. $y'' + y = e^{(1+i)t}$, $y_p(t) = A e^{(1+i)t}$

3.9   The Method of Variation of Parameters

Section 3.8 discussed the method of undetermined coefficients as a technique for finding a particular solution of the constant coefficient equation

$$ay'' + by' + cy = g(t). \tag{1}$$

The method of undetermined coefficients can be applied to equation (1) as long as the nonhomogeneous term, $g(t)$, is one of the types listed in Table 3.1 of Section 3.8 or is a linear combination of types listed in the table.

In this section, we consider the general linear second order nonhomogeneous differential equation

$$y'' + p(t)y' + q(t)y = g(t). \tag{2}$$

Unlike in Section 3.8, we do not insist that this differential equation have constant coefficients or that $g(t)$ belong to some special class of functions. The only restriction we place on differential equation (2) is that the functions $p(t), q(t)$, and $g(t)$ be continuous on the t-interval of interest.

The technique we discuss, the method of variation of parameters, is one that uses a knowledge of the complementary solution of (2) to construct a corresponding particular solution.

Discussion of the Method

Assume that $\{y_1(t), y_2(t)\}$ is a fundamental set of solutions of the homogeneous equation $y'' + p(t)y' + q(t)y = 0$. In other words, the complementary solution is given by

$$y_C(t) = c_1 y_1(t) + c_2 y_2(t).$$

To obtain a particular solution of (2), we "vary the parameters." That is, we replace the constants c_1 and c_2 by functions $u_1(t)$ and $u_2(t)$ and look for a particular solution of (2) having the form

$$y_P(t) = y_1(t)u_1(t) + y_2(t)u_2(t). \tag{3}$$

In (3), there are two functions, $u_1(t)$ and $u_2(t)$, that we are free to specify. One obvious constraint on $u_1(t)$ and $u_2(t)$ is that they must be chosen so that y_P is a solution of nonhomogeneous equation (2). Generally speaking, however, two constraints can be imposed when we want to determine two functions. We will impose a second constraint as we proceed, choosing it so as to simplify the calculations.

Substituting (3) into the left-hand side of equation (2) requires us to compute the first and second derivatives of (3). Computing the first derivative leads to

$$y_P' = [y_1'u_1 + y_2'u_2] + [y_1 u_1' + y_2 u_2'].$$

The grouping of terms in this equation is motivated by the fact that we now impose a constraint on the functions $u_1(t)$ and $u_2(t)$. We require

$$y_1 u_1' + y_2 u_2' = 0. \tag{4}$$

With this constraint, the derivative of y_P becomes

$$y_P' = y_1'u_1 + y_2'u_2, \tag{5}$$

while y_P'' is

$$y_P'' = y_1''u_1 + y_2''u_2 + y_1'u_1' + y_2'u_2'. \tag{6}$$

Notice that the first and second derivatives of y_P have been simplified and y_P'' does not involve u_1'' or u_2''.

Inserting $y_P = y_1 u_1 + y_2 u_2$ into the differential equation $y'' + p(t)y' + q(t)y = g(t)$ and using (5) and (6), we find

$$[y_1''u_1 + y_2''u_2 + y_1'u_1' + y_2'u_2'] + p(t)[y_1'u_1 + y_2'u_2] + q(t)[y_1 u_1 + y_2 u_2] = g(t).$$

Rearranging terms yields

$$[y_1'' + p(t)y_1' + q(t)y_1]u_1 + [y_2'' + p(t)y_2' + q(t)y_2]u_2 + [y_1'u_1' + y_2'u_2'] = g(t). \tag{7}$$

Since y_1 and y_2 are solutions of $y'' + p(t)y' + q(t)y = 0$, equation (7) reduces to

$$y_1'u_1' + y_2'u_2' = g(t). \tag{8}$$

We therefore obtain two constraints, equations (4) and (8), for the two unknown functions $u_1'(t)$ and $u_2'(t)$. We can combine these two equations into the matrix equation

$$\begin{bmatrix} y_1(t) & y_2(t) \\ y_1'(t) & y_2'(t) \end{bmatrix} \begin{bmatrix} u_1'(t) \\ u_2'(t) \end{bmatrix} = \begin{bmatrix} 0 \\ g(t) \end{bmatrix}. \tag{9}$$

If the (2×2) coefficient matrix has an inverse, then we can solve equation (9) for the unknowns $u_1'(t)$ and $u_2'(t)$. Once they are determined, we can find $u_1(t)$ and $u_2(t)$ by computing antiderivatives.

The coefficient matrix in equation (9) is invertible if and only if its determinant is nonzero. Note, however, that the determinant of this matrix is the Wronskian of the functions y_1 and y_2,

$$W(t) = y_1(t)y_2'(t) - y_1'(t)y_2(t).$$

Since $\{y_1(t), y_2(t)\}$ is a fundamental set of solutions, we are assured that the Wronskian is nonzero for all values of t in our interval of interest. Solving equation (9) for u_1' and u_2' gives

$$\begin{bmatrix} u_1'(t) \\ u_2'(t) \end{bmatrix} = \frac{1}{W(t)} \begin{bmatrix} y_2'(t) & -y_2(t) \\ -y_1'(t) & y_1(t) \end{bmatrix} \begin{bmatrix} 0 \\ g(t) \end{bmatrix},$$

or

$$u_1'(t) = \frac{-y_2(t)g(t)}{W(t)} \quad \text{and} \quad u_2'(t) = \frac{y_1(t)g(t)}{W(t)}.$$

Antidifferentiating to obtain $u_1(t)$ and $u_2(t)$, we have a particular solution,

$$y_P(t) = y_1(t)u_1(t) + y_2(t)u_2(t).$$

Explicitly, the particular solution we have obtained is

$$y_P(t) = -y_1(t) \int \frac{y_2(s)g(s)}{W(s)}\, ds + y_2(t) \int \frac{y_1(s)g(s)}{W(s)}\, ds. \tag{10}$$

Once we calculate the two antiderivatives in equation (10), we will have determined a particular function that solves the nonhomogeneous equation (2).

E X A M P L E

1

Find the general solution of the differential equation

$$y'' - 2y' + y = e^t \ln t, \qquad t > 0. \tag{11}$$

Solution: Note that the nonhomogeneous term, $g(t) = e^t \ln t$, does not appear in Table 3.1 as a candidate for the method of undetermined coefficients. Since the method of undetermined coefficients is not applicable, we turn to the method of variation of parameters.

For equation (11), the complementary solution is

$$y_C(t) = c_1 y_1(t) + c_2 y_2(t) = c_1 e^t + c_2 t e^t.$$

Using variation of parameters to find a particular solution, we assume

$$y_P(t) = e^t u_1(t) + t e^t u_2(t).$$

Substituting this expression into the nonhomogeneous differential equation (11) and applying the constraint from equation (4), $e^t u_1'(t) + t e^t u_2'(t) = 0$, we obtain

$$\begin{bmatrix} e^t & t e^t \\ e^t & e^t + t e^t \end{bmatrix} \begin{bmatrix} u_1'(t) \\ u_2'(t) \end{bmatrix} = \begin{bmatrix} 0 \\ e^t \ln t \end{bmatrix}.$$

The determinant of the coefficient matrix is $W(t) = e^{2t}$. Solving this matrix equation gives

$$\begin{bmatrix} u_1'(t) \\ u_2'(t) \end{bmatrix} = \begin{bmatrix} -t \ln t \\ \ln t \end{bmatrix}.$$

Computing antiderivatives yields

$$u_1(t) = \int u_1'(t)\, dt = -\int t \ln t\, dt = -\frac{t^2}{2} \ln t + \frac{t^2}{4} + K_1$$

and

$$u_2(t) = \int u_2'(t)\, dt = \int \ln t\, dt = t \ln t - t + K_2.$$

We can choose the constants of integration to suit our convenience because all we need is *some* particular solution, $y_P = y_1 u_1 + y_2 u_2$. For convenience, we set both K_1 and K_2 equal to zero and obtain

$$y_P(t) = e^t \left[-\frac{t^2}{2} \ln t + \frac{t^2}{4} \right] + te^t[t \ln t - t]$$

$$= \frac{t^2 e^t}{2} \left[\ln t - \frac{3}{2} \right].$$

The general solution of equation (11) is, therefore,

$$y(t) = c_1 e^t + c_2 t e^t + \frac{t^2 e^t}{2} \left[\ln t - \frac{3}{2} \right]. \quad \diamond$$

EXAMPLE

2

Observe that $y_1(t) = t$ is a solution of the homogeneous equation

$$t^2 y'' - t y' + y = 0, \qquad t > 0.$$

Use this observation to solve the nonhomogeneous initial value problem

$$t^2 y'' - t y' + y = t, \qquad y(1) = 1, \qquad y'(1) = 4.$$

Solution: The first step in finding the general solution of the nonhomogeneous equation is determining a fundamental set of solutions $\{y_1, y_2\}$. Thus, we need to find a second solution, $y_2(t)$, to go along with the given solution, $y_1(t) = t$.

The method of reduction of order, described in Section 3.4, can be used. It leads to a second solution,

$$y_2(t) = t \ln t.$$

The functions t and $t \ln t$ can be shown to form a fundamental set of solutions for the homogeneous equation. Therefore, the complementary solution of the nonhomogeneous equation is

$$y_C(t) = c_1 t + c_2 t \ln t, \qquad t > 0.$$

Since the differential equation has variable coefficients, we cannot use the method of undetermined coefficients to find a particular solution. Instead, we

(continued)

(continued)

use the method of variation of parameters. Assume a particular solution of the form

$$y_P(t) = tu_1(t) + [t \ln t]u_2(t).$$

The simplest approach is to substitute this form into the given nonhomogeneous differential equation to determine one equation for u_1' and u_2', then use the constraint

$$tu_1'(t) + [t \ln t]u_2'(t) = 0$$

to form a second equation for u_1' and u_2'.

As an alternative, we could go directly to equation (9) or equation (10) to determine a particular solution. If we proceed in this fashion, however, we have to make certain that the nonhomogeneous equation under consideration is in the standard form given by equation (2). Since the differential equation in this example does *not* have this form, we need to rewrite it as

$$y'' - \frac{1}{t}y' + \frac{1}{t^2}y = \frac{1}{t}.$$

Doing so, we can identify the term $g(t)$ in equations (9) and (10), $g(t) = 1/t$.

Both approaches lead to the following system of equations for u_1' and u_2':

$$\begin{bmatrix} t & t \ln t \\ 1 & 1 + \ln t \end{bmatrix} \begin{bmatrix} u_1' \\ u_2' \end{bmatrix} = \begin{bmatrix} 0 \\ \dfrac{1}{t} \end{bmatrix}.$$

Solving this system, we obtain

$$\begin{bmatrix} u_1' \\ u_2' \end{bmatrix} = \begin{bmatrix} -t^{-1} \ln t \\ t^{-1} \end{bmatrix}.$$

Computing antiderivatives yields

$$u_1(t) = -\int \frac{1}{t} \ln t \, dt = -\frac{(\ln t)^2}{2} + K_1,$$

$$u_2(t) = \int \frac{1}{t} \, dt = \ln t + K_2.$$

We can set both of the arbitrary constants equal to zero, obtaining a particular solution

$$y_P(t) = t\left[-\frac{(\ln t)^2}{2} \right] + [t \ln t] \ln t = \frac{t}{2}(\ln t)^2.$$

The general solution of the nonhomogeneous equation is therefore

$$y(t) = c_1 t + c_2 t \ln t + \frac{t}{2}(\ln t)^2, \qquad t > 0.$$

Imposing the initial conditions shows that the solution of the initial value problem is

$$y(t) = t + 3t \ln t + \frac{t}{2}(\ln t)^2, \qquad t > 0. \; \diamondsuit$$

Figure 3.13 displays the graph of the solution $y(t)$. Note that $\lim_{t \to 0^+} y(t) = 0$. The solution is well behaved near $t = 0$, even though the differential equation has coefficient functions that are not defined at $t = 0$.

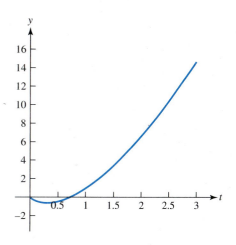

FIGURE 3.13

The solution of the initial value problem in Example 2. Even though some of the coefficient functions are not defined at $t = 0$, the solution $y(t)$ has a limit as t approaches 0 from the right.

EXERCISES

Exercises 1–14:

For the given differential equation,

(a) Determine the complementary solution, $y_C(t) = c_1 y_1(t) + c_2 y_2(t)$.

(b) Use the method of variation of parameters to construct a particular solution. Then form the general solution.

1. $y'' + 4y = 4$

2. $y'' + y = \sec t, \quad -\pi/2 < t < \pi/2$

3. $y'' - [2 + (1/t)]y' + [1 + (1/t)]y = te^t, \quad 0 < t < \infty$. [The functions $y_1(t) = e^t$ and $y_2(t) = t^2 e^t$ are solutions of the homogeneous equation.]

4. $y'' - y = \dfrac{1}{1 + e^t}$

5. $y'' - y = e^t$

6. $y'' - (2/t)y' + (2/t^2)y = t/(1 + t^2), \quad 0 < t < \infty$. [The function $y_1(t) = t^2$ is a solution of the homogeneous equation.]

7. $y'' - 2y' + y = e^t$

8. $y'' + 36y = \csc^3(6t)$

9. $y'' - (2 \cot t)y' + (2 \csc^2 t - 1)y = t \sin t, \quad 0 < t < \pi$. [The functions $y_1(t) = \sin t$ and $y_2(t) = t \sin t$ are both solutions of the homogeneous equation.]

10. $t^2 y'' - ty' + y = t \ln t, \quad 0 < t < \infty$. [The functions $y_1(t) = t$ and $y_2(t) = t \ln t$ are both solutions of the homogeneous equation.]

11. $y'' + [t/(1 - t)]y' - [1/(1 - t)]y = (t - 1)e^t, \, 1 < t < \infty$. [The functions $y_1(t) = t$ and $y_2(t) = e^t$ are both solutions of the homogeneous equation.]

12. $y'' + 4ty' + (2 + 4t^2)y = t^2 e^{-t^2}$. [The functions $y_1(t) = e^{-t^2}$ and $y_2(t) = te^{-t^2}$ are both solutions of the homogeneous equation.]

13. $(t-1)^2 y'' - 4(t-1)y' + 6y = t$, $\quad 1 < t < \infty$. [The function $y_1(t) = (t-1)^2$ is a solution of the homogeneous equation.]

14. $y'' - [2 + (2/t)]y' + [1 + (2/t)]y = e^t$, $\quad 0 < t < \infty$. [The function $y_1(t) = e^t$ is a solution of the homogeneous equation.]

15. Consider the homogeneous differential equation $y'' + p(t)y' + q(t)y = g(t)$. Let $\{y_1, y_2\}$ be a fundamental set of solutions for the corresponding homogeneous equation, and let $W(t)$ denote the Wronskian of this fundamental set. Show that the particular solution that vanishes at $t = t_0$ is given by

$$y_P(t) = \int_{t_0}^{t} [y_1(t)y_2(\lambda) - y_2(t)y_1(\lambda)]\frac{g(\lambda)}{W(\lambda)}\, d\lambda.$$

Exercises 16–18:

The given expression is the solution of the initial value problem

$$y'' + \alpha y' + \beta y = g(t), \qquad y(0) = y_0, \qquad y'(0) = y_0'.$$

Determine the constants α, β, y_0, and y_0'.

16. $y(t) = \dfrac{1}{2}\displaystyle\int_0^t \sin[2(t-\lambda)]g(\lambda)\, d\lambda$

17. $y(t) = e^{-t} + \displaystyle\int_0^t \frac{e^{t-\lambda} - e^{-(t-\lambda)}}{2}g(\lambda)\, d\lambda = e^{-t} + \int_0^t \sinh(t-\lambda)g(\lambda)\, d\lambda$

18. $y(t) = t + \displaystyle\int_0^t (t-\lambda)g(\lambda)\, d\lambda$

3.10   Forced Mechanical Vibrations, Electrical Networks, and Resonance

In this section, we use what we know about solving nonhomogeneous second order linear differential equations to study the behavior of mechanical systems (such as floating objects and spring-mass-dashpot systems) that are subjected to externally applied forces.

  We also consider simple electrical networks containing resistors, inductors, and capacitors (called *RLC* networks) that are driven by voltage and current sources. All of these applications ultimately give rise to the same mathematical problem. And, as we shall see, the physical phenomenon of resonance is an important consideration common to all of these applications.

Forced Mechanical Vibrations

A Buoyant Body   Consider again the problem of the buoyant body discussed in Section 3.1. (See Figure 3.1.) A cylindrical block of cross-sectional area A, height L, and mass density ρ is placed in a liquid having mass density ρ_l. Since we assume $\rho < \rho_l$, the block floats in the liquid. In equilibrium, it sinks a depth Y into the liquid; at this depth, the weight of the block equals the weight of the liquid displaced. The quantity $y(t)$ represents the instantaneous vertical displacement of the block from its equilibrium position, measured positive downward. Suppose now that an externally applied vertical force, $F_a(t)$, acts on the buoyant body. Newton's second law of motion ultimately leads to a nonhomogeneous differential equation for the displacement $y(t)$:

$$\frac{d^2y}{dt^2} + \frac{\rho_l g}{\rho L}y = \frac{1}{\rho AL}F_a(t). \tag{1}$$

In equation (1), g denotes gravitational acceleration and the term ρAL is the block's mass. Defining a radian frequency, $\omega_0 = \sqrt{\rho_l g/\rho L}$, and a force per unit mass, $f_a(t) = (1/\rho AL)F_a(t)$, we can rewrite equation (1) as

$$y'' + \omega_0^2 y = f_a(t). \tag{2}$$

Equation (2) is a second order constant coefficient linear nonhomogeneous differential equation. To uniquely prescribe the bobbing motion of the floating body, we would add initial conditions that specify the initial displacement, $y(t_0) = y_0$, and the initial velocity, $y'(t_0) = y_0'$, of the block.

A Spring-Mass-Dashpot System   As in Section 3.6, we can use Newton's second law of motion to derive a nonhomogenous differential equation for the displacement of a mass suspended from a spring-dashpot connection and acted upon by an applied force. The resulting differential equation is

$$my'' + \gamma y' + ky = F_a(t), \tag{3}$$

where $F_a(t)$ denotes an applied vertical force. The positive constants m, γ, and k represent the mass, damping coefficient, and spring constant of the system, respectively. The dependent variable $y(t)$ measures downward displacement from the equilibrium rest position. (See Figure 3.8.)

If there is no damping (that is, if $\gamma = 0$) and if we define radian frequency $\omega_0 = \sqrt{k/m}$ and force per unit mass $f_a(t) = (1/m)F_a(t)$, equation (3) can be rewritten as

$$y'' + \omega_0^2 y = f_a(t). \tag{4}$$

Note that equation (4), describing a spring-mass system, is identical in structure to equation (2), which describes the bobbing motion of a buoyant body. Both applications lead to the same mathematical problem. As we shall see shortly, other applications (such as *RLC* networks) also lead to differential equations having exactly the same structure as equations (2) and (3). Since these applications all lead to the same mathematical problem, we will discuss equations (2) and (3) in their own right rather than investigate each application separately. The Exercises focus on specific applications.

Oscillatory Applied Forces and Resonance

Examples 1 and 2 concern an important special case of equation (2) where the applied force is a sinusoidally varying force,

$$f_a(t) = F \cos \omega_1 t,$$

where F is a constant. For simplicity, we assume the system is initially at rest in equilibrium. Thus, for the special case under consideration, the corresponding initial value problem is

$$y'' + \omega_0^2 y = F \cos\omega_1 t, \qquad t > 0,$$
$$y(0) = 0, \qquad y'(0) = 0.$$

(5)

The complementary solution of (5) is

$$y_C(t) = c_1 \sin\omega_0 t + c_2 \cos\omega_0 t.$$

(6)

In order to find a particular solution, we need to consider two separate cases, $\omega_1 \neq \omega_0$ and $\omega_1 = \omega_0$. From a mathematical perspective, the method of undetermined coefficients discussed in Section 3.8 leads us to expect two different types of particular solutions. If $\omega_1 \neq \omega_0$, the nonhomogeneous term, $g(t) = F\cos\omega_1 t$, is not a solution of the homogeneous equation; when $\omega_1 = \omega_0$, $g(t)$ is a solution of the homogeneous equation.

This mathematical perspective is consistent with the physics of the problem. We should expect different behavior on purely physical grounds. Radian frequency ω_0 (or, more properly, $f_0 = \omega_0/2\pi$) is called the **natural frequency** of the vibrating system. It represents the frequency at which the system would vibrate if no applied force were present and the system were merely responding to some initial disturbance. The applied force acts on the system with its own applied frequency ω_1. In the special case where the natural and applied frequencies are equal, the applied force pushes and pulls on the system with a frequency precisely equal to that at which the system tends naturally to vibrate. This precise reinforcement leads to the phenomenon of **resonance**. For this reason, the natural frequency of the system is also referred to as its **resonant frequency**.

E X A M P L E

1

Assume $\omega_1 \neq \omega_0$. Solve the initial value problem (5).

Solution: Since $\omega_1 \neq \omega_0$, Table 3.1 in Section 3.8 suggests a particular solution of the form

$$y_P(t) = A\cos\omega_1 t + B\sin\omega_1 t.$$

Substituting this form into the nonhomogeneous equation, we obtain (see Exercise 1)

$$y_P(t) = -\frac{F}{\omega_1^2 - \omega_0^2}\cos\omega_1 t.$$

Imposing the initial conditions on the general solution

$$y(t) = c_1 \sin\omega_0 t + c_2 \cos\omega_0 t - \frac{F}{\omega_1^2 - \omega_0^2}\cos\omega_1 t,$$

we obtain the solution of initial value problem (5),

$$y(t) = \frac{F}{\omega_1^2 - \omega_0^2}[\cos\omega_0 t - \cos\omega_1 t].$$

(7)

Figure 3.14 shows solution (7) for the special case where $\omega_1 = 12\pi$ s^{-1}, $\omega_0 = 10\pi$ s^{-1}, and F is chosen so that $F/(\omega_1^2 - \omega_0^2) = 2$ cm. The example therefore assumes that the natural frequency of the system is 5 Hz while the applied frequency is 6 Hz. (For definiteness, the unit of length is chosen to be the centimeter.)

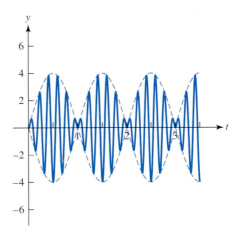

FIGURE 3.14

The solution of initial value problem (5), as given by equation (7). For the case shown, $\omega_1 = 12\pi$, $\omega_0 = 10\pi$, and $F = 2(\omega_1^2 - \omega_0^2)$.

We can better understand Figure 3.14, including the dashed envelope, if we use trigonometric identities to recast equation (7). Suppose we define an average radian frequency, $\overline{\omega}$, and a difference (or **beat**) radian frequency, β, by

$$\overline{\omega} = \frac{\omega_1 + \omega_0}{2}, \qquad \beta = \frac{\omega_1 - \omega_0}{2}.$$

With these definitions, we have $\omega_1 = \overline{\omega} + \beta$ and $\omega_0 = \overline{\omega} - \beta$. Equation (7) becomes

$$y(t) = \frac{F}{4\overline{\omega}\beta}[\cos(\overline{\omega}t - \beta t) - \cos(\overline{\omega}t + \beta t)].$$

Using the trigonometric identity $\cos(\theta_1 \pm \theta_2) = \cos\theta_1 \cos\theta_2 \mp \sin\theta_1 \sin\theta_2$, we note that

$$\cos(\overline{\omega}t - \beta t) - \cos(\overline{\omega}t + \beta t) = 2\sin\overline{\omega}t \sin\beta t.$$

Therefore, we obtain an alternative representation for the solution $y(t)$:

$$y(t) = A(t)\sin\overline{\omega}t, \tag{8}$$

where

$$A(t) = \frac{F\sin\beta t}{2\overline{\omega}\beta}.$$

Using equation (8), we can interpret solution (7) as the product of a variable amplitude term, $A(t)$, and a sinusoidal term, $\sin\overline{\omega}t$. In cases where ω_1 and ω_0 are nearly equal, the amplitude term, whose behavior is governed by the factor $\sin\beta t$, is slowly varying relative to the sinusoidal term, $\sin\overline{\omega}t$ (because $|\beta| \ll \overline{\omega}$). The combination of these disparate rates of variation gives rise to the phenomenon of "beats" seen in Figure 3.14.

Since $\overline{\omega} = 11\pi$ s^{-1} and $\beta = \pi$ s^{-1}, $\sin\beta t = \sin\pi t$ varies much more slowly than $\sin\overline{\omega}t = \sin 11\pi t$. The multiplicative factor

$$A(t) = \frac{F\sin\beta t}{2\overline{\omega}\beta} = 4\sin\pi t$$

(continued)

(continued)

defines a slowly varying sinusoidal envelope for the more rapidly varying $\sin \overline{\omega} t = \sin 11 \pi t$. The dashed envelope shown in Figure 3.14 is defined by the graphs of $y = \pm 4 \sin \pi t$. ❖

E X A M P L E

2

Assume $\omega_1 = \omega_0$. Solve initial value problem (5).

Solution: In this case,

$$y'' + \omega_0^2 y = F \cos \omega_0 t.$$

Since the functions $\cos \omega_0 t$ and $\sin \omega_0 t$ are both solutions of the homogeneous equation, Table 3.1 of Section 3.8 prescribes a trial solution of the form

$$y_P(t) = At \cos \omega_0 t + Bt \sin \omega_0 t.$$

Substituting this form leads (see Exercise 1) to the particular solution

$$y_P(t) = \frac{F}{2\omega_0} t \sin \omega_0 t$$

and the general solution

$$y(t) = c_1 \sin \omega_0 t + c_2 \cos \omega_0 t + \frac{F}{2\omega_0} t \sin \omega_0 t.$$

The initial conditions imply that c_1 and c_2 are zero. The solution of initial value problem (5) is therefore

$$y(t) = \frac{F}{2\omega_0} t \sin \omega_0 t. \tag{9}$$

Figure 3.15 shows solution (9) for the case

$$\omega_0 = 10\pi \text{ s}^{-1} \quad \text{and} \quad \frac{F}{2\omega_0} = 4\pi \text{ cm/s}.$$

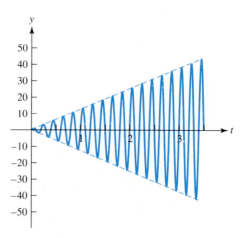

FIGURE 3.15

The solution of initial value problem (5) as given by equation (9). For the case shown, $\omega_0 = 10\pi$ and $F = 8\pi\omega_0$. The solution envelope grows linearly with time, illustrating the phenomenon of resonance.

In this case, the applied force reinforces the natural frequency vibrations of the mechanical system and the envelope of the solution (shown by the dashed lines in Figure 3.15) grows linearly with time. This is the phenomenon of resonance.

Does the solution (9) make sense? The vibration amplitude of a real physical system certainly does not continue to grow indefinitely. Therefore, one would expect equation (9) to describe the behavior of a real system for a limited time at best. Once the vibration amplitude becomes sufficiently large, the assumptions made in deriving the mathematical models cease to be valid. For example, equation (3) and special case (4) for the spring-mass system assume the validity of Hooke's law. When mass displacement amplitude becomes too large, however, the force-displacement relation becomes more complicated than the simple linear relation embodied in Hooke's law. ❖

REMARK: One property of a well-posed problem is continuous dependence upon the data. Roughly speaking, if an initial value problem is to be a reasonable model of reality, its solution should not change uncontrollably when a parameter (such as a coefficient in the differential equation or an initial condition) is changed slightly. Therefore, we might reasonably ask whether it is possible to see resonant solution (9) emerge from nonresonant solution (7) or (8) in the limit as $\omega_1 \to \omega_0$. Note that (8) can be rewritten as

$$y(t) = \frac{F}{2\overline{\omega}}t \left(\frac{\sin \beta t}{\beta t} \right) \sin \overline{\omega} t. \tag{10}$$

Suppose we fix t and let $\omega_1 \to \omega_0$. Then, from their definitions, $\overline{\omega} \to \omega_0$ and $\beta \to 0$. We know from calculus that

$$\lim_{x \to 0} \frac{\sin x}{x} = 1.$$

Therefore, for any fixed value of t, we do indeed obtain resonant solution (9) from nonresonant expression (8) in the limit as $\omega_1 \to \omega_0$.

The Effect of Damping on Resonance

There are no perpetual motion machines. All physical systems have at least some small loss or damping present. Therefore, it is of interest to see what happens if we add damping to an otherwise resonant system. Suppose we consider the initial value problem

$$y'' + 2\delta y' + \omega_0^2 y = F \cos \omega_0 t, \qquad y(0) = 0, \qquad y'(0) = 0, \tag{11a}$$

where δ is a positive constant (the factor 2 is added for convenience). What does the solution look like? In Exercise 11(a), you are asked to show that the solution of this initial value problem is

$$y(t) = \frac{F}{2\delta} \left[\frac{\sin \omega_0 t}{\omega_0} - \frac{e^{-\delta t} \sin \left(\sqrt{\omega_0^2 - \delta^2}\, t \right)}{\sqrt{\omega_0^2 - \delta^2}} \right]. \tag{11b}$$

[In (11), we tacitly assume that $\omega_0^2 > \delta^2$.] As a check, you can show [see Exercise 11(b)] that for any fixed t and ω_0, expression (11b) reduces to (9) as $\delta \to 0$.

Equation (11b) shows the effect of damping on the otherwise resonant system. If we fix δ at some positive value and let $t \to \infty$, the second term in equation (11b) tends to zero. Therefore, with the addition of damping, displacement does not grow indefinitely, as it appears to in Figure 3.15. As time increases, displacement approaches a steady-state behavior given by the first term,

$$\frac{F}{2\delta\omega_0}\sin\omega_0 t.$$

As δ becomes smaller, the amplitude of these steady-state oscillations becomes correspondingly larger. Figure 3.16 shows the variation of displacement for the case

$$\omega_0 = 10\pi \text{ s}^{-1}, \qquad \frac{F}{2\omega_0} = 4\pi \text{ cm/s}, \qquad \delta = 0.5 \text{ s}^{-1}.$$

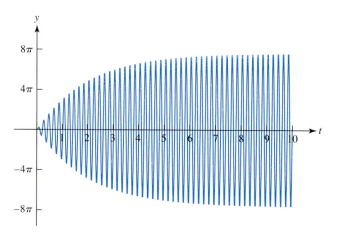

FIGURE 3.16

The solution of $y'' + 2\delta y' + \omega_0^2 y = F\cos\omega_0 t$ for the case $\omega_0 = 10\pi$, $F = 8\pi\omega_0$, and $\delta = 0.5$. As you can see by comparing this graph with the resonant solution graphed in Figure 3.15, the inclusion of damping eliminates the unbounded linear growth in the solution envelope that characterizes the resonant case. As noted, however, the steady-state oscillations, $(F/2\delta\omega_0)\sin\omega_0 t$, have an amplitude proportional to δ^{-1}.

Nonresonant Excitation with Damping Present

Suppose we now change the radian frequency of the applied force in the previous problem to $\omega_1 \neq \omega_0$. In that case, the problem becomes

$$y'' + 2\delta y' + \omega_0^2 y = F\cos\omega_1 t, \qquad y(0) = 0, \qquad y'(0) = 0. \tag{12a}$$

This amounts to the addition of a damping force to the problem defined by equation (5). Again assuming that $\omega_0^2 > \delta^2$, the solution of problem (12a) is

$$y(t) = \frac{F}{(\omega_0^2 - \omega_1^2)^2 + (2\delta\omega_1)^2}[(\omega_0^2 - \omega_1^2)\cos\omega_1 t + 2\delta\omega_1\sin\omega_1 t]$$

$$- \frac{Fe^{-\delta t}}{(\omega_0^2 - \omega_1^2)^2 + (2\delta\omega_1)^2}\left[(\omega_0^2 - \omega_1^2)\cos\left(\sqrt{\omega_0^2 - \delta^2}\,t\right)\right. \tag{12b}$$

$$\left. + \frac{\delta(\omega_0^2 + \omega_1^2)}{\sqrt{\omega_0^2 - \delta^2}}\sin\left(\sqrt{\omega_0^2 - \delta^2}\,t\right)\right].$$

This solution seems relatively complicated. However, two checks can be made. At a fixed value of t, the solution should reduce to (11b) in the limit as $\omega_1 \to \omega_0$. Also, the solution should reduce to (7) if we fix ω_1 but let $\delta \to 0$. Exercise 12 asks you not only to derive this solution, but also to make these checks upon its correctness.

The second term in the solution of equation (12a) represents a transient term, one that tends to zero as time increases. The first term is the steady-state portion of the solution. Figure 3.17 shows the behavior of the solution for the case where

$$\omega_0 = 10\pi \text{ s}^{-1}, \qquad \omega_1 = 12\pi \text{ s}^{-1}, \qquad \frac{F}{2\omega_0} = 4\pi \text{ cm/s}, \qquad \delta = 0.5 \text{ s}^{-1}.$$

Compare this behavior with that exhibited in Figures 3.14 and 3.16. When time t is relatively small, before the effects of damping become pronounced, the solution exhibits a difference frequency modulation envelope that is qualitatively similar to the behavior shown in Figure 3.14. As time progresses, however, damping eventually diminishes the second term in the solution to a negligibly small contribution and the solution becomes essentially the steady-state portion given by the first term. In this respect, the long-term behavior qualitatively resembles the damping-perturbed resonant case exhibited in Figure 3.16.

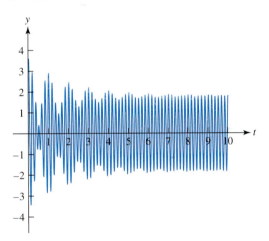

FIGURE 3.17

The solution of equation (12) for representative values of ω_0, ω_1, F, and δ. Initially, for small values of t, damping is not significant and the motion is similar to that shown in Figure 3.14, exhibiting beats. As t grows, damping diminishes the second term in equation (12) and the motion has a period similar to the applied force.

RLC Networks

We now consider networks containing resistors, inductors, and capacitors. The application of Kirchhoff's[7] voltage law and Kirchhoff's current law to these networks leads us to second order differential equations.

---

[7] Gustav Robert Kirchhoff (1824–1887) was a German physicist who made important contributions to network theory, elasticity, and our understanding of blackbody radiation. A lunar crater is named in his honor.

Consider first the series RLC network shown in Figure 3.18. A voltage source $V_S(t)$ having the polarity shown is connected in series with circuit elements having resistance R, inductance L, and capacitance C. A current $I(t)$, assumed positive in the sense shown, flows in the loop. In essence, Kirchhoff's voltage law asserts that the voltage at each point in the network is a well-defined single-valued quantity. Therefore, as we make an excursion around the loop, the sum of the voltage rises must equal the sum of the voltage drops. If we proceed around the loop in Figure 3.18 in a clockwise manner, the voltage rise is the source voltage $V_S(t)$, while the voltage drops are the drops across the three circuit elements. The voltage drop across the resistor is $I(t)R$, the drop across the inductor is $L\,(dI/dt)$, and the drop across the capacitor is $(1/C)Q(t)$, where $Q(t)$ represents the electric charge on the capacitor.

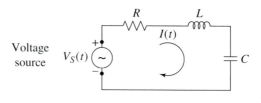

FIGURE 3.18

A series RLC network, with voltage source $V_S(t)$ and loop current $I(t)$.

An application of Kirchhoff's voltage law therefore leads to the equation

$$V_S(t) = RI + L\frac{dI}{dt} + \frac{1}{C}Q(t). \tag{13a}$$

To obtain a differential equation for a single dependent variable, we use the fact that electric current is the rate of change of electric charge with respect to time,

$$I(t) = \frac{dQ}{dt}.$$

One approach is to rewrite equation (13a) as a second order differential equation for the electric charge, obtaining

$$L\frac{d^2Q}{dt^2} + R\frac{dQ}{dt} + \frac{1}{C}Q = V_S(t).$$

This equation, supplemented by initial conditions specifying the charge $Q(t_0)$ and current $Q'(t_0) = I(t_0)$ at some initial time t_0, can be solved for the charge $Q(t)$. Differentiating this solution yields the desired current, $I(t)$. A second approach is simply to differentiate equation (13a), obtaining a second order differential equation for the current $I(t)$. In that case,

$$\frac{d^2I}{dt^2} + \frac{R}{L}\frac{dI}{dt} + \frac{1}{LC}I = \frac{1}{L}\frac{dV_S(t)}{dt}. \tag{13b}$$

Equation (13b) is a nonhomogeneous second order linear differential equation for the unknown loop current $I(t)$. To uniquely prescribe circuit performance, we must add initial conditions $I(t_0) = I_0$ and $I'(t_0) = I'_0$ at some initial time

t_0. [Specifying $I'(t_0)$ is tantamount to specifying the voltage drop across the inductor at time t_0.]

As a second example of an *RLC* network, consider the network shown in Figure 3.19. In this case, the three circuit elements are connected in parallel with a current source $I_S(t)$, whose current output is assumed to flow in the direction shown. This time, the dependent variable of interest is nodal voltage $V(t)$, assumed to have the polarity shown.

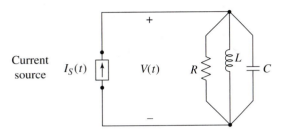

FIGURE 3.19

A parallel *RLC* network, with current source $I_S(t)$ and nodal voltage $V(t)$.

The governing physical principle, Kirchhoff's current law, states that electric current does not accumulate at a circuit node. Therefore, the total current flowing into a node must equal the total current flowing out. Consider the upper node. The current flowing in is the source current, while the current flowing out is the current flowing "down" through each of the circuit elements. The current through the resistor is $(1/R)V(t)$, the current through the capacitor is $C(dV/dt)$, and the current through the inductor is $(1/L) \int V(s)\, ds$ (an antiderivative of the nodal voltage). Applying Kirchhoff's current law to the network in Figure 3.19 leads us to the equation

$$I_S(t) = \frac{1}{R} V + C \frac{dV}{dt} + \frac{1}{L} \int V(s)\, ds.$$

Upon differentiating and rearranging terms, the equation becomes

$$\frac{d^2V}{dt^2} + \frac{1}{RC} \frac{dV}{dt} + \frac{1}{LC} V = \frac{1}{C} \frac{dI_S(t)}{dt}. \tag{14}$$

Specifying V and V' (that is, the currents through the resistor and capacitor) at some initial time t_0 will uniquely determine circuit performance.

REMARK: If we short circuit the resistor (that is, set $R = 0$) in the series circuit (Figure 3.18) or if we open circuit the resistor (that is, let $R \to \infty$) in the parallel circuit (Figure 3.19), we remove the dissipative (damping) element in each case and obtain a lossless *LC* circuit. Such circuits can exhibit resonance. Note that equations (13b) and (14) become identical in structure to equations (2) and (4) with a resonant radian frequency defined by

$$\omega_0^2 = \frac{1}{LC}.$$

EXERCISES

1. Consider the differential equation $y'' + \omega_0^2 y = F \cos \omega t$.

    (a) Determine the complementary solution of this differential equation.

    (b) Use the method of undetermined coefficients to find a particular solution in each of the cases: (i) $\omega = \omega_1 \neq \omega_0$, (ii) $\omega = \omega_0$.

Exercises 2–5:

A 10-kg object suspended from the end of a vertically hanging spring stretches the spring 9.8 cm. At time $t = 0$, the resulting spring-mass system is disturbed from its rest state by the given applied force, $F(t)$. The force $F(t)$ is expressed in newtons and is positive in the downward direction; time is measured in seconds.

(a) Determine the spring constant, k.

(b) Formulate and solve the initial value problem for $y(t)$, where $y(t)$ is the displacement of the object from its equilibrium rest state, measured positive in the downward direction.

(c) Plot the solution and determine the maximum excursion from equilibrium made by the object on the t-interval $0 \leq t < \infty$ or state that there is no such maximum.

2. $F(t) = 20 \cos 10t$

3. $F(t) = 20e^{-t}$

4. $F(t) = 20 \cos 8t$

5. $F(t) = \begin{cases} 20, & 0 \leq t \leq \pi \\ 0, & \pi < t < \infty \end{cases}$

[Hint: Solve Exercise 5 on the t-interval $0 \leq t \leq \pi$ and then use the fact that position $y(t)$ and velocity $y'(t)$ are both continuous at $t = \pi$ to formulate and solve a second initial value problem on the t-interval $\pi < t < \infty$.]

6. Consider the initial value problem $my'' + ky = 20 \cos 8\pi t, y(0) = 0, y'(0) = 0$, modeling the response of a spring-mass system, initially at rest, to an applied force; assume that the unit of force is the newton. Suppose the motion shown in the figure is recorded and can be described mathematically by the formula $y(t) = 0.1 \sin(\pi t) \sin(7\pi t)$ m. What are the values of mass m and spring constant k for this system? [Hint: Recall the identity $\cos(\alpha \pm \beta) = \cos\alpha \cos\beta \mp \sin\alpha \sin\beta$.]

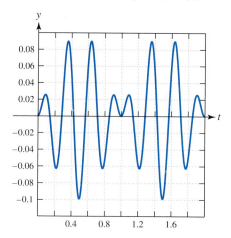

Figure for Exercise 6

Exercises 7–10:

Consider the initial value problem

$$my'' + \gamma y' + ky = F(t), \qquad y(0) = 0, \qquad y'(0) = 0,$$

modeling the motion of a spring-mass-dashpot system initially at rest and subjected to an applied force $F(t)$, where the unit of force is the newton (N). Assume that $m = 2$ kg, $\gamma = 8$ kg/s, and $k = 80$ N/m.

(a) Solve the initial value problem for the given applied force. In Exercise 10, use the fact that the system displacement $y(t)$ and velocity $y'(t)$ remain continuous at times when the applied force is discontinuous.

(b) Determine the long-time behavior of the system. In particular, is $\lim_{t\to\infty} y(t) = 0$? If not, describe in qualitative terms what the system is doing as $t \to \infty$.

7. $F(t) = 20 \cos 8t$

8. $F(t) = 20 e^{-t}$

9. $F(t) = 20 \sin 6t$

10. $F(t) = \begin{cases} 20, & 0 \le t \le \dfrac{\pi}{2} \\[2mm] 0, & \dfrac{\pi}{2} < t < \infty \end{cases}$

11. Consider the initial value problem $my'' + \gamma y' + ky = \overline{F} \cos \sqrt{k/m}\, t$, $y(0) = 0$, $y'(0) = 0$. If we set $\gamma/m = 2\delta$, $\omega_0^2 = k/m$, and $\overline{F}/m = F$, we obtain initial value problem (11a). Assume that $\omega_0^2 > 2\delta$. Note that the radian frequency of the applied force is ω_0; this is the resonant radian frequency of the corresponding undamped system.

(a) Derive equation (11b), showing that the solution of this initial value problem is

$$y(t) = \frac{F}{2\delta} \left[\frac{\sin(\omega_0 t)}{\omega_0} - \frac{e^{-\delta t} \sin\left(\sqrt{\omega_0^2 - \delta^2}\, t\right)}{\sqrt{\omega_0^2 - \delta^2}} \right].$$

(b) Show, for any fixed values $t > 0$ and $\omega_0 > 0$, that

$$\lim_{\delta \to 0^+} \left\{ \frac{F}{2\delta} \left[\frac{\sin(\omega_0 t)}{\omega_0} - \frac{e^{-\delta t} \sin\left(\sqrt{\omega_0^2 - \delta^2}\, t\right)}{\sqrt{\omega_0^2 - \delta^2}} \right] \right\} = \frac{F}{2\omega_0} t \sin(\omega_0 t).$$

This limit is the response of the undamped spring-mass system to resonant frequency excitation.

(c) Suppose that we know the values of mass m and spring constant k (and $\overline{F}$, the amplitude of the applied force). Explain how we might use our knowledge of the solution in part (a) (observed over a long time interval) to estimate the damping constant δ.

12. Consider the initial value problem given in equation (12a),

$$y'' + 2\delta y' + \omega_0^2 y = F \cos \omega_1 t, \qquad y(0) = 0, \qquad y'(0) = 0,$$

where $\omega_0^2 > \delta^2$ and $\omega_1 \ne \omega_0$. The radian frequency of the applied force is therefore not equal to ω_0.

(a) Solve the initial value problem for $y(t)$ and verify that equation (12b) represents the solution.

(b) Assume that $t > 0$ and $\delta > 0$ are fixed. Show that

$$\lim_{\omega_1 \to \omega_0} y(t) = \frac{F}{2\delta} \left[\frac{\sin(\omega_0 t)}{\omega_0} - \frac{e^{-\delta t} \sin\left(\sqrt{\omega_0^2 - \delta^2}\, t\right)}{\sqrt{\omega_0^2 - \delta^2}} \right].$$

(c) Assume now that $t > 0$ and ω_1 are fixed. Show that

$$\lim_{\delta \to 0^+} y(t) = \frac{F}{\omega_1^2 - \omega_0^2} \left[\cos(\omega_0 t) - \cos(\omega_1 t) \right].$$

13. The Great Zacchini, daredevil extraordinaire, is a circus performer whose act consists of being "shot from a cannon" to a safety net some distance away. The "cannon" is a frictionless tube containing a large spring, as shown in the figure. The spring constant is $k = 150$ lb/ft, and the spring is precompressed 10 ft prior to launching the acrobat. Assume that the spring obeys Hooke's law and that Zacchini weighs 150 lb. Neglect the weight of the spring.

(a) Let $x(t)$ represent spring displacement along the tube axis, measured positive in the upward direction. Show that Newton's second law of motion leads to the differential equation $mx'' = -kx - mg\cos(\pi/4), x < 0$, where m is the mass of the daredevil. Specify appropriate initial conditions.

(b) With what speed does he emerge from the tube when the spring is released?

(c) If the safety net is to be placed at the same height as the mouth of the "cannon," how far downrange from the cannon's mouth should the center of the net be placed?

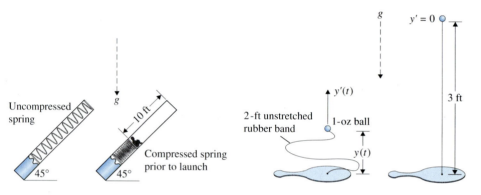

Figure for Exercise 13   Figure for Exercise 14

14. **A Change of Independent Variable (see Section 2.9)**   A popular child's toy consists of a small rubber ball attached to a wooden paddle by a rubber band; see the figure. Assume a 1-oz ball is connected to the paddle by a rubber band having an unstretched length of 2 ft. When the ball is launched vertically upward by the paddle with an initial speed of 30 ft/sec, the rubber band is observed to stretch 1 ft (to a total length of 3 ft) when the ball has reached its highest point. Assume the rubber band behaves like a spring and obeys Hooke's law for this amount of stretching. Our objective is to determine the spring constant k. (Neglect the weight of the rubber band.)

The motion occurs in two phases. Until the ball has risen 2 ft above the paddle, it acts like a projectile influenced only by gravity. Once the height of the ball exceeds 2 ft, the ball acts like a mass on a spring, acted upon by the elastic restoring force of the rubber band and gravity.

(a) Assume the ball leaves the paddle at time $t = 0$. Let t_2 and t_3 represent the times at which the height of the ball is 2 ft and 3 ft, respectively, and let m denote the mass of the rubber ball. Show that an application of Newton's second law of motion leads to the following two-part description of the problem:

(i) $my'' = -mg$, $0 < t < t_2$, $y(0) = 0$, $y'(0) = 30$

(ii) $my'' = -k(y - 2) - mg$, $t_2 < t < t_3$.

Here, y and y' are assumed to be continuous at $t = t_2$. We also know that $y(t_2) = 2$, $y(t_3) = 3$, and $y'(t_3) = 0$.

If we attempt to solve the problem "directly" with time as the independent variable, it is relatively difficult, since the times t_2 and t_3 must be determined as part of the problem. Since height $y(t)$ is an increasing function of time t over the interval of interest, however, we can view time t as a function of height, y, and use y as the independent variable.

(b) Let $v = y' = dy/dt$. If we adopt y as the independent variable, the acceleration becomes $y'' = dv/dt = (dv/dy)(dy/dt) = v(dv/dy)$. Therefore,

(i) $mv \dfrac{dv}{dy} = -mg, \quad 0 < y < 2, \quad v(0) = 30$

(ii) $mv \dfrac{dv}{dy} = -k(y-2) - mg, \quad 2 < y < 3.$

Here, v is continuous at $y = 2$ and $v|_{y=3} = 0$.

Solve these two separable differential equations, impose the accompanying supplementary conditions, and determine the spring constant k.

Network Problems   Use the following consistent set of scaled units (referred to as the Scaled SI Unit System) in Exercises 15–18.

Quantity	Unit	Symbol
Voltage	volt	V
Current	milliampere (mA)	I
Time	millisecond (ms)	t
Resistance	kilohm (kΩ)	R
Inductance	henry (H)	L
Capacitance	microfarad (μF)	C

Exercises 15–16:

Consider the series LC network shown in the figure. Assume that at time $t = 0$, the current and its time rate of change are both zero. For the given source voltage, determine the current $I(t)$.

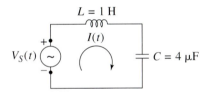

Figure for Exercises 15–16

15. $V_S(t) = 5 \sin 3t$ volts   **16.** $V_S(t) = 10te^{-t}$ volts

Exercises 17–18:

Consider the parallel RLC network shown in the figure. Assume that at time $t = 0$, the voltage $V(t)$ and its time rate of change are both zero. For the given source current, determine the voltage $V(t)$.

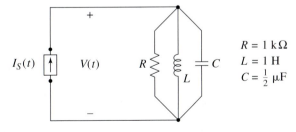

Figure for Exercises 17–18

17. $I_S(t) = 1 - e^{-t}$ mA   **18.** $I_S(t) = 5 \sin t$ mA

3.11 Higher Order Linear Homogeneous Differential Equations

So far in this chapter, we have studied second order linear differential equations

$$y'' + p_1(t)y' + p_0(t)y = g(t),$$

with emphasis on the important special case of constant coefficient equations. We now consider higher order linear equations. An nth order linear differential equation has the form

$$y^{(n)} + p_{n-1}(t)y^{(n-1)} + \cdots + p_2(t)y'' + p_1(t)y' + p_0(t)y = g(t). \tag{1}$$

Our study of equation (1) and the associated initial value problem follows a familiar pattern. We first generalize the existence and uniqueness results of Section 3.1, then examine the homogeneous case of equation (1), and finally discuss nonhomogeneous equations. The basic theory and solution techniques for second order linear equations extend naturally to the higher order case.

Our motivation for studying higher order equations is twofold. The theory of higher order linear differential equations is important in certain applications, and the way it generalizes second order linear theory is aesthetically appealing. Fourth order linear equations arise, for example, in modeling the loading and bending of beams (see the Projects at the end of this chapter).

Existence and Uniqueness

Theorem 3.5 generalizes the existence and uniqueness results presented earlier for initial value problems involving linear differential equations; see Theorem 2.1 (first order problems) and Theorem 3.1 (second order problems).

Theorem 3.5

Let $p_0(t), p_1(t), \ldots, p_{n-1}(t)$ and $g(t)$ be continuous functions defined on the interval $a < t < b$, and let t_0 be in (a, b). Then the initial value problem

$$y^{(n)} + p_{n-1}(t)y^{(n-1)} + \cdots + p_2(t)y'' + p_1(t)y' + p_0(t)y = g(t),$$

$$y(t_0) = y_0, \qquad y'(t_0) = y_0', \qquad y''(t_0) = y_0'', \qquad \ldots, \qquad y^{(n-1)}(t_0) = y_0^{(n-1)}$$

has a unique solution defined on the entire interval (a, b).

Comparing Theorems 2.1, 3.1, and 3.5, we see that the language and conclusions of the three theorems are virtually identical. In fact, we shall see in Chapter 4 that Theorems 2.1, 3.1, and 3.5 can all be viewed as special cases of an overarching existence-uniqueness theorem for systems of first order linear equations (see Theorem 4.1 in Section 4.2).

The Principle of Superposition and Fundamental Sets of Solutions

Consider the nth order linear homogeneous equation

$$y^{(n)} + p_{n-1}(t)y^{(n-1)} + \cdots + p_2(t)y'' + p_1(t)y' + p_0(t)y = 0, \qquad a < t < b. \tag{2}$$

The principle of superposition, stated in Theorem 3.2 for second order linear equations, applies to higher order linear equations as well. In particular, if $y_1(t), y_2(t), \ldots, y_r(t)$ are solutions of equation (2), then the linear combination

$$y(t) = c_1 y_1(t) + c_2 y_2(t) + \cdots + c_r y_r(t)$$

is also a solution of equation (2).

The idea of a fundamental set of solutions for a second order linear homogeneous differential equation also extends to nth order linear homogeneous equations. Let $\{y_1(t), y_2(t), \ldots, y_n(t)\}$ be a set of n solutions of the homogeneous differential equation (2). This set is a **fundamental set of solutions** if every solution of (2) can be represented as a linear combination of the form

$$y(t) = c_1 y_1(t) + c_2 y_2(t) + \cdots + c_n y_n(t), \qquad a < t < b. \tag{3}$$

Constructing Fundamental Sets

Consider the initial value problem

$$y^{(n)} + p_{n-1}(t)y^{(n-1)} + \cdots + p_2(t)y'' + p_1(t)y' + p_0(t)y = 0,$$
$$y(t_0) = y_0, \qquad y'(t_0) = y_0', \qquad y''(t_0) = y_0'', \qquad \ldots, \qquad y^{(n-1)}(t_0) = y_0^{(n-1)}. \tag{4}$$

Note that every solution of homogeneous equation (2) can be viewed as the unique solution of some initial value problem represented by (4). Simply fix a point t_0 in (a, b), and use the values of the function and its first $n - 1$ derivatives at t_0 as initial conditions.

Let $y_0, y_0', \ldots, y_0^{(n-1)}$ be an arbitrary set of n constants, and let $u(t)$ be the corresponding unique solution of initial value problem (4). Assume that $\{y_1, y_2, \ldots, y_n\}$ is a fundamental set of solutions of (2). Since $\{y_1, y_2, \ldots, y_n\}$ is a fundamental set of solutions, there are constants $c_1, c_2, \ldots, c_n$ such that $u(t) = c_1 y_1(t) + c_2 y_2(t) + \cdots + c_n y_n(t), a < t < b$. The initial conditions in (4) lead to a system of equations that can be written in matrix form as

$$\begin{bmatrix} y_1(t_0) & y_2(t_0) & \cdots & y_n(t_0) \\ y_1'(t_0) & y_2'(t_0) & & y_n'(t_0) \\ \vdots & & & \vdots \\ y_1^{(n-1)}(t_0) & y_2^{(n-1)}(t_0) & \cdots & y_n^{(n-1)}(t_0) \end{bmatrix} \begin{bmatrix} c_1 \\ c_2 \\ \vdots \\ c_n \end{bmatrix} = \begin{bmatrix} y_0 \\ y_0' \\ \vdots \\ y_0^{(n-1)} \end{bmatrix}. \tag{5}$$

By Theorem 3.5, initial value problem (4) has a unique solution for any choice of the initial conditions. Therefore, matrix equation (5) has a solution for any choice of $y_0, y_0', \ldots, y_0^{(n-1)}$, and this means that the $(n \times n)$ coefficient matrix has a nonzero determinant. [From linear algebra, if an $(n \times n)$ matrix equation $A\mathbf{x} = \mathbf{b}$ is consistent for every right-hand side $\mathbf{b}$, then the matrix A is invertible. Equivalently, the determinant of A is nonzero.]

As in Section 3.2, the determinant of the coefficient matrix in equation (5) is called the **Wronskian** and is denoted as $W(t)$:

$$W(t) = \begin{vmatrix} y_1(t) & y_2(t) & \cdots & y_n(t) \\ y_1'(t) & y_2'(t) & & y_n'(t) \\ \vdots & & & \vdots \\ y_1^{(n-1)}(t) & y_2^{(n-1)}(t) & \cdots & y_n^{(n-1)}(t) \end{vmatrix}. \tag{6}$$

We have just seen that if $\{y_1(t), y_2(t), \ldots, y_n(t)\}$ is a fundamental set of solutions for (2), then the corresponding Wronskian (6) is nonzero for all $t, a < t < b$. Conversely, if $\{y_1(t), y_2(t), \ldots, y_n(t)\}$ is a set of n solutions of (2) and if the corresponding Wronskian is nonzero on (a, b), then $\{y_1(t), y_2(t), \ldots, y_n(t)\}$ is a fundamental set of solutions. To prove this, let $u(t)$ be any solution of (2), and let $y_0 = u(t_0), y_0' = u'(t_0), \ldots, y_0^{(n-1)} = u^{(n-1)}(t_0)$. Solve equation (5) for $c_1, c_2, \ldots, c_n$, and define $\hat{y}(t) = c_1 y_1(t) + c_2 y_2(t) + \cdots + c_n y_n(t)$. Note that $u(t)$ and $\hat{y}(t)$ are both solutions of initial value problem (4). Therefore, by Theorem 3.5, $u(t) = \hat{y}(t) = c_1 y_1(t) + c_2 y_2(t) + \cdots + c_n y_n(t), a < t < b$.

E X A M P L E

1

Consider the fourth order differential equation

$$\frac{d^4 y}{dt^4} - y = 0, \qquad -\infty < t < \infty.$$

It can be verified that the functions $y_1(t) = e^t$, $y_2(t) = e^{-t}$, $y_3(t) = \cos t$, and $y_4(t) = \sin t$ are solutions of this equation.

(a) Show that these four solutions form a fundamental set of solutions.
(b) Represent the solution $\bar{y}(t) = \sinh t + \sin(t + \pi/3)$ in terms of this fundamental set.

Solution:

(a) Computing the Wronskian, we obtain

$$W(t) = \begin{vmatrix} e^t & e^{-t} & \cos t & \sin t \\ e^t & -e^{-t} & -\sin t & \cos t \\ e^t & e^{-t} & -\cos t & -\sin t \\ e^t & -e^{-t} & \sin t & -\cos t \end{vmatrix} = -8(\cos^2 t + \sin^2 t) = -8.$$

Since the Wronskian is nonzero on $(-\infty, \infty)$, it follows that $\{y_1, y_2, y_3, y_4\}$ is a fundamental set of solutions.

(b) We use the fact that $\sinh t = (e^t - e^{-t})/2$ and the trigonometric identity $\sin(A + B) = \sin A \cos B + \sin B \cos A$ to obtain

$$\bar{y}(t) = \sinh t + \sin\left(t + \frac{\pi}{3}\right) = \frac{1}{2}e^t - \frac{1}{2}e^{-t} + \frac{1}{2}\sin t + \frac{\sqrt{3}}{2}\cos t. \quad ❖$$

Abel's Theorem

Let $\{y_1(t), y_2(t), \ldots, y_n(t)\}$ be a set of solutions of the linear homogeneous equation (2). We have seen that $\{y_1(t), y_2(t), \ldots, y_n(t)\}$ is a fundamental set of solutions if and only if the corresponding Wronskian, $W(t)$, is nonzero on (a, b). Abel's theorem shows that the Wronskian is either zero throughout (a, b) or is never zero in (a, b). This fact allows us to choose a single convenient test point, t_0, and use the value $W(t_0)$ to decide whether $\{y_1(t), y_2(t), \ldots, y_n(t)\}$ is a fundamental set of solutions.

We state Abel's theorem for the general nth order case, but prove it only for $n = 2$.

Theorem 3.6

Let $y_1(t), y_2(t), \ldots, y_n(t)$ denote n solutions of the differential equation

$$y^{(n)} + p_{n-1}(t)y^{(n-1)} + \cdots + p_1(t)y' + p_0(t)y = 0, \qquad a < t < b,$$

where $p_0(t), p_1(t), \ldots, p_{n-1}(t)$ are continuous on (a, b). Let $W(t)$ be the Wronskian of $y_1(t), y_2(t), \ldots, y_n(t)$. Then the function $W(t)$ is a solution of the first order linear differential equation

$$W' = -p_{n-1}(t)W.$$

Therefore, if t_0 is any point in the interval (a, b), it follows that

$$W(t) = W(t_0)e^{-\int_{t_0}^{t} p_{n-1}(s)\,ds}, \qquad a < t < b. \tag{7}$$

● **PROOF:**   We prove Theorem 3.6 for the case $n = 2$. Let y_1 and y_2 be solutions of the second order linear equation

$$y'' + p_1(t)y' + p_0(t)y = 0.$$

The Wronskian is $W = y_1 y_2' - y_2 y_1'$. Differentiating and simplifying, we obtain

$$\begin{aligned}
W' &= y_1 y_2'' - y_2 y_1'' \\
&= y_1[-p_1(t)y_2' - p_0(t)y_2] - y_2[-p_1(t)y_1' - p_0(t)y_1] \\
&= -p_1(t)[y_1 y_2' - y_2 y_1'] \\
&= -p_1(t)W.
\end{aligned}$$

Solving the resulting first order linear equation $W' = -p_1(t)W$ leads to (7). ●

The proof of Theorem 3.6 for general n can be found in most advanced texts on differential equations. The basic argument is similar to that used for the case $n = 2$. The computations are more involved, however, since one needs to compute the derivative of an $(n \times n)$ determinant of functions.

By equation (7), if $W(t_0) \neq 0$, then $W(t) \neq 0$ for all t in (a, b). On the other hand, if $W(t_0) = 0$ then $W(t)$ is also zero for all t in (a, b). The point t_0 is arbitrary; Abel's theorem implies that the Wronskian of a set of solutions of (2) either is zero throughout (a, b) or is never zero in (a, b).

Additional Observations

We conclude by making some additional observations about fundamental sets of solutions of the homogeneous equation (2).

Fundamental Sets Always Exist

When differential equation (2) has constant coefficients, we can explicitly construct fundamental sets of solutions. For the general case of variable coefficients, however, we are usually unable to explicitly construct solutions of (2). In such cases, it is logical to ask whether fundamental sets of solutions do, in fact, exist. The following theorem provides an affirmative answer.

Theorem 3.7

Consider the nth order linear homogeneous differential equation

$$y^{(n)} + p_{n-1}(t)y^{(n-1)} + \cdots + p_1(t)y' + p_0(t)y = 0, \qquad a < t < b,$$

where $p_0(t), p_1(t), \ldots, p_{n-1}(t)$ are continuous on (a, b). A fundamental set of solutions exists for this equation.

● **PROOF:**   We prove Theorem 3.7 for the case $n = 2$. Let t_0 be any point in (a, b), and let y_1 and y_2 be solutions of the initial value problems

$$y_1'' + p_1(t)y_1' + p_0(t)y_1 = 0, \qquad y_1(t_0) = 1, \qquad y_1'(t_0) = 0$$
$$y_2'' + p_1(t)y_2' + p_0(t)y_2 = 0, \qquad y_2(t_0) = 0, \qquad y_2'(t_0) = 1.$$

Existence-uniqueness Theorem 3.5 assures us that each of these initial value problems has a unique solution on (a, b). The fact that $\{y_1, y_2\}$ forms a fundamental set of solutions follows immediately from the observation that

$$W(t_0) = \begin{vmatrix} y_1(t_0) & y_2(t_0) \\ y_1'(t_0) & y_2'(t_0) \end{vmatrix} = \begin{vmatrix} 1 & 0 \\ 0 & 1 \end{vmatrix} = 1. \ ●$$

Fundamental Sets of Solutions Are Linearly Independent

A set of functions defined on a common domain, say $f_1(t), f_2(t), \ldots, f_r(t)$ defined on the interval $a < t < b$, is called a **linearly dependent set** if there exist constants $k_1, k_2, \ldots, k_r$, not all zero, such that

$$k_1 f_1(t) + k_2 f_2(t) + \cdots + k_r f_r(t) = 0, \qquad a < t < b. \tag{8}$$

A set of functions that is not linearly dependent is called **linearly independent**. Thus, a set of functions is linearly independent if equation (8) implies $k_1 = k_2 = \cdots = k_r = 0$. If a set of functions is linearly dependent, then at least one of the functions can be expressed as a linear combination of the others. For example, if $k_1 \neq 0$ in (8), then

$$f_1(t) = -(k_2/k_1)f_2(t) - (k_3/k_1)f_3(t) - \cdots - (k_r/k_1)f_r(t), \qquad a < t < b.$$

Loosely speaking, the functions comprising a linearly independent set are all "basically different," while those forming a linearly dependent set are not. The following theorem, whose proof is left to the exercises, shows that a fundamental set of solutions is a linearly independent set.

Theorem 3.8

Consider the nth order linear homogeneous differential equation

$$y^{(n)} + p_{n-1}(t)y^{(n-1)} + \cdots + p_1(t)y' + p_0(t)y = 0, \qquad a < t < b,$$

where $p_0(t), p_1(t), \ldots, p_{n-1}(t)$ are continuous on (a, b). A fundamental set of solutions for this equation is necessarily a linearly independent set of functions.

How Fundamental Sets Are Related

Fundamental sets of solutions for equation (2) always exist. In fact, this linear homogeneous equation has infinitely many fundamental sets. These fundamental sets must be related to each other, since any given fundamental set can be used to construct all solutions of (2). The following theorem shows how these fundamental sets are related.

Theorem 3.9

Let $\{y_1(t), y_2(t), \ldots, y_n(t)\}$ be a fundamental set of solutions of the nth order linear homogeneous differential equation

$$y^{(n)} + p_{n-1}(t)y^{(n-1)} + \cdots + p_1(t)y' + p_0(t)y = 0, \qquad a < t < b,$$

where $p_0(t), p_1(t), \ldots, p_{n-1}(t)$ are continuous on (a, b). Let $\{\bar{y}_1(t), \bar{y}_2(t), \ldots, \bar{y}_n(t)\}$ be a set of solutions of the differential equation. Then there exists an $(n \times n)$ matrix A such that

$$[\,\bar{y}_1(t), \bar{y}_2(t), \ldots, \bar{y}_n(t)\,] = [y_1(t), y_2(t), \ldots, y_n(t)]A.$$

Moreover, $\{\bar{y}_1(t), \bar{y}_2(t), \ldots, \bar{y}_n(t)\}$ is a fundamental set of solutions if and only if the determinant of A is nonzero.

● **PROOF:**   We prove Theorem 3.9 for the case $n = 2$. Since $\{y_1, y_2\}$ is a fundamental set, we can express $\bar{y}_1$ and $\bar{y}_2$ as linear combinations of y_1 and y_2:

$$\begin{aligned} \bar{y}_1 &= a_{11}y_1 + a_{21}y_2 \\ \bar{y}_2 &= a_{12}y_1 + a_{22}y_2 \end{aligned} \quad \text{or} \quad [\bar{y}_1, \bar{y}_2] = [y_1, y_2]\begin{bmatrix} a_{11} & a_{12} \\ a_{21} & a_{22} \end{bmatrix} = [y_1, y_2]A.$$

Since the equation $[\bar{y}_1, \bar{y}_2] = [y_1, y_2]A$ holds for all t in (a, b), we can differentiate it and obtain the matrix equation

$$\begin{bmatrix} \bar{y}_1 & \bar{y}_2 \\ \bar{y}_1' & \bar{y}_2' \end{bmatrix} = \begin{bmatrix} y_1 & y_2 \\ y_1' & y_2' \end{bmatrix} A.$$

Using the fact that the determinant of the product of two matrices is the product of their determinants, we have

$$\overline{W}(t) = W(t) \det(A),$$

where $\overline{W}(t)$ and $W(t)$ denote the Wronskians of $\{\bar{y}_1, \bar{y}_2\}$ and $\{y_1, y_2\}$, respectively. Since $W(t) \neq 0$ on (a, b), it follows that $\overline{W}(t) \neq 0$ if and only if $\det(A) \neq 0$. ●

EXERCISES

Exercises 1–6:

In each exercise,

(a) Verify that the given functions form a fundamental set of solutions.

(b) Solve the initial value problem.

1. $y''' = 0$; $y(1) = 4$, $y'(1) = 2$, $y''(1) = 0$

$y_1(t) = 2$, $y_2(t) = t - 1$, $y_3(t) = t^2 - 1$

2. $y''' - y' = 0$; $y(0) = 4$, $y'(0) = 1$, $y''(0) = 3$

   $y_1(t) = 1$, $y_2(t) = e^t$, $y_3(t) = e^{-t}$

3. $y^{(4)} + 4y'' = 0$; $y(0) = 0$, $y'(0) = -1$, $y''(0) = -4$, $y'''(0) = 8$

   $y_1(t) = 1$, $y_2(t) = t$, $y_3(t) = \cos 2t$, $y_4(t) = \sin 2t$

4. $y''' + 2y'' = 0$; $y(0) = 0$, $y'(0) = 3$, $y''(0) = -8$

   $y_1(t) = 1$, $y_2(t) = t$, $y_3(t) = e^{-2t}$

5. $ty''' + 3y'' = 0$, $t > 0$; $y(2) = \frac{1}{2}$, $y'(2) = -\frac{5}{4}$, $y''(2) = \frac{1}{4}$

   $y_1(t) = 1$, $y_2(t) = t$, $y_3(t) = t^{-1}$

6. $t^2 y''' + ty'' - y' = 0$, $t < 0$; $y(-1) = 1$, $y'(-1) = -1$, $y''(-1) = -1$

   $y_1(t) = 1$, $y_2(t) = \ln(-t)$, $y_3(t) = t^2$

Exercises 7–10:

Consider the given differential equation on the interval $-\infty < t < \infty$. Assume that the members of a solution set satisfy the initial conditions. Do the solutions form a fundamental set?

7. $y'' + 2ty' + t^2 y = 0$, $y_1(1) = 2$, $y_1'(1) = -1$, $y_2(1) = -4$, $y_2'(1) = 2$

8. $y'' + ty = 0$, $y_1(0) = 0$, $y_1'(0) = 2$, $y_2(0) = -1$, $y_2'(0) = 0$

9. $y''' + (\sin t)y = 0$, $y_1(0) = 1$, $y_1'(0) = -1$, $y_1''(0) = 0$, $y_2(0) = 0$, $y_2'(0) = 0$,

   $y_2''(0) = 2$, $y_3(0) = 2$, $y_3'(0) = -2$, $y_3''(0) = 1$

10. $y''' + e^t y'' + y = 0$, $y_1(1) = 0$, $y_1'(1) = 1$, $y_1''(1) = 1$, $y_2(1) = 1$, $y_2'(1) = -1$,

   $y_2''(1) = 0$, $y_3(1) = -1$, $y_3'(1) = 0$, $y_3''(1) = 0$

Exercises 11–15:

The given differential equation has a fundamental set of solutions whose Wronskian $W(t)$ is such that $W(0) = 1$. What is $W(4)$?

11. $y'' + \dfrac{t}{2}y' + y = 0$ **12.** $y''' + \dfrac{t}{2}y' + y = 0$ **13.** $y''' + y'' + ty = 0$

14. $y^{(4)} - y'' + y = 0$ **15.** $(t^2 + 1)y''' - 2ty'' + y = 0$

Exercises 16–19:

Find a fundamental set $\{\bar{y}_1, \bar{y}_2\}$ satisfying the given initial conditions.

16. $y'' - y = 0$, $\bar{y}_1(0) = 1$, $\bar{y}_1'(0) = 0$, $\bar{y}_2(0) = 0$, $\bar{y}_2'(0) = 1$

17. $y'' + y = 0$, $\bar{y}_1(0) = 1$, $\bar{y}_1'(0) = 1$, $\bar{y}_2(0) = 1$, $\bar{y}_2'(0) = -1$

18. $y'' + 4y' + 5y = 0$, $\bar{y}_1(0) = 1$, $\bar{y}_1'(0) = -1$, $\bar{y}_2(0) = 0$, $\bar{y}_2'(0) = 1$

19. $y'' + 4y' + 4y = 0$, $\bar{y}_1(0) = -1$, $\bar{y}_1'(0) = 2$, $\bar{y}_2(0) = 1$, $\bar{y}_2'(0) = -1$

Exercises 20–22:

In each exercise, $\{y_1, y_2, y_3\}$ is a fundamental set of solutions and $\{\bar{y}_1, \bar{y}_2, \bar{y}_3\}$ is a set of solutions.

(a) Find a (3×3) constant matrix A such that $[\bar{y}_1(t), \bar{y}_2(t), \bar{y}_3(t)] = [y_1(t), y_2(t), y_3(t)]A$.

(b) Determine whether $\{\bar{y}_1, \bar{y}_2, \bar{y}_3\}$ is also a fundamental set by calculating $\det(A)$.

20. $y''' - y' = 0$, $\{y_1(t), y_2(t), y_3(t)\} = \{1, e^t, e^{-t}\}$,

   $\{\bar{y}_1(t), \bar{y}_2(t), \bar{y}_3(t)\} = \{\cosh t, 1 - \sinh t, 2 + \sinh t\}$

21. $y''' - y'' = 0$, $\{y_1(t), y_2(t), y_3(t)\} = \{1, t, e^{-t}\}$,

   $\{\bar{y}_1(t), \bar{y}_2(t), \bar{y}_3(t)\} = \{1 - 2t, t + 2, e^{-(t+2)}\}$

22. $t^2y''' + ty'' - y' = 0, t > 0,$ $\{y_1(t), y_2(t), y_3(t)\} = \{1, \ln t, t^2\},$
$$\{\bar{y}_1(t), \bar{y}_2(t), \bar{y}_3(t)\} = \{2t^2 - 1, 3, \ln t^3\}$$

Exercises 23–26:

The Wronskian formed from a solution set of the given differential equation has the specified value at $t = 1$. Determine $W(t)$.

23. $y''' - 3y'' + 3y' - y = 0;$ $W(1) = 1$

24. $y''' + (\sin t)y'' + (\cos t)y' + 2y = 0;$ $W(1) = 0$

25. $t^3y''' + t^2y'' - 2y = 0,$ $t > 0;$ $W(1) = 3$

26. $t^3y''' - 2y = 0,$ $t > 0;$ $W(1) = 3$

27. Linear Independence   For definiteness, consider Theorem 3.8 in the case $n = 3$. Let $\{y_1, y_2, y_3\}$ be a fundamental set of solutions for $y''' + p_2(t)y'' + p_1(t)y' + p_0(t)y = 0$, where p_0, p_1, p_2 are continuous on $a < t < b$. Show that the fundamental set is linearly independent. [Hint: Consider the equation $k_1y_1(t) + k_2y_2(t) + k_3y_3(t) = 0$ along with the first and second derivatives of this equation, evaluated at some point t_0 in $a < t < b$.]

3.12   Higher Order Homogeneous Constant Coefficient Differential Equations

Consider the nth order linear homogeneous differential equation

$$y^{(n)} + a_{n-1}y^{(n-1)} + \cdots + a_1y' + a_0y = 0, \qquad -\infty < t < \infty, \tag{1}$$

where $a_0, a_1, \ldots, a_{n-1}$ are real constants. The general solution is given by

$$y(t) = c_1y_1(t) + c_2y_2(t) + \cdots + c_ny_n(t),$$

where $\{y_1(t), y_2(t), \ldots, y_n(t)\}$ is a fundamental set of solutions. How do we determine a fundamental set of solutions for equation (1)?

As in Section 3.3, we look for solutions of the form $y(t) = e^{\lambda t}$, where λ is a constant to be determined. Substituting this form into equation (1) leads to

$$(\lambda^n + a_{n-1}\lambda^{n-1} + \cdots + a_1\lambda + a_0)e^{\lambda t} = 0, \qquad -\infty < t < \infty.$$

The exponential function $e^{\lambda t}$ is nonzero for all values of λ (whether real or complex). Therefore, in order for $y(t) = e^{\lambda t}$ to be a solution, we need

$$\lambda^n + a_{n-1}\lambda^{n-1} + \cdots + a_1\lambda + a_0 = 0. \tag{2}$$

The nth degree polynomial in equation (2) is called the **characteristic polynomial**, while equation (2) itself is called the **characteristic equation**. The roots of the characteristic equation define solutions of (1) having the form $y(t) = e^{\lambda t}$.

Roots of the Characteristic Equation

When we considered the case $n = 2$, we were able to list three possibilities for the roots of the characteristic polynomial (two distinct real roots, one repeated real root, or two distinct complex roots). While we cannot list all the possibilities for the general case, we can make some useful observations.

Complex Roots Occur in Complex Conjugate Pairs

Since the coefficients $a_0, a_1, \ldots, a_{n-1}$ are real constants, the complex-valued roots of the characteristic equation always occur in complex conjugate pairs. One simple consequence of this observation is the fact that every characteristic polynomial of odd degree has at least one real root.

We noted in Section 3.5 that if $\lambda = \alpha + i\beta$ and $\bar{\lambda} = \alpha - i\beta$ are a complex conjugate pair of roots of the characteristic equation, then real-valued solutions corresponding to these two roots are

$$y_1(t) = e^{\alpha t} \cos \beta t \quad \text{and} \quad y_2(t) = e^{\alpha t} \sin \beta t. \tag{3}$$

This result is true for the general nth order linear homogeneous equation as well.

E X A M P L E

1

Find the general solution of the third order differential equation

$$y''' + 4y' = 0, \qquad -\infty < t < \infty.$$

Solution: Looking for solutions of the form $e^{\lambda t}$ leads to the characteristic equation $\lambda^3 + 4\lambda = 0$, or

$$\lambda(\lambda^2 + 4) = \lambda(\lambda + 2i)(\lambda - 2i) = 0.$$

The three roots are therefore $\lambda_1 = 0, \lambda_2 = 2i$, and $\lambda_3 = \bar{\lambda}_2 = -2i$. The corresponding real-valued solutions are

$$y_1(t) = 1, \qquad y_2(t) = \cos 2t, \qquad y_3(t) = \sin 2t.$$

To show these three solutions constitute a fundamental set, we calculate the Wronskian and find

$$W(t) = \begin{vmatrix} 1 & \cos 2t & \sin 2t \\ 0 & -2\sin 2t & 2\cos 2t \\ 0 & -4\cos 2t & -4\sin 2t \end{vmatrix} = 8(\sin^2 2t + \cos^2 2t) = 8.$$

Since the Wronskian is nonzero, the three solutions form a fundamental set of solutions. The general solution of the differential equation is

$$y(t) = c_1 + c_2 \cos 2t + c_3 \sin 2t, \qquad -\infty < t < \infty.$$

In this example, we computed the determinant $W(t)$. But, recalling Abel's theorem, we could have simply evaluated $W(t)$ at a convenient point, say $t = 0$. Note, for this equation, that Abel's theorem would also have anticipated that $W(t)$ is a constant function. ❖

If the Characteristic Polynomial Has Distinct Roots, the Corresponding Solutions Form a Fundamental Set

In Example 1, the characteristic equation had three distinct roots: the real number $\lambda_1 = 0$ and the complex conjugate pair $\lambda_2 = 2i$ and $\lambda_3 = -2i$. The corresponding set of solutions formed a fundamental set of solutions. In the exercises, we ask you to show that this result holds in general. That is, if the characteristic equation has n distinct roots $\lambda_1, \lambda_2, \ldots, \lambda_n$, then the set of solutions $\{e^{\lambda_1 t}, e^{\lambda_2 t}, \ldots, e^{\lambda_n t}\}$ forms a fundamental set of solutions. (If some of these distinct roots are complex-valued, we can replace the conjugate pair of complex exponentials with the corresponding real-valued pair of solutions and the conclusion remains the same.)

Roots of the Characteristic Equation May Have Multiplicity Greater than 2 and May Be Complex

For the second order linear homogeneous equations discussed earlier, a repeated root of the corresponding quadratic characteristic equation is, of necessity, real-valued. If λ_1 is a repeated root of the quadratic characteristic equation, then the corresponding solutions

$$y_1(t) = e^{\lambda_1 t} \quad \text{and} \quad y_2(t) = te^{\lambda_1 t}$$

form a fundamental set of solutions.

For higher order differential equations, the situation is more complicated. For example,

(a) *A root can be repeated more than once.*   In particular, if the characteristic polynomial has the form

$$p(\lambda) = \lambda^n + a_{n-1}\lambda^{n-1} + \cdots + a_1\lambda + a_0 = (\lambda - \lambda_1)^r q(\lambda), \tag{4}$$

where q is a polynomial of degree $n - r$ and $q(\lambda_1) \neq 0$, then we say λ_1 is a **root of multiplicity r**. If λ_1 is a root of multiplicity r, then the functions

$$e^{\lambda_1 t}, \qquad te^{\lambda_1 t}, \qquad t^2 e^{\lambda_1 t}, \qquad \ldots, \qquad t^{r-1}e^{\lambda_1 t} \tag{5}$$

form a set of r solutions of the differential equations. The remaining $n - r$ solutions needed to form a fundamental set of solutions can be determined by examining the roots of $q(\lambda) = 0$.

(b) *A repeated root might be complex.*   We recall, however, that complex roots arise in complex conjugate pairs. Therefore, if $\lambda_1 = \alpha + i\beta$ is a root of multiplicity r, then $\overline{\lambda}_1 = \alpha - i\beta$ is also a root of multiplicity r. In such cases, the characteristic polynomial has the form

$$\lambda^n + a_{n-1}\lambda^{n-1} + \cdots + a_1\lambda + a_0 = (\lambda - \lambda_1)^r (\lambda - \overline{\lambda}_1)^r \hat{q}(\lambda),$$

where $\hat{q}(\lambda)$ is a polynomial of degree $n - 2r$ and where $\hat{q}(\lambda_1) \neq 0$ and $\hat{q}(\overline{\lambda}_1) \neq 0$. In this case, the functions

$$\begin{array}{lllll} e^{\alpha t}\cos\beta t, & te^{\alpha t}\cos\beta t, & t^2 e^{\alpha t}\cos\beta t, & \ldots, & t^{r-1}e^{\alpha t}\cos\beta t \\ e^{\alpha t}\sin\beta t, & te^{\alpha t}\sin\beta t, & t^2 e^{\alpha t}\sin\beta t, & \ldots, & t^{r-1}e^{\alpha t}\sin\beta t \end{array} \tag{6}$$

form a set of $2r$ real-valued solutions.

E X A M P L E

2

Find the general solution of

(a) $y^{(6)} + 3y^{(5)} + 3y^{(4)} + y''' = 0$ (b) $y^{(5)} - y^{(4)} + 2y''' - 2y'' + y' - y = 0$

Solution:

(a) The characteristic polynomial is

$$p(\lambda) = \lambda^6 + 3\lambda^5 + 3\lambda^4 + \lambda^3 = \lambda^3(\lambda^3 + 3\lambda^2 + 3\lambda + 1) = \lambda^3(\lambda + 1)^3.$$

(continued)

(continued)

Therefore, $\lambda = 0$ and $\lambda = -1$ are each roots of multiplicity 3. By (5), the functions $\{1, t, t^2, e^{-t}, te^{-t}, t^2 e^{-t}\}$ are solutions. The Wronskian can be shown to be nonzero, and hence the solutions form a fundamental set; the general solution is

$$y(t) = c_1 + c_2 t + c_3 t^2 + c_4 e^{-t} + c_5 t e^{-t} + c_6 t^2 e^{-t}.$$

(b) The characteristic polynomial is

$$p(\lambda) = \lambda^5 - \lambda^4 + 2\lambda^3 - 2\lambda^2 + \lambda - 1 = (\lambda - 1)(\lambda^2 + 1)^2.$$

The characteristic equation has one real root, $\lambda = 1$, and a repeated pair of complex conjugate roots, $\pm i$. The set of solutions $\{e^t, \cos t, \sin t, t \cos t, t \sin t\}$ forms a fundamental set of solutions. The general solution is

$$y(t) = c_1 e^t + c_2 \cos t + c_3 \sin t + c_2 t \cos t + c_3 t \sin t. \quad \diamond$$

It can be shown that solutions (5) and (6), arising from repeated roots, will always form a fundamental set of solutions when they are combined with the solutions that arise from any remaining roots of the characteristic equation.

Solving the Differential Equation $y^{(n)} - ay = 0$

Let a be a real number. The characteristic equation for $y^{(n)} - ay = 0$ is $\lambda^n - a = 0$. Finding the roots of this equation amounts to finding the n different nth roots of the real number a. The first step is to write a in polar form as

$$a = Re^{i\alpha},$$

where $R = |a|$ and where $\alpha = 0$ when $a > 0$ and $\alpha = \pi$ when $a < 0$. Recall Euler's formula from Section 3.5,

$$e^{i\theta} = \cos\theta + i\sin\theta.$$

For any integer k, it follows that $e^{i2k\pi} = 1$. Therefore, we can write $a = Re^{i(\alpha + 2k\pi)}$, $k = 0, \pm 1, \pm 2, \ldots$, and hence

$$a^{1/n} = R^{1/n} e^{i(\alpha + 2k\pi)/n}, \qquad k = 0, \pm 1, \pm 2, \ldots. \tag{7}$$

In equation (7), $R^{1/n}$ is the positive real nth root of $R = |a|$. We generate the n distinct roots of $\lambda^n - a = 0$ by setting $k = 0, 1, \ldots, n-1$ in equation (7). (Other integer values of k simply replicate these values.) Once we determine the n roots, we can use Euler's formula to rewrite them in the form $x + iy$.

E X A M P L E

3

Find the general solution of

$$y^{(4)} + 16y = 0, \qquad -\infty < t < \infty.$$

Solution: The characteristic equation is $\lambda^4 + 16 = 0$. Therefore, in this example, $a = -16$ and $n = 4$. Using equation (7) with $R = 16$ and $\alpha = \pi$, we find the four roots are

$$\lambda_k = 2e^{i(\pi + 2k\pi)/4}, \qquad k = 0, 1, 2, 3,$$

or

$$\lambda_0 = 2e^{i\pi/4} = \sqrt{2} + i\sqrt{2}$$
$$\lambda_1 = 2e^{i3\pi/4} = -\sqrt{2} + i\sqrt{2}$$
$$\lambda_2 = 2e^{i5\pi/4} = -\sqrt{2} - i\sqrt{2}$$
$$\lambda_3 = 2e^{i7\pi/4} = \sqrt{2} - i\sqrt{2}.$$

The general solution, expressed in terms of real-valued solutions, is

$$y(t) = e^{\sqrt{2}t}\left(c_1 \cos\sqrt{2}\,t + c_2 \sin\sqrt{2}\,t\right) + e^{-\sqrt{2}t}\left(c_3 \cos\sqrt{2}\,t + c_4 \sin\sqrt{2}\,t\right). \;❖$$

When a is a real number, the complex roots of $\lambda^n - a = 0$ occur in complex conjugate pairs. If we plot these n roots in the complex plane, we see they all lie on a circle of radius $|a|^{1/n}$ and the angular separation between roots is $2\pi/n$. Figure 3.20 shows the four roots of $\lambda^4 + 16 = 0$ found in Example 3.

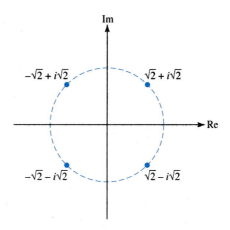

FIGURE 3.20

The four roots of $\lambda^4 + 16 = 0$ lie on a circle of radius $16^{1/4} = 2$ and have an angular separation of $2\pi/4 = \pi/2$ radians. The roots occur in two complex conjugate pairs.

EXERCISES

Exercises 1–18:

In each exercise,

(a) Find the general solution of the differential equation.

(b) If initial conditions are specified, solve the initial value problem.

1. $y''' - 4y' = 0$ **2.** $y''' + y'' - y' - y = 0$ **3.** $y''' + y'' + 4y' + 4y = 0$

4. $16y^{(4)} - 8y'' + y = 0$ **5.** $16y^{(4)} + 8y'' + y = 0$ **6.** $y'' - y = 0$

7. $y''' - 2y'' - y' + 2y = 0$ **8.** $y^{(4)} - y = 0$ **9.** $y''' + 8y = 0$

10. $2y''' - y'' = 0$ **11.** $y''' + y' = 0$ **12.** $y^{(4)} + 2y'' + y = 0$

13. $y^{(6)} - y = 0$ **14.** $y^{(4)} - y''' + y' - y = 0$

15. $y''' + 2y'' + y' = 0,\quad y(0) = 0,\; y'(0) = 0,\; y''(0) = 1$

16. $y''' + 4y' = 0$, $y(0) = 1$, $y'(0) = 6$, $y''(0) = 4$

17. $y''' + 3y'' + 3y' + y = 0$, $y(0) = 0$, $y'(0) = 1$, $y''(0) = 0$

18. $y^{(4)} - y''' = 0$, $y(0) = 0$, $y'(0) = 0$, $y''(0) = 0$, $y'''(0) = 1$

Exercises 19–20:

The given differential equations have repeated roots; α and β are real constants.

(a) Use equation (5) or (6), as appropriate, to determine a solution set.

(b) Show that the solution set found in part (a) is a fundamental set of solutions by evaluating the Wronskian at $t = 0$.

19. $y''' - 3\alpha y'' + 3\alpha^2 y' - \alpha^3 y = 0$ **20.** $y^{(4)} + 2\beta^2 y'' + \beta^4 y = 0$

Exercises 21–25:

In each exercise, you are given the general solution of

$$y^{(4)} + a_3 y''' + a_2 y'' + a_1 y' + a_0 y = 0,$$

where a_3, a_2, a_1, and a_0 are real constants. Use the general solution to determine the constants a_3, a_2, a_1, and a_0. [Hint: Construct the characteristic equation from the given general solution.]

21. $y(t) = c_1 + c_2 t + c_3 \cos 3t + c_4 \sin 3t$

22. $y(t) = c_1 \cos t + c_2 \sin t + c_3 \cos 2t + c_4 \sin 2t$

23. $y(t) = c_1 e^t + c_2 t e^t + c_3 e^{-t} + c_4 t e^{-t}$

24. $y(t) = c_1 e^{-t} \sin t + c_2 e^{-t} \cos t + c_3 e^t \sin t + c_4 e^t \cos t$

25. $y(t) = c_1 e^t + c_2 t e^t + c_3 t^2 e^t + c_4 t^3 e^t$

Exercises 26–30:

Consider the nth order homogeneous linear differential equation

$$y^{(n)} + a_{n-1} y^{(n-1)} + \cdots + a_3 y''' + a_2 y'' + a_1 y' + a_0 y = 0,$$

where $a_0, a_1, a_2, \ldots, a_{n-1}$ are real constants. In each exercise, several functions belonging to a fundamental set of solutions for this equation are given.

(a) What is the smallest value n for which the given functions can belong to such a fundamental set?

(b) What is the fundamental set?

26. $y_1(t) = t$, $y_2(t) = e^t$, $y_3(t) = \cos t$

27. $y_1(t) = e^t$, $y_2(t) = e^t \cos 2t$, $y_3(t) = e^{-t} \cos 2t$

28. $y_1(t) = t^2 \sin t$, $y_2(t) = e^t \sin t$

29. $y_1(t) = t \sin t$, $y_2(t) = t^2 e^t$

30. $y_1(t) = t^2$, $y_2(t) = e^{2t}$

Exercises 31–35:

Consider the nth order differential equation

$$y^{(n)} - ay = 0,$$

where a is a real number. In each exercise, some information is presented about the solutions of this equation. Use the given information to deduce both the order n ($n \geq 1$) of the differential equation and the value of the constant a. (If more than one answer is possible, determine the smallest possible order n and the corresponding value of a.)

31. $|a| = 1$ and $\lim_{t \to \infty} y(t) = 0$ for all solutions $y(t)$ of the equation.

32. $|a| = 2$ and all nonzero solutions of the differential equation are exponential functions.

33. $y(t) = t^3$ is a solution of the differential equation.

34. $|a| = 4$ and all solutions of the differential equation are bounded functions on the interval $-\infty < t < \infty$.

35. Two solutions are $y_1(t) = e^{-t}$ and $y_2(t) = e^{t/2}\sin(\sqrt{3}\,t/2)$.

36. Assume the characteristic equation of $y^{(n)} + a_{n-1}y^{(n-1)} + \cdots + a_1 y' + a_0 y = 0$ has distinct roots $\lambda_1, \lambda_2, \ldots, \lambda_n$. It can be shown that the Vandermonde[8] determinant has the value

$$
\begin{vmatrix}
1 & 1 & \cdots & 1 \\
\lambda_1 & \lambda_2 & & \lambda_n \\
\lambda_1^2 & \lambda_2^2 & & \lambda_n^2 \\
\vdots & & & \vdots \\
\lambda_1^{n-1} & \lambda_2^{n-1} & \cdots & \lambda_n^{n-1}
\end{vmatrix}
= \prod_{\substack{i,j=1 \\ i>j}}^{n} (\lambda_i - \lambda_j).
$$

Use this fact to show that $\{e^{\lambda_1 t}, e^{\lambda_2 t}, \ldots, e^{\lambda_n t}\}$ is a fundamental set of solutions.

3.13   Higher Order Linear Nonhomogeneous Differential Equations

We now consider the nth order linear nonhomogeneous differential equation

$$
y^{(n)} + p_{n-1}(t)y^{(n-1)} + \cdots + p_1(t)y' + p_0(t)y = g(t), \qquad a < t < b. \tag{1}
$$

The arguments made in Section 3.7, establishing the solution structure for second order equations, apply as well in the nth order case. The arguments rely on the fact that the differential equation is linear; they do not depend on the order of the equation. The solution structure of equation (1) can be represented schematically as

The general solution of the nonhomogeneous equation	=	The general solution of the homogeneous equation	+	A particular solution of the nonhomogeneous equation.

The general solution of the homogeneous equation is the **complementary solution**, denoted by $y_C(t)$. The one solution of the nonhomogeneous equation that we have somehow found is the **particular solution**, denoted by $y_P(t)$. The general solution of nonhomogeneous equation (1) has the form

$$
y(t) = y_C(t) + y_P(t).
$$

Finding a Particular Solution

In Sections 3.8 and 3.9, we discussed the method of undetermined coefficients and the method of variation of parameters for constructing particular solutions.

---

[8] Alexandre-Theophile Vandermonde (1735–1796) was a violinist who turned to mathematics when he was 35 years old. His four published mathematical papers made noteworthy contributions to the study of algebraic equations, topology, combinatorics, and determinants. Surprisingly, the determinant that now bears his name appears nowhere in his published works.

Both methods have straightforward generalizations to nth order equations. With respect to undetermined coefficients, Table 3.1 in Section 3.8 applies to higher order equations as well as second order equations. The only change is that the integer r in Table 3.1 can now range as high as the order of the equation, n. We illustrate the method of undetermined coefficients in Example 1.

E X A M P L E

1

Choose an appropriate form for a particular solution of

$$y^{(6)} + 3y^{(5)} + 3y^{(4)} + y''' = t + 2te^{-t} + \sin t, \qquad -\infty < t < \infty.$$

Solution: The first step is to find the complementary solution. The characteristic polynomial is $\lambda^6 + 3\lambda^5 + 3\lambda^4 + \lambda^3 = \lambda^3(\lambda^3 + 3\lambda^2 + 3\lambda + 1) = \lambda^3(\lambda + 1)^3$. Since $\lambda = 0$ and $\lambda = -1$ are repeated roots, the general solution of the homogeneous equation is

$$y_C(t) = c_1 + c_2 t + c_3 t^2 + c_4 e^{-t} + c_5 te^{-t} + c_6 t^2 e^{-t}.$$

Therefore, the method of undetermined coefficients suggests that we look for a particular solution having the form

$$y_P(t) = t^3(A_1 t + A_0) + t^3(B_1 te^{-t} + B_0 e^{-t}) + C\cos t + D\sin t.$$

The t^3 multipliers ensure that no term in the assumed form is a solution of the homogeneous equation. ❖

Variation of Parameters

The method of variation of parameters, discussed in Section 3.9, can be extended to find a particular solution of a linear nonhomogeneous nth order equation. As with second order equations, we assume we know a fundamental set of solutions of the homogeneous equation, $\{y_1(t), y_2(t), \ldots, y_n(t)\}$. The complementary solution is therefore

$$y_C(t) = c_1 y_1(t) + c_2 y_2(t) + \cdots + c_n y_n(t), \qquad a < t < b.$$

We now "vary the parameters," by replacing the constants $c_1, c_2, \ldots, c_n$ with functions $u_1(t), u_2(t), \ldots, u_n(t)$, and assume a particular solution of the form

$$y_P(t) = y_1(t)u_1(t) + y_2(t)u_2(t) + \cdots + y_n(t)u_n(t), \qquad a < t < b. \tag{2}$$

The functions $u_1(t), u_2(t), \ldots, u_n(t)$ must be chosen so that (2) is a solution of nonhomogeneous equation (1). However, since there are n functions in (2), we are free to impose $n - 1$ additional constraints on the n functions. Specifically, we impose the following $n - 1$ constraints:

$$y_1 u_1' + y_2 u_2' + \cdots + y_n u_n' = 0$$
$$y_1' u_1' + y_2' u_2' + \cdots + y_n' u_n' = 0$$
$$y_1'' u_1' + y_2'' u_2' + \cdots + y_n'' u_n' = 0 \tag{3}$$
$$\vdots$$
$$y_1^{(n-2)} u_1' + y_2^{(n-2)} u_2' + \cdots + y_n^{(n-2)} u_n' = 0.$$

The purpose of (3) is to make successive derivatives of $y_P(t)$ [where $y_P(t)$ is defined by equation (2)] have the following simple forms:

$$y'_P = y'_1 u_1 + y'_2 u_2 + \cdots + y'_n u_n$$
$$y''_P = y''_1 u_1 + y''_2 u_2 + \cdots + y''_n u_n$$
$$\vdots \qquad\qquad\qquad\qquad \text{(4)}$$
$$y_P^{(n-1)} = y_1^{(n-1)} u_1 + y_2^{(n-1)} u_2 + \cdots + y_n^{(n-1)} u_n.$$

When we substitute representation (2) for y_P into differential equation (1), use (3), and also use the fact that each of the functions $y_1, y_2, \ldots, y_n$ is a solution of the homogeneous equation, we obtain

$$y_1^{(n-1)} u'_1 + y_2^{(n-1)} u'_2 + \cdots + y_n^{(n-1)} u'_n = g. \qquad \text{(5)}$$

Taken together, equations (3) and (5) form a set of n linear equations for the n unknowns, $u'_1, u'_2, \ldots, u'_n$. In matrix form, this system of equations is

$$\begin{bmatrix} y_1 & y_2 & \cdots & y_n \\ y'_1 & y'_2 & \cdots & y'_n \\ \vdots & & & \vdots \\ y_1^{(n-1)} & y_2^{(n-1)} & \cdots & y_n^{(n-1)} \end{bmatrix} \begin{bmatrix} u'_1 \\ u'_2 \\ \vdots \\ u'_n \end{bmatrix} = \begin{bmatrix} 0 \\ 0 \\ \vdots \\ g \end{bmatrix}, \qquad a < t < b. \qquad \text{(6)}$$

The determinant of the $(n \times n)$ coefficient matrix is the Wronskian of a fundamental set of solutions, $\{y_1, y_2, \ldots, y_n\}$. Therefore, since the Wronskian determinant is nonzero for all values t in the interval (a, b), the system (6) has a unique solution for the unknowns $u'_1, u'_2, \ldots, u'_n$. Once these n functions are determined, we compute $u_1, u_2, \ldots, u_n$ by antidifferentiation and form y_P as prescribed by equation (2).

In principle, the method of variation of parameters is very general. However, the practical limitations noted in Section 3.9 for the second order case also apply to the nth order case. If the differential equation coefficients are not constants, it may be very difficult to determine a fundamental set of solutions of the homogeneous equation. Even if we know a fundamental set, it may be impossible to express the antiderivatives of $u'_1, u'_2, \ldots, u'_n$ in terms of known functions. The following example, however, is one in which the entire computational program of variation of parameters can be performed explicitly.

E X A M P L E

2

Consider the nonhomogeneous differential equation

$$t^3 y''' - 3t^2 y'' + 6ty' - 6y = t, \qquad 0 < t < \infty. \qquad \text{(7)}$$

(a) Verify that the functions $y_1(t) = t, y_2(t) = t^2, y_3(t) = t^3$ form a fundamental set of solutions for the associated homogeneous equation.

(b) Use variation of parameters to find a particular solution of the nonhomogeneous equation.

Solution:

(a) In Section 8.3, we discuss methods of finding solutions for homogeneous equations such as $t^3 y''' - 3t^2 y'' + 6ty' - 6y = 0$. For now, the fact that these

(continued)

(continued)

functions are solutions can be verified by direct substitution. The Wronskian of the solutions $y_1(t) = t$, $y_2(t) = t^2$, $y_3(t) = t^3$ is nonzero on $0 < t < \infty$:

$$W(t) = \begin{vmatrix} t & t^2 & t^3 \\ 1 & 2t & 3t^2 \\ 0 & 2 & 6t \end{vmatrix} = 2t^3.$$

Therefore, the given functions form a fundamental set of solutions.

(b) Assume a particular solution of the form

$$y_P = tu_1(t) + t^2 u_2(t) + t^3 u_3(t).$$

If we want to use equations (5) and (6), we must first divide equation (7) by t^3 to put it in the standard form (1); this step is necessary in order to properly identify the nonhomogeneous term, $g(t)$. Here, the function $g(t)$ is given by $g(t) = t^{-2}$. Using equation (6), we arrive at

$$\begin{bmatrix} t & t^2 & t^3 \\ 1 & 2t & 3t^2 \\ 0 & 2 & 6t \end{bmatrix} \begin{bmatrix} u_1' \\ u_2' \\ u_3' \end{bmatrix} = \begin{bmatrix} 0 \\ 0 \\ t^{-2} \end{bmatrix}. \tag{8}$$

Solving system (8), we find

$$\begin{bmatrix} u_1' \\ u_2' \\ u_3' \end{bmatrix} = \begin{bmatrix} 1/(2t) \\ -1/t^2 \\ 1/(2t^3) \end{bmatrix}.$$

We obtain the functions u_1, u_2, u_3 by computing convenient antiderivatives:

$$u_1(t) = \frac{1}{2}\ln|t| = \frac{1}{2}\ln t, \qquad u_2(t) = \frac{1}{t}, \qquad u_3(t) = -\frac{1}{4t^2},$$

where $0 < t < \infty$. Thus, one particular solution is

$$y_P(t) = \frac{t}{2}\ln t + t^2\left(\frac{1}{t}\right) + t^3\left(\frac{-1}{4t^2}\right) = \frac{t}{2}\ln t + \frac{3t}{4}, \qquad 0 < t < \infty.$$

Since the term $3t/4$ is a solution of the homogeneous equation, we can dispense with it and use a simpler particular solution,

$$y_P(t) = \frac{t}{2}\ln t, \qquad 0 < t < \infty.$$

The general solution is

$$y(t) = y_C(t) + y_P(t) = c_1 t + c_2 t^2 + c_3 t^3 + \frac{t}{2}\ln t, \qquad 0 < t < \infty. \; ❖$$

EXERCISES

Exercises 1–14:

For each differential equation,

(a) Find the complementary solution.

(b) Find a particular solution.

(c) Formulate the general solution.

1. $y''' - y' = e^{2t}$ **2.** $y''' - y' = 4 + 2\cos 2t$ **3.** $y''' - y' = 4t$

4. $y''' - y' = -4e^t$ **5.** $y''' + y'' = 6e^{-t}$ **6.** $y''' - y'' = 4e^{-2t}$

7. $y''' - 2y'' + y' = t + 4e^t$ **8.** $y''' - 3y'' + 3y' - y = 12e^t$

9. $y''' - y = e^t$ **10.** $y''' + y = e^t + \cos t$

11. $y^{(4)} - y = t + 1$ **12.** $y^{(4)} - y = \cos 2t$

13. $y''' + y = t^3$ **14.** $y''' + y'' = 4$

Exercises 15–21:

For each differential equation,

(a) Find the complementary solution.

(b) Formulate the *appropriate form* for the particular solution suggested by the method of undetermined coefficients. You need not evaluate the undetermined coefficients.

15. $y''' - 4y'' + 4y' = t^3 + 4t^2 e^{2t}$ **16.** $y''' - 3y'' + 3y' - y = e^t + 4e^t \cos 3t + 4$

17. $y^{(4)} - 16y = 4t \sin 2t$ **18.** $y^{(4)} + 8y'' + 16y = t \cos 2t$

19. $y^{(4)} - y = te^{-t} + (3t + 4)\cos t$ **20.** $y''' - y'' = t^2 + \cos t$

21. $y^{(4)} + 4y = e^t \sin t$

Exercises 22–24:

Consider the nonhomogeneous differential equation

$$y''' + ay'' + by' + cy = g(t).$$

In each exercise, the general solution of the differential equation is given, where c_1, c_2, and c_3 represent arbitrary constants. Use this information to determine the constants a, b, c and the function $g(t)$.

22. $y = c_1 + c_2 t + c_3 e^{2t} + 4\sin 2t$ **23.** $y = c_1 \sin 2t + c_2 \cos 2t + c_3 e^t + t^2$

24. $y = c_1 + c_2 t + c_3 t^2 - 2t^3$

Exercises 25–26:

Consider the nonhomogeneous differential equation

$$t^3 y''' + at^2 y'' + bty' + cy = g(t), \qquad t > 0.$$

In each exercise, the general solution of the differential equation is given, where c_1, c_2, and c_3 represent arbitrary constants. Use this information to determine the constants a, b, c and the function $g(t)$.

25. $y = c_1 + c_2 t + c_3 t^3 + t^4$ **26.** $y = c_1 t + c_2 t^2 + c_3 t^4 + 2\ln t$

27. (a) Verify that $\{t, t^2, t^4\}$ is a fundamental set of solutions of the differential equation

$$t^3 y''' - 4t^2 y'' + 8ty' - 8y = 0.$$

(b) Find the general solution of

$$t^3 y''' - 4t^2 y'' + 8ty' - 8y = 2\sqrt{t}, \qquad t > 0.$$

[Hint: Cramer's rule can be used to solve the system of equations arising in the method of variation of parameters.]

28. Using the information of Exercise 27(a), find the general solution of

$$t^3 y''' - 4t^2 y'' + 8ty' - 8y = 2t, \qquad t > 0.$$

29. Using the information of Exercise 27(a), find the general solution of

$$t^3 y''' - 4t^2 y'' + 8ty' - 8y = 6t^3, \qquad t > 0.$$

Exercises 30–33:

Find the solution of the differential equation that satisfies the given conditions.

30. $y^{(4)} - y = e^{-t}$, $y(0) = 0$, $\lim_{t \to \infty} y(t) = 0$

31. $y''' + y = \cos t$, $y(0) = 1$, $|y(t)| \le 2$ for all t, $0 \le t < \infty$

32. $y''' + y'' = 4e^{-2t}$, $y(0) = 2$, $\lim_{t \to \infty} y(t) = 1$

33. $y''' + y = e^{-t}$, $y(0) = 1$, $\lim_{t \to \infty} y(t) = 0$

CHAPTER 3 REVIEW EXERCISES

These review exercises provide you with an opportunity to test your understanding of the concepts and solution techniques developed in this chapter. The end-of-section exercises deal with the topics discussed in the section. These review exercises, however, require you to identify an appropriate solution technique before solving the problem.

Exercises 1–30:

In each exercise, determine the general solution. If initial conditions are given, solve the initial value problem.

1. $y'' + 2y' + 2y = 0$

2. $y'' - y = -6 \sin t$

3. $y'' + 4y = 0$, $y(0) = 3$, $y'(0) = 2$

4. $y''' + 9y' = 0$

5. $y'' - 5y' + 6y = 0$

6. $y'' - 4y' + 13y = 0$

7. $y^{(4)} - 81y = 0$

8. $y'' + y = \tan t$, $-\dfrac{\pi}{2} < t < \dfrac{\pi}{2}$

9. $y'' + 9y = 0$

10. $y'' + 5y' + 6y = 6t$, $y(0) = -1$, $y'(0) = 1$

11. $y'' - 2y' + y = t^{-1}e^t$, $t > 0$

12. $y'' + 2y' + 2y = 5 \cos t$

13. $y'' + 2y' + y = 8$, $y(0) = 10$, $y'(0) = 1$

14. $y''' + y' = e^{2t}$

15. $y'' = 6t + 4$

16. $y'' + 2y' + y = 0$

17. $y^{(4)} - y = 4t$

18. $y'' - 3y' + 2y = -3e^t$

19. $y'' + y = \sec t + t$, $-\dfrac{\pi}{2} < t < \dfrac{\pi}{2}$

20. $y''' + y'' - y' - y = 0$

21. $y'' - 6y' + 9y = 0$

22. $y'' + 6y' + 8y = 0$

23. $y'' - y = 6 - t^2$

24. $y'' - 9y = 0$

25. $y'' - 20y' + 100y = 0$

26. $y'' + 4y' + 4y = e^{-2t}$

27. $y'' - 4y' = 0$

28. $y'' - 4y' + 4y = 0$

29. $y^{(4)} - 4y''' + 4y'' = 0$

30. $y'' + y = 2 + t^2$, $y(0) = 2$, $y'(0) = 3$

PROJECTS

Project 1: Modeling Buoyant Motion

Consider the paraboloid of revolution whose cross section is shown in Figure 3.21. The upper radius of this solid body is r, and its height is h. Our initial goal is to derive a differential equation modeling the bobbing motion of such a solid when it is floating in liquid. The respective densities of the solid and liquid are ρ and ρ_l, where we assume

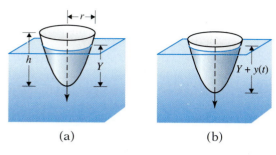

FIGURE 3.21

(a) The equilibrium state. (b) The dynamic state; $y(t)$ is positive downward.

$\rho < \rho_l$ so that the solid floats in the liquid. Assume further that the paraboloid bobs vertically up and down when disturbed from its equilibrium rest state.

1. As shown in Figure 3.21, Y represents the depth to which the paraboloid sinks when it is floating at rest, and $y(t)$ denotes the displacement of the body from its equilibrium state (measured positive in the downward direction). Derive a differential equation for $y(t)$ as follows:

   (a) Compute the weight of the solid. Apply the law of buoyancy (the weight of the solid equals the weight of the liquid displaced) to obtain an expression for the equilibrium depth Y.

   (b) Assume that the solid is displaced from its equilibrium state and apply Newton's law of motion $ma = F$ to derive a differential equation for $y(t)$. The net downward force acting upon the body is the difference between its weight and the upward buoyant force.

   The differential equation you derive should have the form

   $$y''(t) + \alpha y(t) + \beta y^2(t) = 0, \tag{1}$$

   where α and β are positive constants. This is a second order autonomous *nonlinear* differential equation. We have no techniques for solving such an equation. Nevertheless, we can use it to answer the question posed in part 2.

2. Consider the problem illustrated in Figure 3.22. A bullet-like projectile, having the shape of a paraboloid of revolution, is dropped from rest a short distance l above

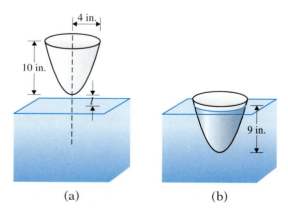

FIGURE 3.22

(a) The initial state when dropped. (b) Maximum penetration into the liquid.

the surface of a liquid. Assume that $\rho = 0.5\rho_l$, $h = 10$ in., $r = 4$ in., and $g = 32$ ft/sec^2. Assume that the tip of the paraboloid penetrates the liquid to a maximum depth of 9 in.

Determine the distance l. Assume that the drag forces in the atmosphere and in the liquid are negligible. [Hint: Divide the problem into two parts. Make a change of independent variable in (1), adopting y as the new independent variable; recall Section 2.9. Use this equation to determine the velocity with which the projectile impacts the liquid surface. Use this information, in turn, to determine l.]

Project 2: A Simple Centrifuge

The mechanical system shown in Figure 3.23 is a simple model of a centrifuge. A particle having mass m is initially positioned in a frictionless tube that rotates horizontally about a fixed pivot. As the tube rotates, the particle's radial distance from the pivot will increase and the particle will eventually exit the tube. Our goal is to analyze this behavior mathematically.

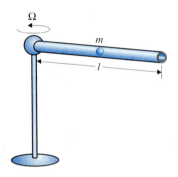

FIGURE 3.23

A simple centrifuge.

The key observation is the fact that at any given instant, the particle experiences no forces in the radial direction. Newton's law of motion tells us, therefore, that the component of the acceleration vector in the radial direction must vanish. Obtain the relevant differential equation as follows:

1. Consider Figure 3.24, where $\mathbf{e}_r$ and $\mathbf{e}_\theta$ are unit vectors in the radial and tangential directions, respectively. Show that

$$\mathbf{e}_r = (\cos\theta)\mathbf{i} + (\sin\theta)\mathbf{j}, \qquad \mathbf{e}_\theta = (-\sin\theta)\mathbf{i} + (\cos\theta)\mathbf{j}, \tag{2}$$

where $\mathbf{i}$ are $\mathbf{j}$ are unit vectors in the x and y directions, respectively.

2. If the angle θ changes with time, the unit vectors $\mathbf{e}_r$ and $\mathbf{e}_\theta$ will likewise change with time since their orientations will change. Use equation (2) and the fact that $\mathbf{i}$ and $\mathbf{j}$ are constant vectors to show that

$$\frac{d\mathbf{e}_r}{dt} = \theta'\mathbf{e}_\theta \quad \text{and} \quad \frac{d\mathbf{e}_\theta}{dt} = -\theta'\mathbf{e}_r, \tag{3}$$

where $\theta' = d\theta/dt$. Next, use equation (3) to derive expressions for $d^2\mathbf{e}_r/dt^2$ and $d^2\mathbf{e}_\theta/dt^2$.

3. The position vector describing the particle's location in the horizontal plane can be represented as $\mathbf{r} = r\mathbf{e}_r$, where r is the (time-varying) radial distance of the particle from the pivot. Differentiate this expression twice with respect to time, using the

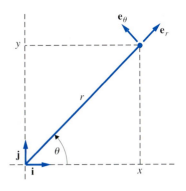

FIGURE 3.24

Diagram for equation (2).

product rule and the relations derived in part 2, to show that

$$\mathbf{a} = \frac{d^2\mathbf{r}}{dt^2} = [r'' - r(\theta')^2]\mathbf{e}_r + [r\theta'' + 2r'\theta']\mathbf{e}_\theta.$$

In our problem, angular velocity $\theta' = \Omega$ is prescribed, whereas the radial acceleration vanishes. The differential equation determining the radial distance of the particle from the pivot is therefore

$$r'' - \Omega^2 r = 0. \tag{4}$$

When angular velocity Ω is constant, equation (4) can be solved using the techniques developed in this chapter.

4. A frictionless tube 2 m in length is rotating with a constant angular velocity of 30 revolutions per minute. At time $t = 0$ the tube is aligned with the positive x-axis, and at that instant a particle is injected into the tube at the pivot with a radial velocity $r'(0) = r_0'$. What should the injection radial velocity r_0' be if we want the particle to exit the tube at the first instant the tube becomes aligned with the negative x-axis? What will be the radial velocity of the particle when it exits the tube?

Project 3: A Glimpse at Linear Two-Point Boundary Value Problems

The problems considered in this chapter are initial value problems, problems in which all supplementary constraints are imposed at a single value of the independent variable. Another important type of problem is a two-point boundary value problem, in which constraints (called boundary conditions) are imposed upon the solution at two different values of the independent variable. In applications, the two points at which constraints are imposed are typically the endpoints of the interval of interest. (Project 4 considers an application of this type.)

The purpose of this exercise is to briefly illustrate some of the differences that exist between initial value problems and boundary value problems. We have seen that definitive statements can be made guaranteeing the existence of unique solutions of initial value problems. For boundary value problems, however, the situation is more complicated. The problems below illustrate that a two-point boundary value problem may have a unique solution, infinitely many solutions, or no solution.

In each case, first obtain the general solution of the differential equation and then impose the boundary conditions.

1. Consider the two-point boundary value problem

$$y'' - y = 0, \qquad y(0) = \alpha, \quad y(T) = \beta.$$

Show that this problem has a unique solution on the interval $0 \le t \le T$ for every choice of the constants α, β, and T, where we assume $T > 0$. Obtain the solution in the special case where $\alpha = 0$, $\beta = 2$, and $T = 1$.

2. Consider the two-point boundary value problem

$$y'' + y = 0, \qquad y(0) = \alpha, \qquad y(T) = \beta.$$

Show that this problem has

(a) A unique solution for all choices of α and β if $T \ne n\pi, n = 0, 1, 2, \ldots$.

(b) Infinitely many solutions if $T = \pi$ and $\beta = -\alpha$.

(c) No solution if $T = \pi$ and $\beta \ne -\alpha$.

In each of the following exercises, obtain the general solution of the differential equation and impose the boundary conditions. State whether the boundary value problem has a unique solution, infinitely many solutions, or no solution. If a solution exists, whether unique or not, determine it.

3. $y'' + y = 0, \quad y'(0) = 1, \quad y(\pi) = 1$
4. $y'' + 2y' + y = 0, \quad y(0) = 0, \quad y(2) = -2$
5. $y'' + 4y = 0, \quad y(0) = 0, \quad y(\pi/2) = 0$
6. $y'' + y = 2, \quad y(0) = 1, \quad y(2\pi) = 0$
7. $y'' - 2y' + 2y = 0, \quad y(0) = 0, \quad y(1) = -2$
8. $y''' - y' = 0, \quad y(0) = 0, \quad y'(0) = 0, \quad y(1) = 1$

Project 4: Vibrations of a Clamped-End Beam

Consider the beam shown in Figure 3.25(a). It is uniform in cross section and composition, has length l, and is clamped at both ends. Assume that a distributed loading or force per unit length, denoted by $w(x, t)$, is applied vertically to the beam. This loading is a function of position x along the beam and time t. In response to this dynamic loading, the beam will flex or deflect. We denote the beam displacement at point x and time t by $y(x, t)$. Both w and y are assumed positive in the downward direction [see Figure 3.25(b)].

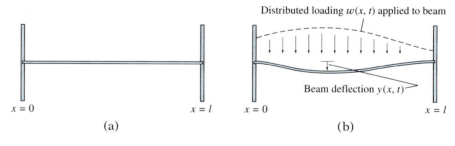

FIGURE 3.25

(a) The beam in an unloaded state. (b) The beam in a loaded state, at time t.

A mathematical description of how the beam deflects under loading is needed. A model frequently used is the following partial differential equation, known as the Euler-

Bernoulli beam equation:

$$\rho\frac{\partial^2 y(x,t)}{\partial t^2} + EI\frac{\partial^4 y(x,t)}{\partial x^4} = w(x,t), \qquad 0 < x < l, \qquad 0 < t < \infty. \tag{5}$$

In equation (5), ρ is the mass per unit length, E is Young's modulus (a constant characterizing the stiffness of the beam material), and I is the area moment of inertia (a constant determined by the beam's cross-sectional geometry).

Because the beam is clamped at its ends, beam displacement and the slope of this displacement must vanish at both $x = 0$ and $x = l$. Therefore, beam displacement $y(x,t)$ must satisfy the following four constraints (known as boundary conditions):

$$y(0,t) = 0, \qquad \frac{\partial y(0,t)}{\partial x} = 0, \qquad y(l,t) = 0, \qquad \frac{\partial y(l,t)}{\partial x} = 0, \qquad 0 \le t < \infty. \tag{6}$$

Consider the case where the loading function is

$$w(x,t) = w_0 \sin^2(\pi x/l)\ \sin(\omega t),$$

where w_0 is a positive constant. Therefore, at any fixed point x along the beam, the loading varies sinusoidally in time with radian frequency $\omega = 2\pi f$. When the factor $\sin(\omega t)$ is positive, the loading is pressing downward; when $\sin(\omega t)$ is negative, the loading is pulling upward. The amplitude or strength of the loading at point x is $w_0 \sin^2(\pi x/l)$. This amplitude is largest at the beam center, and it vanishes at both endpoints.

1. Assume a solution of the form

$$y(x,t) = u(x)\sin(\omega t). \tag{7}$$

We are therefore assuming that the beam deflection has the same $\sin(\omega t)$ time dependence as the applied loading. Substitute (7) into equations (5) and (6). Show that we obtain the following problem for $u(x)$:

$$EI\frac{d^4 u(x)}{dx^4} - \rho\omega^2 u(x) = w_0 \sin^2(\pi x/l), \qquad 0 < x < l$$

$$u(0) = 0, \qquad u'(0) = 0, \qquad u(l) = 0, \qquad u'(l) = 0. \tag{8}$$

Problem (8) consists of a fourth order linear nonhomogeneous ordinary differential equation and four supplementary conditions. Note, however, that the four constraints are not all given at the same value of independent variable x. Problem (8) is not an initial value problem; it is a two-point boundary value problem. The four constraints in (8) are referred to as boundary conditions. Although the theory for such boundary value problems will not be discussed here, this particular problem has a unique solution that we can obtain using the techniques developed in this chapter.

2. Obtain the general solution of the differential equation in (8), assuming that $(\rho\omega^2/EI)^{1/4} \ne 2\pi/l$. [Hint: Use the trigonometric identity $\sin^2\theta = (1 - \cos 2\theta)/2$.]

3. Impose the boundary conditions in (8). This will lead to a system of four equations for the four arbitrary constants in the general solution found in part 2.

4. Use computer software to solve the linear system found in part 3 and obtain the solution $u(x)$ for the parameter values $l = 2$ m, $\omega = 2\pi$ (therefore, $f = 1$ Hz), $\rho = 0.6$ kg/m, $EI = 32$ N·m^2, and $w_0 = 100$ N.

5. Use computer software to plot the solution $u(x)$ and determine the maximum deflection of the beam under this loading.

4

First Order Linear Systems

4.1   Introduction

A linear system of algebraic equations is a familiar concept. For instance, consider this system of three linear equations in three unknowns:

$$
\begin{aligned}
2x_1 + x_2 - 2x_3 &= 3 \\
5x_1 + 2x_2 + 9x_3 &= 7 \\
3x_1 - x_2 + 4x_3 &= 6.
\end{aligned}
$$

Solving this system entails finding all values x_1, x_2, x_3 that simultaneously satisfy each of the three equations. In most cases, such systems of equations cannot be solved "one equation at a time." Rather, we have to deal with the system as a whole, applying techniques from matrix theory to obtain the solution.

Systems of First Order Linear Differential Equations

In this chapter, we consider systems of first order linear differential equations. In general, we will be interested in systems of n first order linear differential equations. When $n = 3$, such a system has the form

$$
\begin{aligned}
y_1' &= p_{11}(t)y_1 + p_{12}(t)y_2 + p_{13}(t)y_3 + g_1(t) \\
y_2' &= p_{21}(t)y_1 + p_{22}(t)y_2 + p_{23}(t)y_3 + g_2(t) \\
y_3' &= p_{31}(t)y_1 + p_{32}(t)y_2 + p_{33}(t)y_3 + g_3(t), \qquad a < t < b.
\end{aligned}
\tag{1}
$$

Solving this problem amounts to finding all functions $y_1(t)$, $y_2(t)$, and $y_3(t)$ that simultaneously satisfy the three differential equations on the interval of interest. Here again, as with a system of algebraic equations, we cannot normally solve a system of differential equations "one equation at a time." We cannot, for example, use the techniques of Chapter 2 to solve the first equation for $y_1(t)$, because the functions $y_2(t)$ and $y_3(t)$ are not known. Instead, we have to develop techniques that deal with the system of equations as a whole. In this regard, techniques from matrix theory will be of central importance.

E X A M P L E

1

A Two-Tank Mixing Problem

Consider the two-tank connection shown in Figure 4.1. As in Chapter 2, the solute and solvent are assumed to be salt and water, respectively, and the solutions in both tanks are "well-stirred." Assume each tank has a capacity of 500 gal. Initially, Tank 1 contains 200 gal of fresh water, while Tank 2 has 50 lb of salt dissolved in 300 gal of water. At time $t = 0$, the flow begins at the rates and concentrations shown in Figure 4.1. Let the amounts of salt in the two tanks at time t be denoted by $Q_1(t)$ and $Q_2(t)$, respectively. The problem is to determine $Q_1(t)$ and $Q_2(t)$ on the time interval that is physically relevant (that is, we will stop the flow if one of the tanks becomes completely filled or completely drained). Time t is in minutes.

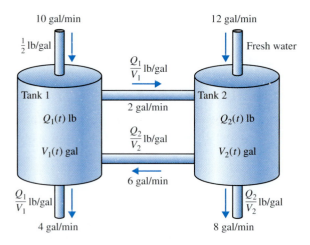

FIGURE 4.1

The two-tank mixing problem described in Example 1.

Solution: As a first step, we determine how the volumes of fluid vary in both tanks. Tank 1 has a total of 16 gal of solution entering per minute and 6 gal leaving per minute. Since Tank 1 contains 200 gal at $t = 0$ and gains 10 gal of fluid per minute, the volume of liquid in Tank 1 is given by $V_1(t) = 200 + 10t$ gal.

Tank 2, on the other hand, gains 14 gal of fluid per minute but also loses 14 gal/min. Therefore, the volume of liquid in Tank 2 remains constant at $V_2(t) = 300$ gal. These considerations of volume determine the t-interval of interest. Since the capacity of Tank 1 is 500 gal, it will be completely filled in 30 minutes. The interval of interest is therefore $0 \le t \le 30$.

We obtain the relevant system of differential equations by applying the principle of "conservation of salt" to each tank: The rate of change of the amount of salt in a tank is equal to the rate at which salt enters the tank minus the rate at which salt leaves the tank. From Figure 4.1, it follows that

$$\frac{dQ_1}{dt} = 5 + 6\left[\frac{Q_2}{V_2}\right] - 6\left[\frac{Q_1}{V_1}\right] = -\frac{6}{200 + 10t}Q_1 + \frac{6}{300}Q_2 + 5,$$

$$\frac{dQ_2}{dt} = 2\left[\frac{Q_1}{V_1}\right] - 14\left[\frac{Q_2}{V_2}\right] = \frac{2}{200 + 10t}Q_1 - \frac{14}{300}Q_2.$$

(2)

The initial value problem, to be solved on the interval $0 \le t \le 30$, consists of these differential equations together with initial conditions $Q_1(0) = 0$, $Q_2(0) = 50$. ❖

The Calculus of Matrix Functions

Systems of differential equations such as (1) and (2) can be rewritten as a single matrix equation, and techniques from matrix theory can be employed to find solutions of such systems. We now discuss some helpful background in the calculus of matrix functions.

Throughout Chapter 4, we will be concerned with $(m \times n)$ matrices whose entries are real-valued functions of the real variable t. Such functions are called **matrix-valued functions** or, simply, **matrix functions**. For example, a (3×2) matrix function has the form

$$A(t) = \begin{bmatrix} a_{11}(t) & a_{12}(t) \\ a_{21}(t) & a_{22}(t) \\ a_{31}(t) & a_{32}(t) \end{bmatrix}.$$

(3)

In (3), the entries $a_{ij}(t)$ are real-valued functions defined on a common interval $a < t < b$.

When a matrix function has a single column, such as the (3×1) matrix function

$$\mathbf{y}(t) = \begin{bmatrix} y_1(t) \\ y_2(t) \\ y_3(t) \end{bmatrix},$$

we usually refer to $\mathbf{y}(t)$ as a **vector-valued function** or, simply, a **vector function**. (In matrix theory, it is common practice to use boldface type to denote column vectors. Therefore, whenever we refer to a vector function, we will use boldface also.)

The Arithmetic of Matrix Functions

For a fixed value of t, a matrix function is a constant matrix, and thus all the familiar rules of matrix arithmetic hold for matrix functions as well. Rather than stating general theorems giving detailed rules for the arithmetic of matrix functions, we simply illustrate the rules in Example 2.

EXAMPLE 2

We consider (2×1) and (2×2) matrix functions for simplicity. All functions are assumed to be defined on some common interval, say $a < t < b$, and all equalities listed hold on that interval.

(a) *Equality:*

$$\begin{bmatrix} a_1(t) \\ a_2(t) \end{bmatrix} = \begin{bmatrix} b_1(t) \\ b_2(t) \end{bmatrix}$$

if and only if $a_1(t) = b_1(t)$ and $a_2(t) = b_2(t)$.

(b) *Addition:*

$$\begin{bmatrix} a_1(t) \\ a_2(t) \end{bmatrix} + \begin{bmatrix} b_1(t) \\ b_2(t) \end{bmatrix} = \begin{bmatrix} a_1(t) + b_1(t) \\ a_2(t) + b_2(t) \end{bmatrix}$$

(c) *Scalar Multiplication:*

$$f(t) \begin{bmatrix} a_1(t) \\ a_2(t) \end{bmatrix} = \begin{bmatrix} f(t)a_1(t) \\ f(t)a_2(t) \end{bmatrix}$$

(d) *Matrix Multiplication:*

$$\begin{bmatrix} a_{11}(t) & a_{12}(t) \\ a_{21}(t) & a_{22}(t) \end{bmatrix} \begin{bmatrix} b_1(t) \\ b_2(t) \end{bmatrix} = \begin{bmatrix} a_{11}(t)b_1(t) + a_{12}(t)b_2(t) \\ a_{21}(t)b_1(t) + a_{22}(t)b_2(t) \end{bmatrix}$$

(e) *Matrix Inversion:* Let $A(t)$ be a (2×2) matrix function,

$$A(t) = \begin{bmatrix} a_{11}(t) & a_{12}(t) \\ a_{21}(t) & a_{22}(t) \end{bmatrix}.$$

Then $A^{-1}(t)$ exists for all t such that $\det[A(t)] \neq 0$. ❖

EXAMPLE 3

Consider the (2×2) matrix function

$$A(t) = \begin{bmatrix} t & 1 \\ 4t & 4t^2 \end{bmatrix}, \qquad -\infty < t < \infty.$$

Determine all values t such that $A(t)$ is invertible, and find $A^{-1}(t)$ for those values t.

Solution: The matrix function $A(t)$ is invertible if and only if $\det[A(t)] \neq 0$. For this matrix,

$$\det[A(t)] = 4t^3 - 4t = 4t(t-1)(t+1).$$

Therefore, $A^{-1}(t)$ exists for all t except $t = -1, 0, 1$. The inverse of the (2×2) matrix

$$\begin{bmatrix} a & b \\ c & d \end{bmatrix}$$

is given by

$$\begin{bmatrix} a & b \\ c & d \end{bmatrix}^{-1} = \frac{1}{ad - bc} \begin{bmatrix} d & -b \\ -c & a \end{bmatrix}, \qquad ad - bc \neq 0.$$

In our case,

$$A^{-1}(t) = \frac{1}{4t(t^2 - 1)} \begin{bmatrix} 4t^2 & -1 \\ -4t & t \end{bmatrix} = \begin{bmatrix} \dfrac{t}{t^2 - 1} & \dfrac{-1}{4t(t^2 - 1)} \\ \dfrac{-1}{t^2 - 1} & \dfrac{1}{4(t^2 - 1)} \end{bmatrix}, \qquad t \neq -1, 0, 1. \; \blacklozenge$$

Limits and Derivatives of Matrix Functions

The concept of the limit of a vector function is familiar from calculus. For example, let $\mathbf{r}(t)$ denote the vector function

$$\mathbf{r}(t) = f(t)\mathbf{i} + g(t)\mathbf{j} + h(t)\mathbf{k},$$

where the three component functions $f(t)$, $g(t)$, and $h(t)$ are defined in an open interval containing the point $t = a$. The limit of $\mathbf{r}(t)$ as t approaches a is

$$\lim_{t \to a} \mathbf{r}(t) = \left[\lim_{t \to a} f(t)\right] \mathbf{i} + \left[\lim_{t \to a} g(t)\right] \mathbf{j} + \left[\lim_{t \to a} h(t)\right] \mathbf{k},$$

provided the limits of the three component functions exist.

We take the same approach in defining limits of a matrix function. Let $A(t)$ be an $(m \times n)$ matrix function having component functions $a_{ij}(t)$, all defined in an open interval containing the point $t = a$. To say that

$$\lim_{t \to a} A(t) = L$$

means

$$\lim_{t \to a} a_{ij}(t) = l_{ij}, \qquad 1 \leq i \leq m, \quad 1 \leq j \leq n.$$

For instance, if $A(t)$ is a (2×2) matrix function, then

$$\lim_{t \to a} A(t) = L$$

if and only if

$$\lim_{t \to a} A(t) = \lim_{t \to a} \begin{bmatrix} a_{11}(t) & a_{12}(t) \\ a_{21}(t) & a_{22}(t) \end{bmatrix} = \begin{bmatrix} \lim\limits_{t \to a} a_{11}(t) & \lim\limits_{t \to a} a_{12}(t) \\ \lim\limits_{t \to a} a_{21}(t) & \lim\limits_{t \to a} a_{22}(t) \end{bmatrix} = \begin{bmatrix} l_{11} & l_{12} \\ l_{21} & l_{22} \end{bmatrix} = L.$$

If one or more of the component function limits do not exist, then the limit of the matrix function does not exist. For example, if

$$B(t) = \begin{bmatrix} t & t^{-1} \\ 0 & e^t \end{bmatrix}, \quad \text{then} \quad \lim_{t \to 2} B(t) = \begin{bmatrix} 2 & \frac{1}{2} \\ 0 & e^2 \end{bmatrix}.$$

However, $\lim_{t \to 0} B(t)$ does not exist.

As in single variable calculus, we say the matrix function $A(t)$ is **continuous at $t = a$** if $A(t)$ is defined in a neighborhood of $t = a$ and if

$$\lim_{t \to a} A(t) = A(a). \tag{4}$$

To define a derivative (an instantaneous rate of change), we are led to a limit of the form

$$A'(t) = \lim_{h \to 0} \frac{1}{h}[A(t+h) - A(t)].$$

For example, let $A(t)$ be a (2×2) matrix function with differentiable component functions. Then

$$A'(t) = \lim_{h \to 0} \frac{1}{h}[A(t+h) - A(t)]$$

$$= \begin{bmatrix} \lim\limits_{h \to 0} \dfrac{a_{11}(t+h) - a_{11}(h)}{h} & \lim\limits_{h \to 0} \dfrac{a_{12}(t+h) - a_{12}(h)}{h} \\ \lim\limits_{h \to 0} \dfrac{a_{21}(t+h) - a_{21}(h)}{h} & \lim\limits_{h \to 0} \dfrac{a_{22}(t+h) - a_{22}(h)}{h} \end{bmatrix}$$

$$= \begin{bmatrix} a'_{11}(t) & a'_{12}(t) \\ a'_{21}(t) & a'_{22}(t) \end{bmatrix}.$$

As this special case suggests, the derivative of a matrix function is the matrix of derivatives of its component functions. In general, let $A(t)$ be an $(m \times n)$ matrix function

$$A(t) = \begin{bmatrix} a_{11}(t) & a_{12}(t) & \cdots & a_{1n}(t) \\ a_{21}(t) & a_{22}(t) & & a_{2n}(t) \\ \vdots & & & \vdots \\ a_{m1}(t) & a_{m2}(t) & \cdots & a_{mn}(t) \end{bmatrix},$$

where each of the component functions is differentiable on the interval (a, b). Then the derivative $A'(t)$ exists and is given by

$$A'(t) = \begin{bmatrix} a'_{11}(t) & a'_{12}(t) & \cdots & a'_{1n}(t) \\ a'_{21}(t) & a'_{22}(t) & & a'_{2n}(t) \\ \vdots & & & \vdots \\ a'_{m1}(t) & a'_{m2}(t) & \cdots & a'_{mn}(t) \end{bmatrix}, \qquad a < t < b. \tag{5}$$

We refer to $A(t)$ as a **differentiable matrix function** or, simply, a **differentiable matrix**.

Representing Linear Systems in Matrix Terms

We can express systems of linear differential equations in matrix terms. For example, recall the (3×3) system (1)

$$y'_1 = p_{11}(t)y_1 + p_{12}(t)y_2 + p_{13}(t)y_3 + g_1(t)$$
$$y'_2 = p_{21}(t)y_1 + p_{22}(t)y_2 + p_{23}(t)y_3 + g_2(t)$$
$$y'_3 = p_{31}(t)y_1 + p_{32}(t)y_2 + p_{33}(t)y_3 + g_3(t), \qquad a < t < b.$$

Define $\mathbf{y}(t)$, $P(t)$, and $\mathbf{g}(t)$ as follows:

$$\mathbf{y}(t) = \begin{bmatrix} y_1(t) \\ y_2(t) \\ y_3(t) \end{bmatrix}, \qquad P(t) = \begin{bmatrix} p_{11}(t) & p_{12}(t) & p_{13}(t) \\ p_{21}(t) & p_{22}(t) & p_{23}(t) \\ p_{31}(t) & p_{32}(t) & p_{33}(t) \end{bmatrix}, \qquad \mathbf{g}(t) = \begin{bmatrix} g_1(t) \\ g_2(t) \\ g_3(t) \end{bmatrix}.$$

Using these definitions, we can express the (3×3) system in matrix terms as

$$\mathbf{y}'(t) = P(t)\mathbf{y}(t) + \mathbf{g}(t). \tag{6}$$

The value of (6) is more than just shorthand. The notation also helps us understand the principles of solving systems of linear differential equations because it takes our eyes away from the details of individual equations and allows us to see the system as a single entity. We can view (6) as a single differential equation involving a matrix-valued dependent variable.

Some Useful Formulas

The fact that the derivative of a matrix function is simply the matrix of derivatives can be used to verify the following familiar-looking formulas.

Let $A(t)$ and $B(t)$ be two differentiable $(m \times n)$ matrix functions. Then

$$[A(t) + B(t)]' = A'(t) + B'(t). \tag{7}$$

Let $A(t)$ be a differentiable $(m \times n)$ matrix function, and let $f(t)$ be a differentiable scalar function. Then

$$[f(t)A(t)]' = f'(t)A(t) + f(t)A'(t). \tag{8}$$

Let $A(t)$ be a differentiable $(m \times n)$ matrix function, and let $B(t)$ be a differentiable $(n \times p)$ matrix function. Then

$$[A(t)B(t)]' = A'(t)B(t) + A(t)B'(t). \tag{9}$$

Formula (9) is the analog of the familiar product formula in calculus. Since the functions involved are matrix functions, however, it is imperative that the order of matrix multiplications be preserved.

Antiderivatives of Matrix Functions

Since the derivative of a matrix function reduces to the matrix of derivatives, it is not surprising that antiderivatives of matrix functions are found by antidifferentiating each component of the matrix function. That is, if $A(t)$ is the $(m \times n)$ matrix function

$$A(t) = \begin{bmatrix} a_{11}(t) & a_{12}(t) & \cdots & a_{1n}(t) \\ a_{21}(t) & a_{22}(t) & & a_{2n}(t) \\ \vdots & & & \vdots \\ a_{m1}(t) & a_{m2}(t) & \cdots & a_{mn}(t) \end{bmatrix},$$

then the antiderivative of $A(t)$ is the $(m \times n)$ matrix function

$$\int A(t)\,dt = \begin{bmatrix} \int a_{11}(t)\,dt & \int a_{12}(t)\,dt & \cdots & \int a_{1n}(t)\,dt \\ \int a_{21}(t)\,dt & \int a_{22}(t)\,dt & & \int a_{2n}(t)\,dt \\ \vdots & & & \vdots \\ \int a_{m1}(t)\,dt & \int a_{m2}(t)\,dt & \cdots & \int a_{mn}(t)\,dt \end{bmatrix}.$$

E X A M P L E

4

Determine the antiderivative of

$$A(t) = \begin{bmatrix} 2e^{2t} & 2t \\ 0 & -1 \end{bmatrix}.$$

Solution: Since the antiderivative of a matrix function is the matrix of antiderivatives,

$$\int A(t)\,dt = \begin{bmatrix} e^{2t}+c_{11} & t^2+c_{12} \\ c_{21} & -t+c_{22} \end{bmatrix} = \begin{bmatrix} e^{2t} & t^2 \\ 0 & -t \end{bmatrix} + \begin{bmatrix} c_{11} & c_{12} \\ c_{21} & c_{22} \end{bmatrix} = \begin{bmatrix} e^{2t} & t^2 \\ 0 & -t \end{bmatrix} + C. ❖$$

As Example 4 illustrates, when we calculate the antiderivative of a matrix function, we need to allow for different arbitrary constants in each component function antidifferentiation. Therefore, the general antiderivative of a matrix function is a matrix of convenient antiderivatives *plus* an arbitrary constant matrix of comparable dimensions.

E X E R C I S E S

Exercises 1–5:

For the given matrix functions $A(t)$, $B(t)$, and $\mathbf{c}(t)$, make the indicated calculations

$$A(t) = \begin{bmatrix} t-1 & t^2 \\ 2 & 2t+1 \end{bmatrix}, \quad B(t) = \begin{bmatrix} t & -1 \\ 0 & t+2 \end{bmatrix}, \quad \mathbf{c}(t) = \begin{bmatrix} t+1 \\ -1 \end{bmatrix}.$$

1. $2A(t) - 3tB(t)$ **2.** $A(t)B(t) - B(t)A(t)$ **3.** $A(t)\mathbf{c}(t)$

4. $\det[tA(t)]$ **5.** $\det[B(t)A(t)]$

Exercises 6–9:

Determine all values t such that $A(t)$ is invertible and, for those t-values, find $A^{-1}(t)$.

6. $A(t) = \begin{bmatrix} t+1 & t \\ t & t+1 \end{bmatrix}$ **7.** $A(t) = \begin{bmatrix} t & 2 \\ 2 & t-3 \end{bmatrix}$

8. $A(t) = \begin{bmatrix} \sin t & -\cos t \\ \sin t & \cos t \end{bmatrix}$ **9.** $A(t) = \begin{bmatrix} e^t & e^{3t} \\ e^{2t} & e^{4t} \end{bmatrix}$

Exercises 10–11:

Find $\lim_{t \to 0} A(t)$ or state that the limit does not exist.

10. $A(t) = \begin{bmatrix} \dfrac{\sin t}{t} & t \cos t & \dfrac{3}{t+1} \\ e^{3t} & \sec t & \dfrac{2t}{t^2-1} \end{bmatrix}$

11. $A(t) = \begin{bmatrix} te^{-t} & \tan t \\ t^2 - 2 & e^{\sin t} \end{bmatrix}$

Exercises 12–13:

Find $A'(t)$ and $A''(t)$. For what values of t are the matrices $A(t)$, $A'(t)$, and $A''(t)$ defined?

12. $A(t) = \begin{bmatrix} \sin t & 3t \\ t^2 + 2 & 5 \end{bmatrix}$

13. $A(t) = \begin{bmatrix} 7 & \ln |t| \\ \sqrt{1-t} & e^{3t} \end{bmatrix}$

Exercises 14–15:

Each of the systems of linear differential equations can be expressed in the form $\mathbf{y}' = P(t)\mathbf{y} + \mathbf{g}(t)$. Determine $P(t)$ and $\mathbf{g}(t)$.

14. $y_1' = t^2 y_1 + 3y_2 + \sec t$
$\quad y_2' = (\sin t)y_1 + ty_2 - 5$

15. $y_1' = t^{-1}y_1 + (t^2 + 1)y_2 + t$
$\quad y_2' = 4y_1 + t^{-1}y_2 + 8t \ln t$

Exercises 16–22:

Determine the general form of $A(t)$ by constructing antiderivatives as needed and imposing any given constraints.

16. $A'(t) = \begin{bmatrix} -1 \\ 2t \end{bmatrix}$

17. $A'(t) = [1 \quad t \quad e^t], \quad A(0) = [-1 \quad 1 \quad 0]$

18. $A'(t) = \begin{bmatrix} 2t & 1 \\ \cos t & 3t^2 \end{bmatrix}, \quad A(0) = \begin{bmatrix} 2 & 5 \\ 1 & -2 \end{bmatrix}$

19. $A'(t) = \begin{bmatrix} t^{-1} & 4t \\ 5 & 3t^2 \end{bmatrix}, \quad A(1) = \begin{bmatrix} 2 & 5 \\ 1 & -2 \end{bmatrix}$

20. $A''(t) = \begin{bmatrix} 2 \\ e^t \end{bmatrix}$

21. $A''(t) = \begin{bmatrix} 2t & \sin t \\ 0 & 0 \end{bmatrix}, \quad A(0) = \begin{bmatrix} 0 & 1 \\ 0 & 0 \end{bmatrix}, \quad A'(0) = \begin{bmatrix} 0 & 0 \\ 1 & 0 \end{bmatrix}$

22. $A''(t) = \begin{bmatrix} 1 & t \\ 0 & 0 \end{bmatrix}, \quad A(0) = \begin{bmatrix} 1 & 1 \\ -2 & 1 \end{bmatrix}, \quad A(1) = \begin{bmatrix} -1 & 2 \\ -2 & 3 \end{bmatrix}$

Exercises 23–24:

Calculate $A(t) = \int_0^t B(s)\,ds$.

23. $B(s) = \begin{bmatrix} 2s & \cos s & 2 \\ 5 & (s+1)^{-1} & 3s^2 \end{bmatrix}$

24. $B(s) = \begin{bmatrix} e^s & 6s \\ \cos 2\pi s & \sin 2\pi s \end{bmatrix}$

25. Let $A(t)$ be an $(n \times n)$ matrix function. We use the notation $A^2(t)$ to mean the matrix function $A(t)A(t)$.

(a) Construct an explicit (2×2) differentiable matrix function to show that

$$\frac{d}{dt}[A^2(t)] \quad \text{and} \quad 2A(t)\frac{d}{dt}[A(t)]$$

are generally not equal.

(b) What is the correct formula relating the derivative of $A^2(t)$ to the matrices $A(t)$ and $A'(t)$?

26. Construct an example of a (2×2) matrix function $A(t)$ such that $A^2(t)$ is a constant matrix but $A(t)$ is not a constant matrix.

27. Let $A(t)$ be an $(n \times n)$ matrix function that is both differentiable and invertible on some t-interval of interest. It can be shown that $A^{-1}(t)$ is likewise differentiable on this interval. Differentiate the matrix identity $A^{-1}(t)A(t) = I$ to obtain the following formula:

$$\frac{d}{dt}[A^{-1}(t)] = -A^{-1}(t)A'(t)A^{-1}(t).$$

[Hint: Recall the product rule, equation (9). Notice that the formula you derive is **not** the same as the power rule of single-variable calculus.]

28. Consider the matrix function

$$A(t) = \begin{bmatrix} t & t^3 \\ 0 & 2t \end{bmatrix}.$$

Explicitly calculate both $(d/dt)[A^{-1}(t)]$ and $-A^{-1}(t)A'(t)A^{-1}(t)$ for this special case to illustrate the formula derived in Exercise 27.

Exercises 29–32:

Consider the two-tank mixing apparatus shown in the figure. Each tank has a capacity of 500 gal and initially contains 100 gal of fresh water. At time $t = 0$, the well-stirred mixing process begins with the specified input concentration and flow rates.

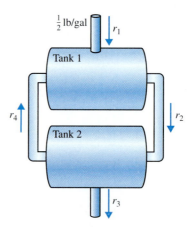

$\frac{1}{2}$ lb/gal

r_1

Tank 1

r_4

r_2

Tank 2

r_3

Figure for Exercises 29–32

(a) Determine the volume of solution in each tank as a function of time.

(b) Determine the time interval of interest. (The process stops when a tank is full or empty.)

(c) Let $Q_1(t)$ and $Q_2(t)$ denote the amounts of salt (in pounds) in Tanks 1 and 2 at time t (in minutes). Derive the initial value problem with Q_1 and Q_2 as dependent variables describing the mixing process.

29. $r_1 = r_3 = 5$ gal/min, $r_2 = r_4 = 10$ gal/min

30. $r_1 = r_3 = 5$ gal/min, $r_2 = 6$ gal/min, $r_4 = 4$ gal/min

31. $r_1 = 5$ gal/min, $r_3 = 0$, $r_2 = r_4 = 5$ gal/min

32. $r_1 = 5$ gal/min, $r_3 = 10$ gal/min, $r_2 = r_4 = 5$ gal/min

4.2   Existence and Uniqueness

Section 4.1 introduced matrix and vector functions and showed how to deal with the calculus of such functions. We now discuss initial value problems involving systems of first order differential equations. Consider the initial value problem

$$
\begin{aligned}
y_1' &= p_{11}(t)y_1 + p_{12}(t)y_2 + \cdots + p_{1n}(t)y_n + g_1(t) \\
y_2' &= p_{21}(t)y_1 + p_{22}(t)y_2 + \cdots + p_{2n}(t)y_n + g_2(t) \\
&\ \ \vdots \\
y_n' &= p_{n1}(t)y_1 + p_{n2}(t)y_2 + \cdots + p_{nn}(t)y_n + g_n(t), \qquad a < t < b, \\
y_1(t_0) &= y_1^0, \qquad y_2(t_0) = y_2^0, \qquad \cdots \qquad , y_n(t_0) = y_n^0,
\end{aligned}
\tag{1}
$$

where $y_1^0, y_2^0, \ldots, y_n^0$ are n constants specified at some point t_0 in the t-interval of interest, (a, b). The n^2 coefficient functions $p_{11}(t), p_{12}(t), \ldots, p_{nn}(t)$ and the n functions $g_1(t), g_2(t), \ldots, g_n(t)$ are given functions, defined on $a < t < b$.

We can recast problem (1) in matrix form as

$$
\mathbf{y}'(t) = P(t)\mathbf{y}(t) + \mathbf{g}(t), \qquad \mathbf{y}(t_0) = \mathbf{y}_0,
\tag{2}
$$

where, for $a < t < b$,

$$
\mathbf{y}(t) = \begin{bmatrix} y_1(t) \\ y_2(t) \\ \vdots \\ y_n(t) \end{bmatrix}, \qquad
P(t) = \begin{bmatrix} p_{11}(t) & p_{12}(t) & \cdots & p_{1n}(t) \\ p_{21}(t) & p_{22}(t) & & p_{2n}(t) \\ \vdots & & & \vdots \\ p_{n1}(t) & p_{n2}(t) & \cdots & p_{nn}(t) \end{bmatrix},
$$

$$
\mathbf{g}(t) = \begin{bmatrix} g_1(t) \\ g_2(t) \\ \vdots \\ g_n(t) \end{bmatrix}, \qquad
\mathbf{y}_0 = \begin{bmatrix} y_1^0 \\ y_2^0 \\ \vdots \\ y_n^0 \end{bmatrix}.
$$

The differential equation in (2) is referred to as a **first order system of linear differential equations** or, more simply, a **first order linear system**. If the $(n \times 1)$ vector function $\mathbf{g}(t)$ is the $(n \times 1)$ zero vector, then the system is called a first order **homogeneous** linear system; if $\mathbf{g}(t)$ is not identically zero on the interval of interest, the system is a first order **nonhomogeneous** linear system.

We sometimes need to distinguish the differential equation in (2), where the dependent variable $\mathbf{y}(t)$ is a vector-valued function, from the differential equations considered in Chapters 1–3, where the dependent variable $y(t)$ is a single real-valued function. We will refer to the differential equations studied in Chapters 1–3 as **scalar differential equations**. In particular, Chapters 2 and 3 dealt with first order, second order, and nth order linear scalar differential equations.

An Existence and Uniqueness Theorem

What are the conditions needed to ensure that initial value problem (2) has a unique solution? A general theme emerged from the existence/uniqueness results of Chapters 2 and 3—namely,

If (a, b) denotes the interval of interest, then continuity of the coefficient functions of a linear differential equation, together with continuity of the nonhomogeneous term $g(t)$, is sufficient to guarantee existence of a unique solution of the initial value problem on the entire interval (a, b).

Theorem 4.1 continues this theme. The theory of linear differential equations—whether scalar equations or first order systems—has an underlying conceptual unity. Theorem 4.1 is an overarching result, including Theorems 2.1, 3.1, and 3.5 as special cases.

Theorem 4.1

Consider the initial value problem

$$\mathbf{y}'(t) = P(t)\mathbf{y}(t) + \mathbf{g}(t), \qquad \mathbf{y}(t_0) = \mathbf{y}_0,$$

where $\mathbf{y}(t)$, $P(t)$, $\mathbf{g}(t)$, and $\mathbf{y}_0$ are defined as in equation (2). Let the n^2 components of $P(t)$ and the n components of $\mathbf{g}(t)$ be continuous on the interval (a, b), and let t_0 be in (a, b). Then the initial value problem has a unique solution that exists on the entire interval (a, b).

E X A M P L E

1

Consider the initial value problem

$$y_1' = (\sin 2t)y_1 + \frac{t}{t^2 - 2t - 8}y_2 + 4, \qquad y_1(1) = 2$$

$$y_2' = (\ln|t + 1|)y_1 + e^{-2t}y_2 + \sec t, \qquad y_2(1) = 0.$$

Determine the largest t-interval on which Theorem 4.1 guarantees the existence of a unique solution of this problem.

Solution: The given problem can be rewritten as

$$\mathbf{y}'(t) = P(t)\mathbf{y}(t) + \mathbf{g}(t), \qquad a < t < b,$$
$$\mathbf{y}(t_0) = \mathbf{y}_0,$$

where

$$\mathbf{y}(t) = \begin{bmatrix} y_1(t) \\ y_2(t) \end{bmatrix}, \qquad P(t) = \begin{bmatrix} \sin 2t & \dfrac{t}{t^2 - 2t - 8} \\ \ln|t + 1| & e^{-2t} \end{bmatrix},$$

$$\mathbf{g}(t) = \begin{bmatrix} 4 \\ \sec t \end{bmatrix}, \qquad t_0 = 1, \qquad \mathbf{y}_0 = \begin{bmatrix} 2 \\ 0 \end{bmatrix}.$$

According to Theorem 4.1, a unique solution is guaranteed to exist on the largest interval (a, b), containing the point $t_0 = 1$, in which the four components of $P(t)$ and the two components of $\mathbf{g}(t)$ are continuous.

The functions $p_{11}(t) = \sin 2t$, $p_{22}(t) = e^{-2t}$, and $g_1(t) = 4$ are continuous for all t, $-\infty < t < \infty$. The function

$$p_{12}(t) = \frac{t}{t^2 - 2t - 8} = \frac{t}{(t - 4)(t + 2)}$$

has discontinuities at $t = -2$ and $t = 4$. Similarly, $p_{21}(t) = \ln|t + 1|$ is discontinuous at $t = -1$, while $g_2(t) = \sec t$ has discontinuities at odd multiples of $\pi/2$. Since $\pi/2 \approx 1.571$, the largest interval containing $t_0 = 1$ on which all six functions are continuous is $-1 < t < \pi/2$. Theorem 4.1 guarantees the existence of a unique solution on the interval $-1 < t < \pi/2$. ❖

Rewriting an *n*th Order Scalar Linear Equation as a First Order Linear System

Theorem 4.1 is an overarching result because it is always possible to rewrite an nth order scalar linear differential equation as a system of n first order linear differential equations. We introduce the technique with a simple example, rewriting a second order scalar linear differential equation as a system of two first order linear equations. Initial conditions can be transformed as well.

Consider the scalar initial value problem

$$y'' - e^t y' + 3y = \sin 2t, \qquad y(0) = 2, \qquad y'(0) = -1. \tag{3}$$

Make the change of variables

$$y_1(t) = y(t), \qquad y_2(t) = y'(t). \tag{4}$$

Define the vector function $\mathbf{y}(t)$ by

$$\mathbf{y}(t) = \begin{bmatrix} y_1(t) \\ y_2(t) \end{bmatrix}.$$

From (3) and (4), we have

$$y_1'(t) = y'(t) = y_2(t)$$
$$y_2'(t) = y''(t) = e^t y'(t) - 3y(t) + \sin 2t = e^t y_2(t) - 3y_1(t) + \sin 2t.$$

Therefore, initial value problem (3) can be rewritten as

$$\begin{bmatrix} y_1'(t) \\ y_2'(t) \end{bmatrix} = \begin{bmatrix} y_2(t) \\ e^t y_2(t) - 3y_1(t) + \sin 2t \end{bmatrix}, \qquad \mathbf{y}(0) = \begin{bmatrix} 2 \\ -1 \end{bmatrix}.$$

This initial value problem has the form $\mathbf{y}'(t) = P(t)\mathbf{y}(t) + \mathbf{g}(t)$, $\mathbf{y}(0) = \mathbf{y}_0$, where

$$P(t) = \begin{bmatrix} 0 & 1 \\ -3 & e^t \end{bmatrix}, \qquad \mathbf{g}(t) = \begin{bmatrix} 0 \\ \sin 2t \end{bmatrix}, \qquad \mathbf{y}_0 = \begin{bmatrix} 2 \\ -1 \end{bmatrix}.$$

In general, consider the nth order scalar linear differential equation

$$y^{(n)} + p_{n-1}(t)y^{(n-1)} + \cdots + p_1(t)y' + p_0(t)y = g(t). \tag{5}$$

Make the change of variables

$$y_1(t) = y(t), \qquad y_2(t) = y'(t), \qquad y_3(t) = y''(t), \qquad \ldots, \qquad y_n(t) = y^{(n-1)}(t). \tag{6}$$

Define

$$\mathbf{y}(t) = \begin{bmatrix} y_1(t) \\ y_2(t) \\ \vdots \\ y_n(t) \end{bmatrix},$$

and calculate the components of $\mathbf{y}'(t)$ using equations (6) and (5). The result is a linear system, $\mathbf{y}'(t) = P(t)\mathbf{y}(t) + \mathbf{g}(t)$. [If there are initial conditions associated with differential equation (5), we can use (6) to express the initial conditions in the form $\mathbf{y}(t_0) = \mathbf{y}_0$.]

EXAMPLE 2

Rewrite the scalar initial value problem as a first order linear system,

$$y''' - t^2 y'' + 3ty' + 5y = e^{-4t}, \qquad y(0) = 1, \qquad y'(0) = 3, \qquad y''(0) = 7.$$

Solution: Define $\mathbf{y}(t)$ using the change of variables (6):

$$\mathbf{y}(t) = \begin{bmatrix} y_1(t) \\ y_2(t) \\ y_3(t) \end{bmatrix} = \begin{bmatrix} y(t) \\ y'(t) \\ y''(t) \end{bmatrix}.$$

Differentiating $\mathbf{y}(t)$, we obtain

$$\mathbf{y}'(t) = \begin{bmatrix} y_1'(t) \\ y_2'(t) \\ y_3'(t) \end{bmatrix} = \begin{bmatrix} y'(t) \\ y''(t) \\ y'''(t) \end{bmatrix} = \begin{bmatrix} y_2(t) \\ y_3(t) \\ t^2 y_3(t) - 3ty_2(t) - 5y_1(t) + e^{-4t} \end{bmatrix}.$$

Therefore, the initial value problem can be written as

$$\begin{bmatrix} y_1'(t) \\ y_2'(t) \\ y_3'(t) \end{bmatrix} = \begin{bmatrix} 0 & 1 & 0 \\ 0 & 0 & 1 \\ -5 & -3t & t^2 \end{bmatrix} \begin{bmatrix} y_1(t) \\ y_2(t) \\ y_3(t) \end{bmatrix} + \begin{bmatrix} 0 \\ 0 \\ e^{-4t} \end{bmatrix}, \qquad \mathbf{y}(0) = \begin{bmatrix} 1 \\ 3 \\ 7 \end{bmatrix}. \ \diamond$$

The ability to rewrite an nth order scalar equation as a first order linear system leads to an important conceptual unity for the theory of differential equations. As we present the theory of first order linear systems in this chapter, we will point out how the results being developed relate to analogous results from Chapters 2 and 3.

What are the practical implications of this relationship? On one hand, the techniques we have seen in prior chapters for solving nth order scalar problems have not been rendered obsolete; they remain as relevant as ever. As we will see, if an initial value problem for a scalar differential equation can be solved using the techniques of Chapter 3, that process is usually easier than rewriting the equation as a first order system and then solving the resulting matrix problem.

On the other hand, the ability to recast higher order scalar problems as first order systems is very useful, for example, in applying numerical algorithms to solve these problems. Consider, for instance, the initial value problem in Example 2. Although Theorem 3.5 assures us that the problem has a unique solution on $-\infty < t < \infty$, we have no method to explicitly construct the solution; the third order nonhomogeneous differential equation, although linear, has variable coefficients. Therefore, we might choose to solve this problem numerically. As we will see in Section 4.9, Euler's method (as well as other more accurate numerical methods) can be extended to apply to initial value problems for first order systems. In fact, when we want to use a numerical method to solve a higher order scalar problem, the first step normally is to rewrite the scalar problem as a first order system.

EXERCISES

Exercises 1–4:

Find the largest interval $a < t < b$ such that a unique solution of the given initial value problem is guaranteed to exist.

1. $y_1' = t^{-1}y_1 + (\tan t)y_2, \quad y_1(3) = 0$
$y_2' = (\ln |t|)y_1 + e^t y_2, \quad y_2(3) = 1$

2. $y_1' = y_1 + (\tan t)y_2 + (t+1)^{-2} \quad y_1(0) = 0$
$y_2' = (t^2 - 2)y_1 + 4y_2, \quad y_2(0) = 0$

3. $t^2 y_1' = (\cos t)y_1 + y_2 + 1, \quad y_1(1) = 0$
$y_2' = 2y_1 + 4ty_2 + \sec t, \quad y_2(1) = 2$

4. $(t+2)y_1' = 3ty_1 + 5y_2, \quad y_1(1) = 0$
$(t-2)y_2' = 2y_1 + 4ty_2, \quad y_2(1) = 2$

Exercises 5–6:

Verify, for any values c_1 and c_2, that the functions $y_1(t)$ and $y_2(t)$ satisfy the given system of linear differential equations.

5. $y_1' = 4y_1 + y_2, \quad y_1(t) = c_1 e^{5t} + c_2 e^{3t}$
$y_2' = y_1 + 4y_2, \quad y_2(t) = c_1 e^{5t} - c_2 e^{3t}$

6. $y_1' = y_1 + y_2, \quad y_1(t) = c_1 e^t \cos t + c_2 e^t \sin t$
$y_2' = -y_1 + y_2, \quad y_2(t) = -c_1 e^t \sin t + c_2 e^t \cos t$

Exercises 7–8:

For each of the exercises,

(a) Rewrite the equations from the given exercise in vector form as $\mathbf{y}'(t) = A\mathbf{y}(t)$, identifying the constant matrix A.

(b) Rewrite the solution of the equations in part (a) in vector form as $\mathbf{y}(t) = c_1\mathbf{y}_1(t) + c_2\mathbf{y}_2(t)$.

7. Exercise 5 **8.** Exercise 6

Exercises 9–10:

Each exercise lists a candidate for the solution, $\mathbf{y}(t)$, of the equation $\mathbf{y}' = A\mathbf{y}$, where A is the given constant matrix. Verify that $\mathbf{y}(t)$ is indeed a solution for any choice of the constants c_1 and c_2. Find values of c_1 and c_2 such that $\mathbf{y}(t)$ solves the given initial value problem. [According to Theorem 4.1, you have found the unique solution of $\mathbf{y}' = A\mathbf{y}$, $\mathbf{y}(0) = \mathbf{y}_0$.]

9. $\mathbf{y}' = A\mathbf{y}, \quad \mathbf{y}(0) = \begin{bmatrix} 4 \\ -3 \end{bmatrix}$, where $A = \begin{bmatrix} 1 & -2 \\ 1 & 4 \end{bmatrix}$ and $\mathbf{y}(t) = c_1 e^{2t}\begin{bmatrix} 2 \\ -1 \end{bmatrix} + c_2 e^{3t}\begin{bmatrix} 1 \\ -1 \end{bmatrix}$

10. $\mathbf{y}' = A\mathbf{y}, \quad \mathbf{y}(0) = \begin{bmatrix} -1 \\ 8 \end{bmatrix}$, where $A = \begin{bmatrix} 3 & 2 \\ 4 & 1 \end{bmatrix}$ and $\mathbf{y}(t) = c_1 e^{5t}\begin{bmatrix} 1 \\ 1 \end{bmatrix} + c_2 e^{-t}\begin{bmatrix} -1 \\ 2 \end{bmatrix}$

Exercises 11–14:

As in Example 2, rewrite the scalar linear differential equation as a system of first order linear differential equations of the form $\mathbf{y}' = P(t)\mathbf{y} + \mathbf{g}(t)$. Identify the matrix function $P(t)$ and the vector function $\mathbf{g}(t)$.

11. $y'' + t^2 y' + 4y = \sin t$

12. $(\cos t)y'' - 3ty' + \sqrt{t}\,y = t^2 + 1$

13. $e^t y''' + 5y'' + t^{-1}y' + (\tan t)y = 1$

14. $2y'' + ty + e^{3t} = y''' + (\cos t)y'$

Exercises 15–17:

Each initial value problem was obtained from an initial value problem for a higher order scalar differential equation. What is the corresponding scalar initial value problem?

15. $\mathbf{y}' = \begin{bmatrix} 0 & 1 \\ -3 & 2 \end{bmatrix}\mathbf{y} + \begin{bmatrix} 0 \\ 2\cos 2t \end{bmatrix}, \quad \mathbf{y}(-1) = \begin{bmatrix} 1 \\ 4 \end{bmatrix}$

16. $\mathbf{y}' = \begin{bmatrix} y_2 \\ y_3 \\ -2y_1 + 4y_3 + e^{3t} \end{bmatrix}, \quad \mathbf{y}(0) = \begin{bmatrix} 1 \\ -2 \\ 3 \end{bmatrix}$

17. $\mathbf{y}' = \begin{bmatrix} y_2 \\ y_3 \\ y_4 \\ y_2 + y_3\sin(y_1) + y_3^2 \end{bmatrix}, \quad \mathbf{y}(1) = \begin{bmatrix} 0 \\ 0 \\ -1 \\ 2 \end{bmatrix}$

Exercises 18–21:

Exercises 11–17 dealt with rewriting a single scalar equation as a first order system. Frequently, however, we need to convert systems of higher order equations into a single first order system. In each exercise, rewrite the given system of two second order equations as a system of four first order linear equations of the form $\mathbf{y}' = P(t)\mathbf{y} + \mathbf{g}(t)$. In each exercise, use the following change of variables and identify $P(t)$ and $\mathbf{g}(t)$:

$$\mathbf{y}(t) = \begin{bmatrix} y_1(t) \\ y_2(t) \\ y_3(t) \\ y_4(t) \end{bmatrix} = \begin{bmatrix} y(t) \\ y'(t) \\ z(t) \\ z'(t) \end{bmatrix}.$$

18. $y'' = tz' + y' + z$
$z'' = y' + z' + 2ty$

19. $y'' = t^{-1}y' + 4y - tz + (\sin t)z' + e^{2t}$
$z'' = y - 5z'$

20. $y'' = 7y' + 4y - 8z + 6z' + t^2$
$z'' = 5z' + 2z - 6y' + 3y - \sin t$

21. $15z + 9y' + 3y'' = 12y - 6z' + 3t^2$
$z' + 5y - z'' = 2z - 6y' + t$

4.3   Homogeneous Linear Systems

Our previous discussions of linear differential equations began with homogenous equations. We use the same approach here. Consider the system of n first order homogeneous linear differential equations,

$$\begin{aligned} y_1' &= p_{11}(t)y_1 + p_{12}(t)y_2 + \cdots + p_{1n}(t)y_n \\ y_2' &= p_{21}(t)y_1 + p_{22}(t)y_2 + \cdots + p_{2n}(t)y_n \\ &\vdots \\ y_n' &= p_{n1}(t)y_1 + p_{n2}(t)y_2 + \cdots + p_{nn}(t)y_n, \qquad a < t < b. \end{aligned} \tag{1}$$

This **first order homogeneous linear system** can be written in matrix form as

$$\mathbf{y}' = P(t)\mathbf{y}, \qquad a < t < b, \tag{2}$$

where

$$y(t) = \begin{bmatrix} y_1(t) \\ y_2(t) \\ \vdots \\ y_n(t) \end{bmatrix} \quad \text{and} \quad P(t) = \begin{bmatrix} p_{11}(t) & p_{12}(t) & \cdots & p_{1n}(t) \\ p_{21}(t) & p_{22}(t) & & p_{2n}(t) \\ \vdots & & & \vdots \\ p_{n1}(t) & p_{n2}(t) & \cdots & p_{nn}(t) \end{bmatrix}.$$

The Principle of Superposition

Theorem 4.2 states a superposition principle for equation (2).

Theorem 4.2

> Let $\mathbf{y}_1(t), \mathbf{y}_2(t), \ldots, \mathbf{y}_r(t)$ be any r solutions of the homogeneous linear equation
>
> $$\mathbf{y}' = P(t)\mathbf{y}, \qquad a < t < b.$$
>
> Then, for any r constants $c_1, c_2, \ldots, c_r$, the linear combination
>
> $$\mathbf{y}(t) = c_1\mathbf{y}_1(t) + c_2\mathbf{y}_2(t) + \cdots + c_r\mathbf{y}_r(t)$$
>
> is also a solution on the t-interval $a < t < b$.

● **PROOF:**   The proof of Theorem 4.2 is conceptually the same as the proof of Theorem 3.2; we simply substitute the expression for $\mathbf{y}(t)$ into differential equation (2), obtaining

$$\mathbf{y}' = (c_1\mathbf{y}_1 + c_2\mathbf{y}_2 + \cdots + c_r\mathbf{y}_r)' = c_1\mathbf{y}_1' + c_2\mathbf{y}_2' + \cdots + c_r\mathbf{y}_r'. \tag{3}$$

Since $\mathbf{y}_i' = P(t)\mathbf{y}_i$ for $1 \le i \le n$, we can rewrite (3) as

$$\mathbf{y}' = c_1 P(t)\mathbf{y}_1 + c_2 P(t)\mathbf{y}_2 + \cdots + c_r P(t)\mathbf{y}_r = P(t)(c_1\mathbf{y}_1 + c_2\mathbf{y}_2 + \cdots + c_r\mathbf{y}_r) = P(t)\mathbf{y}.$$

●

Fundamental Sets and the General Solution

We are mainly concerned with Theorem 4.2 in the case $r = n$. In this context, we again introduce the concept of a fundamental set of solutions.

Let $\{\mathbf{y}_1(t), \mathbf{y}_2(t), \ldots, \mathbf{y}_n(t)\}$ be a set of n solutions of the linear system (2). This set of solutions is called a **fundamental set of solutions** if every solution of (2) can be written as a linear combination of the form $c_1\mathbf{y}_1(t) + c_2\mathbf{y}_2(t) + \cdots + c_n\mathbf{y}_n(t)$. If $\{\mathbf{y}_1(t), \mathbf{y}_2(t), \ldots, \mathbf{y}_n(t)\}$ is a fundamental set of solutions of $\mathbf{y}' = P(t)\mathbf{y}$, then the expression

$$\mathbf{y}(t) = c_1\mathbf{y}_1(t) + c_2\mathbf{y}_2(t) + \cdots + c_n\mathbf{y}_n(t)$$

is called the **general solution.**

EXAMPLE

1

Consider the first order linear homogeneous system $\mathbf{y}' = A\mathbf{y}$, $-\infty < t < \infty$, where

$$A = \begin{bmatrix} 0 & 1 & 0 \\ -6 & 5 & 0 \\ 0 & 0 & 1 \end{bmatrix}.$$

You can verify by direct substitution that the vector functions

$$\mathbf{y}_1(t) = \begin{bmatrix} e^{2t} \\ 2e^{2t} \\ 0 \end{bmatrix}, \qquad \mathbf{y}_2(t) = \begin{bmatrix} e^{3t} \\ 3e^{3t} \\ 0 \end{bmatrix}, \qquad \mathbf{y}_3(t) = \begin{bmatrix} 0 \\ 0 \\ e^t \end{bmatrix}$$

are a solution of $\mathbf{y}' = A\mathbf{y}$. From Theorem 4.2, it follows that the linear combination

$$\mathbf{y}(t) = c_1\mathbf{y}_1(t) + c_2\mathbf{y}_2(t) + c_3\mathbf{y}_3(t) \tag{4}$$

is also a solution for any choice of constants c_1, c_2, c_3. Is (4) the general solution of $\mathbf{y}' = A\mathbf{y}$? We will show in Example 2 that the answer is "Yes." Also, in Section 4.4, we describe how to obtain the three solutions $\mathbf{y}_1(t)$, $\mathbf{y}_2(t)$, and $\mathbf{y}_3(t)$. ❖

Two Important Identities

We take note of a simple but important matrix identity known as the **column form for matrix-vector multiplication**. Let A be an $(m \times n)$ matrix, and let $\mathbf{x}$ be an $(n \times 1)$ vector. We can represent A in **column form** as

$$A = [\mathbf{A}_1, \mathbf{A}_2, \dots, \mathbf{A}_n].$$

[In this column form representation, $\mathbf{A}_1, \mathbf{A}_2, \dots, \mathbf{A}_n$ denote the columns of A; each $\mathbf{A}_i$ is an $(m \times 1)$ vector.] Let the vector $\mathbf{x}$ be given by

$$\mathbf{x} = \begin{bmatrix} x_1 \\ x_2 \\ \vdots \\ x_n \end{bmatrix}.$$

Then the matrix-vector product $A\mathbf{x}$ is equal to the linear combination

$$A\mathbf{x} = x_1\mathbf{A}_1 + x_2\mathbf{A}_2 + \cdots + x_n\mathbf{A}_n. \tag{5}$$

For instance, consider the linear combination $\mathbf{y}(t) = c_1\mathbf{y}_1(t) + c_2\mathbf{y}_2(t) + c_3\mathbf{y}_3(t)$ appearing in equation (4) in Example 1. Using identity (5), we can rewrite this linear combination in the form $\mathbf{y}(t) = \Psi(t)\mathbf{c}$, where

$$\Psi(t) = [\mathbf{y}_1(t), \mathbf{y}_2(t), \mathbf{y}_3(t)] = \begin{bmatrix} e^{2t} & e^{3t} & 0 \\ 2e^{2t} & 3e^{3t} & 0 \\ 0 & 0 & e^t \end{bmatrix} \quad \text{and} \quad \mathbf{c} = \begin{bmatrix} c_1 \\ c_2 \\ c_3 \end{bmatrix}.$$

A related identity involves matrix-matrix multiplication. Let B be an $(m \times n)$ matrix, and let $C = [\mathbf{C}_1, \mathbf{C}_2, \dots, \mathbf{C}_r]$ be an $(n \times r)$ matrix. Then, in column form, the matrix product BC can be written as

$$BC = [B\mathbf{C}_1, B\mathbf{C}_2, \dots, B\mathbf{C}_r]. \tag{6}$$

In other words [see (6)], the ith column of BC is the product of the matrix B and the ith column of C.

The Wronskian

Consider the initial value problem

$$\mathbf{y}' = P(t)\mathbf{y}, \qquad \mathbf{y}(t_0) = \mathbf{y}_0, \qquad a < t < b, \tag{7}$$

where the $(n \times n)$ matrix $P(t)$ is continuous on (a, b) and where t_0 is in (a, b). Let $\{\mathbf{y}_1(t), \mathbf{y}_2(t), \ldots, \mathbf{y}_n(t)\}$ be a fundamental set of solutions for $\mathbf{y}' = P(t)\mathbf{y}$. Therefore, the general solution of $\mathbf{y}' = P(t)\mathbf{y}$ is

$$\mathbf{y}(t) = c_1\mathbf{y}_1(t) + c_2\mathbf{y}_2(t) + \cdots + c_n\mathbf{y}_n(t).$$

The unique solution of initial value problem (7) is found by imposing the initial condition

$$c_1\mathbf{y}_1(t_0) + c_2\mathbf{y}_2(t_0) + \cdots + c_n\mathbf{y}_n(t_0) = \mathbf{y}_0.$$

Using (5), we can write this vector equation as

$$[\mathbf{y}_1(t_0), \mathbf{y}_2(t_0), \ldots, \mathbf{y}_n(t_0)] \begin{bmatrix} c_1 \\ c_2 \\ \vdots \\ c_n \end{bmatrix} = \mathbf{y}_0. \tag{8a}$$

For brevity, let $\Psi(t) = [\mathbf{y}_1(t), \mathbf{y}_2(t), \ldots, \mathbf{y}_n(t)]$, and let $\mathbf{c}$ denote the $(n \times 1)$ vector of constants in equation (8a). Using this notation, we can write equation (8a) as

$$\Psi(t_0)\mathbf{c} = \mathbf{y}_0. \tag{8b}$$

By Theorem 4.1, equation (8b) has a unique solution for every right-hand side $\mathbf{y}_0$ and choice of t_0 in (a, b). Therefore, $\det[\Psi(t)] \neq 0$ for all t in (a, b).

This discussion leads us once more to the definition of a Wronskian. Let $\{\mathbf{y}_1(t), \mathbf{y}_2(t), \ldots, \mathbf{y}_n(t)\}$ be a set of n solutions of $\mathbf{y}' = P(t)\mathbf{y}$, and let $\Psi(t) = [\mathbf{y}_1(t), \mathbf{y}_2(t), \ldots, \mathbf{y}_n(t)]$. The **Wronskian**, $W(t)$, is defined to be

$$W(t) = \det[\Psi(t)].$$

We have seen that if the columns of $\Psi(t)$ form a fundamental set of solutions, then $W(t) \neq 0$ for all t in (a, b). The converse is true as well and can be established with essentially the same arguments used in Section 3.11; see equations (4)–(6) in Section 3.11. Theorem 4.3 gives the resulting characterization of fundamental sets.

Theorem 4.3

Let $\mathbf{y}_1(t), \mathbf{y}_2(t), \ldots, \mathbf{y}_n(t)$ be n solutions of the homogeneous linear equation

$$\mathbf{y}' = P(t)\mathbf{y}, \qquad a < t < b,$$

where $P(t)$ is continuous on (a, b). Let $W(t)$ denote the Wronskian of these solutions. Then $\{\mathbf{y}_1(t), \mathbf{y}_2(t), \ldots, \mathbf{y}_n(t)\}$ is a fundamental set of solutions if and only if $W(t) \neq 0$ on (a, b).

EXAMPLE

2

Consider the initial value problem

$$\mathbf{y}' = A\mathbf{y}, \qquad \mathbf{y}(0) = \mathbf{y}_0, \qquad -\infty < t < \infty,$$

where

$$A = \begin{bmatrix} 0 & 1 & 0 \\ -6 & 5 & 0 \\ 0 & 0 & 1 \end{bmatrix}, \qquad \mathbf{y}_0 = \begin{bmatrix} 3 \\ 7 \\ 4 \end{bmatrix}.$$

From Example 1, we know that three solutions of the differential equation $\mathbf{y}' = A\mathbf{y}$ are

$$\mathbf{y}_1(t) = \begin{bmatrix} e^{2t} \\ 2e^{2t} \\ 0 \end{bmatrix}, \qquad \mathbf{y}_2(t) = \begin{bmatrix} e^{3t} \\ 3e^{3t} \\ 0 \end{bmatrix}, \qquad \mathbf{y}_3(t) = \begin{bmatrix} 0 \\ 0 \\ e^t \end{bmatrix}.$$

(a) Show that $\{\mathbf{y}_1(t), \mathbf{y}_2(t), \mathbf{y}_3(t)\}$ is a fundamental set of solutions.
(b) Solve the initial value problem.

Solution:

(a) The Wronskian is

$$W(t) = \begin{vmatrix} e^{2t} & e^{3t} & 0 \\ 2e^{2t} & 3e^{3t} & 0 \\ 0 & 0 & e^t \end{vmatrix} = e^t(3e^{5t} - 2e^{5t}) = e^{6t}.$$

Since the Wronskian is nonzero for all t, $\{\mathbf{y}_1(t), \mathbf{y}_2(t), \mathbf{y}_3(t)\}$ is a fundamental set of solutions on $-\infty < t < \infty$.

(b) From part (a), the general solution is

$$\mathbf{y}(t) = c_1\mathbf{y}_1(t) + c_2\mathbf{y}_2(t) + c_3\mathbf{y}_3(t) = \Psi(t)\mathbf{c},$$

where

$$\Psi(t) = \begin{bmatrix} e^{2t} & e^{3t} & 0 \\ 2e^{2t} & 3e^{3t} & 0 \\ 0 & 0 & e^t \end{bmatrix}.$$

Imposing the initial condition, $\mathbf{y}(0) = \Psi(0)\mathbf{c} = \mathbf{y}_0$, we obtain

$$\begin{bmatrix} 1 & 1 & 0 \\ 2 & 3 & 0 \\ 0 & 0 & 1 \end{bmatrix} \begin{bmatrix} c_1 \\ c_2 \\ c_3 \end{bmatrix} = \begin{bmatrix} 3 \\ 7 \\ 4 \end{bmatrix}.$$

Therefore, $\mathbf{c}$ is given by

$$\mathbf{c} = \Psi(0)^{-1}\mathbf{y}_0 = \begin{bmatrix} 3 & -1 & 0 \\ -2 & 1 & 0 \\ 0 & 0 & 1 \end{bmatrix} \begin{bmatrix} 3 \\ 7 \\ 4 \end{bmatrix} = \begin{bmatrix} 2 \\ 1 \\ 4 \end{bmatrix}.$$

The solution of the initial value problem is

$$\mathbf{y}(t) = \Psi(t)\mathbf{c} = \begin{bmatrix} 2e^{2t} + e^{3t} \\ 4e^{2t} + 3e^{3t} \\ 4e^{t} \end{bmatrix}. \; \diamondsuit$$

Abel's Theorem

In Section 3.11, we stated Abel's theorem and used it to establish an important dichotomy property for Wronskians formed from solutions of scalar linear homogeneous equations. We now present (without proof) a generalization of Abel's theorem. This generalization, stated in Theorem 4.4, again implies that the Wronskian of a set of solutions either vanishes nowhere or vanishes everywhere on the t-interval of interest. Theorem 4.4 shows, therefore, that we need only establish that the Wronskian is nonzero at some convenient point in order to demonstrate that a solution set is a fundamental set.

Before stating Theorem 4.4, we need to define a new quantity. Let A be an $(n \times n)$ matrix,

$$A = \begin{bmatrix} a_{11} & a_{12} & \cdots & a_{1n} \\ a_{21} & a_{22} & & a_{2n} \\ \vdots & & & \vdots \\ a_{n1} & a_{n2} & \cdots & a_{nn} \end{bmatrix}.$$

The **trace of A**, denoted by $\text{tr}[A]$, is defined to be the sum of the diagonal elements of A,

$$\text{tr}[A] = a_{11} + a_{22} + a_{33} + \cdots + a_{nn}.$$

For instance, the (3×3) matrix A in Examples 1 and 2,

$$A = \begin{bmatrix} 0 & 1 & 0 \\ -6 & 5 & 0 \\ 0 & 0 & 1 \end{bmatrix},$$

has $\text{tr}[A] = 6$. For the matrix function $P(t)$ in equation (2),

$$\text{tr}[P(t)] = p_{11}(t) + p_{22}(t) + p_{33}(t) + \cdots + p_{nn}(t).$$

Having this preliminary definition, we are ready to state Abel's theorem.

Theorem 4.4

Let $\{\mathbf{y}_1(t), \mathbf{y}_2(t), \ldots, \mathbf{y}_n(t)\}$ be a set of solutions of

$$\mathbf{y}' = P(t)\mathbf{y}, \qquad a < t < b,$$

and let $W(t)$ be the Wronskian of these solutions. Then $W(t)$ satisfies the scalar differential equation

$$W'(t) = \text{tr}[P(t)]W(t).$$

Moreover, if t_0 is any point in $a < t < b$, then

$$W(t) = W(t_0)e^{\int_{t_0}^{t} \text{tr}[P(s)]\,ds}, \qquad a < t < b. \tag{9}$$

As we see from equation (9), if $W(t_0) = 0$, then the Wronskian vanishes identically on the t-interval of interest. On the other hand, if $W(t_0) \neq 0$, then the Wronskian is never zero in (a, b).

In the special case where an nth order linear scalar differential equation is recast as a first order linear system, the definition of the Wronskian and the conclusion of Abel's theorem stated for systems reduce precisely to their counterparts in Chapter 3; see Exercise 35.

Additional Observations

We make some additional observations about the linear system $\mathbf{y}' = P(t)\mathbf{y}$ that parallel those made in Section 3.11 for scalar linear homogeneous equations. We leave the verification to the exercises.

1. Fundamental sets always exist; see Exercise 36.
2. Fundamental sets are linearly independent sets of functions; see Exercise 37.
3. Fundamental sets are not unique. Fundamental sets are related as described by Theorem 4.5.

Theorem 4.5

Let $\{\mathbf{y}_1(t), \mathbf{y}_2(t), \ldots, \mathbf{y}_n(t)\}$ be a fundamental set of solutions of

$$\mathbf{y}' = P(t)\mathbf{y}, \qquad a < t < b,$$

where the $(n \times n)$ matrix function $P(t)$ is continuous on the interval (a, b). Let

$$\Psi(t) = \left[\mathbf{y}_1(t), \mathbf{y}_2(t), \ldots, \mathbf{y}_n(t)\right]$$

denote the $(n \times n)$ matrix function formed from the fundamental set. Let $\{\hat{\mathbf{y}}_1(t), \hat{\mathbf{y}}_2(t), \ldots, \hat{\mathbf{y}}_n(t)\}$ be any other set of n solutions of the differential equation, and let

$$\hat{\Psi}(t) = \left[\hat{\mathbf{y}}_1(t), \hat{\mathbf{y}}_2(t), \ldots, \hat{\mathbf{y}}_n(t)\right]$$

denote the $(n \times n)$ matrix formed from this other set of solutions. Then,

(a) There is a unique $(n \times n)$ constant matrix C such that

$$\hat{\Psi}(t) = \Psi(t)C, \qquad a < t < b.$$

(b) Moreover, $\{\hat{\mathbf{y}}_1(t), \hat{\mathbf{y}}_2(t), \ldots, \hat{\mathbf{y}}_n(t)\}$ is also a fundamental set of solutions if and only if the determinant of C is nonzero.

Fundamental Matrices

As we have seen, it is often convenient to introduce an $(n \times n)$ matrix function

$$\Psi(t) = \left[\mathbf{y}_1(t), \mathbf{y}_2(t), \ldots, \mathbf{y}_n(t)\right],$$

where $\Psi(t)$ is formed using a set of solutions as its columns. We refer to such a matrix $\Psi(t)$ as a **solution matrix**. In addition, if the set of solutions forms a fundamental set of solutions, we call $\Psi(t)$ a **fundamental matrix** of $\mathbf{y}' = P(t)\mathbf{y}$.

Part (a) of Theorem 4.5 states that any solution matrix can be expressed as any given fundamental matrix multiplied on the right by an $(n \times n)$ constant matrix C. Part (b) of Theorem 4.5 states that any other fundamental matrix can be expressed as the given fundamental matrix multiplied on the right by an invertible $(n \times n)$ constant matrix C.

EXERCISES

Exercises 1–6:

(a) Rewrite the given system of linear homogeneous differential equations as a homogeneous linear system of the form $\mathbf{y}' = P(t)\mathbf{y}$.

(b) Verify that the given function $\mathbf{y}(t)$ is a solution of $\mathbf{y}' = P(t)\mathbf{y}$.

1. $\begin{aligned} y_1' &= 9y_1 - 4y_2 \\ y_2' &= 15y_1 - 7y_2 \end{aligned}$, $\mathbf{y}(t) = \begin{bmatrix} 2e^{3t} \\ 3e^{3t} \end{bmatrix}$

2. $\begin{aligned} y_1' &= -3y_1 - 2y_2 \\ y_2' &= 4y_1 + 3y_2 \end{aligned}$, $\mathbf{y}(t) = \begin{bmatrix} e^t + e^{-t} \\ -2e^t - e^{-t} \end{bmatrix}$

3. $\begin{aligned} y_1' &= y_1 + 4y_2 \\ y_2' &= -y_1 + y_2 \end{aligned}$, $\mathbf{y}(t) = \begin{bmatrix} 2e^t \cos 2t \\ -e^t \sin 2t \end{bmatrix}$

4. $\begin{aligned} y_1' &= y_2 \\ y_2' &= \frac{2}{t^2}y_1 - \frac{2}{t}y_2 \end{aligned}$, $t > 0$, $\mathbf{y}(t) = \begin{bmatrix} -t^2 + 3t \\ -2t + 3 \end{bmatrix}$

5. $\begin{aligned} y_1' &= y_2 + y_3 \\ y_2' &= -6y_1 - 3y_2 + y_3, \\ y_3' &= -8y_1 - 2y_2 + 4y_3 \end{aligned}$ $\mathbf{y}(t) = \begin{bmatrix} e^t \\ -e^t \\ 2e^t \end{bmatrix}$

6. $\begin{aligned} y_1' &= 2y_1 + y_2 + y_3 \\ y_2' &= y_1 + y_2 + 2y_3, \\ y_3' &= y_1 + 2y_2 + y_3 \end{aligned}$ $\mathbf{y}(t) = \begin{bmatrix} 2e^t + e^{4t} \\ -e^t + e^{4t} \\ -e^t + e^{4t} \end{bmatrix}$

Exercises 7–14:

Determine whether the given functions form a fundamental set of solutions for the linear system.

7. $\mathbf{y}' = \begin{bmatrix} 0 & -1 \\ -1 & 0 \end{bmatrix} \mathbf{y}$, $\mathbf{y}_1(t) = \begin{bmatrix} e^t \\ -e^t \end{bmatrix}$, $\mathbf{y}_2(t) = \begin{bmatrix} e^{-t} \\ e^{-t} \end{bmatrix}$

8. $\mathbf{y}' = \begin{bmatrix} 2 & -2 \\ 0 & 1 \end{bmatrix} \mathbf{y}$, $\mathbf{y}_1(t) = \begin{bmatrix} 3e^{2t} \\ 0 \end{bmatrix}$, $\mathbf{y}_2(t) = \begin{bmatrix} 2e^t \\ e^t \end{bmatrix}$

9. $\mathbf{y}' = \begin{bmatrix} 0 & 1 \\ 0 & 2 \end{bmatrix} \mathbf{y}$, $\mathbf{y}_1(t) = \begin{bmatrix} 1 \\ 0 \end{bmatrix}$, $\mathbf{y}_2(t) = \begin{bmatrix} -2 \\ 0 \end{bmatrix}$

10. $\mathbf{y}' = \begin{bmatrix} 0 & 2 \\ -2 & 0 \end{bmatrix} \mathbf{y}$, $\mathbf{y}_1(t) = \begin{bmatrix} \cos 2t \\ -\sin 2t \end{bmatrix}$, $\mathbf{y}_2(t) = \begin{bmatrix} -2 \sin 2t \\ -2 \cos 2t \end{bmatrix}$

11. $\mathbf{y}' = \begin{bmatrix} 1 & -1 \\ 5 & -1 \end{bmatrix} \mathbf{y}$, $\mathbf{y}_1(t) = \begin{bmatrix} \cos 2t \\ \cos 2t + \sin 2t \end{bmatrix}$, $\mathbf{y}_2(t) = \begin{bmatrix} \sin 2t \\ \sin 2t - 2 \cos 2t \end{bmatrix}$

12. $\mathbf{y}' = \begin{bmatrix} -1 & 2 \\ 0 & -1 \end{bmatrix} \mathbf{y}$, $\mathbf{y}_1(t) = \begin{bmatrix} e^{-t} \\ 0 \end{bmatrix}$, $\mathbf{y}_2(t) = \begin{bmatrix} e^{-t} \\ e^{-t} \end{bmatrix}$

13. $\mathbf{y}' = \begin{bmatrix} 1 & -2 & 1 \\ 0 & -1 & 1 \\ 0 & 0 & 0 \end{bmatrix} \mathbf{y}$, $\quad \mathbf{y}_1(t) = \begin{bmatrix} e^t \\ 0 \\ 0 \end{bmatrix}$, $\quad \mathbf{y}_2(t) = \begin{bmatrix} e^{-t} \\ e^{-t} \\ 0 \end{bmatrix}$, $\quad \mathbf{y}_3(t) = \begin{bmatrix} 1 \\ 1 \\ 1 \end{bmatrix}$

14. $\mathbf{y}' = \begin{bmatrix} 0 & 1 \\ t^{-2} & -t^{-1} \end{bmatrix} \mathbf{y}$, $\quad \mathbf{y}_1(t) = \begin{bmatrix} t \\ 1 \end{bmatrix}$, $\quad \mathbf{y}_2(t) = \begin{bmatrix} t^{-1} \\ -t^{-2} \end{bmatrix}$, $\quad 0 < t < \infty$

Exercises 15–23:

(a) Verify that the given functions are solutions of the homogeneous linear system.

(b) Compute the Wronskian of the solution set. On the basis of this calculation, can you assert that the set of solutions forms a fundamental set?

(c) If the given solutions form a fundamental set, state the general solution of the linear homogeneous system. Express the general solution as the product $\mathbf{y}(t) = \Psi(t)\mathbf{c}$, where $\Psi(t)$ is a square matrix whose columns are the solutions forming the fundamental set and $\mathbf{c}$ is a column vector of arbitrary constants.

(d) If the solutions form a fundamental set, impose the given initial condition and find the unique solution of the initial value problem.

15. $\mathbf{y}' = \begin{bmatrix} 9 & -4 \\ 15 & -7 \end{bmatrix} \mathbf{y}$, $\quad \mathbf{y}(0) = \begin{bmatrix} 1 \\ 1 \end{bmatrix}$; $\quad \mathbf{y}_1(t) = \begin{bmatrix} 2e^{3t} \\ 3e^{3t} \end{bmatrix}$, $\quad \mathbf{y}_2(t) = \begin{bmatrix} 2e^{-t} \\ 5e^{-t} \end{bmatrix}$

16. $\mathbf{y}' = \begin{bmatrix} 9 & -4 \\ 15 & -7 \end{bmatrix} \mathbf{y}$, $\quad \mathbf{y}(0) = \begin{bmatrix} 0 \\ 1 \end{bmatrix}$; $\quad \mathbf{y}_1(t) = \begin{bmatrix} 2e^{3t} - 4e^{-t} \\ 3e^{3t} - 10e^{-t} \end{bmatrix}$, $\quad \mathbf{y}_2(t) = \begin{bmatrix} 4e^{3t} + 2e^{-t} \\ 6e^{3t} + 5e^{-t} \end{bmatrix}$

17. $\mathbf{y}' = \begin{bmatrix} 3 & 2 \\ -4 & -3 \end{bmatrix} \mathbf{y}$, $\quad \mathbf{y}(0) = \begin{bmatrix} 1 \\ 1 \end{bmatrix}$; $\quad \mathbf{y}_1(t) = \begin{bmatrix} e^{-t} \\ -2e^{-t} \end{bmatrix}$, $\quad \mathbf{y}_2(t) = \begin{bmatrix} -3e^{-t} \\ 6e^{-t} \end{bmatrix}$

18. $\mathbf{y}' = \begin{bmatrix} -3 & -5 \\ 2 & -1 \end{bmatrix} \mathbf{y}$, $\quad \mathbf{y}(0) = \begin{bmatrix} 5 \\ 2 \end{bmatrix}$; $\quad \mathbf{y}_1(t) = \begin{bmatrix} -5e^{-2t} \cos 3t \\ e^{-2t}(\cos 3t - 3\sin 3t) \end{bmatrix}$,

$\mathbf{y}_2(t) = \begin{bmatrix} -5e^{-2t} \sin 3t \\ e^{-2t}(3\cos 3t + \sin 3t) \end{bmatrix}$

19. $\mathbf{y}' = \begin{bmatrix} -3 & -2 \\ 4 & 3 \end{bmatrix} \mathbf{y}$, $\quad \mathbf{y}(1) = \begin{bmatrix} 1 \\ -3 \end{bmatrix}$; $\quad \mathbf{y}_1(t) = \begin{bmatrix} e^t \\ -2e^t \end{bmatrix}$, $\quad \mathbf{y}_2(t) = \begin{bmatrix} e^{-t} \\ -e^{-t} \end{bmatrix}$

20. $\mathbf{y}' = \begin{bmatrix} 1 & -1 \\ -2 & 2 \end{bmatrix} \mathbf{y}$, $\quad \mathbf{y}(-1) = \begin{bmatrix} -2 \\ 4 \end{bmatrix}$; $\quad \mathbf{y}_1(t) = \begin{bmatrix} 1 \\ 1 \end{bmatrix}$, $\quad \mathbf{y}_2(t) = \begin{bmatrix} e^{3t} \\ -2e^{3t} \end{bmatrix}$

21. $\mathbf{y}' = \begin{bmatrix} 2t^{-2} & 1 - 2t^{-1} + 2t^{-2} \\ -2t^{-2} & 2t^{-1} - 2t^{-2} \end{bmatrix} \mathbf{y}$, $\quad \mathbf{y}(2) = \begin{bmatrix} -2 \\ 2 \end{bmatrix}$, $\quad t > 0$; $\quad \mathbf{y}_1(t) = \begin{bmatrix} t^2 - 2t \\ 2t \end{bmatrix}$,

$\mathbf{y}_2(t) = \begin{bmatrix} t - 1 \\ 1 \end{bmatrix}$

22. $\mathbf{y}' = \begin{bmatrix} -2 & 0 & 0 \\ 0 & 1 & 4 \\ 0 & -1 & 1 \end{bmatrix} \mathbf{y}$, $\quad \mathbf{y}(0) = \begin{bmatrix} 3 \\ 4 \\ -2 \end{bmatrix}$; $\quad \mathbf{y}_1(t) = \begin{bmatrix} e^{-2t} \\ 0 \\ 0 \end{bmatrix}$, $\quad \mathbf{y}_2(t) = \begin{bmatrix} 0 \\ 2e^t \cos 2t \\ -e^t \sin 2t \end{bmatrix}$,

$\mathbf{y}_3(t) = \begin{bmatrix} 0 \\ 2e^t \sin 2t \\ e^t \cos 2t \end{bmatrix}$

23. $\mathbf{y}' = \begin{bmatrix} -21 & -10 & 2 \\ 22 & 11 & -2 \\ -110 & -50 & 11 \end{bmatrix} \mathbf{y}, \quad \mathbf{y}(0) = \begin{bmatrix} 3 \\ -10 \\ -16 \end{bmatrix}; \quad \mathbf{y}_1(t) = \begin{bmatrix} 5e^t \\ -11e^t \\ 0 \end{bmatrix}, \quad \mathbf{y}_2(t) = \begin{bmatrix} e^t \\ 0 \\ 11e^t \end{bmatrix},$

$$\mathbf{y}_3(t) = \begin{bmatrix} e^{-t} \\ -e^{-t} \\ 5e^{-t} \end{bmatrix}$$

Exercises 24–27:

The given functions are solutions of the homogeneous linear system.

(a) Compute the Wronskian of the solution set and verify that the solution set is a fundamental set of solutions.

(b) Compute the trace of the coefficient matrix.

(c) Verify Abel's theorem by showing that, for the given point t_0,

$$W(t) = W(t_0)e^{\int_{t_0}^{t} \text{tr}[P(s)]ds}.$$

24. $\mathbf{y}' = \begin{bmatrix} 6 & 5 \\ -7 & -6 \end{bmatrix} \mathbf{y}; \quad \mathbf{y}_1(t) = \begin{bmatrix} 5e^{-t} \\ -7e^{-t} \end{bmatrix}, \quad \mathbf{y}_2(t) = \begin{bmatrix} e^t \\ -e^t \end{bmatrix}, \quad t_0 = -1, \quad -\infty < t < \infty$

25. $\mathbf{y}' = \begin{bmatrix} 9 & 5 \\ -7 & -3 \end{bmatrix} \mathbf{y}; \quad \mathbf{y}_1(t) = \begin{bmatrix} 5e^{2t} \\ -7e^{2t} \end{bmatrix}, \quad \mathbf{y}_2(t) = \begin{bmatrix} e^{4t} \\ -e^{4t} \end{bmatrix}, \quad t_0 = 0, \quad -\infty < t < \infty$

26. $\mathbf{y}' = \begin{bmatrix} 1 & t \\ 0 & -t^{-1} \end{bmatrix} \mathbf{y}, \ t \neq 0; \quad \mathbf{y}_1(t) = \begin{bmatrix} -1 \\ t^{-1} \end{bmatrix}, \quad \mathbf{y}_2(t) = \begin{bmatrix} e^t \\ 0 \end{bmatrix}, \quad t_0 = 1, \quad 0 < t < \infty$

27. $\mathbf{y}' = \begin{bmatrix} 2 & 1 & 1 \\ 1 & 1 & 2 \\ 1 & 2 & 1 \end{bmatrix} \mathbf{y}; \quad \mathbf{y}_1(t) = \begin{bmatrix} 2e^t \\ -e^t \\ -e^t \end{bmatrix}, \quad \mathbf{y}_2(t) = \begin{bmatrix} 0 \\ -e^{-t} \\ e^{-t} \end{bmatrix}, \quad \mathbf{y}_3(t) = \begin{bmatrix} e^{4t} \\ e^{4t} \\ e^{4t} \end{bmatrix},$

$t_0 = 0, \quad -\infty < t < \infty$

28. The homogeneous linear system

$$\mathbf{y}' = \begin{bmatrix} 3 & 1 \\ -2 & \alpha \end{bmatrix} \mathbf{y}$$

has a fundamental set of solutions whose Wronskian is constant, $W(t) = 4$, $-\infty < t < \infty$. What is the value α?

Exercises 29–32:

In each exercise,

(a) Verify that the matrix $\Psi(t)$ is a fundamental matrix of the given linear system.

(b) Determine a constant matrix C such that the given matrix $\hat{\Psi}(t)$ can be represented as $\hat{\Psi}(t) = \Psi(t)C$.

(c) Use your knowledge of the matrix C and assertion (b) of Theorem 4.5 to determine whether $\hat{\Psi}(t)$ is also a fundamental matrix or simply a solution matrix.

29. $\mathbf{y}' = \begin{bmatrix} 0 & 1 \\ 1 & 0 \end{bmatrix} \mathbf{y}, \quad \Psi(t) = \begin{bmatrix} e^t & e^{-t} \\ e^t & -e^{-t} \end{bmatrix}, \quad \hat{\Psi}(t) = \begin{bmatrix} \sinh t & \cosh t \\ \cosh t & \sinh t \end{bmatrix}$

30. $\mathbf{y}' = \begin{bmatrix} 0 & 1 \\ 1 & 0 \end{bmatrix} \mathbf{y}, \quad \Psi(t) = \begin{bmatrix} e^t & e^{-t} \\ e^t & -e^{-t} \end{bmatrix}, \quad \hat{\Psi}(t) = \begin{bmatrix} 2e^t - e^{-t} & e^t + 3e^{-t} \\ 2e^t + e^{-t} & e^t - 3e^{-t} \end{bmatrix}$

31. $\mathbf{y}' = \begin{bmatrix} 1 & 1 \\ 0 & -2 \end{bmatrix} \mathbf{y}, \quad \Psi(t) = \begin{bmatrix} e^t & e^{-2t} \\ 0 & -3e^{-2t} \end{bmatrix}, \quad \hat{\Psi}(t) = \begin{bmatrix} 2e^{-2t} & 0 \\ -6e^{-2t} & 0 \end{bmatrix}$

32. $\mathbf{y}' = \begin{bmatrix} 1 & 1 & 1 \\ 0 & -1 & 1 \\ 0 & 0 & 2 \end{bmatrix} \mathbf{y}, \quad \Psi(t) = \begin{bmatrix} e^t & e^{-t} & 4e^{2t} \\ 0 & -2e^{-t} & e^{2t} \\ 0 & 0 & 3e^{2t} \end{bmatrix}, \quad \hat{\Psi}(t) = \begin{bmatrix} e^t + e^{-t} & 4e^{2t} & e^t + 4e^{2t} \\ -2e^{-t} & e^{2t} & e^{2t} \\ 0 & 3e^{2t} & 3e^{2t} \end{bmatrix}$

Exercises 33–34:

The matrix $\Psi(t)$ is a fundamental matrix of the given homogeneous linear system. Find a constant matrix C such that $\hat{\Psi}(t) = \Psi(t)C$ is a fundamental matrix satisfying $\hat{\Psi}(0) = I$, where I is the (2×2) identity matrix.

33. $\mathbf{y}' = \begin{bmatrix} 0 & 1 \\ 1 & 0 \end{bmatrix} \mathbf{y}, \quad \Psi(t) = \begin{bmatrix} e^t & e^{-t} \\ e^t & -e^{-t} \end{bmatrix}$ **34.** $\mathbf{y}' = \begin{bmatrix} 1 & 1 \\ 0 & -2 \end{bmatrix} \mathbf{y}, \quad \Psi(t) = \begin{bmatrix} e^t & e^{-2t} \\ 0 & -3e^{-2t} \end{bmatrix}$

35. Consider the nth order scalar equation $y^{(n)} + p_{n-1}(t)y^{(n-1)} + \cdots + p_1(t)y' + p_0(t)y = 0$. For the special cases $n = 2, n = 3$, and $n = 4$, rewrite the scalar equation as a first order system $\mathbf{y}' = A\mathbf{y}$. Verify that Abel's theorem, as stated in Theorem 4.4 for systems, has the same conclusion as does Abel's theorem stated in Theorem 3.6 for scalar equations.

36. Let $\mathbf{e}_1, \mathbf{e}_2, \ldots, \mathbf{e}_n$ denote the columns of the $(n \times n)$ identity matrix I. Let $P(t)$ be continuous on (a, b), and let t_0 be in (a, b). Let $\mathbf{y}_1(t), \mathbf{y}_2(t), \ldots, \mathbf{y}_n(t)$ denote the solutions of $\mathbf{y}_j' = P(t)\mathbf{y}_j, \mathbf{y}_j(t_0) = \mathbf{e}_j, j = 1, 2, \ldots, n$. Show that $\{\mathbf{y}_1(t), \mathbf{y}_2(t), \ldots, \mathbf{y}_n(t)\}$ is a fundamental set of solutions.

37. Let $\{\mathbf{y}_1(t), \mathbf{y}_2(t), \ldots, \mathbf{y}_n(t)\}$ be a fundamental set of solutions of the linear system $\mathbf{y}' = P(t)\mathbf{y}$, where the matrix function $P(t)$ is continuous on $a < t < b$. Prove that $\{\mathbf{y}_1(t), \mathbf{y}_2(t), \ldots, \mathbf{y}_n(t)\}$ is a linearly independent set of functions on (a, b). [Hint: One approach is to rewrite the equation $k_1\mathbf{y}_1(t) + k_2\mathbf{y}_2(t) + \cdots + k_n\mathbf{y}_n(t) = \mathbf{0}$ as $\Psi(t)\mathbf{k} = \mathbf{0}$, where $\Psi(t) = [\mathbf{y}_1(t), \mathbf{y}_2(t), \ldots, \mathbf{y}_n(t)]$. Now consider $\Psi(t)\mathbf{k} = \mathbf{0}$ at some arbitrary point t_0 in (a, b).]

4.4   Constant Coefficient Homogeneous Systems; the Eigenvalue Problem

Consider the first order homogeneous equation

$$\mathbf{y}' = A\mathbf{y}, \qquad -\infty < t < \infty,$$

where $\mathbf{y}(t)$ is an $(n \times 1)$ vector function and A is an $(n \times n)$ matrix of real-valued constants. The general solution of $\mathbf{y}' = A\mathbf{y}$ has the form

$$\mathbf{y}(t) = c_1\mathbf{y}_1(t) + c_2\mathbf{y}_2(t) + \cdots + c_n\mathbf{y}_n(t), \tag{1}$$

where $\{\mathbf{y}_1(t), \mathbf{y}_2(t), \ldots, \mathbf{y}_n(t)\}$ is a fundamental set of solutions. We now address the problem of finding a fundamental set of solutions for $\mathbf{y}' = A\mathbf{y}$.

The Eigenvalue Problem

In Chapters 2 and 3, we found solutions of the linear homogeneous constant coefficient scalar equation by looking for solutions of the form $y(t) = e^{\lambda t}$. For the present case, $\mathbf{y}' = A\mathbf{y}$, we take a similar approach. This time, however, we must find solutions that are vector functions. Therefore, we look for solutions of the form

$$\mathbf{y}(t) = e^{\lambda t}\mathbf{x}, \tag{2}$$

where λ is a constant (possibly complex) and $\mathbf{x}$ is an $(n \times 1)$ constant vector. To ensure that $\mathbf{y}(t)$ is a nonzero solution, we require that $\mathbf{x}$ be a nonzero vector.

Substituting the trial form $\mathbf{y}(t) = e^{\lambda t}\mathbf{x}$ into the left-hand side of $\mathbf{y}' = A\mathbf{y}$ leads to

$$\mathbf{y}' = (e^{\lambda t}\mathbf{x})' = (e^{\lambda t})'\mathbf{x} = \lambda e^{\lambda t}\mathbf{x} = e^{\lambda t}(\lambda \mathbf{x}). \tag{3}$$

Substituting $\mathbf{y}(t) = e^{\lambda t}\mathbf{x}$ into the right-hand side of $\mathbf{y}' = A\mathbf{y}$ yields

$$A\mathbf{y} = A(e^{\lambda t}\mathbf{x}) = e^{\lambda t}(A\mathbf{x}). \tag{4}$$

Equating expressions (3) and (4) gives

$$e^{\lambda t}(\lambda \mathbf{x}) = e^{\lambda t}(A\mathbf{x}).$$

Canceling the nonzero factor $e^{\lambda t}$ and rearranging, we obtain

$$A\mathbf{x} = \lambda \mathbf{x}, \qquad \mathbf{x} \neq \mathbf{0}. \tag{5}$$

Equation (5) is known as an **eigenvalue problem** and is important in mathematics, science, and engineering. The problem posed by equation (5) is that of finding constants λ, called **eigenvalues**, and corresponding nonzero vectors $\mathbf{x}$, called **eigenvectors**, such that $A\mathbf{x} = \lambda \mathbf{x}$.

The combination of an eigenvalue λ and a corresponding eigenvector $\mathbf{x}$ is referred to as an **eigenpair** and denoted by $(\lambda, \mathbf{x})$. For every eigenpair $(\lambda, \mathbf{x})$ of the matrix A, the associated vector function

$$\mathbf{y}(t) = e^{\lambda t}\mathbf{x} \tag{6}$$

is a solution of $\mathbf{y}' = A\mathbf{y}$.

If $\mathbf{x}$ is an eigenvector of A corresponding to an eigenvalue λ, then so is the vector $a\mathbf{x}$, where a is any nonzero constant. Hence, if $(\lambda, \mathbf{x})$ is an eigenpair for A, then so is $(\lambda, a\mathbf{x})$, $a \neq 0$. Eigenvectors are not unique. In view of equation (6), however, lack of uniqueness is not surprising. That is, if $\mathbf{y}(t) = e^{\lambda t}\mathbf{x}$ is a solution of the homogeneous linear equation $\mathbf{y}' = A\mathbf{y}$, then so is $\tilde{\mathbf{y}}(t) = a\mathbf{y}(t)$ (see Theorem 4.2).

EXAMPLE

1

Consider the homogeneous first order system

$$\begin{aligned} y_1' &= y_1 + 2y_2 \\ y_2' &= 2y_1 + y_2. \end{aligned}$$

We can rewrite this system of equations as $\mathbf{y}' = A\mathbf{y}$, where

$$\mathbf{y}(t) = \begin{bmatrix} y_1(t) \\ y_2(t) \end{bmatrix} \quad \text{and} \quad A = \begin{bmatrix} 1 & 2 \\ 2 & 1 \end{bmatrix}.$$

Let

$$\mathbf{x}_1 = \begin{bmatrix} 1 \\ -1 \end{bmatrix} \quad \text{and} \quad \mathbf{x}_2 = \begin{bmatrix} 1 \\ 1 \end{bmatrix}.$$

(continued)

(continued)

(a) Use equation (5) to show that $\mathbf{x}_1$ and $\mathbf{x}_2$ are eigenvectors of A. Determine the corresponding eigenvalues λ_1 and λ_2.

(b) Use equation (6) to determine two solutions, $\mathbf{y}_1(t)$ and $\mathbf{y}_2(t)$, of $\mathbf{y}' = A\mathbf{y}$.

(c) Calculate the Wronskian, and decide whether $\{\mathbf{y}_1(t), \mathbf{y}_2(t)\}$ is a fundamental set of solutions.

Solution:

(a) Calculating $A\mathbf{x}_1$, we find

$$A\mathbf{x}_1 = \begin{bmatrix} 1 & 2 \\ 2 & 1 \end{bmatrix} \begin{bmatrix} 1 \\ -1 \end{bmatrix} = \begin{bmatrix} -1 \\ 1 \end{bmatrix} = (-1)\,\mathbf{x}_1.$$

Therefore, $(\lambda_1, \mathbf{x}_1) = (-1, \mathbf{x}_1)$ is an eigenpair of A. Similarly, calculating $A\mathbf{x}_2$, we obtain

$$A\mathbf{x}_2 = \begin{bmatrix} 1 & 2 \\ 2 & 1 \end{bmatrix} \begin{bmatrix} 1 \\ 1 \end{bmatrix} = \begin{bmatrix} 3 \\ 3 \end{bmatrix} = 3\mathbf{x}_2.$$

Therefore, $(\lambda_2, \mathbf{x}_2) = (3, \mathbf{x}_2)$ is a second eigenpair of A.

(b) From part (a), we have eigenpairs $(\lambda_1, \mathbf{x}_1)$ and $(\lambda_2, \mathbf{x}_2)$. Using equation (6), we can form two solutions, $\mathbf{y}_1(t) = e^{\lambda_1 t}\mathbf{x}_1$ and $\mathbf{y}_2(t) = e^{\lambda_2 t}\mathbf{x}_2$:

$$\mathbf{y}_1(t) = e^{-t} \begin{bmatrix} 1 \\ -1 \end{bmatrix} = \begin{bmatrix} e^{-t} \\ -e^{-t} \end{bmatrix} \quad \text{and} \quad \mathbf{y}_2(t) = e^{3t} \begin{bmatrix} 1 \\ 1 \end{bmatrix} = \begin{bmatrix} e^{3t} \\ e^{3t} \end{bmatrix}.$$

(c) To determine whether these two solutions form a fundamental set of solutions, we calculate the Wronskian, $W(t)$, of $\mathbf{y}_1$ and $\mathbf{y}_2$. From part (b), our solution matrix $\Psi(t)$ is

$$\Psi(t) = [\mathbf{y}_1(t), \mathbf{y}_2(t)] = \begin{bmatrix} e^{-t} & e^{3t} \\ -e^{-t} & e^{3t} \end{bmatrix}.$$

The Wronskian is

$$W(t) = \det[\Psi(t)] = 2e^{2t}.$$

Since the Wronskian is never zero, we know by Theorem 4.3 that $\{\mathbf{y}_1(t), \mathbf{y}_2(t)\}$ is a fundamental set of solutions of $\mathbf{y}' = A\mathbf{y}$. The general solution is therefore

$$\mathbf{y}(t) = c_1 \begin{bmatrix} e^{-t} \\ -e^{-t} \end{bmatrix} + c_2 \begin{bmatrix} e^{3t} \\ e^{3t} \end{bmatrix} = \begin{bmatrix} e^{-t} & e^{3t} \\ -e^{-t} & e^{3t} \end{bmatrix} \begin{bmatrix} c_1 \\ c_2 \end{bmatrix}. \quad ❖$$

Finding Eigenpairs

Example 1 suggests a procedure to find the general solution of $\mathbf{y}' = A\mathbf{y}$ when A is an $(n \times n)$ constant matrix. Each eigenpair $(\lambda, \mathbf{x})$ gives rise to a solution of the form $\mathbf{y}(t) = e^{\lambda t}\mathbf{x}$. The general solution is the linear combination

$$\mathbf{y}(t) = c_1\mathbf{y}_1(t) + c_2\mathbf{y}_2(t) + \cdots + c_n\mathbf{y}_n(t),$$

where $\{\mathbf{y}_1(t), \mathbf{y}_2(t), \ldots, \mathbf{y}_n(t)\}$ is a fundamental set.

Some obvious questions are

1. Given an $(n \times n)$ constant matrix A, do there always exist eigenpairs $(\lambda, \mathbf{x})$? Is it possible to find n different eigenpairs and thereby form n different solutions? Will these solutions form a fundamental set?
2. How do we find these eigenpairs?

The eigenvalue problem (5) consists of finding scalars λ and nonzero vectors $\mathbf{x}$ such that $A\mathbf{x} = \lambda\mathbf{x}$ or, equivalently, $A\mathbf{x} - \lambda\mathbf{x} = \mathbf{0}$. We can rewrite the equation $A\mathbf{x} - \lambda\mathbf{x} = \mathbf{0}$ as

$$(A - \lambda I)\mathbf{x} = \mathbf{0}, \qquad \mathbf{x} \neq \mathbf{0}, \tag{7}$$

where I denotes the $(n \times n)$ identity matrix.

To solve (7), we use a result from linear algebra stating that the matrix equation $(A - \lambda I)\mathbf{x} = \mathbf{0}$ has a nonzero solution $\mathbf{x}$ if and only if the determinant of $A - \lambda I$ is zero. Therefore, λ is an eigenvalue of A if and only if $\det[A - \lambda I] = 0$; that is,

$$\begin{vmatrix} a_{11} - \lambda & a_{12} & \cdots & a_{1n} \\ a_{21} & a_{22} - \lambda & & a_{2n} \\ \vdots & & & \vdots \\ a_{n1} & a_{n2} & \cdots & a_{nn} - \lambda \end{vmatrix} = 0. \tag{8}$$

Evaluating this determinant (by a cofactor expansion, for instance) shows that (8) is a polynomial equation of the form

$$p(\lambda) = 0, \tag{9}$$

where $p(\lambda)$ is a polynomial of degree n in the variable λ. The polynomial $p(\lambda)$ is called the **characteristic polynomial**, and equation (8) is called the **characteristic equation**. The eigenvalues of A are, therefore, the roots of the characteristic equation. Since an nth degree polynomial has n zeros (counting multiplicity), an $(n \times n)$ matrix A always has n eigenvalues. The eigenvalues may be zero or nonzero, real or complex, and some of them may be repeated (that is, they may have multiplicity greater than one).

Since we are assuming the matrix A is real-valued, the coefficients of the characteristic polynomial are real numbers. Consequently, any complex roots of the characteristic equation occur in complex conjugate pairs. Thus, if $\lambda = \alpha + i\beta$ is an eigenvalue of A, so is $\bar{\lambda} = \alpha - i\beta$.

For each eigenvalue λ, we know the homogeneous system of equations $(A - \lambda I)\mathbf{x} = \mathbf{0}$ has a nontrivial solution. Therefore, an eigenvector is obtained by forming the homogeneous system $(A - \lambda I)\mathbf{x} = \mathbf{0}$ and finding a nontrivial solution.

E X A M P L E

2

(a) Find the eigenpairs of

$$A = \begin{bmatrix} 4 & -2 \\ 1 & 1 \end{bmatrix}.$$

(b) Do the solutions of $\mathbf{y}' = A\mathbf{y}$ created from these eigenpairs form a fundamental set?

(continued)

(continued)

Solution:

(a) We first find the eigenvalues of A by finding the roots of $p(\lambda) = 0$:

$$p(\lambda) = \det(A - \lambda I) = \begin{vmatrix} 4 - \lambda & -2 \\ 1 & 1 - \lambda \end{vmatrix}$$

$$= (4 - \lambda)(1 - \lambda) + 2$$

$$= \lambda^2 - 5\lambda + 6 = (\lambda - 3)(\lambda - 2).$$

The matrix therefore has two real distinct eigenvalues, $\lambda_1 = 3$ and $\lambda_2 = 2$.

For each eigenvalue, we solve the homogeneous system of equations $(A - \lambda I)\mathbf{x} = \mathbf{0}$ and choose a nontrivial solution to serve as the eigenvector. For $\lambda_1 = 3$, the system is $(A - 3I)\mathbf{x} = \mathbf{0}$, or

$$\begin{bmatrix} 1 & -2 \\ 1 & -2 \end{bmatrix} \begin{bmatrix} x_1 \\ x_2 \end{bmatrix} = \begin{bmatrix} 0 \\ 0 \end{bmatrix}.$$

Because the coefficient matrix has a zero determinant, the homogeneous system $(A - 3I)\mathbf{x} = \mathbf{0}$ has nontrivial solutions. One such nontrivial solution is

$$\mathbf{x}_1 = \begin{bmatrix} 2 \\ 1 \end{bmatrix}.$$

We next obtain an eigenvector $\mathbf{x}_2$ corresponding to $\lambda_2 = 2$ by solving $(A - 2I)\mathbf{x} = \mathbf{0}$:

$$\begin{bmatrix} 2 & -2 \\ 1 & -1 \end{bmatrix} \begin{bmatrix} x_1 \\ x_2 \end{bmatrix} = \begin{bmatrix} 0 \\ 0 \end{bmatrix}.$$

A convenient nontrivial solution is

$$\mathbf{x}_2 = \begin{bmatrix} 1 \\ 1 \end{bmatrix}.$$

Eigenpairs of A are

$$\lambda_1 = 3, \mathbf{x}_1 = \begin{bmatrix} 2 \\ 1 \end{bmatrix} \quad \text{and} \quad \lambda_2 = 2, \mathbf{x}_2 = \begin{bmatrix} 1 \\ 1 \end{bmatrix}.$$

(b) Two solutions of $\mathbf{y}' = A\mathbf{y}$ are

$$\mathbf{y}_1(t) = \begin{bmatrix} 2e^{3t} \\ e^{3t} \end{bmatrix} \quad \text{and} \quad \mathbf{y}_2(t) = \begin{bmatrix} e^{2t} \\ e^{2t} \end{bmatrix}.$$

Forming the Wronskian, we obtain

$$W(t) = \begin{vmatrix} 2e^{3t} & e^{2t} \\ e^{3t} & e^{2t} \end{vmatrix} = e^{5t}.$$

Since the Wronskian is nonzero everywhere, the solutions form a fundamental set on $(-\infty, \infty)$. ❖

EXAMPLE
3

(a) Find the eigenpairs of the (3×3) matrix
$$A = \begin{bmatrix} 1 & -7 & 3 \\ -1 & -1 & 1 \\ 4 & -4 & 0 \end{bmatrix}.$$

(b) Do the solutions of $\mathbf{y}' = A\mathbf{y}$ created from these eigenpairs form a fundamental set?

Solution:

(a) We find the eigenvalues by solving the characteristic equation $p(\lambda) = 0$. To calculate the characteristic polynomial $p(\lambda) = \det(A - \lambda I)$, we use a cofactor expansion along the first row:

$$p(\lambda) = \begin{vmatrix} 1 - \lambda & -7 & 3 \\ -1 & -1 - \lambda & 1 \\ 4 & -4 & -\lambda \end{vmatrix}$$

$$= (1 - \lambda) \begin{vmatrix} -1 - \lambda & 1 \\ -4 & -\lambda \end{vmatrix} + 7 \begin{vmatrix} -1 & 1 \\ 4 & -\lambda \end{vmatrix} + 3 \begin{vmatrix} -1 & -1 - \lambda \\ 4 & -4 \end{vmatrix}$$

$$= (1 - \lambda)(\lambda^2 + \lambda + 4) + 7(\lambda - 4) + 3(4\lambda + 8)$$

$$= -\lambda^3 + 16\lambda = -\lambda(\lambda - 4)(\lambda + 4).$$

The eigenvalues of A are $\lambda_1 = 0$, $\lambda_2 = 4$, and $\lambda_3 = -4$. (In this example, $\lambda_1 = 0$ is an eigenvalue. Although eigenvectors must be nonzero, eigenvalues *can* be zero. In fact [see equation (8)], $\lambda = 0$ is an eigenvalue whenever $\det[A] = 0$.)

We now compute the eigenvectors. An eigenvector corresponding to $\lambda_1 = 0$ is a nonzero solution of $A\mathbf{x} = \mathbf{0}$,

$$\begin{bmatrix} 1 & -7 & 3 \\ -1 & -1 & 1 \\ 4 & -4 & 0 \end{bmatrix} \begin{bmatrix} x_1 \\ x_2 \\ x_3 \end{bmatrix} = \begin{bmatrix} 0 \\ 0 \\ 0 \end{bmatrix}. \qquad \textbf{(10)}$$

In this case, unlike the situation in Example 2, we cannot find a solution by inspection. Instead, we solve system (10) using Gaussian elimination. Elementary row operations can be used to row reduce system (10) into the following equivalent system:

$$\begin{bmatrix} 1 & 0 & -\frac{1}{2} \\ 0 & 1 & -\frac{1}{2} \\ 0 & 0 & 0 \end{bmatrix} \begin{bmatrix} x_1 \\ x_2 \\ x_3 \end{bmatrix} = \begin{bmatrix} 0 \\ 0 \\ 0 \end{bmatrix}.$$

The solution of this system is $x_1 = \frac{1}{2}x_3$, $x_2 = \frac{1}{2}x_3$. For convenience, we set $x_3 = 2$, obtaining the eigenvector

$$\mathbf{x}_1 = \begin{bmatrix} 1 \\ 1 \\ 2 \end{bmatrix}.$$

Similarly, an eigenvector corresponding to $\lambda_2 = 4$ is a nonzero solution of

(continued)

(continued)

$(A - 4I)\mathbf{x} = \mathbf{0}$,

$$\begin{bmatrix} -3 & -7 & 3 \\ -1 & -5 & 1 \\ 4 & -4 & -4 \end{bmatrix} \begin{bmatrix} x_1 \\ x_2 \\ x_3 \end{bmatrix} = \begin{bmatrix} 0 \\ 0 \\ 0 \end{bmatrix}.$$

Using Gaussian elimination, we find the equivalent system

$$\begin{bmatrix} 1 & 0 & -1 \\ 0 & 1 & 0 \\ 0 & 0 & 0 \end{bmatrix} \begin{bmatrix} x_1 \\ x_2 \\ x_3 \end{bmatrix} = \begin{bmatrix} 0 \\ 0 \\ 0 \end{bmatrix}.$$

Therefore, $x_1 = x_3$, $x_2 = 0$. Choosing $x_3 = 1$ leads to the eigenvector

$$\mathbf{x}_2 = \begin{bmatrix} 1 \\ 0 \\ 1 \end{bmatrix}.$$

Lastly, we leave it as an exercise to show that

$$\mathbf{x}_3 = \begin{bmatrix} 2 \\ 1 \\ -1 \end{bmatrix}$$

is an eigenvector corresponding to $\lambda_3 = -4$.
  Therefore, we have eigenpairs

$$\lambda_1 = 0, \quad \mathbf{x}_1 = \begin{bmatrix} 1 \\ 1 \\ 2 \end{bmatrix}, \qquad \lambda_2 = 4, \quad \mathbf{x}_2 = \begin{bmatrix} 1 \\ 0 \\ 1 \end{bmatrix}, \qquad \lambda_3 = -4, \quad \mathbf{x}_3 = \begin{bmatrix} 2 \\ 1 \\ -1 \end{bmatrix}.$$

(b) Three solutions of $\mathbf{y}' = A\mathbf{y}$ are

$$\mathbf{y}_1(t) = \begin{bmatrix} 1 \\ 1 \\ 2 \end{bmatrix}, \qquad \mathbf{y}_2(t) = \begin{bmatrix} e^{4t} \\ 0 \\ e^{4t} \end{bmatrix}, \qquad \mathbf{y}_3(t) = \begin{bmatrix} 2e^{-4t} \\ e^{-4t} \\ -e^{-4t} \end{bmatrix}.$$

According to Abel's theorem (Theorem 4.4), we can determine whether these solutions form a fundamental set by evaluating $W(t_0)$ at some convenient choice of t_0. Choosing $t_0 = 0$, we obtain

$$W(0) = \begin{vmatrix} 1 & 1 & 2 \\ 1 & 0 & 1 \\ 2 & 1 & -1 \end{vmatrix} = 4.$$

Therefore, these three solutions form a fundamental set on the interval $-\infty < t < \infty$. ❖

Eigenpair computations, such as those of the previous two examples, have built-in checks available that should be exploited. In computing eigenvectors, the Gaussian elimination process that transforms the coefficient matrix $A - \lambda I$ to echelon form must produce at least one row of zeros. If that does not occur, you should realize you've made a mistake. Another check is simply to compute the product $A\mathbf{x}$ and verify that it equals $\lambda\mathbf{x}$.

EXERCISES

Exercises 1–10:

Rewrite the linear system as a matrix equation $\mathbf{y}' = A\mathbf{y}$, and compute the eigenvalues of the matrix A.

1. $\begin{aligned} y_1' &= 5y_1 + 3y_2 \\ y_2' &= -6y_1 - 4y_2 \end{aligned}$

2. $\begin{aligned} y_1' &= 3y_1 + 2y_2 \\ y_2' &= -4y_1 - 3y_2 \end{aligned}$

3. $\begin{aligned} y_1' &= y_1 + y_2 \\ y_2' &= 2y_1 + 2y_2 \end{aligned}$

4. $\begin{aligned} y_1' &= 2y_1 + y_2 \\ y_2' &= -y_1 \end{aligned}$

5. $\begin{aligned} y_1' &= 2y_2 \\ y_2' &= -2y_1 \end{aligned}$

6. $\begin{aligned} y_1' &= 4y_1 - 2y_2 \\ y_2' &= 5y_1 - 2y_2 \end{aligned}$

7. $\begin{aligned} y_1' &= 5y_1 \\ y_2' &= y_2 + 3y_3 \\ y_3' &= 2y_2 + 2y_3 \end{aligned}$

8. $\begin{aligned} y_1' &= y_2 - 3y_3 \\ y_2' &= -5y_2 - 4y_3 \\ y_3' &= 8y_2 + 7y_3 \end{aligned}$

9. $\begin{aligned} y_1' &= -2y_1 + 3y_2 + y_3 \\ y_2' &= -8y_1 + 13y_2 + 5y_3 \\ y_3' &= 11y_1 - 17y_2 - 6y_3 \end{aligned}$

10. $\begin{aligned} y_1' &= y_1 - 7y_2 + 3y_3 \\ y_2' &= -y_1 - y_2 + y_3 \\ y_3' &= 4y_1 - 4y_2 \end{aligned}$

Exercises 11–17:

In each exercise, λ is an eigenvalue of the given matrix A. Determine an eigenvector corresponding to λ.

11. $A = \begin{bmatrix} -4 & 3 \\ -4 & 4 \end{bmatrix}$, $\quad \lambda = 2$

12. $A = \begin{bmatrix} 5 & 3 \\ -4 & -3 \end{bmatrix}$, $\quad \lambda = -1$

13. $A = \begin{bmatrix} 1 & 1 \\ -4 & 6 \end{bmatrix}$, $\quad \lambda = 5$

14. $A = \begin{bmatrix} 1 & -7 & 3 \\ -1 & -1 & 1 \\ 4 & -4 & 0 \end{bmatrix}$, $\quad \lambda = -4$

15. $A = \begin{bmatrix} 3 & 1 & 1 \\ -1 & 1 & -1 \\ 2 & 1 & 2 \end{bmatrix}$, $\quad \lambda = 2$

16. $A = \begin{bmatrix} 1 & 3 & 1 \\ 2 & 1 & 2 \\ 4 & 3 & -2 \end{bmatrix}$, $\quad \lambda = 5$

17. $A = \begin{bmatrix} -2 & 3 & 1 \\ -8 & 13 & 5 \\ 11 & -17 & -6 \end{bmatrix}$, $\quad \lambda = 0$

Exercises 18–23:

For the given linear system $\mathbf{y}' = A\mathbf{y}$,

(a) Compute the eigenpairs of the coefficient matrix A.

(b) For each eigenpair found in part (a), form a solution of $\mathbf{y}' = A\mathbf{y}$.

(c) Does the set of solutions found in part (b) form a fundamental set of solutions?

18. $\mathbf{y}' = \begin{bmatrix} 4 & 2 \\ -1 & 1 \end{bmatrix} \mathbf{y}$

19. $\mathbf{y}' = \begin{bmatrix} 7 & -3 \\ 16 & -7 \end{bmatrix} \mathbf{y}$

20. $\mathbf{y}' = \begin{bmatrix} 3 & 2 \\ -9 & -6 \end{bmatrix} \mathbf{y}$

21. $\mathbf{y}' = \begin{bmatrix} 0 & 1 \\ 1 & 0 \end{bmatrix} \mathbf{y}$

22. $\mathbf{y}' = \begin{bmatrix} 2 & 1 \\ 0 & -1 \end{bmatrix} \mathbf{y}$

23. $\mathbf{y}' = \begin{bmatrix} -5 & -6 \\ 3 & 4 \end{bmatrix} \mathbf{y}$

Exercises 24–27:

In each exercise, an eigenvalue λ is given for the matrix A.

(a) Find an eigenvector corresponding to the given eigenvalue λ.

(b) Find the other two eigenvalues of the matrix A.

(c) Find eigenvectors corresponding to the eigenvalues found in part (b).

(d) Do the three solutions of $\mathbf{y}' = A\mathbf{y}$ formed from the eigenpairs make up a fundamental set of solutions?

24. $A = \begin{bmatrix} 2 & 1 & 2 \\ 0 & 2 & 2 \\ 0 & 4 & 0 \end{bmatrix}$, $\quad \lambda = -2$ **25.** $A = \begin{bmatrix} 1 & 2 & 0 \\ -4 & 7 & 0 \\ 0 & 0 & 1 \end{bmatrix}$, $\quad \lambda = 1$

26. $A = \begin{bmatrix} 3 & 1 & 0 \\ -6 & -5 & 2 \\ -7 & -8 & 4 \end{bmatrix}$, $\quad \lambda = 2$ **27.** $A = \begin{bmatrix} 3 & 1 & 0 \\ -8 & -6 & 2 \\ -9 & -9 & 4 \end{bmatrix}$, $\quad \lambda = 2$

28. Consider the (2×2) matrix

$$A = \begin{bmatrix} a & b \\ b & d \end{bmatrix},$$

where A has all real entries. Show that A has only real eigenvalues. [Hint: Calculate the characteristic polynomial, and use the quadratic formula.] The matrix A is a **symmetric matrix** since $A = A^T$. The symbol A^T denotes the **transpose of A**, where A^T is obtained by interchanging the rows and columns of A. For example,

$$\text{if} \quad A = \begin{bmatrix} x & y \\ u & v \end{bmatrix}, \quad \text{then} \quad A^T = \begin{bmatrix} x & u \\ y & v \end{bmatrix}.$$

29. Consider the (2×2) matrix

$$A = \begin{bmatrix} a & b \\ -b & a \end{bmatrix},$$

where a and b are real numbers and b is nonzero. Show that the eigenvalues of A are complex.

30. Let A be a (2×2) matrix with eigenvalues λ_1 and λ_2, where $\lambda_1 \neq \lambda_2$. Let $\mathbf{x}_1$ and $\mathbf{x}_2$ be corresponding eigenvectors. Show that $\{\mathbf{x}_1, \mathbf{x}_2\}$ is a linearly independent set. [Hint: Suppose $k_1\mathbf{x}_1 + k_2\mathbf{x}_2 = \mathbf{0}$. Multiply this equation by A and obtain $k_1\lambda_1\mathbf{x}_1 + k_2\lambda_2\mathbf{x}_2 = \mathbf{0}$. Next, multiply $k_1\mathbf{x}_1 + k_2\mathbf{x}_2 = \mathbf{0}$ by λ_1 and obtain $k_1\lambda_1\mathbf{x}_1 + k_2\lambda_1\mathbf{x}_2 = \mathbf{0}$. Argue that $k_1 = k_2 = 0$.]

31. Let A be an $(n \times n)$ matrix with eigenvalue λ and eigenvector $\mathbf{x}$. Let α be any constant. Use the definition, $A\mathbf{x} = \lambda\mathbf{x}$, $\mathbf{x} \neq \mathbf{0}$, to show that $\lambda + \alpha$ is an eigenvalue of the matrix $A + \alpha I$ and that $\alpha\lambda$ is an eigenvalue of αA. Similarly, if A is invertible, show that $\lambda \neq 0$ and that $1/\lambda$ is an eigenvalue of A^{-1}.

32. Let A be an $(n \times n)$ matrix with eigenvalue λ and eigenvector $\mathbf{x}$.

(a) Use the definition, $A\mathbf{x} = \lambda\mathbf{x}$, $\mathbf{x} \neq \mathbf{0}$, to show that λ^2 is an eigenvalue of A^2.

(b) Let A be a (2×2) matrix such that $A\mathbf{x} = \lambda\mathbf{x}$, where $\lambda = -2$ and

$$\mathbf{x} = \begin{bmatrix} 3 \\ -1 \end{bmatrix}.$$

Determine the vector $A^3\mathbf{x}$. [This is a special case of the general result: λ^k is an eigenvalue of A^k.]

4.5   Real Eigenvalues and the Phase Plane

Consider the linear system $\mathbf{y}' = A\mathbf{y}$, where A is an $(n \times n)$ constant matrix. As we saw in the previous section, solutions of $\mathbf{y}' = A\mathbf{y}$ can be found by determining eigenpairs of A. In particular, if $(\lambda, \mathbf{x})$ is an eigenpair, then $\mathbf{y}(t) = e^{\lambda t}\mathbf{x}$ is a solution of $\mathbf{y}' = A\mathbf{y}$.

In order to construct a fundamental set of solutions for $\mathbf{y}' = A\mathbf{y}$, we need to find a set of n solutions $\{\mathbf{y}_1(t), \mathbf{y}_2(t), \ldots, \mathbf{y}_n(t)\}$ such that $\det[\Psi(t)] \neq 0$, where $\Psi(t)$ is the $(n \times n)$ matrix

$$\Psi(t) = [\mathbf{y}_1(t), \mathbf{y}_2(t), \ldots, \mathbf{y}_n(t)].$$

Since every eigenpair $(\lambda, \mathbf{x})$ leads to a solution $\mathbf{y}(t) = e^{\lambda t}\mathbf{x}$ of $\mathbf{y}' = A\mathbf{y}$, the first step in finding a fundamental set of solutions is to calculate the eigenpairs of A. We then ask whether there are n solutions of the form $\mathbf{y}_i(t) = e^{\lambda_i t}\mathbf{x}_i$ such that $\det[\Psi(t)] \neq 0$. As in Chapter 3, we first consider the case where A has real and distinct eigenvalues. Section 4.6 examines the case where A has complex eigenvalues, and then Section 4.7 considers the case of repeated eigenvalues.

This section also introduces a geometric tool known as the phase plane, which allows us to visualize solutions of a two-dimensional constant coefficient system

$$y_1' = a_{11}y_1 + a_{12}y_2$$
$$y_2' = a_{21}y_1 + a_{22}y_2.$$

Recognizing a Fundamental Set of Solutions

Suppose that we are given an $(n \times n)$ constant coefficient system $\mathbf{y}' = A\mathbf{y}$ and that we have found n eigenpairs $(\lambda_1, \mathbf{x}_1), (\lambda_2, \mathbf{x}_2), \ldots, (\lambda_n, \mathbf{x}_n)$ for the coefficient matrix A. Do the corresponding solutions $\mathbf{y}_i(t) = e^{\lambda_i t}\mathbf{x}_i, i = 1, 2, \ldots, n$ form a fundamental set of solutions? One way to answer this question is to calculate the Wronskian, $\det[\Psi(t)]$, where $\Psi(t) = [\mathbf{y}_1(t), \ldots, \mathbf{y}_n(t)]$. However, in certain cases, we can answer the question without having to actually calculate the Wronskian. The basis for this assertion is Theorem 4.6.

Theorem 4.6

Consider the homogeneous linear system $\mathbf{y}' = A\mathbf{y}$, $-\infty < t < \infty$. Let the constant $(n \times n)$ matrix A have eigenpairs $(\lambda_1, \mathbf{x}_1), (\lambda_2, \mathbf{x}_2), \ldots, (\lambda_n, \mathbf{x}_n)$, where the eigenvectors are linearly independent. Then the set of solutions

$$\{e^{\lambda_1 t}\mathbf{x}_1, e^{\lambda_2 t}\mathbf{x}_2, \ldots, e^{\lambda_n t}\mathbf{x}_n\}$$

is a fundamental set of solutions.

Theorem 4.6 follows because $\det[\mathbf{x}_1, \mathbf{x}_2, \ldots, \mathbf{x}_n] \neq 0$ if $\{\mathbf{x}_1, \mathbf{x}_2, \ldots, \mathbf{x}_n\}$ is a set of linearly independent $(n \times 1)$ vectors and because $\det[\Psi(0)] = \det[\mathbf{x}_1, \mathbf{x}_2, \ldots, \mathbf{x}_n]$.

When A is an $(n \times n)$ constant matrix that possesses a set of n linearly independent eigenvectors, we say that A has a **full set of eigenvectors**. Theorem 4.6 shows how to form a fundamental set of solutions for $\mathbf{y}' = A\mathbf{y}$ when A has a full set of eigenvectors. When A fails to have a set of n linearly independent eigenvectors, we say that A is **defective**. Section 4.7 discusses techniques for finding a fundamental set of solutions when A is defective.

There are two cases where we can assert immediately that an $(n \times n)$ constant matrix A has a full set of eigenvectors:

(a) If A has n distinct eigenvalues, then A has a full set of eigenvectors.

(b) If A is a symmetric real matrix, then A has a full set of eigenvectors.

The conclusion in Case (a) follows from Theorem 4.6 and the fact that eigenvectors corresponding to distinct eigenvalues are linearly independent. The fact that distinct eigenvalues have linearly independent eigenvectors is proved in most linear algebra texts; a proof is not given here. However, Exercise 30 in Section 4.4 asks you to establish this fact when $n = 2$. The coefficient matrices in Examples 2 and 3 of Section 4.4 have distinct eigenvalues, and consequently each has a full set of eigenvectors.

An $(n \times n)$ matrix A is symmetric if $A^T = A$, where A^T denotes the transpose of A. It can be shown that all the eigenvalues of a real symmetric matrix are real. In Exercise 28 of Section 4.4, you are asked to prove this when $n = 2$. It also can be shown that a real symmetric matrix always has a full set of eigenvectors; in Exercise 11, you are asked to prove this when $n = 2$.

E X A M P L E

1

Solve the initial value problem

$$y_1' = y_1 - 3y_2, \qquad y_1(0) = 1$$
$$y_2' = y_1 + 5y_2, \qquad y_2(0) = 1.$$

Solution: The coefficient matrix

$$A = \begin{bmatrix} 1 & -3 \\ 1 & 5 \end{bmatrix}$$

has eigenpairs

$$\lambda_1 = 4, \ \mathbf{x}_1 = \begin{bmatrix} -1 \\ 1 \end{bmatrix} \quad \text{and} \quad \lambda_2 = 2, \ \mathbf{x}_2 = \begin{bmatrix} -3 \\ 1 \end{bmatrix}.$$

Since the eigenvalues are distinct, A has a full set of eigenvectors and the general solution is

$$\mathbf{y}(t) = c_1 e^{4t} \begin{bmatrix} -1 \\ 1 \end{bmatrix} + c_2 e^{2t} \begin{bmatrix} -3 \\ 1 \end{bmatrix} = \begin{bmatrix} -e^{4t} & -3e^{2t} \\ e^{4t} & e^{2t} \end{bmatrix} \begin{bmatrix} c_1 \\ c_2 \end{bmatrix}.$$

Imposing the initial condition, we have

$$\mathbf{y}(0) = \begin{bmatrix} -1 & -3 \\ 1 & 1 \end{bmatrix} \begin{bmatrix} c_1 \\ c_2 \end{bmatrix} = \begin{bmatrix} 1 \\ 1 \end{bmatrix} \quad \text{or} \quad \begin{bmatrix} c_1 \\ c_2 \end{bmatrix} = \begin{bmatrix} 2 \\ -1 \end{bmatrix}.$$

Therefore, the solution of the initial value problem is

$$\mathbf{y}(t) = \begin{bmatrix} -e^{4t} & -3e^{2t} \\ e^{4t} & e^{2t} \end{bmatrix} \begin{bmatrix} 2 \\ -1 \end{bmatrix} = \begin{bmatrix} -2e^{4t} + 3e^{2t} \\ 2e^{4t} - e^{2t} \end{bmatrix}. \quad ❖$$

E X A M P L E

2

Find the general solution of

$$\begin{aligned} y_1' &= 2y_1 - y_2 - y_3 \\ y_2' &= -y_1 + 2y_2 - y_3 \\ y_3' &= -y_1 - y_2 + 2y_3. \end{aligned}$$

Solution: The coefficient matrix is

$$A = \begin{bmatrix} 2 & -1 & -1 \\ -1 & 2 & -1 \\ -1 & -1 & 2 \end{bmatrix}.$$

Since A is real and symmetric, we know that A has a full set of eigenvectors. The characteristic polynomial is

$$p(\lambda) = -\lambda^3 + 6\lambda^2 - 9\lambda = -\lambda(\lambda - 3)^2.$$

The matrix A therefore has an eigenvalue $\lambda_1 = 0$ and repeated eigenvalues $\lambda_2 = \lambda_3 = 3$. One eigenpair is

$$\lambda_1 = 0, \qquad \mathbf{x}_1 = \begin{bmatrix} 1 \\ 1 \\ 1 \end{bmatrix}.$$

To find eigenvectors corresponding to the repeated eigenvalue $\lambda = 3$, we solve the system $(A - 3I)\mathbf{x} = \mathbf{0}$,

$$\begin{bmatrix} -1 & -1 & -1 \\ -1 & -1 & -1 \\ -1 & -1 & -1 \end{bmatrix} \begin{bmatrix} x_1 \\ x_2 \\ x_3 \end{bmatrix} = \begin{bmatrix} 0 \\ 0 \\ 0 \end{bmatrix}.$$

This system reduces to $x_1 + x_2 + x_3 = 0$, and hence

$$\mathbf{x} = \begin{bmatrix} x_1 \\ x_2 \\ x_3 \end{bmatrix} = \begin{bmatrix} -x_2 - x_3 \\ x_2 \\ x_3 \end{bmatrix} = x_2 \begin{bmatrix} -1 \\ 1 \\ 0 \end{bmatrix} + x_3 \begin{bmatrix} -1 \\ 0 \\ 1 \end{bmatrix}.$$

Therefore, two linearly independent eigenvectors corresponding to $\lambda = 3$ are

$$\mathbf{x}_2 = \begin{bmatrix} -1 \\ 1 \\ 0 \end{bmatrix} \quad \text{and} \quad \mathbf{x}_3 = \begin{bmatrix} -1 \\ 0 \\ 1 \end{bmatrix}.$$

The general solution of the linear system $\mathbf{y}' = A\mathbf{y}$ is

$$\mathbf{y}(t) = c_1 \begin{bmatrix} 1 \\ 1 \\ 1 \end{bmatrix} + c_2 e^{3t} \begin{bmatrix} -1 \\ 1 \\ 0 \end{bmatrix} + c_3 e^{3t} \begin{bmatrix} -1 \\ 0 \\ 1 \end{bmatrix} = \begin{bmatrix} 1 & -e^{3t} & -e^{3t} \\ 1 & e^{3t} & 0 \\ 1 & 0 & e^{3t} \end{bmatrix} \begin{bmatrix} c_1 \\ c_2 \\ c_3 \end{bmatrix}. \; \diamondsuit$$

The Phase Plane

Given a scalar differential equation $y' = f(t, y)$, we can see how the solutions behave by graphing them in the ty-plane. If explicit solutions are not available, we can construct a direction field in the ty-plane and use it to obtain approximate solution curves.

Now, consider a two-dimensional linear system $\mathbf{y}' = P(t)\mathbf{y} + \mathbf{g}(t)$,

$$y_1' = p_{11}(t)y_1 + p_{12}(t)y_2 + g_1(t)$$
$$y_2' = p_{21}(t)y_1 + p_{22}(t)y_2 + g_2(t).$$

To graph the solutions of this system, we would need three dimensions, since solutions would need to be plotted in (t, y_1, y_2)-space. There is, however, an alternative approach that is very useful for a two-dimensional *autonomous* system such as the constant coefficient system $\mathbf{y}' = A\mathbf{y}$,

$$y_1' = a_{11}y_1 + a_{12}y_2$$
$$y_2' = a_{21}y_1 + a_{22}y_2. \qquad \text{(1)}$$

The alternative approach uses a graphical tool known as the phase plane. In the phase plane approach, we treat the independent variable t as a parameter and plot solutions in the (y_1, y_2)-plane. For any fixed value of t, the solution is represented as a point $(y_1(t), y_2(t))$ in the phase plane. As t varies, the points $(y_1(t), y_2(t))$ trace out a curve in the phase plane. By plotting a collection of such curves, we gain graphical insight into the behavior of solutions.

E X A M P L E

3

Construct a phase plane plot for the system

$$y_1' = 0.6y_1 + 0.8y_2$$
$$y_2' = 0.8y_1 - 0.6y_2. \qquad \text{(2)}$$

Solution: For this system, the coefficient matrix is

$$A = \begin{bmatrix} 0.6 & 0.8 \\ 0.8 & -0.6 \end{bmatrix}.$$

The general solution is

$$\mathbf{y}(t) = c_1 e^t \begin{bmatrix} 2 \\ 1 \end{bmatrix} + c_2 e^{-t} \begin{bmatrix} 1 \\ -2 \end{bmatrix}. \qquad \text{(3)}$$

Figure 4.2 displays a representative collection of phase plane solution curves. These curves correspond to a sampling of starting points $(y_1(0), y_2(0))$. The arrows on each curve indicate the direction in which the solution point moves as t increases. The origin is an equilibrium point.

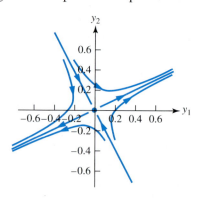

FIGURE 4.2

Phase plane solution curves for the system in Example 3. The general solution (3) shows that, except when $c_1 = 0$, the points $(y_1(t), y_2(t))$ move away from the origin and are asymptotic to the line $y_2 = \frac{1}{2}y_1$.

We can infer the qualitative behavior of these phase plane solution curves from the general solution (3). For example, solutions originating on the line $y_2 = -2y_1$ remain on this line and move inward, toward the origin. This special case corresponds to $c_1 = 0$ in (3). [When $c_1 = 0$, it follows that $y_1(t) = c_2 e^{-t}$ and $y_2(t) = -2c_2 e^{-t} = -2y_1(t)$. The motion is inward, toward the origin, since e^{-t} decreases as t increases.]

Solutions originating on the line $y_2 = \frac{1}{2}y_1$ move away from the origin along this line as t increases. Such solutions correspond to $c_2 = 0$ in (3). [In this special case, $y_1(t) = 2c_1 e^t$ and $y_2(t) = c_1 e^t = \frac{1}{2}y_1(t)$.]

All other nonzero solutions correspond to cases in which both constants c_1 and c_2 are nonzero. Since e^{-t} approaches zero and e^t grows as time increases, these solution curves move away from the origin and approach the line $y_2 = \frac{1}{2}y_1$ as an asymptote.

In Example 3, the general solution (3) gives us all the information necessary to understand the phase plane behavior of solutions of the linear system (2). One virtue of the phase plane, however, is that the situation can be reversed. In a manner analogous to that used with direction fields, we can obtain qualitative information without actually solving the differential equation. At any point in the phase plane, we know from calculus that the slope of a line tangent to the solution curve of system (1) passing through that point is

$$\frac{dy_2}{dy_1} = \frac{dy_2/dt}{dy_1/dt} = \frac{a_{21}y_1 + a_{22}y_2}{a_{11}y_1 + a_{12}y_2}.$$

For system (2) in Example 3, we have

$$\frac{dy_2}{dy_1} = \frac{4y_1 - 3y_2}{3y_1 + 4y_2}. \tag{4}$$

Therefore, if we evaluate the right-hand side of (4) at a grid of phase plane sampling points and place small filaments with slopes equal to these values at the points, we can generate a qualitative picture of the flow of phase plane solution curves. We can also assign a direction to the filaments, indicating the instantaneous direction in which the point $(y_1(t), y_2(t))$ is moving along the phase plane curve. For example, at the point $(y_1, y_2) = (1, 0)$, we find $dy_2/dy_1 = \frac{4}{3}$. Therefore, the filament has slope $\frac{4}{3}$ and is directed upward and to the right, since both dy_1/dt and dy_2/dt are positive at that point. At the point $(0, 1)$, the filament has slope of $-\frac{3}{4}$. Moreover, since $dy_2/dt < 0$ and $dy_1/dt > 0$, the arrow on this filament is directed downward and to the right. Repeating this calculation at each point on a phase plane grid, we obtain a **phase plane direction field** such as that shown in Figure 4.3. (A phase plane direction field is sometimes referred to as a **vector field**, since it attaches a vector to each point in the plane.)

While there are a number of computer software packages that can generate phase plane direction fields, we can obtain some rough qualitative information from equation (4) without making the extensive calculations required for a direction field. On the line $4y_1 - 3y_2 = 0$, the numerator of (4) vanishes and the phase plane filaments are horizontal; they are directed to the right if the denominator $3y_1 + 4y_2 > 0$ and directed to the left if $3y_1 + 4y_2 < 0$. The filaments are vertical on the line $3y_1 + 4y_2 = 0$, where the denominator of (4) vanishes. The arrows on these vertical filaments point upward if $4y_1 - 3y_2 > 0$ and downward if $4y_1 - 3y_2 < 0$. In the phase plane sectors lying between these intersecting lines, we can evaluate the numerator and denominator of (4) and get a sense of the typical directed filament. Figure 4.4 illustrates these ideas.

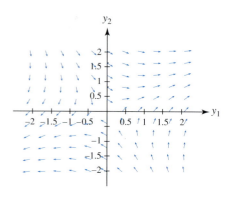

FIGURE 4.3

A phase plane direction field for the system (2) in Example 3. When a phase plane solution curve $(y_1(t), y_2(t))$ passes through a grid point, it is moving in the direction of the arrow attached to that grid point.

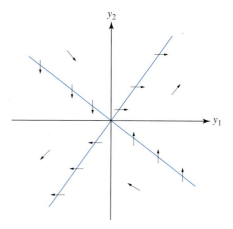

FIGURE 4.4

We can obtain rough qualitative information about the phase plane behavior of solutions of system (2) by dividing the phase plane into four sectors, using the two lines $3y_1 + 4y_2 = 0$ and $4y_1 - 3y_2 = 0$; see equation (4).

EXERCISES

Exercises 1–10:

In each exercise, find the general solution of the homogeneous linear system and then solve the given initial value problem.

1. $y_1' = -y_1 - y_2,$ $y_1(0) = -2$
$y_2' = 6y_1 + 4y_2,$ $y_2(0) = 6$

2. $y_1' = y_1 + y_2,$ $y_1(0) = 3$
$y_2' = -2y_1 - 2y_2,$ $y_2(0) = -4$

3. $\mathbf{y}' = \begin{bmatrix} -5 & -2 \\ 12 & 5 \end{bmatrix} \mathbf{y},$ $\mathbf{y}(1) = \begin{bmatrix} 1 \\ 0 \end{bmatrix}$

4. $\mathbf{y}' = \begin{bmatrix} 1 & 2 \\ 0 & 3 \end{bmatrix} \mathbf{y},$ $\mathbf{y}(0) = \begin{bmatrix} 4 \\ 1 \end{bmatrix}$

5. $\mathbf{y}' = \begin{bmatrix} 1 & 2 \\ 2 & 1 \end{bmatrix} \mathbf{y}, \quad \mathbf{y}(-1) = \begin{bmatrix} 2 \\ 2 \end{bmatrix}$

6. $\begin{aligned} y_1' &= 2y_1 + y_2 + 2y_3, & y_1(0) &= 4 \\ y_2' &= 3y_2 + 2y_3, & y_2(0) &= 3 \\ y_3' &= y_3, & y_3(0) &= -1 \end{aligned}$

7. $\mathbf{y}' = \begin{bmatrix} 3 & 1 & 1 \\ 1 & 3 & 1 \\ 1 & 1 & 3 \end{bmatrix} \mathbf{y}, \quad \mathbf{y}(0) = \begin{bmatrix} -1 \\ 1 \\ 6 \end{bmatrix}$

[For Exercise 7, the characteristic polynomial is $p(\lambda) = -(\lambda - 5)(\lambda - 2)^2$.]

8. $\mathbf{y}' = \begin{bmatrix} 2 & 2 & 2 \\ 2 & 2 & 2 \\ 2 & 2 & 2 \end{bmatrix} \mathbf{y}, \quad \mathbf{y}(0) = \begin{bmatrix} 2 \\ 5 \\ 5 \end{bmatrix}$

[For Exercise 8, the characteristic polynomial is $p(\lambda) = -\lambda^2(\lambda - 6)$.]

9. $\mathbf{y}' = \begin{bmatrix} 7 & 10 & 0 \\ -5 & -8 & 0 \\ 3 & 1 & 1 \end{bmatrix} \mathbf{y}, \quad \mathbf{y}(0) = \begin{bmatrix} -6 \\ 4 \\ -8 \end{bmatrix}$

10. $\mathbf{y}' = \begin{bmatrix} 3 & 1 & 2 \\ 0 & 8 & 15 \\ 0 & -6 & -11 \end{bmatrix} \mathbf{y}, \quad \mathbf{y}(0) = \begin{bmatrix} 2 \\ 5 \\ -2 \end{bmatrix}$

11. Let $A = \begin{bmatrix} a & b \\ b & c \end{bmatrix}$ be a (2×2) real symmetric matrix. In Exercise 28 of Section 4.4, it was shown that such a matrix has only real eigenvalues. We now show that A has a full set of eigenvectors. Note, by Exercise 30 of Section 4.4, that if A has distinct eigenvalues, then A has a full set of eigenvectors. Thus, the only case to consider is the case where A has repeated eigenvalues, $\lambda_1 = \lambda_2$.

(a) If $\lambda_1 = \lambda_2$, show that $a = c$, $b = 0$, and therefore $A = aI$.

(b) Exhibit a pair of linearly independent eigenvectors in this case.

12. Let A be an $(n \times n)$ real symmetric matrix. Show that eigenvectors belonging to distinct eigenvalues are orthogonal. That is, if $A\mathbf{x}_1 = \lambda_1 \mathbf{x}_1$ and $A\mathbf{x}_2 = \lambda_2 \mathbf{x}_2$, where $\lambda_1 \neq \lambda_2$, then $\mathbf{x}_1^T \mathbf{x}_2 = 0$. [Hint: Consider the matrix product $\mathbf{x}_1^T A \mathbf{x}_2$, and use the symmetry of A to show that $(\lambda_1 - \lambda_2)\mathbf{x}_1^T \mathbf{x}_2 = 0$. You will also need to recall that if the matrix product of R and S is defined, then $(RS)^T = S^T R^T$.]

Tank-Flushing Problems Consider the flow systems schematically shown in the figures for Exercises 13 and 14. In each case, a flushing out of the system is initiated at time $t = 0$. Fresh water is pumped into each tank, and well-stirred mixtures flow out. Each flow rate is equal to r gal/min, and we let $Q_j(t)$ represent the amount of solute (in pounds) in the jth tank at time t. Each tank has fluid volume V, remaining constant. All tanks in the system have an identical flow environment.

13. (a) Consider the two-tank flow system shown in the figure on the next page. Apply the "conservation of salt" principle to each tank, and derive the homogeneous linear equation for $Q_j(t)$, $j = 1, 2$.

(b) The fact that the two tanks have identical capacity and experience the same environment reflects itself in the fact that the coefficient matrix is a (real) symmetric matrix. Determine the eigenvalues and corresponding eigenvectors of this matrix, and form the general solution of the homogeneous linear system.

(c) Assume that the initial amounts of solute in Tanks 1 and 2 are $Q_1(0) = Q_0$ and $Q_2(0) = 2Q_0$, respectively, where Q_0 is a positive constant. Assume that $r/V = 0.02 \, \text{sec}^{-1}$. Determine the amount of flushing time required to reduce the amount of salt in each of the two tanks to $0.01Q_0$ or less.

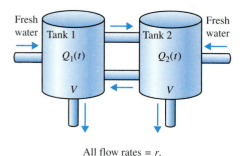

All flow rates = r.

Figure for Exercise 13

14. Consider the three-tank flow system shown in the figure.

(a) As in Exercise 13, derive a homogeneous linear equation for $Q_j(t), j = 1, 2, 3$.

(b) Determine the eigenvalues and eigenvectors of the coefficient matrix. You will find that one eigenvalue appears as a root of multiplicity two of the characteristic equation. You should, however, be able to find two linearly independent eigenvectors corresponding to this repeated eigenvalue. Form the general solution of the homogeneous linear system.

(c) If the initial amounts of salt in the three tanks are $Q_1(0) = Q_0$, $Q_2(0) = 2Q_0$, and $Q_3(0) = 3Q_0$, respectively, determine the solution of the resulting initial value problem. (Your answer will involve the constants r/V and Q_0 as well as time t.)

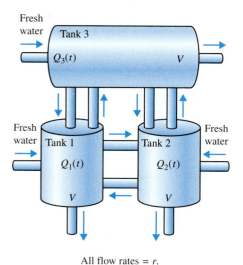

All flow rates = r.

Figure for Exercise 14

15. (a) Let $\mathbf{y}(t)$ denote the solution of the autonomous linear system $\mathbf{y}' = A\mathbf{y}, \mathbf{y}(0) = \mathbf{y}_0$. Show that $\mathbf{y}(t - t_0)$ is the solution of the initial value problem $\mathbf{y}' = A\mathbf{y}, \mathbf{y}(t_0) = \mathbf{y}_0$. (Recall Theorem 2.3 in Section 2.5.)

(b) Let A be a constant (2×2) matrix. Suppose the solution of $\mathbf{y}' = A\mathbf{y}, \mathbf{y}(0) = \mathbf{y}_0$ is given by

$$\mathbf{y}(t) = \begin{bmatrix} e^t - 2e^{-t} \\ 3e^t + e^{-t} \end{bmatrix}.$$

Let $\hat{\mathbf{y}}(t)$ denote the solution of $\mathbf{y}' = A\mathbf{y}, \ \mathbf{y}(-1) = \mathbf{y}_0$. Determine $\hat{\mathbf{y}}(2)$.

16. **Equilibrium Solutions**   Consider the linear system $\mathbf{y}' = A\mathbf{y}$, where

$$A = \begin{bmatrix} 2 & 1 \\ -4 & \alpha \end{bmatrix}.$$

(a) For what values of the constant α is $\mathbf{y} = \mathbf{0}$ the only equilibrium solution?

(b) For what values of α does more than one equilibrium solution exist? In this case, how many are there? Where do these values lie when plotted in the phase plane?

17. Match each linear system with one of the phase plane direction fields.

(a) $\mathbf{y}' = \begin{bmatrix} 1 & -3 \\ 0 & -2 \end{bmatrix} \mathbf{y}$

(b) $\mathbf{y}' = \dfrac{-1}{3} \begin{bmatrix} 4 & 1 \\ 2 & 5 \end{bmatrix} \mathbf{y}$

(c) $\mathbf{y}' = \begin{bmatrix} 3 & 1 \\ 1 & 3 \end{bmatrix} \mathbf{y}$

(d) $\mathbf{y}' = \dfrac{1}{2} \begin{bmatrix} 1 & 3 \\ 3 & 1 \end{bmatrix} \mathbf{y}$

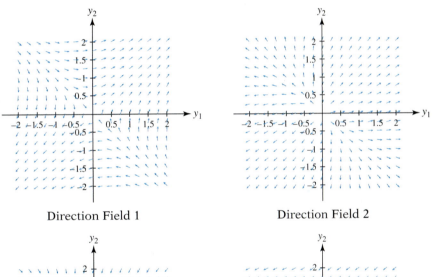

Direction Field 1   Direction Field 2

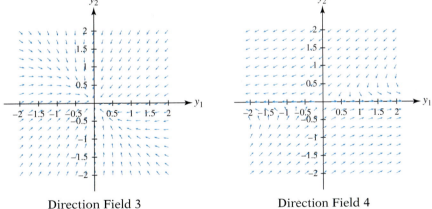

Direction Field 3   Direction Field 4

Figure for Exercise 17

Exercises 18–21:

In each exercise, the general solution of a (2×2) linear system $\mathbf{y}' = A\mathbf{y}$ is given, along with an initial condition. Sketch the phase plane solution trajectory that satisfies the given initial condition.

18. $\mathbf{y}(t) = c_1 e^{-t} \begin{bmatrix} 1 \\ -1 \end{bmatrix} + c_2 e^{-2t} \begin{bmatrix} 1 \\ 1 \end{bmatrix}$, $\mathbf{y}(0) = \begin{bmatrix} 3 \\ -1 \end{bmatrix}$

19. $\mathbf{y}(t) = c_1 e^{2t} \begin{bmatrix} 1 \\ 1 \end{bmatrix} + c_2 e^{t} \begin{bmatrix} 0 \\ 1 \end{bmatrix}$, $\mathbf{y}(0) = \begin{bmatrix} 0 \\ -2 \end{bmatrix}$

20. $\mathbf{y}(t) = c_1 e^{-t} \begin{bmatrix} 1 \\ -1 \end{bmatrix} + c_2 e^{t} \begin{bmatrix} 1 \\ 1 \end{bmatrix}$, $\mathbf{y}(0) = \begin{bmatrix} 2 \\ 0 \end{bmatrix}$

21. $y(t) = c_1 \begin{bmatrix} 2 \\ 1 \end{bmatrix} + c_2 e^{-t} \begin{bmatrix} -1 \\ 1 \end{bmatrix}$, $\mathbf{y}(0) = \begin{bmatrix} 3 \\ 3 \end{bmatrix}$

4.6   Complex Eigenvalues

In this section, we study the differential equation $\mathbf{y}' = A\mathbf{y}$, where A is a real constant ($n \times n$) matrix possessing complex conjugate eigenvalues. As an example, consider the linear system

$$\begin{aligned} y_1' &= y_1 + y_2 \\ y_2' &= -y_1 + y_2. \end{aligned}$$

The coefficient matrix

$$A = \begin{bmatrix} 1 & 1 \\ -1 & 1 \end{bmatrix}$$

has a pair of complex eigenvalues, $\lambda_1 = 1 + i$ and $\lambda_2 = 1 - i$. To find an eigenvector corresponding to λ_1, we seek a nontrivial solution of $(A - \lambda_1 I)\mathbf{x} = \mathbf{0}$. Therefore,

$$\begin{bmatrix} -i & 1 \\ -1 & -i \end{bmatrix} \begin{bmatrix} x_1 \\ x_2 \end{bmatrix} = \begin{bmatrix} 0 \\ 0 \end{bmatrix}$$

or, equivalently,

$$\begin{bmatrix} -i & 1 \\ 0 & 0 \end{bmatrix} \begin{bmatrix} x_1 \\ x_2 \end{bmatrix} = \begin{bmatrix} 0 \\ 0 \end{bmatrix}.$$

A nontrivial solution is given by

$$\mathbf{x}_1 = \begin{bmatrix} 1 \\ i \end{bmatrix}.$$

Similarly, an eigenvector corresponding to $\lambda_2 = 1 - i$ is

$$\mathbf{x}_2 = \begin{bmatrix} 1 \\ -i \end{bmatrix}.$$

For this matrix, the eigenvalues and eigenvectors occur in conjugate pairs, and this property is common to every real matrix with complex eigenvalues. If A is a real matrix, then

1. Complex eigenvalues always occur in conjugate pairs.
2. If λ is a complex eigenvalue with a corresponding eigenvector $\mathbf{x}$, then $\bar{\mathbf{x}}$ is an eigenvector for the eigenvalue $\bar{\lambda}$.

Therefore, if $(\lambda, \mathbf{x})$ is an eigenpair of a real matrix A, then so is $(\bar{\lambda}, \bar{\mathbf{x}})$. Note the computational implications: Once an eigenvector $\mathbf{x}$ corresponding to a complex eigenvalue λ has been determined, we need only form the complex conjugate $\bar{\mathbf{x}}$ to obtain an eigenvector corresponding to $\bar{\lambda}$.

The Real and Imaginary Parts of a Complex-Valued Solution Are Also Solutions

For applications, we often want to convert the complex solutions of $\mathbf{y}' = A\mathbf{y}$ that arise from complex eigenpairs into real-valued solutions. For example, the matrix

$$A = \begin{bmatrix} 1 & 1 \\ -1 & 1 \end{bmatrix}$$

has eigenvalues $\lambda_1 = 1 + i$ and $\lambda_2 = \bar{\lambda}_1 = 1 - i$ and corresponding eigenvectors

$$\mathbf{x}_1 = \begin{bmatrix} 1 \\ i \end{bmatrix} \quad \text{and} \quad \mathbf{x}_2 = \bar{\mathbf{x}}_1 = \begin{bmatrix} 1 \\ -i \end{bmatrix}.$$

Since the eigenvalues are distinct, we conclude that

$$\{\mathbf{y}_1(t), \mathbf{y}_2(t)\} = \left\{ e^{(1+i)t} \begin{bmatrix} 1 \\ i \end{bmatrix}, e^{(1-i)t} \begin{bmatrix} 1 \\ -i \end{bmatrix} \right\} \tag{1}$$

is a fundamental set of solutions for $\mathbf{y}' = A\mathbf{y}$. How do we convert these two solutions into a fundamental set of real-valued solutions?

As in Chapter 3, the key result is that both the real and the imaginary parts of a complex-valued solution are also solutions.

Theorem 4.7

Consider the differential equation $\mathbf{y}' = A\mathbf{y}$, $-\infty < t < \infty$, where A is an $(n \times n)$ real matrix. Let $\mathbf{y}(t) = \mathbf{u}(t) + i\mathbf{v}(t)$ be a complex-valued solution of this differential equation, where $\mathbf{u}(t)$ and $\mathbf{v}(t)$ are each real-valued $(n \times 1)$ vector functions representing the real and imaginary parts of $\mathbf{y}(t)$, respectively. Then $\mathbf{u}(t)$ and $\mathbf{v}(t)$ are each solutions of $\mathbf{y}' = A\mathbf{y}$, $-\infty < t < \infty$.

● **PROOF:**   Substitute $\mathbf{y}(t) = \mathbf{u}(t) + i\mathbf{v}(t)$ into the left-hand side of the differential equation $\mathbf{y}' = A\mathbf{y}$ to obtain

$$[\mathbf{u}(t) + i\mathbf{v}(t)]' = \begin{bmatrix} [u_1(t) + iv_1(t)]' \\ [u_2(t) + iv_2(t)]' \\ \vdots \\ [u_n(t) + iv_n(t)]' \end{bmatrix} = \begin{bmatrix} u_1'(t) \\ u_2'(t) \\ \vdots \\ u_n'(t) \end{bmatrix} + i \begin{bmatrix} v_1'(t) \\ v_2'(t) \\ \vdots \\ v_n'(t) \end{bmatrix} = \mathbf{u}'(t) + i\mathbf{v}'(t). \tag{2}$$

Substituting $\mathbf{y}(t) = \mathbf{u}(t) + i\mathbf{v}(t)$ into the right-hand side of the differential equation $\mathbf{y}' = A\mathbf{y}$, we obtain

$$A[\mathbf{u}(t) + i\mathbf{v}(t)] = A\mathbf{u}(t) + A[i\mathbf{v}(t)] = A\mathbf{u}(t) + iA\mathbf{v}(t). \tag{3}$$

Two complex quantities are equal if and only if their corresponding real and imaginary parts are equal. Therefore, equating expressions (2) and (3), we see that

$$\mathbf{u}'(t) = A\mathbf{u}(t), \qquad \mathbf{v}'(t) = A\mathbf{v}(t), \qquad -\infty < t < \infty. \bullet$$

The following example illustrates the use of Theorem 4.7 to convert a set of complex-valued solutions into a set of real-valued solutions.

EXAMPLE

1

Find a real-valued fundamental set of solutions of $\mathbf{y}' = A\mathbf{y}$, $-\infty < t < \infty$, where

$$A = \begin{bmatrix} 1 & 1 \\ -1 & 1 \end{bmatrix}.$$

Solution: From equation (1),

$$\mathbf{y}(t) = e^{(1+i)t} \begin{bmatrix} 1 \\ i \end{bmatrix} \tag{4}$$

is a complex-valued solution of $\mathbf{y}' = A\mathbf{y}$. From Euler's formula [see equation (9) in Section 3.5],

$$e^{(1+i)t} = e^t e^{it} = e^t(\cos t + i \sin t) = e^t \cos t + i e^t \sin t.$$

Therefore, we can write solution (4) as

$$\mathbf{y}(t) = (e^t \cos t + i e^t \sin t) \begin{bmatrix} 1 \\ i \end{bmatrix}$$

$$= \begin{bmatrix} e^t \cos t \\ -e^t \sin t \end{bmatrix} + i \begin{bmatrix} e^t \sin t \\ e^t \cos t \end{bmatrix}$$

$$= \mathbf{u}(t) + i\mathbf{v}(t).$$

It follows from Theorem 4.7 that the two real functions

$$\mathbf{u}(t) = \begin{bmatrix} e^t \cos t \\ -e^t \sin t \end{bmatrix} \quad \text{and} \quad \mathbf{v}(t) = \begin{bmatrix} e^t \sin t \\ e^t \cos t \end{bmatrix}$$

are also solutions of the differential equation. (You can also verify this claim by direct substitution into the differential equation.)

To show that they form a fundamental set of solutions on $-\infty < t < \infty$, we calculate the Wronskian and find

$$W(t) = \begin{vmatrix} e^t \cos t & e^t \sin t \\ -e^t \sin t & e^t \cos t \end{vmatrix} = e^{2t}(\cos^2 t + \sin^2 t) = e^{2t} \neq 0.$$

Therefore, the general solution of the differential equation is $\mathbf{y}(t) = c_1 \mathbf{u}(t) + c_2 \mathbf{v}(t)$:

$$\mathbf{y}(t) = c_1 \begin{bmatrix} e^t \cos t \\ -e^t \sin t \end{bmatrix} + c_2 \begin{bmatrix} e^t \sin t \\ e^t \cos t \end{bmatrix} = \begin{bmatrix} e^t \cos t & e^t \sin t \\ -e^t \sin t & e^t \cos t \end{bmatrix} \begin{bmatrix} c_1 \\ c_2 \end{bmatrix}. ❖$$

REMARK: According to Theorem 4.5 in Section 4.3, the two fundamental sets [the original set of complex-valued solutions (1) and the real-valued set constructed in Example 1] are related via multiplication on the right by a constant nonsingular matrix. We can illustrate Theorem 4.5 here by noting that

$$\begin{bmatrix} e^t \cos t & e^t \sin t \\ -e^t \sin t & e^t \cos t \end{bmatrix} = \begin{bmatrix} e^{(1+i)t} & e^{(1-i)t} \\ ie^{(1+i)t} & -ie^{(1-i)t} \end{bmatrix} \begin{bmatrix} \dfrac{1}{2} & \dfrac{-i}{2} \\ \dfrac{1}{2} & \dfrac{i}{2} \end{bmatrix},$$

where the constant matrix on the right-hand side is nonsingular since

$$\begin{vmatrix} \dfrac{1}{2} & \dfrac{-i}{2} \\ \dfrac{1}{2} & \dfrac{i}{2} \end{vmatrix} = \dfrac{i}{2} \neq 0.$$

E X A M P L E

2

Solve the initial value problem $\mathbf{y}' = A\mathbf{y}$, $-\infty < t < \infty$,

$$A = \begin{bmatrix} 1 & 2 & -2 \\ 2 & 5 & -2 \\ 4 & 12 & -5 \end{bmatrix}, \qquad \mathbf{y}(0) = \begin{bmatrix} 1 \\ 1 \\ 0 \end{bmatrix}.$$

Solution: Since the characteristic polynomial is a cubic polynomial with real coefficients, there will be at least one real eigenvalue. Using a cofactor expansion along the first row, we obtain

$$p(\lambda) = \det(A - \lambda I) = \begin{vmatrix} 1 - \lambda & 2 & -2 \\ 2 & 5 - \lambda & -2 \\ 4 & 12 & -5 - \lambda \end{vmatrix}$$

$$= (1 - \lambda)\begin{vmatrix} 5 - \lambda & -2 \\ 12 & -5 - \lambda \end{vmatrix} - 2\begin{vmatrix} 2 & -2 \\ 4 & -5 - \lambda \end{vmatrix} - 2\begin{vmatrix} 2 & 5 - \lambda \\ 4 & 12 \end{vmatrix}$$

$$= -(\lambda^3 - \lambda^2 + 3\lambda + 5).$$

Computer software could certainly be used to determine the roots of $p(\lambda) = 0$. In this particular case, however, we see by inspection that $\lambda = -1$ is a root. Therefore, synthetic division and the quadratic formula yields

$$p(\lambda) = -(\lambda + 1)(\lambda^2 - 2\lambda + 5) = -(\lambda + 1)(\lambda - 1 - 2i)(\lambda - 1 + 2i).$$

The eigenvalues are $\lambda_1 = -1$, $\lambda_2 = 1 + 2i$, $\lambda_3 = \bar{\lambda}_2 = 1 - 2i$. We now compute the eigenvectors.

For $\lambda_1 = -1$, we solve $(A - \lambda_1 I)\mathbf{x}_1 = (A + I)\mathbf{x}_1 = \mathbf{0}$, or

$$\begin{bmatrix} 2 & 2 & -2 \\ 2 & 6 & -2 \\ 4 & 12 & -4 \end{bmatrix} \begin{bmatrix} x_1 \\ x_2 \\ x_3 \end{bmatrix} = \begin{bmatrix} 0 \\ 0 \\ 0 \end{bmatrix}.$$

Using elementary row operations, we obtain an equivalent homogeneous system

$$\begin{bmatrix} 1 & 0 & -1 \\ 0 & 1 & 0 \\ 0 & 0 & 0 \end{bmatrix} \begin{bmatrix} x_1 \\ x_2 \\ x_3 \end{bmatrix} = \begin{bmatrix} 0 \\ 0 \\ 0 \end{bmatrix}.$$

(continued)

(continued)

A nontrivial solution is

$$\mathbf{x}_1 = \begin{bmatrix} 1 \\ 0 \\ 1 \end{bmatrix}.$$

For $\lambda_2 = 1 + 2i$, we solve $(A - \lambda_2 I)\mathbf{x} = [A - (1 + 2i)I]\mathbf{x} = \mathbf{0}$, or

$$\begin{bmatrix} -2i & 2 & -2 \\ 2 & 4 - 2i & -2 \\ 4 & 12 & -6 - 2i \end{bmatrix} \begin{bmatrix} x_1 \\ x_2 \\ x_3 \end{bmatrix} = \begin{bmatrix} 0 \\ 0 \\ 0 \end{bmatrix}.$$

In this case, elementary row operations lead to an equivalent system

$$\begin{bmatrix} 1 & 0 & -\dfrac{i}{2} \\ 0 & 1 & -\dfrac{1}{2} \\ 0 & 0 & 0 \end{bmatrix} \begin{bmatrix} x_1 \\ x_2 \\ x_3 \end{bmatrix} = \begin{bmatrix} 0 \\ 0 \\ 0 \end{bmatrix}.$$

A nontrivial solution is

$$\mathbf{x}_2 = \begin{bmatrix} i \\ 1 \\ 2 \end{bmatrix}.$$

Although we do not need it to solve the given initial value problem, we know an eigenvector $\mathbf{x}_3$ corresponding to the eigenvalue $\lambda_3 = \bar{\lambda}_2 = 1 - 2i$ is given by

$$\mathbf{x}_3 = \bar{\mathbf{x}}_2 = \begin{bmatrix} -i \\ 1 \\ 2 \end{bmatrix}.$$

We now develop a real-valued fundamental set of solutions. One solution [corresponding to the eigenpair $(\lambda_1, \mathbf{x}_1)$] is

$$\mathbf{y}_1(t) = e^{-t} \begin{bmatrix} 1 \\ 0 \\ 1 \end{bmatrix} = \begin{bmatrix} e^{-t} \\ 0 \\ e^{-t} \end{bmatrix}.$$

Since $\mathbf{y}' = A\mathbf{y}$ is a system of three first order equations, we know that a fundamental set must consist of three solutions. To obtain the two other solutions needed, we take the complex-valued solution determined by the eigenpair $(\lambda_2, \mathbf{x}_2)$,

$$\mathbf{y}(t) = e^{(1+2i)t} \begin{bmatrix} i \\ 1 \\ 2 \end{bmatrix},$$

and decompose it into the form $\mathbf{u}(t) + i\mathbf{v}(t)$. [The other complex-valued solution, determined by the eigenpair $(\lambda_3, \mathbf{x}_3)$, decomposes into $\mathbf{u}(t) - i\mathbf{v}(t)$, yielding (essentially) the same pair of real-valued solutions.] Decomposing $\mathbf{y}(t)$, we obtain

$$\mathbf{y}(t) = e^{(1+2i)t} \begin{bmatrix} i \\ 1 \\ 2 \end{bmatrix} = e^t (\cos 2t + i \sin 2t) \begin{bmatrix} i \\ 1 \\ 2 \end{bmatrix}$$

$$= \begin{bmatrix} -e^t \sin 2t \\ e^t \cos 2t \\ 2e^t \cos 2t \end{bmatrix} + i \begin{bmatrix} e^t \cos 2t \\ e^t \sin 2t \\ 2e^t \sin 2t \end{bmatrix}.$$

We complete the fundamental set by setting

$$\mathbf{y}_2(t) = \begin{bmatrix} -e^t \sin 2t \\ e^t \cos 2t \\ 2e^t \cos 2t \end{bmatrix} \quad \text{and} \quad \mathbf{y}_3(t) = \begin{bmatrix} e^t \cos 2t \\ e^t \sin 2t \\ 2e^t \sin 2t \end{bmatrix}.$$

The general solution, $\mathbf{y}(t) = c_1 \mathbf{y}_1(t) + c_2 \mathbf{y}_2(t) + c_3 \mathbf{y}_3(t)$, is therefore

$$\mathbf{y}(t) = c_1 \begin{bmatrix} e^{-t} \\ 0 \\ e^{-t} \end{bmatrix} + c_2 \begin{bmatrix} -e^t \sin 2t \\ e^t \cos 2t \\ 2e^t \cos 2t \end{bmatrix} + c_3 \begin{bmatrix} e^t \cos 2t \\ e^t \sin 2t \\ 2e^t \sin 2t \end{bmatrix}$$

$$= \begin{bmatrix} e^{-t} & -e^t \sin 2t & e^t \cos 2t \\ 0 & e^t \cos 2t & e^t \sin 2t \\ e^{-t} & 2e^t \cos 2t & 2e^t \sin 2t \end{bmatrix} \begin{bmatrix} c_1 \\ c_2 \\ c_3 \end{bmatrix}. \tag{5}$$

The fact that our three solutions form a fundamental set can be verified directly by noting that the Wronskian, evaluated at $t = 0$, is

$$W(0) = \begin{vmatrix} 1 & 0 & 1 \\ 0 & 1 & 0 \\ 1 & 2 & 0 \end{vmatrix} = -1 \neq 0.$$

To solve the given initial value problem, we impose the initial condition in equation (5), finding

$$\begin{bmatrix} 1 & 0 & 1 \\ 0 & 1 & 0 \\ 1 & 2 & 0 \end{bmatrix} \begin{bmatrix} c_1 \\ c_2 \\ c_3 \end{bmatrix} = \begin{bmatrix} 1 \\ 1 \\ 0 \end{bmatrix}.$$

The solution is

$$\begin{bmatrix} c_1 \\ c_2 \\ c_3 \end{bmatrix} = \begin{bmatrix} -2 \\ 1 \\ 3 \end{bmatrix},$$

and therefore the unique solution of the initial value problem is

$$\mathbf{y}(t) = -2 \begin{bmatrix} e^{-t} \\ 0 \\ e^{-t} \end{bmatrix} + \begin{bmatrix} -e^t \sin 2t \\ e^t \cos 2t \\ 2e^t \cos 2t \end{bmatrix} + 3 \begin{bmatrix} e^t \cos 2t \\ e^t \sin 2t \\ 2e^t \sin 2t \end{bmatrix} = \begin{bmatrix} -2e^{-t} - e^t \sin 2t + 3e^t \cos 2t \\ e^t \cos 2t + 3e^t \sin 2t \\ -2e^{-t} + 2e^t \cos 2t + 6e^t \sin 2t \end{bmatrix}.$$

❖

Phase Plane Trajectories

The following example develops phase plane trajectories for a pair of (2×2) systems $\mathbf{y}' = A\mathbf{y}$, where A has complex eigenvalues.

E X A M P L E

3

Sketch several phase plane trajectories for

$$\text{(a)}\ \begin{aligned} y_1' &= \quad y_2 \\ y_2' &= -4y_1 \end{aligned} \qquad \text{(b)}\ \begin{aligned} y_1' &= \quad y_1 + y_2 \\ y_2' &= -y_1 + y_2 \end{aligned}$$

Note that the solution of system (b) is given in Example 1.

Solution:

(a) The coefficient matrix has eigenpairs $(\lambda_1, \mathbf{x}_1)$ and $(\lambda_2, \mathbf{x}_2)$, where

$$\lambda_1 = 2i, \ \ \mathbf{x}_1 = \begin{bmatrix} 1 \\ 2i \end{bmatrix} \quad \text{and} \quad \lambda_2 = -2i, \ \ \mathbf{x}_2 = \begin{bmatrix} 1 \\ -2i \end{bmatrix}.$$

Thus, the general solution, expressed in terms of a real-valued fundamental set, is

$$\mathbf{y}(t) = c_1 \begin{bmatrix} \cos 2t \\ -2 \sin 2t \end{bmatrix} + c_2 \begin{bmatrix} \sin 2t \\ 2 \cos 2t \end{bmatrix}.$$

Since the component functions $\cos 2t$ and $\sin 2t$ are periodic with period π, the trajectories are closed curves; as t increases by π units, the phase plane point returns to its original position; see Figure 4.5(a). The phase plane trajectories form a family of concentric ellipses centered at the equilibrium solution $\mathbf{y} = 0$, and they are traversed in a clockwise manner.

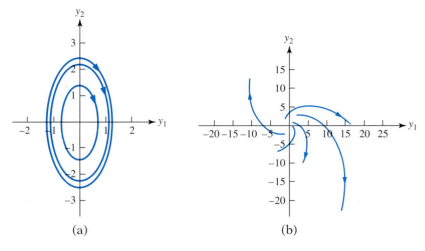

(a)   (b)

FIGURE 4.5

(a) Phase plane trajectories for the system in part (a) of Example 3 are ellipses. These trajectories are traced in a clockwise manner. (b) Phase plane trajectories for the system in part (b) are spirals, traced in a clockwise manner.

To establish that the trajectories are ellipses, we rewrite the arbitrary constants as $c_1 = R\cos\alpha$ and $c_2 = R\sin\alpha$. With this,

$$y_1(t) = R(\cos\alpha\cos 2t + \sin\alpha\sin 2t) = R\cos(2t - \alpha)$$
$$y_2(t) = -2R(\cos\alpha\sin 2t - \sin\alpha\cos 2t) = -2R\sin(2t - \alpha).$$

Therefore, it follows that the phase plane points $(y_1(t), y_2(t))$ lie on the ellipse

$$\frac{y_1^2}{R^2} + \frac{y_2^2}{4R^2} = 1.$$

(b) From Example 1, the general solution in component form is

$$y_1(t) = e^t(c_1\cos t + c_2\sin t)$$
$$y_2(t) = e^t(-c_1\sin t + c_2\cos t).$$

Setting $c_1 = R\cos\alpha$ and $c_2 = R\sin\alpha$ as in part (a), we obtain

$$y_1(t) = Re^t\cos(t - \alpha)$$
$$y_2(t) = -Re^t\sin(t - \alpha).$$

Therefore, the phase plane points $(y_1(t), y_2(t))$ lie on the spiral

$$y_1^2 + y_2^2 = R^2 e^{2t}.$$

As t increases, the phase plane points spiral outward while moving clockwise about the origin; see Figure 4.5(b). ❖

EXERCISES

Exercises 1–10:

Find the eigenvalues and eigenvectors of the given matrix A.

1. $A = \begin{bmatrix} 2 & 1 \\ -1 & 2 \end{bmatrix}$ **2.** $A = \begin{bmatrix} 0 & -9 \\ 1 & 0 \end{bmatrix}$ **3.** $A = \begin{bmatrix} 0 & 1 \\ -2 & -2 \end{bmatrix}$ **4.** $A = \begin{bmatrix} 3 & 2 \\ -5 & -3 \end{bmatrix}$

5. $A = \begin{bmatrix} -5 & -2 \\ 5 & 1 \end{bmatrix}$ **6.** $A = \begin{bmatrix} 3 & 1 \\ -2 & 1 \end{bmatrix}$ **7.** $A = \begin{bmatrix} -1 & -0.5 & 0 \\ 0.5 & -1 & 0 \\ 0 & 0 & 2 \end{bmatrix}$

8. $A = \begin{bmatrix} 0 & 0 & 0 \\ 0 & 3 & -5 \\ 0 & 2 & -3 \end{bmatrix}$ **9.** $A = \begin{bmatrix} 2 & 2 & 9 \\ 1 & -1 & 3 \\ -1 & -1 & -4 \end{bmatrix}$ **10.** $A = \begin{bmatrix} 1 & -4 & -1 \\ 3 & 2 & 3 \\ 1 & 1 & 3 \end{bmatrix}$

Exercises 11–16:

In each exercise, one or more eigenvalues and corresponding eigenvectors are given for a real matrix A. Determine a fundamental set of solutions for $\mathbf{y}' = A\mathbf{y}$, where the fundamental set consists entirely of *real* solutions.

11. A is (2×2) with an eigenvalue $\lambda = 4 + 2i$ and corresponding eigenvector

$$\mathbf{x} = \begin{bmatrix} 4 \\ -1 + i \end{bmatrix}.$$

12. A is (2×2) with an eigenvalue $\lambda = i$ and corresponding eigenvector

$$\mathbf{x} = \begin{bmatrix} -2 + i \\ 5 \end{bmatrix}.$$

13. A is (2×2) with an eigenvalue $\lambda = 2i$ and corresponding eigenvector

$$\mathbf{x} = \begin{bmatrix} -1 - i \\ 1 \end{bmatrix}.$$

14. A is (2×2) with an eigenvalue $\lambda = 1 + i$ and corresponding eigenvector

$$\mathbf{x} - \begin{bmatrix} -1 + i \\ i \end{bmatrix}.$$

15. A is (3×3) with a complex eigenvalue $\lambda = 2 + 3i$ and corresponding eigenvector

$$\mathbf{x} = \begin{bmatrix} -5 + 3i \\ 3 + 3i \\ 2 \end{bmatrix},$$

and a real eigenvalue $\lambda = 2$ and corresponding eigenvector

$$\mathbf{x} = \begin{bmatrix} 1 \\ 0 \\ -1 \end{bmatrix}.$$

16. A is (4×4) and has two different complex eigenvalues: $\lambda = 1 + 5i$ with corresponding eigenvector

$$\mathbf{x} = \begin{bmatrix} i \\ 1 \\ 0 \\ 0 \end{bmatrix},$$

and $\lambda = 1 + 2i$ with corresponding eigenvector

$$\mathbf{x} = \begin{bmatrix} 0 \\ 0 \\ i \\ 1 \end{bmatrix}.$$

Exercises 17–26:

Solve the initial value problem. Eigenpairs of the coefficient matrices were determined in Exercises 1–10.

17. $y_1' = 2y_1 + y_2, \quad y_1(0) = 4$
$\quad\;\; y_2' = -y_1 + 2y_2, \quad y_2(0) = 7$

18. $\mathbf{y}' = \begin{bmatrix} 0 & -9 \\ 1 & 0 \end{bmatrix} \mathbf{y}, \quad \mathbf{y}(0) = \begin{bmatrix} 6 \\ 2 \end{bmatrix}$

19. $\mathbf{y}' = \begin{bmatrix} 0 & 1 \\ -2 & -2 \end{bmatrix} \mathbf{y}, \quad \mathbf{y}(0) = \begin{bmatrix} 2 \\ 2 \end{bmatrix}$

20. $y_1' = 3y_1 + 2y_2, \quad y_1(0) = -1$
$\quad\;\; y_2' = -5y_1 - 3y_2, \quad y_2(0) = 1$

21. $y_1' = -5y_1 - 2y_2, \quad y_1(0) = 0$
$\quad\;\; y_2' = 5y_1 + y_2, \quad y_2(0) = -2$

22. $\mathbf{y}' = \begin{bmatrix} 3 & 1 \\ -2 & 1 \end{bmatrix} \mathbf{y}, \quad \mathbf{y}(0) = \begin{bmatrix} 8 \\ 6 \end{bmatrix}$

23. $\mathbf{y}' = \begin{bmatrix} -1 & -0.5 & 0 \\ 0.5 & -1 & 0 \\ 0 & 0 & 2 \end{bmatrix} \mathbf{y}, \quad \mathbf{y}(0) = \begin{bmatrix} 2 \\ 3 \\ -1 \end{bmatrix}$

24. $y_1' = 0, \quad\quad\quad\;\; y_1(\pi/2) = -1$
$\quad\;\; y_2' = 3y_2 - 5y_3, \quad y_2(\pi/2) = 1$
$\quad\;\; y_3' = 2y_2 - 3y_3, \quad y_3(\pi/2) = 2$

25. $\mathbf{y}' = \begin{bmatrix} 2 & 2 & 9 \\ 1 & -1 & 3 \\ -1 & -1 & -4 \end{bmatrix} \mathbf{y}, \quad \mathbf{y}(0) = \begin{bmatrix} 12 \\ 2 \\ 4 \end{bmatrix}$

26.
$$\mathbf{y}' = \begin{bmatrix} 1 & -4 & -1 \\ 3 & 2 & 3 \\ 1 & 1 & 3 \end{bmatrix} \mathbf{y}, \quad \mathbf{y}(0) = \begin{bmatrix} -1 \\ 9 \\ 4 \end{bmatrix}$$

27. Let A be a real (2×2) matrix having $\lambda = \alpha + i\beta$ as a complex eigenvalue, with β nonzero. Show that any eigenvector $\mathbf{x}$ corresponding to λ must have at least one complex component. [Hint: Assume that $\mathbf{x}$ is a real vector and deduce a contradiction.]

Exercises 28–31:

In each exercise, consider the initial value problem $\mathbf{y}' = A\mathbf{y}, \mathbf{y}(0) = \mathbf{y}_0$ for the given coefficient matrix A. In each exercise, the matrix A contains a real parameter μ.

(a) Determine all values of μ for which A has distinct real eigenvalues and all values of μ for which A has distinct complex eigenvalues.

(b) For what values of μ found in part (a) does $\sqrt{y_1(t)^2 + y_2(t)^2} \to 0$ as $t \to \infty$ for every initial vector $\mathbf{y}_0$?

28. $A = \begin{bmatrix} 1 & 3 \\ \mu & -2 \end{bmatrix}$ **29.** $A = \begin{bmatrix} -2 & \mu \\ 1 & -3 \end{bmatrix}$ **30.** $A = \begin{bmatrix} -3 & -\mu \\ \mu & 1 \end{bmatrix}$ **31.** $A = \begin{bmatrix} -3 & \mu \\ \mu & 1 \end{bmatrix}$

32. Match the phase plane plot of the solution with the appropriate initial value problem given on the next page. (The arrows on the trajectories indicate how the solution point moves as t increases.)

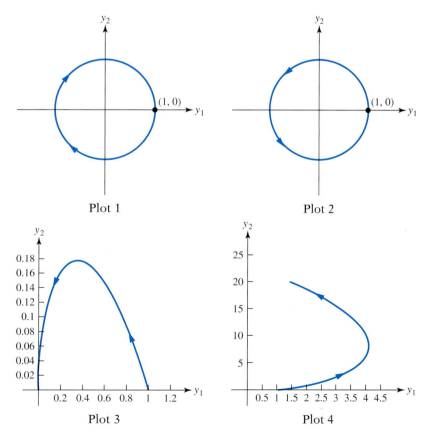

Plot 1   Plot 2

Plot 3   Plot 4

Figure for Exercise 32

$$\text{(a) } \mathbf{y}' = \begin{bmatrix} 1 & -0.5 \\ 0.5 & 1 \end{bmatrix} \mathbf{y}, \quad \mathbf{y}(0) = \begin{bmatrix} 1 \\ 0 \end{bmatrix} \qquad \text{(b) } \mathbf{y}' = \begin{bmatrix} 0 & -1 \\ 1 & 0 \end{bmatrix} \mathbf{y}, \quad \mathbf{y}(0) = \begin{bmatrix} 1 \\ 0 \end{bmatrix}$$

$$\text{(c) } \mathbf{y}' = \begin{bmatrix} 0 & 2 \\ -2 & 0 \end{bmatrix} \mathbf{y}, \quad \mathbf{y}(0) = \begin{bmatrix} 1 \\ 0 \end{bmatrix} \qquad \text{(d) } \mathbf{y}' = \begin{bmatrix} -1 & -0.5 \\ 0.5 & -1 \end{bmatrix} \mathbf{y}, \quad \mathbf{y}(0) = \begin{bmatrix} 1 \\ 0 \end{bmatrix}$$

Exercises 33–36:

A complex solution of the differential equation $\mathbf{y}' = A\mathbf{y}$ is given, where A is a real (2×2) matrix. Let $\mathbf{y}(t)$ denote any solution of $\mathbf{y}' = A\mathbf{y}$, where $\mathbf{y}(0) \neq \mathbf{0}$. As t increases, how will the phase plane trajectory of the solution behave? Will the solution point (a) move around the origin on a circular orbit, (b) move around the origin on an elliptical orbit, (c) spiral inward toward the origin, or (d) spiral outward away from the origin?

33. $\mathbf{y}(t) = e^{it} \begin{bmatrix} i \\ 2 \end{bmatrix}$ **34.** $\mathbf{y}(t) = e^{(-1+i)t} \begin{bmatrix} 1 \\ i \end{bmatrix}$

35. $\mathbf{y}(t) = e^{-2it} \begin{bmatrix} i \\ 2 \end{bmatrix}$ **36.** $\mathbf{y}(t) = e^{(2-i)t} \begin{bmatrix} i \\ -1 \end{bmatrix}$

4.7   Repeated Eigenvalues

In Section 3.4, we discussed the second order homogeneous scalar equation

$$y'' - 2\alpha y' + \alpha^2 y = 0. \tag{1}$$

Looking for solutions of the form $y(t) = e^{\lambda t}$ led us to a characteristic polynomial with repeated roots,

$$\lambda^2 - 2\alpha\lambda + \alpha^2 = (\lambda - \alpha)^2.$$

One solution of equation (1) is $y_1(t) = e^{\alpha t}$. A second solution needed to form a fundamental set of solutions is $y_2(t) = te^{\alpha t}$.

In our present study of the homogeneous first order linear system $\mathbf{y}' = A\mathbf{y}$, an analogous situation arises when the constant coefficient matrix A has repeated eigenvalues. (We say that A has **repeated eigenvalues** whenever the characteristic equation, $\det[A - \lambda I] = 0$, has repeated roots.) The problem of finding a fundamental set of solutions is more complicated when A has repeated eigenvalues than in the repeated root scalar case. In some cases, such as when A is a real symmetric matrix, the presence of repeated eigenvalues presents no new difficulties. We now consider the other cases, where some new ideas are required. An example illustrates the complications that may arise from the presence of repeated eigenvalues.

Complications from Repeated Eigenvalues

If A has repeated eigenvalues, are there enough linearly independent eigenvectors to form a fundamental set of solutions for $\mathbf{y}' = A\mathbf{y}$? For an illustration of the complications that may arise from repeated eigenvalues, consider the linear system $\mathbf{y}' = A\mathbf{y}$, where

$$A = \begin{bmatrix} \alpha & \beta \\ 0 & \alpha \end{bmatrix}. \tag{2}$$

The characteristic polynomial for A is

$$p(\lambda) = (\lambda - \alpha)^2.$$

Thus, $\lambda = \alpha$ is a repeated eigenvalue of A.

If $\beta = 0$ in equation (2), then

$$\mathbf{x}_1 = \begin{bmatrix} 1 \\ 0 \end{bmatrix} \quad \text{and} \quad \mathbf{x}_2 = \begin{bmatrix} 0 \\ 1 \end{bmatrix}$$

are eigenvectors corresponding to the eigenvalue $\lambda = \alpha$. [In fact, the matrix $A - \alpha I$ is the zero matrix, and thus any nonzero (2×1) vector is an eigenvector.] By Abel's theorem, the functions

$$\mathbf{y}_1(t) = e^{\alpha t}\mathbf{x}_1 \quad \text{and} \quad \mathbf{y}_2(t) = e^{\alpha t}\mathbf{x}_2 \tag{3}$$

form a fundamental set of solutions for $\mathbf{y}' = A\mathbf{y}$. However, what happens if $\beta \neq 0$ in equation (2)? In this event, as shown below, there is (essentially) only *one* eigenvector $\mathbf{x}_1$. Hence, we need a further analysis of $\mathbf{y}' = A\mathbf{y}$ in order to find the second solution needed for a fundamental set of solutions.

If $\beta \neq 0$, the eigenvector equation $(A - \alpha I)\mathbf{x} = \mathbf{0}$ is

$$\begin{bmatrix} 0 & \beta \\ 0 & 0 \end{bmatrix} \begin{bmatrix} x_1 \\ x_2 \end{bmatrix} = \begin{bmatrix} 0 \\ 0 \end{bmatrix}.$$

Since $\beta \neq 0$, it follows that $x_2 = 0$. Thus, every eigenvector corresponding to $\lambda = \alpha$ has the form

$$\mathbf{x} = \begin{bmatrix} x_1 \\ x_2 \end{bmatrix} = \begin{bmatrix} c \\ 0 \end{bmatrix} = c \begin{bmatrix} 1 \\ 0 \end{bmatrix} = c\mathbf{x}_1. \tag{4}$$

We now have one member of a fundamental set of solutions, namely

$$\mathbf{y}_1(t) = e^{\alpha t}\mathbf{x}_1 = \begin{bmatrix} e^{\alpha t} \\ 0 \end{bmatrix}.$$

How do we find a second solution, $\mathbf{y}_2(t)$?

Finding a Second Solution When the Value β in Matrix (2) Is Nonzero

For the simple example (2), we can find a second solution $\mathbf{y}_2(t)$ by sequentially solving the component equations. Let

$$\mathbf{y}(t) = \begin{bmatrix} y_1(t) \\ y_2(t) \end{bmatrix}.$$

In component form, the differential equation $\mathbf{y}' = A\mathbf{y}$ is

$$\begin{aligned} y_1' &= \alpha y_1 + \beta y_2 \\ y_2' &= \alpha y_2. \end{aligned} \tag{5}$$

We first solve the second equation, finding $y_2(t) = c_2 e^{\alpha t}$. Next, we substitute this expression for $y_2(t)$ into the first equation, obtaining

$$y_1' = \alpha y_1 + \beta c_2 e^{\alpha t}.$$

Solving this first order linear equation, we arrive at $y_1(t) = c_1 e^{\alpha t} + c_2 \beta t e^{\alpha t}$. Therefore, the general solution of $\mathbf{y}' = A\mathbf{y}$ is

$$\mathbf{y}(t) = \begin{bmatrix} y_1(t) \\ y_2(t) \end{bmatrix} = \begin{bmatrix} c_1 e^{\alpha t} + c_2 \beta t e^{\alpha t} \\ c_2 e^{\alpha t} \end{bmatrix} = \begin{bmatrix} c_1 e^{\alpha t} \\ 0 \end{bmatrix} + \begin{bmatrix} c_2 \beta t e^{\alpha t} \\ c_2 e^{\alpha t} \end{bmatrix}$$

$$= c_1 e^{\alpha t} \begin{bmatrix} 1 \\ 0 \end{bmatrix} + c_2 \left\{ t e^{\alpha t} \begin{bmatrix} \beta \\ 0 \end{bmatrix} + e^{\alpha t} \begin{bmatrix} 0 \\ 1 \end{bmatrix} \right\}$$

$$= c_1 \mathbf{y}_1(t) + c_2 \mathbf{y}_2(t).$$

As in the repeated-root scalar case, we see that the function $t e^{\alpha t}$ enters into the second solution,

$$\mathbf{y}_2(t) = t e^{\alpha t} \begin{bmatrix} \beta \\ 0 \end{bmatrix} + e^{\alpha t} \begin{bmatrix} 0 \\ 1 \end{bmatrix}.$$

We also see, however, that the second solution is not simply of the form $t e^{\alpha t} \mathbf{x}_1$; rather, it has the form $\mathbf{y}_2(t) = t e^{\alpha t} \mathbf{v}_1 + e^{\alpha t} \mathbf{v}_2$, where $\mathbf{v}_1$ and $\mathbf{v}_2$ are nonzero constant vectors. [Note that the vector we are calling $\mathbf{v}_1$ is actually an eigenvector, since $\mathbf{v}_1 = \beta \mathbf{x}_1$; see equation (4).]

To verify, for this example, that $\{\mathbf{y}_1(t), \mathbf{y}_2(t)\}$ is a fundamental set of solutions, let $\Psi(t) = [\mathbf{y}_1(t), \mathbf{y}_2(t)]$. At $t = 0$, the Wronskian is nonzero since

$$W(0) = \det[\Psi(0)] = \det[\mathbf{y}_1(0), \mathbf{y}_2(0)] = \begin{vmatrix} 1 & 0 \\ 0 & 1 \end{vmatrix} = 1.$$

How do we find a second solution when we cannot solve the component equations sequentially as we did in equation (5)? In the next subsection, we will use this example as a guide to develop a procedure for finding a second solution of $\mathbf{y}' = A\mathbf{y}$ in the case where A is a (2×2) constant matrix with a repeated eigenvalue. Later, we comment on the general case where A is an $(n \times n)$ matrix.

Finding a Fundamental Set of Solutions When an Eigenvalue Is Repeated

Consider the differential equation $\mathbf{y}' = A\mathbf{y}$, $-\infty < t < \infty$, where A is a real constant (2×2) matrix and where the characteristic polynomial is $p(\lambda) = (\lambda - \alpha)^2$.

If $\{\mathbf{x}_1, \mathbf{x}_2\}$ is a set of two linearly independent eigenvectors corresponding to the repeated eigenvalue $\lambda = \alpha$, then

$$\mathbf{y}_1(t) = e^{\alpha t} \mathbf{x}_1 \quad \text{and} \quad \mathbf{y}_2(t) = e^{\alpha t} \mathbf{x}_2$$

form a fundamental set of solutions of $\mathbf{y}' = A\mathbf{y}$. Suppose, however, that a set of two linearly independent eigenvectors corresponding to the eigenvalue $\lambda = \alpha$ does not exist. In such a case, $\mathbf{y}_1(t) = e^{\alpha t} \mathbf{x}_1$ is one solution. How do we find a second solution, $\mathbf{y}_2(t)$?

Motivated by the previous example, we look for a second solution of the form

$$\mathbf{y}_2(t) = t e^{\alpha t} \mathbf{v}_1 + e^{\alpha t} \mathbf{v}_2, \tag{6}$$

where $\mathbf{v}_1$ and $\mathbf{v}_2$ are nonzero constant (2×1) vectors to be determined. Substituting this representation into the differential equation, we obtain

$$(e^{\alpha t} + \alpha t e^{\alpha t})\mathbf{v}_1 + \alpha e^{\alpha t}\mathbf{v}_2 = A(te^{\alpha t}\mathbf{v}_1 + e^{\alpha t}\mathbf{v}_2).$$

We can rewrite this equation as

$$te^{\alpha t}(A\mathbf{v}_1 - \alpha\mathbf{v}_1) + e^{\alpha t}(A\mathbf{v}_2 - \alpha\mathbf{v}_2 - \mathbf{v}_1) = \mathbf{0}, \qquad -\infty < t < \infty. \qquad \textbf{(7)}$$

The set $\{te^{\alpha t}, e^{\alpha t}\}$ is a linearly independent set of functions on any t-interval of interest. Therefore, if equation (7) is to hold, each of the constant matrix coefficients must vanish. We obtain, therefore, a pair of matrix equations that the nonzero vectors $\mathbf{v}_1$ and $\mathbf{v}_2$ must satisfy:

$$\begin{aligned} (A - \alpha I)\mathbf{v}_1 &= \mathbf{0} \\ (A - \alpha I)\mathbf{v}_2 &= \mathbf{v}_1. \end{aligned} \qquad \textbf{(8)}$$

The first of these equations is simply the eigenvector equation, so we take $\mathbf{v}_1$ to be an eigenvector corresponding to $\lambda = \alpha$. Consider the second equation,

$$(A - \alpha I)\mathbf{v}_2 = \mathbf{v}_1. \qquad \textbf{(9)}$$

At first glance, equation (9) should give us cause for concern. The coefficient matrix $A - \alpha I$ is singular (that is, noninvertible) since α is an eigenvalue of A. We recall that a nonhomogeneous system of equations having a singular coefficient matrix, such as system (9), has either no solution or infinitely many solutions. In the present case, however, it can be shown that equation (9) always has infinitely many solutions. Selecting a particular solution of (9) determines $\mathbf{v}_2$; having $\mathbf{v}_2$, we can form a second solution, $\mathbf{y}_2(t)$. It can be shown that the pair of solutions obtained, $\{\mathbf{y}_1(t), \mathbf{y}_2(t)\}$, is a fundamental set of solutions. The vector $\mathbf{v}_2$ in equation (8) is called a *generalized eigenvector of order 2*. See Exercises 34–37.

E X A M P L E

1

Solve the initial value problem

$$\mathbf{y}' = \begin{bmatrix} 2 & -1 \\ 1 & 4 \end{bmatrix}\mathbf{y}, \qquad \mathbf{y}(1) = \begin{bmatrix} 1 \\ 3 \end{bmatrix}.$$

Solution: The characteristic polynomial is

$$p(\lambda) = \det[A - \lambda I] = \begin{vmatrix} 2 - \lambda & -1 \\ 1 & 4 - \lambda \end{vmatrix} = \lambda^2 - 6\lambda + 9 = (\lambda - 3)^2.$$

The coefficient matrix, therefore, has $\lambda = 3$ as a repeated eigenvalue. The corresponding eigenvector equation is $(A - 3I)\mathbf{x} = \mathbf{0}$, or

$$\begin{bmatrix} -1 & -1 \\ 1 & 1 \end{bmatrix}\begin{bmatrix} x_1 \\ x_2 \end{bmatrix} = \begin{bmatrix} 0 \\ 0 \end{bmatrix}.$$

This equation reduces to $x_1 + x_2 = 0$, and hence the eigenvectors have the form

$$\mathbf{x} = \begin{bmatrix} x_1 \\ x_2 \end{bmatrix} = \begin{bmatrix} -x_2 \\ x_2 \end{bmatrix} = x_2 \begin{bmatrix} -1 \\ 1 \end{bmatrix}, \qquad x_2 \neq 0.$$

(continued)

(continued)

A convenient choice for an eigenvector is

$$\mathbf{v}_1 = \begin{bmatrix} -1 \\ 1 \end{bmatrix}.$$

Therefore, one solution of $\mathbf{y}' = A\mathbf{y}$ is

$$\mathbf{y}_1(t) = e^{3t}\mathbf{v}_1.$$

The eigenvalue $\lambda = 3$ does not have two linearly independent eigenvectors, since every eigenvector is a nonzero multiple of $\mathbf{v}_1$. Therefore, we look for a second solution having the form $\mathbf{y}_2(t) = te^{3t}\mathbf{v}_1 + e^{3t}\mathbf{v}_2$, where [see equation (9)] $\mathbf{v}_2$ satisfies the equation $(A - 3I)\mathbf{x} = \mathbf{v}_1$, or

$$\begin{bmatrix} -1 & -1 \\ 1 & 1 \end{bmatrix} \begin{bmatrix} x_1 \\ x_2 \end{bmatrix} = \begin{bmatrix} -1 \\ 1 \end{bmatrix}.$$

This equation reduces to $x_1 + x_2 = 1$, and hence the solution is

$$\mathbf{v}_2 = \begin{bmatrix} 1 - x_2 \\ x_2 \end{bmatrix} = \begin{bmatrix} 1 \\ 0 \end{bmatrix} + x_2 \begin{bmatrix} -1 \\ 1 \end{bmatrix}.$$

There are infinitely many choices for x_2, but we only need one solution. Choosing $x_2 = 0$ for convenience, we obtain

$$\mathbf{v}_2 = \begin{bmatrix} 1 \\ 0 \end{bmatrix}.$$

Therefore, a second solution is $\mathbf{y}_2(t) = te^{3t}\mathbf{v}_1 + e^{3t}\mathbf{v}_2$:

$$\mathbf{y}_2(t) = te^{3t} \begin{bmatrix} -1 \\ 1 \end{bmatrix} + e^{3t} \begin{bmatrix} 1 \\ 0 \end{bmatrix} = \begin{bmatrix} -te^{3t} + e^{3t} \\ te^{3t} \end{bmatrix}.$$

Computing the Wronskian of the two solutions, we find

$$W(t) = \begin{vmatrix} -e^{3t} & -te^{3t} + e^{3t} \\ e^{3t} & te^{3t} \end{vmatrix} = -e^{6t} \neq 0.$$

These two solutions form a fundamental set, and the general solution is given by $\mathbf{y}(t) = c_1\mathbf{y}_1(t) + c_2\mathbf{y}_2(t)$:

$$\mathbf{y}(t) = c_1 \begin{bmatrix} -e^{3t} \\ e^{3t} \end{bmatrix} + c_2 \begin{bmatrix} -te^{3t} + e^{3t} \\ te^{3t} \end{bmatrix} = \begin{bmatrix} -e^{3t} & -te^{3t} + e^{3t} \\ e^{3t} & te^{3t} \end{bmatrix} \begin{bmatrix} c_1 \\ c_2 \end{bmatrix}.$$

Imposing the initial condition,

$$\mathbf{y}(1) = \begin{bmatrix} 1 \\ 3 \end{bmatrix} = \begin{bmatrix} -e^3 & 0 \\ e^3 & e^3 \end{bmatrix} \begin{bmatrix} c_1 \\ c_2 \end{bmatrix},$$

leads to $c_1 = -e^{-3}$ and $c_2 = 4e^{-3}$. The solution of the initial value problem is, therefore,

$$\mathbf{y}(t) = \begin{bmatrix} -4te^{3(t-1)} + 5e^{3(t-1)} \\ 4te^{3(t-1)} - e^{3(t-1)} \end{bmatrix}. \quad \diamondsuit$$

Algebraic Multiplicity and Geometric Multiplicity

So far in this section, we have concentrated on the equation $\mathbf{y}' = A\mathbf{y}$ in the case where A is a constant (2×2) matrix with a repeated eigenvalue. The case where A is an $(n \times n)$ matrix with repeated eigenvalues is more complicated, and a comprehensive treatment is beyond the scope of our present discussion. We do, however, give some indications of the underlying ideas in the Exercises. In addition, some observations about the $(n \times n)$ case can be made at this point.

In looking for a fundamental set of solutions for $\mathbf{y}' = A\mathbf{y}$, there is more to consider than simply whether A has repeated eigenvalues. If an eigenvalue is repeated, the question then arises as to whether there exist enough linearly independent eigenvectors. These considerations lead to the following definitions.

Let α be an eigenvalue of an $(n \times n)$ matrix A. The **algebraic multiplicity** of α is the order of α as a root of the characteristic equation, $p(\lambda) = 0$. If

$$p(\lambda) = (\lambda - \alpha)^r q(\lambda),$$

where $q(\lambda)$ is a polynomial of degree $n - r$ and $q(\alpha) \neq 0$, then α is an eigenvalue of algebraic multiplicity r. The **geometric multiplicity** of the eigenvalue α is the number of linearly independent eigenvectors that can be found corresponding to this eigenvalue.

Since the characteristic polynomial has degree n, the algebraic multiplicity of α is an integer r, where $1 \leq r \leq n$. Similarly, a set of $(n \times 1)$ vectors cannot be linearly independent unless the set contains n or fewer vectors. Thus, the geometric multiplicity of α is an integer s, where $1 \leq s \leq n$. As we note later in equation (10), the inequality $s \leq r$ holds for every matrix A.

EXAMPLE 2

In each of the following cases, determine the algebraic and geometric multiplicity of the eigenvalue $\lambda = 2$.

(a) $A_1 = \begin{bmatrix} 2 & 1 & 0 \\ 0 & 2 & 1 \\ 0 & 0 & 2 \end{bmatrix}$ (b) $A_2 = \begin{bmatrix} 2 & 1 & 0 \\ 0 & 2 & 0 \\ 0 & 0 & 2 \end{bmatrix}$ (c) $A_3 = \begin{bmatrix} 2 & 0 & 0 \\ 0 & 2 & 0 \\ 0 & 0 & 2 \end{bmatrix}$

Solution: In each case, the characteristic polynomial is $p(\lambda) = -(\lambda - 2)^3$. Therefore, in each case, the eigenvalue $\lambda = 2$ has algebraic multiplicity 3.

(a) All solutions (see Exercise 13) of the eigenvector equation $(A_1 - 2I)\mathbf{x} = \mathbf{0}$ have the form

$$\mathbf{x} = \begin{bmatrix} a \\ 0 \\ 0 \end{bmatrix} = a \begin{bmatrix} 1 \\ 0 \\ 0 \end{bmatrix} = a\mathbf{x}_1.$$

Therefore, for the matrix A_1, the geometric multiplicity of the eigenvalue $\lambda = 2$ is 1.

(b) All solutions of the eigenvector equation $(A_2 - 2I)\mathbf{x} = \mathbf{0}$ have the following form (see Exercise 13):

$$\mathbf{x} = \begin{bmatrix} b \\ 0 \\ c \end{bmatrix} = b \begin{bmatrix} 1 \\ 0 \\ 0 \end{bmatrix} + c \begin{bmatrix} 0 \\ 0 \\ 1 \end{bmatrix} = b\mathbf{x}_1 + c\mathbf{x}_2,$$

(continued)

(continued)

where $\mathbf{x}_1$ and $\mathbf{x}_2$ are linearly independent. Therefore, for the matrix A_2, the geometric multiplicity of the eigenvalue $\lambda = 2$ is 2.

(c) Finally, since the matrix $A_3 - 2I$ is the (3×3) zero matrix, every nonzero (3×1) vector is an eigenvector of A_3. Therefore, for the matrix A_3, the geometric multiplicity of $\lambda = 2$ is 3. ❖

If we want to solve the equation $\mathbf{y}' = A_1\mathbf{y}$ in Example 2, we need to find two more solutions to form a fundamental set of solutions, as one solution is $\mathbf{y}_1(t) = e^{2t}\mathbf{x}_1$. By contrast, if we want to solve $\mathbf{y}' = A_2\mathbf{y}$, we need find only one additional solution, as we know two linearly independent solutions, $\mathbf{y}_1(t) = e^{2t}\mathbf{x}_1$ and $\mathbf{y}_2(t) = e^{2t}\mathbf{x}_2$. Finally, if we want to solve $\mathbf{y}' = A_3\mathbf{y}$, then (since every three-dimensional vector is an eigenvector of A_3) we can select any three linearly independent vectors $\mathbf{u}$, $\mathbf{v}$, $\mathbf{w}$ and we are assured (by Theorem 4.6) that the solutions $\mathbf{y}_1(t) = e^{2t}\mathbf{u}$, $\mathbf{y}_2(t) = e^{2t}\mathbf{v}$, and $\mathbf{y}_3(t) = e^{2t}\mathbf{w}$ form a fundamental set of solutions.

Defective Matrices

The following inequality is established in advanced texts:

$$\text{The geometric multiplicity of an eigenvalue} \leq \text{The algebraic multiplicity of an eigenvalue.} \tag{10}$$

In determining the structure of a fundamental set of solutions of $\mathbf{y}' = A\mathbf{y}$, the key question is "How does the geometric multiplicity of each eigenvalue relate to its algebraic multiplicity?" If the two multiplicities for a given eigenvalue λ are equal, the corresponding solutions entering into the fundamental set will all be of the form $e^{\lambda t}\mathbf{x}$ (whether the eigenvalue is distinct or repeated). If the geometric multiplicity of an eigenvalue is strictly less than its algebraic multiplicity, we have a "deficiency of eigenvectors" and solutions of a more complicated form become part of the fundamental set.

A matrix that has at least one eigenvalue with a geometric multiplicity that is strictly less than its algebraic multiplicity cannot have a full set of eigenvectors; such a matrix is called **defective**. Thus, in Example 2, the matrices A_1 and A_2 are defective but the matrix A_3 has a full set of eigenvectors.

Phase Plane Trajectories

We examine the phase plane trajectories of the linear system $\mathbf{y}' = A\mathbf{y}$, where A is a real (2×2) constant matrix with repeated eigenvalues $\lambda_1 = \lambda_2 = \alpha$, with $\alpha \neq 0$.

If A has a full set of eigenvectors, then it can be shown that $A = \alpha I$; see Exercise 24. In this case, the general solution has the form

$$\mathbf{y}(t) = c_1 e^{\alpha t} \begin{bmatrix} 1 \\ 0 \end{bmatrix} + c_2 e^{\alpha t} \begin{bmatrix} 0 \\ 1 \end{bmatrix}. \tag{11}$$

From (11), we obtain $c_2 y_1(t) = c_1 y_2(t)$, and thus the phase plane solution points move on rays emanating from the origin. The motion is inward toward the origin if $\alpha < 0$ and outward from the origin if $\alpha > 0$. Figure 4.6 illustrates the case where $\alpha < 0$.

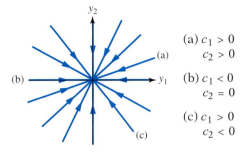

(a) $c_1 > 0$
      $c_2 > 0$

(b) $c_1 < 0$
      $c_2 = 0$

(c) $c_1 > 0$
      $c_2 < 0$

FIGURE 4.6

When $A = \alpha I$, phase plane trajectories move along rays emanating from the origin. Here, where $\alpha < 0$, motion is toward the origin.

If A is defective, the general solution has the form

$$\mathbf{y}(t) = c_1 e^{\alpha t} \mathbf{v}_1 + c_2 e^{\alpha t}(t\mathbf{v}_1 + \mathbf{v}_2).$$

In this case, as in equation (11), motion is toward the origin when $\alpha < 0$ and away from the origin when $\alpha > 0$. Unlike the situation illustrated in Figure 4.6, however, the solution points need not move along rays. Figure 4.7 shows the phase plane direction field for the linear system treated in Example 1. In this example, the general solution was found to be

$$\mathbf{y}(t) = c_1 e^{3t} \begin{bmatrix} -1 \\ 1 \end{bmatrix} + c_2 e^{3t} \begin{bmatrix} -t+1 \\ t \end{bmatrix}. \tag{12}$$

As can be seen from Figure 4.7, all trajectories move outward from the origin. If $c_2 = 0$ in (12), the solution points move outward along the line $y_2 = -y_1$. If $c_2 \neq 0$, then $y_2(t) \approx -y_1(t)$ for large values of t.

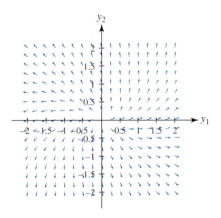

FIGURE 4.7

The phase plane direction field for Example 1. From the general solution in equation (12), we see that phase plane solution trajectories approach the line $y_2 = -y_1$ as t increases.

EXERCISES

Exercises 1–12:

Consider the given initial value problem $\mathbf{y}' = A\mathbf{y}$, $\mathbf{y}(t_0) = \mathbf{y}_0$.

(a) Find the eigenvalues and eigenvectors of the coefficient matrix A.

(b) Construct a fundamental set of solutions.

(c) Solve the initial value problem.

1. $\begin{aligned} y_1' &= 2y_1 + y_2, & y_1(0) &= 3 \\ y_2' &= -y_1, & y_2(0) &= -1 \end{aligned}$

2. $\begin{aligned} y_1' &= y_2, & y_1(0) &= 0 \\ y_2' &= -y_1 - 2y_2, & y_2(0) &= 2 \end{aligned}$

3. $\mathbf{y}' = \begin{bmatrix} -2 & 1 \\ 0 & -2 \end{bmatrix} \mathbf{y}, \quad \mathbf{y}(0) = \begin{bmatrix} 1 \\ -1 \end{bmatrix}$

4. $\mathbf{y}' = \begin{bmatrix} 1 & -1 \\ 4 & 5 \end{bmatrix} \mathbf{y}, \quad \mathbf{y}(0) = \begin{bmatrix} 1 \\ 1 \end{bmatrix}$

5. $\mathbf{y}' = \begin{bmatrix} 6 & 0 \\ 2 & 6 \end{bmatrix} \mathbf{y}, \quad \mathbf{y}(0) = \begin{bmatrix} -2 \\ 0 \end{bmatrix}$

6. $\begin{aligned} y_1' &= y_2, & y_1(1) &= -1 \\ y_2' &= -y_1 + 2y_2, & y_2(1) &= 2 \end{aligned}$

7. $\mathbf{y}' = \begin{bmatrix} -4 & 1 \\ -1 & -2 \end{bmatrix} \mathbf{y}, \quad \mathbf{y}(-1) = \begin{bmatrix} 1 \\ 0 \end{bmatrix}$

8. $\mathbf{y}' = \begin{bmatrix} 6 & 1 \\ -1 & 4 \end{bmatrix} \mathbf{y}, \quad \mathbf{y}(0) = \begin{bmatrix} 4 \\ -4 \end{bmatrix}$

9. $\mathbf{y}' = \begin{bmatrix} 2 & 1 & 0 \\ 0 & 2 & 0 \\ 0 & 0 & 1 \end{bmatrix} \mathbf{y}, \quad \mathbf{y}(0) = \begin{bmatrix} 1 \\ 3 \\ -2 \end{bmatrix}$

10. $\mathbf{y}' = \begin{bmatrix} 1 & -1 & 0 \\ 1 & 3 & 0 \\ 0 & 0 & 2 \end{bmatrix} \mathbf{y}, \quad \mathbf{y}(0) = \begin{bmatrix} -3 \\ -2 \\ 0 \end{bmatrix}$

11. $\begin{aligned} y_1' &= 0, & y_1(0) &= 4 \\ y_2' &= 5y_2 - y_3, & y_2(0) &= 1 \\ y_3' &= 4y_2 + y_3, & y_3(0) &= 1 \end{aligned}$

12. $\mathbf{y}' = \begin{bmatrix} -3 & 0 & -36 \\ 0 & 1 & 0 \\ 1 & 0 & 9 \end{bmatrix} \mathbf{y}, \quad \mathbf{y}(0) = \begin{bmatrix} 0 \\ -1 \\ 2 \end{bmatrix}$

13. Find the eigenvalues and eigenvectors of

(a) $A_1 = \begin{bmatrix} 2 & 1 & 0 \\ 0 & 2 & 1 \\ 0 & 0 & 2 \end{bmatrix}$ (b) $A_2 = \begin{bmatrix} 2 & 1 & 0 \\ 0 & 2 & 0 \\ 0 & 0 & 2 \end{bmatrix}$

14. Consider the homogeneous linear system $\mathbf{y}' = A_1 \mathbf{y}$, where A_1 is given in Exercise 13.

(a) Write the three component differential equations of $\mathbf{y}' = A_1 \mathbf{y}$, and solve these equations sequentially, finding first $y_3(t)$, then $y_2(t)$, and then $y_1(t)$. [For example, the third component equation is $y_3' = 2y_3$. Therefore, $y_3(t) = c_3 e^{2t}$.]

(b) Rewrite the component solutions obtained in part (a) as a single matrix equation of the form $\mathbf{y}(t) = \Psi(t)\mathbf{c}$, where $\Psi(t)$ is a (3×3) solution matrix and

$$\mathbf{c} = \begin{bmatrix} c_1 \\ c_2 \\ c_3 \end{bmatrix}.$$

Show that $\Psi(t)$ is, in fact, a fundamental matrix. [Note that this observation is consistent with the fact that the component solutions obtained in part (a) form the general solution.]

15. Repeat Exercise 14 for the homogeneous linear system $\mathbf{y}' = A_2\mathbf{y}$.

16. The Scalar Repeated-Root Equation Revisited   Consider the homogeneous scalar equation $y'' - 2\alpha y' + \alpha^2 y = 0$, where α is a real constant. Recall from Section 3.4 that the general solution is $y(t) = c_1 e^{\alpha t} + c_2 t e^{\alpha t}$.

(a) Recast $y'' - 2\alpha y' + \alpha^2 y = 0$ as a first order linear system $\mathbf{z}' = A\mathbf{z}$.

(b) Show that the (2×2) matrix A has eigenvalue α with algebraic multiplicity 2 and geometric multiplicity 1.

(c) Obtain the general solution of $\mathbf{z}' = A\mathbf{z}$. As a check, does $z_1(t)$ equal the general solution $y(t) = c_1 e^{\alpha t} + c_2 t e^{\alpha t}$? Is $z_2(t)$ equal to $z_1'(t)$?

Exercises 17–23:

For each matrix A, find the eigenvalues and eigenvectors. Give the geometric and algebraic multiplicity of each eigenvalue. Does A have a full set of eigenvectors?

17. $A = \begin{bmatrix} 5 & 0 & 0 \\ 1 & 5 & 0 \\ 1 & 1 & 5 \end{bmatrix}$
   18. $A = \begin{bmatrix} 5 & 0 & 0 \\ 1 & 5 & 0 \\ 1 & 0 & 5 \end{bmatrix}$
   19. $A = \begin{bmatrix} 5 & 0 & 1 \\ 0 & 5 & 0 \\ 0 & 0 & 5 \end{bmatrix}$

20. $A = \begin{bmatrix} 5 & 0 & 0 \\ 0 & 5 & 0 \\ 0 & 0 & 5 \end{bmatrix}$
   21. $A = \begin{bmatrix} 2 & 0 & 0 & 0 \\ 1 & 2 & 0 & 0 \\ 0 & 0 & 3 & 0 \\ 0 & 0 & 1 & 3 \end{bmatrix}$

22. $A = \begin{bmatrix} 2 & 0 & 0 & 0 \\ 0 & 2 & 0 & 0 \\ 0 & 0 & 2 & 0 \\ 0 & 0 & 1 & 2 \end{bmatrix}$
   23. $A = \begin{bmatrix} 2 & 0 & 0 & 0 \\ 0 & 2 & 0 & 0 \\ 0 & 0 & 2 & 0 \\ 0 & 0 & 1 & 3 \end{bmatrix}$

24. Let A be a real (2×2) matrix having repeated eigenvalue $\lambda_1 = \lambda_2 = \alpha$ and a full set of eigenvectors, $\mathbf{x}_1$ and $\mathbf{x}_2$. Show that $A = \alpha I$. [Hint: Let $T = [\mathbf{x}_1, \mathbf{x}_2]$ be the invertible (2×2) matrix whose columns are the eigenvectors. Show that $AT = \alpha T$.]

Exercises 25–28:

In each exercise, the general solution of the linear system $\mathbf{y}' = A\mathbf{y}$ is given. Determine the coefficient matrix A.

25. $y_1(t) = c_1 e^{-t}(1 + 2t) + 4c_2 t e^{-t}$
$y_2(t) = -c_1 t e^{-t} + c_2 e^{-t}(1 - 2t)$

26. $\mathbf{y}(t) = \begin{bmatrix} c_1 e^{-2t} \\ c_2 e^{-2t} \end{bmatrix}$

27. $\mathbf{y}(t) = c_1 \begin{bmatrix} e^t(1 + t) \\ t e^t \end{bmatrix} + c_2 \begin{bmatrix} -t e^t \\ e^t(1 - t) \end{bmatrix}$

28. $\mathbf{y}(t) = \begin{bmatrix} e^{t/2} & 0 \\ 0 & e^{t/2} \end{bmatrix} \begin{bmatrix} c_1 \\ c_2 \end{bmatrix}$

29. Match the linear system with one of the phase plane direction fields on the next page.

(a) $y_1' = -y_1$
$y_2' = 2y_1 - y_2$

(b) $\mathbf{y}' = \begin{bmatrix} 2 & 0 \\ 0 & 2 \end{bmatrix} \mathbf{y}$

(c) $y_1' = y_1 + y_2$
$y_2' = \quad y_2$

(d) $\mathbf{y}' = \begin{bmatrix} -0.5 & 0 \\ 0 & -0.5 \end{bmatrix} \mathbf{y}$

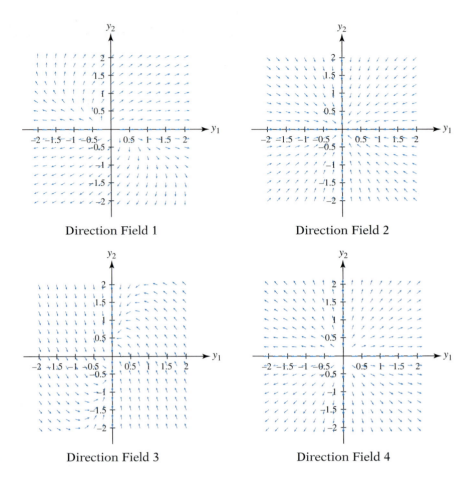

Direction Field 1   Direction Field 2

Direction Field 3   Direction Field 4

Figure for Exercises 29

Exercises 30–33:

Consider the linear system $\mathbf{y}' = A\mathbf{y}$, where A is a real (2×2) constant matrix with repeated eigenvalues. Use the given information to determine the matrix A.

30. Phase plane solution trajectories have horizontal tangents on the line $y_2 = 2y_1$ and vertical tangents on the line $y_1 = 0$. The matrix A has a nonzero repeated eigenvalue and $a_{21} = -1$.

31. All nonzero phase plane solution points move away from the origin on straight line paths as t increases. In addition, $a_{22} = \frac{3}{2}$.

32. Phase plane solution trajectories have horizontal tangents on the line $y_2 = 0$ and vertical tangents on the line $y_2 = 2y_1$. All nonzero phase plane solution points move away from the origin as t increases. In addition, $a_{12} = -1$.

33. All phase plane solution points remain stationary as t increases.

Generalized Eigenvectors   Let A be an $(n \times n)$ matrix. The ideas introduced in equation (8) can be extended. Let $\mathbf{v}_1 \neq \mathbf{0}$ be an eigenvector of A corresponding to the eigenvalue λ, and suppose we can generate the following "chain" of nonzero vectors:

$$(A - \lambda I)\mathbf{v}_1 = \mathbf{0}$$
$$(A - \lambda I)\mathbf{v}_2 = \mathbf{v}_1$$
$$(A - \lambda I)\mathbf{v}_3 = \mathbf{v}_2$$
$$\vdots$$
$$(A - \lambda I)\mathbf{v}_r = \mathbf{v}_{r-1}$$

(13)

In (13), the vector $\mathbf{v}_j$ is called a **generalized eigenvector of order j**. Define

$$\mathbf{y}_k(t) = e^{\lambda t}\left(\mathbf{v}_k + t\,\mathbf{v}_{k-1} + \cdots + \frac{t^{k-1}}{(k-1)!}\mathbf{v}_1\right). \qquad (14)$$

Exercise 36 asks you to show, for $r = 3$ and $k = 1, 2, 3$, that $\mathbf{y}_k(t)$ is a solution of $\mathbf{y}' = A\mathbf{y}$. If λ has algebraic multiplicity m and geometric multiplicity 1, then it can be shown that there is a chain (13) consisting of m different generalized eigenvectors and that these generalized eigenvectors are linearly independent (see Exercise 37). Thus, equation (14) defines a set of m linearly independent solutions of $\mathbf{y}' = A\mathbf{y}$—as many solutions as the multiplicity of the eigenvalue. (If λ has geometric multiplicity 2 or larger, the situation is more complicated.)

Exercises 34–35:

Using equations (13) and (14), find a fundamental set of solutions for the linear system $\mathbf{y}' = A\mathbf{y}$. In each exercise, you are given an eigenvalue λ, where λ has algebraic multiplicity 3 and geometric multiplicity 1 and an eigenvector $\mathbf{v}_1$.

34. $A = \begin{bmatrix} 2 & 1 & 0 \\ 0 & 2 & 1 \\ 0 & 0 & 2 \end{bmatrix}$, $\lambda = 2$, $\mathbf{v}_1 = \begin{bmatrix} 1 \\ 0 \\ 0 \end{bmatrix}$ 35. $A = \begin{bmatrix} 4 & 0 & 0 \\ 2 & 4 & 0 \\ 1 & 3 & 4 \end{bmatrix}$, $\lambda = 4$, $\mathbf{v}_1 = \begin{bmatrix} 0 \\ 0 \\ 1 \end{bmatrix}$

36. Let $\mathbf{v}_1$, $\mathbf{v}_2$, and $\mathbf{v}_3$ be a chain of nonzero vectors, as in equation (13). Show that the vector function $\mathbf{y}_3(t)$ defined in equation (14) is a solution of $\mathbf{y}' = A\mathbf{y}$.

37. Let $\mathbf{v}_1$, $\mathbf{v}_2$, and $\mathbf{v}_3$ be a chain of nonzero vectors, as in equation (13). Show that these vectors form a linearly independent set of vectors. [Hint: Begin with the dependence equation $c_1\mathbf{v}_1 + c_2\mathbf{v}_2 + c_3\mathbf{v}_3 = \mathbf{0}$ and multiply both sides by $A - \lambda I$.]

4.8   Nonhomogeneous Linear Systems

We now address the problem of finding the general solution of a nonhomogeneous first order linear system,

$$\mathbf{y}' = P(t)\mathbf{y} + \mathbf{g}(t), \qquad a < t < b. \qquad (1)$$

In (1), $\mathbf{y}(t)$ is an $(n \times 1)$ vector function, $P(t)$ is an $(n \times n)$ matrix function, and the nonhomogeneous term, $\mathbf{g}(t)$, is an $(n \times 1)$ vector function. The component functions of $P(t)$ and $\mathbf{g}(t)$ are assumed to be continuous on $a < t < b$.

The Structure of the General Solution

In analyzing the structure of the general solution of (1), we return once more to a theme that has permeated all our discussions of nonhomogeneous linear equations. If $\mathbf{y}_1(t)$ and $\mathbf{y}_2(t)$ represent any two solutions of $\mathbf{y}' = P(t)\mathbf{y} + \mathbf{g}(t)$, we ask, "How do they differ?"

To answer this question, we form the difference function, $\mathbf{w}(t) = \mathbf{y}_1(t) - \mathbf{y}_2(t)$. Differentiating $\mathbf{w}(t)$ yields

$$\begin{aligned} \mathbf{w}' &= (\mathbf{y}_1 - \mathbf{y}_2)' \\ &= \mathbf{y}_1' - \mathbf{y}_2' \\ &= [P(t)\mathbf{y}_1 + \mathbf{g}(t)] - [P(t)\mathbf{y}_2 + \mathbf{g}(t)] \\ &= P(t)(\mathbf{y}_1 - \mathbf{y}_2) \\ &= P(t)\mathbf{w}. \end{aligned}$$

Thus, the difference between any two solutions of the nonhomogeneous linear equation is a solution of the homogeneous linear equation. This leads to the familiar decomposition

The general solution of the nonhomogeneous linear system $\mathbf{y}' = P(t)\mathbf{y} + \mathbf{g}(t)$	=	The general solution of the homogeneous linear system $\mathbf{y}' = P(t)\mathbf{y}$	+	A particular solution of the nonhomogeneous linear system $\mathbf{y}' = P(t)\mathbf{y} + \mathbf{g}(t)$.

As before, we refer to the general solution of the homogeneous system, $\mathbf{y}' = P(t)\mathbf{y}$, as the *complementary solution* and denote it by $\mathbf{y}_C(t)$. A solution of the nonhomogeneous system that we have somehow found is called a *particular solution* and is denoted by $\mathbf{y}_P(t)$.

The following theorem, an analog of the superposition principle given in Theorem 3.4, holds for nonhomogeneous linear systems. We leave the proof as an exercise.

Theorem 4.8

Let $\mathbf{u}(t)$ be a solution of

$$\mathbf{y}' = P(t)\mathbf{y} + \mathbf{g}_1(t), \qquad a < t < b,$$

and let $\mathbf{v}(t)$ be a solution of

$$\mathbf{y}' = P(t)\mathbf{y} + \mathbf{g}_2(t), \qquad a < t < b.$$

Let a_1 and a_2 be any constants. Then the vector function $\mathbf{y}_P(t) = a_1\mathbf{u}(t) + a_2\mathbf{v}(t)$ is a particular solution of

$$\mathbf{y}' = P(t)\mathbf{y} + a_1\mathbf{g}_1(t) + a_2\mathbf{g}_2(t), \qquad a < t < b.$$

The following example illustrates Theorem 4.8.

E X A M P L E

1

Find the general solution of

$$\mathbf{y}' = \begin{bmatrix} 1 & 2 \\ 2 & 1 \end{bmatrix} \mathbf{y} + \begin{bmatrix} e^{2t} \\ -2t \end{bmatrix}, \qquad -\infty < t < \infty.$$

Solution: We saw earlier (in Example 1 of Section 4.4) that the general solution of the homogeneous equation,

$$\mathbf{y}' = \begin{bmatrix} 1 & 2 \\ 2 & 1 \end{bmatrix} \mathbf{y}, \qquad -\infty < t < \infty,$$

is

$$\mathbf{y}_C(t) = c_1 \begin{bmatrix} e^{-t} \\ -e^{-t} \end{bmatrix} + c_2 \begin{bmatrix} e^{3t} \\ e^{3t} \end{bmatrix} = \begin{bmatrix} e^{-t} & e^{3t} \\ -e^{-t} & e^{3t} \end{bmatrix} \begin{bmatrix} c_1 \\ c_2 \end{bmatrix}.$$

Having the complementary solution, $\mathbf{y}_C(t)$, we turn our attention to the task of somehow finding a particular solution, $\mathbf{y}_P(t)$, of the nonhomogeneous equation. Note that the nonhomogeneous term,

$$\mathbf{g}(t) = \begin{bmatrix} e^{2t} \\ -2t \end{bmatrix},$$

can be decomposed as follows:

$$\begin{bmatrix} e^{2t} \\ -2t \end{bmatrix} = e^{2t} \begin{bmatrix} 1 \\ 0 \end{bmatrix} + t \begin{bmatrix} 0 \\ -2 \end{bmatrix}.$$

Using the superposition principle in Theorem 4.8, we decompose the differential equation and separately find particular solutions of

$$\mathbf{u}' = \begin{bmatrix} 1 & 2 \\ 2 & 1 \end{bmatrix} \mathbf{u} + e^{2t} \begin{bmatrix} 1 \\ 0 \end{bmatrix} \quad \text{and} \quad \mathbf{v}' = \begin{bmatrix} 1 & 2 \\ 2 & 1 \end{bmatrix} \mathbf{v} + t \begin{bmatrix} 0 \\ -2 \end{bmatrix}. \tag{2}$$

Consider the first equation in (2). Remembering the method of undetermined coefficients, we look for a solution of the form $\mathbf{u}_p(t) = e^{2t}\mathbf{a}$, where $\mathbf{a}$ is a constant (2×1) vector to be determined. Substituting $\mathbf{u}_p(t) = e^{2t}\mathbf{a}$ into the first differential equation in (2) leads to

$$2e^{2t}\mathbf{a} = \begin{bmatrix} 1 & 2 \\ 2 & 1 \end{bmatrix} (e^{2t}\mathbf{a}) + e^{2t}\begin{bmatrix} 1 \\ 0 \end{bmatrix}, \qquad -\infty < t < \infty.$$

Canceling the common e^{2t} factor and rearranging terms, we see that $\mathbf{a}$ must satisfy the condition

$$\begin{bmatrix} 1 & -2 \\ -2 & 1 \end{bmatrix} \mathbf{a} = \begin{bmatrix} 1 \\ 0 \end{bmatrix}, \quad \text{or} \quad \mathbf{a} = \begin{bmatrix} -\frac{1}{3} \\ -\frac{2}{3} \end{bmatrix}.$$

Thus, a particular solution is

$$\mathbf{u}_p(t) = e^{2t}\mathbf{a} = \begin{bmatrix} -\frac{1}{3}e^{2t} \\ -\frac{2}{3}e^{2t} \end{bmatrix}.$$

To find a particular solution of the second equation in (2), we look for a solution having the form $\mathbf{v}_p(t) = t\mathbf{b} + \mathbf{c}$, where $\mathbf{b}$ and $\mathbf{c}$ are constant (2×1) vectors to be determined. Substituting this guess into the differential equation leads to

$$\mathbf{b} = \begin{bmatrix} 1 & 2 \\ 2 & 1 \end{bmatrix} (t\mathbf{b} + \mathbf{c}) + t \begin{bmatrix} 0 \\ -2 \end{bmatrix}$$

or, after collecting like powers of t,

$$t\left(\begin{bmatrix} 1 & 2 \\ 2 & 1 \end{bmatrix} \mathbf{b} + \begin{bmatrix} 0 \\ -2 \end{bmatrix}\right) + \left(\begin{bmatrix} 1 & 2 \\ 2 & 1 \end{bmatrix} \mathbf{c} - \mathbf{b}\right) = \mathbf{0}, \qquad -\infty < t < \infty.$$

Since the set of functions $\{t, 1\}$ is linearly independent on any t-interval, this equation holds only if the coefficients of t and 1 are $\mathbf{0}$—that is,

$$\begin{bmatrix} 1 & 2 \\ 2 & 1 \end{bmatrix} \begin{bmatrix} b_1 \\ b_2 \end{bmatrix} = \begin{bmatrix} 0 \\ 2 \end{bmatrix} \quad \text{and} \quad \begin{bmatrix} 1 & 2 \\ 2 & 1 \end{bmatrix} \begin{bmatrix} c_1 \\ c_2 \end{bmatrix} = \begin{bmatrix} b_1 \\ b_2 \end{bmatrix}. \tag{3}$$

(continued)

(continued)

The two matrix equations in (3) can be solved sequentially, yielding

$$\mathbf{b} = \begin{bmatrix} \frac{4}{3} \\ -\frac{2}{3} \end{bmatrix} \quad \text{and} \quad \mathbf{c} = \begin{bmatrix} -\frac{8}{9} \\ \frac{10}{9} \end{bmatrix}.$$

Therefore, the solution $\mathbf{v}(t) = t\mathbf{b} + \mathbf{c}$ is

$$\mathbf{v}(t) = \begin{bmatrix} \frac{4}{3}t - \frac{8}{9} \\ -\frac{2}{3}t + \frac{10}{9} \end{bmatrix}.$$

Applying Theorem 4.8, we have for a particular solution of the given differential equation

$$\mathbf{y}_P(t) = \mathbf{u}(t) + \mathbf{v}(t) = \begin{bmatrix} -\frac{1}{3}e^{2t} + \frac{4}{3}t - \frac{8}{9} \\ -\frac{2}{3}e^{2t} - \frac{2}{3}t + \frac{10}{9} \end{bmatrix}.$$

The general solution is therefore

$$\mathbf{y}(t) = \mathbf{y}_C(t) + \mathbf{y}_P(t) = c_1 \begin{bmatrix} e^{-t} \\ -e^{-t} \end{bmatrix} + c_2 \begin{bmatrix} e^{3t} \\ e^{3t} \end{bmatrix} + \begin{bmatrix} -\frac{1}{3}e^{2t} + \frac{4}{3}t - \frac{8}{9} \\ -\frac{2}{3}e^{2t} - \frac{2}{3}t + \frac{10}{9} \end{bmatrix}. \quad \text{❖}$$

Comparing Solution Methods

In Chapter 3, the method of undetermined coefficients was seen to be an effective way to find particular solutions when the differential equation had constant coefficients and when the nonhomogeneous term was of a certain form; see Table 3.1. Example 1 provides a simple illustration of how these ideas can be extended and applied to the constant coefficient linear system $\mathbf{y}' = A\mathbf{y} + \mathbf{g}(t)$ when the nonhomogeneous vector function $\mathbf{g}(t)$ has components of a certain form. The Exercises give additional illustrations.

However, in contrast to the scalar problem, the complexity of the matrix problem makes the "educated guesswork" at the core of this method difficult to implement systematically. We shall not discuss the method of undetermined coefficients any further.

The method of variation of parameters (considered in Sections 3.9 and 3.13 for scalar linear equations) extends to linear systems. Therefore, we shall concentrate on the method of variation of parameters. In Chapter 5, we show how Laplace transforms also can be used to solve constant coefficient nonhomogeneous linear first order systems.

As the first step in applying the method of variation of parameters to a nonhomogeneous system of differential equations, we revisit the concept of a fundamental matrix.

Fundamental Matrices

Section 4.3 introduced the concepts of a solution matrix and a fundamental matrix. A *solution matrix*, $\Psi(t)$, is an $(n \times n)$ matrix whose n columns are each solutions of the homogeneous linear first order system $\mathbf{y}' = P(t)\mathbf{y}$, $a < t < b$.

Thus, if $\mathbf{y}_1(t), \mathbf{y}_2(t), \ldots, \mathbf{y}_n(t)$ are solutions of $\mathbf{y}' = P(t)\mathbf{y}$, then

$$\Psi(t) = [\mathbf{y}_1(t), \mathbf{y}_2(t), \ldots, \mathbf{y}_n(t)]$$

is a solution matrix. In addition, if these solutions form a fundamental set of solutions, then the solution matrix $\Psi(t)$ is called a *fundamental matrix*.

When we introduced the concepts of a solution matrix and a fundamental matrix in Section 4.3, our primary focus was on the solutions $\{\mathbf{y}_1, \mathbf{y}_2, \ldots, \mathbf{y}_n\}$. We used solution matrices and fundamental matrices as a way to organize solutions into an array. Initial conditions are conveniently imposed using such arrays. We also use solution matrices to define the Wronskian of the solution set.

At this point, we begin a subtle but important shift of emphasis. We now view solution matrices and fundamental matrices as $(n \times n)$ matrix functions that are mathematical entities in their own right. In particular, solution matrices and fundamental matrices for $\mathbf{y}' = P(t)\mathbf{y}$ can themselves be viewed as solutions of the *matrix* differential equation $\Psi' = P(t)\Psi$. Some important properties of solution matrices and fundamental matrices are summarized in Theorem 4.9.

Theorem 4.9

Consider the homogeneous linear first order system

$$\mathbf{y}' = P(t)\mathbf{y}, \qquad a < t < b,$$

where $\mathbf{y}(t)$ is an $(n \times 1)$ vector function and $P(t)$ is an $(n \times n)$ coefficient matrix, continuous on (a, b).

(a) Let $\Psi(t)$ be any solution matrix of $\mathbf{y}' = P(t)\mathbf{y}$, $a < t < b$. Then $\Psi(t)$ satisfies the matrix differential equation

$$\Psi' = P(t)\Psi, \qquad a < t < b.$$

(b) Let Ψ_0 represent any given constant $(n \times n)$ matrix, and let t_0 be any fixed point in the interval $a < t < b$. Then there is a unique $(n \times n)$ matrix $\Psi(t)$ that solves the initial value problem

$$\Psi' = P(t)\Psi, \qquad \Psi(t_0) = \Psi_0, \qquad a < t < b.$$

Moreover, if the constant matrix Ψ_0 is invertible, then the matrix $\Psi(t)$ is a fundamental matrix of $\mathbf{y}' = P(t)\mathbf{y}$, $a < t < b$.

(c) If $\Psi(t)$ is any fundamental matrix and $\hat{\Psi}(t)$ is any solution matrix of $\mathbf{y}' = P(t)\mathbf{y}$, $a < t < b$, then there exists an $(n \times n)$ constant matrix C such that

$$\hat{\Psi}(t) = \Psi(t)C, \qquad a < t < b.$$

Moreover, the matrix $\hat{\Psi}(t)$ is also a fundamental matrix if and only if $\det[C] \neq 0$.

● **PROOF:**

(a) Express the solution matrix in column form as

$$\Psi(t) = [\mathbf{y}_1(t), \mathbf{y}_2(t), \ldots, \mathbf{y}_n(t)].$$

Recall, from Section 4.1, that $\Psi'(t) = [\mathbf{y}_1'(t), \mathbf{y}_2'(t), \ldots, \mathbf{y}_n'(t)]$. Therefore,

$$\Psi'(t) = [\mathbf{y}_1'(t), \mathbf{y}_2'(t), \ldots, \mathbf{y}_n'(t)]$$
$$= [P(t)\mathbf{y}_1(t), P(t)\mathbf{y}_2(t), \ldots, P(t)\mathbf{y}_n(t)]$$
$$= P(t)[\mathbf{y}_1(t), \mathbf{y}_2(t), \ldots, \mathbf{y}_n(t)]$$
$$= P(t)\Psi(t), \quad a < t < b.$$

(b) Let the constant matrix Ψ_0 be represented in terms of its columns as $\Psi_0 = [\Psi_1, \Psi_2, \ldots, \Psi_n]$. The initial value problem

$$\Psi' = P(t)\Psi, \qquad \Psi(t_0) = \Psi_0, \qquad a < t < b$$

is equivalent to the n separate initial value problems

$$\mathbf{y}_j' = P(t)\mathbf{y}_j, \qquad \mathbf{y}_j(t_0) = \Psi_j, \qquad 1 \leq j \leq n, \qquad a < t < b.$$

By Theorem 4.1, each of these initial value problems has a unique solution. Hence, the solution matrix $\Psi(t)$ is also unique. Moreover, if Ψ_0 is invertible, then $\det[\Psi(t_0)] \neq 0$. By Theorem 4.3, $\Psi(t)$ is a fundamental matrix since the Wronskian, $W(t_0) = \det[\Psi(t_0)]$, is nonzero at t_0.

(c) This result is simply a restatement of Theorem 4.5. ●

Example 2 illustrates several parts of Theorem 4.9.

E X A M P L E

2

Find the unique matrix solution of the initial value problem $\Psi' = A\Psi$, $\Psi(0) = \Psi_0$, $-\infty < t < \infty$, where

$$A = \begin{bmatrix} 1 & 2 \\ 2 & 1 \end{bmatrix}, \qquad \Psi_0 = \begin{bmatrix} 3 & 2 \\ 1 & -4 \end{bmatrix}.$$

Solution: We saw in Example 1 that the two vector functions

$$\mathbf{y}_1(t) = \begin{bmatrix} e^{-t} \\ -e^{-t} \end{bmatrix} \quad \text{and} \quad \mathbf{y}_2(t) = \begin{bmatrix} e^{3t} \\ e^{3t} \end{bmatrix}$$

form a fundamental set of solutions of $\mathbf{y}' = A\mathbf{y}$, $-\infty < t < \infty$. Therefore, the (2×2) matrix function

$$\Psi(t) = \begin{bmatrix} e^{-t} & e^{3t} \\ -e^{-t} & e^{3t} \end{bmatrix}$$

is a fundamental matrix for $\mathbf{y}' = A\mathbf{y}$. By part (a) of Theorem 4.9, the matrix $\Psi(t)$ satisfies the given differential equation, $\Psi' = A\Psi$. However, $\Psi(t)$ does not satisfy the initial condition since

$$\Psi(0) = \begin{bmatrix} 1 & 1 \\ -1 & 1 \end{bmatrix} \neq \Psi_0.$$

By part (c) of Theorem 4.9, we can represent the desired solution, $\hat{\Psi}(t)$, as

$$\hat{\Psi}(t) = \Psi(t)C,$$

where C is a constant nonsingular (2×2) matrix. Imposing the initial conditions, we see that C must satisfy the equation $\hat{\Psi}(0) = \Psi(0)C = \Psi_0$. Solving for C, we find

$$C = [\Psi(0)]^{-1}\Psi_0 \quad \text{or} \quad C = \begin{bmatrix} \frac{1}{2} & -\frac{1}{2} \\ \frac{1}{2} & \frac{1}{2} \end{bmatrix} \begin{bmatrix} 3 & 2 \\ 1 & -4 \end{bmatrix} = \begin{bmatrix} 1 & 3 \\ 2 & -1 \end{bmatrix}.$$

The solution of the initial value problem is $\hat{\Psi}(t) = \Psi(t)C$, or

$$\hat{\Psi}(t) = \begin{bmatrix} e^{-t} & e^{3t} \\ -e^{-t} & e^{3t} \end{bmatrix} \begin{bmatrix} 1 & 3 \\ 2 & -1 \end{bmatrix} = \begin{bmatrix} e^{-t} + 2e^{3t} & 3e^{-t} - e^{3t} \\ -e^{-t} + 2e^{3t} & -3e^{-t} - e^{3t} \end{bmatrix}. \quad \diamondsuit$$

The Variation of Parameters Formula

Consider the nonhomogeneous initial value problem

$$\mathbf{y}' = P(t)\mathbf{y} + \mathbf{g}(t), \qquad \mathbf{y}(t_0) = \mathbf{y}_0, \qquad a < t < b. \tag{4}$$

We assume that the $(n \times n)$ coefficient matrix $P(t)$ and the $(n \times 1)$ vector function $\mathbf{g}(t)$ are continuous on (a, b) and that t_0 is some point lying in this interval.

The method of variation of parameters, as developed in Sections 3.9 and 3.13, is based on an assumed knowledge of the complementary solution. For the linear system (4), the analogous assumption is a knowledge of a fundamental set of solutions of the homogeneous problem; that is, we assume we know a fundamental matrix, $\Psi(t)$, where

$$\Psi' = P(t)\Psi, \qquad a < t < b. \tag{5}$$

The complementary solution of (4) has the form $\mathbf{y}_C(t) = \Psi(t)\mathbf{c}$, where $\mathbf{c}$ is an arbitrary constant $(n \times 1)$ vector. Therefore, we "vary the parameter" and look for a solution of the nonhomogeneous equation (4) of the form $\mathbf{y}(t) = \Psi(t)\mathbf{u}(t)$, where $\mathbf{u}(t)$ is an unknown $(n \times 1)$ matrix function to be determined. Substituting this representation into equation (4) leads to

$$[\Psi(t)\mathbf{u}(t)]' = P(t)[\Psi(t)\mathbf{u}(t)] + \mathbf{g}(t). \tag{6}$$

Differentiating the product on the left-hand side of (6), we obtain

$$\Psi'(t)\mathbf{u}(t) + \Psi(t)\mathbf{u}'(t) = P(t)[\Psi(t)\mathbf{u}(t)] + \mathbf{g}(t). \tag{7}$$

Using the fact that $\Psi'(t) = P(t)\Psi(t)$, we can reduce equation (7) to $\Psi(t)\mathbf{u}'(t) = \mathbf{g}(t)$, or

$$\mathbf{u}'(t) = \Psi^{-1}(t)\mathbf{g}(t). \tag{8}$$

[Note that $\Psi(t)$, being a fundamental matrix, is invertible.]

We can solve for the unknown matrix function $\mathbf{u}(t)$ in equation (8) by antidifferentiation:

$$\mathbf{u}(t) = \mathbf{u}_0 + \int_{t_0}^{t} \Psi^{-1}(s)\mathbf{g}(s)\,ds, \tag{9}$$

where $\mathbf{u}(t_0) = \mathbf{u}_0$ is an arbitrary constant $(n \times 1)$ vector. By including an arbitrary vector $\mathbf{u}_0$ in (9), we have found a representation $\mathbf{y}(t) = \Psi(t)\mathbf{u}(t)$ for the *general solution* of $\mathbf{y}' = P(t)\mathbf{y} + \mathbf{g}(t)$, namely

$$\mathbf{y}(t) = \Psi(t)\mathbf{u}_0 + \Psi(t)\int_{t_0}^{t} \Psi^{-1}(s)\mathbf{g}(s)\,ds. \tag{10}$$

Note that equation (10) has the structure mentioned at the beginning of this section; that is, $\mathbf{y}(t)$ is the sum of the complementary solution, $\mathbf{y}_C(t) = \Psi(t)\mathbf{u}_0$, and a particular solution

$$\mathbf{y}_P(t) = \Psi(t) \int_{t_0}^{t} \Psi^{-1}(s)\mathbf{g}(s)\,ds.$$

(This particular solution is the one that vanishes at $t = t_0$.)

We can solve initial value problem (4) by imposing the initial condition in equation (10). Since $\mathbf{y}(t_0) = \Psi(t_0)\mathbf{u}_0$, we have

$$\mathbf{u}_0 = \Psi^{-1}(t_0)\mathbf{y}_0.$$

Thus, the solution of initial value problem (4) is

$$\mathbf{y}(t) = \Psi(t)\Psi^{-1}(t_0)\mathbf{y}_0 + \Psi(t) \int_{t_0}^{t} \Psi^{-1}(s)\mathbf{g}(s)\,ds. \tag{11}$$

We refer to equation (11) as the **variation of parameters** formula for the solution of the initial value problem.

EXAMPLE 3

Using variation of parameters, solve the initial value problem

$$\mathbf{y}' = \begin{bmatrix} 1 & 2 \\ 2 & 1 \end{bmatrix} \mathbf{y} + \begin{bmatrix} e^{2t} \\ -2t \end{bmatrix}, \qquad \mathbf{y}(0) = \mathbf{0}.$$

In Example 1, the general solution of this differential was found using the method of undetermined coefficients. Verify that imposing the initial conditions on the general solution of Example 1 yields the same solution as variation of parameters.

Solution: In Example 2, we found a fundamental matrix to be

$$\Psi(t) = \begin{bmatrix} e^{-t} & e^{3t} \\ -e^{-t} & e^{3t} \end{bmatrix}.$$

The inverse of this fundamental matrix is

$$\Psi^{-1}(t) = \frac{1}{2} \begin{bmatrix} e^{t} & -e^{t} \\ e^{-3t} & e^{-3t} \end{bmatrix}.$$

By the variation of parameters formula (11),

$$\mathbf{y}(t) = \begin{bmatrix} e^{-t} & e^{3t} \\ -e^{-t} & e^{3t} \end{bmatrix} \int_{0}^{t} \frac{1}{2} \begin{bmatrix} e^{s} & -e^{s} \\ e^{-3s} & e^{-3s} \end{bmatrix} \begin{bmatrix} e^{2s} \\ -2s \end{bmatrix} ds$$

$$= \frac{1}{2} \begin{bmatrix} e^{-t} & e^{3t} \\ -e^{-t} & e^{3t} \end{bmatrix} \int_{0}^{t} \begin{bmatrix} e^{3s} + 2se^{s} \\ e^{-s} - 2se^{-3s} \end{bmatrix} ds.$$

Performing the integration and then the matrix-vector multiplication, we obtain

$$\mathbf{y}(t) = \frac{1}{18} \begin{bmatrix} 7e^{3t} - 6e^{2t} + 15e^{-t} + 24t - 16 \\ 7e^{3t} - 12e^{2t} - 15e^{-t} - 12t + 20 \end{bmatrix}. \tag{12}$$

If we impose the initial condition $\mathbf{y}(t) = \mathbf{0}$ on the general solution derived in Example 1, we obtain $c_1 = \frac{5}{6}$ and $c_2 = \frac{7}{18}$, and solution (12) again ensues. ❖

EXERCISES

Exercises 1–9:

Method of Undetermined Coefficients   Consider the initial value problem $\mathbf{y}' = A\mathbf{y} + \mathbf{g}(t)$, $\mathbf{y}(0) = \mathbf{y}_0$.

(a) Form the complementary solution.

(b) Construct a particular solution by assuming the form suggested and solving for the undetermined constant vectors $\mathbf{a}$, $\mathbf{b}$, and $\mathbf{c}$.

(c) Form the general solution.

(d) Impose the initial condition to obtain the solution of the initial value problem.

1. $\mathbf{y}' = \begin{bmatrix} -2 & 1 \\ 1 & -2 \end{bmatrix} \mathbf{y} + \begin{bmatrix} 1 \\ 1 \end{bmatrix}$, $\mathbf{y}_0 = \begin{bmatrix} 3 \\ 1 \end{bmatrix}$.   Try $\mathbf{y}_P(t) = \mathbf{a}$.

2. $\mathbf{y}' = \begin{bmatrix} 0 & 1 \\ -1 & 0 \end{bmatrix} \mathbf{y} + \begin{bmatrix} 1 \\ 2 \end{bmatrix}$, $\mathbf{y}_0 = \begin{bmatrix} 1 \\ 1 \end{bmatrix}$.   Try $\mathbf{y}_P(t) = \mathbf{a}$.

3. $\mathbf{y}' = \begin{bmatrix} 2 & 1 \\ 1 & 2 \end{bmatrix} \mathbf{y} + \begin{bmatrix} e^{-t} \\ 0 \end{bmatrix}$, $\mathbf{y}_0 = \begin{bmatrix} 0 \\ 0 \end{bmatrix}$.   Try $\mathbf{y}_P(t) = e^{-t}\mathbf{a}$.

4. $\mathbf{y}' = \begin{bmatrix} 0 & 1 \\ -1 & 0 \end{bmatrix} \mathbf{y} + \begin{bmatrix} e^{t} \\ -1 \end{bmatrix}$, $\mathbf{y}_0 = \begin{bmatrix} 1 \\ 0 \end{bmatrix}$.   Try $\mathbf{y}_P(t) = e^{t}\mathbf{a} + \mathbf{b}$.

5. $\mathbf{y}' = \begin{bmatrix} 0 & 1 \\ 1 & 0 \end{bmatrix} \mathbf{y} + \begin{bmatrix} t \\ -1 \end{bmatrix}$, $\mathbf{y}_0 = \begin{bmatrix} 2 \\ -1 \end{bmatrix}$.   Try $\mathbf{y}_P(t) = t\mathbf{a} + \mathbf{b}$.

6. $\mathbf{y}' = \begin{bmatrix} 0 & -1 \\ -1 & 0 \end{bmatrix} \mathbf{y} + \begin{bmatrix} t \\ e^{2t} \end{bmatrix}$, $\mathbf{y}_0 = \begin{bmatrix} 0 \\ 1 \end{bmatrix}$.   Try $\mathbf{y}_P(t) = e^{2t}\mathbf{a} + t\mathbf{b} + \mathbf{c}$.

7. $\mathbf{y}' = \begin{bmatrix} -3 & -2 \\ 4 & 3 \end{bmatrix} \mathbf{y} + \begin{bmatrix} \sin t \\ 0 \end{bmatrix}$, $\mathbf{y}_0 = \begin{bmatrix} 0 \\ 0 \end{bmatrix}$.   Try $\mathbf{y}_P(t) = (\sin t)\mathbf{a} + (\cos t)\mathbf{b}$.

8. $\mathbf{y}' = \begin{bmatrix} 1 & 1 & 0 \\ 0 & -1 & 0 \\ 0 & 0 & 1 \end{bmatrix} \mathbf{y} + \begin{bmatrix} 2 \\ 1 \\ 2 \end{bmatrix}$, $\mathbf{y}_0 = \begin{bmatrix} 0 \\ 0 \\ 0 \end{bmatrix}$.   Try $\mathbf{y}_P(t) = \mathbf{a}$.

9. $\mathbf{y}' = \begin{bmatrix} 1 & 1 & 0 \\ 0 & 1 & 0 \\ 0 & 0 & 1 \end{bmatrix} \mathbf{y} + \begin{bmatrix} 1 \\ t \\ 0 \end{bmatrix}$, $\mathbf{y}_0 = \begin{bmatrix} 1 \\ 0 \\ 2 \end{bmatrix}$.   Try $\mathbf{y}_P(t) = t\mathbf{a} + \mathbf{b}$.

10. As an illustration of the difficulties that may arise in using the method of undetermined coefficients, consider

$$\mathbf{y}' = \begin{bmatrix} 1 & 1 \\ 1 & 1 \end{bmatrix} \mathbf{y} + \begin{bmatrix} e^{2t} \\ 0 \end{bmatrix}.$$

(a) Determine the complementary solution.

(b) Show that seeking a particular solution of the form $\mathbf{y}_P(t) = e^{2t}\mathbf{a}$ does not work.

(c) Since a member of the fundamental set of solutions comprising the complementary solution is a constant vector multiplied by e^{2t}, we might consider $\mathbf{y}_P(t) = te^{2t}\mathbf{a}$ to be a reasonable guess. Show that this guess does not work either. (We obtain the general solution in Exercise 15, using the method of variation of parameters.)

11. Consider the initial value problem

$$\mathbf{y}' = \begin{bmatrix} 0 & 2 \\ -2 & 0 \end{bmatrix} \mathbf{y} + \mathbf{g}(t), \qquad \mathbf{y}\left(\frac{\pi}{2}\right) = \mathbf{y}_0.$$

Suppose we know that

$$\mathbf{y}(t) = \begin{bmatrix} 1 + \sin 2t \\ e^t + \cos 2t \end{bmatrix}$$

is the unique solution. Determine $\mathbf{g}(t)$ and $\mathbf{y}_0$.

12. Consider the initial value problem

$$\mathbf{y}' = \begin{bmatrix} 1 & t \\ t^2 & 1 \end{bmatrix} \mathbf{y} + \mathbf{g}(t), \qquad \mathbf{y}(1) = \begin{bmatrix} 2 \\ -1 \end{bmatrix}.$$

Suppose we know that

$$\mathbf{y}(t) = \begin{bmatrix} t + \alpha \\ t^2 + \beta \end{bmatrix}$$

is the unique solution. Find $\mathbf{g}(t)$ and the constants α and β.

13. Let $P(t)$ be a (2×2) matrix with continuous entries. Consider the differential equation $\mathbf{y}' = P(t)\mathbf{y} + \mathbf{g}(t)$. Suppose we know the solution is $\mathbf{y}_1(t)$ when $\mathbf{g}(t) = \mathbf{g}_1(t)$ and $\mathbf{y}_2(t)$ when $\mathbf{g}(t) = \mathbf{g}_2(t)$. Determine $P(t)$ if

$$\mathbf{g}_1(t) = \begin{bmatrix} -2 \\ 0 \end{bmatrix} \quad \text{and} \quad \mathbf{y}_1(t) = \begin{bmatrix} 1 \\ e^{-t} \end{bmatrix}, \qquad \mathbf{g}_2(t) = \begin{bmatrix} e^t \\ -1 \end{bmatrix} \quad \text{and} \quad \mathbf{y}_2(t) = \begin{bmatrix} e^t \\ -1 \end{bmatrix}.$$

[Hint: Form the matrix equation $[\mathbf{y}_1', \mathbf{y}_2'] = P(t)[\mathbf{y}_1, \mathbf{y}_2] + [\mathbf{g}_1(t), \mathbf{g}_2(t)]$.]

Exercises 14–21:

Method of Variation of Parameters   Use the method of variation of parameters to solve the given initial value problem.

14. $y_1' = -2y_1 + 5y_2 + 1, \quad y_1(0) = 3$
$\quad\;\; y_2' = \quad y_1 + 2y_2 + 1, \quad y_2(0) = 1$

15. $\mathbf{y}' = \begin{bmatrix} 1 & 1 \\ 1 & 1 \end{bmatrix} \mathbf{y} + \begin{bmatrix} e^{2t} \\ 0 \end{bmatrix}, \quad \mathbf{y}(0) = \begin{bmatrix} 0 \\ 0 \end{bmatrix}$

16. $y_1' = \quad y_2 + e^t, \quad y_1(0) = \frac{5}{4}$
$\quad\;\; y_2' = y_1, \qquad\qquad y_2(0) = -\frac{1}{4}$

17. $\mathbf{y}' = \begin{bmatrix} 1 & -1 \\ -1 & 1 \end{bmatrix} \mathbf{y} + \begin{bmatrix} 0 \\ 1 \end{bmatrix}, \quad \mathbf{y}(0) = \begin{bmatrix} 1 \\ -1 \end{bmatrix}$

18. $\mathbf{y}' = \begin{bmatrix} 9 & -4 \\ 15 & -7 \end{bmatrix} \mathbf{y} + \begin{bmatrix} e^t \\ 0 \end{bmatrix}, \quad \mathbf{y}(0) = \begin{bmatrix} 2 \\ 5 \end{bmatrix}$

19. $\mathbf{y}' = \begin{bmatrix} 0 & 1 \\ -1 & 0 \end{bmatrix} \mathbf{y} + \begin{bmatrix} 2 \\ 1 \end{bmatrix}, \quad \mathbf{y}(0) = \begin{bmatrix} 0 \\ 1 \end{bmatrix}$

20. $\mathbf{y}' = \begin{bmatrix} 1 & 1 \\ 0 & 1 \end{bmatrix} \mathbf{y} + \begin{bmatrix} 1 \\ 1 \end{bmatrix}, \quad \mathbf{y}(0) = \begin{bmatrix} 0 \\ 0 \end{bmatrix}$

21. $\mathbf{y}' = \begin{bmatrix} 0 & -2 \\ 2 & 0 \end{bmatrix} \mathbf{y} + \begin{bmatrix} 1 \\ 0 \end{bmatrix}, \quad \mathbf{y}(\pi/2) = \begin{bmatrix} 0 \\ 0 \end{bmatrix}$

Equilibrium Solutions   Consider the linear system $\mathbf{y}' = A\mathbf{y} + \mathbf{b}$, where A is a constant $(n \times n)$ matrix and $\mathbf{b}$ is a constant $(n \times 1)$ vector. An *equilibrium solution*, $\mathbf{y}(t)$, is a constant solution of the differential equation.

22. If the matrix A is invertible, show that $\mathbf{y}' = A\mathbf{y} + \mathbf{b}$ has a unique equilibrium solution. If the matrix A is not invertible, must the differential equation $\mathbf{y}' = A\mathbf{y} + \mathbf{b}$ possess an equilibrium solution? If an equilibrium solution does exist in this case, is it unique?

Exercises 23–27:

In each exercise, determine all equilibrium solutions (if any).

23. $\mathbf{y}' = \begin{bmatrix} 1 & 4 \\ -1 & -3 \end{bmatrix} \mathbf{y} + \begin{bmatrix} 2 \\ 1 \end{bmatrix}$

24. $\mathbf{y}' = \begin{bmatrix} 2 & -1 \\ -1 & 1 \end{bmatrix} \mathbf{y} + \begin{bmatrix} 2 \\ -1 \end{bmatrix}$

25. $\mathbf{y}' = \begin{bmatrix} 1 & -1 \\ -1 & 1 \end{bmatrix} \mathbf{y} + \begin{bmatrix} 2 \\ -2 \end{bmatrix}$

26. $\mathbf{y}' = \begin{bmatrix} 1 & 1 & 0 \\ 0 & -1 & 2 \\ 0 & 0 & 1 \end{bmatrix} \mathbf{y} + \begin{bmatrix} 2 \\ 3 \\ 2 \end{bmatrix}$

27. $\mathbf{y}' = \begin{bmatrix} 1 & 0 & 0 \\ 0 & 1 & 1 \\ 0 & 1 & 1 \end{bmatrix} \mathbf{y} + \begin{bmatrix} -2 \\ 0 \\ 0 \end{bmatrix}$

Exercises 28–32:

Consider the homogeneous linear system $\mathbf{y}' = A\mathbf{y}$. Recall that any associated fundamental matrix satisfies the matrix differential equation $\Psi' = A\Psi$. In each exercise, construct a fundamental matrix that solves the matrix initial value problem $\Psi' = A\Psi$, $\Psi(t_0) = \Psi_0$.

28. $\Psi' = \begin{bmatrix} 1 & -1 \\ -1 & 1 \end{bmatrix} \Psi, \quad \Psi(1) = \begin{bmatrix} 1 & 0 \\ 0 & 1 \end{bmatrix}$

29. $\Psi' = \begin{bmatrix} 0 & 2 \\ -2 & 0 \end{bmatrix} \Psi, \quad \Psi\left(\frac{\pi}{4}\right) = \begin{bmatrix} 1 & -1 \\ 0 & 1 \end{bmatrix}$

30. $\Psi' = \begin{bmatrix} 1 & -1 \\ -1 & 1 \end{bmatrix} \Psi, \quad \Psi(0) = \begin{bmatrix} 1 & 0 \\ 2 & 1 \end{bmatrix}$

31. $\Psi' = \begin{bmatrix} 3 & -4 \\ 2 & -3 \end{bmatrix} \Psi, \quad \Psi(0) = \begin{bmatrix} 1 & 0 \\ 0 & 1 \end{bmatrix}$

32. $\Psi' = \begin{bmatrix} 1 & 4 \\ -1 & 1 \end{bmatrix} \Psi, \quad \Psi\left(\frac{\pi}{4}\right) = \begin{bmatrix} 1 & 0 \\ 0 & 1 \end{bmatrix}$

Exercises 33–35:

The flow system shown in the figure is activated at time $t = 0$. Let $Q_i(t)$ denote the amount of solute present in the ith tank at time t. For simplicity, we assume all the flow rates are a constant 10 gal/min. It follows that volume of solution in each tank remains constant; we assume the volume to be 1000 gal.

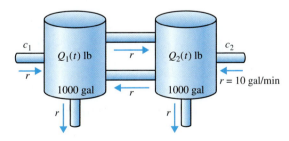

Figure for Exercises 33–35

(a) Derive a linear system of differential equations for $\mathbf{Q}(t) = \begin{bmatrix} Q_1(t) \\ Q_2(t) \end{bmatrix}$.

(b) Solve the initial value problem defined by the given inflow concentrations and initial conditions. Also, determine $\lim_{t \to \infty} \mathbf{Q}(t)$.

(c) In Exercises 33 and 34, the inflow concentrations are constant. Compute the equilibrium solution of the system in part (a). What is the physical significance of this equilibrium solution?

(d) In Exercise 35, the system in part (a) is not autonomous. Graph $Q_1(t)$ and $Q_2(t)$. Determine the maximum amounts of solute in each tank.

33. $c_1 = 0.5$ lb/gal, $c_2 = 0$, $Q_1(0) = Q_2(0) = 0$

34. $c_1 = c_2 = 0.5$ lb/gal, $Q_1(0) = 20$ lb, $Q_2(0) = 0$

35. $c_1 = 0.5e^{-2t/100}$ lb/gal, $c_2 = 0$, $Q_1(0) = 0$, $Q_2(0) = 20$ lb

36. Consider the RL network shown in the figure. Assume that the loop currents I_1 and I_2 are zero until a voltage source $V_S(t)$, having the polarity shown, is turned on at time $t = 0$. Applying Kirchhoff's voltage law to each loop, we obtain the equations

$$-V_S(t) + L_1 \frac{dI_1}{dt} + R_1 I_1 + R_3 (I_1 - I_2) = 0$$

$$R_3 (I_2 - I_1) + R_2 I_2 + L_2 \frac{dI_2}{dt} = 0.$$

(a) Formulate the initial value problem for the loop currents, $\begin{bmatrix} I_1(t) \\ I_2(t) \end{bmatrix}$, assuming that

$$L_1 = L_2 = 0.5 \, H, \qquad R_1 = R_2 = 1 \, k\Omega, \quad \text{and} \quad R_3 = 2 \, k\Omega.$$

(b) Determine a fundamental matrix for the associated linear homogeneous system.

(c) Use the method of variation of parameters to solve the initial value problem for the case where $V_S(t) = 1$ for $t > 0$.

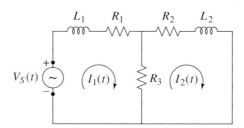

Figure for Exercise 36–37

37. Solve the network of Exercise 36 if the source voltage is $V_S(t) = 2e^{-2t}$ volts.

4.9 Numerical Methods for Systems of Linear Differential Equations

We introduced Euler's method in Section 2.10 as a simple numerical algorithm for approximating the solution of a first order scalar initial value problem,

$$y' = f(t, y), \qquad y(a) = y_0, \qquad a \leq t \leq b. \tag{1}$$

In this section, we begin by extending Euler's method to the first order linear system

$$\mathbf{y}' = P(t)\mathbf{y} + \mathbf{g}(t), \qquad \mathbf{y}(a) = \mathbf{y}_0, \qquad a \leq t \leq b. \tag{2}$$

We also describe how the familiar fourth order Runge-Kutta method [see equations (9a) and (9b) in Section 2.10] can be extended to the initial value problem (2). Later, in Chapter 7, we will discuss methods for systems of first order nonlinear equations $\mathbf{y}' = \mathbf{f}(t, \mathbf{y})$.

Recall the form of Euler's method for a scalar equation. For initial value problem (1), we often partition the interval $[a, b]$ into N subintervals, each having length h, where

$$h = \frac{b - a}{N}.$$

For a given step size h, we define grid points $t_0, t_1, \ldots, t_N$ by the formula[1]

$$t_k = t_0 + kh, \qquad 0 \le k \le N, \quad \text{where} \quad t_0 = a.$$

Note that $a = t_0 < t_1 < t_2 < \ldots < t_N = b$ and that $t_{k+1} = t_k + h$, $0 \le k \le N - 1$. The numerical solution of initial value problem (1) consists of the points (t_k, y_k), $0 \le k \le N$, where the values y_k are numerical approximations to the actual solution values $y(t_k)$.

Euler's Method Is a Finite Difference Method

In Section 2.10, we developed Euler's method from geometric considerations, using the direction field for $y' = f(t, y)$ as a starting point. We now present a different derivation, one that generalizes to systems of first order differential equations.

Let $y(t)$ be the solution of the initial value problem (1), and let $h > 0$ be a given step size. From calculus, we know that the derivative, $y'(t)$, is defined by

$$\lim_{\Delta t \to 0} \frac{y(t + \Delta t) - y(t)}{\Delta t} = y'(t).$$

Therefore, if $y(t)$ is the solution of initial value problem (1) and if the step size h is small, we expect the following approximation to be good:

$$\frac{y(t + h) - y(t)}{h} \approx y'(t) = f(t, y(t)). \tag{3}$$

Evaluating approximation (3) at a grid point $t = t_k$, we obtain

$$y(t_k + h) \approx y(t_k) + hf(t_k, y(t_k)), \qquad 0 \le k \le N - 1.$$

Therefore, once we have an estimate y_k of $y(t_k)$, this approximation leads us to an estimate y_{k+1} of $y(t_{k+1})$:

$$y_{k+1} = y_k + hf(t_k, y_k), \qquad 0 \le k \le N - 1, \qquad y_0 = y(t_0). \tag{4}$$

Equation (4) is Euler's method applied to the scalar initial value problem

$$y' = f(t, y), \qquad y(t_0) = y_0.$$

The approximation (3) is called a **finite difference approximation** to $y'(t)$. Methods derived from finite difference approximations, such as Euler's method,

---

[1]We assume a constant step size h in order to simplify the discussion. As noted in Section 2.10, many implementations of numerical methods use variable-step algorithms rather than a fixed-step algorithm. Such variable-step methods use error estimates to monitor errors as the algorithm proceeds; when errors exceed a prescribed upper level, the steplength is reduced, and when errors are below a prescribed lower level, the steplength is lengthened.

are known as **finite difference methods**. Finite difference methods can be generalized in a natural way to systems of first order differential equations.

Extending Euler's Method to First Order Linear Systems

Consider the initial value problem

$$\mathbf{y}' = P(t)\mathbf{y} + \mathbf{g}(t), \qquad \mathbf{y}(t_0) = \mathbf{y}_0, \qquad a \leq t \leq b, \qquad (5)$$

where the $(n \times n)$ matrix function $P(t)$ and the $(n \times 1)$ vector function $\mathbf{g}(t)$ are continuous on $[a, b]$. Euler's method for initial value problem (5) begins with a generalization of the finite difference approximation (3). In particular, let $\mathbf{y}(t)$ be the unique solution of initial value problem (5), where

$$\mathbf{y}(t) = \begin{bmatrix} y_1(t) \\ \vdots \\ y_n(t) \end{bmatrix}.$$

As we know from Section 4.1,

$$\mathbf{y}'(t) = \begin{bmatrix} y_1'(t) \\ \vdots \\ y_n'(t) \end{bmatrix} = \begin{bmatrix} \lim_{\Delta t \to 0} \dfrac{y_1(t + \Delta t) - y_1(t)}{\Delta t} \\ \vdots \\ \lim_{\Delta t \to 0} \dfrac{y_n(t + \Delta t) - y_n(t)}{\Delta t} \end{bmatrix}$$

$$= \lim_{\Delta t \to 0} \frac{1}{\Delta t} \begin{bmatrix} y_1(t + \Delta t) - y_1(t) \\ \vdots \\ y_n(t + \Delta t) - y_n(t) \end{bmatrix}$$

$$= \lim_{\Delta t \to 0} \frac{1}{\Delta t}[\mathbf{y}(t + \Delta t) - \mathbf{y}(t)].$$

As before, let $h > 0$ be a given step size, where $h = (b - a)/N$, and let

$$t_k = t_0 + kh, \qquad 0 \leq k \leq N, \qquad t_0 = a.$$

Since $\mathbf{y}'(t) = P(t)\mathbf{y}(t) + \mathbf{g}(t)$, we expect (for small h) that

$$\frac{1}{h}[\mathbf{y}(t + h) - \mathbf{y}(t)] \approx P(t)\mathbf{y}(t) + \mathbf{g}(t).$$

Evaluating this approximation at the grid point $t = t_k$, we obtain

$$\mathbf{y}(t_k + h) \approx \mathbf{y}(t_k) + h[P(t_k)\mathbf{y}(t_k) + \mathbf{g}(t_k)].$$

Therefore, once we have an estimate $\mathbf{y}_k$ of $\mathbf{y}(t_k)$, this approximation gives us an estimate $\mathbf{y}_{k+1}$ of $\mathbf{y}(t_{k+1})$. Define

$$\mathbf{y}_{k+1} = \mathbf{y}_k + h[P(t_k)\mathbf{y}_k + \mathbf{g}(t_k)], \qquad 0 \leq k \leq N - 1, \qquad \mathbf{y}_0 = \mathbf{y}(t_0). \qquad (6)$$

Iteration (6) is Euler's method for the initial value problem (5).

There are obvious mathematical questions about the algorithm (similar to those raised in Section 2.10 for the scalar problem) that need to be answered. These will be addressed in Chapter 7.

EXAMPLE 1

Consider the two-tank mixing problem formulated in Section 4.1. The flow schematic is shown in Figure 4.8. The corresponding initial value problem is

$$\frac{d}{dt}\begin{bmatrix} Q_1 \\ Q_2 \end{bmatrix} = \begin{bmatrix} -\dfrac{6}{200+10t} & \dfrac{6}{300} \\[2mm] \dfrac{2}{200+10t} & -\dfrac{14}{300} \end{bmatrix}\begin{bmatrix} Q_1 \\ Q_2 \end{bmatrix} + \begin{bmatrix} 5 \\ 0 \end{bmatrix}, \quad 0 \le t \le 30, \quad \begin{bmatrix} Q_1(0) \\ Q_2(0) \end{bmatrix} = \begin{bmatrix} 0 \\ 50 \end{bmatrix}.$$

$$\text{(7)}$$

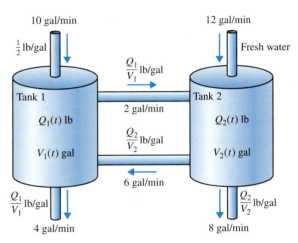

FIGURE 4.8

The two-tank mixing problem discussed in Example 1.

In (7), $Q_1(t)$ and $Q_2(t)$ represent the amounts of salt (in pounds) in Tanks 1 and 2, respectively, at time t (in minutes). Recall that salt solutions enter and leave Tank 1 at different rates, leading to the variable coefficient matrix in (7). At time $t = 30$ min, Tank 1 is filled to capacity and the flow stops. Use Euler's method (6) to estimate $Q_1(t)$ and $Q_2(t)$ on this time interval; use a step size of $h = 0.01$. Plot $Q_1(t)$ and $Q_2(t)$ as functions of t, and also plot the concentration of salt in each tank as a function of t.

Solution: The first order system has the form

$$\mathbf{Q}' = P(t)\mathbf{Q} + \mathbf{g}, \qquad 0 \le t \le 30,$$

where [see equation (7)] $P(t)$ is a (2×2) matrix function and $\mathbf{g}$ is a (2×1) constant vector. Euler's method, applied to this problem, is the iteration

$$\mathbf{Q}_{k+1} = \mathbf{Q}_k + h[P(t_k)\mathbf{Q}_k + \mathbf{g}], \qquad 0 \le k \le N-1, \qquad \mathbf{Q}_0 = \begin{bmatrix} 0 \\ 50 \end{bmatrix},$$

(continued)

(continued)

where $t_k = kh, h = 30/N$. In component form, the algorithm is

$$Q_{1,k+1} = Q_{1,k} + h \left[-\frac{6}{200 + 10t_k} Q_{1,k} + \frac{6}{300} Q_{2,k} + 5 \right]$$

$$Q_{2,k+1} = Q_{2,k} + h \left[\frac{2}{200 + 10t_k} Q_{1,k} - \frac{14}{300} Q_{2,k} \right], \qquad 0 \le k \le N - 1, \tag{8}$$

where $Q_{1,0} = 0, Q_{2,0} = 50$.

The vector $\mathbf{Q}_j$ is an approximation to the exact solution, $\mathbf{Q}(t_j)$, at time $t_j = jh$. Since $h = 0.01$, the number of steps is $N = 3000$.

Figure 4.9(a) shows the result of implementing Euler's method with $h = 0.01$. Figure 4.9(b) displays the Euler's method approximations to the concentrations,

$$c_m(t) = \frac{Q_m(t)}{V_m(t)}, \qquad m = 1, 2.$$

As graphed in Figure 4.9, the answers seem reasonable. We expect the salt concentration in Tank 1 to increase, but not to exceed the maximum inflow concentration of 0.5 lb/gal. Likewise, the 2 gal/min inflow from Tank 1 into Tank 2 mitigates the "flushing out" of Tank 2 that would otherwise occur. The concentration in Tank 2 therefore decreases somewhat slowly over the 30-min interval.

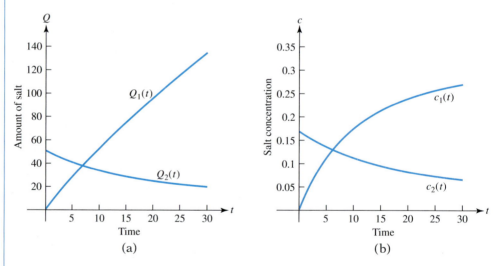

FIGURE 4.9

The results of applying Euler's method to the initial value problem in Example 1. ❖

Runge-Kutta Methods for Systems

Euler's method is a conceptually important but relatively crude numerical algorithm. In order to obtain a high level of accuracy, Euler's method usually demands a very small step size. And, of course, small steps require longer run-times as well as close attention to the accumulation of roundoff errors. One alternative described in Section 2.10, higher order Runge-Kutta methods, usually provides more accuracy for a given step size h and is generally more efficient than Euler's method.

Recall the fourth order Runge-Kutta method described in Section 2.10. For the scalar initial value problem (1), this method has the form

$$y_{k+1} = y_k + \frac{h}{6}[K_1 + 2K_2 + 2K_3 + K_4], \tag{9}$$

where

$$
\begin{aligned}
K_1 &= f(t_k, y_k) \\
K_2 &= f(t_k + h/2, y_k + (h/2)K_1) \\
K_3 &= f(t_k + h/2, y_k + (h/2)K_2) \\
K_4 &= f(t_k + h, y_k + hK_3).
\end{aligned} \tag{10}
$$

Just as Euler's method can be extended to systems, most Runge-Kutta methods can be extended to systems without any loss of accuracy. In particular, consider the initial value problem

$$\mathbf{y}' = \mathbf{f}(t, \mathbf{y}), \qquad \mathbf{y}(t_0) = \mathbf{y}_0. \tag{11}$$

The extension of the Runge-Kutta method (9)–(10) takes the form

$$\mathbf{y}_{k+1} = \mathbf{y}_k + \frac{h}{6}\left[\mathbf{K}_1 + 2\mathbf{K}_2 + 2\mathbf{K}_3 + \mathbf{K}_4\right], \tag{12}$$

where

$$
\begin{aligned}
\mathbf{K}_1 &= \mathbf{f}(t_k, \mathbf{y}_k) \\
\mathbf{K}_2 &= \mathbf{f}(t_k + h/2, \mathbf{y}_k + (h/2)\mathbf{K}_1) \\
\mathbf{K}_3 &= \mathbf{f}(t_k + h/2, \mathbf{y}_k + (h/2)\mathbf{K}_2) \\
\mathbf{K}_4 &= \mathbf{f}(t_k + h, \mathbf{y}_k + h\mathbf{K}_3).
\end{aligned} \tag{13}
$$

E X A M P L E
2

As a test case to illustrate how the Runge-Kutta philosophy can improve accuracy, consider the initial value problem

$$
\begin{aligned}
y_1' &= y_1 - y_2 + 3, & y_1(0) &= 1 \\
y_2' &= -y_1 + y_2 + 2t, & y_2(0) &= 3.
\end{aligned}
$$

(a) Solve this initial value problem mathematically.

(b) Solve this initial value problem numerically on the interval $0 \le t \le 2$, using Euler's method and the Runge-Kutta method (12). Use a constant step size of $h = 0.1$.

(c) Tabulate the exact solution values and both sets of numerical approximations at $t = 0.5$, $t = 1.0$, $t = 1.5$, and $t = 2.0$. Is the Runge-Kutta method more accurate than Euler's method for this test case?

(continued)

(continued)

Solution:

(a) The exact (mathematical) solution is

$$\mathbf{y}(t) = \frac{1}{2} \begin{bmatrix} -e^{2t} + t^2 + 4t + 3 \\ e^{2t} + t^2 + 2t + 5 \end{bmatrix}.$$

(b) For this problem, we have

$$\mathbf{f}(t, \mathbf{y}) = \begin{bmatrix} y_1 - y_2 + 3 \\ -y_1 + y_2 + 2t \end{bmatrix}.$$

We used MATLAB as a programming environment for this example. The MATLAB code that evaluates $\mathbf{f}(t, \mathbf{y})$ is shown in Figure 4.10. MATLAB codes implementing Euler's method and the Runge-Kutta method are shown in Figures 4.11 and 4.12, respectively.

```
function yp=f(t,y)
yp=zeros(2,1);
yp(1)=y(1)-y(2)+3;
yp(2)=-y(1)+y(2)+2*t;
```

FIGURE 4.10

The MATLAB function subprogram to evaluate $\mathbf{f}(t, \mathbf{y})$ in Example 2.

```
%
%   Set the initial conditions for the
%      initial value problem in Example 2
%
t=0;
y=[1,3]';
h=0.1;
output=[t,y(1),y(2)];
%
%
%   Execute Euler's method
%      on the interval [0,2]
%
for i=1:20
    y=y+h*f(t,y);
    t=t+h;
    output=[output;t,y(1),y(2)];
end
```

FIGURE 4.11

A MATLAB script that carries out Euler's method for the initial value problem in Example 2.

```
%
%   Set the initial conditions for the
%      initial value problem in Example 2
%
t=0;
y=[1,3]';
h=0.1;
output=[t,y(1),y(2)];
%
%   Execute the fourth-order Runge-Kutta method
%      on the interval [0,2]
%
for i=1:20
    ttemp=t;
    ytemp=y;
    k1=f(ttemp,ytemp);
    ttemp=t+h/2;
    ytemp=y+(h/2)*k1;
    k2=f(ttemp,ytemp);
    ttemp=t+h/2;
    ytemp=y+(h/2)*k2;
    k3=f(ttemp,ytemp);
    ttemp=t+h;
    ytemp=y+h*k3;
    k4=f(ttemp,ytemp);
    y=y+(h/6)*(k1+2*k2+2*k3+k4);
    t=t+h;
    output=[output;t,y(1),y(2)];
end
```

FIGURE 4.12

A Runge-Kutta code for the initial value problem in Example 2.

(c) Table 4.1 compares the values.

TABLE 4.1

The results of Example 2. Here, y_1^E denotes Euler's method estimates of $y_1(t)$, y_1^{RK} denotes the Runge-Kutta method estimates of $y_1(t)$, and y_1^T denotes the true value of $y_1(t)$. Similar notation is used in the last three columns. Note that the Runge-Kutta estimates are more accurate than the Euler's method estimates.

t	y_1^E	y_1^{RK}	y_1^T	y_2^E	y_2^{RK}	y_2^T
0.5000	1.3558	1.2659	1.2659	4.3442	4.4841	4.4841
1.0000	0.8541	0.3056	0.3055	7.0459	7.6944	7.6945
1.5000	−2.1535	−4.4174	−4.4178	12.7535	15.1674	15.1678
2.0000	−11.7688	−19.7978	−19.7991	25.5688	33.7978	33.7991

E X A M P L E

3

Use the fourth order Runge-Kutta method (12) to estimate the solution of

$$y'' - 2ty' + y = e^{t/2}, \qquad y(1) = 1, \qquad y'(1) = 0, \qquad 1 \le t \le 2.$$

Use a uniform step size of $h = 0.01$.

Solution: The differential equation is scalar, second order, linear, and non-homogeneous, with variable coefficients. None of the analytic techniques described in Chapter 3 are applicable. In order to use the Runge-Kutta method (12), we have to reformulate the second order equation as a system of two first order equations:

$$y'_1 = y_2, \qquad\qquad y_1(1) = 1$$
$$y'_2 = -y_1 + 2ty_2 + e^{t/2}, \qquad y_2(1) = 0.$$

We can use the Runge-Kutta code listed in Figure 4.12, changing the first three lines to read

```
t=1;
y=[1,0]';
h=0.01;
```

We also have to change the for loop to read "for i=1:100" and modify the last two lines of the function m-file in Figure 4.10 to read

```
yp(1)=y(2);
yp(2)=-y(1)+2*t*y(2)+exp(t/2);
```

After making these changes and executing the program, we obtain the results shown in Figure 4.13.

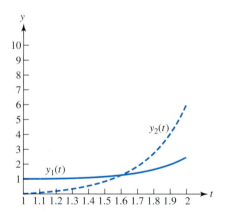

FIGURE 4.13

The results of Example 3. The solid curve is the numerical estimate of the graph of $y_1(t) = y(t)$; the dashed curve is the estimate of the graph of $y_2(t) = y'(t)$.

EXERCISES

Exercises 1–5:

In each exercise, assume that a numerical solution is desired on the interval $t_0 \le t \le t_0 + T$, using a uniform step size h.

(a) As in equation (8), write the Euler's method algorithm in explicit form for the given initial value problem. Specify the starting values t_0 and $\mathbf{y}_0$.

(b) Give a formula for the kth t-value, t_k. What is the range of the index k if we choose $h = 0.01$?

(c) Use a calculator to carry out two steps of Euler's method, finding $\mathbf{y}_1$ and $\mathbf{y}_2$. Use a step size of $h = 0.01$ for the given initial value problem. Hand calculations such as these are used to check the coding of a numerical algorithm.

1. $\mathbf{y}' = \begin{bmatrix} 1 & 2 \\ 2 & 3 \end{bmatrix} \mathbf{y} + \begin{bmatrix} 1 \\ 1 \end{bmatrix}$, $\mathbf{y}(0) = \begin{bmatrix} -1 \\ 1 \end{bmatrix}$, $0 \le t \le 1$

2. $\mathbf{y}' = \begin{bmatrix} 1 & t \\ 2+t & 2 \end{bmatrix} \mathbf{y} + \begin{bmatrix} 1 \\ t \end{bmatrix}$, $\mathbf{y}(1) = \begin{bmatrix} 2 \\ 1 \end{bmatrix}$, $1 \le t \le 1.5$

3. $\mathbf{y}' = \begin{bmatrix} -t^2 & t \\ 2-t & 0 \end{bmatrix} \mathbf{y} + \begin{bmatrix} 1 \\ t \end{bmatrix}$, $\mathbf{y}(1) = \begin{bmatrix} 2 \\ 0 \end{bmatrix}$, $1 \le t \le 4$

4. $\mathbf{y}' = \begin{bmatrix} 1 & 0 & 1 \\ 3 & 2 & 1 \\ 1 & 2 & 0 \end{bmatrix} \mathbf{y} + \begin{bmatrix} 0 \\ 2 \\ t \end{bmatrix}$, $\mathbf{y}(-1) = \begin{bmatrix} 0 \\ 0 \\ 1 \end{bmatrix}$, $-1 \le t \le 0$

5. $\mathbf{y}' = \begin{bmatrix} \dfrac{1}{t} & \sin t \\ 1-t & 1 \end{bmatrix} \mathbf{y} + \begin{bmatrix} 0 \\ t^2 \end{bmatrix}$, $\mathbf{y}(1) = \begin{bmatrix} 0 \\ 0 \end{bmatrix}$, $1 \le t \le 6$

Exercises 6–9:

In each exercise,

(a) As in Example 3, rewrite the given scalar initial value problem as an equivalent initial value problem for a first order system.

(b) Write the Euler's method algorithm, $\mathbf{y}_{k+1} = \mathbf{y}_k + h[P(t_k)\mathbf{y}_k + \mathbf{g}(t_k)]$, in explicit form for the given problem. Specify the starting values t_0 and $\mathbf{y}_0$.

(c) Using a calculator and a uniform step size of $h = 0.01$, carry out two steps of Euler's method, finding $\mathbf{y}_1$ and $\mathbf{y}_2$. What are the corresponding numerical approximations to the solution $y(t)$ at times $t = 0.01$ and $t = 0.02$?

6. $y'' + y = t^{3/2}$, $y(0) = 1$, $y'(0) = 0$

7. $y'' + y' + t^2 y = 2$, $y(0) = 1$, $y'(0) = 1$

8. $y''' + 2y' + ty = t + 1$, $y(0) = 1$, $y'(0) = -1$, $y''(0) = 0$

9. $\dfrac{d}{dt}\left[e^t \dfrac{dy}{dt} \right] + y = 2e^t$, $y(0) = -1$, $y'(0) = 1$

Exercises 10–18:

For the problem in the exercise specified,

(a) Write a program that carries out Euler's method. Use a step size of $h = 0.01$.

(b) Run your program on the interval given.

(c) Check your numerical solution by comparing the first two values, y_1 and y_2, with the hand calculations.

(d) Plot the components of the numerical solution on a common graph over the time interval of interest.

10. Exercise 1   **11.** Exercise 2   **12.** Exercise 3

13. Exercise 4   **14.** Exercise 5   **15.** Exercise 6

16. Exercise 7   **17.** Exercise 8   **18.** Exercise 9

Estimating the Numerical Error of Euler's Method   When solving problems, we should apply all available tests or checks before accepting an answer. In addition to the checks provided by common sense, the physics of the problem being modeled, and the mathematical theory of differential equations, there are additional checks available for testing the accuracy of numerical algorithms. We now describe such a test.

Suppose we apply Euler's method to the initial value problem $y' = P(t)y + g(t)$, $y(t_0) = y_0$, $a \le t \le b$. We observed in Section 2.10 that the error in Euler's method is reduced approximately in half when the step size h is halved (this result will be justified in Chapter 7). This approximate halving of the error leads to a process for estimating the error. In particular, let t^* be a point in $[a, b]$ and choose a step size h by defining $h = (t^* - t_0)/n$, where n is a positive integer. Let y_n denote the Euler's method estimate to $y(t^*)$ obtained using a step size h. Let $\bar{y}_{2n}$ denote the Euler's method estimate to $y(t^*)$ obtained using a step size of $h/2$. We anticipate, by halving the step size, that the error will be (approximately) halved as well:

$$y(t^*) - \bar{y}_{2n} \approx \frac{y(t^*) - y_n}{2}.$$

Some algebraic manipulation leads to the following estimate of the error, $y(t^*) - \bar{y}_{2n}$:

$$y(t^*) - \bar{y}_{2n} \approx \bar{y}_{2n} - y_n. \tag{14}$$

Exercises 19–22:

(a) Compute the error estimate (14) by using your Euler's method program to solve the given initial value problem. In each case, let $t^* = 1$. Use $h = 0.01$ and $h = 0.005$.

(b) Solve the initial value problem mathematically, and determine the exact solution at $t = t^*$.

(c) Compare the actual error, $y(t^*) - \bar{y}_{2n}$, with the estimate of the error $\bar{y}_{2n} - y_n$. [Note that estimate (14) is also applicable at any of the intermediate points $0.01, 0.02, \ldots, 0.99$.]

19. $y'' - y = t$, $y(0) = 2$, $y'(0) = -1$

20. $y'' + 4y = 3 \cos t + 3 \sin t$, $y(0) = 4$, $y'(0) = 5$

21. $y' = \begin{bmatrix} 2 & -1 \\ 1 & 2 \end{bmatrix} y$, $y(0) = \begin{bmatrix} 0 \\ 2 \end{bmatrix}$ **22.** $y' = \begin{bmatrix} -1 & 1 \\ 1 & -1 \end{bmatrix} y$, $y(0) = \begin{bmatrix} 3 \\ -1 \end{bmatrix}$

23. **Draining a Two-Tank System**   Consider the flow system shown in the figure. Tank 1 initially contains 40 lb of salt dissolved in 200 gal of water, while Tank 2 initially contains 40 lb of salt dissolved in 500 gal of water. At time $t = 0$, the draining process is initiated, with well-stirred mixtures flowing at the rates and directions shown. The volumes of fluid in each tank change with time. Note, in particular, that Tank 1 empties in 20 min. Therefore, the flow processes shown in the figure cease to be valid after 20 min. (Tank 2 will still contain 100 gal of fluid after 20 min.)

(a) Let $Q_j(t)$ represent the amount of salt in tank j at time t, $j = 1, 2$. Formulate the initial value problem for

$$\mathbf{Q}(t) = \begin{bmatrix} Q_1(t) \\ Q_2(t) \end{bmatrix},$$

valid for $0 < t < 20$ min.

(b) Solve the initial value problem on the interval $0 \le t \le 19.9$ min, using Euler's method. Use a uniform step size with $h = 0.01$.

(c) Plot the amounts of salt (in pounds) in each tank, $Q_1(t)$ and $Q_2(t)$, on the same graph for $0 \le t \le 19.9$ min.

(d) On what positive t-interval does Theorem 4.1 guarantee a unique solution of the initial value problem formulated in (a)? Should we be on the alert for possible numerical problems as time t increases to 20 min? Explain.

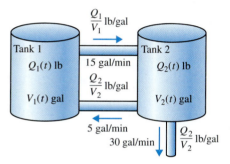

Figure for Exercise 23

24. **A Spring-Mass-Dashpot System with Variable Damping** As we saw in Section 3.6, the differential equation modeling unforced damped motion of a mass suspended from a spring is $my'' + \gamma y' + ky = 0$, where $y(t)$ represents the downward displacement of the mass from its equilibrium position. Assume a mass $m = 1$ kg and a spring constant $k = 4\pi^2$ N/m. Also assume the damping coefficient γ is varying with time:

$$\gamma(t) = 2te^{-t/2} \text{ kg/sec.}$$

Assume, at time $t = 0$, the mass is pulled down 20 cm and released.

(a) Formulate the appropriate initial value problem for the second order scalar differential equation, and rewrite it as an equivalent initial value problem for a first order linear system.

(b) Applying Euler's method, numerically solve this problem on the interval $0 \le t \le 10$ min. Use a step size of $h = 0.005$.

(c) Plot the numerical solution on the time interval $0 \le t \le 10$ min. Explain, in qualitative terms, the effect of the variable damping upon the solution.

Exercises 25–27:

Write a program that applies the Runge-Kutta algorithm (12) to the given problem. Run the program on the interval given, with a constant step size of $h = 0.01$. Plot the components of the solution.

25. $y_1' = -y_1 + y_2 + 2,$ $y_1(0) = 1,$ $0 \le t \le 2$
    $y_2' = -y_1 - y_2,$ $y_2(0) = 0$

26. $y_1' = -y_1 + y_2,$ $y_1(0) = 1,$ $0 \le t \le 1$
    $y_2' = \quad\quad y_2 + t,$ $y_2(0) = 0$

27. $\mathbf{y}' = \begin{bmatrix} 1 & t \\ 0 & 1 \end{bmatrix} \mathbf{y}, \quad \mathbf{y}(1) = \begin{bmatrix} 0 \\ 1 \end{bmatrix}, \quad 1 \le t \le 2$

28. Suppose the Runge-Kutta method (12) is applied to the initial value problem $\mathbf{y}' = A\mathbf{y}, \mathbf{y}(0) = \mathbf{y}_0$, where A is a constant square matrix [thus, $\mathbf{f}(t, \mathbf{y}) = A\mathbf{y}$].

    (a) Express each of the vectors $\mathbf{K}_j$ in terms of h, A, and $\mathbf{y}_k, j = 1, 2, 3, 4$.

    (b) Show that the Runge-Kutta method, when applied to this initial value problem, can be unraveled to obtain

    $$\mathbf{y}_{k+1} = \left(I + hA + \frac{h^2}{2!} A^2 + \frac{h^3}{3!} A^3 + \frac{h^4}{4!} A^4 \right) \mathbf{y}_k. \tag{15}$$

    (c) Use the differential equation $\mathbf{y}' = A\mathbf{y}$ to express the nth derivative, $\mathbf{y}^{(n)}(t)$, in terms of A and $\mathbf{y}(t)$. Express the Taylor series expansion

    $$\mathbf{y}(t + h) = \sum_{n=0}^{\infty} \mathbf{y}^{(n)}(t) \frac{h^n}{n!}$$

    in terms of h, A, and $\mathbf{y}(t)$. Compare the Taylor series with the right-hand side of (15), with $t = t_k$ and $\mathbf{y}(t_k) = \mathbf{y}_k$. How well does (15) replicate the Taylor series?

29. The exact solution of the initial value problem

    $$\mathbf{y}' = \begin{bmatrix} 0.5 & 1 \\ 1 & 0.5 \end{bmatrix} \mathbf{y}, \quad \mathbf{y}(0) = \begin{bmatrix} 1 \\ 0 \end{bmatrix} \quad \text{is given by} \quad \mathbf{y}(t) = \frac{1}{2} \begin{bmatrix} e^{-t/2} + e^{3t/2} \\ -e^{-t/2} + e^{3t/2} \end{bmatrix}.$$

    (a) Write a program that applies the Runge-Kutta method (12) to this problem.

    (b) Run your program on the interval $0 \le t \le 1$, using step size $h = 0.01$.

    (c) Run your program on the interval $0 \le t \le 1$, using step size $h = 0.005$.

    (d) Let $\mathbf{y}_{100}$ and $\bar{\mathbf{y}}_{200}$ denote the numerical approximations to $\mathbf{y}(1)$ computed in parts (b) and (c), respectively. Compute the error vectors $\mathbf{y}(1) - \mathbf{y}_{100}$ and $\mathbf{y}(1) - \bar{\mathbf{y}}_{200}$. By roughly what fractional amount is the error reduced when the step size is halved?

4.10 The Exponential Matrix and Diagonalization

The solution of the scalar initial value problem $y' = ay, y(0) = y_0$ is

$$y(t) = e^{at} y_0.$$

If A is a constant $(n \times n)$ matrix, the solution of $\mathbf{y}' = A\mathbf{y}, \mathbf{y}(0) = \mathbf{y}_0$ is

$$\mathbf{y}(t) = \Phi(t)\mathbf{y}_0, \tag{1}$$

where $\Phi(t)$ is the fundamental matrix that reduces to the identity at $t = 0$. Is it possible to represent the solution (1) in the form

$$\mathbf{y}(t) = e^{At}\mathbf{y}_0? \tag{2}$$

There are two issues to be resolved. First, how do we give meaning to the exponential of a square matrix? Second, if we can give meaning to e^{At}, is expression (2) the unique solution of $\mathbf{y}' = A\mathbf{y}, \mathbf{y}(0) = \mathbf{y}_0$? In other words, is $e^{At} = \Phi(t)$?

The Exponential Matrix

To see how we might define the exponential matrix e^{At}, assume for the present discussion that the solution $\mathbf{y}(t)$ of the initial value problem $\mathbf{y}' = A\mathbf{y}, \mathbf{y}(0) = \mathbf{y}_0$

has the series expansion

$$\mathbf{y}(t) = \sum_{m=0}^{\infty} \frac{t^m}{m!} \mathbf{y}^{(m)}(0). \tag{3}$$

We can use the initial value problem itself to evaluate the vectors $\mathbf{y}^{(m)}(0)$:

$$\mathbf{y}^{(0)}(0) = \mathbf{y}(0) = \mathbf{y}_0$$
$$\mathbf{y}^{(1)}(0) = \mathbf{y}'(0) = A\mathbf{y}(0) = A\mathbf{y}_0$$
$$\mathbf{y}^{(2)}(t) = [A\mathbf{y}(t)]' = A\mathbf{y}'(t) = A^2\mathbf{y}(t); \quad \text{therefore,} \quad \mathbf{y}^{(2)}(0) = A^2\mathbf{y}_0.$$

In general, we obtain $\mathbf{y}^{(m)}(0) = A^m\mathbf{y}_0$, and hence the series (3) becomes

$$\mathbf{y}(t) = \sum_{m=0}^{\infty} \frac{t^m}{m!} A^m \mathbf{y}_0 = \left(I + tA + \frac{t^2}{2!}A^2 + \frac{t^3}{3!}A^3 + \cdots \right) \mathbf{y}_0. \tag{4}$$

The partial sums of the series of matrix powers in equation (4) have the form

$$S_k(t) = \left(I + tA + \frac{t^2}{2!}A^2 + \frac{t^3}{3!}A^3 + \cdots + \frac{t^k}{k!}A^k \right).$$

It can be shown, for any constant $(n \times n)$ matrix A, that

$$\lim_{k \to \infty} S_k(t)$$

exists for all values of t. We define the $(n \times n)$ limit matrix to be the **exponential matrix** and denote it as e^{At}. Therefore,

$$e^{At} = \sum_{m=0}^{\infty} \frac{t^m}{m!} A^m. \tag{5}$$

It can be shown that the matrix exponential is differentiable and that its derivative can be calculated by differentiating the series (5) term by term. Assuming the validity of termwise differentiation, it follows that

$$\frac{d}{dt} e^{At} = A e^{At}. \tag{6}$$

By (6), $\mathbf{y}(t) = e^{At}\mathbf{y}_0$ is a solution of $\mathbf{y}' = A\mathbf{y}$ for any $(n \times 1)$ vector $\mathbf{y}_0$. Using this fact, along with the observation that e^{At} reduces to the identity when $t = 0$ and the fact that the solution of

$$\Phi'(t) = A\Phi(t), \qquad \Phi(0) = I$$

is unique, it follows that $e^{At} = \Phi(t)$.

Properties of the Exponential Matrix

In view of the close correspondence of the series defining e^{At} when A is a matrix and $e^{\alpha t}$ when α is a scalar, it is not surprising that the exponential matrix and the scalar exponential function possess similar properties. For instance,

$$e^{At_1} e^{At_2} = e^{A(t_1 + t_2)} \tag{7a}$$

and

$$(e^{At})^{-1} = e^{-At}. \tag{7b}$$

While properties (7a) and (7b) resemble properties of the scalar exponential function, they also have interpretations in terms of differential equations. In particular, consider the initial value problem $\mathbf{y}' = A\mathbf{y}, \mathbf{y}(0) = \mathbf{y}_0$. We can think of e^{At} as a matrix that propagates solutions forward by t units in time; that is, multiplying the initial state $\mathbf{y}(0) = \mathbf{y}_0$ by e^{At} moves the solution forward in time to $\mathbf{y}(t) = e^{At}\mathbf{y}_0$.

Taking this propagator viewpoint, property (7a) says either we can move the solution forward in one step or we can move it in stages. We can evolve the solution forward in one step from time $t = 0$ to time $t = t_1 + t_2$ by forming $\mathbf{y}(t_1 + t_2) = e^{A(t_1+t_2)}\mathbf{y}_0$. On the other hand, we can achieve the same result by first moving forward t_1 units in time using $\mathbf{y}(t_1) = e^{At_1}\mathbf{y}_0$ and then moving forward an additional t_2 units by forming $\mathbf{y}(t_1 + t_2) = e^{At_2}\mathbf{y}(t_1) = e^{At_2}e^{At_1}\mathbf{y}_0$. In general, it can be shown (see Exercise 20) that

$$\mathbf{y}(t + \Delta t) = e^{A\Delta t}\mathbf{y}(t). \tag{8}$$

In a similar vein, suppose we are given $\mathbf{y}(t) = e^{At}\mathbf{y}_0$. We can recover the initial state $\mathbf{y}_0$ either by multiplying by the inverse matrix, obtaining $\mathbf{y}_0 = (e^{At})^{-1}\mathbf{y}(t)$, or by propagating the solution backwards t units in time by forming $\mathbf{y}_0 = e^{-At}\mathbf{y}(t)$.

E X A M P L E
1

Use the series (5) to calculate the exponential matrix e^{At} for

$$A = \begin{bmatrix} \lambda_1 & 0 \\ 0 & \lambda_2 \end{bmatrix}.$$

Solution: Since A is a diagonal matrix,

$$A^m = \begin{bmatrix} \lambda_1^m & 0 \\ 0 & \lambda_2^m \end{bmatrix}.$$

Therefore, the matrix series (5) becomes

$$e^{tA} = \left(I + t\begin{bmatrix} \lambda_1 & 0 \\ 0 & \lambda_2 \end{bmatrix} + \frac{t^2}{2!}\begin{bmatrix} \lambda_1^2 & 0 \\ 0 & \lambda_2^2 \end{bmatrix} + \frac{t^3}{3!}\begin{bmatrix} \lambda_1^3 & 0 \\ 0 & \lambda_2^3 \end{bmatrix} + \cdots \right)$$

$$= \sum_{m=0}^{\infty} \frac{t^m}{m!} \begin{bmatrix} \lambda_1^m & 0 \\ 0 & \lambda_2^m \end{bmatrix} = \begin{bmatrix} \sum_{m=0}^{\infty} \frac{t^m}{m!}\lambda_1^m & 0 \\ 0 & \sum_{m=0}^{\infty} \frac{t^m}{m!}\lambda_2^m \end{bmatrix}$$

$$= \begin{bmatrix} e^{\lambda_1 t} & 0 \\ 0 & e^{\lambda_2 t} \end{bmatrix}. \; ❖$$

E X A M P L E
2

Use the series (5) to calculate the exponential matrix e^{At} for

$$A = \begin{bmatrix} \lambda & 1 \\ 0 & \lambda \end{bmatrix}.$$

Solution: It can be shown (see Exercise 21) that $A^m = \lambda^m I + m\lambda^{m-1}E$, where E denotes the matrix

$$E = \begin{bmatrix} 0 & 1 \\ 0 & 0 \end{bmatrix}.$$

Therefore,

$$e^{tA} = \sum_{m=0}^{\infty} \frac{t^m}{m!}(\lambda^m I + m\lambda^{m-1}E) = \begin{bmatrix} \sum_{m=0}^{\infty} \frac{t^m}{m!}\lambda^m & \sum_{m=0}^{\infty} \frac{t^m}{(m-1)!}\lambda^{m-1} \\ 0 & \sum_{m=0}^{\infty} \frac{t^m}{m!}\lambda^m \end{bmatrix}$$

$$= \begin{bmatrix} e^{\lambda t} & te^{\lambda t} \\ 0 & e^{\lambda t} \end{bmatrix}. \diamond$$

Similar Matrices

We say that an $(n \times n)$ matrix A is **similar** to an $(n \times n)$ matrix B if there exists an $(n \times n)$ invertible matrix T such that

$$T^{-1}AT = B.$$

If A is similar to B, it follows that B is similar to A since $T^{-1}AT = B$ implies

$$A = TBT^{-1} = (T^{-1})^{-1}B(T^{-1}).$$

Therefore, it is appropriate to refer to the matrices A and B as a pair of **similar matrices**. The act of forming the matrix product $T^{-1}AT$ is often referred to as a **similarity transformation**. Among other things, the concept of similarity is important because

(a) If A and B are similar $(n \times n)$ matrices, then A and B have the same characteristic polynomial and, hence, the same eigenvalues (see Theorem 4.10).
(b) If A and B are similar $(n \times n)$ matrices, then solutions of $\mathbf{w}' = B\mathbf{w}$ are related to solutions of $\mathbf{y}' = A\mathbf{y}$ by the transformation $\mathbf{y}(t) = T\mathbf{w}(t)$.

Theorem 4.10

Let A and B be similar $(n \times n)$ matrices. Then A and B have the same characteristic polynomial.

● **PROOF:**   Since A and B are similar, there is an invertible $(n \times n)$ matrix T such that $T^{-1}AT = B$. Let $p_A(\lambda)$ and $p_B(\lambda)$ denote the characteristic polynomials of matrices A and B, respectively. Observe that

$$p_B(\lambda) = \det[B - \lambda I] = \det[T^{-1}AT - \lambda I] = \det[T^{-1}AT - \lambda T^{-1}T]$$
$$= \det[T^{-1}(A - \lambda I)T] = \det[T^{-1}]\det[A - \lambda I]\det[T]$$
$$= p_A(\lambda).$$

(To obtain the last equality, we used the fact that $\det[T^{-1}]\det[T] = \det[I] = 1$.) ●

Diagonalization

Consider a similarity transformation of the form $T^{-1}AT = D$, where the matrix D is a diagonal matrix,

$$D = \begin{bmatrix} d_{11} & 0 & 0 & \cdots & 0 \\ 0 & d_{22} & 0 & \cdots & 0 \\ 0 & 0 & d_{33} & \cdots & 0 \\ \vdots & & & & \vdots \\ 0 & 0 & 0 & \cdots & d_{nn} \end{bmatrix}.$$

In such cases, we say that matrix A is **similar to a diagonal matrix** or that A is **diagonalizable**. Two questions arise with regard to diagonalization:

1. When is it possible to diagonalize a square matrix A?
2. If it is possible to diagonalize a given matrix A, how do we find the matrix T that accomplishes the diagonalization?

Theorem 4.11 and its corollary address these two questions.

Theorem 4.11

Let A be an $(n \times n)$ matrix similar to a diagonal matrix D. Let T be an invertible matrix such that $T^{-1}AT = D$. Then the diagonal elements of matrix D are the eigenvalues of matrix A, and the columns of matrix T are corresponding eigenvectors of matrix A.

● **PROOF:**   It follows from Theorem 4.10 that matrices A and D have the same eigenvalues. The eigenvalues of diagonal matrix D, however, are its diagonal elements. Therefore, the diagonal elements of D are the eigenvalues of A.

To finish the proof, let D be the diagonal matrix

$$D = \begin{bmatrix} \lambda_1 & 0 & 0 & \cdots & 0 \\ 0 & \lambda_2 & 0 & \cdots & 0 \\ 0 & 0 & \lambda_3 & \cdots & 0 \\ \vdots & & & & \vdots \\ 0 & 0 & 0 & \cdots & \lambda_n \end{bmatrix},$$

and let $\mathbf{t}_1, \mathbf{t}_2, \ldots, \mathbf{t}_n$ denote the n columns of the matrix T so that $T = [\mathbf{t}_1, \mathbf{t}_2, \ldots, \mathbf{t}_n]$. Because $T^{-1}AT = D$, it follows that $AT = TD$:

$$A[\mathbf{t}_1, \mathbf{t}_2, \ldots, \mathbf{t}_n] = [\mathbf{t}_1, \mathbf{t}_2, \ldots, \mathbf{t}_n] \begin{bmatrix} \lambda_1 & 0 & 0 & \cdots & 0 \\ 0 & \lambda_2 & 0 & \cdots & 0 \\ 0 & 0 & \lambda_3 & \cdots & 0 \\ \vdots & & & & \vdots \\ 0 & 0 & 0 & \cdots & \lambda_n \end{bmatrix}. \tag{9}$$

The matrix product on the left-hand side of equation (9) can be rewritten as $[A\mathbf{t}_1, A\mathbf{t}_2, \ldots, A\mathbf{t}_n]$. Similarly, the matrix product on the right-hand side of equation (9) can be rewritten as $[\lambda_1 \mathbf{t}_1, \lambda_2 \mathbf{t}_2, \ldots, \lambda_n \mathbf{t}_n]$. Since matrix equality implies that corresponding columns are equal, we obtain

$$A\mathbf{t}_j = \lambda_j \mathbf{t}_j, \qquad j = 1, 2, \ldots, n. \tag{10}$$

To complete the argument, note that invertibility of the matrix T implies that none of its columns can be the $(n \times 1)$ zero vector. Therefore, $\mathbf{t}_j \neq \mathbf{0}, 1 \leq j \leq n$, and this fact, in conjunction with (10), shows that the columns of T are eigenvectors of A. •

A corollary of Theorem 4.11 characterizes diagonalizable matrices.

Corollary

An $(n \times n)$ matrix A is diagonalizable if and only if it has a set of n linearly independent eigenvectors.

From what we know already, matrices with distinct eigenvalues as well as real symmetric matrices and Hermitian matrices are diagonalizable. In general, A is diagonalizable if and only if A has a full set of eigenvectors.

EXAMPLE 3

As noted, real symmetric matrices are diagonalizable. Let A be the matrix

$$A = \begin{bmatrix} 1 & 2 \\ 2 & 1 \end{bmatrix}.$$

Find an invertible (2×2) matrix T such that $T^{-1}AT = D$.

Solution: We saw in Example 1 of Section 4.4 that eigenpairs of A are

$$\lambda_1 = 3, \quad \mathbf{x}_1 = \begin{bmatrix} 1 \\ 1 \end{bmatrix} \quad \text{and} \quad \lambda_2 = -1, \quad \mathbf{x}_2 = \begin{bmatrix} 1 \\ -1 \end{bmatrix}.$$

Therefore, a matrix T made from the eigenvectors will diagonalize A. So, let T be

$$T = \begin{bmatrix} 1 & 1 \\ 1 & -1 \end{bmatrix}.$$

A direct calculation shows

$$T^{-1}AT = \begin{bmatrix} \frac{1}{2} & \frac{1}{2} \\ \frac{1}{2} & -\frac{1}{2} \end{bmatrix} \begin{bmatrix} 1 & 2 \\ 2 & 1 \end{bmatrix} \begin{bmatrix} 1 & 1 \\ 1 & -1 \end{bmatrix} = \begin{bmatrix} 3 & 0 \\ 0 & -1 \end{bmatrix}. \; ❖$$

The Exponential Matrix Revisited

We again consider the exponential matrix e^{At}, assuming now that A is an $(n \times n)$ diagonalizable matrix. Therefore,

$$A = TDT^{-1}, \tag{11}$$

where D is the $(n \times n)$ diagonal matrix consisting of the eigenvalues of A and T is the $(n \times n)$ matrix formed from the corresponding eigenvectors; recall equations (9) and (10). We now see that equation (11) simplifies the infinite series (5) defining e^{At}.

Observe, by (11), that

$$A^2 = (TDT^{-1})(TDT^{-1}) = TD(T^{-1}T)DT^{-1} = TD^2T^{-1}.$$

In general, it can be shown that

$$A^m = TD^mT^{-1}.$$

Therefore,

$$e^{At} = \sum_{m=0}^{\infty} \frac{t^m}{m!}A^m = \sum_{m=0}^{\infty} \frac{t^m}{m!}TD^mT^{-1} = T\left(\sum_{m=0}^{\infty} \frac{t^m}{m!}D^m\right)T^{-1} = Te^{Dt}T^{-1}. \qquad \text{(12a)}$$

Since D is a diagonal matrix, it follows (as in Example 1) that

$$\sum_{m=0}^{\infty} \frac{t^m}{m!}D^m = \begin{bmatrix} e^{\lambda_1 t} & 0 & \cdots & 0 \\ 0 & e^{\lambda_2 t} & & 0 \\ \vdots & & & \vdots \\ 0 & 0 & \cdots & e^{\lambda_n t} \end{bmatrix}. \qquad \text{(12b)}$$

Using (12b) in (12a), we obtain

$$e^{At} = T\begin{bmatrix} e^{\lambda_1 t} & 0 & \cdots & 0 \\ 0 & e^{\lambda_2 t} & & 0 \\ \vdots & & & \vdots \\ 0 & 0 & \cdots & e^{\lambda_n t} \end{bmatrix}T^{-1}. \qquad \text{(12c)}$$

Decoupling Transformations

There is an alternative approach to solving $\mathbf{y}' = A\mathbf{y}$ when A is diagonalizable. This alternative involves making an appropriate change of dependent variable that transforms $\mathbf{y}' = A\mathbf{y}$ into a collection of decoupled scalar problems. This decoupled system can then be solved using the techniques of Chapters 2 and 3. Example 4 illustrates the ideas, and additional examples are considered in the Exercises.

EXAMPLE 4

Solve the initial value problem

$$\begin{aligned} y_1' &= \; y_1 + 2y_2 + 1, & y_1(0) &= 1 \\ y_2' &= 2y_1 + \; y_2 + t, & y_2(0) &= -1. \end{aligned} \qquad \text{(13)}$$

Solution: This problem has the form $\mathbf{y}' = A\mathbf{y} + \mathbf{g}(t)$, $\mathbf{y}(0) = \mathbf{y}_0$, where

$$A = \begin{bmatrix} 1 & 2 \\ 2 & 1 \end{bmatrix}, \qquad \mathbf{g}(t) = \begin{bmatrix} 1 \\ t \end{bmatrix}, \qquad \mathbf{y}_0 = \begin{bmatrix} 1 \\ -1 \end{bmatrix}.$$

From Example 3, we know that A is diagonalizable. In particular, $T^{-1}AT = D$, where

$$T^{-1} = \frac{1}{2}\begin{bmatrix} 1 & 1 \\ 1 & -1 \end{bmatrix}, \qquad T = \begin{bmatrix} 1 & 1 \\ 1 & -1 \end{bmatrix}, \qquad D = \begin{bmatrix} 3 & 0 \\ 0 & -1 \end{bmatrix}.$$

Let us make the change of variable $\mathbf{z}(t) = T^{-1}\mathbf{y}(t)$ or, equivalently, $\mathbf{y}(t) = T\mathbf{z}(t)$. Since $\mathbf{y}'(t) = T\mathbf{z}'(t)$, the equation $\mathbf{y}' = A\mathbf{y} + \mathbf{g}(t)$ transforms into

$$T\mathbf{z}' = AT\mathbf{z} + \mathbf{g}(t),$$

or

$$\mathbf{z}' = T^{-1}AT\mathbf{z} + T^{-1}\mathbf{g}(t).$$

The initial condition for the transformed system is $\mathbf{z}(0) = T^{-1}\mathbf{y}(0) = T^{-1}\mathbf{y}_0$. Since $T^{-1}AT = D$, the original problem (13) has become $\mathbf{z}' = D\mathbf{z} + T^{-1}\mathbf{g}(t)$, $\mathbf{z}(0) = T^{-1}\mathbf{y}_0$, or

$$\mathbf{z}' = \begin{bmatrix} 3 & 0 \\ 0 & -1 \end{bmatrix}\mathbf{z} + \frac{1}{2}\begin{bmatrix} 1 & 1 \\ 1 & -1 \end{bmatrix}\begin{bmatrix} 1 \\ t \end{bmatrix}, \qquad \mathbf{z}(0) = \frac{1}{2}\begin{bmatrix} 1 & 1 \\ 1 & -1 \end{bmatrix}\begin{bmatrix} 1 \\ -1 \end{bmatrix}.$$

In terms of components, this system has the form

$$\begin{aligned} z_1' &= 3z_1 && + (1+t)/2, & z_1(0) &= 0 \\ z_2' &= && -z_2 + (1-t)/2, & z_2(0) &= 1. \end{aligned} \tag{14}$$

As can be seen, the system (14) is an uncoupled system of first order linear equations of the type studied in Chapter 2. The solutions are

$$z_1(t) = (4e^{3t} - 3t - 4)/18$$
$$z_2(t) = (2 - t)/2.$$

In terms of the original variables, $\mathbf{y}(t) = T\mathbf{z}(t)$, or

$$\mathbf{y}(t) = \begin{bmatrix} 1 & 1 \\ 1 & -1 \end{bmatrix}\mathbf{z}(t) = \frac{1}{18}\begin{bmatrix} 4e^{3t} - 12t + 14 \\ 4e^{3t} + 6t - 22 \end{bmatrix}. \qquad ❖$$

A Warning

Some properties of the scalar exponential function e^{at} generalize to the exponential matrix e^{At}; properties (7a) and (7b) are two such examples. There are limits, however, to the extent to which the scalar properties generalize to e^{At}. If A and B are $(n \times n)$ matrices, then it is generally the case that

$$e^{At}e^{Bt} \neq e^{Bt}e^{At} \quad \text{and} \quad e^{At}e^{Bt} \neq e^{(A+B)t}. \tag{15}$$

The reason the expected results do not materialize is that matrix multiplication is not commutative. That is, it is usually the case that

$$AB \neq BA.$$

EXERCISES

Exercises 1–10:

The given matrix A is diagonalizable.

(a) Find T and D such that $T^{-1}AT = D$.

(b) Using (12c), determine the exponential matrix e^{At}.

1. $A = \begin{bmatrix} 5 & -6 \\ 3 & -4 \end{bmatrix}$ **2.** $A = \begin{bmatrix} 3 & 4 \\ -2 & -3 \end{bmatrix}$ **3.** $A = \begin{bmatrix} 1 & 1 \\ 1 & 1 \end{bmatrix}$ **4.** $A = \begin{bmatrix} 2 & 3 \\ 2 & 3 \end{bmatrix}$

5. $A = \begin{bmatrix} 2 & 3 \\ 3 & 2 \end{bmatrix}$ **6.** $A = \begin{bmatrix} 1 & -2 \\ -2 & 1 \end{bmatrix}$ **7.** $A = \begin{bmatrix} 2 & 0 \\ 1 & 1 \end{bmatrix}$ **8.** $A = \begin{bmatrix} -2 & 2 \\ 0 & 3 \end{bmatrix}$

9. $A = \begin{bmatrix} 0 & 2 \\ -2 & 0 \end{bmatrix}$ **10.** $A = \begin{bmatrix} 1 & 2 \\ -2 & 1 \end{bmatrix}$

Exercises 11–16:

In each exercise, the coefficient matrix A of the given linear system has a full set of eigenvectors and is therefore diagonalizable.

(a) As in Example 4, make the change of variables $\mathbf{z}(t) = T^{-1}\mathbf{y}(t)$, where $T^{-1}AT = D$. Reformulate the given problem as a set of uncoupled problems.

(b) Solve the uncoupled system in part (a) for $\mathbf{z}(t)$, and then form $\mathbf{y}(t) = T\mathbf{z}(t)$ to obtain the solution of the original problem.

11. $\mathbf{y}' = \begin{bmatrix} 6 & -6 \\ 2 & -1 \end{bmatrix} \mathbf{y}, \quad \mathbf{y}(0) = \begin{bmatrix} 1 \\ -3 \end{bmatrix}$ **12.** $\mathbf{y}' = \begin{bmatrix} 1 & 1 \\ 2 & 2 \end{bmatrix} \mathbf{y} + \begin{bmatrix} 0 \\ 3 \end{bmatrix}$

13. $\mathbf{y}' = \begin{bmatrix} -4 & -6 \\ 3 & 5 \end{bmatrix} \mathbf{y} + \begin{bmatrix} e^{2t} \\ -e^{2t} \end{bmatrix}, \quad \mathbf{y}(0) = \begin{bmatrix} 0 \\ 0 \end{bmatrix}$

14. $y_1' = 3y_1 + 2y_2 + 4$
$ y_2' = y_1 + 4y_2 + 1$

15. $y_1' = -9y_1 - 5y_2, \qquad\qquad y_1(0) = 1$
$ y_2' = 8y_1 + 4y_2 + 1, \quad y_2(0) = 0$
$ y_3' = y_3 + 2, \quad y_3(0) = 0$

16. $\begin{bmatrix} 3 & 2 \\ -4 & -3 \end{bmatrix} \mathbf{y}' + \mathbf{y} = \begin{bmatrix} 1 \\ 1 \end{bmatrix}$

17. Consider the differential equation $\mathbf{y}' = \begin{bmatrix} 2 & 1 \\ 0 & 2 \end{bmatrix} \mathbf{y}$. Example 2 shows that the corresponding exponential matrix is $e^{At} = \begin{bmatrix} e^{2t} & te^{2t} \\ 0 & e^{2t} \end{bmatrix}$. Suppose that $\mathbf{y}(1) = \begin{bmatrix} 1 \\ 2 \end{bmatrix}$. Use the propagator property (8) to determine $\mathbf{y}(4)$ and $\mathbf{y}(-1)$.

18. Determine by inspection whether or not the matrix is diagonalizable. Give a reason that supports your conclusion.

(a) $A_1 = \begin{bmatrix} 1 & 1 \\ 0 & 1 \end{bmatrix}$ (b) $A_2 = \begin{bmatrix} 1 & 1 \\ 0 & -1 \end{bmatrix}$ (c) $A_3 = \begin{bmatrix} 1 & 1 \\ 1 & 1 \end{bmatrix}$

19. Let A be a constant $(n \times n)$ diagonalizable matrix. Use the representation (12c) to establish properties (7a) and (7b). That is, show that $e^{At_1}e^{At_2} = e^{A(t_1+t_2)}$ and $(e^{At})^{-1} = e^{-At}$.

20. Use property (7a) to establish the propagator property (8). That is, show that $\mathbf{y}(t + \Delta t) = e^{A\Delta t}\mathbf{y}(t)$.

21. Let $A = \begin{bmatrix} \lambda & 1 \\ 0 & \lambda \end{bmatrix}$, and let $E = \begin{bmatrix} 0 & 1 \\ 0 & 0 \end{bmatrix}$. Use mathematical induction or the binomial formula to show that $A^m = \lambda^m I + m\lambda^{m-1}E$.

22. Consider the differential equation $\mathbf{y}' = A\mathbf{y}$, where $A = \begin{bmatrix} \lambda & 1 & 0 \\ 0 & \lambda & 1 \\ 0 & 0 & \lambda \end{bmatrix}$. Using the infinite

series (5), show that $e^{At} = \begin{bmatrix} e^{\lambda t} & te^{\lambda t} & \frac{1}{2}t^2 e^{\lambda t} \\ 0 & e^{\lambda t} & te^{\lambda t} \\ 0 & 0 & e^{\lambda t} \end{bmatrix}$.

Exercises 23–24:

In each exercise,

(a) Does $AB = BA$?

(b) Calculate the exponential matrices e^{At}, e^{Bt}, and $e^{(A+B)t}$. Does $e^{At}e^{Bt} = e^{(A+B)t}$?

23. $A = \begin{bmatrix} 1 & 1 \\ 0 & -1 \end{bmatrix}$, $B = \begin{bmatrix} 1 & 0 \\ 0 & -1 \end{bmatrix}$ **24.** $A = \begin{bmatrix} 2 & -1 \\ -1 & 2 \end{bmatrix}$, $B = \begin{bmatrix} 1 & 1 \\ 1 & 1 \end{bmatrix}$

Exercises 25–30:

Second Order Linear Systems   We consider systems of second order linear equations. Such systems arise, for instance, when Newton's laws are used to model the motion of coupled spring-mass systems, such as those in Exercises 31–32. In each of Exercises 25–30, let $A = \begin{bmatrix} 2 & 1 \\ 1 & 2 \end{bmatrix}$. Note that the eigenpairs of A are $\lambda_1 = 3, \mathbf{x}_1 = \begin{bmatrix} 1 \\ 1 \end{bmatrix}$ and $\lambda_2 = 1, \mathbf{x}_2 = \begin{bmatrix} 1 \\ -1 \end{bmatrix}$.

(a) Let $T = [\mathbf{x}_1, \mathbf{x}_2]$ denote the matrix of eigenvectors that diagonalizes A. Make the change of variable $\mathbf{z}(t) = T^{-1}\mathbf{y}(t)$, and reformulate the given problem as a set of uncoupled second order linear problems.

(b) Solve the uncoupled problem for $\mathbf{z}(t)$, and then form $\mathbf{y}(t) = T\mathbf{z}(t)$ to solve the original problem.

25. $\mathbf{y}'' + A\mathbf{y} = \mathbf{0}$

26. $y_1'' - 2y_1 - y_2 = 0$, $y_1(0) = 0$, $y_1'(0) = 1$
$y_2'' - y_1 - 2y_2 = 0$, $y_2(0) = 0$, $y_2'(0) = -1$

27. $\mathbf{y}'' + A\mathbf{y} = \begin{bmatrix} 1 \\ 0 \end{bmatrix}$, $\mathbf{y}(0) = \begin{bmatrix} 1 \\ 0 \end{bmatrix}$, $\mathbf{y}'(0) = \begin{bmatrix} 0 \\ 1 \end{bmatrix}$

28. $\mathbf{y}'' + \mathbf{y}' + A\mathbf{y} = \mathbf{0}$ **29.** $y_1'' + 4y_1' + 2y_2' = 0$
$y_2'' + 2y_1' + 4y_2' = 1$

30. $A\mathbf{y}'' + 4\mathbf{y} = \mathbf{0}$

Exercises 31–32:

Consider the spring-mass system shown in the figure on the next page. The system can execute one-dimensional motion on the frictionless horizontal surface. The unperturbed and perturbed systems are labeled (a) and (b) respectively.

31. (a) Show that an application of Newton's second law of motion leads the second order system $M\mathbf{x}'' + K\mathbf{x} = \mathbf{0}$, where

$$M = \begin{bmatrix} m_1 & 0 & 0 \\ 0 & m_2 & 0 \\ 0 & 0 & m_3 \end{bmatrix}, \quad K = \begin{bmatrix} k_1 & -k_1 & 0 \\ -k_1 & k_1 + k_2 & -k_2 \\ 0 & -k_2 & k_2 \end{bmatrix}, \quad \mathbf{x}(t) = \begin{bmatrix} x_1(t) \\ x_2(t) \\ x_3(t) \end{bmatrix}.$$

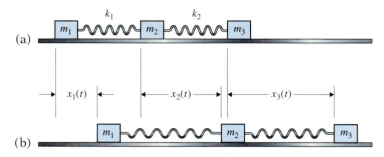

Figure for Exercises 31–32

(b) Let $m_1 = m_2 = m_3 = m$ and $k_1 = k_2 = k$. Determine the eigenpairs of $A = M^{-1}K$.

(c) Obtain the general solution of $\mathbf{x}'' + A\mathbf{x} = \mathbf{0}$.

32. Consider the second order linear system derived in part (a) of Exercise 31.

(a) Show that the matrix K has an eigenvalue $\lambda = 0$. Determine a corresponding eigenvector and denote it as $\mathbf{v}_0$.

(b) Explain the physical significance of the eigenpair $(0, \mathbf{v}_0)$. In particular, what motion does the system execute if the initial conditions are $\mathbf{x}(0) = \mathbf{0}$, $\mathbf{x}'(0) = \mathbf{v}_0$? [Hint: Look for a solution of the form $\mathbf{x}(t) = f(t)\mathbf{v}_0$, where $f(t)$ is a scalar function to be determined.] Describe, in words, how the system is behaving.

33. We know that similar matrices have the same eigenvalues (in fact, they have the same characteristic polynomial). There are many examples that show the converse is not true; that is, there are examples of matrices A and B that have the same characteristic polynomial but are not similar. Show that the following matrices A and B cannot be similar:

$$A = \begin{bmatrix} 1 & 0 \\ 3 & 1 \end{bmatrix} \quad \text{and} \quad B = \begin{bmatrix} 1 & 0 \\ 0 & 1 \end{bmatrix}.$$

34. Drawing on the ideas involved in working Exercise 33, show that if an $(n \times n)$ real matrix A is similar to the $(n \times n)$ identity I, then $A = I$.

35. Give an example that shows that while similar matrices have the same eigenvalues, they may not have the same eigenvectors.

36. Define matrices $P(t)$ and $Q(t)$ as follows:

$$P(t) = \begin{bmatrix} 1 & \cos t \\ 2t & 0 \end{bmatrix}, \qquad Q(t) = \int_0^t P(s)\,ds.$$

Show that $P(t)$ and its derivative $Q(t)$ do not commute. That is, $P(t)Q(t) \neq Q(t)P(t)$.

CHAPTER 4 REVIEW EXERCISES

These review exercises provide you with an opportunity to test your understanding of the concepts and solution techniques developed in this chapter. The end-of-section exercises deal with the topics discussed in the section. These review exercises, however, require you to identify an appropriate solution technique before solving the problem.

Exercises 1–22:

In each exercise, determine the general solution. If initial conditions are given, solve the initial value problem.

1. $\mathbf{y'} = \begin{bmatrix} 3 & -4 \\ 2 & -3 \end{bmatrix} \mathbf{y}$

2. $\mathbf{y'} = \begin{bmatrix} -7 & 6 \\ -9 & 8 \end{bmatrix} \mathbf{y}$

3. $\mathbf{y'} = \begin{bmatrix} 0 & -4 \\ 4 & 0 \end{bmatrix} \mathbf{y}$

4. $\mathbf{y'} = \begin{bmatrix} -2 & -1 \\ 1 & -4 \end{bmatrix} \mathbf{y}$

5. $y_1' = 3y_1 - y_2$
   $y_2' = y_1 + y_2$

6. $y_1' = -2y_1 + y_2$
   $y_2' = -y_1 - 2y_2$

7. $\mathbf{y'} = \begin{bmatrix} 0 & 2 \\ -1 & 3 \end{bmatrix} \mathbf{y} + \begin{bmatrix} 6 \\ 7 \end{bmatrix}$

8. $\mathbf{y'} = \begin{bmatrix} 3 & -2 \\ 2 & -1 \end{bmatrix} \mathbf{y}, \quad \mathbf{y}(0) = \begin{bmatrix} 4 \\ 2 \end{bmatrix}$

9. $y_1' = 2y_1 + y_2 + y_3$
   $y_2' = y_1 + 2y_2 + y_3$
   $y_3' = y_1 + y_2 + 2y_3$

10. $y_1' = 2y_1 + 2y_2$
    $y_2' = 5y_1 + 5y_2$
    $y_3' = \qquad 2y_3$

11. $y_1' = -4y_1 + 6y_2, \quad y_1(0) = 7$
    $y_2' = -3y_1 + 5y_2, \quad y_2(0) = 5$

12. $\mathbf{y'} = \begin{bmatrix} 3 & -5 \\ 1 & -1 \end{bmatrix} \mathbf{y} + \begin{bmatrix} 1 \\ 1 \end{bmatrix}$

13. $\mathbf{y'} = \begin{bmatrix} -6 & 10 \\ -4 & 6 \end{bmatrix} \mathbf{y}, \quad \mathbf{y}(0) = \begin{bmatrix} 5 \\ 5 \end{bmatrix}$

14. $\mathbf{y'} = \begin{bmatrix} -2 & -1 \\ 2 & -4 \end{bmatrix} \mathbf{y}$

15. $\mathbf{y'} = \begin{bmatrix} 2 & -1 & 1 \\ -1 & 2 & 1 \\ 1 & 1 & 2 \end{bmatrix} \mathbf{y}$

16. $\mathbf{y'} = \begin{bmatrix} 1 & 1 & 0 \\ 0 & 1 & 0 \\ 0 & 0 & 1 \end{bmatrix} \mathbf{y}$

17. $\mathbf{y'} = \begin{bmatrix} 1 & 4 \\ -1 & 5 \end{bmatrix} \mathbf{y} + \begin{bmatrix} 7 \\ 11 \end{bmatrix}$

18. $\mathbf{y'} = \begin{bmatrix} 7 & -5 \\ 10 & -7 \end{bmatrix} \mathbf{y}$

19. $\mathbf{y'} = \begin{bmatrix} 0 & 4 \\ -1 & 4 \end{bmatrix} \mathbf{y}$

20. $\mathbf{y'} = \begin{bmatrix} 9 & 20 \\ -4 & -9 \end{bmatrix} \mathbf{y}$

21. $\mathbf{y'} = \begin{bmatrix} -2 & 1 \\ -1 & 0 \end{bmatrix} \mathbf{y}$

22. $\mathbf{y'} = \begin{bmatrix} 11 & -3 \\ 30 & -8 \end{bmatrix} \mathbf{y}$

PROJECTS

Consider the configuration shown in Figure 4.14. The three identical springs of un-stretched length l are assumed to be weightless, and the two identical masses are as-

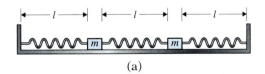

(a)

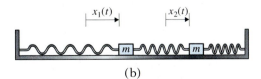

(b)

FIGURE 4.14

A coupled spring-mass system. (a) The equilibrium state.
(b) The perturbed state.

sumed to slide on a frictionless surface. In the vertical direction, the surface exerts a normal force upon each mass equal and opposite to its weight. Therefore, we need only consider motion in the horizontal direction.

Project 1: Derivation of the Equations of Motion

Assume the system is set in motion at time $t = 0$ and there are no externally applied forces. Let $x_1(t)$ and $x_2(t)$ represent the respective horizontal displacements of the two masses from their equilibrium positions, measured positive to the right as shown. Show that the application of Newton's second law of motion to each mass leads to the system of equations

$$\begin{aligned} mx_1'' &= k\,(x_2 - x_1) - kx_1 \\ mx_2'' &= -k\,(x_2 - x_1) - kx_2, \qquad t > 0. \end{aligned}$$

These equations can be rewritten as a second order linear system

$$m\mathbf{x}'' = k \begin{bmatrix} -2 & 1 \\ 1 & -2 \end{bmatrix} \mathbf{x}, \qquad t > 0, \qquad \text{where} \quad \mathbf{x}(t) = \begin{bmatrix} x_1(t) \\ x_2(t) \end{bmatrix}. \tag{1}$$

In addition to the equations of motion (1), we specify the initial position and velocity of each mass by

$$\mathbf{x}(0) = \mathbf{x}_0, \qquad \mathbf{x}'(0) = \mathbf{x}_0'. \tag{2}$$

Project 2: Numerical Solution Using the Exponential Matrix

The initial value problem defined by equation (1) and initial condition (2) can be solved by using the diagonalization techniques of Section 4.10 to transform the problem into two decoupled scalar problems. In this project, however, we will solve the problem by using the exponential matrix to propagate the solution forward in time. Our solution will be numerical in the sense that we will tabulate the solution at discrete times.

The first task is to recast the problem as a first order system. Define

$$y_1 = x_1, \qquad y_2 = x_1', \qquad y_3 = x_2, \qquad y_4 = x_2'.$$

With this, the initial value problem can be rewritten as

$$\mathbf{y}' = A\mathbf{y}, \qquad \mathbf{y}(0) = \mathbf{y}_0, \tag{3}$$

where A is a (4×4) constant matrix and $\mathbf{y}_0$ is a (4×1) vector.

1. Determine A and $\mathbf{y}_0$.
2. Suppose we want to solve (3) on the interval $0 \le t \le T$. Choose a time step $\Delta t = T/N$, where N is a positive integer. The solution of (3) can be tabulated in $0 \le t \le T$ at the time steps $t_j = j\,\Delta t, j = 0, 1, 2, \ldots, N$, using the iteration

$$\mathbf{y}(t_j) = e^{A\Delta t}\mathbf{y}(t_{j-1}), \qquad j = 1, 2, \ldots, N. \tag{4}$$

[The iteration (4) comes from the propagator property of the exponential matrix, $\mathbf{y}(t + \Delta t) = e^{A\Delta t}\mathbf{y}(t)$; recall equation (8) in Section 4.10.] Assume $m = 2$ kg, $k = 8$ N/m, $l = 1$ m, $T = 10$ sec, and $\Delta t = 0.05$ sec. For the initial conditions

$$\mathbf{x}_0 = \begin{bmatrix} -0.25 \\ 0.25 \end{bmatrix} \text{ m}, \qquad \mathbf{x}_0' = \begin{bmatrix} 0 \\ 0 \end{bmatrix} \text{ m/sec},$$

find $\mathbf{y}(t_j)$ using iteration (4), and plot the spring displacements at times $t_0, t_1, \ldots, t_N$. Interpret your results physically, describing how the two masses are moving in the time interval $0 \le t \le T$. [Software such as MATLAB has built-in exponential matrix routines you can call to form $e^{A\Delta t}$. Or you can simply solve the initial value problem $\Phi' = A\Phi$, $\Phi(0) = I$ and use the fact that $\Phi(\Delta t) = e^{A\Delta t}$.]

3. Repeat part (2), but with the initial conditions

$$\mathbf{x}_0 = \begin{bmatrix} 0 \\ 0 \end{bmatrix} \text{ m,} \qquad \mathbf{x}_0' = \begin{bmatrix} 0.25 \\ 0.25 \end{bmatrix} \text{ m/sec.}$$

Project 3: Resonant Behavior of a Coupled Spring-Mass System

Consider the spring-mass system shown in Figure 4.15. The figure shows two springs of negligible weight. The springs have unstretched lengths l_1 and l_2 and have spring constants k_1 and k_2. Masses m_1 and m_2 are attached, and the springs stretch appropriately to achieve the rest configuration shown. The amount of stretching done by each spring is determined by imposing conditions of static equilibrium on each mass. Since the sum of the vertical forces acting on each mass must vanish, we obtain $k_1 Y_1 = m_1 g + k_2 Y_2$ and $m_2 g = k_2 Y_2$. Therefore, $k_1 Y_1 = m_1 g + m_2 g$, where g represents acceleration due to gravity. When the system is disturbed from its equilibrium state, both masses move vertically. Let $y_1(t)$ and $y_2(t)$ represent the displacements of the masses from their equilibrium positions at time t, as shown in Figure 4.15. We assume the system is initially at rest and is set into motion by a force $f(t) = F \sin \omega t$, applied vertically to the mass m_2; see Figure 4.15(c).

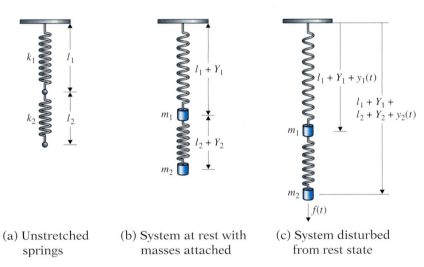

(a) Unstretched   (b) System at rest with   (c) System disturbed
    springs   masses attached   from rest state

FIGURE 4.15

1. Show that Newton's second law of motion leads to the following nonhomogeneous system of second order differential equations:

$$\begin{aligned} m_1 y_1'' &= -(k_1 + k_2)y_1 + k_2 y_2 \\ m_2 y_2'' &= k_2 y_1 - k_2 y_2 + F \sin \omega t. \end{aligned} \tag{5}$$

Since the system is initially at rest, the initial value problem is

$$\mathbf{y}'' + A\mathbf{y} = \mathbf{b} \sin \omega t, \qquad \mathbf{y}(0) = \mathbf{0}, \qquad \mathbf{y}'(0) = \mathbf{0}. \tag{6}$$

Determine the constant matrix A and the constant vector $\mathbf{b}$.

2. Assume that $m_1 = m_2 = m$ and that $k_1 = 2k$ and $k_2 = k$. Denote the eigenpairs of the real symmetric matrix A by $(\lambda_1, \mathbf{x}_1)$ and $(\lambda_2, \mathbf{x}_2)$. Make the change of variable

$\mathbf{z}(t) = T^{-1}\mathbf{y}(t)$, where $T = [\mathbf{x}_1, \mathbf{x}_2]$, and reformulate problem (6) as a pair of decoupled second order initial value problems.

3. Determine the resonant frequencies of the pair of initial value problems derived in part (2). That is, for what values of ω will at least one component of $\mathbf{z}(t)$ have an amplitude that grows linearly with time?

4. Solve the initial value problem derived in part (2) in the case where $m = 1$ kg, $k = 1$ N/m, $l = 1$ m, $F = 0.1$ N, and ω is equal to the largest of the values determined in part (3). Form $\mathbf{y}(t) = T\mathbf{z}(t)$ to obtain the solution of equation (6). Discuss the physical relevance of the solution. Is it physically meaningful for all $t, 0 \le t < \infty$?

Project 4: A Control Problem in Charged Particle Ballistics

Consider a particle, having mass m and electric charge q, moving in a magnetic field. The magnetic field is a vector field $\mathbf{B}$. The motion of the particle in this magnetic field is most conveniently described in terms of vectors.

Let $\mathbf{i}$, $\mathbf{j}$, and $\mathbf{k}$ represent unit vectors in the x, y, and z directions, respectively. The position of the particle is defined by the position vector,

$$\mathbf{r}(t) = x(t)\mathbf{i} + y(t)\mathbf{j} + z(t)\mathbf{k},$$

and its velocity by the corresponding velocity vector,

$$\mathbf{v}(t) = v_x(t)\mathbf{i} + v_y(t)\mathbf{j} + v_z(t)\mathbf{k}.$$

Since $\mathbf{v}(t) = d\mathbf{r}/dt$, the velocity components are

$$v_x(t) = \frac{d}{dt}x(t), \qquad v_y(t) = \frac{d}{dt}y(t), \qquad v_z(t) = \frac{d}{dt}z(t).$$

Figure 4.16 illustrates the problem.

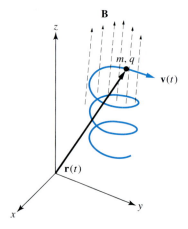

FIGURE 4.16

A charged particle having mass m and electric charge q moves in the magnetic field $\mathbf{B}$. Its motion is described by its position vector $\mathbf{r}(t)$ and velocity vector $\mathbf{v}(t)$.

If the weight of the charged particle is neglected, the only force acting on it is the Lorentz[2] force (the force the magnetic field exerts on the moving charge). The Lorentz force, described using vector notation, is $q\mathbf{v}(t) \times \mathbf{B}$. An application of Newton's second law of motion leads to the vector equation

$$m \frac{d}{dt} \mathbf{v}(t) = q\mathbf{v}(t) \times \mathbf{B}. \tag{7}$$

Suppose the charged particle is launched from the origin at time $t = 0$ with initial velocity $\mathbf{v}(0) = \mathbf{v}_0$. Is it possible to select $\mathbf{v}_0$ so that the particle will be located at a desired location $\mathbf{R}$ at a specific later time $t = \tau > 0$? In other words, can we choose $\mathbf{v}(0) = \mathbf{v}_0$ so that $\mathbf{r}(\tau) = \mathbf{R}$?

1. Let $\mathbf{B} = B\mathbf{k}$ and define $\omega_c = qB/m$. The constant ω_c is a radian frequency known as the **cyclotron frequency**. Rewrite vector equation (7) in the form

$$\mathbf{v}' = A\mathbf{v}, \qquad \mathbf{v}(0) = \mathbf{v}_0.$$

2. Solve the initial value problem, expressing the solution in the form $\mathbf{v}(t) = e^{At}\mathbf{v}_0$.

3. Form $\mathbf{r}(t) = \int_0^t \mathbf{v}(s)\, ds$ and determine conditions under which we can choose $\mathbf{v}_0$ such that $\mathbf{r}(\tau) = \mathbf{R} = \int_0^\tau \mathbf{v}(s)\, ds$.

---

[2]Hendrik Lorentz (1853–1928) was professor of mathematical physics at Leiden University from 1878 until 1912; he thereafter directed research at the Teyler Institute in Haarlem. Lorentz is noted for his studies of atomic structure and of the mathematical transformations (called Lorentz transformations) that form the basis of Einstein's theory of special relativity. Along with his student Pieter Zeeman, Lorentz was awarded the Nobel Prize in 1902.

Laplace Transforms

5.1   Introduction

When you begin to study a new topic such as the Laplace transform, two questions arise: "What is it?" and "Why is it important?" A scientist often uses Laplace[1] transforms to solve a mathematical problem for the same reason that a motorist leaves a congested highway and travels a network of back roads to reach his or her destination. The most easily traveled path between two points is not always the most direct one.

One of the important applications of Laplace transforms is solving constant coefficient linear differential equations that have discontinuous right-hand sides. In particular, many mechanical and electrical systems are driven by sources that switch on and off. Such systems are often modeled by an initial value problem of the form

$$y'' + ay' + by = g(t), \qquad y(t_0) = y_0, \qquad y'(t_0) = y'_0,$$

---

[1] Pierre-Simon Laplace (1749–1827) was a French scientist renowned for his work in mathematics, celestial mechanics, probability theory, and physics. The Laplace transform, the Laplace probability density function, and Laplace's equation (arising in the study of potential theory) are some mathematical entities named in his honor.

where the right-hand side $g(t)$ has discontinuities at those times when the source changes abruptly. Laplace transform techniques are a convenient tool for analyzing such initial value problems.

The philosophy underlying the use of Laplace transforms is illustrated in Figure 5.1. We have a problem to solve—for example, determining the behavior of some mechanical or electrical system. Instead of attacking the problem directly, we transform (or map) the original problem into a new problem. This mapping is accomplished by means of the mathematical operation known as the Laplace transform. The original problem is often referred to as the *time domain problem* since the independent variable for the original problem is usually time. The new problem, resulting from the Laplace transformation, is referred to as the *transform domain problem*. After obtaining this new transform domain problem, we solve it and then transform the solution back into the time domain by performing another mapping, known as the inverse Laplace transform. The inverse mapping thus gives us the solution of the original time domain problem, the problem of interest.

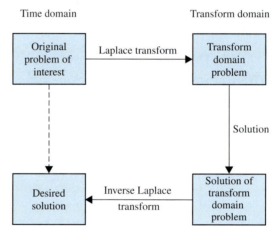

FIGURE 5.1

The philosophy underlying the use of Laplace transforms.

In order for the problem-solving approach diagrammed in Figure 5.1 to be attractive, the following three steps must be easier to implement than the direct solution approach:

1. calculating the Laplace transform,
2. solving the transformed problem, and
3. calculating the inverse Laplace transform.

For many of the problems we treat, this will be the case. Constant coefficient linear *differential* equations will be transformed into *algebraic* equations. The resulting transform domain problem typically entails solving a single linear algebraic equation or a system of linear algebraic equations.

We will consider a variety of constant coefficient linear differential equations (both scalar equations and systems) and show how these problems can be solved using Laplace transforms. We will also consider several problems, such as the parameter identification problem in the following example, that are not so straightforwardly solved using the techniques developed so far.

EXAMPLE

1

Consider a vibrating mechanical system that exists in a "black box," as in Figure 5.2. Assume you are confident that the vibrating system can be modeled as the spring-mass-dashpot arrangement shown, but you do not have the internal access needed to directly measure the spring constant k, the mass m, or the damping constant γ. You can only apply a force at the external access point and measure the resulting displacement.

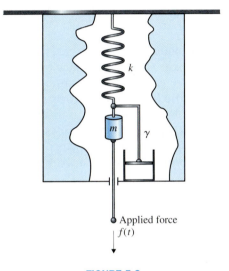

FIGURE 5.2

A cutaway schematic of a "black box" vibrating system.

Mathematically (as we saw in Section 3.10), the mechanical system in Figure 5.2 is described by the initial value problem,

$$my'' + \gamma y' + ky = f(t), \qquad t \geq 0, \qquad y(0) = 0, \qquad y'(0) = 0.$$

The system is at rest at time $t = 0$ when force $f(t)$ is applied. The applied force $f(t)$ is known for $t \geq 0$; the displacement $y(t)$ is monitored and is thus known for $t \geq 0$. The parameters m, γ, and k, however, are unknown.

Assuming we know the input-output relation [that is, $f(t)$ and $y(t)$] for one given applied force and the corresponding measured displacement, we ask the following two questions:

1. Is it possible to predict what the output will be if another input is applied to the system? In other words, if we apply a different external force, $\tilde{f}(t)$, is it possible to predict the resulting displacement, $\tilde{y}(t)$?

2. Is it possible to determine the constants m, γ, and k from a knowledge of the single input-output history given by $f(t)$ and $y(t)$?

We will see in Section 5.4 that the use of Laplace transforms provides a relatively easy way to answer both questions. ❖

The Laplace Transform

The first use of Laplace transforms as an operational tool for solving constant coefficient linear differential equations is often attributed to the British physicist Oliver Heaviside.[2] As noted earlier, the Laplace transform maps a function of t, say $f(t)$, into a new function, $F(s)$, of a new *transform variable s*. [In terms of notation, we generally use lowercase letters to designate time domain functions, such as $f(t)$, and capital letters to denote corresponding transform domain functions, $F(s)$.]

Let $f(t)$ be a function defined on the interval $0 \le t < \infty$. The **Laplace transform of $f(t)$** is defined by the improper integral

$$\mathcal{L}\{f(t)\} = F(s) = \int_0^\infty f(t)e^{-st}\,dt, \tag{1}$$

provided the integral exists. As the notation of equation (1) indicates, we often denote $F(s)$, the Laplace transform of $f(t)$, by the symbol $\mathcal{L}\{f(t)\}$. The new transform variable s is assumed to be a real variable. (In more advanced treatments of the Laplace transform, the variable s is allowed to be a complex variable.)

As we look at equation (1), the first issue to settle is that of identifying the properties $f(t)$ must possess in order for its Laplace transform to exist—that is, in order for the improper integral in equation (1) to converge. Recall from calculus that the improper integral in (1) is defined by

$$\int_0^\infty f(t)e^{-st}\,dt = \lim_{T\to\infty} \int_0^T f(t)e^{-st}\,dt, \tag{2}$$

provided the limit exists. When the limit (2) exists, we say that the improper integral **converges**, and we define the improper integral to be this limit value. If the limit in (2) does not exist, we say that the improper integral **diverges**.

Whether or not limit (2) exists depends on the nature of $f(t)$ and on the value of the transform variable s; note that s plays the role of a parameter in the integral. In this section, we identify a large class of functions that possess Laplace transforms. It is important to realize, however, that not every function has a Laplace transform. The third function considered in Example 2 illustrates this fact.

E X A M P L E

2

Find the Laplace transform, if it exists, of

(a) $f(t) = e^{at}$ (b) $f(t) = t$ (c) $f(t) = e^{t^2}$

Solution:

(a) Applying the definition, we see that

$$\mathcal{L}\{e^{at}\} = \int_0^\infty e^{at}e^{-st}\,dt = \lim_{T\to\infty} \int_0^T e^{-(s-a)t}\,dt, \tag{3}$$

---

[2]Oliver Heaviside (1850–1925) studied electricity and magnetism while employed as a telegrapher. He is remembered for his great simplification of Maxwell's equations, his contributions to vector analysis, and his development of operational calculus. In 1902, Heaviside conjectured that a conducting layer exists in the atmosphere which allows radio waves to follow the curvature of Earth. This layer, now called the Heaviside layer, was detected in 1923.

provided the limit exists. Since

$$\int_0^T e^{-(s-a)t}\,dt = \begin{cases} T, & s = a \\ \dfrac{1 - e^{-(s-a)T}}{s-a}, & s \neq a, \end{cases}$$

the improper integral defined by the limit (3) exists if and only if $s > a$. Therefore,

$$F(s) = \mathcal{L}\{e^{at}\} = \frac{1}{s-a}, \quad s > a. \tag{4}$$

(b) Similarly,

$$\mathcal{L}\{t\} = \int_0^\infty t e^{-st}\,dt = \lim_{T\to\infty} \int_0^T t e^{-st}\,dt, \tag{5}$$

provided the limit exists. Since

$$\int_0^T t e^{-st}\,dt = \begin{cases} \dfrac{T^2}{2}, & s = 0 \\ -\dfrac{e^{-sT}}{s^2}(1 + sT) + \dfrac{1}{s^2}, & s \neq 0, \end{cases}$$

the improper integral defined by the limit (5) exists if and only if $s > 0$. Therefore,

$$F(s) = \mathcal{L}\{t\} = \frac{1}{s^2}, \quad s > 0. \tag{6}$$

(c) Applying the definition gives

$$\mathcal{L}\{e^{t^2}\} = \int_0^\infty e^{t^2} e^{-st}\,dt = \lim_{T\to\infty} \int_0^T e^{t(t-s)}\,dt,$$

provided the limit exists. For any fixed value of s, however, the integrand, $e^{t(t-s)}$, is greater than 1 whenever $t > s$. Therefore, the limit does not exist for any value s and the function $f(t) = e^{t^2}$ does not possess a Laplace transform. ❖

Piecewise Continuous Functions and Exponentially Bounded Functions

We now identify a class of functions that possess Laplace transforms. If $f(t)$ is a member of this class, then its Laplace transform exists for all $s > a$, where a is a constant that depends on the particular function f.

We begin with two definitions. A function $f(t)$ is called **piecewise continuous on the interval** $0 \leq t \leq T$ if

(a) There are at most finitely many points, $0 \leq t_1 < t_2 < \cdots < t_{N_T} \leq T$, at which $f(t)$ is not continuous, and

(b) At any point of discontinuity, t_j, the one-sided limits

$$\lim_{t\to t_j^-} f(t) \quad \text{and} \quad \lim_{t\to t_j^+} f(t)$$

both exist. (If a discontinuity occurs at an endpoint, 0 or T, then only the interior one-sided limits need exist.) These discontinuities are called **jump discontinuities**.

Condition (b) says the only discontinuities allowed for a piecewise continuous function are jump discontinuities. Condition (a) says a function that is piecewise continuous on the interval $[0, T]$ never has more than a finite number of jump discontinuities in $[0, T]$. Note that the number of discontinuity points is a nondecreasing function of the interval length T; that is, if $T_2 > T_1$, then $N_{T_2} \geq N_{T_1}$.

A function defined on $0 \leq t < \infty$ is called **piecewise continuous on the interval** $0 \leq t < \infty$ if it is piecewise continuous on $0 \leq t \leq T$ for all $T > 0$.

An example of a function $f(t)$ that is piecewise continuous on $0 \leq t < \infty$ is the "square wave" shown in Figure 5.3, where

$$f(t) = \begin{cases} 1, & 0 \leq t \leq 1, \\ 0, & 1 < t < 2, \end{cases} \qquad f(t+2) = f(t).$$

Note that $f(t)$ is a periodic function with period 2. Every discontinuity of $f(t)$ is a jump discontinuity. While $f(t)$ has infinitely many discontinuities in $0 \leq t < \infty$, the function never has more than a finite number of discontinuities in a finite interval $0 \leq t \leq T$.

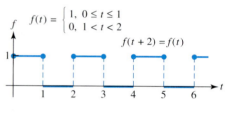

FIGURE 5.3

The function $f(t)$ is called a "square wave." It has infinitely many discontinuities in $0 \leq t < \infty$. They are all jump discontinuities, and there are never more than a finite number in any finite subinterval of $0 \leq t < \infty$. Therefore, $f(t)$ is piecewise continuous on $0 \leq t < \infty$.

Our next definition is concerned with measuring how fast $|f(t)|$ grows as $t \to \infty$. A function $f(t)$ defined on $0 \leq t < \infty$ is called **exponentially bounded** on $0 \leq t < \infty$ if there are constants M and a, with $M \geq 0$, such that

$$|f(t)| \leq Me^{at}, \qquad 0 \leq t < \infty.$$

Figure 5.4 illustrates the nature of this definition. A function $f(t)$ is exponentially bounded if we can find constants M and a such that the graph of $f(t)$ is contained in the region R, where R is bounded above by $y = Me^{at}$ and below by $y = -Me^{at}$.

If a function $f(t)$ is bounded on $0 \leq t < \infty$, then it is also exponentially bounded; that is, if $|f(t)| \leq M, 0 \leq t < \infty$, then $|f(t)| \leq Me^{at}, 0 \leq t < \infty$, with $a = 0$.

Existence of the Laplace Transform

Theorem 5.1 establishes the existence of the Laplace transform for all functions that are piecewise continuous and exponentially bounded on $0 \leq t < \infty$. The proof is given in advanced texts.

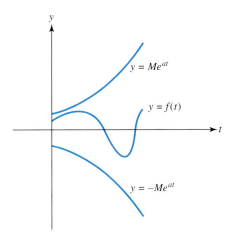

FIGURE 5.4

The function $f(t)$ is exponentially bounded because its graph is bounded above by $y = Me^{at}$ and below by $y = -Me^{at}$. Hence, $|f(t)| \le Me^{at}, 0 \le t < \infty$.

Theorem 5.1

Let $f(t)$ be piecewise continuous and exponentially bounded on $0 \le t < \infty$, where $|f(t)| \le Me^{at}, 0 \le t < \infty$. Then the Laplace transform,

$$F(s) = \int_0^\infty f(t)e^{-st}\,dt,$$

exists for all $s > a$.

In this chapter, we restrict our attention to functions that are piecewise continuous and exponentially bounded on $0 \le t < \infty$. The next theorem, stated without proof, gives some closure properties for this special class of functions, asserting that we can form linear combinations and products of functions in the class and that any new functions produced will also belong to the class (and thus have Laplace transforms).

Theorem 5.2

Let $f_1(t)$ and $f_2(t)$ be piecewise continuous and exponentially bounded on $0 \le t < \infty$, where

$$|f_1(t)| \le M_1 e^{a_1 t} \quad \text{and} \quad |f_2(t)| \le M_2 e^{a_2 t}.$$

(a) Let $f(t) = c_1 f_1(t) + c_2 f_2(t)$, where c_1 and c_2 are arbitrary constants. Then $f(t)$ is also piecewise continuous and exponentially bounded on $0 \le t < \infty$. In fact, $|f(t)| \le Me^{at}$, where $M = |c_1|M_1 + |c_2|M_2$ and $a = \max\{a_1, a_2\}$. Moreover, $F(s) = \mathcal{L}\{f(t)\}$ is given by

$$F(s) = \mathcal{L}\{c_1 f_1(t) + c_2 f_2(t)\} = c_1 \mathcal{L}\{f_1(t)\} + c_2 \mathcal{L}\{f_2(t)\}$$
$$= c_1 F_1(s) + c_2 F_2(s), \qquad s > a.$$

(b) Let $g(t) = f_1(t)f_2(t)$. Then $g(t)$ is also piecewise continuous and exponentially bounded on $0 \le t < \infty$. In fact, $|g(t)| \le Me^{at}$, where $M = M_1 M_2$ and $a = a_1 + a_2$. It follows that $G(s) = \mathcal{L}\{g(t)\}$ exists for $s > a$.

Since the Laplace transform satisfies the formula

$$\mathcal{L}\{c_1 f_1(t) + c_2 f_2(t)\} = c_1 \mathcal{L}\{f_1(t)\} + c_2 \mathcal{L}\{f_2(t)\}, \tag{7}$$

we say that the Laplace transform is a **linear transformation** on the set of piecewise continuous, exponentially bounded functions.

E X A M P L E

3

Determine whether the functions are exponentially bounded and piecewise continuous on $0 \le t < \infty$:

(a) $f(t) = \begin{cases} 1, & 0 \le t \le 1, \\ 0, & 1 < t < 2, \end{cases}$ $f(t) = f(t-2)$ for $t \ge 2$

(b) $g(t) = te^t$, $0 \le t < \infty$

(c) $k(t) = e^{t^2}$, $0 \le t < \infty$

Solution:

(a) The function $f(t)$ is defined on $0 \le t < \infty$ as a periodic function having period 2. This function, whose graph was shown earlier in Figure 5.3, has jump discontinuities at the positive integers. As was noted in Figure 5.3, $f(t)$ is piecewise continuous on $0 \le t < \infty$. Since the function is bounded, it is also exponentially bounded; in fact, we have

$$|f(t)| \le Me^{at}, \quad \text{with} \quad M = 1 \quad \text{and} \quad a = 0.$$

We will calculate the Laplace transform of $f(t)$ in Section 5.4 when we discuss Laplace transforms of periodic functions.

(b) Since $g(t)$ is continuous, it is certainly piecewise continuous on $0 \le t < \infty$. It remains to show that $g(t)$ is exponentially bounded. Let $\alpha > 1$, and consider the function

$$\varphi(t) = g(t)e^{-\alpha t} = te^{-(\alpha - 1)t}, \quad 0 \le t < \infty.$$

It can be shown that

$$0 \le \varphi(t) \le \frac{1}{(\alpha - 1)e}, \quad 0 \le t < \infty.$$

This inequality implies that

$$0 \le g(t) \le \frac{1}{(\alpha - 1)e}e^{\alpha t}, \quad 0 \le t < \infty,$$

and we conclude that $g(t)$ is exponentially bounded on $0 \le t < \infty$, with $M = 1/[(\alpha - 1)e]$ and $a = \alpha > 1$.

(c) If $k(t)$ were exponentially bounded, then there would be constants M and a such that $e^{t^2} \le Me^{at}$ for all nonnegative values t. In turn, this inequality would imply

$$e^{t(t-a)} \le M, \quad 0 \le t < \infty.$$

But, as t grows, the inequality has to fail eventually. Thus, $k(t) = e^{t^2}$ is not exponentially bounded. The function is, however, piecewise continuous since it is continuous. ❖

The Inverse Laplace Transform and Uniqueness

Using the Laplace transform to solve problems involves three separate steps: (1) applying the transform to obtain a new transform domain problem, (2) solving the new transform domain problem, and (3) applying the inverse transform that maps the transform domain solution back to the time domain, resulting in the solution of the problem of interest.

In order to define the inverse mapping (that is, the inverse Laplace transform), we need to know that the Laplace transform operation, when applied to functions that are piecewise continuous and exponentially bounded, possesses an underlying uniqueness property. In particular, given a transform domain function $F(s)$, we want unambiguously to identify a function $f(t)$ that has $F(s)$ as its transform. The following theorem, which we present without proof, addresses the uniqueness question.

Theorem 5.3

Let $f_1(t)$ and $f_2(t)$ be piecewise continuous and exponentially bounded on $0 \leq t < \infty$. Let $F_1(s)$ and $F_2(s)$ represent their respective Laplace transforms. Suppose, for some constant a, that

$$F_1(s) = F_2(s), \qquad s > a.$$

Then $f_1(t) = f_2(t)$ at all points $t \geq 0$ where both functions are continuous.

This theorem gives about the best result we can hope for. As an illustration, consider the function $f_1(t) = e^{-t}, t \geq 0$. We saw in Example 2 that

$$\mathcal{L}\{e^{at}\} = \frac{1}{s-a}, \qquad s > a.$$

Therefore,

$$\mathcal{L}\{e^{-t}\} = F(s) = \frac{1}{s+1}, \qquad s > -1.$$

Suppose we create a new function $f_2(t)$ by simply redefining $f_1(t)$ to be zero at each of the positive integers:

$$f_2(t) = \begin{cases} e^{-t}, & t \text{ not an integer} \\ 0, & t \text{ an integer.} \end{cases} \tag{8}$$

The graph of the function $f_2(t)$ is shown in Figure 5.5. Observe, even though $f_1(t)$ and $f_2(t)$ are different functions, that

$$\int_0^T f_1(t)e^{-st}\, dt = \int_0^T f_2(t)e^{-st}\, dt \tag{9}$$

for every $T > 0$. Therefore, $\mathcal{L}\{f_1(t)\} = \mathcal{L}\{f_2(t)\} = F(s), s > -1$.

As we see from equation (9), the improper integral defining the Laplace transform is insensitive to changes in the value of a function at a finite number of points in $0 \leq t \leq T$. This insensitivity, however, does not pose a serious

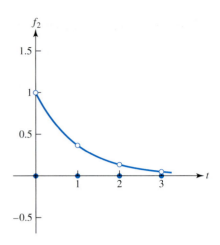

FIGURE 5.5

The graph of the function $f_2(t)$ defined by equation (8). Note that the graph of $f_2(t)$ is identical to the graph of $f_1(t) = e^{-t}$ except at $t = 0, 1, 2, \ldots$. Even though the functions $f_1(t)$ and $f_2(t)$ are different, their Laplace transforms are the same [see equation (9)].

practical problem since we are interested in physically relevant functions. For example, in defining the inverse Laplace transform of

$$F(s) = \frac{1}{s-a},$$

we will choose it to be the continuous function $f(t) = e^{at}, t \geq 0$.

Our approach to determining inverse Laplace transforms will be a tabular one. In the next several sections, we will compute Laplace transforms of functions and build up a library of **Laplace transform pairs**, such as the pair

$$f(t) = e^{at}, \quad t \geq 0 \quad \text{and} \quad F(s) = \frac{1}{s-a}, \quad s > a. \tag{10}$$

Determining an inverse Laplace transform will essentially consist of a simple "table look-up" process. That is, we find the appropriate transform domain function $F(s)$ in the table and then take the corresponding time domain function $f(t)$ in the table to be the inverse transform of $F(s)$. The next example illustrates this approach. (In more advanced treatments, within the theory of complex variables, a more fundamental approach to computing inverse Laplace transforms is developed.)

[Note: We will use the symbol $\mathcal{L}^{-1}\{\ \}$ to denote the operation of taking the inverse Laplace transform.]

E X A M P L E

4

What is the inverse Laplace transform of

$$F(s) = \frac{2s}{s^2 - 1}, \qquad s > 1?$$

That is, for what function $f(t)$ do we have $\mathcal{L}\{f(t)\} = F(s) = 2s/(s^2 - 1)$?

Solution: We first observe that the rational function $F(s)$ has the following partial fraction expansion:

$$\frac{2s}{s^2 - 1} = \frac{1}{s - 1} + \frac{1}{s + 1}.$$

(The topic of partial fractions is reviewed in Section 5.3. For now you can verify this claim by simply recombining the right-hand side.) Since the Laplace transform is a linear transformation, the inverse Laplace transform is likewise a linear transformation. In particular,

$$\mathcal{L}^{-1}\left\{\frac{2s}{s^2 - 1}\right\} = \mathcal{L}^{-1}\left\{\frac{1}{s - 1} + \frac{1}{s + 1}\right\} = \mathcal{L}^{-1}\left\{\frac{1}{s - 1}\right\} + \mathcal{L}^{-1}\left\{\frac{1}{s + 1}\right\}.$$

Recalling the Laplace transform pair listed earlier in equation (10), we obtain

$$\mathcal{L}^{-1}\left\{\frac{1}{s - 1}\right\} = e^t \quad \text{and} \quad \mathcal{L}^{-1}\left\{\frac{1}{s + 1}\right\} = e^{-t}.$$

Therefore,

$$\mathcal{L}^{-1}\left\{\frac{2s}{s^2 - 1}\right\} = e^t + e^{-t} = 2\cosh t, \qquad t \geq 0. \; \diamondsuit$$

EXERCISES

Exercises 1–12:

As in Example 2, use the definition to find the Laplace transform for $f(t)$, if it exists. In each exercise, the given function $f(t)$ is defined on the interval $0 \leq t < \infty$. If the Laplace transform exists, give the domain of $F(s)$. In Exercises 9–12, also sketch the graph of $f(t)$.

1. $f(t) = 1$ **2.** $f(t) = e^{3t}$ **3.** $f(t) = te^{-t}$ **4.** $f(t) = t - 5$

5. $f(t) = te^{t\sqrt{t}}$ **6.** $f(t) = e^{(t-1)^2}$ **7.** $f(t) = |t - 1|$ **8.** $f(t) = (t - 2)^2$

9. $f(t) = \begin{cases} 0, & 0 \leq t < 1 \\ 1, & 1 \leq t \end{cases}$ **10.** $f(t) = \begin{cases} 0, & 0 \leq t < 1 \\ t - 1, & 1 \leq t \end{cases}$

11. $f(t) = \begin{cases} 0, & 0 \leq t < 1 \\ 1, & 1 \leq t < 2 \\ 0, & 2 \leq t \end{cases}$ **12.** $f(t) = \begin{cases} 0, & 0 \leq t < 1 \\ t - 1, & 1 \leq t < 2 \\ 0, & 2 \leq t \end{cases}$

13. Let n be a positive integer. Using integration by parts, establish the reduction formula

$$\int t^n e^{-st}\, dt = -\frac{t^n e^{-st}}{s} + \frac{n}{s}\int t^{n-1} e^{-st}\, dt, \qquad s > 0.$$

14. For $s > 0$ and n a positive integer, evaluate the limits:

(a) $\lim\limits_{t \to 0} t^n e^{-st}$ (b) $\lim\limits_{t \to \infty} t^n e^{-st}$

[Hint: Use L'Hôpital's rule to establish limit (b).]

15. (a) Use Exercises 13 and 14 to derive a reduction formula for the Laplace transform of $f(t) = t^n$,

$$\mathcal{L}\{t^n\} = \frac{n}{s}\mathcal{L}\{t^{n-1}\}, \qquad s > 0. \tag{11}$$

(b) From Example 2, we have $\mathcal{L}\{t\} = 1/s^2, s > 0$. Use this fact, together with reduction formula (11), to calculate $\mathcal{L}\{t^k\}$ for $k = 2, 3, \ldots, 5$.

(c) Formulate a conjecture as to the Laplace transform of $f(t) = t^m$, where m is an arbitrary positive integer.

Exercises 16–21:

From a table of integrals,

$$\int e^{\alpha u} \sin \beta u \, du = e^{\alpha u} \frac{\alpha \sin \beta u - \beta \cos \beta u}{\alpha^2 + \beta^2}$$

$$\int e^{\alpha u} \cos \beta u \, du = e^{\alpha u} \frac{\alpha \cos \beta u + \beta \sin \beta u}{\alpha^2 + \beta^2}.$$

Use these integrals to find the Laplace transform of $f(t)$, if it exists. If the Laplace transform exists, give the domain of $F(s)$.

16. $f(t) = \cos \omega t$ **17.** $f(t) = \sin \omega t$ **18.** $f(t) = \cos[\omega(t-2)]$

19. $f(t) = \sin[\omega(t-2)]$ **20.** $f(t) = e^{3t} \sin t$ **21.** $f(t) = e^{-2t} \cos 4t$

Exercises 22–23:

Use the linearity property (7) along with the transforms found in Example 2,

$$\mathcal{L}\{e^{at}\} = \frac{1}{s-a}, \quad s > a \quad \text{and} \quad \mathcal{L}\{t\} = \frac{1}{s^2}, \quad s > 0,$$

to calculate the Laplace transform $R(s) = \mathcal{L}\{r(t)\}$ of the given function $r(t)$. For what values s does the Laplace transform exist?

22. $r(t) = 2e^{-5t} + 6t$ **23.** $r(t) = 5e^{-7t} + t + 2e^{2t}$

Exercises 24–31:

In each exercise, a function $f(t)$ is given. In Exercises 28 and 29, the symbol $[\![u]\!]$ denotes the **greatest integer function**, $[\![u]\!] = n$ when $n \leq u < n+1$, n an integer, $n = \ldots, -2, -1, 0, 1, 2, \ldots$.

(a) Is $f(t)$ continuous on $0 \leq t < \infty$, discontinuous but piecewise continuous on $0 \leq t < \infty$, or neither?

(b) Is $f(t)$ exponentially bounded on $0 \leq t < \infty$? If so, determine values of M and a such that $|f(t)| \leq Me^{at}, 0 \leq t < \infty$.

24. $f(t) = \tan t$ **25.** $f(t) = e^t \sin t$ **26.** $f(t) = t^2 e^{-t}$ **27.** $f(t) = \cosh 2t$

28. $f(t) = [\![t]\!]$ **29.** $f(t) = [\![e^{2t}]\!]$ **30.** $f(t) = \dfrac{e^{t^2}}{e^{2t} + 1}$ **31.** $f(t) = \dfrac{1}{t}$

Exercises 32–35:

Determine whether the given improper integral converges. If the integral converges, give its value.

32. $\displaystyle\int_0^\infty \frac{1}{1+t^2} \, dt$ **33.** $\displaystyle\int_0^\infty \frac{t}{1+t^2} \, dt$

34. $\displaystyle\int_0^\infty e^{-t} \cos(e^{-t}) \, dt$ **35.** $\displaystyle\int_0^\infty t e^{-t^2} \, dt$

Exercises 36–39:

Suppose that $\mathcal{L}\{f_1(t)\} = F_1(s)$ and $\mathcal{L}\{f_2(t)\} = F_2(s), s > a$. Use the fact that

$$\mathcal{L}^{-1}\{c_1 F_1(s) + c_2 F_2(s)\} = c_1 \mathcal{L}^{-1}\{F_1(s)\} + c_2 \mathcal{L}^{-1}\{F_2(s)\}, \quad a < s$$

to determine the inverse Laplace transform of the given function. Refer to the examples in this section and equation (11) in Exercise 15.

36. $F(s) = \dfrac{3}{s-2}$

37. $F(s) = -\dfrac{2}{s^2} + \dfrac{1}{s+1}$

38. $F(s) = \dfrac{4s}{s^2-4} = \dfrac{2}{s+2} + \dfrac{2}{s-2}$

39. $F(s) = \dfrac{2}{s^2-1} = \dfrac{1}{s-1} - \dfrac{1}{s+1}$

5.2 Laplace Transform Pairs

This section develops a library of Laplace transform pairs that we will use to solve problems. We begin by defining a function known as the *unit step function* or the *Heaviside step function*.

The Unit Step Function

The **unit step function** or **Heaviside step function**, $h(t)$, is the piecewise continuous function defined by

$$h(t) = \begin{cases} 1, & t \geq 0 \\ 0, & t < 0. \end{cases}$$

Figure 5.6 displays graphs of $h(t)$ and its "shifted argument" counterpart, $h(t - \alpha), \alpha > 0$.

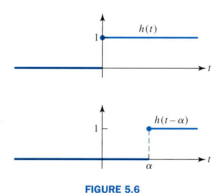

FIGURE 5.6

The graphs of the unit step function, $h(t)$, and the shifted step function, $h(t - \alpha)$.

The Laplace transform of the unit step function, $h(t)$, is given by

$$\mathcal{L}\{h(t)\} = \int_0^\infty h(t)e^{-st}\,dt = \int_0^\infty e^{-st}\,dt = -\frac{e^{-st}}{s}\Big|_0^\infty = \frac{1}{s}, \qquad s > 0. \qquad \textbf{(1a)}$$

In equation (1a), we use a common notation:

$$f(t)\Big|_a^\infty = \lim_{t \to \infty} f(t) - f(a).$$

For the shifted step function, $h(t - \alpha)$, we obtain the Laplace transform

$$\mathcal{L}\{h(t - \alpha)\} = \int_0^\infty h(t - \alpha)e^{-st}\,dt = \int_\alpha^\infty e^{-st}\,dt = -\frac{e^{-st}}{s}\bigg|_\alpha^\infty = \frac{e^{-s\alpha}}{s}, \tag{1b}$$

$$s > 0 \quad \text{and} \quad \alpha \geq 0.$$

Note that the unit step function $h(t)$ and the constant function $f(t) = 1$ are identical on $0 \leq t < \infty$. Therefore, they have the same Laplace transform,

$$\mathcal{L}\{1\} = \frac{1}{s}, \qquad s > 0. \tag{1c}$$

Transforms of Polynomial, Exponential, and Trigonometric Functions

In this subsection, we develop some common transform pairs, starting with the polynomial function $f(t) = t^n$. Also see Exercise 15 in Section 5.1.

The Laplace Transform of f(t) = t^n   For $n = 1$, we use integration by parts to obtain

$$\mathcal{L}\{t\} = \int_0^\infty te^{-st}\,dt = \left[-\frac{te^{-st}}{s} - \frac{e^{-st}}{s^2}\right]\bigg|_0^\infty = \frac{1}{s^2}, \qquad s > 0. \tag{2a}$$

In general, for any positive integer n, integration by parts yields

$$\mathcal{L}\{t^n\} = \int_0^\infty t^n e^{-st}\,dt = -\frac{t^n e^{-st}}{s}\bigg|_0^\infty + \frac{n}{s}\int_0^\infty t^{n-1}e^{-st}\,dt.$$

You can use L'Hôpital's rule to show that $\lim_{t \to \infty} t^n e^{-st} = 0, s > 0$. Therefore, we obtain the following reduction formula for $\mathcal{L}\{t^n\}$:

$$\mathcal{L}\{t^n\} = \frac{n}{s}\mathcal{L}\{t^{n-1}\}, \qquad s > 0. \tag{2b}$$

Applying reduction formula (2b) recursively, we find, for $s > 0$,

$$\mathcal{L}\{t^2\} = \frac{2}{s}\mathcal{L}\{t\} = \frac{2}{s^3}, \qquad \mathcal{L}\{t^3\} = \frac{3}{s}\mathcal{L}\{t^2\} = \frac{3 \cdot 2}{s^4}$$

and, in general,

$$\mathcal{L}\{t^n\} = \frac{n!}{s^{n+1}}, \qquad n = 1, 2, 3, \ldots, \qquad s > 0. \tag{3}$$

The Laplace Transform of f(t) = e^{\alpha t}   We saw in Section 5.1 that

$$\mathcal{L}\{e^{\alpha t}\} = \frac{1}{s - \alpha}, \qquad s > \alpha. \tag{4}$$

The Laplace Transforms of f(t) = sin ωt *and f(t) = cos* ωt Using integration by parts twice yields

$$\mathcal{L}\{\sin \omega t\} = \int_0^\infty e^{-st} \sin \omega t \, dt$$

$$= \left[-\frac{e^{-st} \sin \omega t}{s} - \frac{\omega e^{-st} \cos \omega t}{s^2} \right] \Big|_0^\infty - \frac{\omega^2}{s^2} \int_0^\infty e^{-st} \sin \omega t \, dt$$

$$= \frac{\omega}{s^2} - \frac{\omega^2}{s^2} \mathcal{L}\{\sin \omega t\}, \qquad s > 0.$$

Solving for $\mathcal{L}\{\sin \omega t\}$, we find

$$\mathcal{L}\{\sin \omega t\} = \frac{\omega}{s^2 + \omega^2}, \qquad s > 0. \tag{5a}$$

Similarly,

$$\mathcal{L}\{\cos \omega t\} = \frac{s}{s^2 + \omega^2}, \qquad s > 0. \tag{5b}$$

We know from Section 5.1 that the Laplace transform defines a linear transformation on the set of piecewise continuous and exponentially bounded functions; that is, if $f(t)$ and $g(t)$ are piecewise continuous and exponentially bounded, then

$$\mathcal{L}\{c_1 f(t) + c_2 g(t)\} = c_1 \mathcal{L}\{f(t)\} + c_2 \mathcal{L}\{g(t)\}.$$

We can use this linearity property to extend our library of transforms. For example, combining linearity with the transform for $\mathcal{L}\{t^n\}$ listed in equation (3), we obtain the Laplace transform of any polynomial. The next example provides an illustration.

E X A M P L E

1

Use the transform pairs developed above to find the Laplace transform of

(a) $p(t) = 2t^3 + 5t - 3, \qquad t \geq 0$ (b) $f(t) = 4 \cos^2 3t, \qquad t \geq 0$

Solution:

(a) Using linearity and equation (3), we have

$$\mathcal{L}\{p(t)\} = \mathcal{L}\{2t^3 + 5t - 3\} = 2\mathcal{L}\{t^3\} + 5\mathcal{L}\{t\} - 3\mathcal{L}\{1\}$$

$$= 2\frac{3!}{s^4} + 5\frac{1}{s^2} - 3\frac{1}{s} = \frac{12 + 5s^2 - 3s^3}{s^4}, \qquad s > 0.$$

(b) We have no transform pair directly involving $\cos^2 \omega t$. However, we can use a trigonometric identity to rewrite $f(t) = 4 \cos^2 3t$ as $f(t) = 2 + 2 \cos 6t$. Using linearity and equation (5b) yields

$$\mathcal{L}\{f(t)\} = 2\mathcal{L}\{1\} + 2\mathcal{L}\{\cos 6t\}$$

$$= 2\frac{1}{s} + 2\frac{s}{s^2 + 36} = \frac{4s^2 + 72}{s(s^2 + 36)}, \qquad s > 0. \; ❖$$

Two Shift Theorems

The next two results, established in Theorem 5.4, are often referred to as the **first and second shift theorems**. Like the linearity property illustrated in Example 1, the shift theorems increase the number of functions for which we can easily find Laplace transforms.

Theorem 5.4

> Let $f(t)$ be piecewise continuous and exponentially bounded on $0 \le t < \infty$, where $|f(t)| \le Me^{at}$, $0 \le t < \infty$. Let $F(s) = \mathcal{L}\{f(t)\}$, and let $h(t)$ denote the unit step function. Then
>
> (a) $\mathcal{L}\{e^{\alpha t}f(t)\} = F(s - \alpha), \qquad s > a + \alpha$
>
> (b) $\mathcal{L}\{f(t - \alpha)h(t - \alpha)\} = e^{-\alpha s}F(s), \qquad \alpha \ge 0, \qquad s > a.$

Since $h(t - \alpha) = 0$ when $t < \alpha$ (see Figure 5.6),

$$f(t - \alpha)h(t - \alpha) = 0, \qquad t < \alpha.$$

The graph of $f(t - \alpha)h(t - \alpha)$ looks just like the graph of $f(t)$ except for the fact that it is shifted to the right and remains zero until $t = \alpha$. Figure 5.7 provides an example.

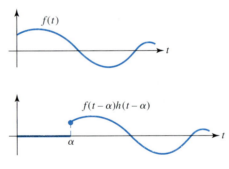

FIGURE 5.7

The graphs of $f(t)$ and $f(t - \alpha)h(t - \alpha)$. Note that the graph of $f(t - \alpha)h(t - \alpha)$ looks like the graph of $f(t)$ except that it is shifted α units to the right.

- **PROOF (of Theorem 5.4):**

(a) The following calculation establishes part (a):

$$\mathcal{L}\{e^{\alpha t}f(t)\} = \int_0^\infty e^{\alpha t}f(t)e^{-st}\, dt = \int_0^\infty f(t)e^{-(s-\alpha)t}\, dt = F(s - \alpha),$$

$$s > a + \alpha.$$

(b) To establish the second shift theorem, we begin with

$$\mathcal{L}\{f(t-\alpha)h(t-\alpha)\} = \int_0^\infty f(t-\alpha)h(t-\alpha)e^{-st}\,dt = \int_\alpha^\infty f(t-\alpha)e^{-st}\,dt.$$

Making the change of variable $\tau = t - \alpha$, we have

$$\mathcal{L}\{f(t-\alpha)h(t-\alpha)\} = \int_\alpha^\infty f(t-\alpha)e^{-st}\,dt = \int_0^\infty f(\tau)e^{-s(\tau+\alpha)}\,d\tau$$

$$= e^{-s\alpha}\int_0^\infty f(\tau)e^{-s\tau}\,d\tau = e^{-s\alpha}F(s), \qquad s > a. \;\bullet$$

Note that parts (a) and (b) of Theorem 5.4 possess a certain duality. Roughly speaking, multiplying a function by an exponential function in the time domain shifts the argument of its Laplace transform. Likewise, shifting the argument in the time domain leads to an exponential multiplicative factor in the transform domain.

E X A M P L E
2

Find

(a) $\mathcal{L}\{e^{2t}t^4\}$ (b) $\mathcal{L}\{e^{\alpha t}\cos\omega t\}$ (c) $\mathcal{L}^{-1}\left\{\dfrac{e^{-5s}}{s^2}\right\}$ (d) $\mathcal{L}^{-1}\left\{\dfrac{e^{-\alpha s}}{s^2+1}\right\}$

Solution:

(a) By Theorem 5.4, multiplying $f(t)$ by $e^{\alpha t}$ shifts the argument of its transform, $F(s)$. That is,

$$\mathcal{L}\{e^{2t}t^4\} = \mathcal{L}\{t^4\}\Big|_{s\to s-2} = \frac{4!}{(s-2)^5}, \qquad s > 2.$$

(b) As in part (a),

$$\mathcal{L}\{e^{\alpha t}\cos\omega t\} = \mathcal{L}\{\cos\omega t\}|_{s\to s-\alpha} = \frac{s-\alpha}{(s-\alpha)^2+\omega^2}, \qquad s > \alpha.$$

(c) We know that $\mathcal{L}\{t\} = 1/s^2, t \ge 0$. By the second shift theorem,

$$\frac{e^{-5s}}{s^2} = \mathcal{L}\{(t-5)h(t-5)\}.$$

Therefore,

$$\mathcal{L}^{-1}\left\{\frac{e^{-5s}}{s^2}\right\} = (t-5)h(t-5) = \begin{cases} 0, & 0 \le t < 5 \\ t-5, & 5 \le t < \infty. \end{cases}$$

The graph of this inverse transform is shown in Figure 5.8.

(d) We know that $\mathcal{L}\{\sin t\} = 1/(s^2+1), t \ge 0$. Using the second shift theorem yields

$$\frac{e^{-\alpha s}}{s^2+1} = \mathcal{L}\{[\sin(t-\alpha)]h(t-\alpha)\}.$$

(continued)

(continued)

Therefore,

$$\mathcal{L}^{-1}\left\{\frac{e^{-\alpha s}}{s^2+1}\right\} = [\sin(t-\alpha)]\,h(t-\alpha) = \begin{cases} 0, & 0 \le t < \alpha \\ \sin(t-\alpha), & \alpha \le t < \infty. \end{cases}$$

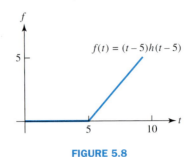

FIGURE 5.8

The graph of $f(t) = (t-5)h(t-5)$.   ❖

The Laplace Transform of Derivatives and Antiderivatives

The utility of Laplace transforms as a tool to solve problems involving constant coefficient linear differential equations is due in large part to the transform pairs in the next theorem that relate the Laplace transform of derivatives and integrals of a function to the Laplace transform of the function itself. We will make extensive use of these results in Sections 5.4–5.7.

Theorem 5.5

(a) Let $f(t)$ be continuous on $0 \le t < \infty$, and let $f'(t)$ exist as a piecewise continuous, exponentially bounded function on $0 \le t < \infty$, where $|f'(t)| \le Me^{at}$, $0 \le t < \infty$. Then

$$\mathcal{L}\{f'(t)\} = s\mathcal{L}\{f(t)\} - f(0) = sF(s) - f(0), \qquad s > \max\{a, 0\}.$$

(b) Let $f'(t)$ be continuous on $0 \le t < \infty$, and let $f''(t)$ exist as a piecewise continuous, exponentially bounded function on $0 \le t < \infty$, where $|f''(t)| \le Me^{at}$, $0 \le t < \infty$. Then

$$\mathcal{L}\{f''(t)\} = s\mathcal{L}\{f'(t)\} - f'(0) = s\,[sF(s) - f(0)] - f'(0)$$

$$= s^2 F(s) - sf(0) - f'(0), \qquad s > \max\{a, 0\}.$$

(c) Let $f(t)$ be piecewise continuous and exponentially bounded on $0 \le t < \infty$, where $|f(t)| \le Me^{at}$, $0 \le t < \infty$. Then

$$\mathcal{L}\left\{\int_0^t f(u)\,du\right\} = \frac{\mathcal{L}\{f(t)\}}{s} = \frac{F(s)}{s}, \qquad s > \max\{a, 0\}.$$

● **PROOF:**   The proof of part (a) is presented to illustrate the relevant ideas. By hypothesis, the function $f(t)$ is continuous. We now show it is also exponentially bounded, and thus $f(t)$ has a Laplace transform. We have

$$|f(t)| = \left| f(0) + \int_0^t f'(u)\, du \right| \leq |f(0)| + \left| \int_0^t f'(u)\, du \right| \leq |f(0)| + \int_0^t Me^{au}\, du,$$

$$t \geq 0.$$

Therefore, for $0 \leq t < \infty$,

$$|f(t)| \leq \begin{cases} |f(0)| + \dfrac{M}{a}(e^{at} - 1) \leq \left[|f(0)| + \dfrac{M}{a} \right] e^{at}, & a > 0 \\[2ex] |f(0)| + Mt, & a = 0 \\[2ex] |f(0)| + \dfrac{M}{|a|}(1 - e^{at}) \leq |f(0)| + \dfrac{M}{|a|}, & a < 0. \end{cases} \qquad (6)$$

From these inequalities, we are able to conclude that $\mathcal{L}\{f(t)\} = F(s)$ exists for $s > \max\{a, 0\}$.

To obtain (a), consider the interval $0 \leq t \leq T$ for some arbitrary $T > 0$. Let $t_1 < t_2 < \cdots < t_N$ represent the points of discontinuity of $f'(t)$ on this interval. Then

$$\int_0^T f'(t)e^{-st}\, dt = \int_0^{t_1} f'(t)e^{-st}\, dt$$

$$+ \int_{t_1}^{t_2} f'(t)e^{-st}\, dt + \cdots + \int_{t_{N-1}}^{t_N} f'(t)e^{-st}\, dt + \int_{t_N}^T f'(t)e^{-st}\, dt.$$

Performing integration by parts on each of these integrals yields

$$\int_0^T f'(t)e^{-st}\, dt = f(t)e^{-st}\Big|_0^{t_1} + f(t)e^{-st}\Big|_{t_1}^{t_2} + \cdots + f(t)e^{-st}\Big|_{t_{N-1}}^{t_N} + f(t)e^{-st}\Big|_{t_N}^T$$

$$+ s\left[\int_0^{t_1} f(t)e^{-st}\, dt + \int_{t_1}^{t_2} f(t)e^{-st}\, dt + \cdots + \int_{t_{N-1}}^{t_N} f(t)e^{-st}\, dt + \int_{t_N}^T f(t)e^{-st}\, dt \right].$$

Since $f(t)$ is continuous, the sum of the endpoint evaluations reduces to $f(T)e^{-sT} - f(0)$. Similarly, the sum of integrals on the right-hand side can be expressed as a single integral from 0 to T. Therefore, we obtain

$$\int_0^T f'(t)e^{-st}\, dt = f(T)e^{-sT} - f(0) + s\int_0^T f(t)e^{-st}\, dt.$$

Now let $T \to \infty$, while assuming that $s > \max\{a, 0\}$. For these values of s, $\lim_{T \to \infty} f(T)e^{-sT} = 0$ and $\lim_{T \to \infty} \int_0^T f(t)e^{-st}\, dt = \mathcal{L}\{f(t)\} = F(s)$. Therefore, the result follows. ●

Note that differentiation in the time domain corresponds, roughly speaking, to multiplication by s in the transform domain, while antidifferentiation in the time domain corresponds to division by s in the transform domain.

Solving Initial Value Problems

The next example, while quite simple, illustrates how we can use Laplace transforms to solve initial value problems. Following a review of the method of partial fractions in Section 5.3, we give a more detailed discussion.

EXAMPLE

3

Consider the initial value problem

$$y' - 3y = g(t), \qquad y(0) = 1,$$

where $g(t)$ is the step function given by

$$g(t) = \begin{cases} 0, & 0 \le t < 2 \\ 6, & 2 \le t < \infty. \end{cases}$$

Let $Y(s)$ denote the Laplace transform of $y(t)$, where $y(t)$ is the unique solution of this initial value problem. Using Theorem 5.5, derive an equation for $Y(s)$ and, taking the inverse Laplace transform, find $y(t)$.

Solution: The nonhomogeneous term $g(t)$ can be represented as $g(t) = 6h(t - 2)$, where $h(t - 2)$ denotes the shifted Heaviside step function. (See Figure 5.6.) Taking Laplace transforms of

$$y'(t) - 3y(t) = 6h(t - 2), \qquad 0 \le t < \infty,$$

we have

$$\mathcal{L}\{y'(t)\} - 3\mathcal{L}\{y(t)\} = 6\mathcal{L}\{h(t - 2)\}.$$

Using part (a) of Theorem 5.5 to evaluate $\mathcal{L}\{y'(t)\}$ and part (b) of Theorem 5.4 to evaluate $\mathcal{L}\{h(t - 2)\}$, we find

$$[sY(s) - y(0)] - 3Y(s) = \frac{6e^{-2s}}{s}.$$

Solving for $Y(s)$ and using the fact that $y(0) = 1$, we obtain

$$Y(s) = \frac{1}{s - 3} + \frac{6e^{-2s}}{s(s - 3)}, \qquad s > 3.$$

Using a partial fraction expansion for the second term on the right-hand side gives

$$Y(s) = \frac{1}{s - 3} + e^{-2s}\left(\frac{2}{s - 3} - \frac{2}{s}\right), \qquad s > 3.$$

Therefore,

$$y(t) = \mathcal{L}^{-1}\left\{\frac{1}{s - 3}\right\} + 2\mathcal{L}^{-1}\left\{e^{-2s}\frac{1}{s - 3}\right\} - 2\mathcal{L}^{-1}\left\{e^{-2s}\frac{1}{s}\right\}.$$

Using the second shifting theorem, we see that $y(t) = e^{3t} + 2h(t - 2)[e^{3t-6} - 1]$, $t \ge 0$, is the unique solution of the initial value problem. In particular, $y(t)$ is the piecewise-defined function

$$y(t) = \begin{cases} e^{3t}, & 0 \le t < 2 \\ e^{3t} + 2[e^{3t-6} - 1], & 2 \le t < \infty. \end{cases}$$

Note that $y(t)$ is continuous at $t = 2$, but it is not differentiable at $t = 2$. ❖

We have restricted our attention in Theorem 5.5 to transform relations for first and second derivatives, since these derivatives appear most frequently in applications. It should be clear that, with appropriate hypotheses, the arguments establishing Theorem 5.5 can be extended and we can obtain similar formulas for higher derivatives. In general,

$$\mathcal{L}\{f^{(n)}(t)\} = s^n F(s) - s^{n-1}f(0) - s^{n-2}f'(0) - \cdots - sf^{(n-2)}(0) - f^{(n-1)}(0),$$

$$n = 1, 2, 3, \ldots.$$

The Laplace transform pairs and relations that we have developed so far are summarized in Table 5.1 at the end of this section.

The importance of Theorem 5.5 cannot be overemphasized. As we have seen, derivatives and antiderivatives of a function $f(t)$ transform into algebraic expressions involving the function's transform $F(s)$. This is the key to the problem simplification achieved by working in the transform domain. Example 4, solving for the response of a simple *RLC* network, illustrates how Laplace transforms achieve such simplifications. The example further illustrates the important role that partial fraction expansions play in the use of Laplace transforms.

Changing Our Point of View

Until now, we have carefully stated the values of s for which the improper integral

$$\mathcal{L}\{f(t)\} = \int_0^\infty f(t)e^{-st}\,dt$$

converges and thus defines the domain of the Laplace transform $F(s)$. Each entry in our table of transform pairs (Table 5.1 at the end of this section) includes the domain of $F(s)$.

It is important to understand and appreciate the underlying mathematical issues. However, when we begin to use Laplace transforms to actually solve initial value problems, we will no longer be so attentive to these details. Part of the reason for this change is that we will be "computing" the Laplace transform, $Y(s)$, of the *unknown* solution, $y(t)$. Since the solution, $y(t)$, is unknown, we cannot easily determine the domain of its transform, $Y(s)$.

Instead, we are going to use the Laplace transform as an operational tool. We will simply assume that the unknown solution is a piecewise continuous and exponentially bounded function whose Laplace transform exists for $s > a$ for some value a. After we formally execute the three steps—Laplace transformation, solution of the transformed problem, and inverse Laplace transformation—we will obtain what we can regard as a *candidate* for the solution of our initial value problem. If we can directly verify that the "candidate solution" obtained by the use of transforms is, in fact, the unique solution of the original problem of interest, then we are done. Example 4 illustrates these points and the three-step Laplace transform solution procedure, using it to analyze a network problem.

The series *RLC* network shown in Figure 5.9 is assumed to be initially quiescent; that is, the current and the charge on the capacitor are both zero for $t \leq 0$. At time $t = 0$, a voltage source $v(t) = v_0 \, te^{-\alpha t}$, having the polarity shown, is turned on. Determine the current $i(t)$ for $t \geq 0$.

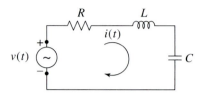

FIGURE 5.9

The *RLC* network analyzed in Example 4.

Solution: Recall that the underlying principle for describing our problem mathematically is Kirchhoff's voltage law (see Section 3.10). As we traverse the network in a clockwise manner, the voltage rise through the source must equal the sum of the voltage drops through the resistor R, inductor L, and capacitor C. The resulting equation, along with the accompanying supplementary conditions, is

$$v(t) = Ri(t) + L\,\frac{di(t)}{dt} + \frac{1}{C}\int_0^t i(u)\,du, \qquad i(0) = 0, \qquad t \geq 0,$$

or

$$\frac{di(t)}{dt} + \frac{R}{L}i(t) + \frac{1}{LC}\int_0^t i(u)\,du = \frac{1}{L}v(t), \qquad i(0) = 0, \qquad t \geq 0. \qquad \textbf{(7)}$$

When we considered this problem in Section 3.10, we differentiated equation (7) to obtain a second order differential equation for the current $i(t)$. Now, however, we will work directly with equation (7), which is an integro-differential equation for the unknown current, $i(t)$.

The first step is to compute the Laplace transform of both sides of equation (7), obtaining

$$\mathcal{L}\left\{\frac{di}{dt}\right\} + \frac{R}{L}\mathcal{L}\{i\} + \frac{1}{LC}\mathcal{L}\left\{\int_0^t i(u)\,du\right\} = \frac{1}{L}\mathcal{L}\{v(t)\}.$$

This equation can be written as

$$sI(s) - i(0) + \frac{R}{L}I(s) + \frac{1}{LC}\frac{I(s)}{s} = \frac{1}{L}V(s). \qquad \textbf{(8)}$$

Notice that the supplementary condition involving $i(0)$ enters directly into the transformed equation (8). In our case, $i(0) = 0$. Since $v(t) = v_0 te^{-\alpha t}$, we have

$$V(s) = v_0\mathcal{L}\{te^{-\alpha t}\} = v_0\mathcal{L}\{t\}\big|_{s \to s+\alpha} = v_0\frac{1}{(s+\alpha)^2}.$$

The transform domain problem is therefore entirely defined by the algebraic equation:

$$sI(s) + \frac{R}{L}I(s) + \frac{1}{LC}\frac{I(s)}{s} = \frac{v_0}{L}\frac{1}{(s+\alpha)^2}. \tag{9}$$

The second step is to solve transform domain problem (9). We find

$$I(s) = \frac{v_0}{L}\frac{s}{(s+\alpha)^2\left(s^2 + \frac{R}{L}s + \frac{1}{LC}\right)}. \tag{10}$$

The third step is to find the inverse Laplace transform of $I(s)$. To accomplish this, we use a partial fraction expansion to decompose rational function (10) into a sum of terms, each of whose inverse Laplace transform is known.

For an illustration of this third step with a specific case, suppose that $I(s)$ in equation (10) is given by

$$I(s) = \frac{50s}{(s+1)^2(s^2 + 4s + 13)}. \tag{11}$$

Expression (11) has the partial fraction expansion

$$\begin{aligned}
I(s) &= \frac{6}{s+1} - \frac{5}{(s+1)^2} - \frac{6s+13}{(s+2)^2+9} \\
&= \frac{6}{s+1} - \frac{5}{(s+1)^2} - 6\frac{s+2}{(s+2)^2+9} - \frac{1}{3}\frac{3}{(s+2)^2+9},
\end{aligned} \tag{12}$$

where we have used the fact that $s^2 + 4s + 13 = (s+2)^2 + 9$. The algebraic manipulations leading to the last expression in (12) were done in anticipation of the inverse Laplace transform computation.

Applying the inverse Laplace transform to $I(s)$ yields

$$\begin{aligned}
i(t) &= \mathcal{L}^{-1}\{I(s)\} \\
&= 6\mathcal{L}^{-1}\left\{\frac{1}{s+1}\right\} - 5\mathcal{L}^{-1}\left\{\frac{1}{(s+1)^2}\right\} \\
&\quad - 6\mathcal{L}^{-1}\left\{\frac{s+2}{(s+2)^2+9}\right\} - \frac{1}{3}\mathcal{L}^{-1}\left\{\frac{3}{(s+2)^2+9}\right\}.
\end{aligned} \tag{13}$$

The required inverse transforms can be obtained from Table 5.1 at the end of this section. When these inverse transforms are used in equation (13), it follows that

$$i(t) = 6e^{-t} - 5te^{-t} - 6e^{-2t}\cos 3t - \tfrac{1}{3}e^{-2t}\sin 3t, \qquad t \geq 0.$$

As a final check, one should verify that this expression for the network current is, in fact, the desired solution. ❖

The network current is plotted in Figure 5.10; its behavior seems reasonable. Since the source voltage is proportional to te^{-t}, one would expect the current to exhibit a transient behavior followed by an approach to zero as $t \to \infty$.

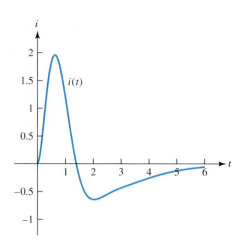

FIGURE 5.10

The network current found in Example 4 for the *RLC* network of Figure 5.9. Since the source voltage is proportional to te^{-t}, we expect the current to consist of an initial transient variation followed by an approach to zero as time increases.

TABLE 5.1

A Table of Laplace Transform Pairs

Time Domain Function $f(t)$, $t \geq 0$	Laplace Transform $F(s)$
1. $h(t) = \begin{cases} 1, & t \geq 0 \\ 0, & t < 0 \end{cases}$	$\dfrac{1}{s}$, $\quad s > 0$
2. t^n, $\quad n = 1, 2, 3, \ldots$	$\dfrac{n!}{s^{n+1}}$, $\quad s > 0$
3. $e^{\alpha t}$	$\dfrac{1}{s - \alpha}$, $\quad s > \alpha$
4. $\sin \omega t$	$\dfrac{\omega}{s^2 + \omega^2}$, $\quad s > 0$
5. $\cos \omega t$	$\dfrac{s}{s^2 + \omega^2}$, $\quad s > 0$
6. $\sinh \alpha t$	$\dfrac{\alpha}{s^2 - \alpha^2}$, $\quad s > |\alpha|$
7. $\cosh \alpha t$	$\dfrac{s}{s^2 - \alpha^2}$, $\quad s > |\alpha|$
8. $e^{\alpha t} f(t)$, with $|f(t)| \leq Me^{at}$	$F(s - \alpha)$, $\quad s > \alpha + a$

(9)–(12) are four special cases of (8):

9. $e^{\alpha t} h(t)$	$\dfrac{1}{s - \alpha}$, $\quad s > \alpha$
10. $e^{\alpha t} t^n$, $\quad n = 1, 2, 3, \ldots$	$\dfrac{n!}{(s - \alpha)^{n+1}}$, $\quad s > \alpha$
11. $e^{\alpha t} \sin \omega t$	$\dfrac{\omega}{(s - \alpha)^2 + \omega^2}$, $\quad s > \alpha$
12. $e^{\alpha t} \cos \omega t$	$\dfrac{(s - \alpha)}{(s - \alpha)^2 + \omega^2}$, $\quad s > \alpha$

TABLE 5.1 (continued)	
13. $f(t - \alpha)h(t - \alpha)$, $\alpha \geq 0$, with $\lvert f(t) \rvert \leq Me^{at}$	$e^{-\alpha s}F(s)$, $s > a$
(14) is a special case of (13):	
14. $h(t - \alpha)$, $\alpha \geq 0$	$\dfrac{e^{-\alpha s}}{s}$, $s > 0$
15. $f'(t)$, with $f(t)$ continuous and $\lvert f'(t) \rvert \leq Me^{at}$	$sF(s) - f(0)$, $s > \max\{a, 0\}$
16. $f''(t)$, with $f'(t)$ continuous and $\lvert f''(t) \rvert \leq Me^{at}$	$s^2F(s) - sf(0) - f'(0)$, $s > \max\{a, 0\}$
17. $f^{(n)}(t)$, with $f^{(n-1)}(t)$ continuous and $\lvert f^{(n)}(t) \rvert \leq Me^{at}$	$s^nF(s) - s^{n-1}f(0) - \cdots$ $- sf^{(n-2)}(0) - f^{(n-1)}(0)$, $s > \max\{a, 0\}$, $n = 1, 2, 3, \ldots$
18. $\displaystyle\int_0^t f(u)\,du$, with $\lvert f(t) \rvert \leq Me^{at}$	$\dfrac{F(s)}{s}$, $s > \max\{a, 0\}$
19. $\dfrac{1}{2\omega^3}(\sin \omega t - \omega t \cos \omega t)$	$\dfrac{1}{(s^2 + \omega^2)^2}$, $s > 0$
20. $\dfrac{t}{2\omega}\sin \omega t$	$\dfrac{s}{(s^2 + \omega^2)^2}$, $s > 0$
21. $tf(t)$	$-F'(s)$

EXERCISES

Exercises 1–12:

Use Table 5.1 to find $\mathcal{L}\{f(t)\}$ for the given function $f(t)$ defined on the interval $t \geq 0$.

1. $f(t) = 3t^2 + 2t + 1$ **2.** $f(t) = 2e^t + 5$ **3.** $f(t) = 1 + \sin 3t$

4. $f(t) = e^{3t-3}h(t - 1)$ **5.** $f(t) = (t - 1)^2 h(t - 1)$ **6.** $f(t) = \sin^2 \omega t$

7. $f(t) = 2te^{-2t}$ **8.** $f(t) = \sin 3t \cos 3t$ **9.** $f(t) = 2th(t - 2)$

10. $f(t) = e^{2t}\cos 3t$ **11.** $f(t) = e^{3t}h(t - 1)$ **12.** $f(t) = e^{4t}(t^2 + 3t + 5)$

Exercises 13–21:

Use Table 5.1 to find $\mathcal{L}^{-1}\{F(s)\}$ for the given $F(s)$.

13. $F(s) = \dfrac{3}{s} + \dfrac{24}{s^4}$ **14.** $F(s) = \dfrac{10}{s^2 + 25} + \dfrac{4}{s - 3}$

15. $F(s) = \dfrac{2s - 4}{(s - 2)^2 + 9}$ **16.** $F(s) = \dfrac{5}{(s - 3)^4}$

17. $F(s) = e^{-2s}\dfrac{3}{s^2 + 9}$ **18.** $F(s) = \dfrac{e^{-2s}}{s - 9}$

19. $F(s) = \dfrac{4s - 6}{s^2 - 2s + 10}$ **20.** $F(s) = \dfrac{e^{-3s}(2s + 7)}{s^2 + 16}$

21. $F(s) = \dfrac{48(e^{-3s} + 2e^{-5s})}{s^5}$

Exercises 22–33:

Combinations of Shifted Heaviside Step Functions Exercises 22–33 deal with combinations of Heaviside step functions. As the two examples below show, we can use combinations of shifted Heaviside step functions to represent pulses.

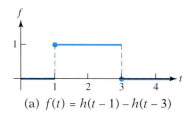

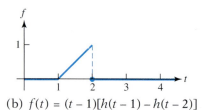

(a) $f(t) = h(t - 1) - h(t - 3)$ (b) $f(t) = (t - 1)[h(t - 1) - h(t - 2)]$

In each exercise, graph the function $f(t)$ for $0 \leq t < \infty$, and use Table 5.1 to find the Laplace transform of $f(t)$.

22. $f(t) = h(t - 1) + h(t - 3)$ **23.** $f(t) = \sin(t - 2\pi)h(t - 2\pi)$

24. $f(t) = t[h(t - 1) - h(t - 3)]$ **25.** $f(t) = h(t) - h(t - 3)$

26. $f(t) = 3[h(t - 1) - h(t - 4)]$ **27.** $f(t) = (2 - t)[h(t - 1) - h(t - 3)]$

28. $f(t) = |2 - t|[h(t - 1) - h(t - 3)]$

29. $f(t) = [h(t - 1) - h(t - 2)] - [h(t - 2) - h(t - 3)]$

30. $h(2 - t)$ **31.** $e^{-2t}h(1 - t)$

32. $h(t - 1) + h(4 - t)$ **33.** $h(t - 2) - h(3 - t)$

Exercises 34–37:

In each exercise, the graph of $f(t)$ is given. Represent $f(t)$ using a combination of Heaviside step functions, and use Table 5.1 to calculate the Laplace transform of $f(t)$.

34.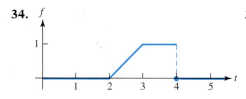

35.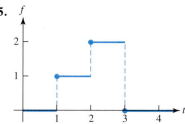

36.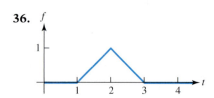

37.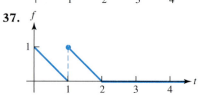

An Introduction to the Method of Partial Fractions We will present a review of the method of partial fractions in Section 5.3. By way of introduction, however, we consider here a special case of the method and use it to solve some initial value problems, as in Examples 3 and 4. Suppose $F(s) = 1/Q(s)$, where $Q(s) = (s - r_1)(s - r_2) \cdots (s - r_n)$ and where the roots $r_1, r_2, \ldots, r_n$ are real and distinct. In this case, there are constants $A_1, A_2, \ldots, A_n$

such that

$$F(s) = \frac{1}{(s - r_1)(s - r_2) \cdots (s - r_n)} = \frac{A_1}{s - r_1} + \frac{A_2}{s - r_2} + \cdots + \frac{A_n}{s - r_n}. \qquad (14)$$

One way to determine the constants $A_1, A_2, \ldots, A_n$ is to recombine the right-hand side into a single rational function and equate the resulting numerator to 1.

Exercises 38–41:

Using a partial fraction expansion, find $\mathcal{L}^{-1}\{F(s)\}$. In Exercise 40, compare your answer with (6) in Table 5.1.

38. $F(s) = \dfrac{12}{(s - 3)(s + 1)}$

39. $F(s) = \dfrac{4}{s(s + 2)}$

40. $F(s) = \dfrac{24e^{-5s}}{s^2 - 9}$

41. $F(s) = \dfrac{10e^{-s}}{s^2 - 5s + 6}$

Exercises 42–45:

As in Examples 3 and 4, use Laplace transform techniques to solve the initial value problem.

42. $y' + 4y = g(t), \quad y(0) = 2, \quad g(t) = \begin{cases} 0, & 0 \le t < 1 \\ 12, & 1 \le t < 3 \\ 0, & 3 \le t < \infty \end{cases}$

43. $y' - y = g(t), \quad y(0) = 1, \quad g(t) = \begin{cases} 0, & 0 \le t < 4 \\ e^{3t}, & 4 \le t < \infty \end{cases}$

44. $y'' - 4y = e^{3t}, \quad y(0) = 0, \quad y'(0) = 0$

45. $y'' - 2y' - 8y = e^t, \quad y(0) = 0, \quad y'(0) = 1$

46. Let $f(t)$ be piecewise continuous and exponentially bounded on the interval $0 \le t < \infty$, and let $F(s)$ denote the Laplace transform of $f(t)$. It is shown in advanced calculus[3] that it is possible to differentiate under the integral sign with respect to the parameter s. That is,

$$\frac{d}{ds} F(s) = \frac{d}{ds} \int_0^\infty e^{-st} f(t) \, dt = \int_0^\infty \frac{d}{ds} [e^{-st} f(t)] \, dt.$$

(a) Use this result to show that $\mathcal{L}\{tf(t)\} = -F'(s)$.

(b) Use the result of part (a) to establish formula (20) in Table 5.1.

Exercises 47–48:

Obtain the Laplace transform of the given function in terms of $\mathcal{L}\{f(t)\} = F(s)$. For Exercise 48, note that $\int_a^t f(\lambda) \, d\lambda = \int_0^t f(\lambda) \, d\lambda - \int_0^a f(\lambda) \, d\lambda$.

47. $\displaystyle\int_0^t \int_0^\lambda f(\sigma) \, d\sigma \, d\lambda$

48. $\displaystyle\int_2^t f(\lambda) \, d\lambda$, given that $\displaystyle\int_0^2 f(\lambda) \, d\lambda = 3$

49. Consider the functions f and g defined on $0 \le t < \infty$,

$$f(t) = h(t)h(3 - t) \quad \text{and} \quad g(t) = h(t) - h(t - 3).$$

(a) Are the two functions identical?

(b) Determine $F(s) = \mathcal{L}\{f(t)\}$ and $G(s) = \mathcal{L}\{g(t)\}$. Is $F(s) = G(s)$?

---

[3] David V. Widder, *Advanced Calculus*, 2nd ed. (Englewood Cliffs, NJ: Prentice Hall, 1961).

5.3   The Method of Partial Fractions

When we solve a problem in the transform domain, the solution is often a rational function of s; this was the case in Examples 3 and 4 of Section 5.2. The function $F(s)$ is called a **rational function** if it has the form

$$F(s) = \frac{N(s)}{D(s)}, \tag{1}$$

where $N(s)$ and $D(s)$ are polynomials.

In order to find $\mathcal{L}^{-1}\{F(s)\}$, we proceed as in Examples 3 and 4 of Section 5.2, using the method of partial fractions to decompose the rational function (1) into a sum of simpler expressions whose inverse transform can be recognized from a table of transform pairs.

The method of partial fractions is usually studied in calculus when antiderivatives of rational functions are computed. Refer back to your calculus text for a comprehensive discussion of the technique. The goal of this section is simply to review the underlying ideas.

Using the Method of Partial Fractions

The starting point for the method of partial fractions is a rational function in which *the degree of the numerator polynomial is strictly less than the degree of the denominator polynomial*. The rational functions we will encounter in the transform domain will have this form.

Let $F(s) = N(s)/D(s)$, where $N(s)$ and $D(s)$ are polynomials having real coefficients and where the degree of $N(s)$ is less than the degree of $D(s)$. The form of the partial fraction expansion is totally determined by the factorization of the denominator polynomial, $D(s)$. Table 5.2 lists the possible factors of the denominator polynomial. For each of the factors in the left-hand column, we need to include the terms in the right-hand column in the partial fraction expansion. The complete partial fraction expansion is the sum of the contributions from all of the denominator factors. This expansion contains constants that must subsequently be determined.

Since the denominator polynomial, $D(s)$, has real coefficients, any complex zeros will occur in complex conjugate pairs. Therefore, irreducible quadratic factors (which correspond to complex conjugate pairs of zeros) will have the forms listed in Table 5.2. Cases 3 and 4 are special versions of cases 5 and 6 that correspond to $\alpha = 0$. Since the numerator polynomial also has real coefficients, the constants in the partial fraction expansion will likewise be real valued. These constants, denoted by capital letters on the right-hand side of Table 5.2, must be determined.

When looking for the inverse transform of a term that arises in case 5 or 6, we usually rewrite the term. By completing the square, we can rewrite an irreducible quadratic factor of the form $s^2 + 2\alpha s + \beta^2$, $\beta^2 > \alpha^2$ as

$$(s^2 + 2\alpha s + \alpha^2) + (\beta^2 - \alpha^2) = (s + \alpha)^2 + \omega^2.$$

This form is associated with the Laplace transforms of $e^{-\alpha t}\sin\omega t$ and $e^{-\alpha t}\cos\omega t$; see the transform pairs (11) and (12) in Table 5.1.

TABLE 5.2

Denominator Polynomial Factors and Their Corresponding Terms
in the Partial Fraction Expansion

Denominator Factor	Partial Fraction Expansion Term
1. Simple real root $s - \alpha$	$\dfrac{A}{s - \alpha}$
2. Repeated real root $(s - \alpha)^n$	$\dfrac{A_n}{(s - \alpha)^n} + \dfrac{A_{n-1}}{(s - \alpha)^{n-1}} + \cdots + \dfrac{A_1}{s - \alpha}$
3. Irreducible quadratic factor $s^2 + \omega^2$	$\dfrac{Bs + C}{s^2 + \omega^2}$
4. Repeated irreducible quadratic factor $(s^2 + \omega^2)^n$	$\dfrac{B_n s + C_n}{(s^2 + \omega^2)^n} + \dfrac{B_{n-1} s + C_{n-1}}{(s^2 + \omega^2)^{n-1}} + \cdots + \dfrac{B_1 s + C_1}{s^2 + \omega^2}$
5. Irreducible quadratic factor $s^2 + 2\alpha s + \beta^2, \quad \beta^2 > \alpha^2$	$\dfrac{Bs + C}{s^2 + 2\alpha s + \beta^2}$
6. Repeated irreducible quadratic factor $(s^2 + 2\alpha s + \beta^2)^n, \quad \beta^2 > \alpha^2$	$\dfrac{B_n s + C_n}{(s^2 + 2\alpha s + \beta^2)^n} + \cdots + \dfrac{B_1 s + C_1}{s^2 + 2\alpha s + \beta^2}$

EXAMPLE 1

Find $\mathcal{L}^{-1}\{F(s)\}$, where

$$F(s) = \frac{s^2 + 4}{s^4 - s^2}.$$

Solution: The function $F(s)$ is a rational function in which the degree of numerator $N(s) = s^2 + 4$ is less than that of denominator $D(s) = s^4 - s^2$. The denominator factors into

$$D(s) = s^2(s^2 - 1) = s^2(s - 1)(s + 1).$$

Therefore, the denominator has $s = 0$ as a repeated real root and $s = \pm 1$ as simple real roots. According to Table 5.2, $F(s)$ has a partial fraction expansion of the form

$$F(s) = \frac{s^2 + 4}{s^4 - s^2} = \frac{A_2}{s^2} + \frac{A_1}{s} + \frac{B}{s - 1} + \frac{C}{s + 1}, \tag{2}$$

where the four constants A_1, A_2, B, and C must be determined.

One way to evaluate the unknown constants is to recombine the right-hand side of (2), obtaining

$$\frac{s^2 + 4}{s^4 - s^2} = \frac{(A_1 + B + C)s^3 + (A_2 + B - C)s^2 - A_1 s - A_2}{s^4 - s^2}. \tag{3}$$

(continued)

(continued)

Since the two expressions in (3) must be equal, the two numerator polynomials must be identical. We therefore obtain four equations for the four unknown constants:

$$A_1 \quad\quad + B + C = 0$$
$$A_2 + B - C = 1$$
$$-A_1 \quad\quad\quad\quad = 0$$
$$-A_2 \quad\quad\quad\quad = 4.$$

This system is easily solved, and we obtain $A_1 = 0, A_2 = -4, B = \frac{5}{2}$, and $C = -\frac{5}{2}$. Thus, we have the partial fraction expansion

$$F(s) = -\frac{4}{s^2} + \frac{\frac{5}{2}}{s-1} - \frac{\frac{5}{2}}{s+1}. \tag{4}$$

From Table 5.1 in the previous section,

$$\mathcal{L}^{-1}\{F(s)\} = -4\mathcal{L}^{-1}\left\{\frac{1}{s^2}\right\} + \frac{5}{2}\mathcal{L}^{-1}\left\{\frac{1}{s-1}\right\} - \frac{5}{2}\mathcal{L}^{-1}\left\{\frac{1}{s+1}\right\}$$

$$= -4t + \frac{5}{2}e^t - \frac{5}{2}e^{-t} = -4t + 5\sinh t, \quad\quad t \geq 0. \; ❖$$

Alternative Approaches for Determining the Constants in a Partial Fraction Expansion

We now consider two alternative approaches for determining the constants in a partial fraction expansion. We explain the first approach below by re-working the expansion in Example 1. We explain the second approach later, in Example 2.

Consider equation (2) in Example 1,

$$\frac{s^2 + 4}{s^2(s-1)(s+1)} = \frac{A_2}{s^2} + \frac{A_1}{s} + \frac{B}{s-1} + \frac{C}{s+1}.$$

If we multiply both sides by $(s-1)$ and cancel common factors, we obtain

$$\frac{s^2 + 4}{s^2(s+1)} = A_2\frac{(s-1)}{s^2} + A_1\frac{(s-1)}{s} + B + C\frac{(s-1)}{(s+1)}. \tag{5}$$

We now determine B by setting $s = 1$ on both sides of expression (5), finding

$$B = \left.\frac{s^2+4}{s^2(s+1)}\right|_{s=1} = \frac{5}{2}.$$

Similarly, multiplying both sides by $(s+1)$ leads to

$$\frac{s^2+4}{s^2(s-1)} = A_2\frac{(s+1)}{s^2} + A_1\frac{(s+1)}{s} + B\frac{(s+1)}{(s-1)} + C,$$

and we find

$$C = \left.\frac{s^2+4}{s^2(s-1)}\right|_{s=-1} = -\frac{5}{2}.$$

So far, we have determined the constants B and C in the expansion

$$\frac{s^2 + 4}{s^4 - s^2} = \frac{A_2}{s^2} + \frac{A_1}{s} + \frac{B}{s - 1} + \frac{C}{s + 1}.$$

We next multiply both sides by s^2, obtaining

$$\frac{s^2 + 4}{s^2 - 1} = A_2 + A_1 s + B\frac{s^2}{s - 1} + C\frac{s^2}{s + 1}. \tag{6}$$

We now evaluate the constant A_2 by setting $s = 0$:

$$A_2 = \left.\frac{s^2 + 4}{s^2 - 1}\right|_{s=0} = -4.$$

To determine A_1, we differentiate both sides of equation (6), finding

$$\frac{-10s}{(s^2 - 1)^2} = A_1 + B\frac{s^2 - 2s}{(s - 1)^2} + C\frac{s^2 + 2s}{(s + 1)^2}.$$

Setting $s = 0$ leads to $A_1 = 0$. In this way, we once more obtain expansion (4) for $F(s)$.

EXAMPLE

2

Find $\mathcal{L}^{-1}\{F(s)\}$, where

$$F(s) = \frac{s^2 + s - 1}{(s^2 + 4s + 4)(s^2 + 2s + 5)}.$$

Solution: Since the degree of the numerator is less than the degree of the denominator, we proceed with the partial fraction expansion. The first quadratic factor in the denominator can be factored as $s^2 + 4s + 4 = (s + 2)^2$. The second quadratic factor, $s^2 + 2s + 5$, is irreducible. Therefore, the partial fraction expansion of $F(s)$ has the form

$$\frac{s^2 + s - 1}{(s + 2)^2(s^2 + 2s + 5)} = \frac{A_2}{(s + 2)^2} + \frac{A_1}{s + 2} + \frac{Bs + C}{s^2 + 2s + 5}. \tag{7}$$

We can certainly recombine the right-hand side of (7), obtaining four equations for the four unknown constants. However, as an alternative strategy, we first determine A_1 and A_2 as explained above and then use another approach to determine B and C. Multiplying equation (7) by $(s + 2)^2$, we obtain A_2:

$$A_2 = \left.\frac{s^2 + s - 1}{s^2 + 2s + 5}\right|_{s=-2} = \frac{1}{5}.$$

Next, we find A_1:

$$A_1 = \left.\frac{d}{ds}\left[\frac{s^2 + s - 1}{s^2 + 2s + 5}\right]\right|_{s=-2} = \left.\frac{s^2 + 12s + 7}{(s^2 + 2s + 5)^2}\right|_{s=-2} = -\frac{13}{25}.$$

Using these values in (7), we have

$$\frac{s^2 + s - 1}{(s + 2)^2(s^2 + 2s + 5)} = \frac{\frac{1}{5}}{(s + 2)^2} - \frac{\frac{13}{25}}{s + 2} + \frac{Bs + C}{s^2 + 2s + 5}. \tag{8}$$

(continued)

(continued)

We now determine the constants B and C by selecting two convenient values of s and evaluating (8) at these values. Setting $s = 0$ in (8) leads to

$$-\frac{1}{20} = \frac{1}{20} - \frac{13}{50} + \frac{C}{5}, \quad \text{or} \quad C = \frac{4}{5}.$$

Similarly, setting $s = -1$ in (8) gives

$$-\frac{1}{4} = \frac{1}{5} - \frac{13}{25} + \frac{-B + \frac{4}{5}}{4}, \quad \text{or} \quad B = \frac{13}{25}.$$

Therefore, the partial fraction expansion is

$$F(s) = \frac{\frac{1}{5}}{(s+2)^2} - \frac{\frac{13}{25}}{s+2} + \frac{\frac{13}{25}s + \frac{4}{5}}{(s+1)^2 + 4}. \tag{9}$$

From Table 5.1,

$$\mathcal{L}^{-1}\left\{\frac{\frac{1}{5}}{(s+2)^2}\right\} = \frac{1}{5}te^{-2t} \quad \text{and} \quad \mathcal{L}^{-1}\left\{-\frac{\frac{13}{25}}{s+2}\right\} = -\frac{13}{25}e^{-2t}, \quad t \geq 0.$$

To obtain the inverse transform of the third expression on the right-hand side of (9), we use formulas 11 and 12 of Table 5.1. To apply these formulas, we first rewrite the term:

$$\frac{\frac{13}{25}s + \frac{4}{5}}{(s+1)^2 + 4} = \frac{\frac{13}{25}(s+1) + \frac{7}{25}}{(s+1)^2 + 4} = \frac{13}{25}\frac{s+1}{(s+1)^2 + 4} + \frac{7}{50}\frac{2}{(s+1)^2 + 4}.$$

From formulas 11 and 12, we conclude that

$$\mathcal{L}^{-1}\left\{\frac{\frac{13}{25}s + \frac{4}{5}}{(s+1)^2 + 4}\right\} = \frac{13}{25}\mathcal{L}^{-1}\left\{\frac{s+1}{(s+1)^2 + 4}\right\} + \frac{7}{50}\mathcal{L}^{-1}\left\{\frac{2}{(s+1)^2 + 4}\right\}$$

$$= \frac{13}{25}e^{-t}\cos 2t + \frac{7}{50}e^{-t}\sin 2t, \quad t \geq 0.$$

Combining these results, we obtain the final answer:

$$f(t) = \mathcal{L}^{-1}\left\{\frac{s^2 + s - 1}{(s^2 + 4s + 4)(s^2 + 2s + 5)}\right\}$$

$$= \frac{1}{5}te^{-2t} - \frac{13}{25}e^{-2t} + \frac{13}{25}e^{-t}\cos 2t + \frac{7}{50}e^{-t}\sin 2t, \quad t \geq 0. \ \diamond$$

EXAMPLE 3

Use Laplace transforms to solve the initial value problem

$$y'' + 4y = 4t + 8, \qquad y(0) = 4, \qquad y'(0) = -1.$$

Solution: Let $y(t)$ denote the solution, and let $Y(s) = \mathcal{L}\{y(t)\}$. Taking Laplace transforms of both sides of $y'' + 4y = 4t + 8$, we obtain

$$\mathcal{L}\{y''\} + 4\mathcal{L}\{y\} = \frac{4}{s^2} + \frac{8}{s}. \tag{10}$$

From Theorem 5.5,

$$\mathcal{L}\{y''\} = s^2 Y(s) - sy(0) - y'(0),$$

and therefore

$$s^2 Y(s) - 4s + 1 + 4Y(s) = \frac{4}{s^2} + \frac{8}{s}. \tag{11}$$

Note that both initial conditions enter into equation (11). The solution of the problem in the transform domain is

$$Y(s) = \frac{4s^3 - s^2 + 8s + 4}{s^2(s^2 + 4)},$$

which has partial fraction expansion

$$Y(s) = \frac{1}{s^2} + \frac{2}{s} + \frac{2s}{s^2 + 4} - \frac{2}{s^2 + 4}.$$

We obtain the solution of the original initial value problem by taking the inverse Laplace transform,

$$y(t) = \mathcal{L}^{-1}\left\{\frac{1}{s^2}\right\} + 2\mathcal{L}^{-1}\left\{\frac{1}{s}\right\} + 2\mathcal{L}^{-1}\left\{\frac{s}{s^2 + 4}\right\} - \mathcal{L}^{-1}\left\{\frac{2}{s^2 + 4}\right\}$$

$$= t + 2 + 2\cos 2t - \sin 2t, \qquad t \geq 0. \; \diamond$$

EXERCISES

Exercises 1–8:

Give the *form* of the partial fraction expansion for the given rational function $F(s)$. You need not evaluate the constants in the expansion. However, if the denominator of $F(s)$ contains irreducible quadratic factors of the form $s^2 + 2\alpha s + \beta^2$, $\beta^2 > \alpha^2$, complete the square and rewrite this factor in the form $(s + \alpha)^2 + \omega^2$.

1. $F(s) = \dfrac{2s + 3}{(s - 1)(s - 2)^2}$

2. $F(s) = \dfrac{s^3 + 3s + 1}{(s - 1)^3(s - 2)^2}$

3. $F(s) = \dfrac{s^2 + 1}{s^2(s^2 + 2s + 10)}$

4. $F(s) = \dfrac{s^2 + 5s - 3}{(s^2 + 16)(s - 2)}$

5. $F(s) = \dfrac{s^2 - 1}{(s^2 - 9)^2}$

6. $F(s) = \dfrac{s^3 - 1}{(s^2 + 1)^2(s + 4)^2}$

7. $F(s) = \dfrac{s^2 + s + 2}{(s^2 + 8s + 17)(s^2 + 6s + 13)}$

8. $F(s) = \dfrac{s^4 + 5s^2 + 2s - 9}{(s^2 + 8s + 17)^2(s - 2)^2}$

Exercises 9–17:

Find the inverse Laplace transform.

9. $F(s) = \dfrac{2}{s - 3}$

10. $F(s) = \dfrac{1}{(s + 1)^3}$

11. $F(s) = \dfrac{4s + 5}{s^2 + 9}$

12. $F(s) = \dfrac{2s - 3}{s^2 - 3s + 2}$

13. $F(s) = \dfrac{3s + 7}{s^2 + 4s + 3}$

14. $F(s) = \dfrac{4s^2 + s + 1}{s^3 + s}$

15. $F(s) = \dfrac{3s^2 + s + 8}{s^3 + 4s}$

16. $F(s) = \dfrac{s^2 + 6s + 8}{s^4 + 8s^2 + 16}$

17. $F(s) = \dfrac{s}{s^3 - 3s^2 + 3s - 1}$

Exercises 18–29:

Use the Laplace transform to solve the initial value problem.

18. $y' + 2y = 26\sin 3t, \; y(0) = 3$

19. $y' - 3y = 13\cos 2t, \; y(0) = 1$

20. $y' + 2y = 4t, \; y(0) = 3$

21. $y' - 3y = e^{3t}, \; y(0) = 1$

22. $y'' + 3y' + 2y = 6e^{-t}$, $y(0) = 1$, $y'(0) = 2$

23. $y'' + 4y = 8t$, $y(0) = 2$, $y'(0) = 6$ **24.** $y'' + 4y = \cos 2t$, $y(0) = 1$, $y'(0) = 1$

25. $y'' + 4y = \sin 2t$, $y(0) = 1$, $y'(0) = 0$

26. $y'' - 2y' + y = e^{2t}$, $y(0) = 0$, $y'(0) = 0$

27. $y'' + 2y' + y = e^{-t}$, $y(0) = 0$, $y'(0) = 1$

28. $y'' + 9y = g(t)$, $y(0) = 1$, $y'(0) = 3$, $g(t) = \begin{cases} 6, & 0 \le t < \pi \\ 0, & \pi \le t < \infty \end{cases}$

29. $y'' + y = g(t)$, $y(0) = 1$, $y'(0) = 0$, $g(t) = \begin{cases} t, & 0 \le t < 2 \\ 0, & 2 \le t < \infty \end{cases}$

Exercises 30–32:

Consider the initial value problem $y'' + \alpha y' + \beta y = 0, y(0) = y_0, y'(0) = y_0'$. The Laplace transform of the solution, $Y(s) = \mathcal{L}\{y(t)\}$, is given. Determine the constants α, β, y_0, and y_0'.

30. $Y(s) = \dfrac{2s - 1}{s^2 + s + 2}$ **31.** $Y(s) = \dfrac{3}{s^2 - 4}$ **32.** $Y(s) = \dfrac{s}{(s + 1)^2}$

5.4   Laplace Transforms of Periodic Functions and System Transfer Functions

In many applications, the nonhomogeneous term in a linear differential equation is a periodic function. We now derive a formula for the Laplace transform of such periodic functions.

Theorem 5.6

Let $f(t)$ be a piecewise continuous periodic function defined on $0 \le t < \infty$, where $f(t)$ has period T. Then

$$\mathcal{L}\{f(t)\} = \frac{\displaystyle\int_0^T e^{-st} f(t)\, dt}{1 - e^{-sT}}, \qquad s > 0.$$

● **PROOF:**   Since $f(t)$ is piecewise continuous, it is bounded on the interval $0 \le t \le T$. Since the function is also periodic, it follows that $f(t)$ is bounded on $0 \le t < \infty$. Therefore, by Theorem 5.1, its Laplace transform exists for $s > 0$. Computing the transform, we find

$$F(s) = \mathcal{L}\{f(t)\} = \int_0^\infty f(t)e^{-st}\, dt = \sum_{n=0}^\infty \int_{nT}^{(n+1)T} f(t)e^{-st}\, dt. \qquad \text{(1)}$$

[In the last step of equation (1), we have decomposed the improper integral into a sum of integrals over the constituent periods.] Consider a representative integral in (1),

$$\int_{nT}^{(n+1)T} f(t)e^{-st}\, dt,$$

where n is an arbitrary but fixed integer. Making the change of variables $\tau = t - nT$ yields

$$\int_{nT}^{(n+1)T} f(t)e^{-st}\,dt = \int_0^T f(\tau + nT)e^{-s(\tau+nT)}\,d\tau = e^{-snT}\int_0^T f(\tau)e^{-s\tau}\,d\tau,$$

where the last equality follows from the periodicity of f. Thus, equation (1) reduces to

$$F(s) = \sum_{n=0}^{\infty} e^{-snT}\int_0^T f(\tau)e^{-s\tau}\,d\tau = \left[\int_0^T f(\tau)e^{-s\tau}\,d\tau\right]\sum_{n=0}^{\infty} e^{-snT}. \qquad \textbf{(2)}$$

Since $s > 0$, it follows that $0 < e^{-sT} < 1$. Therefore, the infinite series in equation (2) is a convergent geometric series,

$$\sum_{n=0}^{\infty} e^{-snT} = \sum_{n=0}^{\infty} (e^{-sT})^n = \frac{1}{1 - e^{-sT}},$$

and Theorem 5.6 follows. ●

E X A M P L E

1

Let T be a positive constant, and consider the square wave

$$f(t) = \begin{cases} 1, & 0 \le t \le \dfrac{T}{2}, \\[2mm] 0, & \dfrac{T}{2} < t < T, \end{cases} \qquad f(t+T) = f(t), \qquad t \ge 0.$$

The graph of $f(t)$ is shown in Figure 5.11. Use Theorem 5.6 to determine the Laplace transform of $f(t)$.

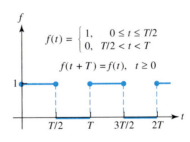

FIGURE 5.11

The graph of the square wave $f(t)$ treated in Example 1. Note that $f(t)$ is periodic with period T and is piecewise continuous on $0 \le t < \infty$.

Solution: By Theorem 5.6,

$$\mathcal{L}\{f(t)\} = \frac{\displaystyle\int_0^T e^{-st}f(t)\,dt}{1 - e^{-sT}} = \frac{\displaystyle\int_0^{T/2} e^{-st}\,dt}{1 - e^{-sT}}, \qquad s > 0.$$

Evaluating this last integral, we find

$$\mathcal{L}\{f(t)\} = \frac{\left(1 - e^{-sT/2}\right)s^{-1}}{1 - e^{-sT}} = \frac{1}{s\left(1 + e^{-sT/2}\right)}. \quad ❖$$

E X A M P L E

2

Find the inverse transform of

$$F(s) = \frac{1}{s^2} - \frac{e^{-s}}{s(1 - e^{-s})}, \qquad s > 0.$$

Solution: Applying the inverse transform operation yields

$$\mathcal{L}^{-1}\{F(s)\} = \mathcal{L}^{-1}\left\{\frac{1}{s^2}\right\} - \mathcal{L}^{-1}\left\{\frac{e^{-s}}{s(1 - e^{-s})}\right\} = t - \mathcal{L}^{-1}\left\{\frac{e^{-s}}{s(1 - e^{-s})}\right\}.$$

Expanding the factor $(1 - e^{-s})^{-1}$ as a geometric series and applying the inverse operation termwise to the convergent series, we obtain

$$\mathcal{L}^{-1}\left\{\frac{e^{-s}}{s(1 - e^{-s})}\right\} = \mathcal{L}^{-1}\left\{\frac{e^{-s}}{s}[1 + e^{-s} + e^{-2s} + e^{-3s} + \cdots]\right\}$$

$$= \mathcal{L}^{-1}\left\{\frac{1}{s}[e^{-s} + e^{-2s} + e^{-3s} + \cdots]\right\}$$

$$= \mathcal{L}^{-1}\left\{\frac{e^{-s}}{s}\right\} + \mathcal{L}^{-1}\left\{\frac{e^{-2s}}{s}\right\} + \mathcal{L}^{-1}\left\{\frac{e^{-3s}}{s}\right\} + \cdots$$

$$= h(t - 1) + h(t - 2) + h(t - 3) + \cdots.$$

The function $g(t) = h(t - 1) + h(t - 2) + h(t - 3) + \cdots$ has the staircase-like graph shown in Figure 5.12(a). Thus, the inverse transform of $F(s)$ is the sawtooth wave function $f(t) = t - g(t)$ whose graph is shown in Figure 5.12(b). [Note: For any fixed value of t, $g(t) = h(t - 1) + h(t - 2) + h(t - 3) + \cdots$ is actually a finite sum, since $h(t - \alpha) = 0$ when $t < \alpha$. For instance, if $t = 2.5$, then $g(2.5) = 2$.]

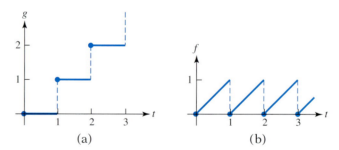

(a)   (b)

FIGURE 5.12

(a) The graph of $g(t) = h(t - 1) + h(t - 2) + h(t - 3) + \cdots$ resembles a staircase. (b) The inverse transform of $F(s)$ in Example 2 is $f(t) = t - g(t)$; this graph is often called a sawtooth wave. ❖

Solution of Parameter Identification Problems and the System Transfer Function

Example 1 of Section 5.1 posed the problem of studying a "black box" that housed a spring-mass-dashpot mechanical system (see Figure 5.2). Two specific questions are

1. If we subject the initially quiescent system to a known force $f(t)$, starting at $t = 0$, and measure the subsequent displacement $y(t)$ for $t \geq 0$, can we

use our measurements to predict what the displacement $\tilde{y}(t)$ would be if a different force $\tilde{f}(t)$ were applied?

2. Can we use our knowledge of the input-output relation [that is, our knowledge of $f(t)$ and $y(t)$] to determine the mass m, the spring constant k, and the damping coefficient γ of the unknown mechanical system?

The relevant mathematical problem is

$$my'' + \gamma y' + ky = f(t), \qquad t > 0$$
$$y(0) = 0, \qquad y'(0) = 0. \tag{3}$$

We now use Laplace transforms to provide affirmative answers to these two questions.

Taking Laplace transforms of both sides of equation (3) and noting the zero initial conditions, we have

$$ms^2 Y(s) + \gamma s Y(s) + k Y(s) = F(s),$$

or

$$Y(s) = \left[\frac{1}{ms^2 + \gamma s + k} \right] F(s). \tag{4}$$

Although the computations in (4) are simple, the result is important. In the time domain, we obtain the output $y(t)$ from the input $f(t)$ by solving an initial value problem. In the transform domain, however, we obtain the output $Y(s)$ from the input $F(s)$ by multiplying $F(s)$ by the function

$$\Phi(s) = \frac{1}{ms^2 + \gamma s + k}. \tag{5}$$

Note that the function $\Phi(s)$ in (5) depends only on the mechanical system; it is sometimes referred to as the **system transfer function**. If we know $\Phi(s)$, we can use multiplication to determine the output $Y(s)$ arising from a given input $F(s)$. Conversely, if we know some input-output pair $F(s)$ and $Y(s)$, we can determine the system transfer function $\Phi(s)$ by forming the quotient

$$\Phi(s) = \frac{Y(s)}{F(s)}.$$

The role of the system transfer function is shown schematically in Figure 5.13.

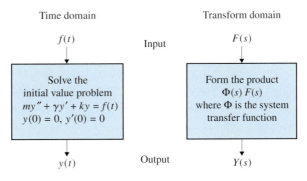

FIGURE 5.13

There are two ways to analyze the mechanical spring-mass-dashpot system. In the time domain, solve the initial value problem given by equation (3). In the transform domain, form the product of the system transfer function $\Phi(s)$ and the input $F(s)$, as in equation (4).

Example 3 illustrates how these transform domain ideas can be used to answer the two questions posed above.

Suppose we know that the response of an initially quiescent mechanical system to an applied force can be modeled as the solution of the spring-mass-dashpot initial value problem

$$my'' + \gamma y' + y = f(t), \qquad t > 0$$
$$y(0) = 0, \qquad y'(0) = 0.$$

Assume, for a known applied force $f(t)$, we can measure the resulting displacement $y(t)$ for $t \geq 0$. We are, however, unable to directly determine the parameters m, γ, and k.

In particular, suppose when we apply a unit step force $f(t) = h(t)$, the displacement is

$$y(t) = -\tfrac{1}{2}e^{-t}\cos t - \tfrac{1}{2}e^{-t}\sin t + \tfrac{1}{2}, \qquad t \geq 0.$$

Use this information to

(a) Predict the displacement should the force $\tilde{f}(t) = e^{-2t}, t \geq 0$ be applied.
(b) Determine the parameters m, γ, and k.

Solution: To solve the problem, we first compute Laplace transforms of the applied force $f(t) = h(t)$ and the ensuing response $y(t) = -\tfrac{1}{2}e^{-t}\cos t - \tfrac{1}{2}e^{-t}\sin t + \tfrac{1}{2}$. We obtain

$$F(s) = \frac{1}{s}$$

$$Y(s) = -\frac{1}{2}\frac{s+1}{(s+1)^2 + 1} - \frac{1}{2}\frac{1}{(s+1)^2 + 1} + \frac{1}{2s} = \frac{1}{s(s^2 + 2s + 2)}.$$

In the transform domain, $Y(s) = \Phi(s)F(s)$, where $\Phi(s)$ is the system transfer function. Therefore, the system transfer function is given by

$$\Phi(s) = \frac{Y(s)}{F(s)} = \frac{1}{s^2 + 2s + 2}. \tag{6}$$

Once we know the system transfer function, we can readily predict the output corresponding to any input.

(a) Suppose the applied force is $\tilde{f}(t) = e^{-2t}$. The Laplace transform of the applied force is $\tilde{F}(s) = 1/(s+2)$. We can find the transform of the displacement from the relationship $\tilde{Y}(s) = \Phi(s)\tilde{F}(s)$:

$$\tilde{Y}(s) = \left[\frac{1}{s^2 + 2s + 2}\right]\frac{1}{s+2} = \frac{1}{(s^2 + 2s + 2)(s+2)}$$

$$= \frac{1}{2}\frac{1}{s+2} - \frac{1}{2}\frac{s+1}{(s+1)^2 + 1} + \frac{1}{2}\frac{1}{(s+1)^2 + 1}.$$

The corresponding time domain output is thus

$$\tilde{y}(t) = \mathcal{L}^{-1}\{\tilde{Y}(s)\} = \tfrac{1}{2}e^{-2t} - \tfrac{1}{2}e^{-t}\cos t + \tfrac{1}{2}e^{-t}\sin t, \qquad t \geq 0.$$

(b) The problem posed in question (b) can be solved by comparing the transfer function

$$\Phi(s) = \frac{1}{s^2 + 2s + 2}$$

with the previously determined form of the transfer function in equation (5),

$$\Phi(s) = \frac{1}{ms^2 + \gamma s + k}.$$

Comparing coefficients, we conclude that

$$m = 1, \qquad \gamma = 2, \qquad k = 2. \; \diamond$$

In the preceding discussion, we assumed that the mechanical system was initially at rest and an applied force then activated the system at time $t = 0$. If the initial conditions are nonzero but known, the same general approach can be used. These ideas are developed in the Exercises.

EXERCISES

Exercises 1–8:

Find the Laplace transform of the periodic function whose graph is shown.

1.
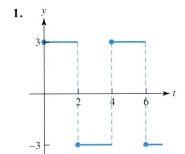

2.
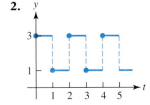

3.
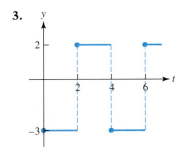

4.
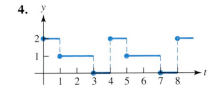

5.
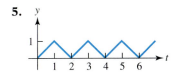

6.
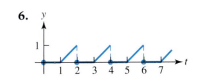

7.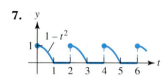

8.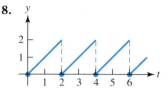

Exercises 9–12:

Sketch the graph of $f(t)$, state the period of $f(t)$, and find $\mathcal{L}\{f(t)\}$.

9. $f(t) = |\sin 2t|$

10. $f(t) = \begin{cases} \sin t, & 0 \le t < \pi, \\ 0, & \pi \le t < 2\pi, \end{cases} \quad f(t + 2\pi) = f(t)$

11. $f(t) = e^{-t}, \quad 0 \le t < 1, \quad f(t+1) = f(t)$

12. $f(t) = 1 - e^{-t}, \quad 0 \le t < 2, \quad f(t+2) = f(t)$

13. Let α be a positive constant. As in Example 2, show that

$$\mathcal{L}^{-1}\left\{ \frac{e^{-\alpha s}}{s(1 - e^{-\alpha s})} \right\} = h(t - \alpha) + h(t - 2\alpha) + h(t - 3\alpha) + \cdots.$$

Sketch the graph of $g(t) = h(t - \alpha) + h(t - 2\alpha) + h(t - 3\alpha) + \cdots$ for $\alpha = 1$ and $0 \le t < 5$.

Exercises 14–15:

In each exercise, use linearity of the inverse transformation and Exercise 13 to find $f(t) = \mathcal{L}^{-1}\{F(s)\}$ for the given transform $F(s)$. Sketch the graph of $f(t)$ for $0 \le t < 5$ in Exercise 14 and $0 \le t < 10$ in Exercise 15.

14. $F(s) = \dfrac{s-1}{s^2} + \dfrac{e^{-s}}{s(1 - e^{-s})}$

15. $F(s) = \dfrac{3}{s^2} - \dfrac{3e^{-2s}}{s(1 - e^{-2s})}$

16. As in Example 2, find $f(t) = \mathcal{L}^{-1}\{F(s)\}$ for $F(s) = 1/2s^2 - (1/s^2)[e^{-2s}/(1 + e^{-2s})]$. Sketch the graph of $f(t)$ for $0 \le t < 12$.

Exercises 17–19:

One-Dimensional Motion with Drag and Periodic Thrust   Assume a body of mass m moves along a horizontal surface in a straight line with velocity $v(t)$. The body is subject to a frictional force proportional to velocity and is propelled forward with a periodic propulsive force $f(t)$. Applying Newton's second law, we obtain the following initial value problem:

$$mv' + kv = f(t), \qquad t \ge 0, \qquad v(0) = v_0.$$

Assume that $m = 1$ kg, $k = 1$ kg/s, and $v_0 = 1$ m/s.

(a) Use Laplace transform methods to determine $v(t)$ for the propulsive force $f(t)$, where $f(t)$ is given in newtons.

(b) Plot $v(t)$ for $0 \le t \le 10$ [this time interval spans the first five periods of $f(t)$]. In Exercise 17, explain why $v(t)$ is constant on the interval $0 \le t \le 1$.

17. $f(t) = \begin{cases} 1, & 0 \le t \le 1, \\ 0, & 1 < t < 2, \end{cases} \quad f(t+2) = f(t)$

18. $f(t) = \begin{cases} 0, & 0 \le t \le 1, \\ 1, & 1 < t < 2, \end{cases} \quad f(t+2) = f(t)$

19. $f(t) = t/2, \quad 0 \le t < 2, \quad f(t+2) = f(t)$

20. An object having mass m is initially at rest on a frictionless horizontal surface. At time $t = 0$, a periodic force is applied horizontally to the object, causing it to move in the positive x-direction. The force, in newtons, is given by

$$f(t) = \begin{cases} f_0, & 0 \leq t \leq T/2, \\ 0, & T/2 < t < T, \end{cases} \qquad f(t+T) = f(t).$$

The initial value problem for the horizontal position, $x(t)$, of the object is $mx''(t) = f(t), x(0) = 0, x'(0) = 0.$

(a) Use Laplace transforms to determine the velocity, $v(t) = x'(t)$, and the position, $x(t)$, of the object.

(b) Let $m = 1$ kg, $f_0 = 1$ N, and $T = 1$ s. What are the velocity v and position x of the object at $t = 1.25$ s?

21. A lake containing 50 million gal of fresh water has a stream flowing through it. Water enters the lake at a constant rate of 5 million gal/day and leaves at the same rate. At some initial time, an upstream manufacturer begins to discharge pollutants into the feeder stream. Each day, during the hours from 8 A.M. to 8 P.M., the stream has a pollutant concentration of 1 mg/gal (10^{-6} kg/gal); at other times, the stream feeds in fresh water. Assume that a well-stirred mixture leaves the lake and that the manufacturer operates seven days per week.

(a) Let $t = 0$ denote the instant that pollutants first enter the lake. Let $q(t)$ denote the amount of pollutant (in kilograms) present in the lake at time t (in days). Use a "conservation of pollutant" principle (rate of change = rate in − rate out) to formulate the initial value problem satisfied by $q(t)$.

(b) Apply Laplace transforms to the problem formulated in (a) and determine $Q(s) = \mathcal{L}\{q(t)\}$.

(c) Determine $q(t) = \mathcal{L}^{-1}\{Q(s)\}$, using the ideas of Example 2. In particular, what is $q(t)$ for $1 \leq t < 2$, the second day of manfacturing?

22. Consider the RL and RC networks shown, with the associated equations for the current $i(t)$.

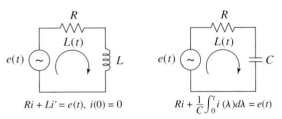

Figure for Exercise 22

Assume that the network element values are $R = 1$ kΩ, $L = 1$ H, $C = 1\mu$F and that $e(t)$, given in volts, is

$$e(t) = \begin{cases} 0, & 0 \leq t \leq 0.5, \\ 1, & 0.5 < t < 1, \end{cases} \qquad e(t+1) = e(t).$$

The associated units of current and time are milliamperes and milliseconds, respectively.

(a) Determine $i(t)$ for the RL network.

(b) Determine $i(t)$ for the RC network.

Transfer Function Problems   Consider the initial value problem

$$ay'' + by' + cy = f(t), \qquad 0 < t < \infty$$
$$y(0) = 0, \qquad y'(0) = 0,$$

(7)

where a, b, and c are constants and $f(t)$ is a known function. We can view problem (7) as defining a linear system, as shown schematically in the figure, where $f(t)$ is a known input and the corresponding solution $y(t)$ is the output. As we have seen, Laplace transforms of the input and output functions satisfy the multiplicative relation, $Y(s) = \Phi(s)F(s)$, where $\Phi(s)$ is the system transfer function.

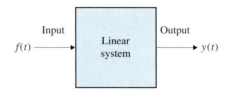

Input   Output

$f(t)$ ———→ | Linear system | ———→ $y(t)$

Figure for Exercises 23–26

23. Show that the term "linear system" is appropriate. In particular, show that if an input $f_1(t)$ produces an output $y_1(t)$ and an input $f_2(t)$ produces an output $y_2(t)$, then the input $f(t) = c_1 f_1(t) + c_2 f_2(t)$ produces the output $y(t) = c_1 y_1(t) + c_2 y_2(t)$. [Hint: Use the superposition principle discussed in Section 3.7.]

24. Suppose that the transfer function for linear system (7) is $\Phi(s) = 1/(2s^2 + 5s + 2)$.

(a) What are the constants a, b, and c?

(b) If $f(t) = e^{-t}$, determine $F(s)$, $Y(s)$, and $y(t)$.

25. Suppose an input $f(t) = t$, when applied to linear system (7), produces the output $y(t) = 2(e^{-t} - 1) + t(e^{-t} + 1), t \geq 0$. What is the system transfer function, $\Phi(s)$?

26. Suppose an input $f(t) = t$, when applied to linear system (7), produces the output $y(t) = 2(e^{-t} - 1) + t(e^{-t} + 1), t \geq 0$. What will be the output if a Heaviside unit step input $f(t) = h(t)$ is applied to the system?

Exercises 27–31:

For the linear system defined by the given initial value problem,

(a) Determine the system transfer function, $\Phi(s)$.

(b) Determine the Laplace transform of the output, $Y(s)$, corresponding to the specified input, $f(t)$.

27. $y'' + 4y = f(t)$, $y(0) = 0$, $y'(0) = 0$; $f(t) = t^2$

28. $y'' + y' + y = f(t)$, $y(0) = 0$, $y'(0) = 0$; $f(t) = \begin{cases} 1, & 0 \leq t \leq 1, \\ -1, & 1 < t < 2, \end{cases}$ $f(t+2) = f(t)$

29. $y'' + 4y' + 4y = f(t)$, $y(0) = 0$, $y'(0) = 0$; $f(t) = t$, $0 \leq t < 1$, $f(t+1) = f(t)$

30. $y''' - 4y = f(t)$, $y(0) = 0$, $y'(0) = 0$, $y''(0) = 0$; $f(t) = e^t + t$

31. $y''' + 4y' = f(t)$, $y(0) = 0$, $y'(0) = 0$, $y''(0) = 0$; $f(t) = \cos 2t$

Exercises 32–33:

We now allow the initial values to be nonzero. Consider the initial value problem

$$y'' + by' + cy = f(t), \qquad 0 < t < \infty$$
$$y(0) = y_0, \qquad y'(0) = y_0'.$$

The input function, $f(t)$, and the Laplace transform of the output function, $Y(s)$, are given. Determine the constants b, c, y_0, and y_0'.

32. $f(t) = h(t)$, the Heaviside unit step function; $Y(s) = (s^2 + 2s + 1)/(s^3 + 3s^2 + 2s)$

33. $f(t) = e^{-t}$; $Y(s) = (s^2 + s + 1)/[(s + 1)(s^2 + 4)]$

5.5   Solving Systems of Differential Equations

In this section, we extend the definition of the Laplace transform to matrix-valued functions and take note of some simple consequences of the extension. We then see how to use Laplace transforms to solve problems involving systems of differential equations.

Laplace Transforms of Matrix-Valued Functions

As we saw in Section 4.1, the integral of a matrix-valued function is simply the matrix of integrals. Similarly, the Laplace transform of a matrix-valued function is the matrix of Laplace transforms. Consider the vector-valued function

$$\mathbf{y}(t) = \begin{bmatrix} y_1(t) \\ y_2(t) \\ \vdots \\ y_n(t) \end{bmatrix}, \tag{1}$$

where each of the component functions is piecewise continuous and exponentially bounded on $0 \leq t < \infty$. The Laplace transform, $\mathcal{L}\{\mathbf{y}(t)\}$, is

$$\mathcal{L}\{\mathbf{y}(t)\} = \int_0^\infty \mathbf{y}(t)e^{-st}\,dt$$

$$= \int_0^\infty \begin{bmatrix} y_1(t) \\ y_2(t) \\ \vdots \\ y_n(t) \end{bmatrix} e^{-st}\,dt = \begin{bmatrix} \int_0^\infty y_1(t)e^{-st}\,dt \\ \int_0^\infty y_2(t)e^{-st}\,dt \\ \vdots \\ \int_0^\infty y_n(t)e^{-st}\,dt \end{bmatrix} = \begin{bmatrix} Y_1(s) \\ Y_2(s) \\ \vdots \\ Y_n(s) \end{bmatrix} = \mathbf{Y}(s). \tag{2}$$

We will use uppercase bold letters to denote the Laplace transform of a vector-valued function.

Similarly, the Laplace transform of an $(m \times n)$ matrix-valued function is the $(m \times n)$ matrix consisting of the Laplace transforms of the component functions. In general, if each component function of a matrix-valued function is Laplace transformable, we say that the matrix function itself is **Laplace transformable**.

EXAMPLE 1

Compute $\mathcal{L}\{\mathbf{y}(t)\}$, where

$$\mathbf{y}(t) = \begin{bmatrix} t \\ -1 \\ e^t \end{bmatrix}.$$

(continued)

(continued)

Solution: Using Table 5.1, we have

$$\mathbf{Y}(s) = \begin{bmatrix} \mathcal{L}\{t\} \\ \mathcal{L}\{-1\} \\ \mathcal{L}\{e^t\} \end{bmatrix} = \begin{bmatrix} \dfrac{1}{s^2} \\ -\dfrac{1}{s} \\ \dfrac{1}{s-1} \end{bmatrix}, \qquad s > 1.$$

Note that the domain of $\mathbf{Y}(s)$ is the intersection of the domains of the component functions. ❖

Some Useful Matrix Formulas

The following results can be established by taking the Laplace transform of each component function and then reassembling the components into a single expression.

1. Let A be a constant $(n \times n)$ matrix, and let $\mathbf{y}(t)$ be an $(n \times p)$ Laplace transformable matrix function. Then

$$\mathcal{L}\{A\mathbf{y}(t)\} = A\mathcal{L}\{\mathbf{y}(t)\} = A\mathbf{Y}(s). \tag{3}$$

2. If each component function satisfies the appropriate hypotheses of Theorem 5.5, then

$$\mathcal{L}\{\mathbf{y}'(t)\} = s\mathbf{Y}(s) - \mathbf{y}(0)$$
$$\mathcal{L}\{\mathbf{y}''(t)\} = s^2\mathbf{Y}(s) - s\mathbf{y}(0) - \mathbf{y}'(0)$$
$$\mathcal{L}\left\{\int_0^t \mathbf{y}(u)\,du\right\} = \frac{1}{s}\mathbf{Y}(s). \tag{4}$$

Solution of the Initial Value Problem for a Nonhomogeneous System

Consider the initial value problem

$$\mathbf{y}' = A\mathbf{y} + \mathbf{g}(t), \qquad t > 0, \qquad \mathbf{y}(0) = \mathbf{y}_0, \tag{5}$$

where $\mathbf{y}(t)$ is the $(n \times 1)$ vector of unknowns and A is a real-valued $(n \times n)$ constant matrix. We also assume the nonhomogeneous term,

$$\mathbf{g}(t) = \begin{bmatrix} g_1(t) \\ g_2(t) \\ \vdots \\ g_n(t) \end{bmatrix},$$

is a Laplace transformable vector function.

Using formulas (3) and (4), we can take the Laplace transform of system (5) and work directly with the matrices rather than dealing with the component equations. We obtain

$$s\mathbf{Y}(s) - \mathbf{y}(0) = A\mathbf{Y}(s) + \mathbf{G}(s),$$

or

$$(sI - A)\mathbf{Y}(s) = \mathbf{y}_0 + \mathbf{G}(s),$$

where $\mathbf{G}(s) = \mathcal{L}\{\mathbf{g}(t)\}$. The solution of the transform domain problem is therefore

$$\mathbf{Y}(s) = (sI - A)^{-1}[\mathbf{y}_0 + \mathbf{G}(s)]. \tag{6}$$

To compute the desired time domain solution, $\mathbf{y}(t) = \mathcal{L}^{-1}\{\mathbf{Y}(s)\}$, we compute the inverse Laplace transform of each component of $\mathbf{Y}(s)$. [Note that $(sI - A)^{-1}$ does not exist when s is an eigenvalue of A.]

EXAMPLE

2

Solve the initial value problem

$$\mathbf{y}' = A\mathbf{y} + \mathbf{g}(t), \qquad 0 \le t < \infty, \qquad \mathbf{y}(0) = \mathbf{y}_0,$$

where

$$A = \begin{bmatrix} 1 & 2 \\ 2 & 1 \end{bmatrix}, \qquad \mathbf{g}(t) = \begin{bmatrix} e^{2t} \\ -2t \end{bmatrix}, \qquad \mathbf{y}_0 = \begin{bmatrix} 1 \\ -2 \end{bmatrix}.$$

(We computed the general solution of this nonhomogeneous linear first order system earlier, in Example 1 of Section 4.8.)

Solution: Taking Laplace transforms and using equations (3) and (4), we obtain [as in equation (6)]

$$\mathbf{Y}(s) = (sI - A)^{-1}[\mathbf{y}_0 + \mathbf{G}(s)],$$

where

$$\mathbf{G}(s) = \begin{bmatrix} \dfrac{1}{s - 2} \\[2mm] -\dfrac{2}{s^2} \end{bmatrix}.$$

Note that

$$(sI - A)^{-1} = \begin{bmatrix} s - 1 & -2 \\ -2 & s - 1 \end{bmatrix}^{-1} = \frac{1}{(s + 1)(s - 3)} \begin{bmatrix} s - 1 & 2 \\ 2 & s - 1 \end{bmatrix}.$$

Therefore, the transform domain solution is

$$\mathbf{Y}(s) = \frac{1}{(s + 1)(s - 3)} \begin{bmatrix} s - 1 & 2 \\ 2 & s - 1 \end{bmatrix} \begin{bmatrix} 1 + \dfrac{1}{s - 2} \\[2mm] -2 - \dfrac{2}{s^2} \end{bmatrix} = \begin{bmatrix} \dfrac{s^4 - 6s^3 + 9s^2 - 4s + 8}{s^2(s + 1)(s - 2)(s - 3)} \\[4mm] \dfrac{-2s^4 + 8s^3 - 8s^2 + 6s - 4}{s^2(s + 1)(s - 2)(s - 3)} \end{bmatrix}.$$

(continued)

(continued)

To obtain the time domain solution, we need to determine the inverse Laplace transform of each component of $\mathbf{Y}(s)$. Using a partial fraction expansion, we write $Y_1(s)$ and $Y_2(s)$ as

$$Y_1(s) = \frac{s^4 - 6s^3 + 9s^2 - 4s + 8}{s^2(s+1)(s-2)(s-3)} = \frac{4}{3}\frac{1}{s^2} - \frac{8}{9}\frac{1}{s} + \frac{7}{3}\frac{1}{s+1} - \frac{1}{3}\frac{1}{s-2} - \frac{1}{9}\frac{1}{s-3}$$

$$Y_2(s) = \frac{-2s^4 + 8s^3 - 8s^2 + 6s - 4}{s^2(s+1)(s-2)(s-3)} = -\frac{2}{3}\frac{1}{s^2} + \frac{10}{9}\frac{1}{s} - \frac{7}{3}\frac{1}{s+1} - \frac{2}{3}\frac{1}{s-2} - \frac{1}{9}\frac{1}{s-3}.$$

Therefore,

$$y_1(t) = \mathcal{L}^{-1}\{Y_1(s)\} = \tfrac{4}{3}t - \tfrac{8}{9} + \tfrac{7}{3}e^{-t} - \tfrac{1}{3}e^{2t} - \tfrac{1}{9}e^{3t}$$

$$y_2(t) = \mathcal{L}^{-1}\{Y_2(s)\} = -\tfrac{2}{3}t + \tfrac{10}{9} - \tfrac{7}{3}e^{-t} - \tfrac{2}{3}e^{2t} - \tfrac{1}{9}e^{3t}, \qquad t \geq 0.$$

We can regroup these terms into the following matrix solution:

$$\mathbf{y}(t) = t\begin{bmatrix} \tfrac{4}{3} \\ -\tfrac{2}{3} \end{bmatrix} + \begin{bmatrix} -\tfrac{8}{9} \\ \tfrac{10}{9} \end{bmatrix} + e^{-t}\begin{bmatrix} \tfrac{7}{3} \\ -\tfrac{7}{3} \end{bmatrix} + e^{2t}\begin{bmatrix} -\tfrac{1}{3} \\ -\tfrac{2}{3} \end{bmatrix} + e^{3t}\begin{bmatrix} -\tfrac{1}{9} \\ -\tfrac{1}{9} \end{bmatrix}, \qquad t \geq 0. \qquad \textbf{(7)}$$

As a check, you can compare solution (7) with the general solution obtained in Section 4.8, Example 1. What values of c_1 and c_2 are needed in Example 1 of Section 4.8 in order to replicate solution (7)? ❖

The System Transfer Function

The preceding discussion indicates that we can identify a system transfer function for a linear constant coefficient system. Consider, in particular, the transform domain solution given by equation (6),

$$\mathbf{Y}(s) = (sI - A)^{-1}[\mathbf{y}_0 + \mathbf{G}(s)].$$

The vector $\mathbf{y}_0 + \mathbf{G}(s)$ is the sum of the initial condition and the transformed nonhomogeneous term; this sum represents the system input. The system output, $\mathbf{Y}(s)$, is obtained by premultiplying the input by the square matrix $(sI - A)^{-1}$. Therefore, the matrix $(sI - A)^{-1}$ is the system transfer function (also called the system transfer matrix). Note that the system transfer matrix for $\mathbf{y}' = A\mathbf{y} + \mathbf{g}(t)$ depends only on the coefficient matrix A.

We now show that the system transfer matrix, $(sI - A)^{-1}$, is actually the Laplace transform of the exponential matrix, e^{tA}. To see why, consider the matrix initial value problem

$$\Phi' = A\Phi, \qquad \Phi(0) = I, \qquad \textbf{(8)}$$

where A is a constant $(n \times n)$ matrix and I is the $(n \times n)$ identity matrix. As we saw in Section 4.10, the solution of initial value problem (8) is

$$\Phi(t) = e^{tA}.$$

However, when we take the Laplace transform of equation (8), we obtain

$$s\mathcal{L}\{\Phi\} - I = A\mathcal{L}\{\Phi\}.$$

Solving for $\mathcal{L}\{\Phi\}$, we find $\mathcal{L}\{\Phi\} = (sI - A)^{-1}$. But, since $\Phi(t) = e^{tA}$, we are led to

$$\mathcal{L}\{e^{tA}\} = (sI - A)^{-1}.$$

This equation is an elegant generalization of the familiar formula

$$\mathcal{L}\{e^{\alpha t}\} = (s - \alpha)^{-1}.$$

A Network Example

Laplace transforms provide a convenient tool for analyzing networks having a more complicated structure than the single loop or single node networks we have studied thus far. As an example, consider the two-loop network shown in Figure 5.14. We assume that the network is initially quiescent; that is, both loop currents are zero at time $t = 0$, and the capacitor has no initial charge. At time $t = 0$, the voltage source $v(t)$ is turned on.

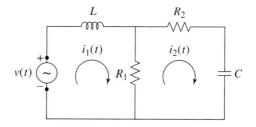

FIGURE 5.14

A two-loop network. The loop currents, $i_1(t)$ and $i_2(t)$, are found by solving the linear system (9).

The mathematical description of this network's behavior is obtained by applying Kirchhoff's voltage law to each loop:

$$v(t) = L \frac{di_1}{dt} + R_1 i_1 - R_1 i_2$$

$$0 = R_1 i_2 - R_1 i_1 + R_2 i_2 + \frac{1}{C} \int_0^t i_2(\lambda) \, d\lambda \qquad \textbf{(9)}$$

$$i_1(0) = i_2(0) = 0.$$

Taking the Laplace transform in equation (9), we obtain

$$V(s) = sLI_1(s) + R_1 I_1(s) - R_1 I_2(s)$$

$$0 = R_1 I_2(s) - R_1 I_1(s) + R_2 I_2(s) + \frac{1}{Cs} I_2(s).$$

This system can be written in matrix form as

$$\begin{bmatrix} R_1 + sL & -R_1 \\ -R_1 & R_1 + R_2 + \dfrac{1}{Cs} \end{bmatrix} \begin{bmatrix} I_1(s) \\ I_2(s) \end{bmatrix} = \begin{bmatrix} V(s) \\ 0 \end{bmatrix}, \qquad \textbf{(10)}$$

where $V(s)$ is the Laplace transform of the known voltage $v(t)$. Note that equation (10) incorporates the initial conditions $i_1(0) = i_2(0) = 0$. We obtain the transform domain solution

$$\begin{bmatrix} I_1(s) \\ I_2(s) \end{bmatrix} = \frac{sV(s)}{(R_1 + R_2)Ls^2 + \left(R_1 R_2 + \dfrac{L}{C}\right)s + \dfrac{R_1}{C}} \begin{bmatrix} R_1 + R_2 + \dfrac{1}{Cs} \\ R_1 \end{bmatrix}.$$

As a convenient particular case, we'll assume the following network element values:

$$R_1 = R_2 = 1 \text{ k}\Omega, \qquad L = 0.5 \text{ H}, \qquad C = 0.5 \text{ } \mu\text{F}.$$

Likewise, we assume that the input voltage is $v(t) = h(t)$, where $h(t)$ is the unit step function; in other words, a 1-volt DC voltage source is switched on at time $t = 0$. Given these values, the transform domain solutions in (10) become

$$I_1(s) = \frac{2(s+1)}{s(s^2 + 2s + 2)} = \frac{1}{s} - \left[\frac{s+1}{(s+1)^2 + 1} - \frac{1}{(s+1)^2 + 1} \right]$$

$$I_2(s) = \frac{1}{s^2 + 2s + 2} = \frac{1}{(s+1)^2 + 1}.$$

Therefore, the resulting time domain network loop currents are

$$i_1(t) = 1 - e^{-t}[\cos t - \sin t]$$

$$i_2(t) = e^{-t} \sin t, \qquad t \geq 0,$$

(11)

where the units of current and time are milliamperes and milliseconds, respectively.

The loop currents behave qualitatively as one would expect. In particular, as $t \to \infty$, the current in Loop 1 approaches a constant unit value and the current in Loop 2 tends to zero. In the limit, the inductor voltage tends to zero and the unit current produces a voltage drop across resistor R_1 equal to the source voltage of 1 volt. In Loop 2, the capacitor voltage,

$$\frac{1}{C} \int_0^t i_2(\lambda) \, d\lambda = 2 \int_0^t e^{-\lambda} \sin \lambda \, d\lambda = 1 - e^{-t}(\sin t + \cos t),$$

tends to unity as $t \to \infty$. In this loop, the voltage across resistor R_2 tends to zero; the buildup of charge and accompanying voltage drop across the capacitor ultimately balance the voltage across resistor R_1. Graphs of the loop currents, $i_1(t)$ and $i_2(t)$, are displayed in Figure 5.15.

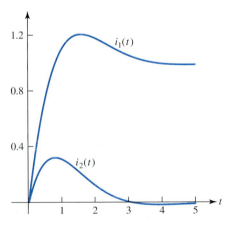

FIGURE 5.15

Graphs of the time domain loop currents given in equation (11).

EXERCISES

Exercises 1–5:

Compute the Laplace transform of the given matrix-valued function $\mathbf{y}(t)$.

1. $\mathbf{y}(t) = \begin{bmatrix} \cos t \\ t \\ te^t \end{bmatrix}$

2. $\mathbf{y}(t) = \dfrac{d}{dt} \begin{bmatrix} e^{-t}\cos 2t \\ 0 \\ t + e^t \end{bmatrix}$

3. $\mathbf{y}(t) = \begin{bmatrix} 1 & -1 \\ 0 & 2 \end{bmatrix} \begin{bmatrix} 2t \\ h(t-2) \end{bmatrix}$

4. $\mathbf{y}(t) = \displaystyle\int_0^t \begin{bmatrix} 1 \\ \lambda \\ e^{-\lambda} \end{bmatrix} d\lambda$

5. $\mathbf{y}(t) = \begin{bmatrix} h(t-1)\sin(t-1) & 0 \\ e^{t-1} & t \end{bmatrix} \begin{bmatrix} 1 \\ -2 \end{bmatrix}$

Exercises 6–8:

Compute the inverse Laplace transform of the given matrix function $\mathbf{Y}(s)$.

6. $\mathbf{Y}(s) = \begin{bmatrix} \dfrac{1}{s} \\[2mm] \dfrac{2}{s^2 + 2s + 2} \\[2mm] \dfrac{1}{s^2 + s} \end{bmatrix}$

7. $\mathbf{Y}(s) = e^{-s} \begin{bmatrix} 1 & -1 \\ 0 & 2 \end{bmatrix} \begin{bmatrix} \dfrac{1}{s} \\[2mm] \dfrac{1}{s^2 + 1} \end{bmatrix}$

8. $\mathbf{Y}(s) = \begin{bmatrix} e^{-s} & -1 & 2 \\ 2 & 0 & 3 \\ 1 & -2 & 1/s \end{bmatrix} \begin{bmatrix} \mathcal{L}\{t^3\} \\ \mathcal{L}\{e^{2t}\} \\ \mathcal{L}\{\sin t\} \end{bmatrix}$

Exercises 9–20:

Use Laplace transforms to solve the given initial value problem.

9. $\mathbf{y}' = \begin{bmatrix} 5 & -4 \\ 5 & -4 \end{bmatrix} \mathbf{y}, \quad \mathbf{y}(0) = \begin{bmatrix} 5 \\ 6 \end{bmatrix}$

10. $\mathbf{y}' = \begin{bmatrix} 5 & -4 \\ 5 & -4 \end{bmatrix} \mathbf{y} + \begin{bmatrix} 0 \\ 1 \end{bmatrix}, \quad \mathbf{y}(0) = \begin{bmatrix} 0 \\ 0 \end{bmatrix}$

11. $\mathbf{y}' = \begin{bmatrix} 5 & -4 \\ 3 & -2 \end{bmatrix} \mathbf{y} + \begin{bmatrix} t \\ 1 \end{bmatrix}, \quad \mathbf{y}(0) = \begin{bmatrix} 0 \\ 0 \end{bmatrix}$

12. $\mathbf{y}' = \begin{bmatrix} 5 & -4 \\ 3 & -2 \end{bmatrix} \mathbf{y}, \quad \mathbf{y}(0) = \begin{bmatrix} 3 \\ 2 \end{bmatrix}$

13. $\mathbf{y}' = \begin{bmatrix} 1 & 4 \\ -1 & 1 \end{bmatrix} \mathbf{y}, \quad \mathbf{y}(0) = \begin{bmatrix} 2 \\ 0 \end{bmatrix}$

14. $\mathbf{y}' = \begin{bmatrix} 1 & 4 \\ -1 & 1 \end{bmatrix} \mathbf{y} + \begin{bmatrix} 0 \\ 3e^t \end{bmatrix}, \quad \mathbf{y}(0) = \begin{bmatrix} 3 \\ 0 \end{bmatrix}$

15. $\mathbf{y}' = \begin{bmatrix} 6 & -3 \\ 8 & -5 \end{bmatrix} \mathbf{y}, \quad \mathbf{y}(1) = \begin{bmatrix} 5 \\ 10 \end{bmatrix}$ [Hint: Make the change of variable $\tau = t - 1$.]

16. $\mathbf{y}'' = \begin{bmatrix} -3 & -2 \\ 4 & 3 \end{bmatrix} \mathbf{y}, \quad \mathbf{y}(0) = \begin{bmatrix} 1 \\ 0 \end{bmatrix}, \quad \mathbf{y}'(0) = \begin{bmatrix} 0 \\ 1 \end{bmatrix}$

17. $\mathbf{y}'' = \begin{bmatrix} 1 & -1 \\ 1 & -1 \end{bmatrix} \mathbf{y} + \begin{bmatrix} t \\ 1 \end{bmatrix}, \quad \mathbf{y}(0) = \begin{bmatrix} 0 \\ 0 \end{bmatrix}, \quad \mathbf{y}'(0) = \begin{bmatrix} 0 \\ 0 \end{bmatrix}$

18. $\mathbf{y}'' = \begin{bmatrix} 1 & -1 \\ 1 & -1 \end{bmatrix} \mathbf{y} + \begin{bmatrix} 2 \\ 1 \end{bmatrix}, \quad \mathbf{y}(0) = \begin{bmatrix} 0 \\ 1 \end{bmatrix}, \quad \mathbf{y}'(0) = \begin{bmatrix} 0 \\ 0 \end{bmatrix}$

19. $\mathbf{y}' = \begin{bmatrix} 6 & 5 & 0 \\ -7 & -6 & 0 \\ 0 & 0 & -2 \end{bmatrix} \mathbf{y}, \quad \mathbf{y}(0) = \begin{bmatrix} 2 \\ -4 \\ -1 \end{bmatrix}$

20. $\mathbf{y}' = \begin{bmatrix} 1 & 0 & 0 \\ 0 & -1 & 1 \\ 0 & 0 & 2 \end{bmatrix} \mathbf{y} + \begin{bmatrix} e^t \\ 1 \\ -2t \end{bmatrix}, \quad \mathbf{y}(0) = \begin{bmatrix} 0 \\ 0 \\ 0 \end{bmatrix}$

21. The Laplace transform was applied to the initial value problem $\mathbf{y}' = A\mathbf{y}, \mathbf{y}(0) = \mathbf{y}_0$, where $\mathbf{y}(t) = \begin{bmatrix} y_1(t) \\ y_2(t) \end{bmatrix}$, A is a (2×2) constant matrix, and $\mathbf{y}_0 = \begin{bmatrix} y_{1,0} \\ y_{2,0} \end{bmatrix}$. The following transform domain solution was obtained:

$$\mathcal{L}\{\mathbf{y}(t)\} = \mathbf{Y}(s) = \frac{1}{s^2 - 9s + 18} \begin{bmatrix} s-2 & -1 \\ 4 & s-7 \end{bmatrix} \begin{bmatrix} y_{1,0} \\ y_{2,0} \end{bmatrix}.$$

(a) What are the eigenvalues of the coefficient matrix A?

(b) What is the coefficient matrix A?

22. A System Cascade   Consider the linear system defined as follows:

$$\mathbf{y}_1' = A\mathbf{y}_1 + \mathbf{g}(t), \qquad \mathbf{y}_1(0) = \mathbf{0}$$
$$\mathbf{y}_2' = A\mathbf{y}_2 + \mathbf{y}_1(t), \qquad \mathbf{y}_2(0) = \mathbf{0},$$

where $\mathbf{y}_1(t), \mathbf{y}_2(t)$, and $\mathbf{g}(t)$ are (2×1) vector functions and A is a (2×2) constant matrix. A schematic of the system is shown in the figure. It consists of two identical stages connected in cascade. The input $\mathbf{g}(t)$ is applied to the first stage, producing an output $\mathbf{y}_1(t)$. This output is then used as input to the second stage, producing an output $\mathbf{y}_2(t)$. We can view this cascade connection as forming an overall linear system determined by input $\mathbf{g}(t)$ and output $\mathbf{y}_2(t)$. Let $\mathbf{Y}_1(s)$, $\mathbf{Y}_2(s)$, and $\mathbf{G}(s)$ denote the Laplace transforms of $\mathbf{y}_1(t), \mathbf{y}_2(t)$, and $\mathbf{g}(t)$, respectively.

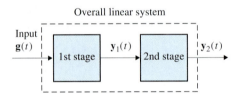

Figure for Exercise 22

(a) Show that $\mathbf{Y}_2(s)$ and $\mathbf{G}(s)$ are related by an equation of the form $\mathbf{Y}_2(s) = \Omega(s)\mathbf{G}(s)$, where $\Omega(s)$ is a (2×2) matrix transfer function for the cascade system. How is $\Omega(s)$ related to the coefficient matrix A?

(b) Suppose

$$A = \begin{bmatrix} 1 & -1 \\ 1 & -1 \end{bmatrix} \quad \text{and} \quad \mathbf{g}(t) = \begin{bmatrix} 1 \\ t \end{bmatrix}.$$

Determine $\Omega(s)$ and $y_2(t), t \geq 0$.

Exercises 23–24:

System Identification   We consider a system analog of the parameter identification problem studied in Section 5.4. Assume that a linear system can be modeled by the initial value problem $\mathbf{y}' = A\mathbf{y} + \mathbf{g}(t), \mathbf{y}(0) = \mathbf{y}_0$.

Assume we can select the input $\mathbf{g}(t)$ and the initial state $\mathbf{y}_0$ and can measure the output $\mathbf{y}(t)$, but we have no direct way of measuring the coefficient matrix A. The task is to determine A by exciting the system with an appropriate selection of inputs and/or initial states and measuring the corresponding outputs. Exercises 23–24 treat particular two-dimensional cases.

In each exercise, use the given input-output information to determine the coefficient matrix A. One approach is to use Laplace transforms. Let $\mathbf{Y}(s)$ and $\mathbf{G}(s)$ represent the Laplace transforms of $\mathbf{y}(t)$ and $\mathbf{g}(t)$, respectively. Then we know that $\mathbf{Y}(s) = (sI - A)^{-1}[\mathbf{y}_0 + \mathbf{G}(s)]$. Form (2×2) matrices

$$[\mathbf{Y}_1(s), \mathbf{Y}_2(s)] \quad \text{and} \quad [\mathbf{y}_{1,0} + \mathbf{G}_1(s), \mathbf{y}_{2,0} + \mathbf{G}_2(s)],$$

using the transformed information as columns. We can obtain an equation relating these two (2×2) matrices and use this equation to determine A.

23. When $\mathbf{g}(t) = \mathbf{0}$ and $\mathbf{y}_0 = \begin{bmatrix} 1 \\ 1 \end{bmatrix}$, the observed output is $\mathbf{y}(t) = \begin{bmatrix} e^{3t} \\ e^{3t} \end{bmatrix}$. When $\mathbf{g}(t) = \mathbf{0}$ and $\mathbf{y}_0 = \begin{bmatrix} 0 \\ 5 \end{bmatrix}$, the observed output is $\mathbf{y}(t) = \begin{bmatrix} 3e^{-2t} - 3e^{3t} \\ 8e^{-2t} - 3e^{3t} \end{bmatrix}$. Determine coefficient matrix A.

24. When $\mathbf{g}(t) = \mathbf{0}$ and $\mathbf{y}_0 = \begin{bmatrix} 0 \\ 1 \end{bmatrix}$, the observed output is $\mathbf{y}(t) = \begin{bmatrix} te^{-2t} \\ e^{-2t} \end{bmatrix}$. When $\mathbf{g}(t) = \begin{bmatrix} 2 \\ 0 \end{bmatrix}$ and $\mathbf{y}_0 = \begin{bmatrix} 1 \\ 0 \end{bmatrix}$, the observed output is $\mathbf{y}(t) = \begin{bmatrix} 1 \\ 0 \end{bmatrix}$. Determine the coefficient matrix A.

25. For the network shown, initially both loop currents are zero and no charge is present on the capacitor. At time $t = 0$, both voltage sources are turned on. An application of Kirchhoff's voltage law, equating the algebraic sum of the voltage drops in a clockwise traversal of each loop to zero, leads to the system of equations

$$-v_1(t) + R_1 i_1 + L\frac{di_1}{dt} + R_2(i_1 - i_2) = 0, \qquad i_1(0) = 0$$

$$R_2(i_2 - i_1) + \frac{1}{C}\int_0^t i_2(\lambda)\, d\lambda + R_3 i_2 + v_2(t) = 0, \qquad i_2(0) = 0.$$

(a) Apply the Laplace transform to this system of equations. Solve the transformed system of equations for the (2×1) vector of transformed loop currents,

$$\begin{bmatrix} I_1(s) \\ I_2(s) \end{bmatrix}.$$

(b) For simplicity, let $R_1 = R_2 = R_3 = 1\,k\Omega$, $L = 1\,H$, and $C = 1\,\mu F$; let $v_1(t) = v_2(t) = te^{-t}$ volts. Solve for the currents, $i_1(t)$ and $i_2(t)$, $t > 0$ (the units being milliamperes).

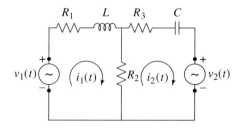

Figure for Exercise 25

5.6 Convolution

When we use Laplace transforms, we often need to find the inverse transform of a product,

$$\mathcal{L}^{-1}\{F(s)G(s)\}.$$

For example, we have seen that the Laplace transform of a system output is the product of the system transfer function and the Laplace transform of the system input. To obtain the time domain output, we must determine the inverse Laplace transform of this product. It is clear from the integral definition of the Laplace transform that the inverse transform of a product of transforms is *not* the product of the inverse transforms. What, then, is it?

This section introduces a mathematical operation known as *convolution*. The convolution operation, denoted by the symbol $*$, starts with two functions $f(t)$ and $g(t)$ defined on $0 \le t < \infty$ and creates a new function $f * g$, also defined on $0 \le t < \infty$. After we define the convolution operation, we will state the convolution theorem; this theorem shows that the Laplace transform of the newly created function $f * g$ is, in fact, the product of the Laplace transforms of the two original functions. Thus, by the convolution theorem,

$$\mathcal{L}^{-1}\{F(s)G(s)\} = (f * g)(t),$$

where $\mathcal{L}\{f(t)\} = F(s)$ and $\mathcal{L}\{g(t)\} = G(s)$.

Although the terminology is new, convolution is an operation we have already encountered several times in our study of linear constant coefficient differential equations.

The Convolution Integral

Let $f(t)$ and $g(t)$ be two functions defined on $0 \le t < \infty$. The **convolution of** $f(t)$ **and** $g(t)$, denoted $f * g$, is the function defined by

$$(f * g)(t) = \int_0^t f(t - \lambda)g(\lambda) \, d\lambda, \qquad 0 \le t < \infty, \tag{1}$$

provided the integral exists. It can be shown that integral (1) exists whenever $f(t)$ and $g(t)$ are piecewise continuous on $0 \le t < \infty$. Moreover, the function $(f * g)(t)$ is piecewise continuous and exponentially bounded on $0 \le t < \infty$ if both $f(t)$ and $g(t)$ possess these properties.

As equation (1) indicates, we use the notation $(f * g)(t)$ to denote the newly created function of t. When we want to designate the convolution of specific functions such as $f(t) = e^{-t}$ and $g(t) = \sin 2t$, we may simply write

$$e^{-t} * \sin 2t.$$

E X A M P L E

1

Calculate the convolution $f * g$, where $f(t) = t$ and $g(t) = e^{-t}$.

Solution: According to definition (1),

$$t * e^{-t} = \int_0^t (t - \lambda)e^{-\lambda} \, d\lambda = t \int_0^t e^{-\lambda} \, d\lambda - \int_0^t \lambda e^{-\lambda} \, d\lambda.$$

Evaluating these integrals, we find

$$t * e^{-t} = t \left[-e^{-\lambda} \right]_0^t - \left[e^{-\lambda}(-\lambda - 1) \right]_0^t = t + e^{-t} - 1. \; ❖$$

The next example illustrates convolution from a geometric point of view.

E X A M P L E

2

Calculate the convolution $f * g$, where

$$f(t) = \begin{cases} t, & 0 \leq t < 1 \\ 0, & 1 \leq t < \infty, \end{cases} \qquad g(t) = \begin{cases} 0, & 0 \leq t < 2 \\ 1, & 2 \leq t < 3 \\ 0, & 3 \leq t < \infty. \end{cases}$$

Solution: The piecewise definition of these two functions provides an opportunity to illustrate the graphical aspects of the convolution operation,

$$(f * g)(t) = \int_0^t f(t - \lambda)g(\lambda) \, d\lambda.$$

The functions in the integrand, $f(\lambda)$ and $g(\lambda)$, are shown in Figure 5.16(a) on the next page. Forming $f(t - \lambda)$ reverses the orientation of the right triangle and translates (or slides) the triangle so that the intersection point of its hypotenuse with the λ-axis occurs at $\lambda = t$ [see Figure 5.16(b)]. As t increases, we can envision this triangle as translating to the right and passing through the region $2 \leq \lambda < 3$, where $g(\lambda) \neq 0$ [see Figures 5.16(c) and 5.16(d)]. At each value of t, the integrand is nonzero only in the overlap region of the right triangle [the graph of $f(t - \lambda)$] and the rectangle [the graph of $g(\lambda)$]. For t in the interval $(2, 4)$, the value $(f * g)(t)$ is equal to the area of the overlap region. Therefore, the convolution integral $\int_0^t f(t - \lambda)g(\lambda) \, d\lambda$ can be evaluated graphically, as is shown in Figures 5.16(b) through 5.16(e). We find

$$(f * g)(t) = \begin{cases} 0, & 0 \leq t < 2 \\ \frac{1}{2}(t - 2)^2, & 2 \leq t < 3 \\ \frac{1}{2}(4 - t)(t - 2), & 3 \leq t < 4 \\ 0, & 4 \leq t < \infty. \end{cases}$$

The graph of the resulting function, $(f * g)(t)$, is given in Figure 5.16(f). ❖

Algebraic Properties of the Convolution Operation

Let f, g, and k be three scalar functions defined on $0 \leq t < \infty$, and let c_1 and c_2 represent arbitrary constants. It can be shown that

$$f * (c_1 g + c_2 k) = c_1(f * g) + c_2(f * k) \tag{2a}$$

$$f * g = g * f \tag{2b}$$

$$(f * g) * k = f * (g * k). \tag{2c}$$

The distributive property, equation (2a), says that the convolution of a function f with a linear combination of functions equals the linear combination of the convolutions. Property (2b) asserts that the convolution operation is commutative; that is, the order in which we choose the two functions doesn't matter

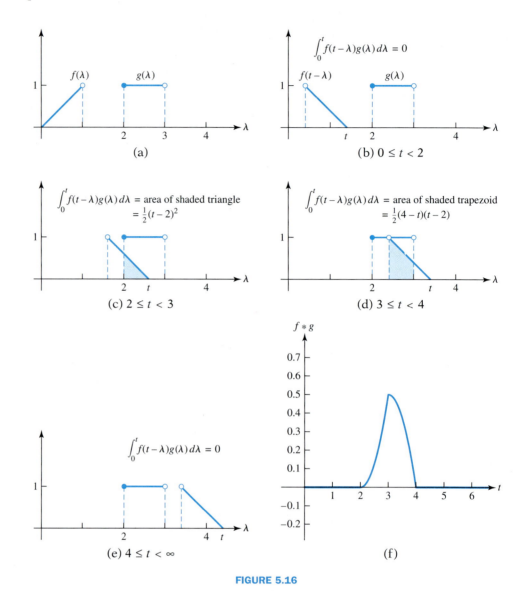

FIGURE 5.16

A graphical interpretation of the calculation $(f * g)(t)$, where $f(t)$ and $g(t)$ are the functions shown in (a); the details are given in Example 2. A graph of the function $(f * g)(t)$ is shown in (f). Note that $(f * g)(t)$ is continuous even though $f(t)$ and $g(t)$ have jump discontinuities.

(see Exercise 1). The associative property, equation (2c), says the convolution of three functions can be done in any order. Therefore, parentheses are unnecessary in (2c), and we can simply write $f * g * k$.

Some Remarks about Convolution

While we have defined the convolution integral only for scalar functions $f(t)$ and $g(t)$, it should be clear that the definition can be extended to compatibly dimensioned matrix-valued functions. For example, if $\mathbf{f}(t)$ is an $(m \times n)$ matrix function and $\mathbf{g}(t)$ is an $(n \times p)$ matrix function, then $(\mathbf{f} * \mathbf{g})(t)$ is the $(m \times p)$

matrix function defined by

$$(\mathbf{f} * \mathbf{g})(t) = \int_0^t \mathbf{f}(t - \lambda)\mathbf{g}(\lambda) \, d\lambda, \qquad 0 \leq t < \infty. \tag{3}$$

Note that we have encountered the convolution integral in previous chapters even though we did not use the term "convolution" there. For example:

1. In Chapter 2, we saw that the solution of the initial value problem $y' = \alpha y + g(t), y(0) = y_0$ is given by

$$y(t) = e^{\alpha t} y_0 + \int_0^t e^{\alpha(t - \lambda)} g(\lambda) \, d\lambda.$$

   The integral term is a convolution integral; therefore, we can interpret the solution as

$$y(t) = e^{\alpha t} y_0 + e^{\alpha t} * g(t). \tag{4a}$$

2. In the discussion of first order constant coefficient linear systems in Section 4.8, we developed the variation of parameters formula for the solution of the initial value problem

$$\mathbf{y}' = A\mathbf{y} + \mathbf{g}(t), \qquad \mathbf{y}(0) = \mathbf{y}_0.$$

   The solution can be represented as

$$\mathbf{y}(t) = \Phi(t)\mathbf{y}_0 + \int_0^t \Phi(t - \lambda)\mathbf{g}(\lambda) \, d\lambda,$$

   where $\Phi(t)$ is the fundamental matrix that reduces to the identity matrix at $t = 0$. Therefore,

$$\mathbf{y}(t) = \Phi(t)\mathbf{y}_0 + \Phi(t) * \mathbf{g}(t). \tag{4b}$$

   In Section 4.10, we saw that $\Phi(t) = e^{tA}$. Therefore, we can also write the solution as

$$\mathbf{y}(t) = e^{tA}\mathbf{y}_0 + e^{tA} * \mathbf{g}(t).$$

The Convolution Theorem

Equations (4a) and (4b) show two instances where the solution of an initial value problem can be related to convolution of functions in the time domain. Theorem 5.7 (the convolution theorem) establishes the connection between convolution of functions in the time domain and multiplication of Laplace transforms in the transform domain.

Theorem 5.7

Let $f(t)$ and $g(t)$ be piecewise continuous and exponentially bounded functions defined on $0 \leq t < \infty$. Let $F(s)$ and $G(s)$ denote their respective Laplace transforms. Then $(f * g)(t)$ is a Laplace transformable function, and its Laplace transform equals the product of $F(s)$ and $G(s)$; that is,

$$\mathcal{L}\{f * g\} = F(s)G(s). \tag{5}$$

● **PROOF:**   We first show that $f * g$ is Laplace transformable. Then we establish the result in equation (5).

The function $f * g$ is actually continuous on $0 \le t < \infty$ whenever $f(t)$ and $g(t)$ are piecewise continuous on $0 \le t < \infty$. Thus (see Theorem 5.1), to show that $f * g$ is Laplace transformable, we need only show that $f * g$ is exponentially bounded on $0 \le t < \infty$. From our hypotheses, we know that $|f(t)| \le M_1 e^{a_1 t}$ and $|g(t)| \le M_2 e^{a_2 t}$. Therefore,

$$|(f * g)(t)| = \left| \int_0^t f(t - \lambda)g(\lambda)\, d\lambda \right| \le \int_0^t \left| f(t - \lambda)g(\lambda) \right| d\lambda$$

$$\le \int_0^t M_1 e^{a_1(t-\lambda)} M_2 e^{a_2 \lambda}\, d\lambda = \begin{cases} M_1 M_2 t e^{a_1 t}, & a_1 = a_2 \\ M_1 M_2 \dfrac{e^{a_2 t} - e^{a_1 t}}{a_2 - a_1}, & a_1 \neq a_2, \end{cases}$$

and it follows that $f * g$ is exponentially bounded.

To complete the argument, we need to establish relation (5). From the definition,

$$\mathcal{L}\{(f * g)(t)\} = \int_0^\infty \left[\int_0^t f(t - \lambda)g(\lambda)\, d\lambda \right] e^{-st}\, dt$$

$$= \int_0^\infty \int_0^t f(t - \lambda)g(\lambda) e^{-st}\, d\lambda\, dt, \tag{6}$$

where we view the integral in (6) as a double integral over the portion of the λt-plane shown in Figure 5.17.

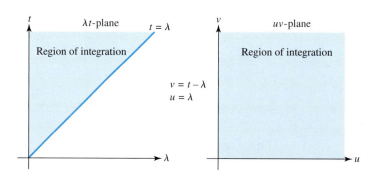

FIGURE 5.17

The regions of integration for the double integrals in equations (6) and (7).

We now introduce the change of variables $u = \lambda, v = t - \lambda$. The boundary lines $\lambda = 0$ and $\lambda = t$ transform into the lines $u = 0$ and $v = 0$, respectively. Note that the Jacobian determinant of this transformation is equal to 1. Therefore, we can rewrite the integral in equation (6) as

$$\mathcal{L}\{f * g\} = \int_0^\infty \int_0^\infty f(v)g(u) e^{-s(u+v)}\, du\, dv$$

$$= \left(\int_0^\infty f(v) e^{-sv}\, dv \right) \left(\int_0^\infty g(u) e^{-su}\, du \right) = F(s)G(s). \ ● \tag{7}$$

As noted earlier, Theorem 5.7 can be used to find $\mathcal{L}^{-1}\{F(s)G(s)\}$:

$$\mathcal{L}^{-1}\{F(s)G(s)\} = \int_0^t f(t-\lambda)g(\lambda)\,d\lambda. \qquad (8)$$

EXAMPLE 3

Use equation (8) to find

$$\mathcal{L}^{-1}\left\{\frac{1}{s^2(s+1)}\right\}.$$

Solution: Applying equation (8), $F(s) = 1/s^2$, and $G(s) = 1/(s+1)$, we have

$$\mathcal{L}^{-1}\left\{\frac{1}{s^2(s+1)}\right\} = \mathcal{L}^{-1}\left\{\frac{1}{s^2}\right\} * \mathcal{L}^{-1}\left\{\frac{1}{s+1}\right\} = t * e^{-t} = t + e^{-t} - 1.$$

(Recall that the convolution $t * e^{-t}$ was computed earlier, in Example 1.) ❖

Multiple Convolutions

In some applications, such as a cascade connection of linear systems, the solution of the problem of interest is a *multiple convolution*. Suppose $f_1(t), f_2(t), \ldots, f_n(t)$ are Laplace transformable functions with Laplace transforms $F_1(s), F_2(s), \ldots, F_n(s)$, respectively. From a repeated application of the convolution theorem, it follows that

$$\mathcal{L}\{f_1 * f_2 * \cdots * f_n\} = F_1(s)F_2(s)\cdots F_n(s).$$

Our next example treats such an application.

EXAMPLE 4

Consider the serial connection of n identical tanks shown in Figure 5.18. Each tank contains V gallons of fresh water. At time $t = 0$, a solution having a concentration of c pounds of salt per gallon flows into Tank 1 at a rate of r gallons per minute, and the well-stirred mixture flows out of Tank 1 and into Tank 2 at the same rate. The well-stirred mixture in Tank 2, in turn, flows into Tank 3 at the same rate. This behavior is replicated throughout the cascade. Since the inflow and outflow rates are the same for each tank, the volume of fluid in each tank remains constant and equal to V. Determine the outflow concentration, $c_n(t)$, of Tank n as a function of time.

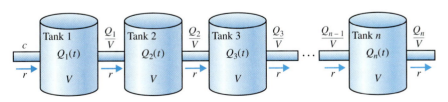

FIGURE 5.18

The n-tank cascade described in Example 4.

Solution: As in Section 2.3, we apply the "conservation of salt" principle to each tank. Let $Q_j(t), j = 1, 2, \ldots, n$ represent the amount of salt (in pounds) in

(continued)

(continued)

the jth tank at time t (in minutes). The following system of initial value problems models this process:

$$Q_1' = rc - r\frac{Q_1}{V}, \qquad Q_1(0) = 0$$

$$Q_j' = r\frac{Q_{j-1}}{V} - r\frac{Q_j}{V}, \qquad Q_j(0) = 0, \qquad j = 2, 3, \ldots, n.$$

The solutions of these differential equations can be obtained recursively. We use a convolution representation for each of the solutions. From equation (4a),

$$Q_1(t) = e^{-(r/V)t} * rc$$

$$Q_j(t) = e^{-(r/V)t} * \frac{r}{V}Q_{j-1}(t), \qquad j = 2, 3, \ldots, n.$$

Therefore, the outflow concentration of the nth tank can be represented as the following multiple convolution:

$$
\begin{aligned}
c_n(t) \;&=\; \frac{1}{V}Q_n(t) \\[4pt]
&=\; \frac{1}{V}e^{-(r/V)t} * \frac{r}{V}Q_{n-1}(t) \\[4pt]
&=\; \frac{1}{V}e^{-(r/V)t} * \frac{r}{V}e^{-(r/V)t} * \frac{r}{V}Q_{n-2}(t) = \cdots \\[4pt]
&=\; \frac{1}{V}e^{-(r/V)t} * \overbrace{\frac{r}{V}e^{-(r/V)t} * \frac{r}{V}e^{-(r/V)t} * \cdots * \frac{r}{V}e^{-(r/V)t}}^{n-1 \text{ functions}} * rc \\[4pt]
&=\; c\left(\frac{r}{V}\right)^n \overbrace{e^{-(r/V)t} * e^{-(r/V)t} * \cdots * e^{-(r/V)t}}^{n \text{ functions}} * 1.
\end{aligned}
$$

By the convolution theorem,

$$\mathcal{L}\{c_n(t)\} = c\left(\frac{r}{V}\right)^n \left[\mathcal{L}\left\{e^{-(r/V)t}\right\}\right]^n \mathcal{L}\{1\}$$

$$= c\left(\frac{r}{V}\right)^n \left[\frac{1}{\left(s + \dfrac{r}{V}\right)^n}\right] \frac{1}{s}.$$

We can recover $c_n(t)$ by taking the inverse transform. From equations (10) and (18) in Table 5.1,

$$\mathcal{L}^{-1}\left\{\frac{1}{s}F(s)\right\} = \int_0^t f(u)\,du \quad \text{and} \quad \mathcal{L}^{-1}\left\{\frac{1}{\left(s + \dfrac{r}{V}\right)^n}\right\} = \frac{t^{n-1}}{(n-1)!}e^{-(r/V)t}.$$

Therefore,

$$c_n(t) = c\left(\frac{r}{V}\right)^n \int_0^t \frac{u^{n-1}}{(n-1)!}e^{-(r/V)u}\,du.$$

We can simplify this integral by making the change of variable $w = (r/V)u$, obtaining

$$c_n(t) = \frac{c}{(n-1)!} \int_0^{rt/V} w^{n-1}e^{-w} \, dw. \tag{9}$$

This final expression can be evaluated for modestly large n using integration-by-parts or using computer software. Figure 5.19 shows a plot of normalized concentration c_n/c vs. rt/V for $n = 3$ and $n = 10$. As we would expect, both normalized concentrations approach a horizontal asymptote of unity. As time evolves, the concentration in all tanks in the cascade builds up to the inflow concentration, c. Figure 5.19 shows, as one would expect, that the concentration in the last tank of the three-tank cascade builds up to this limiting value more rapidly than does the concentration of the last tank in the ten-tank cascade.

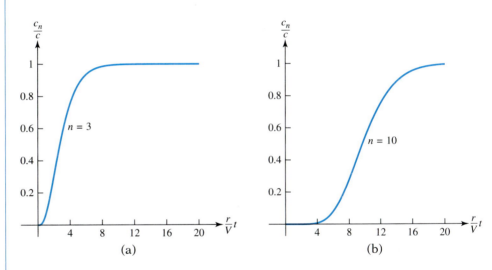

FIGURE 5.19

Graphs of c_n/c vs. rt/V for $n = 3$ and $n = 10$.
[See equation (9) in Example 4.]   ❖

EXERCISES

1. Show that $f * g = g * f$. That is, show that $\int_0^t f(t-\lambda)g(\lambda) \, d\lambda = \int_0^t g(t-\sigma)f(\sigma) \, d\sigma$. [Hint: Use the change of integration variable $\sigma = t - \lambda$. This exercise shows that the convolution operation is commutative.]

Exercises 2–7:

For the given functions $f(t)$ and $g(t)$ defined on $0 \le t < \infty$, compute $f * g$ in two different ways:

(a) by directly evaluating the integral

(b) by computing $\mathcal{L}^{-1}\{F(s)G(s)\}$, where $F(s) = \mathcal{L}\{f(t)\}$ and $G(s) = \mathcal{L}\{g(t)\}$

2. $f(t) = g(t) = h(t)$ 3. $f(t) = t, \ g(t) = t^2$

4. $f(t) = e^t, \ g(t) = e^{-2t}$ 5. $f(t) = t, \ g(t) = \sin t$

6. $f(t) = \sin t, \ g(t) = \cos t$ 7. $f(t) = t, \ g(t) = h(t) - h(t-1)$

Exercises 8–9:

In each exercise, use Laplace transforms to compute the convolution.

8. $P * \mathbf{y}$, where $P(t) = \begin{bmatrix} h(t) & e^t \\ 0 & t \end{bmatrix}$ and $\mathbf{y}(t) = \begin{bmatrix} h(t) \\ e^{-t} \end{bmatrix}$

9. $t * \begin{bmatrix} t \\ \cos t \end{bmatrix}$

Exercises 10–12:

Compute and graph $f * g$.

10. $f(t) = h(t)$, $g(t) = t[h(t) - h(t-2)]$

11. $f(t) = g(t) = h(t-1) - h(t-2)$

12. $f(t) = h(t) - h(t-1)$, $g(t) = h(t-1) - 2h(t-2)$

Exercises 13–15:

Compute the given multiple convolution. (Convolution operations, particularly multiple convolutions, have important applications in probability theory—for example, in computing the probability density function for a sum of independent random variables.[4])

13. $t * t * t$ **14.** $h(t) * e^{-t} * e^{-2t}$ **15.** $t * e^{-t} * e^t$

16. Suppose it is known that $\overbrace{h(t) * h(t) * \cdots * h(t)}^{n \text{ functions}} = Ct^8$. Determine the constant C and the positive integer n.

17. Suppose it is known that $\overbrace{e^{-t} * e^{-t} * \cdots * e^{-t}}^{n \text{ functions}} = Ct^4 e^{\alpha t}$. Determine the constants C and α and the positive integer n.

Exercises 18–26:

The following equations are called **integral equations** because the unknown dependent variable appears within an integral. When the equation also contains derivatives of the dependent variable, it is referred to as an **integro-differential equation**. In each exercise, the given equation is defined for $t \geq 0$. Use Laplace transforms to obtain the solution.

18. $\displaystyle\int_0^t \sin(t-\lambda)y(\lambda)\, d\lambda = t^2$ **19.** $t^2 e^{-t} = \displaystyle\int_0^t \cos(t-\lambda)y(\lambda)\, d\lambda$

20. $y(t) - \displaystyle\int_0^t e^{t-\lambda}y(\lambda)\, d\lambda = t$

21. $\displaystyle\int_0^t y(t-\lambda)y(\lambda)\, d\lambda = 6t^3$. Is the solution $y(t)$ unique? If not, find all possible solutions.

22. $t * y(t) = t^2(1 - e^{-t})$ **23.** $\dfrac{dy}{dt} + \displaystyle\int_0^t y(t-\lambda)e^{-2\lambda}\, d\lambda = 1$, $y(0) = 0$

24. $\mathbf{y}' = h(t) * \mathbf{y}$, $\mathbf{y}(0) = \begin{bmatrix} 1 \\ 2 \end{bmatrix}$

25. $y'' + h(t) * y = 0$, $y(0) = 1$, $y'(0) = 0$

26. $y'' - h(t) * y = 0$, $y(0) = 0$, $y'(0) = 1$

---

[4]Walter C. Giffin, *Transform Techniques for Probability Modeling* (New York: Academic Press, 1975).

Exercises 27–28:

Solve the given initial value problem.

27. $\dfrac{dy}{dt} = t * t, \ y(0) = 1$

28. $y' - y = \displaystyle\int_0^t (t - \lambda)e^\lambda \, d\lambda, \ y(0) = -1$

5.7   The Delta Function and Impulse Response

We often need to determine the behavior of a linear system that is suddenly subjected to an input of short duration and large amplitude. In an electrical network, such an input might be a large applied voltage spike. In a mechanical system, the input might be a very sharp applied force.

System excitations of this sort cause a system response that approximates what is known as the *impulse response* of the linear system. In this section, we discuss the concept of an impulse response and show that it is equal to the inverse Laplace transform of the system transfer function.

An Example of Impulse Response

To introduce the idea of impulse response, we begin with a mass-spring-dashpot system. An example of the "short duration/large amplitude" scenario we want to examine is the following initial value problem:

$$my'' + \gamma y' + ky = p_\varepsilon(t), \qquad t > 0$$
$$y(0) = 0, \qquad y'(0) = 0. \tag{1a}$$

In (1a), we assume that ε is a small positive parameter and that

$$p_\varepsilon(t) = \begin{cases} \dfrac{1}{\varepsilon}, & 0 \le t \le \varepsilon \\[2mm] 0, & \text{otherwise.} \end{cases} \tag{1b}$$

Since ε is small, the applied force p_ε is a pulse of short duration and large amplitude; Figure 5.20 shows the graph of a typical pulse. Note that the applied force p_ε has "unit strength" in the sense that the area under the graph in Figure 5.20 is equal to 1 for any choice of ε.[5] By choosing ε smaller and smaller, we can use the pulse p_ε to model applied forces having larger and larger amplitudes over shorter and shorter periods. Therefore, it is natural to ask the question

> What happens to the system behavior as we make the applied force progressively "sharper" and "stronger"?

In other words, what happens to the solution of initial value problem (1a) as we let $\varepsilon \to 0$?

It can be shown that the solution of initial value problem (1a) is

$$y_\varepsilon(t) = \int_0^t \phi(t - \lambda)p_\varepsilon(\lambda) \, d\lambda, \tag{2a}$$

---

[5]In physics, the linear impulse produced by a constant force is the product of the force times the duration of its application. Therefore, the applied force p_ε has a linear impulse of unity for all ε.

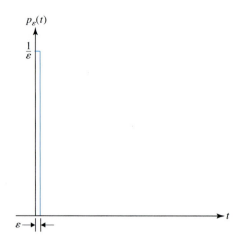

FIGURE 5.20

The function $p_\varepsilon(t)$ is a pulse; see equation (1b).
Note that $\int_{-\infty}^{\infty} p_\varepsilon(t)\, dt = 1$.

where, if the system is underdamped,

$$\phi(t) = \frac{e^{-(\gamma/2m)t} \sin\sqrt{\dfrac{k}{m} - \dfrac{\gamma^2}{4m^2}}\, t}{m\sqrt{\dfrac{k}{m} - \dfrac{\gamma^2}{4m^2}}}. \tag{2b}$$

We use the subscript ε in equation (2a) to denote the fact that the solution $y_\varepsilon(t)$ depends on the parameter ε. For $t \geq \varepsilon$, we see from equation (2a) that

$$y_\varepsilon(t) = \frac{1}{\varepsilon}\int_0^\varepsilon \phi(t-\lambda)\, d\lambda, \qquad t \geq \varepsilon. \tag{3}$$

Since $\phi(t)$ is continuous for all t, we can use the mean value theorem for integrals in equation (3), obtaining

$$y_\varepsilon(t) = \phi(t - \xi), \tag{4}$$

where ξ is some value in the interval $0 \leq \lambda \leq \varepsilon$. As is typical with mean value theorems, the value ξ is known to be sandwiched between 0 and ε but is otherwise unknown. Because of this sandwiching, ξ must approach zero as $\varepsilon \to 0^+$. Since ϕ is continuous,

$$\lim_{\varepsilon \to 0^+} y_\varepsilon(t) = \phi(t).$$

Therefore, as we make ε progressively smaller (that is, as we make the applied force both shorter in duration and correspondingly larger in amplitude), the system response approaches $\phi(t)$, where $\phi(t)$ is the function given in equation (2b). This limiting response is called the **impulse response** of the linear system.

Figure 5.21 shows the impulse response $\phi(t)$ of an underdamped spring-mass-dashpot system with parameters $m = 1$, $\gamma = 2$, and $k = 5$. For these parameters, the function $\phi(t)$ is given by

$$\phi(t) = 0.5e^{-t}\sin 2t. \tag{5}$$

From a heuristic point of view, $\phi(t)$ represents the response of the mechanical system to an *impulsive force*—a force having essentially zero duration and infinite amplitude, but unit area.

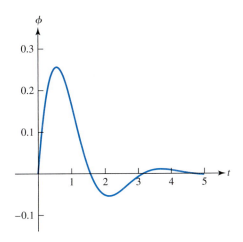

The impulse response function $\phi(t)$ in equation (5).

The Delta Function

From the point of view of applications, it would be nice to have a function $\delta(t)$ that we could use to model an impulsive force. That is, we would like to be able to write

$$
\lim_{\varepsilon \to 0^+} y_\varepsilon(t) = \lim_{\varepsilon \to 0} \int_0^t \phi(t-\lambda) p_\varepsilon(\lambda) \, d\lambda
$$

$$
= \int_0^t \phi(t-\lambda) \delta(\lambda) \, d\lambda \tag{6}
$$

$$
= \phi(t).
$$

The role of the function $\delta(\lambda)$ in (6) would be to evaluate the integrand at $\lambda = 0$. It is important to appreciate, however, that we cannot simply obtain $\delta(\lambda)$ as a limit of $p_\varepsilon(\lambda)$ as $\varepsilon \to 0^+$; that is, we cannot interchange the operations of limit and integration in the first line of (6) because (see Figure 5.20)

$$
\lim_{\varepsilon \to 0^+} p_\varepsilon(\lambda) = \begin{cases} 0, & \lambda \neq 0 \\ \infty, & \lambda = 0. \end{cases}
$$

The **delta function**, denoted by $\delta(t)$, is actually given precise mathematical meaning as a "generalized function" within a branch of mathematics known as the theory of distributions. For our purposes, we will define the delta function, $\delta(t)$, by the limit

$$
\int_a^b f(t) \delta(t-t_0) \, dt = \lim_{\varepsilon \to 0^+} \int_a^b f(t) p_\varepsilon(t-t_0) \, dt, \tag{7a}
$$

whenever $f(t)$ is a function defined and continuous on $[a, b]$. That is, $\delta(t)$ has the property that

$$\int_a^b f(t)\delta(t - t_0)\, dt = \begin{cases} f(t_0), & a \le t_0 < b \\ 0, & \text{otherwise.} \end{cases} \tag{7b}$$

REMARK: The delta function is sometimes referred to as the **Dirac delta function**.[6] Our definition of the delta function in equation (7a) follows directly from our original definition of the pulse function, p_ε. We therefore obtain the value $f(a)$ when $t_0 = a$ and 0 when $t_0 = b$. Other references use a pulse function that is an even function (having value $1/\varepsilon$ in the interval $-\varepsilon/2 \le t \le \varepsilon/2$) as the basis for their definition. In that case, the values in (7b) obtained when $t_0 = a$ and $t_0 = b$ will differ from ours. The reader should always check the definition used by the reference being consulted.

The Laplace Transform of the Delta Function

Equation (7b) can be used as the basis for defining the Laplace transform of $\delta(t)$. We obtain

$$\mathcal{L}\{\delta(t - t_0)\} = \int_0^\infty e^{-st}\delta(t - t_0)\, dt = e^{-st_0}, \qquad t_0 \ge 0. \tag{8}$$

As a special case, when $t_0 = 0$, we have

$$\mathcal{L}\{\delta(t)\} = 1.$$

The Delta Function as a Formal Modeling Tool

It is important to be aware that the delta function is different from the usual functions encountered in calculus. Nevertheless, in many applications people have found it convenient to ignore this distinction; the delta function is often viewed and formally treated as an ordinary function, usually modeling an impulsive input. The solution of the problem of interest typically is given as a convolution integral involving the delta function, and so the answer obtained makes physical sense and can be interpreted as the system response to an idealized impulsive input. The following example illustrates such a formal use of the delta function.

EXAMPLE 1

A body of mass m is at the origin at time $t = 0$, moving in the positive x-direction with velocity v_0. Assume that a frictional force, proportional to the velocity with proportionality constant k, acts to retard the motion. At a time $t_0 > 0$, an impulsive force of strength F_0 acts on the moving body in the direction of the motion. Find the velocity and position of the body as a function of time t.

[6]Paul Adrien Maurice Dirac (1902–1984) was an English mathematical physicist who held the Lucasian Professorship of Mathematics at Cambridge University from 1932 until 1969. After retiring, he moved to Florida, where he continued his research. Dirac is known for his many contributions to quantum theory, particularly the unification theories of quantum mechanics and special relativity.

Solution: We can use the delta function to formally model the impulsive force as

$$F_0\delta(t-t_0), \qquad t>0.$$

Given this model of the impulsive force, Newton's laws of motion lead to the following initial value problem:

$$mv' + kv = F_0\delta(t-t_0), \qquad t>0$$
$$v(0) = v_0. \tag{9}$$

Once we know $v(t)$, the position of the body is given by

$$x(t) = \int_0^t v(\lambda)\,d\lambda. \tag{10}$$

We will use Laplace transforms to solve the problem. Let

$$V(s) = \mathcal{L}\{v(t)\} \quad \text{and} \quad X(s) = \mathcal{L}\{x(t)\}.$$

Noting equation (8), we have for the Laplace transform of equation (9)

$$m[sV(s) - v_0] + kV(s) = F_0 e^{-st_0}.$$

Therefore,

$$V(s) = \frac{v_0}{s+\dfrac{k}{m}} + \frac{F_0}{m}\frac{e^{-st_0}}{\left(s+\dfrac{k}{m}\right)},$$

and hence

$$v(t) = v_0 e^{-(k/m)t} + \frac{F_0}{m}e^{-(k/m)(t-t_0)}h(t-t_0), \qquad t\geq 0. \tag{11}$$

We can find position $x(t)$ by computing the antiderivative of velocity $v(t)$, as in equation (10). Alternatively, we can use the fact that

$$X(s) = \frac{1}{s}V(s)$$

to obtain

$$X(s) = v_0\frac{m}{k}\left[\frac{1}{s} - \frac{1}{s+\dfrac{k}{m}}\right] + \frac{F_0}{k}e^{-st_0}\left[\frac{1}{s} - \frac{1}{s+\dfrac{k}{m}}\right].$$

Taking inverse transforms, we find

$$x(t) = v_0\frac{m}{k}\left[1 - e^{-(k/m)t}\right] + \frac{F_0}{k}\left[1 - e^{-(k/m)(t-t_0)}\right]h(t-t_0), \qquad t\geq 0. \;\; ❖ \tag{12}$$

The solid curves in Figure 5.22 are graphs of velocity and position of the body for the parameter values

$$m = 5 \text{ kg}, \qquad k = 0.5 \text{ kg/s}, \qquad v_0 = 20 \text{ m/s}, \qquad F_0 = 500 \text{ N}, \qquad t_0 = 3 \text{ s}. \tag{13}$$

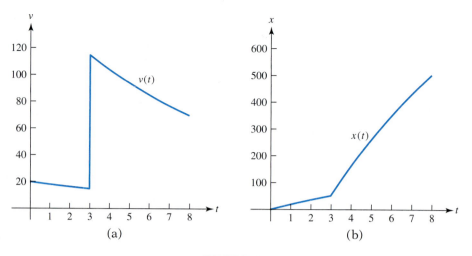

FIGURE 5.22

The results from Example 1. (a) The graph of velocity, $v(t)$, as given in equation (11). The discontinuity is the result of an impulsive force applied at $t = 3$. (b) The graph of position, $x(t)$, as given in equation (12).

As the graph illustrates, application of the impulsive force creates a jump discontinuity in the velocity. This jump is the idealization of the very rapid velocity transition that would occur if the applied force were a very narrow pulse of integrated strength 500 newton-seconds.

The dotted curves in Figure 5.23 show the velocity and position that would result from a force of 5000 N being applied during the interval $3 \leq t \leq 3.1$ sec.

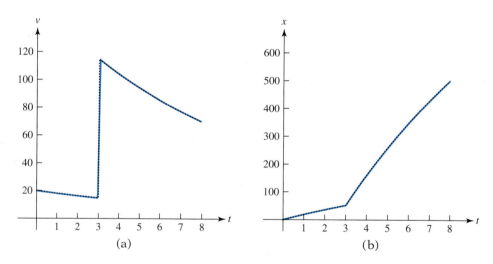

FIGURE 5.23

The graphs of (a) velocity and (b) position for the problems described by equations (9) and (14). Equation (9) models the idealized problem, using the delta function. Equation (14) models the problem using a large (but finite) pulse applied over a t-interval of small (but nonzero) duration. As you can see, the graphs are qualitatively similar.

In other words, the dotted curves arise from solving the initial value problem

$$mv' + kv = F_0 p_\varepsilon(t - t_0), \qquad t > 0$$
$$v(0) = v_0,$$

(14)

with $\varepsilon = 0.1$ sec and all other parameter values as given by (13). The comparison of these graphs illustrates the "idealizing nature" of using the delta function in modeling applications. [The solution of problem (14) is outlined in the Exercises.]

The Impulse Response and the System Transfer Function

The formal use of the delta function as an impulsive source leads to the fact that the impulse response and the system transfer function form a Laplace transform pair. For example, consider the initial value problem

$$a_n \frac{d^n y}{dt^n} + a_{n-1} \frac{d^{n-1} y}{dt^{n-1}} + a_{n-2} \frac{d^{n-2} y}{dt^{n-2}} + \cdots + a_1 \frac{dy}{dt} + a_0 y = \delta(t)$$

(15)

$$y^{(n-1)}(0) = 0, \qquad y^{(n-2)}(0) = 0, \qquad \ldots, \qquad y'(0) = 0, \qquad y(0) = 0,$$

where we use the delta function to model an impulsive nonhomogeneous term. The solution of (15) is the impulse response of an nth order linear system.

Taking Laplace transforms of (15) and using equation (8), we find

$$(a_n s^n + a_{n-1} s^{n-1} + a_{n-2} s^{n-2} + \cdots + a_1 s + a_0) Y(s) = 1,$$

and therefore

$$Y(s) = \frac{1}{a_n s^n + a_{n-1} s^{n-1} + a_{n-2} s^{n-2} + \cdots + a_1 s + a_0}.$$

(16)

The right-hand side of equation (16) is the Laplace transform of the impulse response and is equal to the system transfer function.

EXERCISES

1. Evaluate

   (a) $\displaystyle\int_0^3 (1 + e^{-t}) \delta(t - 2) \, dt$

   (b) $\displaystyle\int_{-2}^1 (1 + e^{-t}) \delta(t - 2) \, dt$

   (c) $\displaystyle\int_{-1}^2 \begin{bmatrix} \cos 2t \\ te^{-t} \end{bmatrix} \delta(t) \, dt$

   (d) $\displaystyle\int_{-3}^2 (e^{2t} + t) \begin{bmatrix} \delta(t + 2) \\ \delta(t - 1) \\ \delta(t - 3) \end{bmatrix} dt$

2. Let $f(t)$ be a function defined and continuous on $0 \le t < \infty$. Determine

   $$f * \delta = \int_0^t f(t - \lambda) \delta(\lambda) \, d\lambda.$$

3. Determine a value of the constant t_0 such that $\int_0^1 \sin^2[\pi(t - t_0)] \delta(t - \frac{1}{2}) \, dt = \frac{3}{4}$.

4. If $\int_1^5 t^n \delta(t - 2) \, dt = 8$, what is the exponent n?

5. Sketch the graph of the function $f(t)$ defined by $f(t) = \int_0^t \delta(\lambda - 1) \, d\lambda$, $0 \le t < \infty$. Can the graph obtained be characterized in terms of a Heaviside step function?

6. Sketch the graph of the function $g(t)$ that is defined by $g(t) = \int_0^t \int_0^\lambda \delta(\sigma - 1)\, d\sigma\, d\lambda$, $0 \le t < \infty$.

7. Sketch the graph of the function $k(t) = \int_0^t [\delta(\lambda - 1) - \delta(\lambda - 2)]\, d\lambda$, $0 \le t < \infty$. Can the graph be characterized in terms of a Heaviside step function or Heaviside step functions?

8. The graph of the function $g(t) = \int_0^t e^{\alpha t} \delta(t - t_0)\, dt$, $0 \le t < \infty$ is shown. Determine the constants α and t_0.

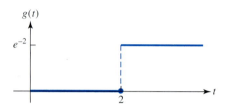

Figure for Exercise 8

Exercises 9–11:

In each exercise, a function $g(t)$ is given.

(a) Solve the initial value problem $y' - y = g(t)$, $y(0) = 0$, using the techniques developed in Chapter 2.

(b) Use Laplace transforms to determine the transfer function $\phi(t)$,

$$\phi' - \phi = \delta(t), \qquad \phi(0) = 0.$$

(c) Evaluate the convolution integral $\phi * g = \int_0^t \phi(t - \lambda) g(\lambda)\, d\lambda$, and compare the resulting function with the solution obtained in part (a).

9. $g(t) = h(t)$ **10.** $g(t) = e^t$ **11.** $g(t) = t$

Exercises 12–20:

Solve the given initial value problem, in which inputs of large amplitude and short duration have been idealized as delta functions. Graph the solution that you obtain on the indicated interval. (In Exercises 19 and 20, plot the two components of the solution on the same graph.)

12. $y' + y = 2 + \delta(t - 1)$, $y(0) = 0$, $0 \le t \le 6$

13. $y' + y = \delta(t - 1) - \delta(t - 2)$, $y(0) = 0$, $0 \le t \le 6$

14. $y'' = \delta(t - 1) - \delta(t - 3)$, $y(0) = 0$, $y'(0) = 0$, $0 \le t \le 6$

15. $y'' + 4\pi^2 y = 2\pi \delta(t - 2)$, $y(0) = 0$, $y'(0) = 0$, $0 \le t \le 6$

16. $y'' - 2y' = \delta(t - 1)$, $y(0) = 1$, $y'(0) = 0$, $0 \le t \le 2$

17. $y'' + 2y' + 2y = \delta(t - 1)$, $y(0) = 0$, $y'(0) = 0$, $0 \le t \le 6$

18. $y'' + 2y' + y = \delta(t - 2)$, $y(0) = 0$, $y'(0) = 1$, $0 \le t \le 6$

19. $\dfrac{d}{dt}\begin{bmatrix} y_1 \\ y_2 \end{bmatrix} = \begin{bmatrix} 1 & 1 \\ 1 & 1 \end{bmatrix}\begin{bmatrix} y_1 \\ y_2 \end{bmatrix} + \delta(t - 1)\begin{bmatrix} 1 \\ 0 \end{bmatrix}$, $\begin{bmatrix} y_1(0) \\ y_2(0) \end{bmatrix} = \begin{bmatrix} 0 \\ 0 \end{bmatrix}$, $0 \le t \le 2$

20. $\dfrac{d}{dt}\begin{bmatrix} y_1 \\ y_2 \end{bmatrix} = \begin{bmatrix} 2 & 1 \\ 0 & 1 \end{bmatrix}\begin{bmatrix} y_1 \\ y_2 \end{bmatrix} + \begin{bmatrix} 0 \\ 1 \end{bmatrix} - \delta(t - 1)\begin{bmatrix} 1 \\ 0 \end{bmatrix}$, $\begin{bmatrix} y_1(0) \\ y_2(0) \end{bmatrix} = \begin{bmatrix} 0 \\ 0 \end{bmatrix}$, $0 \le t \le 2$

PROJECTS

Project 1: Periodic Pinging of a Spring-Mass System

In Chapter 3, we considered the response of a spring-mass system to a periodic applied force. In particular, we saw that the solution of the initial value problem

$$y'' + \omega_0^2 y = F \cos \omega_1 t, \qquad y(0) = 0, \qquad y'(0) = 0$$

has an envelope that grows linearly with time when $\omega_1 = \omega_0$. When $\omega_1 \neq \omega_0$, on the other hand, the solution envelope remains bounded with time.

Suppose that, instead of applying a continuous force, we ping the spring-mass system periodically. In other words, we apply a force of very short duration at equally spaced time intervals. Can we achieve an analogous resonant growth in the solution envelope if the time interval between pings is properly chosen?

Consider the initial value problem

$$y'' + \omega_0^2 y = F \sum_{m=1}^{\infty} \delta(t - mT), \qquad y(0) = 0, \qquad y'(0) = 0,$$

where we use the delta function to model applied pings of short duration. The positive constant T represents the time interval between successive pings.

1. Solve the initial value problem using Laplace transforms. Assume that formal manipulations, such as interchanging the order of inverse Laplace transformation and infinite summation, are valid.
2. Consider the case where the time interval is $T = 2\pi/\omega_0$. In this case, the interval between pings equals the resonant period of the vibrating system. Discuss the qualitative behavior of the solution. Does the solution envelope exhibit some form of resonant growth? As a specific case, assume $\omega_0 = 2\pi$ and $F = 2\pi$. Plot the solution over the time interval $0 \leq t < 10$.
3. Now consider the case where $T = \pi/\omega_0$. (This interval between pings is half the resonant period.) Again, assume that $\omega_0 = 2\pi$ and $F = 2\pi$. Plot the solution over the time interval $0 \leq t < 10$. Does the solution envelope remain bounded or grow with time? Provide a physical rationale for the observed behavior of the solution.

Project 2: Curing Sick Fish

Assume that the tropical fish in a 100-gal aquarium have contracted an ailment and that a soluble medication must be administered to combat the illness. The medicine is packaged in 800-mg doses, and one dose is to be administered daily. Assume that the following facts are known:

(i) A "well-stirred" approximation is valid; that is, the medicine dissolves and disperses itself throughout the tank very rapidly.

(ii) The medicine loses potency at a rate proportional to the amount of medicine present. In fact, the half-life of the medicine (the time span over which the potency is reduced to one-half its initial strength) is one day.

(iii) In order to effectively combat the illness, the concentration of medicine in the tank must be maintained at a level greater than or equal to 5 mg per gallon for a period of 7 days.

Let $q(t)$ denote the amount of potent medicine (in milligrams) in the tank at time t (in days). Assume that the illness is detected at time $t = 0$ and that the first dose of medicine is administered at time $t = 1$. If N doses are administered on consecutive days, the problem to be solved is

$$q'(t) + kq(t) = 800 \sum_{n=1}^{N} \delta(t - n), \qquad q(0) = 0.$$

Because of the well-stirred assumption, we can use the delta function to model the administration of each dose. Moreover, the concentration is $c(t) = q(t)/100$ mg/gal.

1. Determine the constant k and solve the initial value problem using Laplace transforms.

2. Determine the minimum number of doses needed to effectively combat the ailment.

3. What would happen if we continued to administer the medicine (that is, if N became arbitrarily large)? Would the maximum amount of medicine in the tank continue to grow, or would $q(t)$ undergo an initial transient phase and then settle into a periodic, steady-state behavior as time increased?

4. Suppose we define

$$\bar{q}(N) = \int_{N-1}^{N} q(\lambda)\, d\lambda, \qquad N = 1, 2, 3, \ldots .$$

Thus, $\bar{q}(N)$ is the average amount of medicine present in the tank during the Nth day. Show, from the differential equation itself, that if $q(t)$ does settle into a periodic behavior as time increases, then

$$\lim_{N \to \infty} \bar{q}(N) = \frac{800}{k}.$$

Project 3: Locating a Transmission Line Fault

Laplace transformation is an operational tool that can be used to map a given problem into a simpler "transformed problem." We have seen how problems involving ordinary differential equations can be transformed into problems involving simpler algebraic equations. We now consider a problem where Laplace transforms can be used to transform a problem involving partial differential equations into a simpler problem involving ordinary differential equations. The steps outlined in Figure 5.1 remain the same; we first solve this simpler problem and then use the inverse Laplace transform to find the desired solution.

The problem considered is a simple application of the idea of echo location. Knowing how fast sound travels in air, we can determine the distance to a reflection point by measuring the time separation between when a sound is emitted and when its echo is heard. This basic idea can be used to determine where a transmission line fault or disruption is located.

A transmission line is an example of a distributed network. A transmission line is unlike the networks considered earlier in that the voltage and current are functions of both space and time. Consider Figure 5.24, where the transmission line is represented by two parallel cables. The variable x measures distance along the line, with a voltage source or generator positioned at $x = 0$ and the fault (assumed to be an open circuit) located at $x = l$. We assume that the location of the fault is unknown; our goal is to locate it by sending a short pulse down the line and measuring the two-way transit time—the time it takes for the pulse reflected by the fault to return to the source.

As shown in Figure 5.24, we represent the voltage across the line and the current along the line at position x and time t by $v(x, t)$ and $i(x, t)$, respectively. The voltage

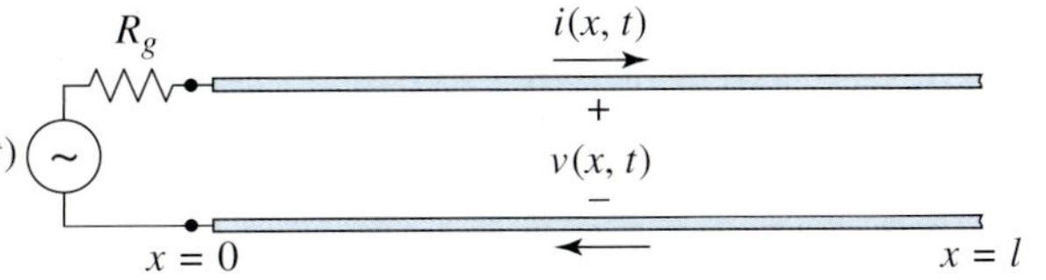

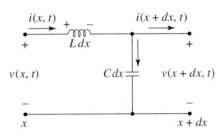

FIGURE 5.24

A transmission line network. A voltage generator is connected at $x = 0$, and an open circuit is assumed to exist at the unknown fault location, $x = l$.

generator is assumed to have an internal resistance R_g. Figure 5.25 depicts a snapshot of a differential transmission line segment taken at some time t. As the figure indicates, the transmission line itself is characterized by a series inductance L per unit length and a shunt capacitance C per unit length. To determine how the transmission line voltage and current behave as functions of space and time, we apply Kirchhoff's voltage and current laws to this differential segment of line. The voltage drop across the inductance is

$$L\frac{\partial i(x, t)}{\partial t}\, dx,$$

while the current flow through the capacitance is

$$C\frac{\partial v(x, t)}{\partial t}\, dx.$$

FIGURE 5.25

Differential transmission line segment equivalent circuit.

If we apply Kirchhoff's voltage law to the circuit in Figure 5.25, we obtain

$$v(x, t) - L\frac{\partial i(x, t)}{\partial t}dx - v(x + dx, t) = 0.$$

Similarly, applying Kirchhoff's current law leads to

$$i(x, t) - C\frac{\partial v(x, t)}{\partial t}dx - i(x + dx, t) = 0.$$

If we divide by dx and let $dx \to 0$, we obtain a pair of partial differential equations

$$\frac{\partial v(x, t)}{\partial x} = -L\frac{\partial i(x, t)}{\partial t}$$

$$\frac{\partial i(x, t)}{\partial x} = -C\frac{\partial v(x, t)}{\partial t}.$$

(1)

We assume that the transmission line is quiescent for $t \leq 0$. That is, we assume

$$i(x, 0) = 0, \qquad v(x, 0) = 0, \qquad 0 \leq x \leq l.$$

(2)

At time $t = 0$, the voltage generator is turned on, emitting a signal $e_g(t), t > 0$. Applying Kirchhoff's voltage law at the generator leads to

$$e_g(t) - i(0, t)R_g - v(0, t) = 0, \qquad t > 0. \tag{3}$$

Lastly, the assumption that an open circuit exists at fault location $x = l$ leads us to the constraint

$$i(l, t) = 0, \qquad t > 0. \tag{4}$$

Equations (1)–(4) constitute the mathematical problem of interest. We are free to select the generator voltage $e_g(t)$. Our goal is to determine a formula for $v(0, t), t > 0$. As we will see, this formula contains l as a parameter. Since $v(0, t)$ is a quantity that can be measured, we will use a voltage measurement to determine the distance l to the fault. Once the location of the fault is known, appropriate repairs can be made.

Since $0 < t < \infty$ is the time interval of interest, we can define the Laplace transforms:

$$V(x, s) \equiv \int_0^\infty v(x, t)e^{-st}dt, \qquad I(x, s) \equiv \int_0^\infty i(x, t)e^{-st}dt. \tag{5}$$

In (5), the variable x is treated as a parameter.

1. Apply the Laplace transform (5) to both sides of equations (1)–(4). Assume that the order of operations can be interchanged. For example,

$$\int_0^\infty \frac{\partial v(x, t)}{\partial x}e^{-st}dt = \frac{\partial}{\partial x}\left[\int_0^\infty v(x, t)e^{-st}dt\right] = \frac{\partial V(x, s)}{\partial x}.$$

Show that an application of the Laplace transform leads to the following transformed problem:

$$\frac{\partial V(x, s)}{\partial x} = -sLI(x, s) \tag{6a}$$

$$\frac{\partial I(x, s)}{\partial x} = -sCV(x, s) \tag{6b}$$

$$E_g(s) - I(0, s)R_g = V(0, s) \tag{6c}$$

$$I(l, s) = 0, \tag{6d}$$

where $E_g(s)$ denotes the Laplace transform of $e_g(t)$.

Note that problem (6) is, in fact, simpler. The only differentiation performed in (6) is with respect to the spatial variable x. If we view the transform variable s as a parameter, then equations (6a)–(6b) are essentially a linear system of ordinary differential equations. A problem involving partial differential equations has been transformed into one that *de facto* involves only ordinary differential equations. Note that problem (6) is not an initial value problem. It is a two-point boundary value problem; the spatial domain is $0 \le x \le l$, and the supplementary conditions (6c)–(6d) are prescribed at the two endpoints.

2. Obtain the general solution of equations (6a)–(6b), viewed as a linear system of ordinary differential equations. Note that since transform variable s is being viewed as a parameter, the two arbitrary constants appearing in the general solution will generally be functions of s.

The quantity $\sqrt{L/C}$ has the dimensions of resistance; it is called the **characteristic impedance** of the transmission line and is often denoted by the symbol Z_0. Assume that $R_g = Z_0$. When this condition holds, the voltage generator is said to be "matched to the transmission line." Impose constraints (6c)–(6d) and show that

$$V(0, s) = \frac{E_g(s)}{2}\left[1 + e^{-s(2l\sqrt{LC})}\right]. \tag{7}$$

3. Determine $v(0, t), t > 0$ by computing the inverse Laplace transform of (7). The product $l\sqrt{LC}$ has the dimensions of time. Assume that the generator voltage $e_g(t)$ is a very short pulse, say

$$e_g(t) = \begin{cases} 10, & 0 < t \leq 0.1 \\ 0, & 0.1 < t < \infty, \end{cases}$$

and that $l\sqrt{LC} = 5$. Graph $v(0, t)$ as a function of time t for $t > 0$.

Explain the physical significance of the two terms comprising $v(0, t)$. Suppose, for example, we know the properties of the transmission line; specifically, suppose we know L and C and, therefore, $\sqrt{LC}$. Explain how your solution can be used to determine the unknown distance l.

Nonlinear Systems

6.1   Introduction

In this chapter, we consider systems of nonlinear differential equations

$$
\begin{aligned}
y_1' &= f_1(t, y_1, y_2, \ldots, y_n) \\
y_2' &= f_2(t, y_1, y_2, \ldots, y_n) \\
&\vdots \\
y_n' &= f_n(t, y_1, y_2, \ldots, y_n), \qquad a < t < b.
\end{aligned}
\tag{1}
$$

To formulate an initial value problem, we specify n initial conditions

$$
y_1(t_0) = y_1^0, \qquad y_2(t_0) = y_2^0, \qquad \ldots, \qquad y_n(t_0) = y_n^0,
\tag{2}
$$

where t_0 is some point belonging to the interval $a < t < b$. The special case of $n = 1$ reduces to the scalar nonlinear problem $y' = f(t, y), y(t_0) = y_0$, treated in Chapter 2.

The Vector Form for a Nonlinear System

We can express the nonlinear system (1) in a compact fashion using vector notation. In particular, define the vector functions

$$\mathbf{y}(t) = \begin{bmatrix} y_1(t) \\ y_2(t) \\ \vdots \\ y_n(t) \end{bmatrix}, \qquad \mathbf{f}(t, \mathbf{y}) = \begin{bmatrix} f_1(t, y_1, y_2, \ldots, y_n) \\ f_2(t, y_1, y_2, \ldots, y_n) \\ \vdots \\ f_n(t, y_1, y_2, \ldots, y_n) \end{bmatrix}, \qquad \mathbf{y}_0 = \begin{bmatrix} y_1^0 \\ y_2^0 \\ \vdots \\ y_n^0 \end{bmatrix}.$$

With this notation, we can write the initial value problem as

$$\mathbf{y}'(t) = \mathbf{f}(t, \mathbf{y}(t)), \qquad a < t < b$$
$$\mathbf{y}(t_0) = \mathbf{y}_0. \tag{3}$$

The linear systems considered in Chapter 4 correspond to a special case of equation (3) where $\mathbf{f}(t, \mathbf{y}) = A(t)\mathbf{y} + \mathbf{g}(t)$.

Autonomous Systems

An important special case occurs when none of the n functions appearing on the right-hand side of system (1) is an explicit function of the independent variable t. In this case, system (1) has the form

$$y_1' = f_1(y_1, y_2, \ldots, y_n)$$
$$y_2' = f_2(y_1, y_2, \ldots, y_n)$$
$$\vdots$$
$$y_n' = f_n(y_1, y_2, \ldots, y_n). \tag{4}$$

System (4) is called an **autonomous** system. In the autonomous case, initial value problems have the form

$$\mathbf{y}' = \mathbf{f}(\mathbf{y})$$
$$\mathbf{y}(t_0) = \mathbf{y}_0. \tag{5}$$

An important feature of solutions of autonomous systems is the nature of their dependence on the independent variable t and the initial value t_0. In Section 2.5, we argued that the solution of a scalar autonomous equation is a function of the *time difference* $t - t_0$; what matters is the value of time t measured relative to the starting time t_0. The same basic argument can be applied to the autonomous system (5), showing that solutions are functions of the difference variable $t - t_0$.

Two-Species Population Models

Modeling the interaction of different species of organisms is important in biological and ecological studies. Consider two species coexisting in some confined environment, say a lake or an island. In some cases, the two species may interact benignly with each other except for the fact that they both compete for the same limited food supply. In other cases, one species may act as a predator and depend on the second species (the prey) as its food supply. Not surprisingly, these two models are referred to as the competing species model and the predator-prey model, respectively.

The ideas underlying the Verhulst population model discussed in Section 2.8 can be extended to describe two-species interactions. We now have two dependent variables, the populations $P_1(t)$ and $P_2(t)$, and their interaction is often modeled by the autonomous nonlinear system

$$
\begin{aligned}
P_1' &= r_1(1 - \alpha_1 P_1 - \beta_1 P_2)P_1 \\
P_2' &= r_2(1 - \beta_2 P_1 - \alpha_2 P_2)P_2,
\end{aligned}
\tag{6}
$$

where the constants r_1, r_2, α_1, and α_2 are positive. The relative birth rates per unit population are $r_1(1 - \alpha_1 P_1 - \beta_1 P_2)$ and $r_2(1 - \beta_2 P_1 - \alpha_2 P_2)$, respectively.

The nonlinear terms having β_1 and β_2 as coefficients are the interaction terms that couple population dynamics. When β_1 and β_2 are positive, an increase in either population decreases the relative birth rate of both populations, since any population increase puts additional stress on the available resources needed by both. If β_1 and β_2 are both zero in equation (6), then the two populations evolve independently of each other; in fact, the two differential equations uncouple, and each population satisfies a separate logistic equation of the type discussed in Section 2.8.

We will treat the predator-prey model in Section 6.7. For now, we leave it as an exercise for you to decide how the competing species model (6) should be modified if one population, say P_1, is a population of predators that depends on the second population, P_2, for its food supply.

The Pendulum

Some problems, such as the motion of the pendulum in Figure 6.1, give rise to second order scalar nonlinear differential equations. Such equations can be recast as first order nonlinear systems and studied as such; this will be our approach in the present chapter.

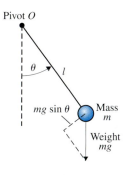

FIGURE 6.1

The pendulum.

Consider the pendulum shown in Figure 6.1. A mass m is attached to the end of a rigid rod of length l. We neglect the weight of the rod and assume the pivot is frictionless. Because of the constraining action of the rod, the (assumed planar) motion of the pendulum mass occurs on the circumference of a circle of radius l centered at the pivot.

The equation of motion for the pendulum can be obtained by equating the sum of the moments about pivot O to the product of the pendulum's moment of inertia and angular acceleration. The resulting formula,

$$\sum M_O = I_O \alpha,$$

can be viewed as a rotational analog of Newton's second law of motion, $F = ma$.

The moment of inertia of the pendulum about pivot O is ml^2. With the counterclockwise direction taken as positive, the moment sum is

$$\sum M_O = -mgl \sin \theta,$$

while the angular acceleration is $\alpha = \theta''$. Therefore, we obtain $-mgl \sin \theta = ml^2 \theta''$, or

$$\theta'' + \frac{g}{l} \sin \theta = 0. \tag{7}$$

To study pendulum motion, we typically specify pendulum position and angular velocity at some initial time, say $t = 0$. Nonlinear differential equation (7), together with the initial conditions $\theta(0) = \theta_0$, $\theta'(0) = \theta_0'$, forms the initial value problem of interest.

We can recast this initial value problem as an initial value problem for a first order nonlinear system by using the ideas introduced in Section 4.2 for linear problems. In particular, let $y_1(t) = \theta(t)$, $y_2(t) = \theta'(t)$, and

$$\mathbf{y}(t) = \begin{bmatrix} y_1(t) \\ y_2(t) \end{bmatrix}.$$

Under this change of variables, we have

$$y_1' = \theta' = y_2 \quad \text{and} \quad y_2' = \theta'' = -\frac{g}{l} \sin \theta = -\frac{g}{l} \sin y_1.$$

Therefore, equation (7) can be rewritten as the first order nonlinear system

$$y_1' = y_2$$
$$y_2' = -\frac{g}{l} \sin y_1.$$

Note that this first order system is autonomous. In vector form, the associated initial value problem is

$$\mathbf{y}' = \mathbf{f}(\mathbf{y}), \qquad \mathbf{y}(0) = \mathbf{y}_0,$$

where

$$\mathbf{f}(\mathbf{y}) = \begin{bmatrix} y_2 \\ -\frac{g}{l} \sin y_1 \end{bmatrix}, \qquad \mathbf{y}_0 = \begin{bmatrix} \theta_0 \\ \theta_0' \end{bmatrix}.$$

Being able to rewrite an initial value problem for a higher order scalar differential equation as an initial value problem for a first order system has several important consequences. For example, in Theorem 6.1 we give the basic existence-uniqueness theory for first order nonlinear systems; this theory

generalizes the scalar results of Section 2.5 in much the same way as Theorem 4.1 generalized the scalar results of Section 2.1 to linear systems. Since systems of higher order nonlinear differential equations can be recast as first order nonlinear systems, Theorem 6.1 will accommodate initial value problems for these scalar higher order equations as well.

In many cases, it is not possible to explicitly solve a nonlinear differential equation, and numerical methods are needed to obtain quantitative information about the solutions. Chapter 7 discusses the development of numerical algorithms, building on the ideas introduced in Section 2.10 and Section 4.9.

Existence and Uniqueness

Consider the initial value problem

$$\mathbf{y}' = \mathbf{f}(t, \mathbf{y}), \qquad \mathbf{y}(t_0) = \mathbf{y}_0, \tag{8}$$

where

$$\mathbf{y}(t) = \begin{bmatrix} y_1(t) \\ y_2(t) \\ \vdots \\ y_n(t) \end{bmatrix} \quad \text{and} \quad \mathbf{f}(t, \mathbf{y}) = \begin{bmatrix} f_1(t, y_1, y_2, \ldots, y_n) \\ f_2(t, y_1, y_2, \ldots, y_n) \\ \vdots \\ f_n(t, y_1, y_2, \ldots, y_n) \end{bmatrix}. \tag{9}$$

Because there are n dependent variables, we consider initial value problem (8) in the $(n+1)$-dimensional open rectangular region R defined by the inequalities

$$a < t < b, \quad \alpha_1 < y_1 < \beta_1, \quad \alpha_2 < y_2 < \beta_2, \quad \ldots, \quad \alpha_n < y_n < \beta_n. \tag{10}$$

Assume the initial condition point $(t_0, \mathbf{y}_0)$ lies in the region R. Theorem 6.1 asserts that continuity of the n component functions of $\mathbf{f}(t, \mathbf{y})$ in equation (9), along with continuity of the n^2 partial derivatives

$$\frac{\partial f_1(t, y_1, y_2, \ldots, y_n)}{\partial y_1}, \quad \frac{\partial f_1(t, y_1, y_2, \ldots, y_n)}{\partial y_2}, \quad \ldots, \quad \frac{\partial f_1(t, y_1, y_2, \ldots, y_n)}{\partial y_n}$$

$$\frac{\partial f_2(t, y_1, y_2, \ldots, y_n)}{\partial y_1}, \quad \frac{\partial f_2(t, y_1, y_2, \ldots, y_n)}{\partial y_2}, \quad \ldots, \quad \frac{\partial f_2(t, y_1, y_2, \ldots, y_n)}{\partial y_n}$$

$$\vdots \qquad\qquad\qquad \vdots \tag{11}$$

$$\frac{\partial f_n(t, y_1, y_2, \ldots, y_n)}{\partial y_1}, \quad \frac{\partial f_n(t, y_1, y_2, \ldots, y_n)}{\partial y_2}, \quad \ldots, \quad \frac{\partial f_n(t, y_1, y_2, \ldots, y_n)}{\partial y_n},$$

is sufficient to ensure the existence of a unique solution of the initial value problem on some interval $c < t < d$ containing t_0. As in the scalar case, however, Theorem 6.1 gives no insight into the size of the interval $c < t < d$.

Theorem 6.1

Consider the initial value problem

$$\mathbf{y}' = \mathbf{f}(t, \mathbf{y}), \qquad \mathbf{y}(t_0) = \mathbf{y}_0,$$

where the initial value point $(t_0, \mathbf{y}_0)$ lies in the region R defined by the inequalities in (10). Let $\mathbf{f}(t, \mathbf{y})$ and the partial derivatives in (11) be continuous in R. Then the initial value problem has a unique solution $\mathbf{y}(t)$ that exists on some t-interval (c, d) containing t_0.

The following example illustrates the application of Theorem 6.1 to a pendulum problem similar to the one shown in Figure 6.1.

EXAMPLE 1

Consider the following initial value problem for a forced pendulum:

$$ml\theta'' + mg\sin\theta = F_0\sin\omega t$$
$$\theta(0) = 0, \qquad \theta'(0) = 0.$$

(12)

This equation describes a sinusoidal tangential force, having amplitude F_0 and radian frequency ω, applied to the pendulum. At time $t = 0$, the pendulum is in the vertically downward position with no initial angular velocity. What can we conclude from Theorem 6.1 about solutions of (12)?

Solution: In order to apply Theorem 6.1, we write the second order differential equation as a first order system. Let $y_1(t) = \theta(t)$, $y_2(t) = \theta'(t)$, and

$$\mathbf{y}(t) = \begin{bmatrix} y_1(t) \\ y_2(t) \end{bmatrix}.$$

We obtain the first order system $\mathbf{y}' = \mathbf{f}(t, \mathbf{y})$, where

$$\mathbf{f}(t, \mathbf{y}) = \begin{bmatrix} y_2 \\ -\dfrac{g}{l}\sin y_1 + \dfrac{F_0}{ml}\sin\omega t \end{bmatrix}.$$

According to Theorem 6.1, we need to examine continuity of the functions

$$f_1(t, y_1, y_2) = y_2 \quad \text{and} \quad f_2(t, y_1, y_2) = -\frac{g}{l}\sin y_1 + \frac{F_0}{ml}\sin\omega t$$

and the four partial derivatives

$$\frac{\partial f_1(t, y_1, y_2)}{\partial y_1} = 0, \qquad \frac{\partial f_1(t, y_1, y_2)}{\partial y_2} = 1,$$

$$\frac{\partial f_2(t, y_1, y_2)}{\partial y_1} = -\frac{g}{l}\cos y_1, \qquad \frac{\partial f_2(t, y_1, y_2)}{\partial y_2} = 0.$$

The functions f_1 and f_2, along with the four partial derivatives, are continuous for all values (t, y_1, y_2) in $t\mathbf{y}$-space. Therefore, applying Theorem 6.1, we can take R to be any open three-dimensional rectangular region in $t\mathbf{y}$-space that contains the initial condition point $(t_0, \mathbf{y}_0) = (0, 0, 0)$. Theorem 6.1 concludes that a unique solution of the initial value problem (12) exists on *some* t-interval containing $t = 0$. ❖

Example 1 not only illustrates the application of the existence-uniqueness theorem but also highlights its shortcomings. The theorem concludes that there is some t-interval of existence-uniqueness but, unlike the linear system case, gives no insight into how large this interval might be. On the one hand, a theorem such as Theorem 6.1 that deals with a very general class of nonlinear systems cannot be expected to do more; it cannot give precise results for particular cases. As shown in Chapter 2, nonlinear initial value problems can have solutions exhibiting a wide variety of behavior. On the other hand, our everyday experience with pendulums suggests that the particular initial value problem considered in Example 1 should have a unique solution on an arbitrarily large t-interval; that is, we don't expect such a mechanical system to behave catastrophically.

For initial value problems involving nonlinear systems, there are virtually no techniques for finding explicit or implicit representations of solutions; we must look for other ways to understand the behavior of these solutions. Our attention, therefore, will be focused in two directions: on determining qualitative information by graphical means and on obtaining quantitative information from numerical methods.

EXERCISES

Exercises 1–9:

In each exercise,

(a) Rewrite the given nth order scalar initial value problem as $\mathbf{y}' = \mathbf{f}(t, \mathbf{y})$, $\mathbf{y}(t_0) = \mathbf{y}_0$, by defining $y_1(t) = y(t), y_2(t) = y'(t), \ldots, y_n(t) = y^{(n-1)}(t)$ and

$$\mathbf{y}(t) = \begin{bmatrix} y_1(t) \\ y_2(t) \\ \vdots \\ y_n(t) \end{bmatrix}.$$

(b) Compute the n^2 partial derivatives $\partial f_i(t, y_1, \ldots, y_n)/\partial y_j$, $i, j = 1, \ldots, n$.

(c) For the system obtained in part (a), determine where in $(n + 1)$-dimensional $t\mathbf{y}$-space the hypotheses of Theorem 6.1 are *not* satisfied. In other words, at what points $(t, y_1, \ldots, y_n)$, if any, does at least one component function $f_i(t, y_1, \ldots, y_n)$ and/or at least one partial derivative function $\partial f_i(t, y_1, \ldots, y_n)/\partial y_j$, $i, j = 1, \ldots, n$ fail to be continuous? What is the largest open rectangular region R where the hypotheses of Theorem 6.1 hold?

1. $y'' + ty' + 2y = 0$, $\quad y(0) = 1$, $\quad y'(0) = 2$

2. $y'' + e^t y = \ln |t|$, $\quad y(-1) = 0$, $\quad y'(-1) = -1$

3. $y'' + ty = \sin y'$, $\quad y(0) = 0$, $\quad y'(0) = 1$

4. $y'' + (y')^3 + y^{1/3} = \tan(t/2)$, $\quad y(1) = 1$, $\quad y'(1) = -2$

5. $ty'' + \dfrac{1}{1 + y + 2y'} = e^{-t}$, $\quad y(2) = 2$, $\quad y'(2) = 1$

6. $y''' + t^2 y'' = \sin t$, $\quad y(1) = 0$, $\quad y'(1) = 1$, $\quad y''(1) = -1$

7. $y''' + y' + y^2 = 0$, $\quad y(-1) = 0$, $\quad y'(-1) = 1$, $\quad y''(-1) = 0$

8. $y''' + \cos(ty') = t(y'')^2$, $\quad y(0) = 1$, $\quad y'(0) = 1$, $\quad y''(0) = -2$

9. $y''' + \dfrac{2t^{1/3}}{(y - 2)(y'' + 2)} = 0$, $\quad y(0) = 0$, $\quad y'(0) = 2$, $\quad y''(0) = 2$

Exercises 10–13:

In each exercise, an initial value problem for a first order nonlinear system is given. Rewrite the problem as an equivalent initial value problem for a higher order nonlinear scalar differential equation.

10. $\dfrac{d}{dt} \begin{bmatrix} y_1 \\ y_2 \end{bmatrix} = \begin{bmatrix} y_2 \\ t\cos^2(y_2) - 3y_1 + t^4 \end{bmatrix}$, $\quad \begin{bmatrix} y_1(2) \\ y_2(2) \end{bmatrix} = \begin{bmatrix} 1 \\ -1 \end{bmatrix}$

11. $\dfrac{d}{dt}\begin{bmatrix} y_1 \\ y_2 \end{bmatrix} = \begin{bmatrix} y_2 \\ y_2 \tan(y_1) + e^{y_2} \end{bmatrix}$, $\begin{bmatrix} y_1(0) \\ y_2(0) \end{bmatrix} = \begin{bmatrix} 0 \\ 1 \end{bmatrix}$

12. $\dfrac{d}{dt}\begin{bmatrix} y_1 \\ y_2 \\ y_3 \end{bmatrix} = \begin{bmatrix} y_2 \\ y_3 \\ y_1 y_2 + y_3^2 \end{bmatrix}$, $\begin{bmatrix} y_1(-1) \\ y_2(-1) \\ y_3(-1) \end{bmatrix} = \begin{bmatrix} -1 \\ 2 \\ -4 \end{bmatrix}$

13. $\dfrac{d}{dt}\begin{bmatrix} y_1 \\ y_2 \\ y_3 \end{bmatrix} = \begin{bmatrix} y_2 \\ y_3 \\ \sqrt{y_2 y_3 + t^2} \end{bmatrix}$, $\begin{bmatrix} y_1(1) \\ y_2(1) \\ y_3(1) \end{bmatrix} = \begin{bmatrix} 1 \\ \frac{1}{2} \\ 3 \end{bmatrix}$

14. Consider the initial value problem

$$\frac{d}{dt}\begin{bmatrix} y_1 \\ y_2 \end{bmatrix} = \begin{bmatrix} \frac{5}{4}y_1^{1/5} + y_2^2 \\ 3y_1 y_2 \end{bmatrix}, \qquad \begin{bmatrix} y_1(0) \\ y_2(0) \end{bmatrix} = \begin{bmatrix} 0 \\ 0 \end{bmatrix}.$$

For the given autonomous system, the two functions $f_1(y_1, y_2) = \frac{5}{4}y_1^{1/5} + y_2^2$ and $f_2(y_1, y_2) = 3y_1 y_2$ are continuous functions for all (y_1, y_2).

(a) Show by direct substitution that

$$y_1(t) = \begin{cases} 0, & -\infty < t \le c, \\ (t - c)^{5/4}, & c < t < \infty, \end{cases} \qquad y_2(t) = 0$$

is a solution of this initial value problem on $-\infty < t < \infty$ for any positive constant c.

(b) Since c is an arbitrary positive constant, the solution of the given initial value problem is clearly not unique. Does this example contradict Theorem 6.1? Explain your answer.

15. Consider the initial value problem $y'' + y^2 = t, y(0) = y_0,\ y'(0) = y_0'$. Can Laplace transforms be used to solve this initial value problem? Explain your answer.

Exercises 16–17:

Give an example of a two-dimensional nonlinear first order system for which the hypotheses of Theorem 6.1 are not satisfied at precisely the specified points in $ty_1 y_2$-space.

16. The points satisfying $1 + t + y_1 + 3y_2 = 0$

17. The points $(t, y_1, y_2) = (1, n\pi, 2), n = 0, \pm 1, \pm 2, \ldots$

18. **Nonlinear Spring-Mass Systems** Hooke's law assumes the restoring force exerted by a spring under tension or compression is proportional to the displacement (the distance stretched or foreshortened). This assumption cannot be valid for large displacements since there are limits to the amount a spring can be stretched or compressed. Suppose we assume that the restoring force $F_R(x)$ is related to spring displacement x by

$$F_R(x) = -\frac{2k\delta}{\pi} \tan\left(\frac{\pi x}{2\delta}\right).$$

In this model, the restoring force has vertical asymptotes at $x = \pm\delta$; the value δ represents the maximum amount the spring can be stretched or compressed. Consider the figure, illustrating a mass m attached to such a spring. Assume that the mass moves on a frictionless horizontal surface and that the spring has unstretched length l. Newton's second law of motion leads to the nonlinear differential equation

$$mx'' + \frac{2k\delta}{\pi} \tan\left(\frac{\pi x}{2\delta}\right) = 0. \tag{13}$$

Nonlinear
spring   Mass

(a) Unstretched state   (b) Spring stretched
a distance $x(t)$

Figure for Exercise 18

(a) Consider $\tan(\pi x/2\delta)$ as a function of x defined on $-\delta < x < \delta$. Expand this function in a Maclaurin series. Show that if we assume $|\pi x/2\delta|$ is small and approximate $\tan(\pi x/2\delta)$ by the first nonvanishing term in this series, we obtain the linear differential equation found previously when we assumed Hooke's law to be valid.

(b) Show that if the first two nonvanishing terms of the Maclaurin expansion are retained, we obtain the differential equation

$$mx'' + k\left[x + \frac{1}{3}\left(\frac{\pi}{2\delta}\right)^2 x^3\right] = 0. \tag{14}$$

Equation (14) is often used to model the onset of nonlinear effects and is referred to as modeling a spring-mass system with cubic nonlinearity.

(c) Rewrite differential equations (13) and (14) as equivalent first order systems.

(d) For each nonlinear system obtained in part (c), determine the points, if any, where the hypotheses of Theorem 6.1 are not satisfied.

19. **Chemical Reactions**   Nonlinear systems often arise when chemical reactions are modeled. One example is described in the reaction diagram in the figure. In the reaction shown, substance A interacts reversibly with enzyme E to form complex C. Complex C, in turn, decomposes irreversibly into the reaction product B and the original enzyme E. The reaction rates k_1, k_1' and k_2 (assumed to be constant) are shown in the figure. With lowercase symbols used to designate concentrations, the governing differential equations are

$$\frac{da}{dt} = -k_1 ae + k_1' c$$

$$\frac{db}{dt} = k_2 c$$

$$\frac{dc}{dt} = k_1 ae - (k_1' + k_2)c \tag{15}$$

$$\frac{de}{dt} = -k_1 ae + (k_1' + k_2)c.$$

Typical initial conditions are $a(0) = a_0$, $b(0) = 0$, $c(0) = 0$, $e(0) = e_0$.

$$A + E \; \underset{k_1'}{\overset{k_1}{\rightleftharpoons}} \; C \qquad C \xrightarrow{k_2} B + E$$

Figure for Exercise 19

(a) Show that the differential equations (15) imply that $d[c(t) + e(t)]/dt = 0$, which implies that $c(t) + e(t) = c(0) + e(0) = e_0$.

(b) Use the observation made in part (a) to eliminate $e(t)$ in (15) and obtain a two-dimensional nonlinear system for the dependent variables $a(t)$ and $c(t)$.

(c) For the two-dimensional system obtained in part (b), at what points in *tac*-space are the hypotheses of Theorem 6.1 satisfied?

6.2   Equilibrium Solutions and Direction Fields

In this section, we extend the concepts of direction fields and equilibrium solutions to systems of autonomous equations. This extension provides a large-scale overview of the qualitative behavior of solutions of autonomous systems.

Equilibrium Solutions

Consider a system of n autonomous differential equations

$$y'_1 = f_1(y_1, y_2, \ldots, y_n)$$
$$y'_2 = f_2(y_1, y_2, \ldots, y_n)$$
$$\vdots$$
$$y'_n = f_n(y_1, y_2, \ldots, y_n)$$

or, in vector terms,

$$\mathbf{y}' = \mathbf{f}(\mathbf{y}). \tag{1}$$

Let $\mathbf{y}_e$ be a constant $(n \times 1)$ vector such that $\mathbf{f}(\mathbf{y}_e) = \mathbf{0}$. The constant vector-valued function $\mathbf{y}(t) = \mathbf{y}_e$, $-\infty < t < \infty$ is called an **equilibrium solution** of the autonomous system (1).

E X A M P L E
1

Find the equilibrium solutions for the pendulum equation

$$y'_1 = y_2$$
$$y'_2 = -\frac{g}{l} \sin y_1.$$

Solution: In vector form, the equation is $\mathbf{y}' = \mathbf{f}(\mathbf{y})$, where

$$\mathbf{y} = \begin{bmatrix} y_1 \\ y_2 \end{bmatrix} \quad \text{and} \quad \mathbf{f}(\mathbf{y}) = \begin{bmatrix} y_2 \\ -\frac{g}{l} \sin y_1 \end{bmatrix}. \tag{2}$$

From (2), the equation $\mathbf{f}(\mathbf{y}) = \mathbf{0}$ requires $y_2 = 0$ and $y_1 = m\pi, m = 0, \pm 1, \pm 2, \ldots$. Therefore, the equilibrium solutions are

$$\mathbf{y}_m(t) = \begin{bmatrix} m\pi \\ 0 \end{bmatrix}, \qquad m = 0, \pm 1, \pm 2, \ldots.$$

These constant solutions of the pendulum equation have a simple physical interpretation. Recall that $y_1 = \theta$ and $y_2 = \theta'$ (see Figure 6.1). Thus, for an even value of m, the pendulum is at rest (since it has zero angular velocity) and is positioned so that it hangs downward. For an odd value of m, the pendulum is also at rest, but it is positioned in the vertically upward position. ❖

EXAMPLE 2

Find the equilibrium solutions of the competing species model

$$P_1' = r_1(1 - \alpha_1 P_1 - \beta_1 P_2)P_1$$
$$P_2' = r_2(1 - \beta_2 P_1 - \alpha_2 P_2)P_2.$$

Solution: Setting both right-hand sides simultaneously equal to zero leads to four equilibrium solutions. One of the equilibrium solutions is the trivial one,

$$\text{(i) } P_1 = P_2 = 0, \quad \text{or} \quad \mathbf{P}_1^{(e)} = \mathbf{0}.$$

[This equilibrium solution corresponds to the absence of both species from the colony.]

Two additional equilibrium solutions are

$$\text{(ii) } P_1 = 0, \qquad P_2 = \frac{1}{\alpha_2}, \quad \text{or} \quad \mathbf{P}_2^{(e)} = \begin{bmatrix} 0 \\ \dfrac{1}{\alpha_2} \end{bmatrix}$$

$$\text{(iii) } P_1 = \frac{1}{\alpha_1}, \qquad P_2 = 0, \quad \text{or} \quad \mathbf{P}_3^{(e)} = \begin{bmatrix} \dfrac{1}{\alpha_1} \\ 0 \end{bmatrix}.$$

[Equilibrium solutions (ii) and (iii) correspond to the absence of one species. The remaining species has the equilibrium value of the corresponding scalar logistic equation (see Section 2.8).]

If neither P_1 nor P_2 is zero, we obtain a fourth equilibrium solution,

$$\text{(iv) } P_1 = \frac{\alpha_2 - \beta_1}{\alpha_1 \alpha_2 - \beta_1 \beta_2}, \qquad P_2 = \frac{\alpha_1 - \beta_2}{\alpha_1 \alpha_2 - \beta_1 \beta_2}, \quad \text{or} \quad \mathbf{P}_4^{(e)} = \begin{bmatrix} \dfrac{\alpha_2 - \beta_1}{\alpha_1 \alpha_2 - \beta_1 \beta_2} \\ \dfrac{\alpha_1 - \beta_2}{\alpha_1 \alpha_2 - \beta_1 \beta_2} \end{bmatrix}.$$

In (iv), we tacitly assume that $\alpha_1 \alpha_2 - \beta_1 \beta_2 \neq 0$. Since populations are nonnegative quantities, equilibrium solution (iv) is physically meaningful only if the constants α_1, α_2, β_1, and β_2 are such that each component of $\mathbf{P}_4^{(e)}$ is positive. In that case, equilibrium solution (iv) corresponds to a state where both populations are present and coexist at constant levels within the colony. ❖

Two-Dimensional Autonomous Systems and the Phase Plane

We now consider a special case—the two-dimensional autonomous system

$$y_1' = f_1(y_1, y_2)$$
$$y_2' = f_2(y_1, y_2). \tag{3}$$

The qualitative behavior of solutions of system (3) can be described and studied graphically. Solution trajectories are plotted in a two-dimensional setting known as the **phase plane**.

The phase plane was introduced in Section 4.5 for studying linear homogeneous constant coefficient systems. In Example 3, we review the main ideas. As noted in Section 4.5, it is natural to graph solutions of a scalar equation $y' = f(t, y)$ in the two-dimensional ty-plane. However, graphing solution curves $\mathbf{y}(t)$ of system (3) requires three-dimensional $ty_1 y_2$-space. As an alternative, we can view the solution components $y_1(t)$ and $y_2(t)$ as defining a parameterized curve in the y_1y_2-plane, or phase plane.

EXAMPLE

3

Consider the initial value problem

$$y_1' = -y_1 - 6y_2$$
$$y_2' = 6y_1 - y_2$$
$$y_1(0) = 1, \qquad y_2(0) = 0.$$

In matrix terms, this autonomous initial value problem has the form $\mathbf{y}' = A\mathbf{y}$, $\mathbf{y}(0) = \mathbf{y}_0$, where

$$\mathbf{y} = \begin{bmatrix} y_1 \\ y_2 \end{bmatrix}, \qquad A = \begin{bmatrix} -1 & -6 \\ 6 & -1 \end{bmatrix}, \quad \text{and} \quad \mathbf{y}_0 = \begin{bmatrix} 1 \\ 0 \end{bmatrix}. \qquad \text{(4)}$$

The solution of initial value problem (4) is

$$\begin{bmatrix} y_1(t) \\ y_2(t) \end{bmatrix} = \begin{bmatrix} e^{-t}\cos 6t \\ e^{-t}\sin 6t \end{bmatrix}. \qquad \text{(5)}$$

Figure 6.2(a) shows the solution graphed in three-dimensional ty_1y_2-space. The graph evolves in a "shrinking" helical or screwlike manner as time increases.

The two-dimensional phase-plane representation is found by graphing the parameterized curve defined by (5). Figure 6.2(b) shows the graph of (5) for $0 \le t \le 3$. The phase-plane graph is a spiral; it is simply the projection of the three-dimensional "helical" trajectory of Figure 6.2(a) on the two-dimensional y_1y_2-plane. As time increases, the solution point spirals counterclockwise inward toward the phase-plane origin (0, 0).

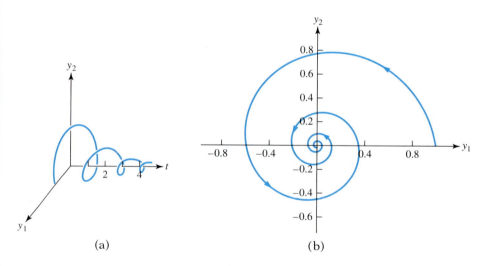

(a) (b)

FIGURE 6.2

(a) The solution of equation (4), graphed in ty_1y_2-space. (b) The solution of equation (4) projected onto the y_1y_2-plane. The arrows show how the phase-plane trajectory is traversed as time increases. Graph (b) corresponds to what an observer would see standing behind the plane $t = 0$ in graph (a) and looking in the direction of increasing t. ❖

Using the Phase Plane to Gather Qualitative Information about Solutions

In studying the geometric aspects of two-dimensional systems, it is often desirable to change the notation. In particular, we can drop the subscripts and denote the dependent variables as $x(t)$ and $y(t)$. With this change of notation, the general nonlinear autonomous two-dimensional system has the form

$$x' = f(x, y)$$
$$y' = g(x, y). \tag{6}$$

The phase plane is now simply the xy-plane.

As illustrated by Figure 6.2(b), we can think of the solution components $x(t)$ and $y(t)$ as defining the coordinates of a point, $(x(t), y(t))$, that is moving in the phase plane; we refer to this point as the **solution point**. [For equilibrium solutions, the components $x(t)$ and $y(t)$ are constant for all t. Therefore, the solution point corresponding to an equilibrium solution is often called an **equilibrium point**.] The plane curve traced out by a solution point is called a **solution curve**. The question we now address is "What qualitative information can we obtain about the motion of a solution point without actually solving the system of differential equations?"

As we know from vector calculus, the vector $\mathbf{v}(t)$ given by

$$\mathbf{v}(t) = x'(t)\mathbf{i} + y'(t)\mathbf{j} \tag{7}$$

is tangent to the solution curve at the point $(x(t), y(t))$ (see Figure 6.3). In particular, if we think of the solution point $(x(t), y(t))$ as moving along the solution curve, then $\mathbf{v}(t)$ is the velocity vector and points in the direction of instantaneous motion at time t.

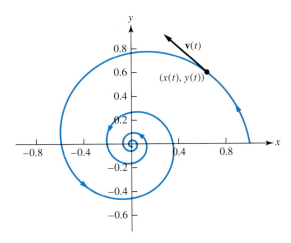

FIGURE 6.3

The solution point $(x(t), y(t))$ lies on a solution curve of system (6). The velocity vector $\mathbf{v}(t) = x'(t)\mathbf{i} + y'(t)\mathbf{j}$ is tangent to the solution curve at $(x(t), y(t))$ and points in the direction the solution point is moving.

Suppose we select an arbitrary point in the phase plane, say $(\bar{x}, \bar{y})$. Assume that a solution curve passes through this point at some time $t = \bar{t}$; that is,

$$(\bar{x}, \bar{y}) = (x(\bar{t}), y(\bar{t})).$$

At the point $(\bar{x}, \bar{y})$, the velocity vector is given by

$$\bar{\mathbf{v}} = x'(\bar{t})\mathbf{i} + y'(\bar{t})\mathbf{j}$$
$$= f(\bar{x}, \bar{y})\mathbf{i} + g(\bar{x}, \bar{y})\mathbf{j}. \tag{8}$$

As illustrated in Figure 6.3, the vector $\bar{\mathbf{v}}$ is tangent to the solution curve at $(\bar{x}, \bar{y})$ and points in the direction of motion. Therefore, by simply evaluating the two right-hand sides of system (6) at a point $(\bar{x}, \bar{y})$ in the phase plane, we can deduce the direction of motion of the solution point at the instant it passes through $(\bar{x}, \bar{y})$. In particular, we see from (8) that the slope m of the line tangent to the solution curve at $(\bar{x}, \bar{y})$ is given by

$$m = \frac{g(\bar{x}, \bar{y})}{f(\bar{x}, \bar{y})}, \qquad f(\bar{x}, \bar{y}) \neq 0. \tag{9}$$

[If $f(\bar{x}, \bar{y}) = 0$ but $g(\bar{x}, \bar{y}) \neq 0$, then the solution curve has a vertical tangent at the point $(\bar{x}, \bar{y})$. If the numerator and denominator both vanish, then $(\bar{x}, \bar{y})$ is an equilibrium point.]

EXAMPLE 4

Consider the autonomous system

$$x' = \tfrac{1}{2}\left(1 - \tfrac{1}{2}x - \tfrac{1}{2}y\right)x$$
$$y' = \tfrac{1}{4}\left(1 - \tfrac{1}{3}x - \tfrac{2}{3}y\right)y.$$

Let $(x(t), y(t))$ denote a solution curve in the phase plane. Determine the velocity vector when the solution curve passes through the given point (x, y).

(a) $(x, y) = (2, 2)$ (b) $(x, y) = \left(\tfrac{1}{2}, \tfrac{3}{2}\right)$

Solution: This system is the competing species autonomous system discussed in Example 2, with the dependent variables renamed and specific values assigned to the constants.

(a) Since $f(2, 2) = -1$ and $g(2, 2) = -\tfrac{1}{2}$, the velocity vector at the point $(2, 2)$ is

$$\mathbf{v} = -\mathbf{i} - \tfrac{1}{2}\mathbf{j}.$$

[At the instant a solution point passes through $(2, 2)$, it is moving downward and to the left. The slope of the line tangent to the phase-plane trajectory at $(2, 2)$ is $\tfrac{1}{2}$.]

(b) At this point, $f\left(\tfrac{1}{2}, \tfrac{3}{2}\right) = 0$ and $g\left(\tfrac{1}{2}, \tfrac{3}{2}\right) = -\tfrac{1}{16}$. Therefore, the velocity vector at $\left(\tfrac{1}{2}, \tfrac{3}{2}\right)$ is

$$\mathbf{v} = -\tfrac{1}{16}\mathbf{j}.$$

[At the instant a solution point passes through $\left(\tfrac{1}{2}, \tfrac{3}{2}\right)$, it is moving vertically downward.] ❖

Phase-Plane Direction Fields

Using equations (8) and (9), we can determine the tangent to a solution curve at a point $(\bar{x}, \bar{y})$ in the phase plane. As in Example 4, we can also determine the instantaneous direction of motion of the solution point when it passes through $(\bar{x}, \bar{y})$. We can use this information to deduce the qualitative behavior of solutions.

A **phase-plane direction field** is constructed by first choosing a suitably dense grid of sampling points in the phase plane. At each grid point, we draw a small arrow, directed along the velocity vector at the grid point. In this way, we generate a qualitative picture of how the solution point moves as time increases. Figure 6.4 shows a direction field for the competing species system treated in Example 4. In Figure 6.4, every arrow is drawn with the same length; this is the construction referred to as a *direction field*. When the arrows at each phase-plane point are drawn so that their lengths are proportional to the speed at that point, the construction is referred to as a **vector field**. Since a vector field contains more information, giving both magnitude and direction of velocity, it might seem that such constructions would always be more desirable. However, vector fields sometimes lead to information overload because the vector field graphs may have intersecting arrows and be visually confusing.

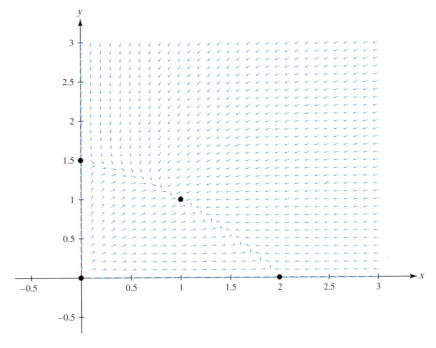

FIGURE 6.4

A portion of the direction field for the autonomous system discussed in Example 4. The arrow at a grid point $(\bar{x}, \bar{y})$ indicates the direction of motion of the solution point as it moves along a solution curve passing through $(\bar{x}, \bar{y})$.

Figure 6.4 provides a good overview of the qualitative behavior of solutions of the autonomous system discussed in Example 4. Recall that the variables x and y correspond to the two populations P_1 and P_2. On the coordinate axes, one or the other population is zero and the direction field arrows point toward the single-species equilibrium points, $(2, 0)$ and $(0, \frac{3}{2})$. Therefore, in the case where only one population is present in the colony, population tends toward the appropriate nonzero equilibrium value, $x = 2$ or $y = \frac{3}{2}$, as time increases. In the general case, where both populations are initially nonzero, the direction field graphed in Figure 6.4 suggests that the populations tend toward the equilibrium point $(1, 1)$ as time evolves.

Putting the Pieces Together

There are computer packages available that generate detailed direction fields such as the one shown in Figure 6.4. However, it is often possible to combine, by hand, the ideas discussed in this section and develop a less detailed phase-plane picture that nevertheless displays the essential features of the system's behavior. For an illustration, consider the competing species model discussed in Example 4,

$$x' = \tfrac{1}{2}\left(1 - \tfrac{1}{2}x - \tfrac{1}{2}y\right)x$$
$$y' = \tfrac{1}{4}\left(1 - \tfrac{1}{3}x - \tfrac{2}{3}y\right)y. \tag{10}$$

First note that

$$f(x, y) = \tfrac{1}{2}\left(1 - \tfrac{1}{2}x - \tfrac{1}{2}y\right)x$$

vanishes on the phase-plane lines $x = 0$ and $x + y = 2$. Likewise,

$$g(x, y) = \tfrac{1}{4}\left(1 - \tfrac{1}{3}x - \tfrac{2}{3}y\right)y$$

vanishes on the lines $y = 0$ and $x + 2y = 3$. Such phase-plane curves, where one of the right-side functions vanishes, are called **nullclines**. Equilibrium points can occur only at a place where two nullclines intersect (although not every intersection point leads to an equilibrium point). For autonomous system (10), there are four equilibrium points:

$$(0, 0), \qquad (2, 0), \qquad \left(0, \tfrac{3}{2}\right), \qquad \text{and} \quad (1, 1).$$

The nullclines and equilibrium points for system (10) are shown in Figure 6.5(a). The nullclines divide the first quadrant of the phase plane into regions where the functions f and g are either positive or negative. Figure 6.5(b) shows the four phase-plane regions defined by the nullclines of system (10) and the corresponding algebraic signs of f and g. From equation (8), we can see that the signs of f and g determine the orientation of the direction field arrows. In region 1, for instance, $f > 0$ and $g > 0$. Therefore, all the direction field arrows in region 1 point upward and to the right.

Figure 6.6 shows, in schematic form, the general orientation of direction field arrows in each of the four open regions. It also shows the orientation of the direction field arrows on the nullclines; these arrows are either horizontal (if $f \neq 0$ and $g = 0$) or vertical (if $f = 0$ and $g \neq 0$). The information in Figure 6.6 is sufficient to deduce the general qualitative behavior of phase-plane trajectories. For example, if x and y are both initially nonzero, then solutions will tend toward the equilibrium state $x = y = 1$ as time increases.

While quite useful, a rough graph such as the one in Figure 6.6 may not be detailed enough to yield a good qualitative picture of the phase-plane trajectories. The next example illustrates this point.

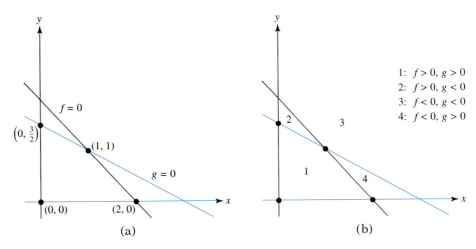

FIGURE 6.5

(a) The lines denote the nullclines for system (10). (b) The nullclines divide the direction field of Figure 6.4 into regions where arrows all have the same general orientation.

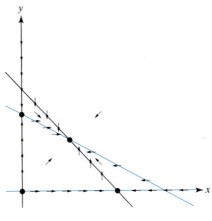

FIGURE 6.6

Along a nullcline, the arrows are either vertical or horizontal. In an open region bounded by nullclines, the single arrow indicates the general orientation of the solution curves in that region. This figure *suggests* that any solution curve starting in one of the open regions will move toward the equilibrium point (1, 1).

E X A M P L E

5

Consider the pendulum equation treated in Example 1, where $g/l = 1$:

$$x' = y$$
$$y' = -\sin x.$$

(a) Sketch the nullclines, as in Figure 6.5(a), marking the equilibrium points with a heavy dot. Then, as in Figure 6.6, add arrows to indicate the flow of solution curves. Does this sketch have enough detail to predict the qualitative nature of phase-plane trajectories?

(continued)

(continued)

(b) Using a computer, sketch a portion of the direction field for $-8 \leq x \leq 8$ and $-6 \leq y \leq 6$. Using your sketch, describe the two different types of phase-plane trajectories and give a physical interpretation for each type. (Recall that x denotes angular displacement from the pendulum's downward-hanging equilibrium position and y denotes angular velocity.)

Solution:

(a) The nullclines consist of the x-axis and the infinite set of vertical lines $x = m\pi, m = 0, \pm 1, \pm 2, \ldots$. A portion of the phase plane is shown in Figure 6.7. This figure also shows the direction field arrows on the nullclines as well as the general "sense of direction" arrows within the vertical phase-plane strips. The arrows are vertical on the nullcline $y = 0$ (the x-axis) and are horizontal on the nullclines $x = m\pi, m = 0, \pm 1, \pm 2, \ldots$. For gaining a good qualitative picture of the phase-plane trajectories, however, this level of description is inadequate.

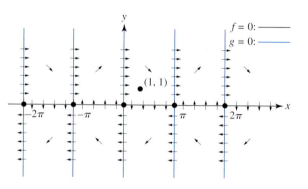

FIGURE 6.7

A sketch showing the main features of the direction field for the pendulum equation in Example 5. Note that we cannot tell from this sketch whether a solution curve passing through the point $(1, 1)$ will continue down until it is below the x-axis or whether it will remain above the x-axis.

   For example, consider a solution curve passing through the point $(1, 1)$. As we see from Figure 6.7, the solution point is moving downward and to the right when it passes through $(1, 1)$. But is the curve falling fast enough that it will cross the x-axis and continue to move down but now to the left? Or is its rate of descent slowing enough that it will intersect the line $x = \pi$ in the upper half of the phase plane and then move up and to the right? We need a more detailed direction field in order to give a reasonable assessment.

(b) Figure 6.8 presents a more detailed direction field plot. This plot indicates that there are two basic types of trajectories. Near the x-axis, there appear to be closed phase-plane trajectories; as time increases, the solution point seems to make clockwise orbits around these closed curves. Far from the x-axis, the trajectories no longer appear to be closed curves. Rather, they appear to be undulating curves that are basically horizontally oriented and that tend to become flatter as distance from the x-axis increases.

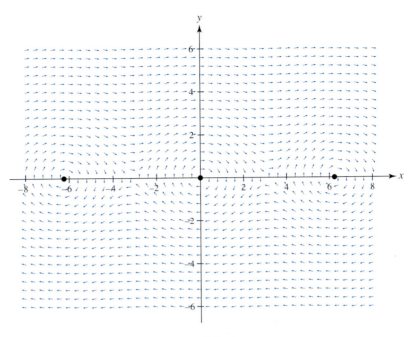

FIGURE 6.8

A portion of the direction field for the pendulum equation in Example 5. Some phase-plane trajectories appear to be closed curves, centered at equilibrium points on the x-axis. Other trajectories appear to be undulating curves where the solution point moves basically in one direction (to the right above the x-axis and to the left below the x-axis). The typical closed trajectory corresponds to a pendulum swinging back and forth, motion in which the pendulum never reaches the vertically upward position. The undulating trajectories correspond to a pendulum that continues to rotate in one direction (counterclockwise if $y = \theta'$ is positive, clockwise if $y = \theta'$ is negative).   ❖

We can understand Figure 6.8 in terms of its physical interpretation. Recall that $x = \theta$ represents the angular displacement of the pendulum from its downward hanging equilibrium position and $y = \theta'$ represents the instantaneous angular velocity of the pendulum. The closed orbits close to the x-axis therefore correspond to motion in which the pendulum swings back and forth. For example, consider a closed trajectory centered at the equilibrium point $(0, 0)$. On such a trajectory, the maximum excursion of $x = \theta$ from zero is less than π. This maximum displacement is reached when the trajectory intersects the x-axis—that is, when angular velocity y is zero. In such a motion, the pendulum never reaches the vertically upward position. It swings up to some maximum angular displacement less than π and then swings back the same amount in the other direction. The continual orbiting of the solution point around a closed trajectory corresponds to this continual back-and-forth swing of the pendulum.

The horizontally configured undulating trajectories correspond to motion in which the pendulum continually rotates about its pivot. In the upper half-plane, $y = \theta'$ is positive and the pendulum is always rotating in the counterclockwise direction. In the lower half-plane, $y = \theta'$ is negative and the pendulum is always rotating in the clockwise direction. Since $y = \theta'$ is never zero, the pendulum never stops and consequently never changes direction.

The pendulum is an example of a conservative system (that is, a system in which energy is conserved). For such systems, and for a more general two-dimensional class of autonomous systems known as *Hamiltonian systems*, we can derive equations for the phase-plane trajectories. We consider such systems in the next section.

EXERCISES

Exercises 1–10:

Find all equilibrium points of the autonomous system.

1. $x' = -x + xy$
   $y' = y - xy$

2. $x' = y(x + 3)$
   $y' = (x - 1)(y - 2)$

3. $x' = (x - 2)(y + 1)$
   $y' = x^2 - 4x + 3$

4. $x' = xy - y + x - 1$
   $y' = xy - 2y$

5. $x' = x^2 - 2xy$
   $y' = 3xy - y^2$

6. $x' = y^2 - xy$
   $y' = 2xy + x^2$

7. $x' = x^2 + y^2 - 8$
   $y' = x^2 - y^2$

8. $x' = x^2 + 2y^2 - 3$
   $y' = 2x^2 + y^2 - 3$

9. $x' = y - 1$
   $y' = xy + x^2$
   $z' = 2y - yz$

10. $x' = z^2 - 1$
   $y' = z - 2xz + yz$
   $z' = -(1 - x - y)^2$

Exercises 11–15:

Rewrite the given scalar differential equation as a first order system, and find all equilibrium points of the resulting system.

11. $y'' + y + y^3 = 0$

12. $y'' + e^y y' + \sin^2 \pi y = 1$

13. $y'' + \dfrac{2y'}{1 + y^4} + y^2 = 1$

14. $y''' - y'' + 2\sin y = 1$

15. $y''' - (y')^2 + \dfrac{4 - y^2}{2 + (y')^2} = 0$

Exercises 16–19:

Use the information provided to determine the unspecified constants.

16. The system

$$x' = x + \alpha xy + \beta$$
$$y' = \gamma y - 3xy + \delta$$

has equilibrium points at $(x, y) = (0, 0)$ and $(2, 1)$. Is $(-2, -2)$ also an equilibrium point?

17. The system

$$x' = \alpha x + \beta xy + 2$$
$$y' = \gamma x + \delta y^2 - 1$$

has equilibrium points at $(x, y) = (1, 1)$ and $(2, 0)$.

18. Consider the system

$$x' = x + \alpha y^3$$
$$y' = x + \beta y + y^4.$$

The slopes of the phase-plane trajectories passing through the points $(x, y) = (2, 1)$ and $(1, -1)$ are 1 and 0, respectively.

19. Consider the system

$$x' = \alpha x^2 + \beta y + 1$$
$$y' = x + \gamma y + y^2.$$

The slopes of the phase-plane trajectories passing through the points $(x, y) = (1, 1)$ and $(1, -1)$ are 0 and 4, respectively. The phase-plane trajectory passing through the point $(x, y) = (0, -1)$ has a vertical tangent.

20. Consider the system

$$x' = x + y^2 - xy^n$$
$$y' = -x + y^{-1}.$$

The slope of the phase-plane trajectory passing through the point $(x, y) = (1, 2)$ is $\frac{1}{6}$. Determine the exponent n.

21. The scalar differential equation $y'' - y' + 2y^2 = \alpha$, when rewritten as a first order system, results in a system having an equilibrium point at $(x, y) = (2, 0)$. Determine the constant α.

22. For the given system, compute the velocity vector $\mathbf{v}(t) = x'(t)\mathbf{i} + y'(t)\mathbf{j}$ at the point $(x, y) = (2, 3)$.

(a) $x' = -x + xy$ (b) $x' = y(x + 3)$ (c) $x' = (x - 2)(y + 1)$

   $y' = y - xy$ $y' = (x - 1)(y - 2)$ $y' = x^2 - 4x + 3$

23. Let A be a (2×2) constant matrix, and let $(\lambda, \mathbf{u})$ be an eigenpair for A. Assume that λ is real, $\lambda \neq 0$, and

$$\mathbf{u} = \begin{bmatrix} u_1 \\ u_2 \end{bmatrix}.$$

Consider the phase plane for the autonomous linear system $\mathbf{y}' = A\mathbf{y}$. We can define a phase-plane line through the origin by the parametric equations $x = \tau u_1, y = \tau u_2$, $-\infty < \tau < \infty$. Let P be any point on this line, say $P = (\tau_0 u_1, \tau_0 u_2)$ for some $\tau_0 \neq 0$.

(a) Show that at the point P, $x' = \tau_0 \lambda u_1$ and $y' = \tau_0 \lambda u_2$.

(b) How is the velocity vector $\mathbf{v}(t) = x'(t)\mathbf{i} + y'(t)\mathbf{j}$ at point P oriented relative to the line?

Exercises 24–27:

In each exercise, a matrix A is given. For each matrix, the vectors

$$\mathbf{u}_1 = \begin{bmatrix} 1 \\ 1 \end{bmatrix} \quad \text{and} \quad \mathbf{u}_2 = \begin{bmatrix} 1 \\ -1 \end{bmatrix}$$

are eigenvectors of A. As discussed in Exercise 23, these eigenvectors are associated with the phase-plane lines $y = x$ and $y = -x$. Solution points of $\mathbf{y}' = A\mathbf{y}$ originating on these two lines remain on these lines as time evolves. Match the given matrix A to one of the four direction fields shown (on the next page) for $\mathbf{y}' = A\mathbf{y}$.

24. $A = \begin{bmatrix} -9 & 1 \\ 1 & -9 \end{bmatrix}$ **25.** $A = \begin{bmatrix} -1 & -3 \\ -3 & -1 \end{bmatrix}$ **26.** $A = \begin{bmatrix} -4 & 6 \\ 6 & -4 \end{bmatrix}$ **27.** $A = \begin{bmatrix} 4 & 2 \\ 2 & 4 \end{bmatrix}$

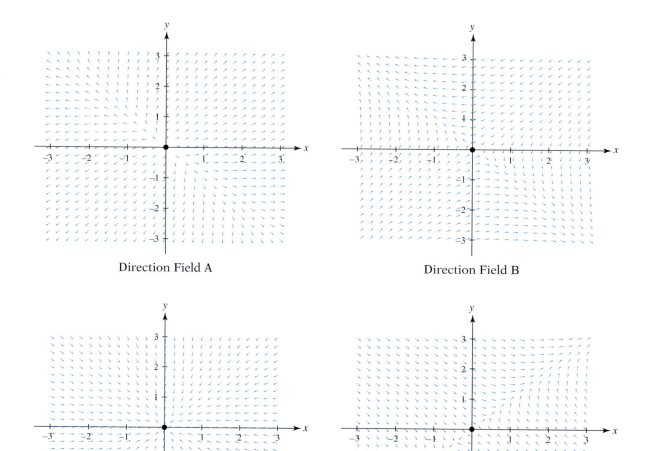

Direction Field A

Direction Field B

Direction Field C

Direction Field D

Figure for Exercises 24–27

28. Suppose that the nonlinear second order equation $y'' + f(y) = 0$ is recast as an autonomous first order system. Show that the nullclines for the resulting system are the horizontal line $y = 0$ and vertical lines of the form $x = \alpha$, where α is a root of $f(y) = 0$. For each such root, what is the nature of the phase-plane point $(x, y) = (\alpha, 0)$?

Exercises 29–31:

(a) Rewrite the given second order equation as an equivalent first order system.

(b) Graph the nullclines of the autonomous system and locate all equilibrium points.

(c) As in Figure 6.6, sketch direction field arrows on the nullclines. Also, sketch an arrow in each open region that suggests the direction in which a solution point is moving when it is in that region.

29. $y'' + y + y^3 = 0$ **30.** $y'' + y(1 - y^2) = 0$ **31.** $y'' + 2\sin^2 y = 1$

Exercises 32–36:

In each exercise,

(a) Graph the nullclines of the autonomous system and locate all equilibrium points.

(b) As in Figure 6.6, sketch direction field arrows on the nullclines. Also, sketch an arrow in each open region that suggests the direction in which a solution point is moving when it is in that region.

32. $x' = 3x - y - 2$ **33.** $x' = -x - y + 2$ **34.** $x' = (2x - y - 2)(4 - x - y)$

  $y' = x - y$ $y' = x - y$ $y' = x - 2y$

35. $x' = (2x - y - 6)(x - y)$ **36.** $x' = x^2 + y - 1$

  $y' = x + y$ $y' = -x^2 + y + 1$

6.3   Conservative Systems

A mathematical model of a physical system often neglects effects such as friction or electrical resistance if they are small enough. We have already encountered several of these idealized mathematical models in our discussion of spring-mass systems, buoyant bodies, and pendulums.

As a consequence of such assumptions, these idealized models obey what is usually called a *conservation law*. In particular, a conservation law means that a quantity, such as energy, remains constant. For example, consider an idealized pendulum. On its upswing, as the bob elevates and simultaneously slows down, energy is converted from kinetic energy to potential energy. On the downswing, potential energy is, in turn, transformed back into kinetic energy. We will show, for this idealized pendulum model, that total pendulum energy (the sum of kinetic and potential energy) remains constant in time. Thus, total pendulum energy is a conserved quantity in the idealized pendulum model.

In general, consider a second order scalar differential equation

$$y'' = f(t, y, y'),$$

and let $y(t)$ be a solution of this differential equation. If there is a function of two variables $H(u, v)$ such that $H(y(t), y'(t))$ remains constant in time, then we call H a **conserved quantity** and say that the differential equation $y'' = f(t, y, y')$ possesses a **conservation law**. We use the same terms to describe the general case of an n-dimensional system with solution components $y_1(t), y_2(t), \ldots, y_n(t)$ for which some function $H(y_1(t), y_2(t), \ldots, y_n(t))$ remains constant.

In this section, we are interested in the following questions:

1. Given a mathematical model, how can we determine (from the structure of the differential equation itself) whether or not it satisfies a conservation law?

2. If the model does possess a conserved quantity, how can we explicitly describe this quantity and use its mathematical description to better understand the system's dynamics?

An Important Class of Second Order Scalar Equations

Consider the differential equation

$$y'' + f(y) = 0. \tag{1}$$

Such differential equations often arise when we apply Newton's laws to a body in one-dimensional motion. In such applications, y'' corresponds to acceleration and the term $-f(y)$ corresponds to the applied force per unit mass. In fact, three of the mathematical models we have discussed—the undamped mass-spring system, the buoyant body, and the pendulum—all have the structure of equation (1).

We now show that equation (1) possesses a conservation law. To see why, and to obtain a description of this law, we first multiply the equation by y', obtaining

$$y'y'' + f(y)y' = 0. \tag{2}$$

Consider the terms in (2). Recalling the chain rule of calculus, we see that the first term on the left-hand side can be written as

$$y'(t)y''(t) = \frac{d}{dt}\left[\tfrac{1}{2}(y'(t))^2\right]. \tag{3}$$

If $F(y)$ denotes an antiderivative of $f(y)$, the chain rule allows us to express the second term in (2) as

$$f(y(t))y'(t) = \frac{d}{dt} F(y(t)). \tag{4}$$

Using (3) and (4), we can rewrite (2) in the form

$$\frac{d}{dt}\left[\tfrac{1}{2}(y'(t))^2 + F(y(t))\right] = 0. \tag{5}$$

Therefore,

$$\tfrac{1}{2}y'(t)^2 + F(y(t)) = C. \tag{6}$$

Equation (6) is the underlying conservation law. For instance, if $y(t)$ represents displacement, then the term $\tfrac{1}{2}y'(t)^2$ is kinetic energy per unit mass. The other term, $F(y(t))$, is potential energy per unit mass (measured relative to some reference value that depends on the particular antiderivative F chosen). The constant C can be interpreted as the (constant) total energy per unit mass.

Phase-Plane Interpretation

Differential equation (1) can be recast as the first order autonomous system

$$\begin{aligned} x' &= y \\ y' &= -f(x), \end{aligned}$$

where x and y now play the roles of $y(t)$ and $y'(t)$, respectively. Thus, the conservation law (6) takes the form

$$\tfrac{1}{2}(y)^2 + F(x) = C. \tag{7}$$

The family of curves obtained by graphing this equation for different values of C is a set of phase-plane trajectories describing the motion. The next example develops these ideas for the pendulum.

EXAMPLE

1

Consider the pendulum equation (recall Example 5, Section 6.2)

$$x' = y$$
$$y' = -\sin x.$$

From (7), the corresponding conservation law is

$$\tfrac{1}{2}y^2 - \cos x = C.$$

If we revert to the original variables, θ and θ', and use E to denote the constant energy of the system, the conservation law has the form[1]

$$\tfrac{1}{2}(\theta')^2 - \cos\theta = E. \tag{8}$$

The term $-\cos\theta$ represents the potential energy of the pendulum bob, measured relative to a zero reference at the horizontal position. Equation (8) is graphed in Figure 6.9 for various energy levels. The direction of solution point motion on these trajectories is indicated with arrows. (The direction field arrows in Figure 6.8 of Section 6.2 are tangent to these phase-plane trajectories.) The entire phase plane is simply a periodic repetition of the portion shown; every 2π increment in $x = \theta$ brings the pendulum back to the same physical configuration.

To understand Figure 6.9, observe from equation (8) that the equilibrium point $(\theta, \theta') = (0, 0)$ (with the bob at rest hanging vertically downward) corresponds to the energy level $E = -1$. In a similar fashion, the equilibrium points $(\theta, \theta') = (\pi, 0)$ and $(\theta, \theta') = (-\pi, 0)$ (with the bob at rest and positioned vertically upward) correspond to the energy level $E = 1$.

The energy value $E = 1$ is a delineating, or "separating," value. Phase-plane trajectories for $-1 < E < 1$ are the closed curves in Figure 6.9 centered at $(0, 0)$. These trajectories correspond to motion in which the pendulum continuously swings back and forth; it does not have enough energy to reach the vertical upward position. The pendulum swings upward to some $\theta_{\max} < \pi$, stops, and then swings downward, achieving the same maximum elevation on its backswing. The two closed curves, labeled (b) and (c) in Figure 6.9, correspond to energy levels $E = -\tfrac{1}{2}$ and $E = \tfrac{1}{2}$. The maximum angular displacements achieved by the pendulum in these two cases are $|\theta_{\max}| = \pi/3$ and $|\theta_{\max}| = 2\pi/3$, respectively. These values correspond to the θ-axis intercepts of the curves.

Energy levels $E > 1$ correspond to motion in which the system possesses enough energy to permit the pendulum to reach the vertical upward position and continue to rotate. Since energy is conserved in this idealized model, the pendulum continues to rotate forever. For each energy level greater than 1, two trajectories are possible. These are not closed trajectories, since total angular displacement increases or decreases monotonically. The trajectories in the upper half-plane (where $\theta' > 0$) correspond to counterclockwise pendulum rotation, while the counterpart trajectories in the lower half-plane (where $\theta' < 0$) represent clockwise pendulum rotation. The eight such trajectories shown in Figure 6.9 correspond to $E = 2$, $E = 3$, $E = 4$, and $E = 5$.

(continued)

---

[1] In Section 2.9, Exercise 22, we derived this conservation law in a different way. A change of independent variable was used to transform $\theta'' + (g/l)\sin\theta = 0$ into a first order separable differential equation. The implicit solution of this equation yielded the conservation law (8).

(continued)

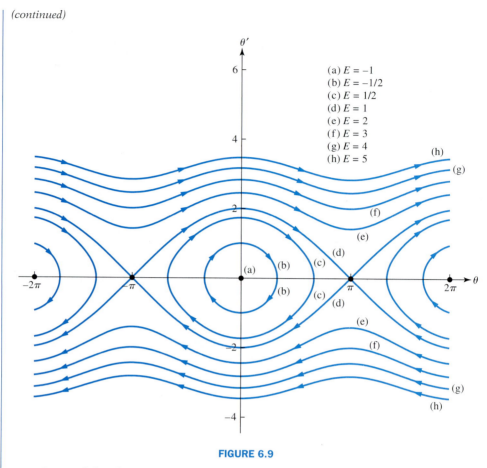

FIGURE 6.9

Some of the phase-plane trajectories for the pendulum equation discussed in Example 1. These curves are graphs of equation (8) for various energy levels E.

The two trajectories corresponding to $E = 1$ appear to connect the equilibrium points $(-\pi, 0)$ and $(\pi, 0)$. These trajectories are called **separatrices**; they separate the closed trajectory curves from the open ones. On the upper separatrix, the solution point approaches the equilibrium point $(\pi, 0)$ as $t \to \infty$. The pendulum swings upward in the counterclockwise direction, slowing down as it approaches the vertical upward position. The pendulum bob approaches this inverted position in the limit as $t \to \infty$. On the lower separatrix, the solution point approaches the equilibrium point $(-\pi, 0)$ as $t \to \infty$. In this case, the pendulum swings upward in the clockwise direction, again slowing down and approaching the inverted position in the limit as $t \to \infty$. ❖

Hamiltonian Systems

We now discuss a class of autonomous first order systems, called *Hamiltonian systems*, that always satisfy a conservation law. We restrict our attention to two-dimensional systems; the Exercises show how the underlying principle can be extended to higher dimensional systems. Hamiltonian systems include the second order scalar equation (1) as a special case.

As a first step, recall the following chain rule from calculus. Assume that a function $H(x, y)$, viewed as a function of two independent variables x and y, is continuous and has continuous first and second partial derivatives (it will be apparent later why continuous second partial derivatives are required). We now form a composition, replacing the variables x and y with differentiable functions of t; we refer to these two functions as $x(t)$ and $y(t)$. The resulting composite function,

$$H(x(t), y(t)),$$

is a differentiable function of t, and its derivative can be found by the chain rule:

$$\frac{d}{dt} H(x(t), y(t)) = \frac{\partial H(x(t), y(t))}{\partial x} \frac{dx}{dt} + \frac{\partial H(x(t), y(t))}{\partial y} \frac{dy}{dt}. \tag{9}$$

Consider now the two-dimensional autonomous system

$$x'(t) = f(x(t), y(t))$$
$$y'(t) = g(x(t), y(t)). \tag{10}$$

System (10) is called a **Hamiltonian system**[2] if there exists a function of two variables $H(x, y)$ that is continuous, with continuous first and second partial derivatives, and such that

$$\frac{\partial H(x, y)}{\partial x} = -g(x, y)$$

$$\frac{\partial H(x, y)}{\partial y} = f(x, y). \tag{11}$$

The function $H(x, y)$ is called the **Hamiltonian function** (or simply the **Hamiltonian**) of the system.

If system (10) is a Hamiltonian system, then the composition $H(x(t), y(t))$ is a conserved quantity of the system. To see why, note that

$$\frac{d}{dt} H(x(t), y(t)) = \frac{\partial H(x(t), y(t))}{\partial x} \frac{dx(t)}{dt} + \frac{\partial H(x(t), y(t))}{\partial y} \frac{dy(t)}{dt}$$

$$= [-g(x(t), y(t))]f(x(t), y(t)) + \left[f(x(t), y(t))\right] g(x(t), y(t))$$

$$= 0.$$

It follows, therefore, that

$$H(x(t), y(t)) = C.$$

Two important questions about Hamiltonian systems are "How can we determine whether a given autonomous system is a Hamiltonian system?" and

---

[2] Sir William Rowan Hamilton (1805–1865) was an Irish mathematician noted for his contributions to optics and dynamics and for the development of the theory of quaternions. Shortly before his death, he was elected the first foreign member of the United States National Academy of Sciences.

"If a system is known to be a Hamiltonian system, how do we determine the conserved quantity H?" We will address both of these questions after the next example, which shows that the second order scalar equation (1), when recast as an autonomous first order system, is a Hamiltonian system.

E X A M P L E
2

If the second order scalar equation $y'' + v(y) = 0$ is rewritten as a first order system, we obtain the autonomous system

$$x' = y$$
$$y' = -v(x).$$

Show that this system is a Hamiltonian system with $H(x, y) = \frac{1}{2}y^2 + V(x)$, where $V(x)$ is any antiderivative of $v(x)$.

Solution: With the notation of (10) and (11), $f(x, y) = y$ and $g(x, y) = -v(x)$. Calculating the partial derivatives of $H(x, y) = \frac{1}{2}y^2 + V(x)$, we find

$$\frac{\partial H(x, y)}{\partial x} = \frac{dV(x)}{dx} = v(x) = -g(x, y)$$

$$\frac{\partial H(x, y)}{\partial y} = y = f(x, y).$$

Thus, from equation (11), the system is a Hamiltonian system. ❖

Recognizing a Hamiltonian System

The following discussion about identifying Hamiltonian systems and constructing Hamiltonians closely parallels the discussion in Section 2.7 about identifying exact differential equations and constructing solutions of exact equations.

In particular, suppose $H(x, y)$ is a Hamiltonian for system (10). Then, from equation (11),

$$\frac{\partial H(x, y)}{\partial x} = -g(x, y) \quad \text{and} \quad \frac{\partial H(x, y)}{\partial y} = f(x, y). \tag{12}$$

From calculus, if the second partial derivatives of $H(x, y)$ exist and are continuous, then the second mixed partial derivatives are equal; that is,

$$\frac{\partial^2 H(x, y)}{\partial x \partial y} = \frac{\partial^2 H(x, y)}{\partial y \partial x}.$$

Therefore, if system (10) is a Hamiltonian system, it necessarily follows that

$$\frac{\partial f(x, y)}{\partial x} = -\frac{\partial g(x, y)}{\partial y}.$$

The following theorem, stated without proof, asserts that this condition is both necessary and sufficient.

Theorem 6.2

Consider the two-dimensional autonomous system

$$x' = f(x, y)$$
$$y' = g(x, y).$$

Assume that $f(x, y)$ and $g(x, y)$ are continuous in the xy-plane. Assume as well that the partial derivatives

$$\frac{\partial f}{\partial x}, \quad \frac{\partial f}{\partial y}, \quad \frac{\partial g}{\partial x}, \quad \frac{\partial g}{\partial y}$$

exist and are continuous in the xy-plane. Then the system is a Hamiltonian system if and only if

$$\frac{\partial f(x, y)}{\partial x} = -\frac{\partial g(x, y)}{\partial y} \tag{13}$$

for all (x, y).

Constructing Hamiltonians

Once a system is known to be a Hamiltonian system, we can construct the Hamiltonian function by the same process of anti-partial-differentiation we used to solve exact differential equations in Section 2.7. We illustrate the ideas in the next example by constructing a Hamiltonian function for a Hamiltonian system.

EXAMPLE 3

Consider the autonomous system

$$x' = y^2 + \cos x$$
$$y' = 2x + 1 + y \sin x.$$

(a) Use Theorem 6.2 to show that this system is a Hamiltonian system.

(b) Find a Hamiltonian function for the system.

Solution:

(a) Calculating the partial derivatives required by test (13), we find

$$\frac{\partial f}{\partial x} = -\sin x \quad \text{and} \quad \frac{\partial g}{\partial y} = \sin x.$$

Since $\partial f / \partial x = -\partial g / \partial y$, we know the system is a Hamiltonian system.

(b) Since the given system is a Hamiltonian system, there must be a function $H(x, y)$ such that

$$\frac{\partial H(x, y)}{\partial x} = -g(x, y) = -2x - 1 - y \sin x$$

$$\frac{\partial H(x, y)}{\partial y} = f(x, y) = y^2 + \cos x. \tag{14}$$

(continued)

(continued)

Choose one of these equations, say the first, and compute an anti-partial-derivative, obtaining

$$H(x, y) = -x^2 - x + y\cos x + q(y), \tag{15}$$

where $q(y)$ is an arbitrary differentiable function of y. [Note: Since the variable y is held fixed when partial differentiation is performed with respect to x, we must allow for this arbitrary function of y when reversing the operation.] From equations (14) and (15), we find

$$\frac{\partial H(x, y)}{\partial y} = y^2 + \cos x = \cos x + \frac{dq(y)}{dy}.$$

Therefore,

$$\frac{dq(y)}{dy} = y^2,$$

and $q(y) = y^3/3 + C$. We can drop the additive arbitrary constant and obtain a Hamiltonian

$$H(x, y) = -x^2 - x + y\cos x + \frac{y^3}{3}. \tag{16}$$

Figure 6.10 shows some phase-plane trajectories [that is, graphs of $H(x, y) = C$] for a few representative values of the constant C.

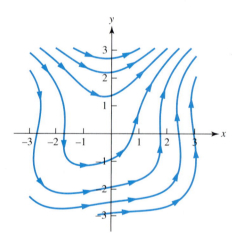

FIGURE 6.10

Some phase-plane trajectories for the autonomous system in Example 3. These curves are level curves of the Hamiltonian (16). ❖

EXERCISES

In each exercise,

(a) As in Example 1, derive a conservation law for the given autonomous equation $x'' + u(x) = 0$. (Your answer should contain an arbitrary constant and therefore define a one-parameter family of conserved quantities.)

(b) Rewrite the given autonomous equation as a first order system of the form

$$x' = f(x, y)$$
$$y' = g(x, y)$$

by setting $y(t) = x'(t)$. The phase plane is then the xy-plane. Express the family of conserved quantities found in (a) in terms of x and y. Determine the equation of the particular conserved quantity whose graph passes through the phase-plane point $(x, y) = (1, 1)$.

(c) Plot the phase-plane graph of the conserved quantity found in part (b), using a computer if necessary. Determine the velocity vector $\mathbf{v} = f(x, y)\mathbf{i} + g(x, y)\mathbf{j}$ at the phase-plane point $(1, 1)$. Add this vector to your graph with the initial point of the vector at $(1, 1)$. What is the geometric relation of this velocity vector to the graph? What information does the velocity vector give about the direction in which the solution point traverses the graph as time increases?

(d) For the solution whose phase-plane trajectory passes through $(1, 1)$, determine whether the solution $x(t)$ is bounded. If the solution is bounded, use the phase-plane plot to estimate the maximum value attained by $|x(t)|$.

1. $x'' + 4x = 0$ 2. $x'' - x = 1$ 3. $x'' + x^3 = 0$

4. $x'' - x^3 = \pi \sin(\pi x)$ 5. $x'' + x^2 = 0$ 6. $x'' + \dfrac{x}{1 + x^2} = 0$

The conservation law for an autonomous second order scalar differential equation having the form $x'' + f(x) = 0$ is given (where y corresponds to x'). Determine the differential equation.

7. $y^2 + x^2 \cos x = C$ 8. $y^2 - e^{-x^2} = C$

9. Consider the differential equation $x'' + x + x^3 = 0$. It has the same structure as the equation used to model the cubically nonlinear spring.

(a) Rewrite the differential equation as a first order system. On the xy-phase plane, sketch the nullclines and locate any equilibrium point(s). Place direction field arrows on the nullclines, indicating the direction in which the solution point traverses the nullclines.

(b) Compute the velocity vector $\mathbf{v} = x'\mathbf{i} + y'\mathbf{j}$ at the four phase-plane points $(x, y) = (\pm 1, \pm 1)$. Locate these points, and draw the velocity vectors on your phase-plane sketch. Use this information, together with the information obtained in part (a), to draw a rough sketch of some phase-plane solution trajectories. Indicate the direction in which the solution point moves on these trajectories.

(c) Determine the conservation law satisfied by solutions of the given differential equation. Determine the equation of the conserved quantity whose graph passes through the phase-plane point $(x, y) = (1, 1)$. Plot the graph of this equation on your phase plane, using computational software as appropriate. Does the graph pass through the other three points, $(-1, 1)$, $(-1, -1)$, and $(1, -1)$, as well? Is the sketch made in part (b) consistent with this graph of the conserved quantity?

10. Each figure shows a phase-plane graph of a conserved quantity for the autonomous differential equation $x'' + \alpha x = 0$, where α is a real constant.

(a) Determine the value of the constant α in each case. What is the equation whose phase-plane graph is shown?

(b) Indicate the direction in which the solution point travels along these phase-plane curves as time increases.

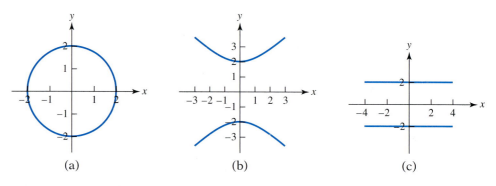

(a)   (b)   (c)

Figure for Exercise 10

In (b), the asymptotes are $y = \pm x$.

11. Consider the autonomous third order scalar equation $y''' + f(y') = 0$, where f is a continuous function. Does this differential equation have a conservation law? If so, obtain the equation of the family of conserved quantities.

12. Consider the equation $mx'' + kx = 0$. We saw in Chapter 3 that this equation models the vibrations of a spring-mass system. The conserved quantity $\frac{1}{2}m(x')^2 + \frac{1}{2}kx^2 = E$ is the (constant) total energy of the system. The first term, $\frac{1}{2}m(x')^2$, is the kinetic energy, while the second term, $\frac{1}{2}kx^2$, is the elastic potential energy. Suppose that damping is now added to the system. The differential equation $mx'' + \gamma x' + kx = 0$ now models the motion (with γ a positive constant). Define $E(t) = \frac{1}{2}m(x')^2 + \frac{1}{2}kx^2$.

(a) Show, in the case of damping, that $E(t)$ is no longer constant. Show, rather, that $dE(t)/dt \le 0$.

(b) Discuss the physical relevance of the observation made in part (a).

Exercises 13–20:

For the given system,

(a) Use Theorem 6.2 to show that the system is a Hamiltonian system.

(b) Find a Hamiltonian function for the system.

(c) Use computational software to graph the phase-plane trajectory passing through $(1, 1)$. Also, indicate the direction of motion for the solution point.

13. $x' = 2x$
  $y' = -2y$

14. $x' = 2xy$
  $y' = -y^2$

15. $x' = x - x^2 + 1$
  $y' = -y + 2xy + 4x$

16. $x' = -8y$
  $y' = 2x$

17. $x' = 2y \cos x$
  $y' = y^2 \sin x$

18. $x' = 2y - x + 3$
  $y' = y + 4x^3 - 2x$

19. $x' = -2y$
  $y' = 3x^2$

20. $x' = xe^{xy}$
  $y' = -2x - ye^{xy}$

Exercises 21–26:

Use Theorem 6.2 to decide whether the given system is a Hamiltonian system. If it is, find a Hamiltonian function for the system.

21. $x' = x^3 + 3\sin(2x + 3y)$
$\quad\;\; y' = -3x^2y - 2\sin(2x + 3y)$

22. $x' = e^{xy} + y^3$
$\quad\;\; y' = -e^{xy} - x^3$

23. $x' = -\sin(2xy) - x$
$\quad\;\; y' = \sin(2xy) + y$

24. $x' = -3x^2 + xe^y$
$\quad\;\; y' = 6xy + 3x - e^y$

25. $x' = y$
$\quad\;\; y' = x - x^2$

26. $x' = x + 2y$
$\quad\;\; y' = x^3 - 2x + y$

Exercises 27–30:

Let $f(u)$ and $g(u)$ be defined and continuously differentiable on the interval $-\infty < u < \infty$, and let $F(u)$ and $G(u)$ be antiderivatives for $f(u)$ and $g(u)$, respectively. In each exercise, use Theorem 6.2 to show that the given system is Hamiltonian. Determine a Hamiltonian function for the system, expressed in terms of F and/or G.

27. $x' = f(y)$
$\quad\;\; y' = g(x)$

28. $x' = f(y) + 2y$
$\quad\;\; y' = g(x) + 6x$

29. $x' = 3f(y) - 2xy$
$\quad\;\; y' = g(x) + y^2 + 1$

30. $x' = f(x - y) + 2y$
$\quad\;\; y' = f(x - y)$

31. A Generalized Hamiltonian System   Consider the two-dimensional autonomous system

$$x' = f(x, y)$$
$$y' = g(x, y).$$

Suppose there exist two functions $K(x, y)$ and $\mu(x, y)$ satisfying

$$\frac{\partial K(x, y)}{\partial x} = -\mu(x, y)g(x, y)$$

$$\frac{\partial K(x, y)}{\partial y} = \mu(x, y)f(x, y).$$

Does the given autonomous system have a conserved quantity? If so, what is the conserved quantity?

32. Higher-Dimensional Autonomous Systems   The ideas underlying Hamiltonian systems extend to higher-dimensional systems. For example, consider the three-dimensional autonomous system

$$x' = f(x, y, z)$$
$$y' = g(x, y, z) \tag{17}$$
$$z' = h(x, y, z).$$

(a) Use the chain rule to show that autonomous system (17) has a conserved quantity if there exists a function $H(x, y, z)$ for which

$$\frac{\partial}{\partial x} H(x, y, z)f(x, y, z) + \frac{\partial}{\partial y} H(x, y, z)g(x, y, z) + \frac{\partial}{\partial z} H(x, y, z)h(x, y, z) = 0.$$

(b) Show that $H(x, y, z) = \cos^2(x) + ye^z$ is a conserved quantity for the system

$$x' = ye^z$$
$$y' = y\cos x \sin x$$
$$z' = \cos x \sin x.$$

6.4   Stability

Differential equations can model different physical behavior at different equilibrium points. For instance, consider the pendulum. The equilibrium points are $(\theta, \theta') = (m\pi, 0)$, where m is an integer. Equilibrium points with m an even integer correspond to the pendulum bob hanging vertically downward, while equilibrium points for m an odd integer correspond to the pendulum bob resting in the inverted position. Suppose a pendulum bob, initially in an equilibrium state, is subjected to a slight perturbation; in other words, it is given a slight displacement and/or a very small angular velocity. If the pendulum is initially hanging downward, we expect the perturbation to remain small—the bob will swing back and forth, making small excursions from the vertical. If the pendulum is initially in the inverted position, however, we expect dramatic changes. The pendulum bob, displaced from its precarious equilibrium state, will fall, ultimately making large departures from its initial equilibrium position.

In everyday language, we might describe the pendulum's downward rest position as a "stable" configuration and the inverted rest position as an "unstable" configuration. Mathematicians have taken these everyday terms and given them precise definitions consistent with our intuitive notion of stable and unstable. In this section, we present and discuss these mathematical definitions. The next section introduces the technique of linearization, which, in many cases, enables us to study and characterize equilibrium point stability by analyzing a simpler associated linear system.

The pendulum example illustrates the question of primary concern: If perturbed slightly from an equilibrium state, will a system exhibit a markedly different behavior? In the case of mechanical systems, instability often means vibrations that grow in amplitude, leading to possible system failure.

Stable and Unstable Equilibrium Points

Consider the autonomous system

$$
\begin{aligned}
y_1' &= f_1(y_1, y_2, \ldots, y_n)\\
y_2' &= f_2(y_1, y_2, \ldots, y_n)\\
&\vdots\\
y_n' &= f_n(y_1, y_2, \ldots, y_n),
\end{aligned}
$$

which we write in vector form as

$$\mathbf{y}' = \mathbf{f}(\mathbf{y}). \tag{1}$$

Assume that the constant vector function

$$\mathbf{y}(t) = \mathbf{y}_e = \begin{bmatrix} y_{1e} \\ y_{2e} \\ \vdots \\ y_{ne} \end{bmatrix}$$

is an equilibrium solution of the system; that is, $\mathbf{f}(\mathbf{y}_e) = \mathbf{0}$.

In order to define precisely what it means for the equilibrium point $\mathbf{y}_e$ to be stable or unstable, we need to be able to compute the distance between points in n-dimensional space. Let $\mathbf{u}$ and $\mathbf{v}$ denote two points in n-dimensional space,

$$\mathbf{u} = \begin{bmatrix} u_1 \\ u_2 \\ \vdots \\ u_n \end{bmatrix} \quad \text{and} \quad \mathbf{v} = \begin{bmatrix} v_1 \\ v_2 \\ \vdots \\ v_n \end{bmatrix}.$$

We define the **norm** of $\mathbf{u}$, denoted by $\|\mathbf{u}\|$, by

$$\|\mathbf{u}\| = \sqrt{u_1^2 + u_2^2 + \cdots + u_n^2}.$$

The **distance** between $\mathbf{u}$ and $\mathbf{v}$, denoted by $\|\mathbf{u} - \mathbf{v}\|$, is the size (the norm) of the difference $\mathbf{u} - \mathbf{v}$:

$$\|\mathbf{u} - \mathbf{v}\| = \sqrt{(u_1 - v_1)^2 + (u_2 - v_2)^2 + \cdots + (u_n - v_n)^2}. \qquad \textbf{(2)}$$

Now, let $\mathbf{y}_e$ be an equilibrium point of the autonomous system $\mathbf{y}' = \mathbf{f}(\mathbf{y})$. We say that the equilibrium point $\mathbf{y}_e$ is **stable** if

> Given any $\varepsilon > 0$, there exists a corresponding $\delta > 0$ such that every solution satisfying $\|\mathbf{y}(0) - \mathbf{y}_e\| < \delta$ also satisfies $\|\mathbf{y}(t) - \mathbf{y}_e\| < \varepsilon$ for all $t \geq 0$.

If an equilibrium point of $\mathbf{y}' = \mathbf{f}(\mathbf{y})$ is not stable, it is called **unstable**.

Interpreting Stability in the Phase Plane

When $n = 2$, we can use the phase plane to give a graphical interpretation of stability. Consider the autonomous system

$$x' = f(x, y)$$
$$y' = g(x, y)$$

having equilibrium solution

$$\mathbf{y}_e = \begin{bmatrix} x_e \\ y_e \end{bmatrix}.$$

We have adopted the notation

$$\mathbf{y}(t) = \begin{bmatrix} x(t) \\ y(t) \end{bmatrix} \quad \text{and} \quad \mathbf{f}(\mathbf{y}) = \begin{bmatrix} f(x, y) \\ g(x, y) \end{bmatrix}$$

so that we can speak of the phase plane as the xy-plane.

We can identify $\mathbf{y}_e$ as the point in the phase plane having coordinates (x_e, y_e). The set of all phase-plane points $\mathbf{y}$ satisfying $\|\mathbf{y} - \mathbf{y}_e\| < r$ is the set of all points lying within a circle of radius r centered at (x_e, y_e).

Consider now the definition of stability. It involves two circles centered at (x_e, y_e), one of radius ε and the other of radius δ (see Figure 6.11). The stability criterion requires that all solutions lying within the circle of radius δ at the initial time $t = 0$ remain within the circle of radius ε for all subsequent time.

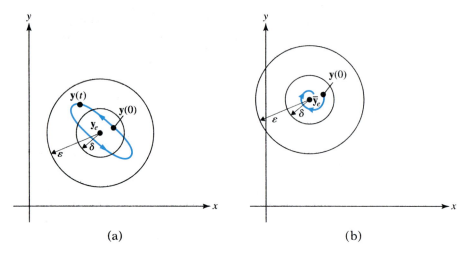

FIGURE 6.11

Two examples of behavior near a stable equilibrium point. In each case, $\mathbf{y}(t) = (x(t), y(t))$ represents a typical solution trajectory near an equilibrium point of the autonomous system $\mathbf{y}' = \mathbf{f}(\mathbf{y})$. (a) When the initial point $\mathbf{y}(0)$ is sufficiently close to $\mathbf{y}_e$, the solution trajectory is a closed curve surrounding $\mathbf{y}_e$. (b) When $\mathbf{y}(0)$ is sufficiently close to $\bar{\mathbf{y}}_e$, the solution trajectory spirals in toward $\bar{\mathbf{y}}_e$.

This situation must hold for all possible choices of $\varepsilon > 0$; whenever we are given an $\varepsilon > 0$, we must be able to find a corresponding $\delta > 0$ that "works." The real test of the definition occurs as we consider smaller and smaller $\varepsilon > 0$. Can we continue to find corresponding values $\delta > 0$ that work? If so, the equilibrium point is stable; if not, it is unstable.

For higher order systems, the same geometrical ideas hold. However, instead of circles in the phase plane, we must consider n-dimensional spheres. We illustrate the concept of stability with two examples involving autonomous linear systems for which explicit general solutions are known.

E X A M P L E

1

Consider the two-dimensional autonomous linear system $\mathbf{y}' = A\mathbf{y}$, where

$$A = \begin{bmatrix} -2 & 1 \\ 1 & -2 \end{bmatrix}. \tag{3}$$

Show that $\mathbf{y}_e = \mathbf{0}$ is the only equilibrium point, and determine whether it is stable or unstable.

Solution: In this case, $\mathbf{f}(\mathbf{y}) = A\mathbf{y}$. The matrix A is invertible, and therefore solving $A\mathbf{y} = \mathbf{0}$ leads to a single equilibrium point, $\mathbf{y}_e = \mathbf{0}$.

To determine the stability properties of this equilibrium point, we apply the stability definition directly to the general solution of this first order linear system. Using the methods of Chapter 4, we find the general solution is

$$\mathbf{y}(t) = c_1 e^{-t} \begin{bmatrix} 1 \\ 1 \end{bmatrix} + c_2 e^{-3t} \begin{bmatrix} 1 \\ -1 \end{bmatrix} = \begin{bmatrix} c_1 e^{-t} + c_2 e^{-3t} \\ c_1 e^{-t} - c_2 e^{-3t} \end{bmatrix}. \tag{4}$$

In order to apply the definition of stability, we need to relate the following two quantities:

1. the distance $\|\mathbf{y}(0) - \mathbf{y}_e\|$, between the initial point and the equilibrium point, and
2. the distance $\|\mathbf{y}(t) - \mathbf{y}_e\|$, between $\mathbf{y}(t)$ and $\mathbf{y}_e$ for $t \geq 0$.

Since $\mathbf{y}_e = \mathbf{0}$, it follows from (4) that

$$\|\mathbf{y}(0) - \mathbf{y}_e\| = \|\mathbf{y}(0)\| = \sqrt{(c_1 + c_2)^2 + (c_1 - c_2)^2} = \sqrt{2(c_1^2 + c_2^2)}.$$

Similarly, for $t \geq 0$,

$$\begin{aligned}
\|\mathbf{y}(t) - \mathbf{y}_e\| = \|\mathbf{y}(t)\| &= \sqrt{(c_1 e^{-t} + c_2 e^{-3t})^2 + (c_1 e^{-t} - c_2 e^{-3t})^2} \\
&= \sqrt{2(c_1^2 e^{-2t} + c_2^2 e^{-6t})} = \sqrt{2(c_1^2 + c_2^2 e^{-4t})e^{-2t}} \\
&\leq \sqrt{2(c_1^2 + c_2^2)e^{-2t}} = \sqrt{2(c_1^2 + c_2^2)}e^{-t} \\
&= \|\mathbf{y}(0)\|e^{-t}.
\end{aligned} \tag{5}$$

Now consider the definition of stability where some value $\varepsilon > 0$ is given. We need to determine a corresponding value $\delta > 0$ such that

If $\mathbf{y}(t)$ is a solution of $\mathbf{y}' = A\mathbf{y}$ and if $\|\mathbf{y}(0)\| < \delta$, then $\|\mathbf{y}(t)\| < \varepsilon$ for all $t \geq 0$.

By (5), we know that

$$\|\mathbf{y}(t)\| \leq \|\mathbf{y}(0)\|e^{-t} \leq \|\mathbf{y}(0)\|, \qquad t \geq 0.$$

Therefore, we can guarantee that $\|\mathbf{y}(t)\| < \varepsilon$ by taking $\delta \leq \varepsilon$. This shows the equilibrium point $\mathbf{y}_e = \mathbf{0}$ is a stable equilibrium point of the system $\mathbf{y}' = A\mathbf{y}$. ❖

EXAMPLE 2

Consider the two-dimensional autonomous linear system $\mathbf{y}' = A\mathbf{y}$, where

$$A = \begin{bmatrix} -1 & -2 \\ -2 & -1 \end{bmatrix}. \tag{6}$$

Show that $\mathbf{y}_e = \mathbf{0}$ is the only equilibrium point, and determine whether it is stable or unstable.

Solution: As in Example 1, $\mathbf{f}(\mathbf{y}) = A\mathbf{y}$, and we see that the matrix A is invertible. Therefore, solving $A\mathbf{y} = \mathbf{0}$ leads to a single equilibrium point, $\mathbf{y}_e = \mathbf{0}$.
   The general solution of $\mathbf{y}' = A\mathbf{y}$ is

$$\mathbf{y}(t) = c_1 e^t \begin{bmatrix} 1 \\ -1 \end{bmatrix} + c_2 e^{-3t} \begin{bmatrix} 1 \\ 1 \end{bmatrix} = \begin{bmatrix} c_1 e^t + c_2 e^{-3t} \\ -c_1 e^t + c_2 e^{-3t} \end{bmatrix}. \tag{7}$$

In Example 1, the coefficient matrix had two negative eigenvalues. In this example, however, the coefficient matrix has a positive eigenvalue, $\lambda = 1$. Since

(continued)

(continued)

$\lim_{t \to \infty} e^t = \infty$, we anticipate that any solution (7) with $c_1 \neq 0$ will become unboundedly large in norm as t increases. Therefore, we anticipate that $\mathbf{y}_e = \mathbf{0}$ is an unstable equilibrium point of this system.

To prove that $\mathbf{y}_e = \mathbf{0}$ is an unstable equilibrium point, we show that, for some $\varepsilon > 0$, there is no δ that works. That is, for every $\delta > 0$, there is at least one solution $\mathbf{y}(t)$ that originates in the circle of radius δ but that eventually gets outside the circle of radius ε. In particular, solutions given by (7) with $c_2 = 0$ and $c_1 \neq 0$ have the form

$$\mathbf{y}(t) = c_1 e^t \begin{bmatrix} 1 \\ -1 \end{bmatrix}, \qquad t \geq 0.$$

Moreover,

$$\|\mathbf{y}(0)\| = \sqrt{2}|c_1| \quad \text{and} \quad \|\mathbf{y}(t)\| = \sqrt{2}|c_1|e^t.$$

This particular family of solutions has phase-plane trajectories that lie on the line $y = -x$ (see Figure 6.12 and Exercise 23 in Section 6.2). Since $\|\mathbf{y}(t)\| = \sqrt{2}|c_1|e^t$, the solution moves away from the origin along this line, growing in norm as t increases. No matter what value $\delta > 0$ we take, we can always choose $|c_1| \neq 0$ but sufficiently small that $\mathbf{y}(0)$ is within the circle of radius δ. But, as long as $|c_1| \neq 0$, the solution $\mathbf{y}(t)$ eventually exits the circle of radius ε. Therefore, $\mathbf{y}_e = \mathbf{0}$ is an unstable equilibrium point of the autonomous system $\mathbf{y}' = A\mathbf{y}$.

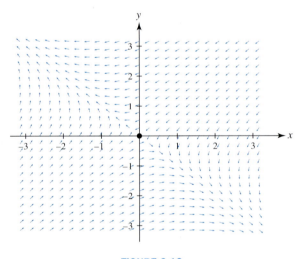

FIGURE 6.12

The direction field for the system in Example 2. The arrows indicate the direction the solution point moves as time increases.   ❖

In Examples 1 and 2, it was relatively straightforward to determine the stability properties of the equilibrium point $\mathbf{y}_e = \mathbf{0}$ because we had an explicit representation of the general solution. The obvious question is "How do we analyze the stability properties of equilibrium points of nonlinear problems, such as the pendulum and competing species problems, when explicit solutions are not attainable?" We address this issue in the next section.

Asymptotic Stability

Example 1 has an interesting aspect. The equilibrium point $\mathbf{y}_e = \mathbf{0}$ not only is stable but also has the feature that all solutions approach it in the limit as $t \to \infty$. This additional feature, wherein all solutions originating sufficiently close to a stable equilibrium point actually approach the equilibrium point as $t \to \infty$, is important enough to warrant its own definition.

Let $\mathbf{y}_e$ be an equilibrium point of the autonomous system $\mathbf{y}' = \mathbf{f}(\mathbf{y})$. We say that $\mathbf{y}_e$ is an **asymptotically stable equilibrium point** if

(a) it is a stable equilibrium point and
(b) there exists a $\delta_0 > 0$ such that $\lim_{t \to \infty} \mathbf{y}(t) = \mathbf{y}_e$ for all solutions initially satisfying $\|\mathbf{y}(0) - \mathbf{y}_e\| < \delta_0$.

Roughly speaking, all solutions starting close enough to a *stable* equilibrium point remain close to it for all subsequent time. All solutions starting sufficiently close to an *asymptotically stable* equilibrium point not only remain close to it for all subsequent time but, in fact, approach it in the limit as $t \to \infty$. Note that asymptotic stability implies stability. However, as the next example shows, an equilibrium point can be stable but not asymptotically stable.

E X A M P L E

3

Consider the two-dimensional autonomous linear system $\mathbf{y}' = A\mathbf{y}$, where

$$A = \begin{bmatrix} 0 & 1 \\ -1 & 0 \end{bmatrix}.$$

Show that $\mathbf{y}_e = \mathbf{0}$ is the only equilibrium point, and determine whether it is asymptotically stable.

Solution: As in Examples 1 and 2, $\mathbf{f}(\mathbf{y}) = A\mathbf{y}$, where the matrix A is invertible. Therefore, solving $A\mathbf{y} = \mathbf{0}$ leads to a single equilibrium point, $\mathbf{y}_e = \mathbf{0}$.

Eigenpairs of the matrix A are

$$\lambda_1 = i, \quad \mathbf{u}_1 = \begin{bmatrix} 1 \\ i \end{bmatrix} \quad \text{and} \quad \lambda_2 = \overline{\lambda_1} = -i, \quad \mathbf{u}_2 = \overline{\mathbf{u}}_1 = \begin{bmatrix} 1 \\ -i \end{bmatrix}.$$

Using the ideas of Section 4.6, we find a real-valued general solution of this system to be

$$\mathbf{y}(t) = c_1 \begin{bmatrix} \cos t \\ -\sin t \end{bmatrix} + c_2 \begin{bmatrix} \sin t \\ \cos t \end{bmatrix}. \tag{8}$$

It follows from (8) that

$$\|\mathbf{y}(t) - \mathbf{y}_e\| = \|\mathbf{y}(t)\| = \sqrt{(c_1 \cos t + c_2 \sin t)^2 + (-c_1 \sin t + c_2 \cos t)^2}$$
$$= \sqrt{c_1^2 + c_2^2} = \|\mathbf{y}(0)\|, \qquad t \geq 0. \tag{9}$$

Therefore, the distance of a solution from the phase-plane origin remains constant in time—the phase-plane trajectories are circles centered at the origin.

(continued)

(continued)

Thus, the equilibrium point $\mathbf{y}_e = \mathbf{0}$ is stable (given $\varepsilon > 0$, we simply take $\delta = \varepsilon$). The equilibrium point is not asymptotically stable, however. In particular, no nonzero solution approaches the equilibrium point as $t \to \infty$. ❖

Stability Characteristics of $\mathbf{y}' = A\mathbf{y}$

The three examples previously considered illustrate the following general stability result, which we present without proof.

Theorem 6.3

Let A be a real invertible (2×2) matrix. Then the autonomous linear system $\mathbf{y}' = A\mathbf{y}$ has a unique equilibrium point $\mathbf{y}_e = \mathbf{0}$. This equilibrium point is

(a) asymptotically stable if all eigenvalues of A have negative real parts. (In other words, the eigenvalues can be real and negative, or they can be a complex conjugate pair with negative real parts.)

(b) stable but not asymptotically stable if the eigenvalues of A are purely imaginary.

(c) unstable if at least one eigenvalue of A has a positive real part. (In other words, the eigenvalues can be real with at least one being positive, or they can be a complex conjugate pair with positive real parts.)

Equilibrium points are sometimes characterized as being isolated or not isolated. A phase-plane equilibrium point is called an **isolated point** if it is the center of some small disk whose interior contains no other equilibrium points. The equilibrium point in Theorem 6.3 (as in all the examples in this section) is the *only* equilibrium point. It is, therefore, an isolated equilibrium point.

We also note, with respect to Theorem 6.3, that the assumption that A is invertible implies $\det(A) \neq 0$. Therefore, $\lambda = 0$ is not an eigenvalue of A. However, if A is not invertible, then $\mathbf{y}_e = \mathbf{0}$ is not the only equilibrium point (nor is it isolated). For example, suppose

$$A = \begin{bmatrix} \alpha & \beta \\ c\alpha & c\beta \end{bmatrix},$$

with α and β not both zero. Thus, $\det(A) = 0$, but A is not the zero matrix. One can show (see Exercise 33) that every point on the phase-plane line $\alpha x + \beta y = 0$ is an equilibrium point.

Theorem 6.3 applies equally well whether the (2×2) matrix A has distinct or repeated eigenvalues. Recall that if A has a repeated (real) eigenvalue $\lambda_1 = \lambda_2 = \lambda$, then solutions involving the function $te^{\lambda t}$, as well as $e^{\lambda t}$, are possible. If $\lambda < 0$, however, we know from calculus that $te^{\lambda t}$ is bounded for $t \geq 0$ and that $\lim_{t \to \infty} te^{\lambda t} = 0$. It follows that $\mathbf{y}_e = \mathbf{0}$ will always be an asymptotically stable equilibrium point when $\lambda < 0$.

In the higher-dimensional case, stability characterization is somewhat more complicated. If A is a real invertible $(n \times n)$ matrix, then the unique equilibrium point $\mathbf{y}_e = \mathbf{0}$ of $\mathbf{y}' = A\mathbf{y}$ is asymptotically stable if all eigenvalues of A have negative real parts. The equilibrium point is unstable if at least one eigenvalue of A has a positive real part. For $n \geq 4$, repeated complex conjugate pairs of purely imaginary eigenvalues, say $\lambda = \pm i\omega$, are possible. In this case, solutions of the form $t \cos \omega t$ and $t \sin \omega t$ are possible when the matrix A does not have n linearly independent eigenvectors; in that event, $\mathbf{y}_e = \mathbf{0}$ is an unstable equilibrium point.

EXERCISES

1. Assume that a two-dimensional autonomous system has an isolated equilibrium point at the origin and that the phase-plane solution curves consist of the family of concentric ellipses $x^2/4 + y^2 = C$, $C \geq 0$.

   (a) Apply the definition to show that the origin is a stable equilibrium point. In particular, given an $\varepsilon > 0$, determine a corresponding $\delta > 0$ so that all solutions starting within a circle of radius δ centered at the origin stay within the circle of radius ε centered at the origin for all $t \geq 0$. (The δ you determine should be expressed in terms of ε.)

   (b) Is the origin an asymptotically stable equilibrium point? Explain.

2. Assume that a two-dimensional autonomous system has an isolated equilibrium point at the origin and that the phase-plane solution curves consist of the family of hyperbolas $-x^2 + y^2 = C, C \geq 0$. Is the equilibrium point stable or unstable? Explain.

3. Consider the differential equation $x'' + \gamma x' + x = 0$, where γ is a real constant.

   (a) Rewrite the given scalar equation as a first order system, defining $y = x'$.

   (b) Determine the values of γ for which the system is (i) asymptotically stable, (ii) stable but not asymptotically stable, (iii) unstable.

Exercises 4–15:

Each exercise lists a linear system $\mathbf{y}' = A\mathbf{y}$, where A is a real constant invertible (2×2) matrix. Use Theorem 6.3 to determine whether the equilibrium point $\mathbf{y}_e = \mathbf{0}$ is asymptotically stable, stable but not asymptotically stable, or unstable.

4. $\mathbf{y}' = \begin{bmatrix} -3 & -2 \\ 4 & 3 \end{bmatrix} \mathbf{y}$ 5. $\mathbf{y}' = \begin{bmatrix} 5 & -14 \\ 3 & -8 \end{bmatrix} \mathbf{y}$ 6. $\mathbf{y}' = \begin{bmatrix} 0 & -2 \\ 2 & 0 \end{bmatrix} \mathbf{y}$

7. $\mathbf{y}' = \begin{bmatrix} 1 & 4 \\ -1 & 1 \end{bmatrix} \mathbf{y}$

8. $x' = -7x - 3y$ 9. $x' = 9x + 5y$
   $y' = 5x + y$ $y' = -7x - 3y$

10. $x' = -3x - 5y$ 11. $x' = 9x - 4y$ 12. $x' = -13x - 8y$
    $y' = 2x - y$ $y' = 15x - 7y$ $y' = 15x + 9y$

13. $x' = 3x - 2y$ 14. $x' = x - 5y$ 15. $x' = -3x + 3y$
    $y' = 5x - 3y$ $y' = x - 3y$ $y' = x - 5y$

Exercises 16–23:

Each exercise lists the general solution of a linear system of the form

$$x' = a_{11}x + a_{12}y$$
$$y' = a_{21}x + a_{22}y,$$

where $a_{11}a_{22} - a_{12}a_{21} \neq 0$. Determine whether the equilibrium point $\mathbf{y}_e = \mathbf{0}$ is asymptotically stable, stable but not asymptotically stable, or unstable.

16. $x = c_1 e^{-2t} + c_2 e^{3t}$ **17.** $x = c_1 e^{2t} + c_2 e^{3t}$ **18.** $x = c_1 e^{-2t} + c_2 e^{-4t}$

      $y = c_1 e^{-2t} + 2c_2 e^{3t}$ $y = c_1 e^{2t} + 2c_2 e^{3t}$ $y = c_1 e^{-2t} + 2c_2 e^{-4t}$

19. $x = c_1 e^t \cos 2t + c_2 e^t \sin 2t$ **20.** $x = c_1 \cos 2t + c_2 \sin 2t$

      $y = -c_1 e^t \sin 2t + c_2 e^t \cos 2t$ $y = -c_1 \sin 2t + c_2 \cos 2t$

21. $x = c_1 e^{-2t} \cos 2t + c_2 e^{-2t} \sin 2t$ **22.** $x = c_1 e^{-2t} + c_2 e^{3t}$

      $y = -c_1 e^{-2t} \sin 2t + c_2 e^{-2t} \cos 2t$ $y = c_1 e^{-2t} - c_2 e^{3t}$

23. $x = c_1 e^{-2t} + c_2 e^{-3t}$

      $y = c_1 e^{-2t} - c_2 e^{-3t}$

24. Consider the nonhomogeneous linear system $\mathbf{y}' = A\mathbf{y} + \mathbf{g}_0$, where A is a real invertible (2×2) matrix and $\mathbf{g}_0$ is a real (2×1) constant vector.

     (a) Determine the unique equilibrium point, $\mathbf{y}_e$, of this system.

     (b) Show how Theorem 6.3 can be used to determine the stability properties of this equilibrium point. [Hint: Adopt the change of dependent variable $\mathbf{z}(t) = \mathbf{y}(t) - \mathbf{y}_e$.]

Exercises 25–28:

Locate the unique equilibrium point of the given nonhomogeneous system, and determine the stability properties of this equilibrium point. Is it asymptotically stable, stable but not asymptotically stable, or unstable?

25. $\mathbf{y}' = \begin{bmatrix} -2 & 1 \\ 1 & -2 \end{bmatrix} \mathbf{y} + \begin{bmatrix} -4 \\ 2 \end{bmatrix}$ **26.** $x' = y + 2$

                                      $y' = -x + 1$

27. $\mathbf{y}' = \begin{bmatrix} 3 & 2 \\ -4 & -3 \end{bmatrix} \mathbf{y} + \begin{bmatrix} -2 \\ 2 \end{bmatrix}$ **28.** $x' = -x + y + 1$

                                        $y' = -10x + 5y + 2$

Exercises 29–32:

Higher Dimensional Systems   In each exercise, locate all equilibrium points for the given autonomous system. Determine whether the equilibrium point or points are asymptotically stable, stable but not asymptotically stable, or unstable.

29. $y_1' = 2y_1 + y_2 + y_3$ **30.** $\dfrac{d}{dt}\begin{bmatrix} y_1 \\ y_2 \\ y_3 \end{bmatrix} = \begin{bmatrix} 1 & -1 & 0 \\ 0 & -1 & 2 \\ 0 & 0 & -1 \end{bmatrix} \begin{bmatrix} y_1 \\ y_2 \\ y_3 \end{bmatrix} + \begin{bmatrix} 2 \\ 0 \\ 3 \end{bmatrix}$

      $y_2' = y_1 + y_2 + 2y_3$

      $y_3' = y_1 + 2y_2 + y_3$

31. $\dfrac{d}{dt}\begin{bmatrix} y_1 \\ y_2 \\ y_3 \\ y_4 \end{bmatrix} = \begin{bmatrix} -3 & -5 & 0 & 0 \\ 2 & -1 & 0 & 0 \\ 0 & 0 & 0 & 2 \\ 0 & 0 & -2 & 0 \end{bmatrix} \begin{bmatrix} y_1 \\ y_2 \\ y_3 \\ y_4 \end{bmatrix}$

32. $y_1' = y_2 - 1$

      $y_2' = y_1 + 2$

      $y_3' = -y_3 + 1$

      $y_4' = -y_4$

33. Let A be a real (2×2) matrix. Assume that A has eigenvalues λ_1 and λ_2, and consider the linear homogeneous system $\mathbf{y}' = A\mathbf{y}$.

     (a) Prove that if λ_1 and λ_2 are both nonzero, then $\mathbf{y}_e = \mathbf{0}$ is an isolated equilibrium point.

(b) Suppose that eigenvalue $\lambda_1 \neq 0$ but that $\lambda_2 = 0$ with corresponding eigenvector $\begin{bmatrix} \beta \\ -\alpha \end{bmatrix}$. Show that all points on the phase-plane line $\alpha x + \beta y = 0$ are equilibrium points. (In this case, $\mathbf{y}_e = \mathbf{0}$ is not an isolated equilibrium point.)

34. Consider the linear system

$$\mathbf{y}' = \begin{bmatrix} -1 & \alpha \\ \alpha & -1 \end{bmatrix} \mathbf{y},$$

where α is a real constant.

(a) What information can be obtained about the eigenvalues of the coefficient matrix simply by examining its structure?

(b) For what value(s) of the constant α is the equilibrium point $\mathbf{y}_e = \mathbf{0}$ an isolated equilibrium point? For what value(s) of the constant α is the equilibrium point $\mathbf{y}_e = \mathbf{0}$ not isolated?

(c) In the case where $\mathbf{y}_e = \mathbf{0}$ is not an isolated equilibrium point, what is the equation of the phase-plane line of equilibrium points?

(d) Is it possible in this example for $\mathbf{y}_e = \mathbf{0}$ to be an isolated equilibrium point that is stable but not asymptotically stable? Explain.

(e) For what values of the constant α, if any, is the equilibrium point $\mathbf{y}_e = \mathbf{0}$ an isolated asymptotically stable equilibrium point? For what values of the constant α, if any, is the equilibrium point $\mathbf{y}_e = \mathbf{0}$ an unstable equilibrium point?

35. Let $A = \begin{bmatrix} 1 & a_{12} \\ a_{21} & a_{22} \end{bmatrix}$ be a real (2×2) matrix. Assume that

$$A \begin{bmatrix} 1 \\ 2 \end{bmatrix} = \begin{bmatrix} 1 & a_{12} \\ a_{21} & a_{22} \end{bmatrix} \begin{bmatrix} 1 \\ 2 \end{bmatrix} = 2 \begin{bmatrix} 1 \\ 2 \end{bmatrix}$$

and that the origin is *not* an isolated equilibrium point of the system $\mathbf{y}' = A\mathbf{y}$. Determine the constants a_{12}, a_{21}, and a_{22}.

6.5 Linearization and the Local Picture

We now consider nonlinear autonomous systems and a technique, known as linearization, for investigating the stability properties of such systems. The stability results cited in Theorem 6.3 for the linear system $\mathbf{y}' = A\mathbf{y}$ will be useful for the nonlinear equations treated in this section. In Section 6.6, we will examine the phase-plane geometry of the linear two-dimensional system $\mathbf{y}' = A\mathbf{y}$ in more detail.

Nonlinear Systems

Although we are going to focus on the case $n = 2$, the ideas are applicable to general n-dimensional autonomous systems. Let

$$\mathbf{y}_e = \begin{bmatrix} x_e \\ y_e \end{bmatrix}$$

be an equilibrium solution of $\mathbf{y}' = \mathbf{f}(\mathbf{y})$. In component form, $\mathbf{y}' = \mathbf{f}(\mathbf{y})$ is given by

$$
\begin{aligned}
x' &= f(x, y) \\
y' &= g(x, y).
\end{aligned}
\tag{1}
$$

For a nonlinear system such as (1), it is usually impossible to obtain explicit solutions. In the absence of explicit solutions, we look for approximations or simplifications that provide qualitative insight into the stability properties of the equilibrium point $\mathbf{y}_e$.

Linearization

From the definition, we know that the issue of equilibrium point stability is ultimately determined by the behavior of solutions very close to the equilibrium point. If any nearby solutions diverge from the equilibrium point, it is an unstable equilibrium point. If all nearby solutions can be suitably confined, the equilibrium point is stable (and perhaps asymptotically stable).

To address the issue of equilibrium point stability, we begin with the observation that if the point (x, y) is near the equilibrium point (x_e, y_e), then the first few terms of the Taylor series expansions of $f(x, y)$ and $g(x, y)$ will yield good approximations to their values near the equilibrium point:

$$
\begin{aligned}
f(x, y) &= f(x_e, y_e) + \frac{\partial f(x_e, y_e)}{\partial x}(x - x_e) + \frac{\partial f(x_e, y_e)}{\partial y}(y - y_e) + \cdots \\
g(x, y) &= g(x_e, y_e) + \frac{\partial g(x_e, y_e)}{\partial x}(x - x_e) + \frac{\partial g(x_e, y_e)}{\partial y}(y - y_e) + \cdots.
\end{aligned}
\tag{2}
$$

We make the following observations:

1. Since (x_e, y_e) is an equilibrium point, $f(x_e, y_e) = g(x_e, y_e) = 0$. Thus, the first term on the right-hand side of each equation in (2) vanishes.

2. The error made in truncating the series [retaining only the linear terms shown on the right-hand sides of (2)] can usually be bounded by a multiple of $\|\mathbf{y} - \mathbf{y}_e\|^2 = (x - x_e)^2 + (y - y_e)^2$.

If the Taylor expansion (2) is used in differential equation (1), we can write the system in matrix form as

$$
\mathbf{y}'(t) =
\begin{bmatrix}
\dfrac{\partial f(x_e, y_e)}{\partial x} & \dfrac{\partial f(x_e, y_e)}{\partial y} \\[2ex]
\dfrac{\partial g(x_e, y_e)}{\partial x} & \dfrac{\partial g(x_e, y_e)}{\partial y}
\end{bmatrix}
(\mathbf{y}(t) - \mathbf{y}_e) + \cdots.
\tag{3}
$$

Note that the (2×2) coefficient matrix in (3) is a *constant* matrix since the partial derivatives are evaluated at the equilibrium point. [In vector calculus, the matrix of first order partial derivatives in (3) is called the **Jacobian matrix** of $\mathbf{f}(\mathbf{y})$.]

Linearization is based on the following two ideas:

1. Since $\mathbf{y}_e$ is a constant vector, the term $\mathbf{y}'(t)$ on the left-hand side of (3) can be replaced by $[\mathbf{y}(t) - \mathbf{y}_e]'$.

2. If we consider solutions close enough to the equilibrium point for the purpose of determining its stability characteristics, the higher order terms in (3) are typically small relative to the linear term, $\mathbf{y}(t) - \mathbf{y}_e$. We will neglect these higher order terms.

Introduce a new dependent variable, $\mathbf{z}(t) = \mathbf{y}(t) - \mathbf{y}_e$. Retaining only the linear term in (3) leads to the corresponding **linearized system**,

$$\mathbf{z}'(t) = \begin{bmatrix} \dfrac{\partial f(x_e, y_e)}{\partial x} & \dfrac{\partial f(x_e, y_e)}{\partial y} \\[3mm] \dfrac{\partial g(x_e, y_e)}{\partial x} & \dfrac{\partial g(x_e, y_e)}{\partial y} \end{bmatrix} \mathbf{z}(t). \tag{4}$$

Note that equation (4) is a homogeneous constant coefficient linear system and that $\mathbf{z} = \mathbf{0}$ is an equilibrium point of the linear system. The stability properties of $\mathbf{z} = \mathbf{0}$ are easy to analyze since we can explicitly find the general solution of equation (4); these properties are summarized in Theorem 6.3.

The underlying premise of linearization is that the stability properties of $\mathbf{z} = \mathbf{0}$ for the linear system (4) should be the same as the stability properties of $\mathbf{y} = \mathbf{y}_e$ for the original nonlinear system (1). We illustrate linearization in the next example. Then we address the question "When does linearization work?"

EXAMPLE 1

Develop the linearized-system approximation for each of the equilibrium points of the nonlinear autonomous system

$$x' = \tfrac{1}{2}\left(1 - \tfrac{1}{2}x - \tfrac{1}{2}y\right)x$$

$$y' = \tfrac{1}{4}\left(1 - \tfrac{1}{3}x - \tfrac{2}{3}y\right)y.$$

Also, determine the stability characteristics of the linearized system in each case.

Solution: This is the competing species model considered in Example 4 of Section 6.2. Recall that this system has four equilibrium solutions,

$$\mathbf{y}_e^{(1)} = \begin{bmatrix} 0 \\ 0 \end{bmatrix}, \qquad \mathbf{y}_e^{(2)} = \begin{bmatrix} 0 \\ \tfrac{3}{2} \end{bmatrix}, \qquad \mathbf{y}_e^{(3)} = \begin{bmatrix} 2 \\ 0 \end{bmatrix}, \qquad \mathbf{y}_e^{(4)} = \begin{bmatrix} 1 \\ 1 \end{bmatrix}.$$

The Jacobian matrix is given by

$$\begin{bmatrix} \dfrac{\partial f(x, y)}{\partial x} & \dfrac{\partial f(x, y)}{\partial y} \\[3mm] \dfrac{\partial g(x, y)}{\partial x} & \dfrac{\partial g(x, y)}{\partial y} \end{bmatrix} = \begin{bmatrix} \tfrac{1}{2} - \tfrac{1}{2}x - \tfrac{1}{4}y & -\tfrac{1}{4}x \\[3mm] -\tfrac{1}{12}y & \tfrac{1}{4} - \tfrac{1}{12}x - \tfrac{1}{3}y \end{bmatrix}. \tag{5}$$

The Jacobian matrix (5) must be evaluated at each of the equilibrium points in order to obtain the coefficient matrix of the appropriate linearized system. At

(continued)

(continued)

the equilibrium point $(0, 0)$, the coefficient matrix is

$$
\begin{bmatrix} \frac{1}{2} - \frac{1}{2}x - \frac{1}{4}y & -\frac{1}{4}x \\ -\frac{1}{12}y & \frac{1}{4} - \frac{1}{12}x - \frac{1}{3}y \end{bmatrix}\Bigg|_{x=y=0} = \begin{bmatrix} \frac{1}{2} & 0 \\ 0 & \frac{1}{4} \end{bmatrix}.
$$

The linearized system is therefore

$$
\mathbf{z}' = \begin{bmatrix} \frac{1}{2} & 0 \\ 0 & \frac{1}{4} \end{bmatrix} \mathbf{z}.
$$

[Observe, for this case, that $\mathbf{z}(t) = \mathbf{y}(t) - \mathbf{y}_e^{(1)} = \mathbf{y}(t)$.] Since the eigenvalues of the coefficient matrix ($\lambda = \frac{1}{2}$ and $\lambda = \frac{1}{4}$) are positive, we conclude from Theorem 6.3 that $\mathbf{z} = \mathbf{0}$ is an unstable equilibrium solution of the linearized system. Therefore, we anticipate that $\mathbf{y}_e^{(1)}$ will also be an unstable equilibrium point of the nonlinear system.

The remaining three equilibrium points can be analyzed in the same manner. The results for all four equilibrium points are summarized in Table 6.1.

TABLE 6.1

Equilibrium Point	$\mathbf{z}(t)$	Linearized System Coefficient Matrix	Eigenvalues of the Coefficient Matrix	Stability Properties of the Linearized System
$(0, 0)$	$\begin{bmatrix} x(t) \\ y(t) \end{bmatrix}$	$\begin{bmatrix} \frac{1}{2} & 0 \\ 0 & \frac{1}{4} \end{bmatrix}$	$\frac{1}{2}, \frac{1}{4}$	Unstable
$(0, \frac{3}{2})$	$\begin{bmatrix} x(t) \\ y(t) - \frac{3}{2} \end{bmatrix}$	$\begin{bmatrix} \frac{1}{8} & 0 \\ -\frac{1}{8} & -\frac{1}{4} \end{bmatrix}$	$\frac{1}{8}, -\frac{1}{4}$	Unstable
$(2, 0)$	$\begin{bmatrix} x(t) - 2 \\ y(t) \end{bmatrix}$	$\begin{bmatrix} -\frac{1}{2} & -\frac{1}{2} \\ 0 & \frac{1}{12} \end{bmatrix}$	$-\frac{1}{2}, \frac{1}{12}$	Unstable
$(1, 1)$	$\begin{bmatrix} x(t) - 1 \\ y(t) - 1 \end{bmatrix}$	$\begin{bmatrix} -\frac{1}{4} & -\frac{1}{4} \\ -\frac{1}{12} & -\frac{1}{6} \end{bmatrix}$	$\frac{-5 - \sqrt{13}}{24}, \frac{-5 + \sqrt{13}}{24}$	Asymptotically stable

❖

The stability properties obtained by studying the linearized systems in Table 6.1 are consistent with the phase-plane direction field portrait of Figure 6.4 in Section 6.2. The three equilibria designated as unstable in Table 6.1 are the ones that appear to have direction field arrows pointing away from them in Figure 6.4. The fourth equilibrium point, designated as asymptotically stable in Table 6.1, appears to have all trajectories moving toward it in Figure 6.4.

When Does Linearization Work?

We restrict our attention to the two-dimensional autonomous system $\mathbf{y}' = \mathbf{f}(\mathbf{y})$. Let $\mathbf{y}_e$ be an equilibrium solution,

$$\mathbf{y}_e = \begin{bmatrix} x_e \\ y_e \end{bmatrix}.$$

We'll proceed as before by introducing a new dependent variable that shifts the equilibrium point to $\mathbf{0}$ and by rewriting $\mathbf{y}' = \mathbf{f}(\mathbf{y})$ in a form that explicitly exhibits the linearized part.

Let $\mathbf{z}(t) = \mathbf{y}(t) - \mathbf{y}_e$, and let A denote the Jacobian matrix evaluated at $\mathbf{y}_e$,

$$A = \begin{bmatrix} \dfrac{\partial f(x_e, y_e)}{\partial x} & \dfrac{\partial f(x_e, y_e)}{\partial y} \\ \dfrac{\partial g(x_e, y_e)}{\partial x} & \dfrac{\partial g(x_e, y_e)}{\partial y} \end{bmatrix}. \tag{6}$$

The system $\mathbf{y}' = \mathbf{f}(\mathbf{y})$ can be rewritten (without any approximation) as

$$\mathbf{z}'(t) = \mathbf{f}(\mathbf{z}(t) + \mathbf{y}_e) = A\mathbf{z}(t) + [\mathbf{f}(\mathbf{z}(t) + \mathbf{y}_e) - A\mathbf{z}(t)]. \tag{7}$$

Define $\mathbf{g}(\mathbf{z}(t)) = \mathbf{f}(\mathbf{z}(t) + \mathbf{y}_e) - A\mathbf{z}(t)$. With this, equation (7) becomes

$$\mathbf{z}'(t) = A\mathbf{z}(t) + \mathbf{g}(\mathbf{z}(t)). \tag{8}$$

Written in this form, we see that $\mathbf{g}(\mathbf{z}(t))$ represents a nonlinear perturbation of the linearized system $\mathbf{z}'(t) = A\mathbf{z}(t)$. The linearization approximation amounts to discarding the nonlinear term, $\mathbf{g}(\mathbf{z}(t))$.

If the behavior of the linearized system $\mathbf{z}' = A\mathbf{z}$ is going to be qualitatively similar to that of the nonlinear system (8) near the equilibrium point $\mathbf{z} = \mathbf{0}$, it seems clear that $\mathbf{g}(\mathbf{z})$ must be suitably "small" near $\mathbf{z} = \mathbf{0}$. In other words, the linear part of (8), $A\mathbf{z}$, must control the basic behavior of solutions near the equilibrium point. We will now describe such a class of nonlinear systems.

A two-dimensional autonomous system $\mathbf{y}' = \mathbf{f}(\mathbf{y})$ is called an **almost linear system at an equilibrium point** $\mathbf{y}_e$ if

(a) $\mathbf{f}(\mathbf{y})$ is a continuous vector-valued function whose component functions possess continuous partial derivatives in an open region of the phase plane containing the equilibrium point, $\mathbf{y}_e$.

(b) The matrix

$$A = \begin{bmatrix} \dfrac{\partial f(x_e, y_e)}{\partial x} & \dfrac{\partial f(x_e, y_e)}{\partial y} \\ \dfrac{\partial g(x_e, y_e)}{\partial x} & \dfrac{\partial g(x_e, y_e)}{\partial y} \end{bmatrix}$$

is invertible.

(c) The perturbation function $\mathbf{g}(\mathbf{z}) = \mathbf{f}(\mathbf{z} + \mathbf{y}_e) - A\mathbf{z}$ is such that

$$\lim_{\|\mathbf{z}\| \to 0} \frac{\|\mathbf{g}(\mathbf{z})\|}{\|\mathbf{z}\|} = 0. \tag{9}$$

REMARKS:

1. The perturbation function $\mathbf{g}(\mathbf{z}) = \mathbf{f}(\mathbf{z} + \mathbf{y}_e) - A\mathbf{z}$ inherits the continuity and differentiability properties assumed for $\mathbf{f}$. Thus, $\mathbf{g}(\mathbf{z})$ is continuous with continuous partial derivatives in an open region of the $\mathbf{z}$-plane containing the origin, $\mathbf{z} = \mathbf{0}$.

2. Limit (9) establishes how "small" the nonlinear perturbation must be near the equilibrium point; the norm of the perturbation must tend to zero faster than $\|\mathbf{z}\|$ as $\mathbf{z}$ approaches the origin.

3. Since matrix A is invertible, $\mathbf{z} = \mathbf{0}$ is the only equilibrium point of the linearized problem $\mathbf{z}' = A\mathbf{z}$. It is also clear that $\mathbf{g}(\mathbf{0}) = \mathbf{0}$ [since $\mathbf{f}(\mathbf{y}_e) = \mathbf{0}$]. Therefore, $\mathbf{z} = \mathbf{0}$ is an equilibrium point of the nonlinear system $\mathbf{z}' = A\mathbf{z} + \mathbf{g}(\mathbf{z})$. In addition, it can be shown that the assumptions made in (a)–(c) imply that $\mathbf{z} = \mathbf{0}$ is an isolated equilibrium point of $\mathbf{z}' = A\mathbf{z} + \mathbf{g}(\mathbf{z})$. Therefore, $\mathbf{y} = \mathbf{y}_e$ is an isolated equilibrium point of the original system, $\mathbf{y}' = \mathbf{f}(\mathbf{y})$.

Theorem 6.4

Let $\mathbf{y}' = \mathbf{f}(\mathbf{y})$ be a two-dimensional autonomous system that is almost linear at an equilibrium point $\mathbf{y} = \mathbf{y}_e$. Let $\mathbf{z}' = A\mathbf{z}$ be the corresponding linearized system.

(a) If $\mathbf{z} = \mathbf{0}$ is an asymptotically stable equilibrium point of $\mathbf{z}' = A\mathbf{z}$, then $\mathbf{y} = \mathbf{y}_e$ is an asymptotically stable equilibrium point of $\mathbf{y}' = \mathbf{f}(\mathbf{y})$.

(b) If $\mathbf{z} = \mathbf{0}$ is an unstable equilibrium point of $\mathbf{z}' = A\mathbf{z}$, then $\mathbf{y} = \mathbf{y}_e$ is an unstable equilibrium point of $\mathbf{y}' = \mathbf{f}(\mathbf{y})$.

(c) If $\mathbf{z} = \mathbf{0}$ is a stable (but not asymptotically stable) equilibrium point of $\mathbf{z}' = A\mathbf{z}$, then no conclusions can be drawn about the stability properties of equilibrium point $\mathbf{y} = \mathbf{y}_e$.

The proof of Theorem 6.4 can be found in more advanced treatments of differential equations, such as Coddington and Levinson.[3] However, the assertions made in Theorem 6.4 should strike you as reasonable. When the linearized system is asymptotically stable, both eigenvalues of the coefficient matrix A have negative real parts. We know, therefore, that the norm of the solution of the linearized system, $\|\mathbf{z}(t)\|$, is exponentially decreasing to zero. If the nonlinear perturbation is sufficiently weak, we might expect this qualitative behavior to persist in the nonlinear system (8).

Similarly, if the linearized system is unstable, at least one of the two eigenvalues of A has a positive real part. The linearized system, therefore, has some solutions that grow exponentially in norm. In this case, we might expect instability to persist when the nonlinear perturbation $\mathbf{g}$ is sufficiently weak. Finally, if the linearized system is stable but not asymptotically stable, then the eigenvalues of A form a purely imaginary complex conjugate pair. In this case, the linearized system is sitting on the fence between stability and instability. It is

---

[3] Earl A. Coddington and Norman Levinson, *Theory of Ordinary Differential Equations* (Malabar, FL: R. E. Krieger, 1984).

possible for the nonlinear perturbation (however small) to tip the balance either way—causing the nonlinear system to be stable or causing it to be unstable. Example 3, found later in this section, illustrates this last point [and thereby proves condition (c) of Theorem 6.4].

E X A M P L E

2

Consider again the nonlinear system discussed in Example 1,

$$x' = \tfrac{1}{2}\left(1 - \tfrac{1}{2}x - \tfrac{1}{2}y\right)x$$

$$y' = \tfrac{1}{4}\left(1 - \tfrac{1}{3}x - \tfrac{2}{3}y\right)y. \tag{10}$$

Use Theorem 6.4 to determine the stability properties of equilibrium point $(1, 1)$ for the nonlinear system (10).

Solution: We begin by making the change of dependent variable $\mathbf{z}(t) = \mathbf{y}(t) - \mathbf{y}_e$:

$$z_1(t) = x(t) - 1 \quad \text{and} \quad z_2(t) = y(t) - 1.$$

With this change of variables, we can rewrite system (10) as

$$\mathbf{z}' = \begin{bmatrix} -\tfrac{1}{4} & -\tfrac{1}{4} \\ -\tfrac{1}{12} & -\tfrac{1}{6} \end{bmatrix} \mathbf{z} + \begin{bmatrix} -\tfrac{1}{4}(z_1^2 + z_1 z_2) \\ -\tfrac{1}{12}(z_1 z_2 + 2z_2^2) \end{bmatrix}. \tag{11}$$

We also know (see Table 6.1) that the linearized system

$$\mathbf{z}' = \begin{bmatrix} -\tfrac{1}{4} & -\tfrac{1}{4} \\ -\tfrac{1}{12} & -\tfrac{1}{6} \end{bmatrix} \mathbf{z}$$

has an asymptotically stable equilibrium point at $\mathbf{z} = \mathbf{0}$.

Theorem 6.4 asserts that $\mathbf{z} = \mathbf{0}$ will be an asymptotically stable equilibrium point of nonlinear system (11) (and therefore $\mathbf{y}_e$ is an asymptotically stable equilibrium point of the original system) if we can show that the system is almost linear at the equilibrium point $\mathbf{y}_e$. Note that the first two conditions of the definition are clearly satisfied. So, to apply Theorem 6.4, all we need to do is establish the limit (9):

$$\frac{\|\mathbf{g}(\mathbf{z})\|}{\|\mathbf{z}\|} \to 0 \quad \text{as} \quad \|\mathbf{z}\| \to 0,$$

where [see equation (11)]

$$\mathbf{g}(\mathbf{z}) = \begin{bmatrix} -\tfrac{1}{4}(z_1^2 + z_1 z_2) \\ -\tfrac{1}{12}(z_1 z_2 + 2z_2^2) \end{bmatrix}.$$

In order to calculate the quotient $\|\mathbf{g}(\mathbf{z})\|/\|\mathbf{z}\|$, it is convenient to introduce polar coordinates $z_1 = r\cos\theta$ and $z_2 = r\sin\theta$. Under this change of variables, we see that $\|\mathbf{z}\| = r$ and

$$\mathbf{g}(\mathbf{z}) = \begin{bmatrix} -\dfrac{r^2}{4}(\cos^2\theta + \sin\theta\cos\theta) \\ -\dfrac{r^2}{12}(\sin\theta\cos\theta + 2\sin^2\theta) \end{bmatrix}.$$

(continued)

(continued)

Thus,

$$\frac{\|\mathbf{g}(\mathbf{z})\|}{\|\mathbf{z}\|} = \frac{\|\mathbf{g}(\mathbf{z})\|}{r}$$

$$= r\sqrt{\left(-\frac{1}{4}(\cos^2\theta + \sin\theta\cos\theta)\right)^2 + \left(-\frac{1}{12}(\sin\theta\cos\theta + 2\sin^2\theta)\right)^2}$$

$$< r\sqrt{\left(\frac{1}{4}(1+1)\right)^2 + \left(\frac{1}{12}(1+2)\right)^2} = \frac{\sqrt{5}}{4}r.$$

Since the right-hand side of this inequality vanishes as $r \to 0$, limit (9) is verified and we conclude that the nonlinear system is almost linear at equilibrium point (1, 1). By Theorem 6.4, this equilibrium point is an asymptotically stable equilibrium point of the given nonlinear system. ❖

Polar Coordinates as Dependent Variables

Polar coordinates are often useful in studying the stability properties of two-dimensional systems. Define new dependent variables $r(t)$ and $\theta(t)$ by means of the relations

$$z_1(t) = r(t)\cos[\theta(t)], \qquad z_2(t) = r(t)\sin[\theta(t)] \tag{12a}$$

$$r(t) = \sqrt{z_1^2(t) + z_2^2(t)}, \qquad \tan[\theta(t)] = \frac{z_2(t)}{z_1(t)}. \tag{12b}$$

We can transform system (8), $\mathbf{z}'(t) = A\mathbf{z}(t) + \mathbf{g}(\mathbf{z}(t))$, into a new system of differential equations in these polar variables.

The motivation for this transformation is the fact that stability properties of the equilibrium point $\mathbf{y} = \mathbf{y}_e$ (or, equivalently, $\mathbf{z} = \mathbf{0}$) depend on how the distance of the solution point $\mathbf{z}(t)$ from the origin varies with time. The radial variable, $r(t) = \|\mathbf{z}(t)\|$, is this distance.

Let the Jacobian matrix A be represented as

$$A = \begin{bmatrix} a_{11} & a_{12} \\ a_{21} & a_{22} \end{bmatrix}.$$

Then, substituting (12a) into $\mathbf{z}'(t) = A\mathbf{z}(t) + \mathbf{g}(\mathbf{z}(t))$, we obtain the matrix equation

$$\begin{bmatrix} \cos\theta & -r\sin\theta \\ \sin\theta & r\cos\theta \end{bmatrix}\begin{bmatrix} r' \\ \theta' \end{bmatrix} = \begin{bmatrix} a_{11} & a_{12} \\ a_{21} & a_{22} \end{bmatrix}\begin{bmatrix} r\cos\theta \\ r\sin\theta \end{bmatrix} + \begin{bmatrix} g_1(r\cos\theta, r\sin\theta) \\ g_2(r\cos\theta, r\sin\theta) \end{bmatrix}. \tag{13}$$

The desired system of differential equations for the polar variables is obtained by multiplying both sides of (13) by

$$\begin{bmatrix} \cos\theta & -r\sin\theta \\ \sin\theta & r\cos\theta \end{bmatrix}^{-1} = \begin{bmatrix} \cos\theta & \sin\theta \\ -r^{-1}\sin\theta & r^{-1}\cos\theta \end{bmatrix}.$$

We illustrate the use of the polar coordinates transformation in the next example. [As previously noted, this example establishes condition (c) of Theorem 6.4.]

EXAMPLE 3

Consider the nonlinear autonomous system

$$x' = y + \alpha x(x^2 + y^2)$$
$$y' = -x + \alpha y(x^2 + y^2),$$

(14)

where α is a constant. This system has $\mathbf{y}_e = \mathbf{0}$ as an equilibrium point for any choice of the parameter α.

Although $\mathbf{y}_e = \mathbf{0}$ is a stable equilibrium point of the linearized system, it is not asymptotically stable. We will show that if $\alpha = 1$, $\mathbf{y}_e = \mathbf{0}$ is an unstable equilibrium point of the nonlinear system, whereas if $\alpha = -1$, $\mathbf{y}_e = \mathbf{0}$ is a stable equilibrium point of the nonlinear system.

Set $\mathbf{z}(t) = \mathbf{y}(t) - \mathbf{y}_e = \mathbf{y}(t)$, and note that the system can be written as $\mathbf{z}' = A\mathbf{z} + \mathbf{g}(\mathbf{z})$, where

$$A = \begin{bmatrix} 0 & 1 \\ -1 & 0 \end{bmatrix}, \qquad \mathbf{g}(\mathbf{z}) = \begin{bmatrix} \alpha x(x^2+y^2) \\ \alpha y(x^2+y^2) \end{bmatrix} = \alpha r^3 \begin{bmatrix} \cos\theta \\ \sin\theta \end{bmatrix}.$$

(15)

The eigenvalues of matrix A are $\pm i$, and we know from Example 3 in Section 6.4 that the associated linearized system is stable but not asymptotically stable at $\mathbf{z} = \mathbf{0}$. Note that the nonlinear system is almost linear at $\mathbf{0}$. In particular, it follows from (15) that the quotient,

$$\frac{\|\mathbf{g}(\mathbf{z})\|}{\|\mathbf{z}\|} = \frac{\|\mathbf{g}(\mathbf{z})\|}{r} = |\alpha| r^2,$$

vanishes as $r \to 0$.

When polar coordinate variables $r(t)$ and $\theta(t)$ are introduced as in (12), it follows (see Exercise 27) that equation (14) transforms into the following simple decoupled system of equations for the new dependent variables:

$$r' = \alpha r^3$$
$$\theta' = -1.$$

(16)

Both of these equations are easy to solve; the differential equation for the radial variable $r(t)$ is a first order separable differential equation. The differential equation for the angular variable $\theta(t)$ can be solved by antidifferentiation. Initial conditions for the original system (14) of the form

$$\begin{bmatrix} x(0) \\ y(0) \end{bmatrix} = \begin{bmatrix} x_0 \\ y_0 \end{bmatrix}$$

transform to corresponding initial conditions for system (16):

$$r(0) = r_0 = \sqrt{x_0^2 + y_0^2}$$
$$\theta(0) = \tan^{-1}\left(\frac{y_0}{x_0}\right).$$

We are interested in the case where $r_0 \neq 0$. The solution of the transformed system (16) is

$$r(t) = \frac{r_0}{\sqrt{1 - 2r_0^2 \alpha t}}, \qquad \theta(t) = -t + \theta_0.$$

(17)

(continued)

(continued)

Consider the two cases $\alpha = 1$ and $\alpha = -1$. For the case $\alpha = 1$, $r(t)$ becomes unbounded as t approaches $\frac{1}{2}r_0^{-2}$. Therefore, since all nonzero solutions eventually move unboundedly far from the equilibrium point, $\mathbf{z} = \mathbf{0}$, the origin is an unstable equilibrium point of the nonlinear system. Therefore, the linearized system is stable (but not asymptotically stable), and the nonlinear system is unstable at the origin.

If $\alpha = -1$, however, we see from (17) that

$$0 \leq r(t) \leq r_0 \quad \text{and} \quad \lim_{t \to \infty} r(t) = 0. \tag{18}$$

In this case, the origin is an asymptotically stable equilibrium point of the nonlinear system. The linearized system is stable (but not asymptotically stable), and the nonlinear system is asymptotically stable at $\mathbf{y} = \mathbf{0}$.

Therefore, we have established assertion (c) of Theorem 6.4: If the linearized system is stable (but not asymptotically stable), nothing can be inferred about the stability properties of the nonlinear system.

In both cases, as time increases and the radial variable changes, the angular variable decreases at a constant rate. The solution point is moving clockwise around the origin with unit angular velocity. Solution point behaviors corresponding to the two cases, $\alpha = \pm 1$, are illustrated in Figure 6.13.

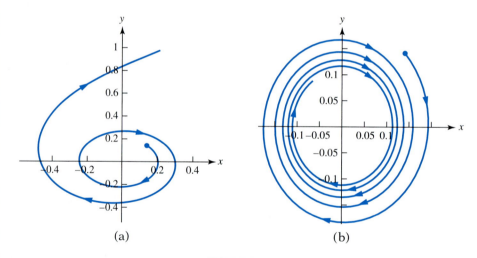

(a)   (b)

FIGURE 6.13

(a) The phase-plane plot of the solution $\mathbf{z}(t)$ in Example 3 with $\alpha = 1$, $r_0 = 0.2$, $\theta_0 = \pi/4$, and $0 \leq t \leq 12$. For this case, $\mathbf{z} = \mathbf{0}$ is an unstable equilibrium point for the nonlinear system (16); the solution point moves away from $\mathbf{z} = \mathbf{0}$ as t increases. (b) The phase-plane plot of the solution $\mathbf{z}(t)$ in Example 3 with $\alpha = -1$ and $0 \leq t \leq 30$. In this case, $\mathbf{z} = \mathbf{0}$ is an asymptotically stable equilibrium point for the nonlinear system (16); the solution point moves toward $\mathbf{z} = \mathbf{0}$ as t increases. ❖

The Pendulum Revisited

As a final example, we consider the stability properties of the pendulum in the context of the ideas developed earlier in this chapter. Recall that, with $g = l$, the pendulum is described by the nonlinear system

$$x' = y$$
$$y' = -\sin x. \tag{19}$$

There are basically two equilibrium configurations,

$$\mathbf{y}_e = \begin{bmatrix} 0 \\ 0 \end{bmatrix} \quad \text{and} \quad \mathbf{y}_e = \begin{bmatrix} \pi \\ 0 \end{bmatrix}.$$

The first corresponds to the pendulum resting in the vertically downward position, and the second corresponds to the pendulum resting in an inverted position. It can be shown (Exercise 11) that system (19) is an almost linear system at both equilibrium points; thus, we can apply Theorem 6.4.

At the equilibrium point $(0, 0)$, the coefficient matrix of the linearized system is

$$A = \begin{bmatrix} 0 & 1 \\ -1 & 0 \end{bmatrix}.$$

The eigenvalues of this matrix are $\pm i$, and thus the linearized system is stable but not asymptotically stable at $(0, 0)$. Therefore, Theorem 6.4 provides no information about stability of the pendulum equation at $(0, 0)$.

At $(\pi, 0)$, the coefficient matrix of the linearized system is

$$A = \begin{bmatrix} 0 & 1 \\ 1 & 0 \end{bmatrix}.$$

The eigenvalues of this matrix are ± 1, and so the linearized system is unstable (since one of the eigenvalues is positive). Therefore, Theorem 6.4 tells us that $(\pi, 0)$ is an unstable equilibrium point for the pendulum equation.

There is another way of deducing stability information. In Section 6.3, we saw that the pendulum leads to a conservative system, and we derived an explicit formula,

$$\tfrac{1}{2}y^2 - \cos x = E, \tag{20}$$

for a pendulum trajectory having energy per unit mass equal to E [see equation (8) in Section 6.3]. Figure 6.14(a) is identical to Figure 6.9 of Section 6.3, showing representative trajectories obtained by graphing equation (20) for various values of E, $E \geq -1$. As the value of the constant E decreases toward the equilibrium point value of $E = -1$, the corresponding phase-plane trajectories form a nested family of "ellipse-like" closed curves. The uniqueness property of solutions prevents these closed curves from intersecting. We expect, therefore, that the origin is a stable equilibrium point of the nonlinear system.

A mathematical argument verifying that $(0, 0)$ is a stable equilibrium point can be developed along the following lines. Given any $\varepsilon > 0$, construct a circle of radius ε centered at the origin. For a value of energy E sufficiently close to -1, we can find a closed trajectory lying entirely within this circle. Let this value of energy be E_ε. Now we choose $\delta > 0$ sufficiently small that a circle of radius δ lies within this closed trajectory. This choice of δ will work as far as satisfying the stability definition is concerned—all solutions originating within the circle of radius δ will remain within the closed trajectory of energy E_ε since solutions

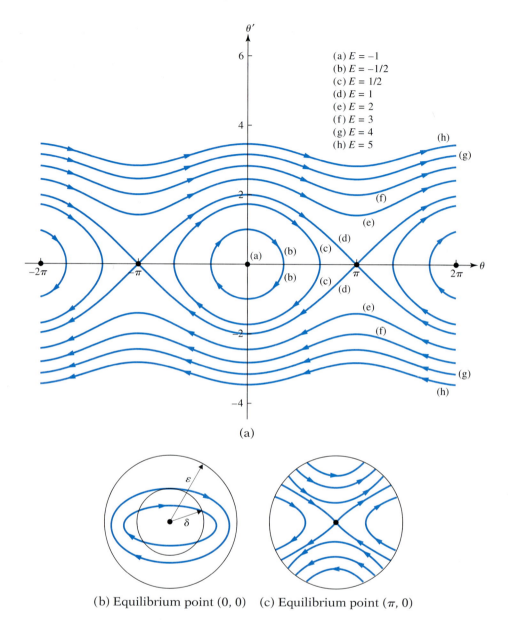

(a) $E = -1$
(b) $E = -1/2$
(c) $E = 1/2$
(d) $E = 1$
(e) $E = 2$
(f) $E = 3$
(g) $E = 4$
(h) $E = 5$

(a)

(b) Equilibrium point $(0, 0)$ (c) Equilibrium point $(\pi, 0)$

FIGURE 6.14

(a) Some of the phase-plane trajectories for the pendulum equation (19).
Note that the separatrices approach the unstable equilibrium point $(\pi, 0)$.
(b) Any trajectory originating within the circle of radius δ must stay within
the trajectory having energy E_ε and must therefore remain within the
circle of radius ε. (c) Every small circle centered at $(\pi, 0)$ has at least one
trajectory that eventually exits the circle.

cannot intersect. Consequently, the trajectories will remain within the circle of
radius ε. Figure 6.14(b) illustrates these ideas.

These geometrical ideas also provide another way of seeing that $(\pi, 0)$ is
an unstable equilibrium point of the nonlinear system. In particular [see Figure 6.14(c)], any circle of radius $\delta > 0$ centered at $(\pi, 0)$ must contain portions of

the separatrices and portions of some of the other trajectories that correspond to large-scale motions of the pendulum. Solution points originating on the latter trajectories eventually exit any small circle of radius ε centered at $(\pi, 0)$. Therefore, the definition fails, and the equilibrium point $(\pi, 0)$ is unstable.

EXERCISES

Exercises 1–9:

In each exercise, the given system is an almost linear system at each of its equilibrium points.

(a) Find the (real) equilibrium points of the given system.

(b) As in Example 2, find the corresponding linearized system $\mathbf{z}' = A\mathbf{z}$ at each equilibrium point.

(c) What, if anything, can be inferred about the stability properties of the equilibrium point(s) by using Theorem 6.4?

1. $x' = x^2 + y^2 - 32$ **2.** $x' = x^2 + 9y^2 - 9$ **3.** $x' = 1 - x^2$

  $y' = y - x$ $y' = x$ $y' = x^2 + y^2 - 2$

4. $x' = x - y - 1$ **5.** $x' = (x - 2)(y - 3)$ **6.** $x' = (x - y)(y + 1)$

  $y' = x^2 - y^2 + 1$ $y' = (x + 2y)(y - 1)$ $y' = (x + 2)(y - 4)$

7. $x' = (x - 2y)(y + 4)$ **8.** $x' = xy - 1$ **9.** $x' = y^2 - x$

  $y' = 2x - y$ $y' = (x + 4y)(x - 1)$ $y' = x^2 - y$

10. Perform a stability analysis of the competing species model at the equilibrium point $(0, 0)$:

$$x' = \tfrac{1}{2}\left(1 - \tfrac{1}{2}x - \tfrac{1}{2}y\right)x$$

$$y' = \tfrac{1}{4}\left(1 - \tfrac{1}{3}x - \tfrac{2}{3}y\right)y.$$

Specifically, repeat the analysis of Example 2 to determine the stability properties of the nonlinear system at this point.

11. Consider the system encountered in the study of pendulum motion,

$$x' = y$$

$$y' = -\sin x,$$

at its equilibrium points $(0, 0)$ and $(\pi, 0)$.

(a) Let $z_1 = x, z_2 = y$. Show that the system becomes

$$\mathbf{z}' = \begin{bmatrix} 0 & 1 \\ -1 & 0 \end{bmatrix} \mathbf{z} + \begin{bmatrix} 0 \\ z_1 - \sin z_1 \end{bmatrix}.$$

(b) Let $z_1 = x - \pi, z_2 = y$. Show that the system becomes

$$\mathbf{z}' = \begin{bmatrix} 0 & 1 \\ 1 & 0 \end{bmatrix} \mathbf{z} - \begin{bmatrix} 0 \\ z_1 - \sin z_1 \end{bmatrix}.$$

(c) Show that the system is almost linear at both equilibrium points. [Hint: One approach is to use Taylor's theorem and polar coordinates.]

Exercises 12–20:

Each exercise lists a nonlinear system $\mathbf{z}' = A\mathbf{z} + \mathbf{g}(\mathbf{z})$, where A is a constant (2×2) invertible matrix and $\mathbf{g}(\mathbf{z})$ is a (2×1) vector function. In each of the exercises, $\mathbf{z} = \mathbf{0}$ is an equilibrium point of the nonlinear system.

(a) Identify A and $\mathbf{g}(\mathbf{z})$.

(b) Calculate $\|\mathbf{g}(\mathbf{z})\|$.

(c) Is $\lim_{\|\mathbf{z}\| \to 0} \|\mathbf{g}(\mathbf{z})\| / \|\mathbf{z}\| = 0$? Is $\mathbf{z}' = A\mathbf{z} + \mathbf{g}(\mathbf{z})$ an almost linear system at $\mathbf{z} = \mathbf{0}$?

(d) If the system is almost linear, use Theorem 6.4 to choose one of the three statements:

    (i) $\mathbf{z} = \mathbf{0}$ is an asymptotically stable equilibrium point.

    (ii) $\mathbf{z} = \mathbf{0}$ is an unstable equilibrium point.

    (iii) No conclusion can be drawn by using Theorem 6.4.

12. $z_1' = 9z_1 - 4z_2 + z_2^2$
$z_2' = 15z_1 - 7z_2$

13. $z_1' = 5z_1 - 14z_2 + z_1z_2$
$z_2' = 3z_1 - 8z_2 + z_1^2 + z_2^2$

14. $z_1' = -3z_1 + z_2 + z_1^2 + z_2^2$
$z_2' = 2z_1 - 2z_2 + (z_1^2 + z_2^2)^{1/3}$

15. $z_1' = -z_1 + 3z_2 + z_2 \cos \sqrt{z_1^2 + z_2^2}$
$z_2' = -z_1 - 5z_2 + z_1 \cos \sqrt{z_1^2 + z_2^2}$

16. $z_1' = -2z_1 + 2z_2 + z_1z_2 \cos z_2$
$z_2' = z_1 - 3z_2 + z_1z_2 \sin z_2$

17. $z_1' = 2z_2 + z_2^2$
$z_2' = -2z_1 + z_1z_2$

18. $z_1' = -3z_1 - 5z_2 + z_1 e^{-\sqrt{z_1^2 + z_2^2}}$
$z_2' = 2z_1 - z_2 + z_2 e^{-\sqrt{z_1^2 + z_2^2}}$

19. $z_1' = 9z_1 + 5z_2 + z_1z_2$
$z_2' = -7z_1 - 3z_2 + z_1^2$

20. $z_1' = 2z_1 + 2z_2$
$z_2' = -5z_1 - 2z_2 + z_1^2$

21. Consider the autonomous system

$$x' = -x + xy + y$$
$$y' = x - xy - 2y.$$

This is the reduced system for the chemical reaction discussed in Exercise 19 of Section 6.1 with $a(t) = x(t)$, $c(t) = y(t)$, $e_0 = 1$, and all rate constants set equal to 1.

(a) Show that this system has a single equilibrium point, $(x_e, y_e) = (0, 0)$.

(b) Determine the linearized system $\mathbf{z}' = A\mathbf{z}$, and analyze its stability properties.

(c) Show that the system is an almost linear system at equilibrium point $(0, 0)$.

(d) Use Theorem 6.4 to determine the equilibrium properties of the given nonlinear system at $(0, 0)$.

22. Consider the nonlinear scalar differential equation $x'' = 1 - (1 + x)^{3/2}$. An equation having this structure arises in modeling the bobbing motion of a floating parabolic trough.

(a) Let $y = x'$ and rewrite the given scalar equation as an equivalent first order system.

(b) Show that the system has a single equilibrium point at $(x_e, y_e) = (0, 0)$.

(c) Determine the linearized system $\mathbf{z}' = A\mathbf{z}$, and analyze its stability properties.

(d) Assume that the system is an almost linear system at equilibrium point $(0, 0)$. Does Theorem 6.4 provide any information about the stability properties of the nonlinear system obtained in part (a)? Explain.

23. Consider again the differential equation of Exercise 22, $x'' = 1 - (1 + x)^{3/2}$.

(a) In Section 6.3, equations having this structure were shown to have a conservation law. Derive this conservation law; it will have the form $y^2/2 + F(x) = C$, where $y = x'$.

(b) Let $C = \frac{1}{2}, \frac{3}{4}$, and 1 in the conservation law derived in part (a). Plot the corresponding phase-plane solution trajectories, using computational software. Are these phase-plane trajectories consistent with a bobbing motion? Assuming these plots typify the behavior of solution trajectories near the origin, do they suggest that the origin is a stable or unstable equilibrium point for the nonlinear system? Explain.

24. Each of the autonomous nonlinear systems fails to satisfy the hypotheses of Theorem 6.4 at the equilibrium point $(0, 0)$. Explain why.

(a) $x' = x - y + xy$

   $y' = -x + y + 2x^2 y^2$

(b) $x' = x - 2y - x^{2/3}$

   $y' = x + y + 2y^{1/3}$

25. **Polar Coordinates**   Consider the system $\mathbf{z}' = A\mathbf{z} + \mathbf{g}(\mathbf{z})$, where

$$A = \begin{bmatrix} a_{11} & a_{12} \\ a_{21} & a_{22} \end{bmatrix} \quad \text{and} \quad \mathbf{g}(\mathbf{z}) = \begin{bmatrix} g_1(\mathbf{z}) \\ g_2(\mathbf{z}) \end{bmatrix}.$$

Show that adopting polar coordinates $z_1(t) = r(t)\cos[\theta(t)]$ and $z_2(t) = r(t)\sin[\theta(t)]$ transforms the system $\mathbf{z}' = A\mathbf{z} + \mathbf{g}(\mathbf{z})$ into

$$r' = r[a_{11}\cos^2\theta + a_{22}\sin^2\theta + (a_{12} + a_{21})\sin\theta\cos\theta] + [g_1\cos\theta + g_2\sin\theta]$$

$$\theta' = [a_{21}\cos^2\theta - a_{12}\sin^2\theta + (a_{22} - a_{11})\sin\theta\cos\theta] + r^{-1}[-g_1\sin\theta + g_2\cos\theta].$$

26. Use the polar equations derived in Exercise 25 to show that if

$$a_{11} = a_{22}, \quad a_{21} = -a_{12}, \quad g_1(\mathbf{z}) = z_1 h\left(\sqrt{z_1^2 + z_2^2}\right), \quad g_2(\mathbf{z}) = z_2 h\left(\sqrt{z_1^2 + z_2^2}\right)$$

for some function h, then the polar equations uncouple into

$$r' = a_{11}r + rh(r)$$

$$\theta' = a_{21}.$$

Note that the radial equation is a separable differential equation and the angle equation can be solved by antidifferentiation.

27. Consider the system $x' = y + \alpha x(x^2 + y^2), y' = -x + \alpha y(x^2 + y^2)$. Introduce polar coordinates and use the results of Exercises 25 and 26 to derive differential equations for $r(t)$ and $\theta(t)$. Solve these differential equations, and then form $x(t)$ and $y(t)$.

Exercises 28–29:

Introduce polar coordinates and transform the given initial value problem into an equivalent initial value problem for the polar variables. Solve the polar initial value problem, and use the polar solution to obtain the solution of the original initial value problem. If the solution exists at time $t = 1$, evaluate it. If not, explain why.

28. $x' = x + x\sqrt{x^2 + y^2}, \quad x(0) = 1$

   $y' = y + y\sqrt{x^2 + y^2}, \quad y(0) = \sqrt{3}$

29. $x' = y - x\ln[x^2 + y^2], \quad x(0) = e/\sqrt{2}$

   $y' = -x - y\ln[x^2 + y^2], \quad y(0) = e/\sqrt{2}$

6.6 Two-Dimensional Linear Systems

We continue studying the phase-plane behavior of solutions of the linear system $\mathbf{y}' = A\mathbf{y}$, where A is a (2×2) real invertible matrix. Since A is invertible, $\mathbf{y} = \mathbf{0}$ is the only equilibrium solution of $\mathbf{y}' = A\mathbf{y}$.

As we have seen, there are two principal reasons for studying this phase-plane behavior. First, $\mathbf{y}' = A\mathbf{y}$ is an important and intrinsically interesting system. Second, such systems arise whenever we linearize about an equilibrium point, zooming in to study the behavior of a nonlinear autonomous system close to an equilibrium point.

By studying the eigenvalues and the phase-plane geometry of the associated eigenvectors at an equilibrium point, we can often sketch a good local picture—one that gives a qualitative description of the nonlinear system behavior near the equilibrium point. Such local pictures complement the large-scale overview provided by the direction field. Taken together, they provide a good overall view of system behavior.

To illustrate the ideas, we consider the competing species problem that has served as a vehicle for discussion throughout this chapter.

EXAMPLE 1

We use linearized system approximations to develop local pictures of system behavior near each of the equilibrium points of the nonlinear system

$$x' = \tfrac{1}{2}\left(1 - \tfrac{1}{2}x - \tfrac{1}{2}y\right)x$$
$$y' = \tfrac{1}{4}\left(1 - \tfrac{1}{3}x - \tfrac{2}{3}y\right)y. \tag{1}$$

System (1) has four equilibrium points, $(0, 0)$, $(0, \tfrac{3}{2})$, $(2, 0)$, and $(1, 1)$. We focus on the equilibrium point $(0, \tfrac{3}{2})$ in order to illustrate the basic ideas. Using

$$\mathbf{z}(t) = \begin{bmatrix} x(t) \\ y(t) - \tfrac{3}{2} \end{bmatrix},$$

we have for the linearized system at $(0, \tfrac{3}{2})$

$$\mathbf{z}' = \begin{bmatrix} \tfrac{1}{8} & 0 \\ -\tfrac{1}{8} & -\tfrac{1}{4} \end{bmatrix} \mathbf{z}.$$

The general solution of this linear system is

$$\mathbf{z}(t) = c_1 e^{t/8} \begin{bmatrix} 3 \\ -1 \end{bmatrix} + c_2 e^{-t/4} \begin{bmatrix} 0 \\ 1 \end{bmatrix}. \tag{2}$$

We can use the eigenpair information in (2) to sketch the qualitative behavior of solution trajectories of $\mathbf{z}' = A\mathbf{z}$. In turn, these sketches provide qualitative information about solutions of the original nonlinear system near the equilibrium point $(0, \tfrac{3}{2})$.

To begin, consider the special case where $c_2 = 0$ and $c_1 \neq 0$. In this case,

$$z_1(t) = 3c_1 e^{t/8} \quad \text{and} \quad z_2(t) = -c_1 e^{t/8}. \tag{3}$$

From (3), these solutions lie on the **z**-plane line

$$z_2 = -\tfrac{1}{3}z_1.$$

Since $e^{t/8}$ increases as t increases, we also see from (3) that both $z_1(t)$ and $z_2(t)$ increase in magnitude as t increases. Therefore, solution points originating on this line remain on this line and move away from the origin as t increases, as shown in the direction field plot in Figure 6.15(a), on the next page. Similarly, consider the companion case where $c_1 = 0$ and $c_2 \neq 0$ in equation (2). In this case, solution points lie on the phase-plane line $z_1 = 0$ and approach the origin as t increases, since $e^{-t/4}$ decreases as t increases [see Figure 6.15(a)].

Solution point behavior in these two special cases enables us to determine qualitatively the general phase-plane characteristics of (2). Consider the general case where $c_1 \neq 0$, $c_2 \neq 0$. For sufficiently small values of t, both exponential functions are roughly comparable in size and both terms in the general solution influence solution point behavior. However, as t increases, the term

$$c_1 e^{t/8} \begin{bmatrix} 3 \\ -1 \end{bmatrix}$$

becomes increasingly dominant and all solution trajectories approach the phase-plane line $z_2 = -\tfrac{1}{3}z_1$ as an asymptote. Therefore, we obtain the phase-plane behavior shown in Figure 6.15(a).

A similar analysis can be used to study behavior near the other equilibrium points, $(0, 0)$, $(2, 0)$, and $(1, 1)$. In all cases, the eigenvalues of the linearized system coefficient matrix are real and distinct. The eigenvectors determine lines through the **z**-plane origin on which solution points travel either toward the origin if the corresponding eigenvalue is negative or away from the origin if the eigenvalue is positive. Using this behavior as a guide, we can sketch in the qualitative behavior of solution points originating elsewhere in the plane. This qualitative behavior is shown in Figures 6.15(b)–(d). ❖

The four **z**-plane-phase portraits in Figure 6.15, when positioned at the corresponding **y**-plane equilibrium points, provide local pictures that are complementary to and consistent with the large-scale overview developed in Section 6.2. This is illustrated in Figure 6.16, where the local equilibrium pictures from Figure 6.15 have been superimposed on Figure 6.6 from Section 6.2. Attention is restricted to the first quadrant, since the dependent variable $\mathbf{y}(t)$ has components that represent (nonnegative) populations.

Classifying Equilibrium Points

In Example 1, the coefficient matrix of the linearized system at each of the four equilibrium points had real, distinct eigenvalues. In two cases, the eigenvalues had the same sign; in the other two, the eigenvalues had opposite signs. If A is an invertible (2×2) matrix, other possibilities exist for the (nonzero) pair of eigenvalues. The eigenvalues might be real and repeated. They might be a complex conjugate pair with nonzero real parts, or they might be a purely imaginary complex conjugate pair. Table 6.2 enumerates the various possibilities and the names assigned to them.

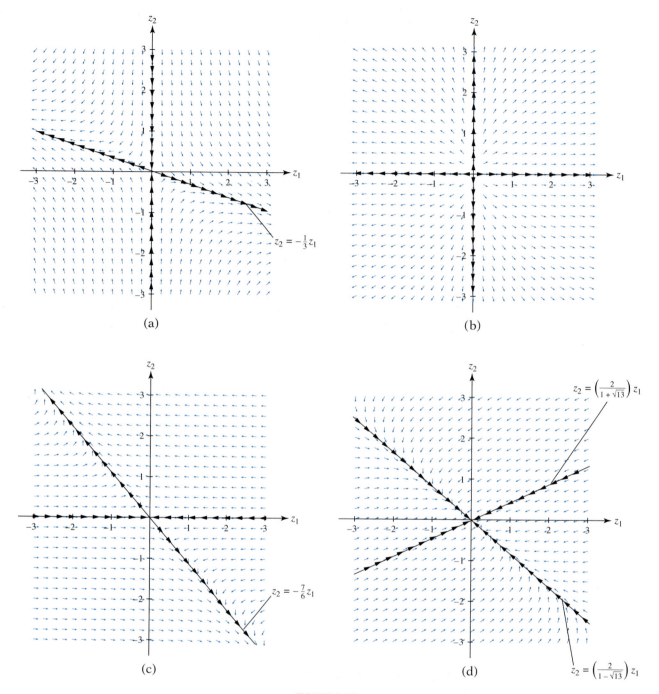

FIGURE 6.15

Direction fields for the various linearizations $\mathbf{z}' = A\mathbf{z}$ associated with the nonlinear system (2). Each $\mathbf{z}$-plane direction field corresponds to an equilibrium point of the nonlinear system. The equilibrium points are (a) $(0, \frac{3}{2})$, (b) $(0, 0)$, (c) $(2, 0)$, and (d) $(1, 1)$.

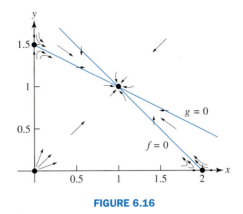

FIGURE 6.16

The local equilibrium pictures from Figure 6.15, superimposed on the qualitative picture developed in Section 6.2 (see Figure 6.6 of Section 6.2).

TABLE 6.2

Classification of the Equilibrium Point at the Origin for $\mathbf{y}' = A\mathbf{y}$

Eigenvalues of A	Type of Equilibrium Point	Stability Characteristics of the Linear System
Real eigenvalues λ_1, λ_2 where		
$\quad \lambda_1 \leq \lambda_2 < 0$	Node	Asymptotically stable
$\quad 0 < \lambda_1 \leq \lambda_2$	Node	Unstable
Real eigenvalues λ_1, λ_2 where		
$\quad \lambda_1 < 0 < \lambda_2$	Saddle point	Unstable
Complex eigenvalues where		
$\quad \lambda_{1,2} = \alpha \pm i\beta, \quad \alpha < 0$	Spiral point	Asymptotically stable
$\quad \lambda_{1,2} = \alpha \pm i\beta, \quad \alpha > 0$	Spiral point	Unstable
Complex eigenvalues where		
$\quad \lambda_{1,2} = \pm i\beta, \quad \beta \neq 0$	Center	Stable but not asymptotically stable

The "node" designation is often divided into two subcategories. If matrix A has two equal (real) eigenvalues and is a scalar multiple of the (2×2) identity matrix, the equilibrium point is called a **proper node**. In all other cases (when A has equal real eigenvalues but only one linearly independent eigenvector or when A has unequal real eigenvalues of the same sign), the equilibrium point is called an **improper node**.

Figures 6.15 and 6.17–6.19 provide some examples of phase-plane behavior at nodes and saddle points. The following three examples illustrate typical behavior at a proper node, a spiral point, and a center.

Proper Node

Consider the linear system

$$\mathbf{y}' = \begin{bmatrix} \alpha & 0 \\ 0 & \alpha \end{bmatrix} \mathbf{y}, \qquad \alpha \neq 0.$$

The origin is a proper node since the eigenvalues are real and equal ($\lambda_1 = \lambda_2 = \alpha$), and the coefficient matrix is a nonzero multiple of the identity matrix. The phase-plane behavior of trajectories is easily recognized if we adopt polar coordinates. Let

$$x = r \cos\theta \quad \text{and} \quad y = r \sin\theta.$$

With this change of variables, the component equations transform into the differential equations for the polar variables,

$$r' = \alpha r$$
$$\theta' = 0.$$

As time increases, the solution points move on rays, since the polar angle θ remains constant. If $\alpha < 0$, solutions approach the origin and the origin is an asymptotically stable equilibrium point. If $\alpha > 0$, solution points move outward along the rays and the origin is an unstable equilibrium point. Note that the rays (the trajectories themselves) are independent of the value of α. The parameter α governs only how quickly solution points move inward or outward along the rays. Figure 6.17 depicts behavior for the case $\alpha > 0$.

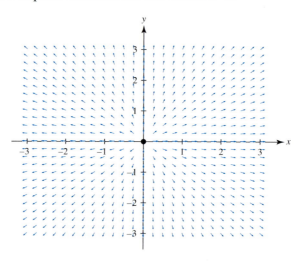

FIGURE 6.17

The origin is a proper node for the system in Example 2. Since $\alpha > 0$ in this example, the origin is an unstable equilibrium point.   ❖

Spiral Point

Consider the linear system

$$\mathbf{y}' = \begin{bmatrix} -1 & -1 \\ 1 & -1 \end{bmatrix} \mathbf{y}.$$

The eigenvalues of the coefficient matrix are the complex conjugate pair $\lambda_1 = -1 + i$, $\lambda_2 = -1 - i$. According to Table 6.2, the origin is an asymptotically stable spiral point. The behavior of solutions, as well as the reason for the terminology "spiral point," can be clearly seen when we change to polar coordinates. For this system, we obtain the following pair of differential equations for the polar variables:

$$r' = -r$$
$$\theta' = 1.$$

Let the initial conditions be $r(0) = r_0$ and $\theta(0) = \theta_0$. Then the solutions are

$$r(t) = r_0 e^{-t}, \qquad \theta(t) = t + \theta_0.$$

Thus, as time increases, a solution point spirals inward toward the origin. Its distance from the origin decreases at an exponential rate while it moves counterclockwise about the origin. This behavior is shown in Figure 6.18.

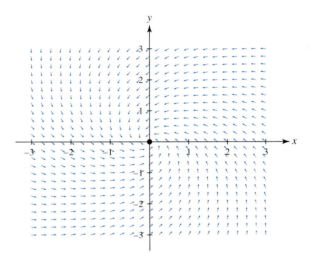

FIGURE 6.18

The origin is a spiral point for the system in Example 3. Since the eigenvalues of A have real part -1, the origin is asymptotically stable. A solution point follows a trajectory that spirals in toward the origin.   ❖

EXAMPLE
4

Center

Consider the linear system

$$\mathbf{y}' = \begin{bmatrix} -4 & 5 \\ -5 & 4 \end{bmatrix} \mathbf{y}.$$

The eigenvalues of the coefficient matrix are the purely imaginary complex conjugate pair $\lambda_1 = 3i$, $\lambda_2 = -3i$. According to Table 6.2, the origin is classified as a stable center. Phase-plane behavior is shown in Figure 6.19.

One way to derive the equations for the elliptical trajectories in Figure 6.19 is to change to polar coordinates. For this linear system, the differential equations for the polar variables are

$$r' = -4r \cos 2\theta$$
$$\theta' = -5 + 4 \sin 2\theta.$$

(continued)

(continued)

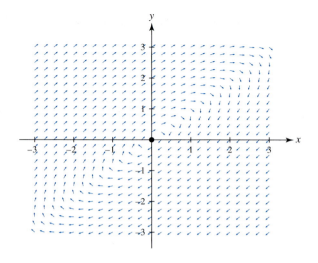

FIGURE 6.19

The origin is a center for the system in Example 4 since the eigenvalues of A are purely imaginary. The origin is a stable equilibrium point, but it is not asymptotically stable. The solution points follow elliptical trajectories about the origin.

Notice that θ is a decreasing function of t. Therefore, an inverse function exists, and we can view r as a function of θ. Using the chain rule, we have

$$\frac{dr}{d\theta} = \frac{dr}{dt}\frac{dt}{d\theta} = -4r\cos 2\theta \frac{1}{-5 + 4\sin 2\theta},$$

or

$$\frac{dr}{d\theta} = \frac{4\cos 2\theta}{5 - 4\sin 2\theta}r.$$

This equation is a first order linear differential equation. Assuming an initial condition of $r = r_0$ when $\theta = 0$, we find the solution

$$r = \frac{r_0}{\sqrt{1 - 0.8\sin 2\theta}}.$$

Note that r is a periodic function of θ, with period π. Since θ is a decreasing function of t, the solution points (r, θ) move clockwise around the closed elliptical trajectories, as shown in Figure 6.19. ❖

An alternative derivation of the trajectory equations is outlined in Exercise 31. This approach leads to equations in terms of the original x, y phase-plane variables.

EXERCISES

Exercises 1–5:

In each exercise, the eigenpairs of a (2×2) matrix A are given where both eigenvalues are real. Consider the phase-plane solution trajectories of the linear system $\mathbf{y}' = A\mathbf{y}$, where

$$\mathbf{y}(t) = \begin{bmatrix} x(t) \\ y(t) \end{bmatrix}.$$

(a) Use Table 6.2 to classify the type and stability characteristics of the equilibrium point at $\mathbf{y} = \mathbf{0}$.

(b) Sketch the two phase-plane lines defined by the eigenvectors. If an eigenvector is $\begin{bmatrix} u_1 \\ u_2 \end{bmatrix}$, the line of interest is $u_2 x - u_1 y = 0$. Solution trajectories originating on such a line stay on the line; they move toward the origin as time increases if the corresponding eigenvalue is negative or away from the origin if the eigenvalue is positive.

(c) Sketch appropriate direction field arrows on both lines. Use this information to sketch a representative trajectory in each of the four phase-plane regions having these lines as boundaries. Indicate the direction of motion of the solution point on each trajectory.

1. $\lambda_1 = 2$, $\mathbf{x}_1 = \begin{bmatrix} 1 \\ 1 \end{bmatrix}$; $\quad \lambda_2 = -1$, $\mathbf{x}_2 = \begin{bmatrix} 1 \\ -1 \end{bmatrix}$

2. $\lambda_1 = 1$, $\mathbf{x}_1 = \begin{bmatrix} 1 \\ 2 \end{bmatrix}$; $\quad \lambda_2 = 2$, $\mathbf{x}_2 = \begin{bmatrix} 2 \\ -1 \end{bmatrix}$

3. $\lambda_1 = 2$, $\mathbf{x}_1 = \begin{bmatrix} 2 \\ 0 \end{bmatrix}$; $\quad \lambda_2 = 1$, $\mathbf{x}_2 = \begin{bmatrix} 0 \\ 2 \end{bmatrix}$

4. $\lambda_1 = -2$, $\mathbf{x}_1 = \begin{bmatrix} 1 \\ 0 \end{bmatrix}$; $\quad \lambda_2 = -1$, $\mathbf{x}_2 = \begin{bmatrix} 1 \\ 1 \end{bmatrix}$

5. $\lambda_1 = 1$, $\mathbf{x}_1 = \begin{bmatrix} 1 \\ 0 \end{bmatrix}$; $\quad \lambda_2 = -1$, $\mathbf{x}_2 = \begin{bmatrix} 2 \\ 1 \end{bmatrix}$

Exercises 6–20:

In each exercise, consider the linear system $\mathbf{y}' = A\mathbf{y}$. Since A is a constant invertible (2×2) matrix, $\mathbf{y} = \mathbf{0}$ is the unique (isolated) equilibrium point.

(a) Determine the eigenvalues of the coefficient matrix A.

(b) Use Table 6.2 to classify the type and stability characteristics of the equilibrium point at the phase-plane origin. If the equilibrium point is a node, designate it as either a proper node or an improper node.

6. $\mathbf{y}' = \begin{bmatrix} 1 & -6 \\ 1 & -4 \end{bmatrix} \mathbf{y}$

7. $\mathbf{y}' = \begin{bmatrix} 6 & -10 \\ 2 & -3 \end{bmatrix} \mathbf{y}$

8. $\mathbf{y}' = \begin{bmatrix} -6 & 14 \\ -2 & 5 \end{bmatrix} \mathbf{y}$

9. $\mathbf{y}' = \begin{bmatrix} 1 & 2 \\ -5 & -1 \end{bmatrix} \mathbf{y}$

10. $\mathbf{y}' = \begin{bmatrix} -1 & 1 \\ -1 & -1 \end{bmatrix} \mathbf{y}$

11. $\mathbf{y}' = \begin{bmatrix} 1 & -6 \\ 2 & -6 \end{bmatrix} \mathbf{y}$

12. $\mathbf{y}' = \begin{bmatrix} 2 & -3 \\ 3 & 2 \end{bmatrix} \mathbf{y}$

13. $\mathbf{y}' = \begin{bmatrix} -2 & -4 \\ 5 & 2 \end{bmatrix} \mathbf{y}$

14. $\mathbf{y}' = \begin{bmatrix} 7 & -24 \\ 2 & -7 \end{bmatrix} \mathbf{y}$

15. $\mathbf{y}' = \begin{bmatrix} -1 & 8 \\ -1 & 5 \end{bmatrix} \mathbf{y}$

16. $\mathbf{y}' = \begin{bmatrix} -2 & 1 \\ -1 & -2 \end{bmatrix} \mathbf{y}$

17. $\mathbf{y}' = \begin{bmatrix} 2 & 4 \\ -4 & -6 \end{bmatrix} \mathbf{y}$

18. $\mathbf{y}' = \begin{bmatrix} 3 & 0 \\ 0 & 3 \end{bmatrix} \mathbf{y}$

19. $\mathbf{y}' = \begin{bmatrix} 1 & 2 \\ -8 & 1 \end{bmatrix} \mathbf{y}$

20. $\mathbf{y}' = \begin{bmatrix} -1 & -2 \\ 2 & 3 \end{bmatrix} \mathbf{y}$

21. Consider the linear system $\mathbf{y}' = A\mathbf{y}$. Four direction fields are shown. Determine which of the four coefficient matrices listed corresponds to each of the direction fields shown.

(a) $A_1 = \begin{bmatrix} -2 & 1 \\ 1 & -2 \end{bmatrix}$ (b) $A_2 = \begin{bmatrix} 1 & 2 \\ -2 & -1 \end{bmatrix}$

(c) $A_3 = \begin{bmatrix} 2 & 1 \\ -1 & -2 \end{bmatrix}$ (d) $A_4 = \begin{bmatrix} 1 & 2 \\ -2 & 1 \end{bmatrix}$

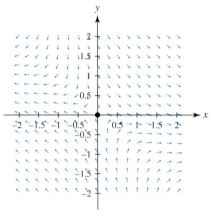

Direction Field 1

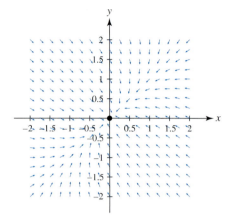

Direction Field 2

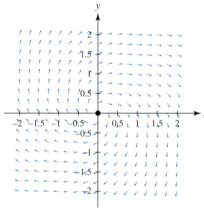

Direction Field 3

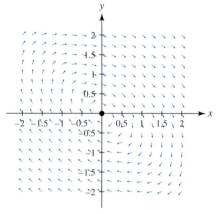

Direction Field 4

Figure for Exercise 21

Exercises 22–25:

Use the information given about the nature of the equilibrium point at the origin to determine the value or range of permissible values for the unspecified entry in the coefficient matrix.

22. The origin is a center for the linear system $\mathbf{y}' = \begin{bmatrix} 2 & 3 \\ -3 & \alpha \end{bmatrix} \mathbf{y}$; determine α.

23. Given $\mathbf{y}' = \begin{bmatrix} -4 & \alpha \\ -2 & 2 \end{bmatrix} \mathbf{y}$, for what values of α (if any) can the origin be an asymptotically stable spiral point?

24. The origin is an asymptotically stable proper node of $\mathbf{y}' = \begin{bmatrix} -2 & 0 \\ \alpha & -2 \end{bmatrix} \mathbf{y}$; determine the value(s) of α.

25. Given $\mathbf{y}' = \begin{bmatrix} 4 & -2 \\ \alpha & -4 \end{bmatrix} \mathbf{y}$, for what values of α (if any) can the origin be an (unstable) saddle point?

Exercises 26–29:

Locate the equilibrium point of the given nonhomogeneous linear system $\mathbf{y}' = A\mathbf{y} + \mathbf{g}_0$. [Hint: Introduce the change of dependent variable $\mathbf{z}(t) = \mathbf{y}(t) - \mathbf{y}_0$, where $\mathbf{y}_0$ is chosen so that the equation can be rewritten as $\mathbf{z}' = A\mathbf{z}$.] Use Table 6.2 to classify the type and stability characteristics of the equilibrium point.

26. $\mathbf{y}' = \begin{bmatrix} 1 & 4 \\ -1 & 1 \end{bmatrix} \mathbf{y} + \begin{bmatrix} 3 \\ 2 \end{bmatrix}$

27. $\mathbf{y}' = \begin{bmatrix} 6 & 5 \\ -7 & -6 \end{bmatrix} \mathbf{y} + \begin{bmatrix} 4 \\ -6 \end{bmatrix}$

28. $x' = 5x - 14y + 2$
$y' = 3x - 8y + 1$

29. $x' = -x + 2$
$y' = 2y - 4$

30. Let

$$A = \begin{bmatrix} a_{11} & a_{12} \\ a_{21} & a_{22} \end{bmatrix}$$

be a real invertible matrix, and consider the system $\mathbf{y}' = A\mathbf{y}$.

(a) What conditions must the matrix entries a_{ij} satisfy to make the equilibrium point $\mathbf{y}_e = \mathbf{0}$ a center?

(b) Assume that the equilibrium point at the origin is a center. Show that the system $\mathbf{y}' = A\mathbf{y}$ is a Hamiltonian system.

(c) Is the converse of the statement in part (b) true? In other words, if the system $\mathbf{y}' = A\mathbf{y}$ is a Hamiltonian system, does it necessarily follow that $\mathbf{y}_e = \mathbf{0}$ is a center? Explain.

31. Consider the linear system of Example 4,

$$\mathbf{y}' = \begin{bmatrix} -4 & 5 \\ -5 & 4 \end{bmatrix} \mathbf{y}.$$

The coefficient matrix has eigenvalues $\lambda_1 = 3i$, $\lambda_2 = -3i$; the equilibrium point at the origin is a center.

(a) Show that the linear system is a Hamiltonian system. Either use the results of Exercise 30 or apply the criterion directly to this example.

(b) Derive the conservation law for this system. The result, $\frac{5}{2}x^2 - 4xy + \frac{5}{2}y^2 = C > 0$, defines a family of ellipses. These ellipses are the trajectories on which the solution point moves as time changes.

(c) Plot the ellipses found in part (b) for $C = \frac{1}{4}, \frac{1}{2}$, and 1. Indicate the direction in which the solution point moves on these ellipses.

Exercises 32–34:

A linear system is given in each exercise.

(a) Determine the eigenvalues of the coefficient matrix A.

(b) Use Table 6.2 to classify the type and stability characteristics of the equilibrium point at $\mathbf{y} = \mathbf{0}$.

(c) The given linear system is a Hamiltonian system. Derive the conservation law for this system.

32. $\mathbf{y}' = \begin{bmatrix} -2 & 1 \\ 5 & 2 \end{bmatrix} \mathbf{y}$

33. $x' = x + 3y$
$y' = -3x - y$

34. $\mathbf{y}' = \begin{bmatrix} 2 & 1 \\ 0 & -2 \end{bmatrix} \mathbf{y}$

6.7 Predator-Prey Population Models

By way of introduction, we pose a question. This question involves the familiar problem of a predatory population being introduced into a colony, either by accident or by design. If the predator is undesirable, our goal may be to remove it from the colony. If the predator is desirable, our goal may be to establish a coexisting ecological balance between the predators and their prey. Which scenarios lead to predator eradication, and which scenarios lead to predator-prey coexistence? And what are the factors responsible for the desired outcome?

Mathematical Modeling

We develop a mathematical model of two-species predator-prey interaction to gain insight into the question posed above.[4] Let $P_1(t)$ and $P_2(t)$ represent the populations of predators and prey (respectively) in a colony at time t. Both populations change with time because of births, deaths, and harvesting. A "conservation of population" principle of the type discussed in Section 2.4 leads to differential equations having the following general structure:

$$\frac{dP_1}{dt} = R_1 P_1 - \mu_1 P_1$$
$$\frac{dP_2}{dt} = R_2 P_2 - \mu_2 P_2. \tag{1}$$

In equation (1), the term R_j represents the net birth rate per unit population of the jth species. We assume that R_1 and R_2 are functions of the populations P_1 and P_2 (but not explicit functions of time t). The nonnegative constants μ_1 and μ_2 represent harvesting rates per unit population. Applying the ideas underlying the logistic population model developed in Section 2.8, we assume the rate functions R_1 and R_2 have the form

$$R_1 = r_1(-1 - \alpha_1 P_1 + \beta_1 P_2)$$
$$R_2 = r_2(1 - \beta_2 P_1 - \alpha_2 P_2), \tag{2}$$

where the parameters r_j, α_j, and β_j are nonnegative constants, $j = 1, 2$.

What are we actually assuming in (2)? In the absence of prey for food (that is, if $\beta_1 = 0$), the predator rate function would be $R_1 = -r_1(1 + \alpha_1 P_1) < 0$ and the predator population would continually decrease. The $\beta_1 P_2$ term embodies the beneficial aspects of the prey food supply on the predator growth rate. The $-\alpha_1 P_1$ term allows for competition among the predators for the available food.

Consider now the prey rate function, R_2. In the absence of predators and limitations on the prey's food supply (that is, if $\beta_2 = 0$ and $\alpha_2 = 0$), the rate function would be $R_2 = r_2 > 0$. In that case, the prey population would grow exponentially whenever $r_2 > \mu_2$. The terms $-\beta_2 P_1$ and $-\alpha_2 P_2$ account for the

---

[4]Important early work in developing and applying such models was done by Volterra. Vito Volterra (1860–1940) was an Italian mathematician and scientist noted for his work on functional calculus, partial differential equations, integral equations and mathematical biology. During his career, he held distinguished positions at the universities of Pisa, Turin, and Rome. In 1931, he was forced to leave the University of Rome after refusing to take an oath of allegiance to the Fascist government. He left Italy the following year and spent the rest of his life abroad.

negative effects of predation and limits on the prey's food supply, respectively. The predator-prey equations we use as our basic model are therefore

$$\frac{dP_1}{dt} = r_1(-1 - \alpha_1 P_1 + \beta_1 P_2)P_1 - \mu_1 P_1$$

$$\frac{dP_2}{dt} = r_2(1 - \beta_2 P_1 - \alpha_2 P_2)P_2 - \mu_2 P_2.$$

(3)

Managing a Predator Population

We now take up the question posed at the beginning of this section: How do we manage a predator population that has been introduced into a colony, whether by accident or by design?

We assume the colony has resource limitations that exert a constraining influence on each of the predator and prey populations. We allow for the possibility of harvesting the predator population but not the prey population. Under these assumptions, system (3) becomes

$$\frac{dP_1}{dt} = r_1 \left(-1 - \frac{\mu}{r_1} - \alpha_1 P_1 + \beta_1 P_2 \right) P_1$$

$$\frac{dP_2}{dt} = r_2 \left(1 - \beta_2 P_1 - \alpha_2 P_2 \right) P_2,$$

(4)

where all the constants on the right-hand side of (4) are assumed positive with the possible exception of the harvesting rate μ, which we may allow to be zero.

From an ecological point of view, we want to know what combination of harvesting strategies and environmental factors will cause the predator population to

(a) die out or

(b) achieve a coexisting balance with the prey population as time evolves.

Rephrasing in mathematical terms, we want to discover which relations among the constants in (4) will cause all solutions to

(a) converge to an equilibrium value $(0, P_{2e})$, where $P_{2e} > 0$, or

(b) converge to an equilibrium value (P_{1e}, P_{2e}), where $P_{1e} > 0, P_{2e} > 0$.

Autonomous system (4) has at most three equilibrium points in the first quadrant of the phase plane:

$$\mathbf{P}_e^{(1)} = \begin{bmatrix} 0 \\ 0 \end{bmatrix}, \qquad \mathbf{P}_e^{(2)} = \begin{bmatrix} 0 \\ \dfrac{1}{\alpha_2} \end{bmatrix}, \qquad \mathbf{P}_e^{(3)} = \begin{bmatrix} \dfrac{-\alpha_2 \left(1 + \dfrac{\mu}{r_1} \right) + \beta_1}{\alpha_1 \alpha_2 + \beta_1 \beta_2} \\[4ex] \dfrac{\beta_2 \left(1 + \dfrac{\mu}{r_1} \right) + \alpha_1}{\alpha_1 \alpha_2 + \beta_1 \beta_2} \end{bmatrix}.$$

The equilibrium point $\mathbf{P}_e^{(1)}$, where $P_{1e} = P_{2e} = 0$, corresponds to neither species being present. The equilibrium point $\mathbf{P}_e^{(2)}$, where

$$P_{1e} = 0, \qquad P_{2e} = \frac{1}{\alpha_2}, \tag{5}$$

corresponds to a complete absence of predators. The third equilibrium point, $\mathbf{P}_e^{(3)}$, is found by solving the system of equations

$$-1 - \frac{\mu}{r_1} - \alpha_1 P_{1e} + \beta_1 P_{2e} = 0$$

$$1 - \beta_2 P_{1e} - \alpha_2 P_{2e} = 0.$$

This equilibrium point is given by

$$P_{1e} = \frac{-\alpha_2 \left(1 + \dfrac{\mu}{r_1}\right) + \beta_1}{\alpha_1 \alpha_2 + \beta_1 \beta_2}, \qquad P_{2e} = \frac{\beta_2 \left(1 + \dfrac{\mu}{r_1}\right) + \alpha_1}{\alpha_1 \alpha_2 + \beta_1 \beta_2}. \tag{6}$$

By definition, the two species coexist when both P_{1e} and P_{2e} are positive. Thus, model (4) predicts that both species can coexist in equilibrium only if

$$\beta_1 > \alpha_2 \left(1 + \frac{\mu}{r_1}\right). \tag{7}$$

If inequality (7) does not hold, then the two species cannot coexist in equilibrium and the only nontrivial equilibrium solution in the first quadrant is (5), wherein predators are absent.

Suppose our goal is to eradicate the predators. We see from (7) that we can eliminate the possibility of equilibrium predator-prey coexistence by sufficiently increasing $\alpha_2[1 + (\mu/r_1)]$ and/or by decreasing β_1. Does this make sense? Increasing μ corresponds to increasing the harvesting rate of predators, while decreasing β_1 corresponds to somehow reducing the beneficial effects of the prey as food for the predators. It seems reasonable that either of these two strategies would be harmful to the predators.

What about increasing α_2, however? Recall that α_2 is the parameter modeling the constraining effects of the available colony resources on the prey population. In the absence of predators, the equilibrium prey population $P_{2e} = 1/\alpha_2$ decreases as α_2 increases. Is it reasonable to conclude that we can adversely impact the predator population by indirectly constraining its food supply?

We want to focus on the role of the parameter α_2 in controlling the population. Toward that end, the parameters $r_1, r_2, \alpha_1, \beta_1, \beta_2$, and μ will all be assigned the value 1, leading to the system

$$\frac{dP_1}{dt} = (-2 - P_1 + P_2)P_1$$

$$\frac{dP_2}{dt} = (1 - P_1 - \alpha_2 P_2)P_2. \tag{8}$$

Table 6.3 summarizes the information we can deduce from linearizing system (8). In either case (whether equilibrium coexistence is possible or impossible), the origin is an unstable saddle point. If equilibrium coexistence is possible, then the equilibrium point $(0, 1/\alpha_2)$ is an unstable saddle point. If coexistence is

TABLE 6.3

Equilibrium Point	Linearized System $\mathbf{z}' = A\mathbf{z}$ $[\mathbf{z}(t) = \mathbf{P}(t) - \mathbf{P}_e]$	Eigenvalues	Stability Properties of System (4)
$P_{1e} = 0$ $P_{2e} = 0$	$A = \begin{bmatrix} -2 & 0 \\ 0 & 1 \end{bmatrix}$	$\lambda_1 = -2$ $\lambda_2 = 1$	Unstable
$P_{1e} = 0$ $P_{2e} = 1/\alpha_2$	$A = \begin{bmatrix} -2 + 1/\alpha_2 & 0 \\ -1/\alpha_2 & -1 \end{bmatrix}$	$\lambda_1 = -2 + 1/\alpha_2$ $\lambda_2 = -1$	Unstable if $\alpha_2 < \frac{1}{2}$, asymptotically stable if $\alpha_2 > \frac{1}{2}$
$P_{1e} = \dfrac{1 - 2\alpha_2}{\alpha_2 + 1}$ $P_{2e} = \dfrac{3}{\alpha_2 + 1}$	$A = \begin{bmatrix} -\dfrac{1 - 2\alpha_2}{\alpha_2 + 1} & \dfrac{1 - 2\alpha_2}{\alpha_2 + 1} \\ -\dfrac{3}{\alpha_2 + 1} & -\dfrac{3\alpha_2}{\alpha_2 + 1} \end{bmatrix}$	$\lambda_{1,2} = -\dfrac{1}{2} \pm \dfrac{1}{2}\sqrt{1 - \dfrac{12(1 - \alpha_2 - 2\alpha_2^2)}{(1 + \alpha_2)^2}}$	Asymptotically stable if $\alpha_2 < \frac{1}{2}$

impossible, however, this equilibrium point becomes an asymptotically stable node.

The third equilibrium point in Table 6.3, corresponding to the coexistence of predators and prey, requires [see equation (7)] that the numerator of P_{1e} be positive; that is,

$$\alpha_2 < \tfrac{1}{2}.$$

When $\alpha_2 < \frac{1}{2}$, this third equilibrium point is either an asymptotically stable spiral point or a node.

The three phase-plane plots in Figure 6.20 correspond to different values of α_2 in system (8). In Figure 6.20(a), $\alpha_2 = 0$. In this case, no resource limitations constrain prey growth, and predator-prey coexistence is possible. All solution trajectories having both species initially present spiral in toward the asymptotically stable equilibrium point $(1, 3)$. In Figure 6.20(b), α_2 has been increased to $\alpha_2 = \frac{9}{20}$. In this case, the equilibrium point of the linearized system is a stable node. Here again, all solution trajectories of the nonlinear system having both species initially present approach the asymptotically stable equilibrium point at $(\frac{2}{29}, \frac{60}{29})$. Lastly [see Figure 6.20(c)], when $\alpha_2 = 1$, coexistence is not possible. All solution trajectories having both species initially present approach the asymptotically stable equilibrium point $(0, 1)$ as $t \to \infty$. In this last case, the predator population tends toward extinction as time progresses. These direction field observations support our previous conclusions.

On one hand, the parametric study illustrated in Figure 6.20 indicates that our interpretation of the model's behavior seems correct. On the other hand, a model such as (4) is at best a gross simplification of reality. The trade-off in

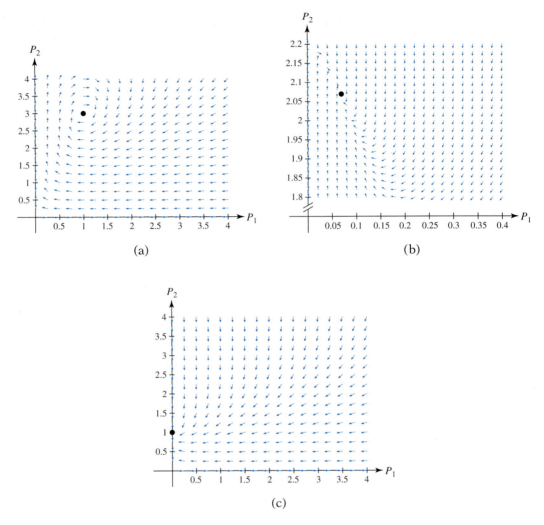

FIGURE 6.20

Portions of the direction field for the predator-prey equations (8), with different choices of α_2: (a) $\alpha_2 = 0$, (b) $\alpha_2 = \frac{9}{20}$, (c) $\alpha_2 = 1$.

modeling is always one of reducing a problem to its essential features without "throwing away" reality. In particular, when model predictions seem counterintuitive, we need to proceed with a healthy skepticism—both scrutinizing and refining the model to gain further confidence and insight.

EXERCISES

Exercises 1–4:

Assume the given autonomous system models the population dynamics of two species, x and y, within a colony.

(a) For each of the two species, answer the following questions.

(i) In the absence of the other species, does the remaining population continuously grow, decline toward extinction, or approach a nonzero equilibrium value as time evolves?

(ii) Is the presence of the other species in the colony beneficial, harmful, or a matter of indifference?

(b) Determine all equilibrium points lying in the first quadrant of the phase plane (including any lying on the coordinate axes).

(c) The given system is an almost linear system at the equilibrium point $(x, y) = (0, 0)$. Determine the stability properties of the system at $(0, 0)$.

1. $x' = x - x^2 - xy$

$y' = y - 3y^2 - \frac{1}{2}xy$

2. $x' = -x - x^2$

$y' = -y + xy$

3. $x' = x - x^2 - xy$

$y' = -y - y^2 + xy$

4. $x' = x - x^2 + xy$

$y' = y - y^2 + xy$

5. A scientist adopted the following mathematical model for a colony containing two species, x and y:

$$x' = r_1(1 + \alpha_1 x + \beta_1 y)x$$
$$y' = r_2(1 + \beta_2 x + \alpha_2 y)y.$$

The following information is known:

(i) If only species x is present in the colony, any initial amount will vary with time as shown in graph (a). If only species y is present, any initial amount will vary as shown in graph (b).

(ii) If both species are initially present, $(x_e, y_e) = (2, 3)$ is an equilibrium point.

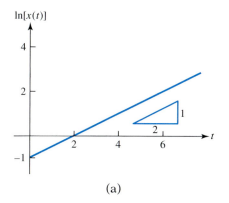

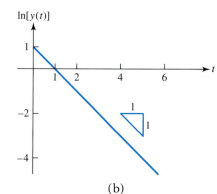

(a)

(b)

Figure for Exercise 5

(a) Determine the six constants r_1, α_1, β_1, r_2, α_2, and β_2.

(b) How do the two populations relate to each other? Is population x beneficial, harmful, or indifferent to population y? Is population y beneficial, harmful, or indifferent to population x?

Exercises 6–7:

Two Competing Species   These exercises explore the question "When one of two species in a colony is desirable and the other is undesirable, is it better to use resources to nurture the growth of the desirable species or to harvest the undesirable one?"

Let $x(t)$ and $y(t)$ represent the populations of two competing species, with $x(t)$ the desirable species. Assume that if resources are invested in promoting the growth of the desirable species, the population dynamics are given by

$$x' = r(1 - \alpha x - \beta y)x + \mu x$$
$$y' = r(1 - \alpha y - \beta x)y. \tag{9}$$

If resources are invested in harvesting the undesirable species, the dynamics are

$$x' = r(1 - \alpha x - \beta y)x$$
$$y' = r(1 - \alpha y - \beta x)y - \mu y. \tag{10}$$

In (10), r, α, β, and μ are positive constants. For simplicity, we assume the same parameter values for both species. For definiteness, assume that $\alpha > \beta > 0$.

6. Consider system (9), which describes the strategy in which resources are invested into nurturing the desirable species.

   (a) Determine the four equilibrium points for the system.

   (b) Show that it is possible, by investing sufficient resources (that is, by making μ large enough), to prevent equilibrium coexistence of the two species.

   (c) Assume that μ is large enough to preclude equilibrium coexistence of the two species. Compute the linearized system at each of the three physically relevant equilibrium points. Determine the stability characteristics of the linearized system at each of these equilibrium points.

   (d) System (9) can be shown to be an almost linear system at each of the equilibrium points. Use this fact and the results of part (c) to infer the stability properties of system (9) at each of the three equilibrium points of interest.

   (e) Sketch the direction field. Will a sufficiently aggressive nurturing of species x ultimately drive undesirable species y to extinction? If so, what is the limiting population of species x?

7. Consider system (10), which describes the strategy in which resources are invested in harvesting the undesirable species. Again assume that $\alpha > \beta > 0$.

   (a) Determine the four equilibrium points for the system.

   (b) Show that it is possible, by investing sufficient resources (that is, by making μ large enough), to prevent equilibrium coexistence of the two species. In fact, if $\mu > r$, show that there are only two physically relevant equilibrium points.

   (c) Assume $\mu > r$. Compute the linearized system at each of the two physically relevant equilibrium points. Determine the stability characteristics of the linearized system at each of these equilibrium points.

   (d) System (10) can be shown to be an almost linear system at each of the equilibrium points. Use this fact and the results of part (c) to infer the stability properties of system (10) at each of the two equilibrium points of interest.

   (e) Sketch the direction field. Will sufficiently aggressive harvesting of species y ultimately drive undesirable species y to extinction? If so, what is the limiting population of species x?

8. Compare the conclusions reached in Exercises 6 and 7. Assume we have sufficient resources to implement either strategy. Which strategy will result in the larger number of desirable species x eventually being present: promoting the desirable species or harvesting the undesirable one? Could this answer have been anticipated by assuming that both strategies will lead to the eventual extinction of species y? Will comparing the resulting one-species equilibrium values for x provide the answer?

9. Three species, designated as x, y, and z, inhabit a colony. Species x and y are two mutually competitive varieties of prey, while z is a predator that depends on x and y for sustenance. In the absence of the other two species, both x and y are known to evolve toward a nonzero equilibrium value as time increases. Species z, however, decreases exponentially toward extinction when both species of prey are absent. Assume that a mathematical model having the following structure is adopted to describe the population dynamics:

$$x' = \pm a_1 x \pm b_1 x^2 \pm c_1 xy \pm d_1 xz$$
$$y' = \pm a_2 y \pm b_2 y^2 \pm c_2 xy \pm d_2 yz$$
$$z' = \pm a_3 z \pm c_3 xz \pm d_3 yz.$$

If we want the constants $a_1, b_1, c_1, a_2, \ldots, d_3$ to be nonnegative, use the information given to select the correct (plus or minus) sign in the model.

Exercises 10–11:

Infectious Disease Dynamics   Consider a colony in which an infectious disease (such as the common cold) is present. The population consists of three "species" of individuals. Let s represent the susceptibles—healthy individuals capable of contracting the illness. Let i denote the infected individuals, and let r represent those who have recovered from the illness. Assume that those who have recovered from the illness are not permanently immunized but can become susceptible again. Also assume that the rate of infection is proportional to si, the product of the susceptible and infected populations. We obtain the model

$$s' = -\alpha si + \gamma r$$
$$i' = \alpha si - \beta i \tag{11}$$
$$r' = \beta i - \gamma r,$$

where α, β, and γ are positive constants.

10. (a) Show that the system of equations (11) describes a population whose size remains constant in time. In particular, show that $s(t) + i(t) + r(t) = N$, a constant.

    (b) Modify (11) to model a situation where those who recover from the disease are permanently immunized. Is $s(t) + i(t) + r(t)$ constant in this case?

    (c) Suppose that those who recover from the disease are permanently immunized but that the disease is a serious one and some of the infected individuals perish. How does the system of equations you formulated in part (b) have to be further modified? Is $s(t) + i(t) + r(t)$ constant in this case?

11. (a) Consider system (11). Use the fact that $s(t) + i(t) + r(t) = N$ to obtain a reduced system of two differential equations for the two dependent variables $s(t)$ and $i(t)$.

    (b) For simplicity, set $\alpha = \beta = \gamma = 1$ and $N = 9$. Determine the equilibrium points of the reduced two-dimensional system.

    (c) Determine the linearized system at each of the equilibrium points found in part (b). Use Table 6.2 to analyze the stability characteristics of each of these linearized systems.

    (d) Show that the nonlinear system is an almost linear system at each of the equilibrium points found in part (b). What are the stability characteristics of the nonlinear system at these points?

Projects

Project 1: A Bobbing Sphere

Consider Figure 6.21. Assume a sphere of radius R weighs half as much as an equivalent volume of water. In its equilibrium state, the sphere floats half-submerged, as shown in Figure 6.21(a). The sphere is disturbed from equilibrium at some instant. Its position is as shown in Figure 6.21(b), with displacement $y(t)$ measured positive downward.

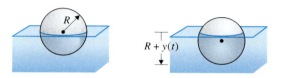

FIGURE 6.21

(a) The equilibrium state of a floating sphere whose weight is one half the weight of an equal volume of water. (b) The perturbed state of the sphere, with $y(t) > 0$ as shown.

1. Compute the volume of the submerged portion of the sphere at the instant when the displacement from equilibrium is $y(t)$. (Archimedes' law of buoyancy states that the upward force acting on the sphere is the weight of the water displaced at that instant.)

2. Apply Newton's second law of motion to obtain the differential equation governing the bobbing motion of the sphere. Considering only the weight and buoyant force, equate my'' to the net downward force (sphere weight minus upward buoyant force). Show that the resulting equation is

$$y'' + \frac{g}{2}\left(\frac{3y}{R} - \frac{y^3}{R^3}\right) = 0. \tag{1}$$

In (1), g denotes the acceleration due to gravity. For what range of values of $y(t)$ is the differential equation (1) physically relevant?

3. Equation (1) defines a conservative system. Derive the corresponding conservation law, and use it to answer the following questions. Assume $R = 0.5$ m and $g = 9.8$ m/s^2.

   (a) If the sphere is raised 10 cm above its equilibrium position and released from rest, what is the maximum vertical speed attained by the sphere in its bobbing motion?

   (b) If the sphere is set into motion with initial conditions $y(0) = 0$, $y'(0) = 1$ m/s, what is the maximum depth that the sphere center will reach as it bobs?

4. Rewrite (1) as an equivalent two-dimensional system of first order equations, where $z_1 = y$ and $z_2 = y'$. Show that the nonlinear system is an almost linear system at its only equilibrium point, $\mathbf{z}_e = \mathbf{0}$.

5. Perform a stability analysis of the linearized system in part 4 at $\mathbf{z} = \mathbf{0}$. Can we use this analysis to infer the stability properties of the corresponding nonlinear system? Explain.

Project 2: Introduction of an Infectious Disease into a Predator-Prey Ecosystem

In this chapter, we have discussed a model of a predator-prey ecosystem,

$$x' = -ax + bxy$$
$$y' = cy - dxy. \tag{2}$$

In (2), $x(t)$ and $y(t)$ represent the populations of predators and prey, respectively, at time t. The terms a, b, c, and d are positive constants. The terms bxy and $-dxy$ account for the beneficial and detrimental impacts of predation on the predator and prey populations.

We also discussed a model for the dynamics of an infectious disease within a population (see Exercise 10 of Section 6.7). In the present discussion, we will assume that infected individuals, when they have recovered, immediately become susceptible again. Therefore, we need not consider a "recovered" population as a separate entity. With this assumption, the infectious disease model considered in Section 6.7 simplifies to

$$s' = -\alpha si + \beta i$$
$$i' = \alpha si - \beta i. \tag{3}$$

In (3), $s(t)$ and $i(t)$ are the populations of susceptible and infected individuals at time t, while α and β are positive constants. In the model (3), the total population, $s(t) + i(t)$, remains constant in time.

We will combine the ideas embodied in these two problems to model a situation where an infectious disease has been introduced into a predator-prey colony. The goal is to determine the behavior of the colony.

Assume the following additional facts.

(i) The disease is benign to the prey; that is, the prey are "carriers." The relative birth rate for infected prey remains the same as that for healthy, susceptible prey.

(ii) The disease is debilitating and ultimately fatal for predators. Once a predator is infected, it can basically be assumed to be deceased. Therefore, we need only consider one population of predators—those that are susceptible.

(iii) The disease is spread among the prey by contact. We assume the rate of infection to be proportional to the product of susceptible and infected populations.

(iv) The predators make no distinction between susceptible and infected individuals in their consumption of prey.

(v) The predators contract the disease only by consumption of prey. The rate of predator infection is proportional to the product of infected prey and susceptible predators.

We obtain a model by dividing the prey population into two subgroups: healthy, susceptible prey and infected prey. Let $s(t)$ and $i(t)$ represent the populations of susceptible and infected prey, respectively, at time t. Let $x(t)$ denote the population of healthy, susceptible predators. The autonomous system that will model the ecosystem is

$$x' = -ax + bxs - \delta xi$$
$$s' = cs - dsx - \alpha si + \beta i \tag{4}$$
$$i' = ci - dxi + \alpha si - \beta i,$$

where the constants $a, b, c, d, \alpha, \beta$, and δ are all positive.

1. Explain the modeling role played by each term in the three differential equations. (For example, the term $-ax$ accounts for the fact that, in the absence of prey, the predator population would decrease and exponentially approach extinction.)

2. As usual, we assume that the variables x, s, and i have been scaled so that one unit of population corresponds to a large number of actual individuals. Assume the following

values for the constants in equation (4):

$$a = 1, \quad b = 1, \quad c = 1, \quad d = 1, \quad \alpha = \tfrac{1}{2}, \quad \beta = 1, \quad \delta = 1.$$

With this, equation (4) becomes

$$x' = -x + xs - xi$$
$$s' = s - sx - \tfrac{1}{2}si + i \tag{5}$$
$$i' = i - xi + \tfrac{1}{2}si - i = -xi + \tfrac{1}{2}si.$$

Show that autonomous system (5) has just one equilibrium point in the first octant of xsi-space, where all three components are strictly positive. What is this equilibrium point?

3. Linearize system (5) about the equilibrium point found in part 2. Let A denote the (3×3) constant coefficient matrix of the linearized system. Without performing any further calculations, answer the following questions:

(a) Must the matrix A have at least one real eigenvalue?

(b) Is it possible for A to have exactly two real eigenvalues?

(c) Is the matrix A real and symmetric? Does it possess any special structure to suggest that it must have three real eigenvalues?

Now use computer software to determine the three eigenvalues of A.

It can be shown (see Coddington and Levinson[5]) that the (isolated) equilibrium point $(x_e, s_e, i_e) = (1, 2, 1)$ of nonlinear system (5) is

(i)   asymptotically stable if all three eigenvalues have negative real parts.

(ii)   unstable if at least one eigenvalue has a positive real part.

Can either of these results be applied in this case to determine the stability properties of the equilibrium point? If so, describe the stability properties of this equilibrium point.

Project 3: Chaos and the Lorenz Equations

In the early 1960s, Edward N. Lorenz,[6] an MIT meteorologist, formulated and studied a system of three nonlinear differential equations that today bear his name. These equations, arising from a model of fluid convection, were analyzed by Lorenz to gain insight into the feasibility of long-range weather forecasting. His findings, published in a classic 1963 paper,[7] illustrate what is now called **deterministic chaos**, a phenomenon wherein even a small number of nonlinear differential equations can exhibit behavior that is highly complicated and extremely sensitive to perturbations in the initial conditions. This sensitivity is sometimes called the "butterfly effect," an allusion to the notion that the flapping of a butterfly's wings on one continent can, after a time, influence the weather on another continent. This project uses a Runge-Kutta method to solve the Lorenz equations numerically and illustrate solution complexity and the "butterfly effect."

[5]Earl A. Coddington and Norman Levinson, *Theory of Ordinary Differential Equations* (Malabar, FL: R. E. Krieger, 1984).
[6]Edward N. Lorenz, professor emeritus, Department of Earth, Atmospheric and Planetary Sciences, MIT, received the 1983 Crafoord Prize "for fundamental contributions to the field of geophysical hydrodynamics, which in a unique way have contributed to a deeper understanding of the large-scale motions of the atmosphere and the sea."
[7]Edward N. Lorenz, "Deterministic Nonperiodic Flow," *Journal of the Atmospheric Sciences*, Vol. 20, March 1963, pp. 130–141. An interesting account of the research activity culminating in these results appears in the book *Chaos* by James Gleick (Viking Press, 1987).

The Lorenz equations are

$$X' = -\sigma X + \sigma Y$$
$$Y' = -XZ + rX - Y$$
$$Z' = XY - bZ,$$

where $\sigma, r,$ and b are positive constants. We will refer to the three dependent variables $X(t), Y(t),$ and $Z(t)$ as "coordinates." In reality, however, they are variables that characterize the intensity and temperature variations of the fluid convective motion.

1. Write a computer program to solve an initial value problem for the Lorenz equations using the fourth order Runge-Kutta method given in equations (11)–(13) of Section 4.9. (Section 4.9 also provides an example of a MATLAB code for the algorithm.)

2. Use your program to solve the Lorenz equations numerically on the time interval $0 \le t \le 50$. Use a step size of $h = 0.01$, and use the following parameter values and initial conditions:

$$\sigma = 10, \qquad b = \tfrac{8}{3}, \qquad r = 28$$
$$X(0) = 0, \qquad Y(0) = 1, \qquad Z(0) = 0.$$

3. We now introduce a very small perturbation of the initial conditions. Repeat the computations of step 2 using the same parameter values and step size, but with the following (different) initial conditions:

$$X(0) = 0.0005, \qquad Y(0) = 0.9999, \qquad Z(0) = 0.0001.$$

Let the "unperturbed" numerical solution obtained in step 2 be denoted by $X(t), Y(t),$ $Z(t)$ and the "perturbed" solution of step 3 by $X_p(t), Y_p(t), Z_p(t)$.

4. Since there are three dependent variables, parametric solution curves are three-dimensional space curves. Such curves are said to exist in phase space (in contrast to the two-dimensional phase plane). Instead of plotting such space curves, we will display solution curve complexity by plotting their projections onto each of the three coordinate planes. We will use computer software to create three separate parametric plots,

$$X(t) \text{ vs. } Y(t), \qquad X(t) \text{ vs. } Z(t), \qquad Y(t) \text{ vs. } Z(t), \qquad 0 \le t \le 50.$$

Illustrate the "butterfly effect" as follows:

(a) Create three separate graphs, displaying each of the "unperturbed" and corresponding "perturbed" pairs of coordinates on the same graph as a function of time. What happens as time progresses? Is there a time beyond which the corresponding coordinate curves bear virtually no resemblance to each other?

(b) To see this chaotic behavior from a different point of view, select one pair of "unperturbed" and "perturbed" coordinates, say $X(t)$ and $X_p(t)$. Now create the parametric plot

$$X(t) \text{ vs. } X_p(t), \qquad 0 \le t \le 50.$$

If $X(t)$ and $X_p(t)$ were to remain close in value to each other for all time, what would this parametric plot look like? Does the plot actually obtained look anything like this?

Numerical Methods

7.1 Introduction

A simple numerical method, Euler's method, was introduced in Section 2.10. In Section 4.9, we extended Euler's method to linear systems. We also described a fourth order Runge-Kutta method that served as the basis for an improved, more accurate algorithm. In this chapter, we discuss the ideas underlying a systematic development of more accurate algorithms.

We begin with the first order scalar initial value problem

$$y' = f(t, y), \qquad y(t_0) = y_0, \qquad t_0 \le t \le t_0 + T.$$

We assume that this problem has a unique solution on the given t-interval. Our goal is to develop algorithms that generate accurate approximations to the solution $y(t)$.

A numerical method frequently begins by imposing a partition of the form $t_0 < t_1 < t_2 < \cdots < t_{N-1} < t_N = t_0 + T$ on the t-interval $[t_0, t_0 + T]$. Often this partition is uniformly spaced—that is, the partition points are defined by

$$t_n = t_0 + nh, \qquad n = 0, 1, 2, \ldots, N, \qquad \text{where} \quad h = \frac{T}{N}.$$

The partition spacing, $h = T/N$, is called the **step length** or the **step size**. At each partition point, t_n, the numerical algorithm generates an approximation, y_n, to the exact solution value, $y(t_n)$. A **numerical solution** of the differential equation consists of the points $\{(t_0, y_0), (t_1, y_1), \ldots, (t_N, y_N)\}$, where

$$y_n \approx y(t_n), \qquad n = 0, 1, \ldots, N.$$

Note that the initial condition provides us with an exact starting point (t_0, y_0). A "good" numerical algorithm is one that generates points (t_n, y_n) that lie "close" to their exact solution counterparts, $(t_n, y(t_n))$ for $n = 1, 2, \ldots, N$. The terms "good" and "close," while intuitively clear, will be made precise later.

Figure 7.1 displays the exact solution of the initial value problem $y' = y^2, y(0) = 1$ on the interval $0 \le t \le 0.95$ and a pair of numerical approximations corresponding to different step lengths h. [The exact solution, $y(t) = (1 - t)^{-1}$, does not exist for $t \ge 1$.]

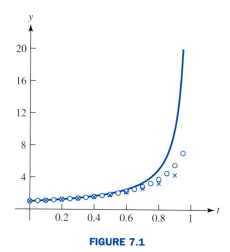

FIGURE 7.1

The initial value problem $y' = y^2, y(0) = 1$ has solution $y(t) = (1 - t)^{-1}, t < 1$. The solid curve is the graph of $y(t)$ for $0 \le t \le 0.95$. The points marked by an ∘ represent a numerical solution with step length $h = 0.05$, and the points marked by an × represent a numerical solution with $h = 0.1$. The numerical solutions were generated by Euler's method. As is usually the case, the smaller step length generates approximations that are more accurate.

Numerical Solutions for Systems of Differential Equations

Focusing our attention on scalar first order initial value problems may seem to be overly restrictive, but that is not the case. The algorithms we develop for first order scalar problems extend directly to first order systems. And, as you have seen, first order systems basically encompass all the differential equations we have considered so far.

We concentrate on first order scalar problems because they possess the virtues of relative simplicity and ease of visualization. In particular, we can graph and compare the exact solution and the numerical solution. To further simplify the development, we restrict our discussions to uniformly spaced partitions of step size h.

The computational aspects of Euler's method were treated earlier. The algorithm was introduced and applied to first order scalar problems in Section 2.10 and extended to first order linear systems in Section 4.9. Euler's method serves as our starting point in the next section, where we briefly review the method and explore ways of improving it.

7.2   Euler's Method, Heun's Method, and the Modified Euler's Method

Euler's Method

As we saw in Section 2.10, Euler's method develops a numerical solution of the initial value problem

$$y' = f(t, y), \qquad y(t_0) = y_0 \tag{1}$$

using the algorithm

$$y_{n+1} = y_n + hf(t_n, y_n), \qquad n = 0, 1, 2, \ldots, N - 1. \tag{2}$$

There are several different ways to derive Euler's method. In Section 2.10, we used a geometric approach based on direction fields. We now discuss two other ways of looking at Euler's method. While they are variations on the basic theme, they provide useful insights as we look for ways to improve Euler's method.

Approximating the Integral Equation

Let $y(t)$ denote the exact solution of initial value problem (1). For now, we restrict our attention to the interval $t_n \leq t \leq t_{n+1}$. Assume that we do not know the exact solution, $y(t)$, but that we have already calculated approximations, y_k, of $y(t_k)$ for $k = 0, 1, \ldots, n$ (see Figure 7.2). Our goal is to find the next approximation, y_{n+1}, of $y(t_{n+1})$.

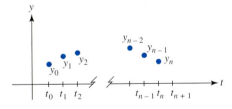

FIGURE 7.2

Let $y(t)$ denote the exact solution of initial value problem (1). While we do not know $y(t)$, we assume we have calculated approximations, y_k, to $y(t_k)$ for $k = 0, 1, \ldots, n$. Our goal is to find the next approximation, y_{n+1}, to $y(t_{n+1})$.

Consider differential equation (1) and its exact solution, $y(t)$. Integrating both sides of equation (1) over the interval $[t_n, t_{n+1}]$, we obtain

$$\int_{t_n}^{t_{n+1}} y'(s) \, ds = \int_{t_n}^{t_{n+1}} f(s, y(s)) \, ds.$$

By the fundamental theorem of calculus, the left-hand integral is $y(t_{n+1}) - y(t_n)$. Therefore, we obtain an equation for $y(t_{n+1})$:

$$y(t_{n+1}) = y(t_n) + \int_{t_n}^{t_{n+1}} f(s, y(s)) \, ds. \tag{3}$$

We cannot use equation (3) computationally because we do not know $y(s)$, $t_n \leq s \leq t_{n+1}$. Suppose, however, that the step length h is small enough that $f(s, y(s))$ is nearly constant over the interval $t_n \leq s \leq t_{n+1}$. In this case, we can approximate the integral by the left Riemann sum,

$$\int_{t_n}^{t_{n+1}} f(s, y(s)) \, ds \approx hf(t_n, y(t_n)). \tag{4}$$

Using approximation (4) in equation (3), we obtain

$$y(t_{n+1}) \approx y(t_n) + hf(t_n, y(t_n)).$$

Replacing $y(t_n)$ by the previously calculated estimate y_n, we are led to Euler's method,

$$y(t_{n+1}) \approx y_n + hf(t_n, y_n) = y_{n+1}.$$

In other words, we can view Euler's method as a left Riemann sum approximation of integral equation (3).

Heun's Method

Looked at in this light, Euler's method might be improved by asking "Are there better numerical integration schemes than approximation (4)?" The trapezoidal rule is one such numerical integration scheme. Using the trapezoidal rule, we can approximate the integral in (3) by

$$\int_{t_n}^{t_{n+1}} f(s, y(s)) \, ds \approx \frac{h}{2} \left[f(t_n, y(t_n)) + f(t_{n+1}, y(t_{n+1})) \right].$$

Using this integral approximation in (3), we obtain

$$y(t_{n+1}) \approx y(t_n) + \frac{h}{2} \left[f(t_n, y(t_n)) + f(t_{n+1}, y(t_{n+1})) \right].$$

Replacing $y(t_n)$ by its estimate y_n leads to

$$y(t_{n+1}) \approx y_n + \frac{h}{2} \left[f(t_n, y_n) + f(t_{n+1}, y(t_{n+1})) \right]. \tag{5}$$

At first glance, it appears we have made matters worse since the unknown $y(t_{n+1})$ appears on the right-hand side of (5), in the term $f(t_{n+1}, y(t_{n+1}))$. Approximation (5), if used as it stands, leads to an implicit algorithm with a nonlinear equation that has to be solved for $y(t_{n+1})$. Suppose, however, that we use Euler's method to approximate the unknown value $y(t_{n+1})$ on the right-hand side of (5):

$$y(t_{n+1}) \approx y_n + \frac{h}{2} \left[f(t_n, y_n) + f(t_{n+1}, y_n + hf(t_n, y_n)) \right].$$

This yields the explicit iteration

$$y_{n+1} = y_n + \frac{h}{2}\left[f(t_n, y_n) + f(t_{n+1}, y_n + hf(t_n, y_n))\right], \qquad n = 0, 1, \ldots, N-1. \quad \textbf{(6)}$$

Algorithm (6) is often called **Heun's method** or the **improved Euler's method**.

The Modified Euler's Method

Another simple numerical integration scheme is the modified Euler's method, in which the integrand is approximated over the interval $t_n \le t \le t_{n+1}$ by its midpoint value. If we use the midpoint rule to approximate the integral in (3) and again use Euler's method to approximate the unknown value $y(t)$ at the midpoint $t = t_n + h/2$, we obtain the algorithm

$$y_{n+1} = y_n + hf\left(t_n + \frac{h}{2}, y_n + \frac{h}{2}f(t_n, y_n)\right), \qquad n = 0, 1, \ldots, N-1. \quad \textbf{(7)}$$

Algorithm (7) is known as the **modified Euler's method**. [There is no universal agreement on the names of algorithms (6) and (7).[1]]

Although algorithms (6) and (7) appear somewhat complicated, they are relatively easy to implement, since computers can readily evaluate compositions of functions. However, you may rightly ask whether algorithm (6) is, as one of its names implies, an improvement on Euler's method. If so, how do we quantitatively describe this improvement? The same question applies to (7), and we address it in Section 7.3. For now, we content ourselves with an example that compares Euler's method with algorithms (6) and (7) for a particular initial value problem.

EXAMPLE 1

Consider the initial value problem

$$y' = y^2, \qquad y(0) = 1, \qquad 0 \le t \le 0.95.$$

Using a step length of $h = 0.05$, compare the results of Euler's method (2), Heun's method (6), and the modified Euler's method (7). [The exact solution is $y(t) = 1/(1-t), t < 1.$]

Solution: For this example, $t_0 = 0$, $T = 0.95$, $N = T/h = 19$, and $f(t, y) = y^2$. Table 7.1 on the next page lists the results. Note that algorithms (6) and (7) do, in fact, represent an improvement over Euler's method. ❖

The relationship between numerical methods for solving differential equations and numerical integration schemes is a reciprocal one. Every numerical integration technique suggests an algorithm for the initial value problem—this is the approach we used in obtaining algorithms (6) and (7) from equation (3). Conversely, an algorithm for the initial value problem gives rise to a corresponding numerical integration scheme. To see why, consider the initial value problem $y' = f(t), y(t_0) = 0$. The solution is simply $y(t) = \int_{t_0}^{t} f(s)\, ds$. Therefore, any numerical method used to solve this initial value problem gives rise to an

---

[1]We are using names found in Peter Henrici, *Discrete Variable Methods in Ordinary Differential Equations* (New York: Wiley, 1962).

TABLE 7.1

The Results of Example 1

As is usually the case, algorithms (6) and (7) give better approximations to $y(t_n)$ than does Euler's method. As is also typical, algorithms (6) and (7) have comparable accuracy. [Note: For this particular initial value problem, Heun's method (6) yields slightly better approximations than the modified Euler's method (7). For other examples, (7) may give slightly better approximations than (6).]

t_n	Euler's Method	Heun's Method	Modified Euler's Method	Exact Solution
0.0000	1.0000	1.0000	1.0000	1.0000
0.0500	1.0500	1.0526	1.0525	1.0526
0.1000	1.1051	1.1109	1.1109	1.1111
0.1500	1.1662	1.1762	1.1761	1.1765
0.2000	1.2342	1.2495	1.2493	1.2500
0.2500	1.3104	1.3326	1.3323	1.3333
0.3000	1.3962	1.4275	1.4271	1.4286
0.3500	1.4937	1.5370	1.5363	1.5385
0.4000	1.6052	1.6645	1.6636	1.6667
0.4500	1.7341	1.8151	1.8137	1.8182
0.5000	1.8844	1.9954	1.9934	2.0000
0.5500	2.0620	2.2153	2.2124	2.2222
0.6000	2.2745	2.4894	2.4850	2.5000
0.6500	2.5332	2.8402	2.8333	2.8571
0.7000	2.8541	3.3049	3.2935	3.3333
0.7500	3.2614	3.9488	3.9289	4.0000
0.8000	3.7932	4.8975	4.8597	5.0000
0.8500	4.5126	6.4264	6.3449	6.6667
0.9000	5.5308	9.2615	9.0471	10.0000
0.9500	7.0603	15.9962	15.2001	20.0000

approximation of the integral. In particular, Euler's method, Heun's method, and the modified Euler's method, when applied to the initial value problem $y' = f(t), y(t_0) = 0$, reduce to a left Riemann sum, the trapezoidal rule, and the midpoint rule, respectively.

Approximating the Taylor Series Expansion

This subsection presents another derivation of Euler's method. Since $y(t)$ is the solution of initial value problem (1) and is assumed to exist on the interval $t_0 \leq t \leq t_0 + T$, we know $y(t)$ is differentiable on that interval. Assume for now not only that the solution is differentiable, but that it can be expanded in a Taylor series at $t = t_n$, where the Taylor series converges in an interval containing $[t_n, t_n + h]$. Therefore, we can express $y(t_{n+1})$ as

$$y(t_{n+1}) = y(t_n + h) = y(t_n) + y'(t_n)h + \frac{y''(t_n)}{2!}h^2 + \frac{y'''(t_n)}{3!}h^3 + \cdots . \qquad (8)$$

Truncating the series (8) after two terms, we obtain the approximation

$$y(t_{n+1}) \approx y(t_n) + y'(t_n)h. \tag{9}$$

Since $y'(t_n) = f(t_n, y(t_n))$, we can rewrite approximation (9) as

$$y(t_{n+1}) \approx y(t_n) + f(t_n, y(t_n))h.$$

Replacing $y(t_n)$ in this approximation by its estimate y_n, we are once more led to Euler's method:

$$y(t_{n+1}) \approx y_n + f(t_n, y_n)h = y_{n+1}.$$

Thus, we obtain Euler's method by truncating the Taylor series (8) after two terms. Viewed in this light, Euler's method might be improved by retaining more terms of the Taylor series—truncating after three, four, or more terms. We investigate this possibility in Section 7.3.

EXERCISES

Most exercises in this chapter require a computer or programmable calculator.

Exercises 1–5:

In each exercise,

(a) Solve the initial value problem analytically, using an appropriate solution technique.

(b) For the given initial value problem, write the Heun's method algorithm,

$$y_{n+1} = y_n + \frac{h}{2}[f(t_n, y_n) + f(t_{n+1}, y_n + hf(t_n, y_n))].$$

(c) For the given initial value problem, write the modified Euler's method algorithm,

$$y_{n+1} = y_n + hf\left(t_n + \frac{h}{2}, y_n + \frac{h}{2}f(t_n, y_n)\right).$$

(d) Use a step size $h = 0.1$. Compute the first three approximations, y_1, y_2, y_3, using the method in part (b).

(e) Use a step size $h = 0.1$. Compute the first three approximations, y_1, y_2, y_3, using the method in part (c).

(f) For comparison, calculate and list the exact solution values, $y(t_1), y(t_2), y(t_3)$.

1. $y' = 2t - 1$, $y(1) = 0$ **2.** $y' = -y$, $y(0) = 1$ **3.** $y' = -ty$, $y(0) = 1$

4. $y' = -y + t$, $y(0) = 0$ **5.** $y^2y' + t = 0$, $y(0) = 1$

Exercises 6–9:

In each exercise,

(a) Find the exact solution of the given initial value problem.

(b) As in Example 1, use a step size of $h = 0.05$ for the given initial value problem. Compute 20 steps of Euler's method, Heun's method, and the modified Euler's method. Compare the numerical values obtained at $t = 1$ by calculating the error $|y(1) - y_{20}|$.

6. $y' = 1 + y^2$, $y(0) = -1$ **7.** $y' = -\dfrac{t}{y}$, $y(0) = 3$

8. $y' + 2y = 4$, $y(0) = 3$ **9.** $y' + 2ty = 0$, $y(0) = 2$

Exercises 10–14:

In each exercise, the given iteration is the result of applying Euler's method, Heun's method, or the modified Euler's method to an initial value problem of the form

$$y' = f(t, y), \qquad y(t_0) = y_0, \qquad t_0 \le t \le t_0 + T.$$

Identify the numerical method, and determine t_0, T, and $f(t, y)$.

10. $y_{n+1} = y_n + h(y_n + t_n^2 y_n^3), \quad y_0 = 1$
$t_n = 2 + nh, \quad h = 0.02, \quad n = 0, 1, 2, \dots, 49$

11. $y_{n+1} = y_n + \dfrac{h}{2}[t_n y_n^2 + 2 + (t_n + h)(y_n + h(t_n y_n^2 + 1))^2], \quad y_0 = 2$
$t_n = 1 + nh, \quad h = 0.05, \quad n = 0, 1, 2, \dots, 99$

12. $y_{n+1} = y_n + h\left(t_n + \dfrac{h}{2}\right)\sin^2\left(y_n + \dfrac{h}{2}t_n \sin^2(y_n)\right), \quad y_0 = 1$

$t_n = nh, \quad h = 0.01, \quad n = 0, 1, 2, \dots, 199$

13. $y_{n+1} = y_n\left(1 + \dfrac{h}{t_n^2 + y_n^2}\right), \quad y_0 = -1$

$t_n = 2 + nh, \quad h = 0.01, \quad n = 0, 1, 2, \dots, 99$

14. $y_{n+1} = y_n + h\left[\sin\left(t_n + \dfrac{h}{2} + y_n + \dfrac{h}{2}\sin(t_n + y_n)\right)\right], \quad y_0 = 1$

$t_n = -1 + nh, \quad h = 0.05, \quad n = 0, 1, 2, \dots, 199$

15. Let h be a fixed positive step size, and let λ be a nonzero constant. Suppose we apply Heun's method or the modified Euler's method to the initial value problem $y' = \lambda y, y(t_0) = y_0$, using this step size h. Show, in either case, that

$$y_k = \left(1 + h\lambda + \frac{(h\lambda)^2}{2!}\right)y_{k-1} \quad \text{and hence} \quad y_k = \left(1 + h\lambda + \frac{(h\lambda)^2}{2!}\right)^k y_0, \qquad k = 1, 2, \dots.$$

Exercises 16–17:

Assume a tank having a capacity of 200 gal initially contains 90 gal of fresh water. At time $t = 0$, a salt solution begins to flow into the tank at a rate of 6 gal/min and the well-stirred mixture flows out at a rate of 1 gal/min. Assume that the inflow concentration fluctuates in time, with the inflow concentration given by $c(t) = 2 - \cos(\pi t)$ lb/gal, where t is in minutes. Formulate the appropriate initial value problem for $Q(t)$, the amount of salt (in pounds) in the tank at time t. Our objective is to approximately determine the amount of salt in the tank when the tank contains 100 gal of liquid.

16. (a) Formulate the initial value problem.

(b) Obtain a numerical solution, using Heun's method with a step size $h = 0.05$.

(c) What is your estimate of $Q(t)$ when the tank contains 100 gal?

17. (a) Formulate the initial value problem.

(b) Obtain a numerical solution, using the modified Euler's method with a step size $h = 0.05$.

(c) What is your estimate of $Q(t)$ when the tank contains 100 gal?

Exercises 18–19:

Let $P(t)$ denote the population of a certain colony, measured in millions of members. Assume that $P(t)$ is the solution of the initial value problem

$$P' = 0.1\left(1 - \frac{P}{3}\right)P + M(t), \qquad P(0) = P_0,$$

where time t is measured in years. Let $M(t) = e^{-t}$. Therefore, the colony experiences a migration influx that is initially strong but soon tapers off. Let $P_0 = \frac{1}{2}$; that is, the colony had 500,000 members at time $t = 0$. Our objective is to estimate the colony size after two years.

18. Obtain a numerical solution of this problem, using Heun's method with a step size $h = 0.05$. What is your estimate of colony size at the end of two years?

19. Obtain a numerical solution of this problem, using the modified Euler's method with a step size $h = 0.05$. What is your estimate of colony size at the end of two years?

20. **Error Estimation** In most applications of numerical methods, as in Exercises 16–19, an exact solution is unavailable to use as a benchmark. Therefore, it is natural to ask, "How accurate is our numerical solution?" For example, how accurate are the solutions obtained in Exercises 16–19 using the step size $h = 0.05$? This exercise provides some insight.

    Suppose we apply Heun's method or the modified Euler's method to the initial value problem $y' = f(t, y), y(t_0) = y_0$ and we use a step size h. It can be shown, for most initial value problems and for h sufficiently small, that the error at a fixed point $t = t^*$ is proportional to h^2. That is, let n be a positive integer, let $h = (t^* - t_0)/n$, and let y_n denote the method's approximation to $y(t^*)$ using step size h. Then

$$\lim_{\substack{h \to 0 \\ t^* \text{ fixed}}} \frac{y(t^*) - y_n}{h^2} = C, \qquad C \neq 0.$$

As a consequence of this limit, reducing a sufficiently small step size by $\frac{1}{2}$ will reduce the error by approximately $\frac{1}{4}$. In particular, let $\hat{y}_{2n}$ denote the method's approximation to $y(t^*)$ using step size $h/2$. Then, for most initial value problems, we expect that $y(t^*) - \hat{y}_{2n} \approx [y(t^*) - y_n]/4$. Rework Example 1, using Heun's method and step sizes of $h = 0.05$, $h = 0.025$, and $h = 0.0125$.

(a) Compare the three numerical solutions at $t = 0.05, 0.10, 0.15, \ldots, 0.95$. Are the errors reduced by about $\frac{1}{4}$ when the step size is reduced by $\frac{1}{2}$? (Since the solution becomes unbounded as t approaches 1 from the left, the expected error reduction may not materialize near $t = 1$.)

(b) Suppose the exact solution is not available. How can the Heun's method solutions obtained using different step sizes be used to estimate the error? [Hint: Assuming that

$$y(t^*) - \hat{y}_{2n} \approx \frac{[y(t^*) - y_n]}{4},$$

derive an expression for $y(t^*) - \hat{y}_{2n}$ that involves only $\hat{y}_{2n}$ and y_n.]

(c) Test the error monitor derived in part (b) on the initial value problem in Example 1.

7.3 Taylor Series Methods

In Section 7.2, we saw that we could obtain Euler's method by truncating the Taylor series for the solution $y(t)$ after the first two terms of the expansion. We therefore anticipate that Euler's method can be improved by retaining more terms of the Taylor series.

In this section, we describe how such an improvement of Euler's method is carried out. In addition, we use the Taylor series expansion as a basis for quantifying the accuracy of numerical algorithms. We begin with some preliminaries:

- First, we state Theorem 7.1. This theorem gives conditions guaranteeing that the solution of an initial value problem has a convergent Taylor series expansion.
- We then present Theorem 7.2, Taylor's theorem. This theorem from calculus enables us to measure the error that arises when we truncate a Taylor series.

Once these preliminary results are in place, we can use Taylor series as a basis for systematically developing algorithms of increasing accuracy. These Taylor series algorithms can, in principle, be made as accurate as we wish. They are not, however, computationally friendly. We combine accuracy with ease of implementation in Section 7.4, when we discuss Runge-Kutta methods.

Preliminaries

We begin with two definitions and then present a theorem guaranteeing that the solution of initial value problem (1),

$$y' = f(t, y), \qquad y(t_0) = y_0, \tag{1}$$

can be expanded in a Taylor series that converges in a neighborhood of the point t_0.

A function $y(t)$, defined on an open interval containing the point $\bar{t}$, is said to be **analytic** at $t = \bar{t}$ if it has a Taylor series expansion

$$y(t) = \sum_{n=0}^{\infty} a_n (t - \bar{t})^n \tag{2a}$$

that converges in an interval $\bar{t} - \delta < t < \bar{t} + \delta$, where $\delta > 0$. It is shown in calculus that if $y(t)$ is analytic at $t = \bar{t}$, then $y(t)$ has derivatives of all orders in the interval $(\bar{t} - \delta, \bar{t} + \delta)$. Moreover, the coefficients of the Taylor series are given by

$$a_n = \frac{y^{(n)}(\bar{t})}{n!}, \qquad n = 0, 1, 2, \ldots. \tag{2b}$$

In general, a function $y(t)$ is said to be **analytic in the interval** $a < t < b$ if it is analytic at every point $\bar{t}$ in this interval.

Consider the function $f(t, y)$ appearing on the right-hand side of differential equation (1). In the context of differential equation (1), $f(t, y)$ is understood to represent $f(t, y(t))$, where $y(t)$ is the unknown solution of interest. In the next definition, however, we view f as a function of two independent variables, t and y.

Let $f(t, y)$ be a function defined in an open region R of the ty-plane containing the point $(\bar{t}, \bar{y})$. The function $f(t, y)$ is said to be **analytic** at $(\bar{t}, \bar{y})$ if it has a two-variable Taylor series expansion

$$f(t, y) = \sum_{m=0}^{\infty} \sum_{n=0}^{\infty} b_{mn} (t - \bar{t})^m (y - \bar{y})^n \tag{3a}$$

that converges in a neighborhood N_ρ of $(\bar{t}, \bar{y})$,

$$N_\rho = \left\{ (t, y) : \sqrt{(t - \bar{t})^2 + (y - \bar{y})^2} < \rho \right\}.$$

We say that $f(t, y)$ is **analytic in a region** R if it is analytic at every point $(\bar{t}, \bar{y})$ in R. The coefficients b_{mn} can be evaluated in terms of the function f and its partial derivatives, evaluated at $(\bar{t}, \bar{y})$; the two-variable Taylor series expansion has the form

$$f(t, y) = f(\bar{t}, \bar{y}) + f_t(\bar{t}, \bar{y})(t - \bar{t}) + f_y(\bar{t}, \bar{y})(y - \bar{y})$$

$$+ \frac{1}{2} \left[f_{tt}(\bar{t}, \bar{y})(t - \bar{t})^2 + 2f_{ty}(\bar{t}, \bar{y})(t - \bar{t})(y - \bar{y}) + f_{yy}(\bar{t}, \bar{y})(y - \bar{y})^2 \right] + \cdots.$$

(3b)

The Existence of Analytic Solutions

It is natural to ask whether analyticity of $f(t, y)$ guarantees analyticity of the solution of initial value problem (1). An affirmative answer is contained in Theorem 7.1, which can be regarded as a refinement of Theorem 2.2. A proof of Theorem 7.1 can be found in Birkhoff and Rota.[2]

Theorem 7.1

Let R denote the rectangle defined by $a < t < b, \alpha < y < \beta$. Let $f(t, y)$ be a function defined and analytic in R, and suppose that (t_0, y_0) is a point in R. Then there is a t-interval (c, d) containing t_0 in which there exists a unique analytic solution of the initial value problem

$$y' = f(t, y), \qquad y(t_0) = y_0.$$

EXAMPLE 1

Consider the initial value problem

$$y' = y^2 + t^2, \qquad y(t_0) = y_0.$$

Here, the function $f(t, y) = y^2 + t^2$ is a polynomial in the variables t and y and is therefore analytic in the entire ty-plane. Hence, the region R can be assumed to be the entire ty-plane. Theorem 7.1 guarantees the existence of a unique analytic solution $y(t)$ in an interval of the form $t_0 - \delta < t < t_0 + \delta$ for some $\delta > 0$. Note that the theorem does not tell us the value of δ, only that such a positive δ exists.

Since the solution is analytic in $t_0 - \delta < t < t_0 + \delta$, we know $y(t)$ has the form

$$y(t) = \sum_{n=0}^{\infty} \frac{y^{(n)}(t_0)}{n!} (t - t_0)^n, \qquad t_0 - \delta < t < t_0 + \delta. ❖$$

We assume throughout this chapter that the hypotheses of Theorem 7.1 are satisfied. This theorem assures us that an analytic solution $y(t)$ exists on some interval of the form $t_0 - \delta < t < t_0 + \delta$. As noted in Example 1, however,

---

[2]Garrett Birkhoff and Gian-Carlo Rota, *Ordinary Differential Equations*, 4th ed. (New York: Wiley, 1989).

Theorem 7.1 does not tell us the size of δ. Since we are interested in generating a numerical solution on an interval of the form $t_0 \le t \le t_0 + T$, we shall also assume that the interval of interest, $[t_0, t_0 + T]$, lies within the existence interval, $(t_0 - \delta, t_0 + \delta)$. Given this assumption, we can expand solution $y(t)$ in a Taylor series about any point $\bar{t}$ lying in the interval of interest. It is important to realize, however, that in practical computations involving nonlinear differential equations there is no *a priori* guarantee that the solution exists on a designated interval of interest, $[t_0, t_0 + T]$.

Using the Differential Equation to Compute the Taylor Series Coefficients

When Euler's method was discussed in Section 2.10, we based the development on the fact that the differential equation determines the direction field. In particular, if we evaluate f at a point $(\bar{t}, \bar{y})$ in the ty-plane, then the value $f(\bar{t}, \bar{y})$ tells us the slope of the solution curve passing through $(\bar{t}, \bar{y})$.

We now show that the differential equation determines much more. In particular, suppose that a solution curve $y(t)$ passes through the point $(\bar{t}, \bar{y})$. We will see that $f(t, y)$ and its partial derivatives, evaluated at $(\bar{t}, \bar{y})$, can be used to calculate all the derivatives of $y(t)$. In turn [see equations (2a) and (2b)], these derivative evaluations completely determine the Taylor series expansion of the solution $y(t)$.

In particular, we know the identity $y'(t) = f(t, y(t))$ holds for t in a neighborhood of $\bar{t}$. Therefore,

$$y'(\bar{t}) = f(\bar{t}, y(\bar{t}))$$
$$= f(\bar{t}, \bar{y}).$$

We find higher derivatives by differentiating the identity $y'(t) = f(t, y(t))$. For example,

$$y''(t) = \frac{d}{dt} y'(t) = \frac{d}{dt} f(t, y(t)). \tag{4a}$$

We use the chain rule to calculate the derivative in equation (4a),

$$\frac{df}{dt} = \frac{\partial f}{\partial t} + \frac{\partial f}{\partial y} \frac{dy}{dt} = \frac{\partial f}{\partial t} + \frac{\partial f}{\partial y} f = f_t + f_y f. \tag{4b}$$

Once the partial derivatives in equation (4b) are computed, we substitute the function $y(t)$ for the second independent variable y, obtaining

$$y''(t) = f_t(t, y(t)) + f_y(t, y(t)) f(t, y(t)).$$

Using the fact that $y(\bar{t}) = \bar{y}$, we have

$$y''(\bar{t}) = f_t(\bar{t}, \bar{y}) + f_y(\bar{t}, \bar{y}) f(\bar{t}, \bar{y}). \tag{5}$$

Equation (5) determines the concavity of the solution curve at the point $(\bar{t}, \bar{y})$, just as $y'(\bar{t}) = f(\bar{t}, \bar{y})$ determines the slope of the solution curve at $(\bar{t}, \bar{y})$.

This differentiation process can be continued to compute higher derivatives of the solution at $(\bar{t}, \bar{y})$. To simplify the notation, we continue to use subscripts to denote partial derivatives and do not explicitly indicate their ultimate evaluation at $(\bar{t}, \bar{y})$. Thus,

$$y'' = f_t + f_y f$$

$$y''' = \frac{d}{dt}\left[f_t + f_y f\right] = \left[f_{tt} + f_{ty}f + (f_{yt} + f_{yy}f)f + f_y(f_t + f_y f)\right]. \tag{6}$$

It is possible, in principle, to continue this differentiation process and compute as many derivatives of $y(t)$ at $t = \bar{t}$ as desired. It is clear from (6), however, that the computations can quickly become cumbersome.

The next example illustrates, however, that when the differential equation has a simple structure, it may be relatively easy to calculate higher derivatives.

E X A M P L E

2

Consider the initial value problem

$$y' = y^2, \qquad y(0) = 1.$$

Evaluate the derivatives $y'(0), y''(0), y'''(0)$, and $y^{(4)}(0)$.

Solution: In this case, $f(t, y) = y^2$ is a polynomial in y. Therefore, Theorem 7.1 applies, and we know the solution $y(t)$ is an analytic function of t in the open interval $(-\delta, \delta)$ for some $\delta > 0$. Since $y'(t) = y^2(t)$, the chain rule yields

$$y''(t) = [y^2(t)]' = 2y(t)y'(t) = 2y^3(t)$$

$$y'''(t) = [2y^3(t)]' = 6y^2(t)y'(t) = 6y^4(t)$$

$$y^{(4)}(t) = [6y^4(t)]' = 24y^3(t)y'(t) = 24y^5(t).$$

Therefore, $y(0) = 1$, $y'(0) = 1$, $y''(0) = 2$, $y'''(0) = 6 = 3!$, and $y^{(4)}(0) = 24 = 4!$. Given these derivative values, the first few terms in the Taylor series expansion of $y(t)$ are

$$y(t) = y(0) + y'(0)t + \frac{y''(0)}{2!}t^2 + \frac{y'''(0)}{3!}t^3 + \frac{y^{(4)}(0)}{4!}t^4 + \cdots$$

$$= 1 + t + t^2 + t^3 + t^4 + \cdots.$$

We recognize this expansion as a geometric series that converges to the exact solution,

$$y(t) = \frac{1}{1-t}, \qquad -1 < t < 1.$$

For this initial value problem, we find (after the fact) that $\delta = 1$. ❖

Taylor Series Methods

The preceding discussion shows how to calculate higher derivatives of the solution $y(t)$ of the initial value problem

$$y' = f(t, y), \qquad y(t_0) = y_0.$$

We can use these ideas to improve Euler's method. Let y_n be an approximation to $y(t_n)$, where t_n and $t_{n+1} = t_n + h$ are in the interval $t_0 \le t \le t_N$. As in equation (8) of Section 7.2, we have

$$y(t_{n+1}) = y(t_n) + y'(t_n)h + \frac{y''(t_n)}{2!}h^2 + \frac{y'''(t_n)}{3!}h^3 + \cdots.$$

Truncating this expansion after p terms, we obtain the approximation

$$y(t_{n+1}) \approx y(t_n) + y'(t_n)h + \frac{y''(t_n)}{2!}h^2 + \frac{y'''(t_n)}{3!}h^3 + \cdots + \frac{y^{(p)}(t_n)}{p!}h^p. \quad (7)$$

As we saw in equation (6), the Taylor series coefficients, $y'(t_n), y''(t_n), y'''(t_n), \ldots$, can be expressed in terms of f and its partial derivatives evaluated at $(t_n, y(t_n))$. For instance, with $p = 1$, (7) becomes

$$y(t_{n+1}) \approx y(t_n) + f(t_n, y(t_n))h.$$

Similarly, for $p = 2$, we obtain from (7)

$$y(t_{n+1}) \approx y(t_n) + f(t_n, y(t_n))h + \left[f_t(t_n, y(t_n)) + f_y(t_n, y(t_n))f(t_n, y(t_n)) \right] \frac{h^2}{2!}.$$

We find similar approximations when $p \geq 3$. In order to use these approximations for computations, we replace $y(t_n)$ by its estimate, y_n. The algorithms we obtain in this manner are collectively referred to as **Taylor series methods**. We use the term **Taylor series method of order p** to identify the Taylor series method obtained from approximation (7). The Taylor series methods of orders 1, 2, and 3 are as follows:

Taylor Series Method of Order 1 (Euler's Method)

$$y_{n+1} = y_n + hf(t_n, y_n), \qquad n = 0, 1, \ldots, N - 1 \quad (8a)$$

Taylor Series Method of Order 2

$$y_{n+1} = y_n + hf(t_n, y_n) + \frac{h^2}{2!} \left[f_t(t_n, y_n) + f_y(t_n, y_n)f(t_n, y_n) \right], \qquad n = 0, 1, \ldots, N - 1 \quad (8b)$$

Taylor Series Method of Order 3

$$y_{n+1} = y_n + hf(t_n, y_n) + \frac{h^2}{2!} \left[f_t(t_n, y_n) + f_y(t_n, y_n)f(t_n, y_n) \right]$$
$$+ \frac{h^3}{3!} \left[f_{tt}(t_n, y_n) + 2f_{ty}(t_n, y_n)f(t_n, y_n) + f_{yy}(t_n, y_n)f^2(t_n, y_n) \right.$$
$$\left. + f_y(t_n, y_n)f_t(t_n, y_n) + f_y^2(t_n, y_n)f(t_n, y_n) \right], \qquad n = 0, 1, \ldots, N - 1. \quad (8c)$$

It is cumbersome to write out all the terms of the general pth order Taylor series method. In order to simplify the notation when discussing Taylor series methods, it is common to denote a pth order Taylor series method as

$$y_{n+1} = y_n + hy_n' + \frac{h^2}{2!}y_n'' + \frac{h^3}{3!}y_n''' + \cdots + \frac{h^p}{p!}y_n^{(p)}, \qquad n = 0, 1, \ldots, N - 1. \quad (9)$$

We are using the name "pth order Taylor series method" to denote method (9). The term *order* has a precise meaning that is given later in this section. Once we state the formal definition of order, however, we will see that method (9) is properly named.

EXAMPLE

3

Consider the initial value problem

$$y' = y^2, \qquad y(0) = 1.$$

Using $h = 0.05$, execute 19 steps of the Taylor series method of order p for $p = 1, 2, 3,$ and 4. Do the results improve as p increases?

Solution: As we saw in Example 2, $y'' = 2y^3, y''' = 6y^4,$ and $y^{(4)} = 24y^5$. The Taylor series methods of orders 1, 2, 3, and 4 are, respectively,

$$y_{n+1} = y_n + hy_n^2,$$
$$y_{n+1} = y_n + hy_n^2 + h^2 y_n^3$$
$$y_{n+1} = y_n + hy_n^2 + h^2 y_n^3 + h^3 y_n^4$$
$$y_{n+1} = y_n + hy_n^2 + h^2 y_n^3 + h^3 y_n^4 + h^4 y_n^5.$$

Table 7.2 illustrates how the Taylor series method estimates improve as the order increases.

TABLE 7.2

In this table, we designate the results of the pth order Taylor series method as "order p" for $p = 1, 2, 3, 4$ and the value of the exact solution at $t = t_n$ as $y(t_n)$. As anticipated, the results improve when we retain more terms in the Taylor series expansion—that is, as the order p increases.

Taylor Series Methods

t_n	Order 1	Order 2	Order 3	Order 4	$y(t_n)$
0.0000	1.0000	1.0000	1.0000	1.0000	1.0000
0.0500	1.0500	1.0525	1.0526	1.0526	1.0526
0.1000	1.1051	1.1108	1.1111	1.1111	1.1111
0.1500	1.1662	1.1759	1.1764	1.1765	1.1765
0.2000	1.2342	1.2491	1.2500	1.2500	1.2500
0.2500	1.3104	1.3320	1.3333	1.3333	1.3333
0.3000	1.3962	1.4266	1.4285	1.4286	1.4286
0.3500	1.4937	1.5357	1.5383	1.5385	1.5385
0.4000	1.6052	1.6626	1.6664	1.6666	1.6667
0.4500	1.7341	1.8123	1.8178	1.8182	1.8182
0.5000	1.8844	1.9914	1.9994	2.0000	2.0000
0.5500	2.0620	2.2095	2.2212	2.2221	2.2222
0.6000	2.2745	2.4805	2.4984	2.4999	2.5000
0.6500	2.5332	2.8264	2.8543	2.8569	2.8571
0.7000	2.8541	3.2822	3.3281	3.3328	3.3333
0.7500	3.2614	3.9093	3.9894	3.9987	4.0000
0.8000	3.7932	4.8227	4.9756	4.9963	5.0000
0.8500	4.5126	6.2661	6.5980	6.6536	6.6667
0.9000	5.5308	8.8443	9.7297	9.9301	10.0000
0.9500	7.0603	14.4850	17.8859	19.1273	20.0000

Example 3 illustrates (for the special case of the differential equation $y' = y^2$) how the Taylor series method of order p becomes more accurate as p increases. We are now ready to make the concept of order precise and to discuss why we expect that higher order methods are usually more accurate than lower order methods.

Taylor's Theorem

We consider the error made when we truncate a Taylor series. Theorem 7.2, known as Taylor's theorem, gives a convenient way of estimating the resulting truncation error. A proof of Taylor's theorem can be found in most calculus books.

Theorem 7.2

Let $y(t)$ be analytic at $t = \bar{t}$, where the Taylor series expansion (2) converges in the interval $\bar{t} - \delta < t < \bar{t} + \delta$. Let m be a positive integer, and let t be in the interval $(\bar{t} - \delta, \bar{t} + \delta)$. Then

$$y(t) = y(\bar{t}) + y'(\bar{t})(t - \bar{t}) + \frac{y''(\bar{t})}{2!}(t - \bar{t})^2 + \cdots$$
$$+ \frac{y^{(m)}(\bar{t})}{m!}(t - \bar{t})^m + \frac{y^{(m+1)}(\xi)}{(m+1)!}(t - \bar{t})^{m+1}, \tag{10}$$

where ξ is some point lying between $\bar{t}$ and t.

In Theorem 7.2, the polynomial

$$P_m(t) = y(\bar{t}) + y'(\bar{t})(t - \bar{t}) + \frac{y''(\bar{t})}{2!}(t - \bar{t})^2 + \cdots + \frac{y^{(m)}(\bar{t})}{m!}(t - \bar{t})^m$$

is referred to as the **Taylor polynomial of degree m**. The term

$$R_m(t) = \frac{y^{(m+1)}(\xi)}{(m+1)!}(t - \bar{t})^{m+1}$$

is the **remainder**, and it measures the error made in approximating $y(t)$ by the Taylor polynomial, $P_m(t)$. When we consider the errors of a numerical method, the role of $\bar{t}$ is typically played by t_n and the generic point t lies in the interval $t_n \leq t \leq t_{n+1}$.

One-Step Methods and the Local Truncation Error

The methods we have considered thus far (Euler's method, Heun's method, the modified Euler's method, and Taylor series methods) are classified as one-step methods. In general, a **one-step method** has the form

$$y_{n+1} = y_n + h\phi(t_n, y_n; h), \qquad n = 0, 1, 2, \ldots, N - 1. \tag{11}$$

These methods are called one step because they use only the most recently computed point, (t_n, y_n), to compute the next point, (t_{n+1}, y_{n+1}). [By contrast, a multistep method uses multiple back values, $(t_n, y_n), (t_{n-1}, y_{n-1}), (t_{n-2}, y_{n-2}), \ldots, (t_{n-k}, y_{n-k})$, to compute (t_{n+1}, y_{n+1}).[3] We restrict our consideration to one-step methods.]

In equation (11), the term $\phi(t_n, y_n; h)$ is called an **increment function**. Different increment functions define different one-step methods. For instance, Euler's method, $y_{n+1} = y_n + hf(t_n, y_n)$, is a one-step method with increment function

$$\phi(t_n, y_n; h) = f(t_n, y_n).$$

Heun's method is a one-step method with increment function

$$\phi(t_n, y_n; h) = \tfrac{1}{2} \left[f(t_n, y_n) + f(t_n + h, y_n + hf(t_n, y_n)) \right].$$

E X A M P L E

4

Write the second order Taylor series method in the form of a one-step method, and identify the increment function $\phi(t_n, y_n; h)$.

Solution: From equation (8b), the second order Taylor series method has the form

$$y_{n+1} = y_n + h \left(f(t_n, y_n) + \frac{h}{2!} \left[f_t(t_n, y_n) + f_y(t_n, y_n)f(t_n, y_n) \right] \right).$$

Thus,

$$\phi(t_n, y_n; h) = f(t_n, y_n) + \frac{h}{2!} \left[f_t(t_n, y_n) + f_y(t_n, y_n)f(t_n, y_n) \right]. \quad \diamond$$

A quantity known as the local truncation error is one of the keys to understanding and assessing the accuracy of one-step methods. Let $y(t)$ denote the solution of the initial value problem $y' = f(t, y), y(t_0) = y_0$, and assume $y(t)$ exists on the interval of interest, $[t_0, t_0 + T]$. Let t_n and $t_{n+1} = t_n + h$ lie in the interval $[t_0, t_0 + T]$. For a given one-step method (11), we define the quantity T_{n+1} by

$$y(t_{n+1}) = y(t_n) + h\phi(t_n, y(t_n); h) + T_{n+1}. \tag{12}$$

The quantities $T_{n+1}, n = 0, 1, \ldots, N - 1$ are called local truncation errors. A **local truncation error**[4] measures how much a single step of the numerical method misses the true solution value, $y(t_{n+1})$, given that the numerical method starts on the solution curve at the point $(t_n, y(t_n))$.

---

[3] John D. Lambert, *Numerical Methods for Ordinary Differential Systems* (Chichester, England: Wiley, 1991).

[4] There is no universal agreement about the definition of local truncation errors. Some texts express the quantity T_{n+1} in equation (12) as $h\tau_{n+1} = T_{n+1}$ and refer to τ_{n+1} as a local truncation error. However, no matter how local truncation errors are defined, there is universal agreement on the definition of the "order" of a one-step method, as given in the next subsection in equation (15).

EXAMPLE

5

Derive an expression for the local truncation errors of Euler's method.

Solution: Since Euler's method is given by $y_{n+1} = y_n + hf(t_n, y_n)$, the local truncation errors are defined by

$$y(t_{n+1}) = y(t_n) + hf(t_n, y(t_n)) + T_{n+1}, \tag{13}$$

where $y(t)$ is the unique solution of the initial value problem $y' = f(t, y), y(t_0) = y_0$. However, $f(t_n, y(t_n)) = y'(t_n)$, and so equation (13) can be expressed as

$$y(t_{n+1}) = y(t_n) + y'(t_n)h + T_{n+1}. \tag{14a}$$

By Taylor's theorem, we can also write

$$y(t_{n+1}) = y(t_n) + y'(t_n)h + \frac{y''(\xi)}{2!}h^2, \tag{14b}$$

where $t_n < \xi < t_{n+1}$. Comparing (14a) and (14b), we see that

$$T_{n+1} = \frac{y''(\xi)}{2!}h^2, \tag{14c}$$

where ξ is some point in the t-interval $t_n < t < t_{n+1}$.

For later use, we note from (14c) that

$$\max_{0 \le n \le N-1} |T_{n+1}| \le Kh^2, \tag{14d}$$

where $K = \max_{t_0 \le t \le t_0 + T} |y''(t)|/2!$. ❖

The Order of a Numerical Method

We now define the order of a numerical method and show that the terminology "Taylor series method of order p" is appropriate. We say that a one-step method has **order p** if there are positive constants K and h_0 such that

For any point t_n in the interval $[t_0, t_0 + T - h_0]$ and any step size h satisfying $0 < h \le h_0$, we have

$$|T_{n+1}| \le Kh^{p+1}. \tag{15}$$

Note that, in inequality (15), the constant K does not depend on the index n. From inequality (14d) of Example 5, we see that Euler's method has order $p = 1$. Similar arguments show that the Taylor series methods (8b) and (8c) have orders 2 and 3, respectively. In general, the Taylor series method

$$y_{n+1} = y_n + hy'_n + \frac{h^2}{2!}y''_n + \frac{h^3}{3!}y'''_n + \cdots + \frac{h^p}{p!}y_n^{(p)}$$

has order p; this is consistent with our prior use of the term.

The order of a numerical method is a measure of how well the method replicates the Taylor expansion of the solution. A numerical method of order p has local truncation errors that satisfy $|T_{n+1}| \le Kh^{p+1}$. From Taylor's theorem, therefore, it follows that a pth order one-step method correctly replicates the Taylor series up to and including the term of order h^p.

The Global Error

The size of the local truncation error for a numerical method tells us how far we would deviate from $y(t_{n+1})$ if we were to take a single step of the method starting on the solution curve at the point $(t_n, y(t_n))$. However, except for the first step of the method [when we start at the initial point $(t_0, y(t_0)) = (t_0, y_0)$], we do not expect to take steps that begin on the solution curve. In this sense, the local truncation error is not a quantity that we can calculate without knowing the true solution of the initial value problem. We are using the concept of the local truncation error to define the order of a numerical method and to establish the convergence of numerical methods.

In practical computations, we are primarily interested in the **global errors**,

$$y(t_n) - y_n \quad \text{for} \quad n = 0, 1, \ldots, N, \tag{16}$$

where $y(t)$ is the true solution of the initial value problem and y_n is the numerical method's estimate to $y(t_n)$.

In discussing local truncation errors and global errors, it is convenient to use the **"Big O" order symbol** (also known as the **Landau symbol**). This symbol is frequently used to characterize inequalities such as (15). We use the notation

$$q(h) = O(h^r), \quad h \to 0 \quad \text{or simply} \quad q(h) = O(h^r)$$

to mean there exists some positive constant K such that $|q(h)| \le Kh^r$ for all positive, sufficiently small h. Thus, inequality (15) can be written as

$$T_{n+1} = O(h^{p+1}).$$

Note that the order of a numerical method, p, is one integer less than the order of the local truncation error. For example, from equation (14d), the local truncation error of Euler's method is $O(h^2)$, and therefore we say that Euler's method is a first order method.

In an appendix to Section 7.4, we state a theorem that shows how (for the types of problems and numerical methods we are considering) the order of the numerical method and the size of the global errors are related. In particular, there is a positive constant M such that the global errors for a pth order method satisfy the inequality

$$\max_{0 \le n \le N} |y(t_n) - y_n| \le Mh^p. \tag{17}$$

Inequality (17) tells us how the global errors are reduced when h is reduced. If we are using a pth order method and if we reduce the step size h by $\frac{1}{2}$, then we anticipate that the global errors will be reduced by about $\left(\frac{1}{2}\right)^p$.

E X A M P L E

6

We again consider the example

$$y' = y^2, \qquad y(0) = 1, \qquad 0 \le t \le 0.95.$$

Use Euler's method to generate numerical solutions, first using step size $h_1 = 0.05$ and then using step size $h_2 = 0.025$. From (17) with $p = 1$, we expect the global errors to be reduced by approximately $\frac{1}{2}$ when h is reduced by $\frac{1}{2}$. Compare the global errors at $t = 0.05, 0.10, 0.15, \ldots, 0.95$. Does it appear

(continued)

(continued)

that the errors resulting from the smaller step size are about half the size of the errors of the larger step?

Solution: The results are listed in Table 7.3. The column headed E_1 gives the global errors $y(t_k) - y_k$, made using $h_1 = 0.05$. Similarly, the column headed E_2 lists the global errors, at the same values of t, made using $h_2 = 0.025$. As predicted by (17), the ratios of E_2 to E_1 (given in the column headed E_2/E_1) are close to 0.5 for smaller values of t. The ratios tend to deviate from 0.5 as the values t_k approach $t = 1$, where the exact solution has a vertical asymptote.

TABLE 7.3

The Results of Example 6
Note, as predicted by (17), that $E_2 \approx E_1/2$.

t_k	E_1 (h = 0.05)	E_2 (h = 0.025)	E_2/E_1
0.0500	0.0026	0.0014	0.5191
0.1000	0.0060	0.0031	0.5206
0.1500	0.0103	0.0054	0.5223
0.2000	0.0158	0.0083	0.5242
0.2500	0.0230	0.0121	0.5264
0.3000	0.0324	0.0171	0.5290
0.3500	0.0448	0.0238	0.5320
0.4000	0.0614	0.0329	0.5355
0.4500	0.0841	0.0454	0.5396
0.5000	0.1156	0.0630	0.5446
0.5500	0.1603	0.0883	0.5507
0.6000	0.2255	0.1259	0.5583
0.6500	0.3239	0.1840	0.5680
0.7000	0.4793	0.2782	0.5805
0.7500	0.7386	0.4412	0.5973
0.8000	1.2068	0.7491	0.6207
0.8500	2.1541	1.4111	0.6551
0.9000	4.4692	3.1700	0.7093
0.9500	12.9397	10.4052	0.8041

❖

The Need for Computationally Friendly Algorithms

Taylor series expansions provide a clear blueprint for how to improve the accuracy of a numerical algorithm. The Exercises develop such algorithms for a variety of problems. In specific cases, as in Examples 1 and 2, the computations are not overly difficult. In other cases, as the order of the algorithm increases, the computations rapidly become unwieldy and the possibility of mistakes in programming the numerical method grows as well. Moreover, Taylor series methods are problem specific; the various partial derivatives of $f(t, y)$ must be recomputed every time we are given a new differential equation.

For these reasons, a Taylor series method is not very attractive as an all-purpose method for solving initial value problems. The challenge is to develop algorithms that replicate the desired number of terms in the Taylor series expansion (thereby achieving the desired accuracy) but do not require calculation of partial derivatives. In particular, we want algorithms that require only evaluations of the function f.

Heun's method and the modified Euler's method, developed in Section 7.2, provide insight into how these goals might be achieved using compositions of functions. Computers can evaluate functions with relative ease, and compositions of functions, while they might look formidable to us, are also evaluated with relative ease on a computer. Nested compositions of functions, such as those used in Heun's method and the modified Euler's method, form the basis of Runge-Kutta methods that are discussed in Section 7.4. Runge-Kutta methods achieve the accuracy of Taylor series methods, but in a computationally friendly way.

EXERCISES

Exercises 1–10:

Assume, for the given differential equation, that $y(0) = 1$.

(a) Use the differential equation itself to determine the values $y'(0), y''(0), y'''(0), y^{(4)}(0)$ and form the Taylor polynomial

$$P_4(t) = y(0) + y'(0)t + \frac{y''(0)}{2!}t^2 + \frac{y'''(0)}{3!}t^3 + \frac{y^{(4)}(0)}{4!}t^4.$$

(b) Verify that the given function is the solution of the initial value problem consisting of the differential equation and initial condition $y(0) = 1$.

(c) Evaluate both the exact solution $y(t)$ and $P_4(t)$ at $t = 0.1$. What is the error $E(0.1) = y(0.1) - P_4(0.1)$? [Note that $E(0.1)$ is the local truncation error incurred in using a Taylor series method of order 4 to step from $t_0 = 0$ to $t_1 = 0.1$ using step size $h = 0.1$.]

1. $y' = -y + 2$; $y(t) = 2 - e^{-t}$

2. $y' = 2ty$; $y(t) = e^{t^2}$

3. $y' = ty^2$; $y(t) = \left(1 - \frac{t^2}{2}\right)^{-1}$

4. $y' = t^2 + y$; $y(t) = 3e^t - (t^2 + 2t + 2)$

5. $y' = y^{1/2}$; $y(t) = \left(1 + \frac{t}{2}\right)^2$

6. $y' = ty^{-1}$; $y(t) = \sqrt{1 + t^2}$

7. $y' = y + \sin t$; $y(t) = \dfrac{3e^t - \cos t - \sin t}{2}$

8. $y' = y^{3/4}$; $y(t) = \left(1 + \frac{t}{4}\right)^4$

9. $y' = 1 + y^2$; $y(t) = \tan\left(t + \frac{\pi}{4}\right)$

10. $y' = -4t^3 y$; $y(t) = e^{-t^4}$

Results analogous to Theorem 7.1 guaranteeing the existence of analytic solutions can be established for higher order scalar problems and first order systems. The development of higher order numerical methods for such problems will be addressed in Section 7.4. Exercises 11–14 illustrate how a series expansion of the solution of a higher order scalar problem can be obtained from the differential equation itself. For example, consider the initial value problem $y'' = f(t, y, y'), y(t_0) = y_0, y'(t_0) = y_0'$. From the equation, we have $y''(t_0) = f(t_0, y_0, y_0')$. Differentiating the identity $y''(t) = f(t, y(t), y'(t))$ allows us to obtain $y'''(t_0)$ and then $y^{(4)}(t_0)$ and so forth.

Exercises 11–14:

In each exercise, for the given t_0,

(a) Obtain the fifth degree Taylor polynomial approximation of the solution,

$$P_5(t) = y(t_0) + y'(t_0)(t - t_0) + \frac{y''(t_0)}{2!}(t - t_0)^2 + \cdots + \frac{y^{(5)}(t_0)}{5!}(t - t_0)^5.$$

(b) If the exact solution is given, calculate the error at $t = t_0 + 0.1$.

11. $y'' - 3y' + 2y = 0$, $y(0) = 1$, $y'(0) = 0$; $t_0 = 0$.
   The exact solution is $y(t) = 2e^t - e^{2t}$.

12. $y'' - y' = 0$, $y(1) = 1$, $y'(1) = 2$; $t_0 = 1$.
   The exact solution is $y(t) = -1 + 2e^{(t-1)}$.

13. $y''' - y' = 0$, $y(0) = 1$, $y'(0) = 2$, $y''(0) = 0$; $t_0 = 0$.
   The exact solution is $y(t) = 1 + e^t - e^{-t}$.

14. $y'' + y + y^3 = 0$, $y(0) = 1$, $y'(0) = 0$; $t_0 = 0$

Exercises 15–18:

In each exercise, determine the largest positive integer r such that $q(h) = O(h^r)$. [Hint: Determine the first nonvanishing term in the Maclaurin expansion of q.]

15. $q(h) = \sin 2h$ **16.** $q(h) = 2h + h^3$

17. $q(h) = 1 - \cos h$ **18.** $q(h) = e^h - (1 + h)$

19. Give an example of functions f and g such that $f(h) = O(h), g(h) = O(h)$ but $(f + g)(h) = O(h^2)$.

Exercises 20–23:

For the given initial value problem,

(a) Execute 20 steps of the Taylor series method of order p for $p = 1, 2, 3$. Use step size $h = 0.05$.

(b) In each exercise, the exact solution is given. List the errors of the Taylor series method calculations at $t = 1$.

20. $y' = \dfrac{t}{y + 1}$, $y(0) = 1$.   The exact solution is $y(t) = -1 + \sqrt{t^2 + 4}$.

21. $y' = 2ty^2$, $y(0) = -1$.   The exact solution is $y(t) = \dfrac{-1}{1 + t^2}$.

22. $y' = \dfrac{1}{2y}$, $y(0) = 1$.   The exact solution is $y(t) = \sqrt{1 + t}$.

23. $y' = \dfrac{1 + y^2}{1 + t}$, $y(0) = 0$.   The exact solution is $y(t) = \tan[\ln(1 + t)]$.

Exercises 24–27:

Assume that a pth order Taylor series method is used to solve an initial value problem. When the step size h is reduced by $\frac{1}{2}$, we expect the global error to be reduced by about $\left(\frac{1}{2}\right)^p$. Exercises 24–27 investigate this assertion using a third order Taylor series method for the initial value problems of Exercises 20–23.

Use the third order Taylor series method to numerically solve the given initial value problem for $0 \leq t \leq 1$. Let E_1 denote the global error at $t = 1$ with step size $h = 0.05$ and E_2 the error at $t = 1$ when $h = 0.025$. Calculate the error ratio E_2/E_1. Is the ratio close to $1/8$?

24. $y' = \dfrac{t}{y+1}$, $y(0) = 1$

25. $y' = 2ty^2$, $y(0) = -1$

26. $y' = \dfrac{1}{2y}$, $y(0) = 1$

27. $y' = \dfrac{1+y^2}{1+t}$, $y(0) = 0$

7.4 Runge-Kutta Methods

In this section, we discuss Runge-Kutta methods as a way of numerically solving the initial value problem

$$y' = f(t, y), \qquad y(t_0) = y_0. \tag{1}$$

Runge-Kutta methods are based on Taylor series methods, but they use nested compositions of function evaluations instead of the partial derivatives of $f(t, y)$ required by a Taylor series method. In theory, one can achieve any desired level of accuracy using the Runge-Kutta approach.

Heun's method and the modified Euler's method are two familiar algorithms that use the Runge-Kutta philosophy of evaluating compositions of functions. For instance, Heun's method has the form

$$y_{n+1} = y_n + \frac{h}{2}\left[f(t_n, y_n) + f(t_n + h, y_n + hf(t_n, y_n))\right], \qquad n = 0, 1, 2, \ldots, N-1.$$

Heun's method is easy to implement—in order to take a step, we need only evaluate the function $f(t, y)$ at the current estimate (t_n, y_n) and at the point $(t_n + h, y_n + hf(t_n, y_n))$. Moreover, as is shown in Example 1, Heun's method is a second order method. In contrast, a second order Taylor series method requires the calculation of two partial derivatives, $f_t(t_n, y_n)$ and $f_y(t_n, y_n)$, in order to make a step with comparable second order accuracy.

EXAMPLE 1

Calculate the order of Heun's method,

$$y_{n+1} = y_n + \frac{h}{2}\left[f(t_n, y_n) + f(t_n + h, y_n + hf(t_n, y_n))\right]. \tag{2}$$

Solution: Let $y(t)$ denote the unique solution of the initial value problem (1). To determine the order of the one-step method (2), we need to find an expression for the local truncation errors, T_{n+1} [recall equation (12) in Section 7.3].

Assume that we apply Heun's method starting on the exact solution curve at t_n—that is, with $y_n = y(t_n)$. To determine the local truncation error, we must first unravel the composition $f(t_n + h, y_n + hf)$ [where functions without arguments will be assumed to be evaluated at (t_n, y_n)]. Expanding $f(t_n + h, y_n + hf)$ in a Taylor series about (t_n, y_n), we obtain

$$f(t_n + h, y_n + hf) = f + (f_t + f_y f)h + \tfrac{1}{2}(f_{tt} + 2f_{ty}f + f_{yy}f^2)h^2 + O(h^3). \tag{3}$$

(continued)

(continued)

Using this expansion in (2) yields

$$y_{n+1} = y_n + \frac{h}{2}\left[f + f + (f_t + f_y f)h + \frac{1}{2}(f_{tt} + 2f_{ty}f + f_{yy}f^2)h^2 + O(h^3)\right]$$

$$= y_n + fh + (f_t + f_y f)\frac{h^2}{2} + (f_{tt} + 2f_{ty}f + f_{yy}f^2)\frac{h^3}{4} + O(h^4).$$
(4)

We compare this expansion with the Taylor series of the exact solution, $y(t_{n+1})$. Using the fact that $y(t_n) = y_n$ and using the expressions for $y'(t_n), y''(t_n)$, $y'''(t_n)$ derived in Section 7.3, we have

$$y(t_{n+1}) = y_n + fh + (f_t + f_y f)\frac{h^2}{2} + (f_{tt} + 2f_{ty}f + f_{yy}f^2 + f_y f_t + f_y^2 f)\frac{h^3}{6} + O(h^4).$$
(5)

Comparing expansions (4) and (5), we see that they agree up to and including the $O(h^2)$ terms but that the $O(h^3)$ term in the Heun method expansion does not correctly replicate the $O(h^3)$ term in the Taylor series of the exact solution. Therefore, the local truncation error of Heun's method is $T_{n+1} = O(h^3)$, and Heun's method is second order. ❖

Second Order Runge-Kutta Methods

To generalize the approach suggested by Heun's method, we choose a set of points $(\theta_i, \gamma_i), i = 1, 2, \ldots, k$ that lie in the ty-plane, in the vertical strip bounded by the lines $t = t_n$ and $t = t_{n+1}$. As Figure 7.3 suggests, these points sample the direction field in the vicinity of the point (t_n, y_n). To formalize this idea of sampling the direction field, consider a one-step method

$$y_{n+1} = y_n + h\phi(t_n, y_n; h),$$
(6)

where the increment function is defined by

$$\phi(t_n, y_n; h) = A_1 f(\theta_1, \gamma_1) + A_2 f(\theta_2, \gamma_2) + \cdots + A_k f(\theta_k, \gamma_k).$$
(7)

The constants $A_1, A_2, \ldots, A_k$ are the **weights** of the method (6). Thus, the increment function is a weighted sum of direction field slopes. For a fixed integer k,

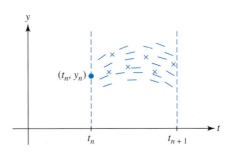

FIGURE 7.3

A portion of the direction field for $y' = f(t, y)$ near our latest estimate (t_n, y_n). We use a weighted sum of direction field evaluations at the points marked "×" to evolve the numerical solution from t_n to t_{n+1}.

the local truncation error is reduced by selecting weights, $\{A_i\}_{i=1}^k$, and direction field sampling points, $\{(\theta_i, \gamma_i)\}_{i=1}^k$, so that the method (6) replicates as many terms in the Taylor series expansion of the local solution as possible.

When $k = 2$, method (6) has the form

$$y_{n+1} = y_n + h[A_1 f(\theta_1, \gamma_1) + A_2 f(\theta_2, \gamma_2)]. \tag{8}$$

We need to choose the sampling points (θ_1, γ_1) and (θ_2, γ_2) and weights A_1 and A_2. Since we are viewing the term $A_1 f(\theta_1, \gamma_1) + A_2 f(\theta_2, \gamma_2)$ as an average slope, we want the sampling points to be near (t_n, y_n) and to be representative of the direction field between $t = t_n$ and $t = t_{n+1}$. A reasonable choice for one of the sampling points is $(\theta_1, \gamma_1) = (t_n, y_n)$. For a second point, our previous study suggests that we might sample somewhere along the "Euler line"—the line of slope $f(t_n, y_n)$ that passes through the point (t_n, y_n). Thus, as a second sampling point, we choose

$$(\theta_2, \gamma_2) = (t_n + \alpha h, y_n + \alpha h f(t_n, y_n)),$$

where α is a constant, $0 < \alpha \le 1$. See Figure 7.4.

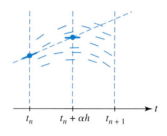

FIGURE 7.4

Given the two-sample method (8), we generally choose one sample at (t_n, y_n) and the second somewhere along the "Euler line," at $(t_n + \alpha h, y_n + \alpha h f(t_n, y_n))$, where α is a constant, $0 < \alpha \le 1$.

With the choices shown in Figure 7.4, method (8) has the form

$$y_{n+1} = y_n + h\left[A_1 f(t_n, y_n) + A_2 f(t_n + \alpha h, y_n + \alpha h f(t_n, y_n))\right]. \tag{9}$$

We now need to select constants A_1, A_2, and α. Since the right-hand side of equation (9) is a function of h, it makes sense to expand the right-hand side in powers of h, with the objective of choosing the constants so that y_{n+1} matches a Taylor series method through as many powers of h as possible.

Expanding the right-hand side of (9) gives

$$y_{n+1} = y_n + h[A_1 f(t_n, y_n) + A_2 f(t_n + \alpha h, y_n + \alpha h f(t_n, y_n))]$$
$$= y_n + h[A_1 f(t_n, y_n) + A_2\{f(t_n, y_n) + f_t(t_n, y_n)\alpha h$$
$$+ f_y(t_n, y_n)\alpha h f(t_n, y_n) + O(h^2)\}]$$
$$= y_n + h(A_1 + A_2)f(t_n, y_n) + h^2\alpha A_2\left[f_t(t_n, y_n) + f_y(t_n, y_n)f(t_n, y_n)\right] + O(h^3).$$

We now attempt to match this expansion with the Taylor series method of order 2,

$$y_{n+1} = y_n + hf(t_n, y_n) + \frac{h^2}{2}\left[f_t(t_n, y_n) + f_y(t_n, y_n)f(t_n, y_n)\right].$$

Our objective is to select the parameters at our disposal, A_1, A_2, and α, so as to maximize the agreement between the expansions. Comparing the two expansions, we see that we can obtain agreement through terms of order h^2 if A_1, A_2, and α satisfy the equations

$$A_1 + A_2 = 1$$

$$\alpha A_2 = \tfrac{1}{2}. \tag{10}$$

Once we satisfy these constraints, the method (9) matches the second order Taylor series method up through terms of order h^2 and therefore, like the second order Taylor series method, has an $O(h^3)$ local truncation error. [This is the best we can do with method (9). It is impossible to select A_1, A_2, and α to match the terms of the third order Taylor series method.]

In (10), we have a system of two (nonlinear) equations in three unknowns. This system has infinitely many solutions,

$$A_2 = \frac{1}{2\alpha} \quad \text{and} \quad A_1 = 1 - \frac{1}{2\alpha}, \tag{11}$$

with $0 < \alpha \le 1$. Since α represents the fraction of the step we move along the Euler line to the second sampling point,

$$(\theta_2, \gamma_2) = (t_n + \alpha h, y_n + \alpha h f(t_n, y_n)),$$

there are two "natural" choices for α, namely $\alpha = \tfrac{1}{2}$ and $\alpha = 1$. If $\alpha = 1$ in equation (11), then $A_1 = \tfrac{1}{2}$ and $A_2 = \tfrac{1}{2}$. With this choice, method (9) reduces to Heun's method,

$$y_{n+1} = y_n + \frac{h}{2}\left[f(t_n, y_n) + f(t_n + h, y_n + hf(t_n, y_n))\right].$$

If $\alpha = \tfrac{1}{2}$ in equation (11), then $A_1 = 0$ and $A_2 = 1$. With this choice, method (9) reduces to the modified Euler's method,

$$y_{n+1} = y_n + hf\left(t_n + \frac{h}{2}, y_n + \frac{hf(t_n, y_n)}{2}\right).$$

R-stage Runge-Kutta Methods

In general, a Runge-Kutta method has the form

$$y_{n+1} = y_n + h\phi(t_n, y_n; h), \tag{12a}$$

where the increment function, $\phi(t_n, y_n; h)$, is given by

$$\phi(t_n, y_n; h) = \sum_{j=1}^{R} A_j K_j(t_n, y_n). \tag{12b}$$

In (12b), the terms A_j are constants (the weights) and the terms $K_j(t_n, y_n)$ are direction field samples, usually called **stages**. The stages are defined sequentially as follows:

$$K_1(t_n, y_n) = f(t_n, y_n)$$

$$K_j(t_n, y_n) = f(t_n + \alpha_j h, y_n + h \sum_{i=1}^{j-1} \beta_{ji} K_i(t_n, y_n)), \qquad j = 2, 3, \ldots, R, \tag{12c}$$

where $0 < \alpha_j \le 1$ and where $\beta_{j,1} + \beta_{j,2} + \cdots + \beta_{j,j-1} = \alpha_j$.

Method (12) is called an **R-stage Runge-Kutta method**. A Runge-Kutta method can be viewed as a "staged" sampling process. That is, for each j, we choose a value α_j that determines the t-coordinate of the jth sampling point. Then [see (12c)] the y-coordinate of the jth sampling point is determined using the prior stages. In this sense, the sampling process is recursive. In (12c), the constraint $0 < \alpha_j \le 1$ means that all sampling points lie between $t = t_n$ and $t = t_{n+1}$. While this description of an R-stage Runge-Kutta method may seem complicated, the format of equation (12) makes programming a Runge-Kutta method very simple (see Figures 7.7 and 7.8 on pages 501 and 502).

An example of a three-stage Runge-Kutta method is

$$y_{n+1} = y_n + \frac{h}{6}(K_1 + 4K_2 + K_3)$$

$$K_1 = f(t_n, y_n)$$

$$K_2 = f\left(t_n + \frac{h}{2}, y_n + \frac{h}{2}K_1\right) \tag{13}$$

$$K_3 = f(t_n + h, y_n - hK_1 + 2hK_2).$$

It is not difficult to show that method (13) has order 3; it matches the third order Taylor series method up through terms of order h^3 but not the order h^4 term.

As we saw in equations (9) and (11), there are infinitely many two-stage, second order Runge-Kutta methods. Similarly, there is an infinite two-parameter family of three-stage, third order Runge-Kutta methods (see Exercises 31–34). Likewise, when the parameters in (12) are chosen properly, there are four-stage, fourth order Runge-Kutta methods. One of the most popular fourth order Runge-Kutta methods (recall Sections 2.10 and 4.9) is

$$y_{n+1} = y_n + \frac{h}{6}(K_1 + 2K_2 + 2K_3 + K_4)$$

$$K_1 = f(t_n, y_n)$$

$$K_2 = f\left(t_n + \frac{h}{2}, y_n + \frac{h}{2}K_1\right) \tag{14}$$

$$K_3 = f\left(t_n + \frac{h}{2}, y_n + \frac{h}{2}K_2\right)$$

$$K_4 = f(t_n + h, y_n + hK_3).$$

Viewing algorithm (14) geometrically, we can envision it as being formed in the following way. First, we calculate K_1, the slope of the tangent line at starting point (t_n, y_n). We proceed a half-step along this tangent line to locate the direction field point at which slope K_2 is evaluated. We use this new slope K_2 to define another line through (t_n, y_n). Proceeding, in turn, a half step along this new line locates the point that determines slope K_3. Finally, we proceed a full step from (t_n, y_n) along the line having slope K_3 to determine the point at which slope K_4 is evaluated. The appropriately weighted average of these four slopes defines the algorithm.

Runge-Kutta Methods for Systems

The discussion in this chapter has focused on the scalar initial value problem

$$y' = f(t, y), \qquad y(t_0) = y_0.$$

As mentioned earlier, the ideas developed and the ensuing methods extend naturally to first order systems. Consider the initial value problem

$$\mathbf{y}' = \mathbf{f}(t, \mathbf{y}), \qquad \mathbf{y}(t_0) = \mathbf{y}_0, \tag{15}$$

where

$$\mathbf{y}(t) = \begin{bmatrix} y_1(t) \\ y_2(t) \\ \vdots \\ y_m(t) \end{bmatrix}, \quad \mathbf{y}_0 = \begin{bmatrix} y_0^{(1)} \\ y_0^{(2)} \\ \vdots \\ y_0^{(m)} \end{bmatrix}$$

and

$$\mathbf{f}(t, \mathbf{y}) = \begin{bmatrix} f_1(t, \mathbf{y}) \\ f_2(t, \mathbf{y}) \\ \vdots \\ f_m(t, \mathbf{y}) \end{bmatrix} = \begin{bmatrix} f_1(t, y_1, y_2, \ldots, y_m) \\ f_2(t, y_1, y_2, \ldots, y_m) \\ \vdots \\ f_m(t, y_1, y_2, \ldots, y_m) \end{bmatrix}.$$

The concept of an analytic function developed in Section 7.3 can be extended to the vector-valued functions $\mathbf{y}(t)$ and $\mathbf{f}(t, \mathbf{y})$. Theorem 7.1 can be extended to give analogous conditions sufficient for the existence of an analytic solution of (15) on an interval of the form $t_0 - \delta < t < t_0 + \delta$ for some $\delta > 0$.

We saw in Section 4.9 how Euler's method and higher order Runge-Kutta methods extend naturally to initial value problems such as (15). For example, the system counterpart of algorithm (14) is

$$\mathbf{y}_{n+1} = \mathbf{y}_n + \frac{h}{6} \left(\mathbf{K}_1 + 2\mathbf{K}_2 + 2\mathbf{K}_3 + \mathbf{K}_4 \right)$$

$$\mathbf{K}_1 = \mathbf{f}(t_n, \mathbf{y}_n)$$

$$\mathbf{K}_2 = \mathbf{f}\left(t_n + \frac{h}{2}, \mathbf{y}_n + \frac{h}{2}\mathbf{K}_1 \right) \tag{16}$$

$$\mathbf{K}_3 = \mathbf{f}\left(t_n + \frac{h}{2}, \mathbf{y}_n + \frac{h}{2}\mathbf{K}_2 \right)$$

$$\mathbf{K}_4 = \mathbf{f}(t_n + h, \mathbf{y}_n + h\mathbf{K}_3).$$

The Damped Pendulum

The next example illustrates how we can apply a Runge-Kutta method to a first order system of the form

$$\mathbf{y}' = \mathbf{f}(t, \mathbf{y}), \qquad \mathbf{y}(t_0) = \mathbf{y}_0.$$

EXAMPLE

2

Consider a pendulum whose motion is influenced not only by its weight but also by a resistive or damping force. The mathematical formulation of this problem leads to an initial value problem involving a scalar second order nonlinear differential equation. We rewrite this scalar second order problem as an equivalent problem for a first order (nonlinear) system and then use the fourth order Runge-Kutta method (16), with a step size of $h = 0.05$, to obtain a numerical solution.

Problem Formulation: The pendulum is formed by a mass m attached to a rod of length l (see Figure 7.5). We neglect the mass of the rod. As the pendulum moves, it is acted on by the force of gravity and also by a damping force, which acts to retard the pendulum motion. We assume this damping force is proportional to the angular velocity of the pendulum and acts in the tangential direction to retard the motion. We obtain

$$ml^2\theta'' = -mgl\sin\theta - \kappa l\theta', \quad \text{or} \quad \theta'' + \frac{\kappa}{ml}\theta' + \frac{g}{l}\sin\theta = 0,$$

where κ is a positive damping constant. We complete the formulation by specifying both θ and θ' at the initial time of interest, say $t = 0$. These two constants give the initial position and initial angular velocity of the pendulum. We adopt the numerical values

$$\frac{\kappa}{ml} = 0.2 \text{ s}^{-1}, \qquad \frac{g}{l} = 1 \text{ s}^{-2}, \qquad \theta(0) = 0 \text{ rad}, \qquad \theta'(0) = 3 \text{ rad/s},$$

and the initial value problem of interest becomes

$$\theta'' + 0.2\theta' + \sin\theta = 0, \qquad \theta(0) = 0, \qquad \theta'(0) = 3.$$

The differential equation is recast as a first order system by defining

$$y_1(t) = \theta(t), \qquad y_2(t) = \theta'(t), \quad \text{and} \quad \mathbf{y}(t) = \begin{bmatrix} y_1(t) \\ y_2(t) \end{bmatrix}.$$

The initial value problem becomes

$$\mathbf{y}' = \begin{bmatrix} y_2 \\ -\sin y_1 - 0.2y_2 \end{bmatrix}, \qquad \mathbf{y}(0) = \begin{bmatrix} 0 \\ 3 \end{bmatrix}. \tag{17}$$

We will solve initial value problem (17) numerically using algorithm (16).

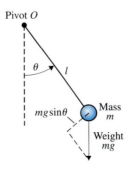

Pivot O

θ

l

$mg\sin\theta$

Mass m

Weight mg

FIGURE 7.5

The damped pendulum described in Example 2.

(continued)

(continued)

What Should We Expect?   Before embarking on a numerical solution, it's usually worthwhile to bring to bear any available physical insights that will help determine what to expect. We know that, generally, solutions of nonlinear initial value problems do not exist on arbitrarily large time intervals. However, because of the nature of the pendulum motion it describes, we expect the exact solution of (17) to exist on an arbitrarily large time interval.

We saw in Chapter 6 that, in the absence of damping, a pendulum starting at $\theta(0) = 0$ with $\theta'(0) = 2$ has just enough energy to reach the inverted position (in the limit as $t \to \infty$). In our case, the initial angular velocity is greater, since $\theta'(0) = 3$. Damping, however, retards the motion and causes the pendulum to lose energy. If damping is not too large, we expect the pendulum to go past the inverted position at least once. If damping is large enough, however, the accompanying loss of energy will more than offset the increase in initial energy and the pendulum will not reach the inverted position. It's not clear at the outset which possibility will occur. In any event, the pendulum eventually will have insufficient energy to reach the inverted position, and it will simply swing back and forth with decreasing amplitude as time increases. Based on these observations, what do you expect the graphs of $\theta(t)$ and $\theta'(t)$ to look like?

Interpreting the Results   Figure 7.6 shows the results of the numerical computation. Note that the graph of $y_1(t) = \theta(t)$ increases from zero to a maximum of about 8.29 rad. Since $2\pi \approx 6.28$, the graph tells us that the pendulum makes one complete counterclockwise revolution, rotating an additional 2 rad $\approx 115°$ beyond the vertically downward position before falling back, beginning to swing back and forth with decreasing amplitude as time progresses. The graph has a horizontal asymptote of 2π, since the pendulum approaches the vertically downward rest position as $t \to \infty$.

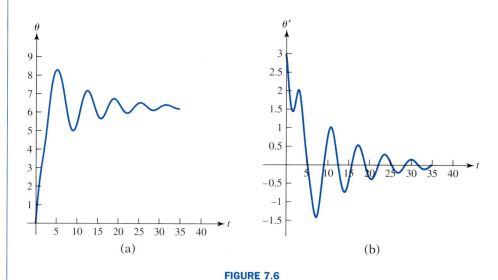

FIGURE 7.6

(a) The graph of $y_1(t) = \theta(t)$. (b) The graph of $y_2(t) = \theta'(t)$.

Is the graph of $y_2(t) = \theta'(t)$ consistent with this physical interpretation? What do the initial minimum and maximum (for $t > 0$) of this graph correspond to? Should they occur while the dependent variable (angular velocity) is positive? Should the zero crossings of this graph occur at the critical points of $y_1(t) = \theta(t)$? Should the maxima of $y_2(t) = \theta'(t)$ occur when the pendulum is in the vertically downward position? Should the graph of $y_2(t) = \theta'(t)$ have a horizontal asymptote of zero? Subjecting your numerical solution to simple common-sense checks such as these is an important final step. ❖

Coding a Runge-Kutta Method

We conclude this section with a short discussion about the practical aspects of writing a program to implement a Runge-Kutta method. Figures 7.7 and 7.8 list the program used to generate the numerical solution of Example 2. This particular code was written in MATLAB, but the principles are the same for any programming language.

```
%
%   Set the initial conditions for the
%      initial value problem of Example 2
%
t=0;
y=[0,3]';
h=0.05;
output=[t,y(1),y(2)];
%
%
%   Execute the fourth order Runge-Kutta method
%      on the interval [0, 30]
%
for i=1:600
    ttemp=t;
    ytemp=y;
    k1=f(ttemp,ytemp);
    ttemp=t+h/2;
    ytemp=y+(h/2)*k1;
    k2=f(ttemp,ytemp);
    ttemp=t+h/2;
    ytemp=y+(h/2)*k2;
    k3=f(ttemp,ytemp);
    ttemp=t+h;
    ytemp=y+h*k3;
    k4=f(ttemp,ytemp);
    y=y+(h/6)*(k1+2*k2+2*k3+k4);
    t=t+h;
    output=[output;t,y(1),y(2)];
end
```

FIGURE 7.7

A Runge-Kutta code for the initial value problem in Example 2.

```
function yp=f(t,y)
yp=zeros(2,1);
yp(1)=y(2);
yp(2)=-sin(y(1))-0.2*y(2);
```

FIGURE 7.8

A function subprogram that evaluates $\mathbf{f}(t, \mathbf{y})$ for the differential equation of Example 2.

Note first that no matter what numerical method we decide to use for the initial value problem

$$\mathbf{y}' = \mathbf{f}(t, \mathbf{y}), \qquad \mathbf{y}(t_0) = \mathbf{y}_0,$$

we need to write a subprogram (or module) that evaluates $\mathbf{f}(t, \mathbf{y})$. Such a module is listed in Figure 7.8 for the initial value problem of Example 2. Figure 7.7 lists a MATLAB program that executes 600 steps of the fourth order Runge-Kutta method (16) for the initial value problem of Example 2.

The code listed in Figure 7.7 stays as close as possible to the notation and format of the fourth order Runge-Kutta method (16). It is always a good idea to use variable names (such as k_1 and k_2) that match the names in the algorithm. Beyond the choice of variable names, the code in Figure 7.8 also mimics the steps of algorithm (16) as closely as possible. Adhering to such conventions makes programs easier to read and debug.

EXERCISES

Exercises 1–10:

We reconsider the initial value problems studied in the Exercises of Section 7.3. The solution of the differential equation satisfying initial condition $y(0) = 1$ is given.

(a) Carry out one step of the third order Runge-Kutta method (13) using a step size $h = 0.1$, obtaining a numerical approximation of the exact solution at $t = 0.1$.

(b) Carry out one step of the fourth order Runge-Kutta method (14) using a step size $h = 0.1$, obtaining a numerical approximation of the exact solution at $t = 0.1$.

(c) Examine the exact solution. Should either or both of the Runge-Kutta methods, in principle, yield an exact answer for the particular problem being considered? Explain.

(d) Compare the numerical values obtained in parts (a) and (b) with the exact solution evaluated at $t = 0.1$. Are the results consistent with your answer in part (c)? Is the error incurred using the four-stage algorithm less than the error for the three-stage calculation?

1. $y' = -y + 2$; $y(t) = 2 - e^{-t}$

2. $y' = 2ty$; $y(t) = e^{t^2}$

3. $y' = ty^2$; $y(t) = \dfrac{2}{2 - t^2}$

4. $y' = t^2 + y$; $y(t) = 3e^t - (t^2 + 2t + 2)$

5. $y' = \sqrt{y}$; $y(t) = \left(1 + \dfrac{t}{2}\right)^2$

6. $y' = \dfrac{t}{y}$; $y(t) = \sqrt{1 + t^2}$

7. $y' = y + \sin t$; $y(t) = \dfrac{3e^t - \cos t - \sin t}{2}$

8. $y' = y^{3/4}; \quad y(t) = \left(1 + \dfrac{t}{4}\right)^4$ **9.** $y' = 1 + y^2; \quad y(t) = \tan\left(t + \dfrac{\pi}{4}\right)$

10. $y' = -4t^3 y; \quad y(t) = e^{-t^4}$

Exercises 11–16:

For the given initial value problem, an exact solution in terms of familiar functions is not available for comparison. If necessary, rewrite the problem as an initial value problem for a first order system. Implement one step of the fourth order Runge-Kutta method (14), using a step size $h = 0.1$, to obtain a numerical approximation of the exact solution at $t = 0.1$.

11. $y'' + ty' + y = 0, \quad y(0) = 1, \quad y'(0) = -1$

12. $\dfrac{d}{dt}\left(e^t \dfrac{dy}{dt}\right) + ty = 1, \quad y(0) = 1, \quad y'(0) = 2$

13. $\mathbf{y}' = \begin{bmatrix} 0 & t \\ e^t & 0 \end{bmatrix}\mathbf{y} + \begin{bmatrix} 1 \\ t \end{bmatrix}, \quad \mathbf{y}(0) = \begin{bmatrix} 2 \\ 1 \end{bmatrix}$ **14.** $\mathbf{y}' = \begin{bmatrix} -1 & t \\ 2 & 0 \end{bmatrix}\mathbf{y}, \quad \mathbf{y}(0) = \begin{bmatrix} -1 \\ 1 \end{bmatrix}$

15. $y''' - ty = 0, \quad y(0) = 1, \quad y'(0) = 0, \quad y''(0) = -1$

16. $y'' + z + ty = 0$

     $z' - y = t, \quad y(0) = 1, \quad y'(0) = 2, \quad z(0) = 0$

Exercises 17–18:

One differential equation for which we can explicitly demonstrate the order of the Runge-Kutta algorithm is the linear homogeneous equation $y' = \lambda y$, where λ is a constant.

17. (a) Verify that the exact solution of $y' = \lambda y, y(t_0) = y_0$ is $y(t) = y_0 e^{\lambda(t-t_0)}$.

     (b) Show, for the three-stage Runge-Kutta method (13), that

$$y(t_n) + h\phi(t_n, y(t_n); h) = y(t_n)\left[1 + \lambda h + \frac{(\lambda h)^2}{2!} + \frac{(\lambda h)^3}{3!}\right].$$

     (c) Show that $y(t_{n+1}) = y(t_n)e^{\lambda h}$.

     (d) What is the order of the local truncation error?

18. Repeat the calculations of Exercise 17 using the four-stage Runge-Kutta method (14). In this case, show that the local truncation error is $O(h^5)$.

Exercises 19–22:

In these exercises, we ask you to use the fourth order Runge-Kutta method (14) to solve the problems in Exercises 20–23 of Section 7.3.

(a) For the given initial value problem, execute 20 steps of the method (14); use step size $h = 0.05$.

(b) The exact solution is given. Compare the numerical approximation y_{20} and the exact solution $y(t_{20}) = y(1)$.

19. $y' = \dfrac{t}{y+1}, \quad y(0) = 1.$ The exact solution is $y(t) = -1 + \sqrt{t^2 + 4}$.

20. $y' = 2ty^2, \quad y(0) = -1.$ The exact solution is $y(t) = \dfrac{-1}{1+t^2}$.

21. $y' = \dfrac{1}{2y}, \quad y(0) = 1.$ The exact solution is $y(t) = \sqrt{1+t}$.

22. $y' = \dfrac{1+y^2}{1+t}$, $y(0) = 0$.   The exact solution is $y(t) = \tan[\ln(1+t)]$.

Exercises 23–25:

In each exercise,

(a) Verify that the given function is the solution of the initial value problem posed. If the initial value problem involves a higher order scalar differential equation, rewrite it as an equivalent initial value problem for a first order system.

(b) Execute the fourth order Runge-Kutta method (16) over the specified t-interval, using step size $h = 0.1$, to obtain a numerical approximation of the exact solution. Tabulate the components of the numerical solution with their exact solution counterparts at the endpoint of the specified interval.

23. $y'' + 2y' + 2y = -2$, $y(0) = 0$, $y'(0) = 1$; $y(t) = e^{-t}(\cos t + 2\sin t) - 1$; $0 \le t \le 2$

24. $\mathbf{y}' = \begin{bmatrix} -1 & \frac{1}{2} \\ \frac{1}{2} & -1 \end{bmatrix} \mathbf{y}$, $\mathbf{y}(0) = \begin{bmatrix} 2 \\ 0 \end{bmatrix}$; $\mathbf{y}(t) = \begin{bmatrix} e^{-t/2} + e^{-3t/2} \\ e^{-t/2} - e^{-3t/2} \end{bmatrix}$; $0 \le t \le 1$

25. $t^2 y'' - t y' + y = t^2$, $y(1) = 2$, $y'(1) = 2$; $y(t) = t(t + 1 - \ln t)$; $1 \le t \le 2$

Exercises 26–30:

These exercises ask you to use numerical methods to study the behavior of some scalar second order initial value problems. In each exercise, use the fourth order Runge-Kutta method (16) and step size $h = 0.05$ to solve the problem over the given interval.

26. $y'' + 4(1 + 3\tanh t)y = 0$, $y(0) = 1$, $y'(0) = 0$; $0 \le t \le 10$.

This problem might model the motion of a spring-mass system in which the mass is released from rest with a unit initial displacement at $t = 0$ and with the spring stiffening as the motion progresses in time. Plot the numerical solutions for $y(t)$ and $y'(t)$. Since $\tanh t$ approaches 1 for large values of t, we might expect the solution to approximate a solution of $y'' + 16y = 0$ for time t sufficiently large. Do your graphs support this conjecture?

27. $y'' + y + y^3 = 0$, $y(0) = 0$, $y'(0) = 1$; $0 \le t \le 10$.

A nonlinear differential equation having this structure arose in modeling the motion of a nonlinear spring. We are interested in assessing the impact of the nonlinear y^3 term on the motion. Plot the numerical solution for $y(t)$. If the nonlinear term were not present, the initial value problem would have solution $y(t) = \sin t$. On the same graph, plot the function $\sin t$. Does the nonlinearity increase or decrease the period of the motion? How do the amplitudes of the motion differ?

28. $\theta'' + \sin\theta = 0.2\sin t$, $\theta(0) = 0$, $\theta'(0) = 0$; $0 \le t \le 50$.

This nonlinear differential equation is used to model the forced motion of a pendulum initially at rest in the vertically downward position. For small angular displacements, the approximation $\sin\theta \approx \theta$ is often used in the differential equation. Note, however, that the solution of the resulting initial value problem $\theta'' + \theta = 0.2\sin t$, $\theta(0) = 0$, $\theta'(0) = 0$ is given by $\theta(t) = -0.1(\sin t - t\cos t)$, leading to pendulum oscillations that continue to grow in amplitude as time increases. Our goal is to determine how the nonlinear $\sin\theta$ term affects the motion. Plot the numerical solutions for $\theta(t)$ and $\theta'(t)$. Describe in simple terms what the pendulum is doing on the time interval considered.

29. $\theta'' + \sin\theta = 0, \quad \theta(0) = 0, \quad \theta'(0) = 2; \quad 0 \le t \le 20.$

This problem models pendulum motion when the pendulum is initially in the vertically downward position with an initial angular velocity of 2 rad/s. For this conservative system, it was shown in Chapter 6 that $(\theta')^2 - 2\cos\theta = 2$. Therefore, the initial conditions have been chosen so that the pendulum will rotate upward in the positive (counterclockwise) direction, slowing down and approaching the vertically upward position as $t \to \infty$. The phase-plane solution point is moving on the separatrix; thus, loosely speaking, the exact solution is "moving on a knife's edge." If the initial velocity is slightly less, the pendulum will not reach the upright position but will reach a maximum value less than π and then proceed to swing back and forth. If the initial velocity is slightly greater, the pendulum will pass through the vertically upright position and continue to rotate counterclockwise. What happens if we solve this problem numerically? Plot the numerical solutions for $\theta(t)$ and $\theta'(t)$. Interpret in simple terms what the numerical solution is saying about the pendulum motion on the time interval considered. Does the numerical solution conserve energy?

30. $mx'' + \dfrac{2k\delta}{\pi}\tan\left(\dfrac{\pi x}{2\delta}\right) = F(t), \quad x(0) = 0, \quad x'(0) = 0; \quad 0 \le t \le 15.$

This problem was used to model a nonlinear spring-mass system (see Exercise 18 in Section 6.1). The motion is assumed to occur on a frictionless horizontal surface. In this equation, m is the mass of the object attached to the spring, $x(t)$ is the horizontal displacement of the mass from the unstretched equilibrium position, and δ is the length that the spring can contract or elongate. The spring restoring force has vertical asymptotes at $x = \pm\delta$. Time t is in seconds.

Let $m = 100$ kg, $\delta = 0.15$ m, and $k = 100$ N/m. Assume that the spring-mass system is initially at rest with the spring at its unstretched length. At time $t = 0$, a force of large amplitude but short duration is applied:

$$F(t) = \begin{cases} F_0 \sin\pi t, & 0 \le t \le 1 \\ 0, & 1 < t < 15 \end{cases} \text{ newtons.}$$

Solve the problem numerically for the two cases $F_0 = 4$ N and $F_0 = 40$ N. Plot the corresponding displacements on the same graph. How do they differ?

Exercises 31–34:

Third Order Runge-Kutta Methods   As given in equation (12), the form of a three-stage Runge-Kutta method is

$$y_{n+1} = y_n + h[A_1 K_1(t_n, y_n) + A_2 K_2(t_n, y_n) + A_3 K_3(t_n, y_n)], \tag{18a}$$

where

$$\begin{aligned} K_1(t_n, y_n) &= f(t_n, y_n) \\ K_2(t_n, y_n) &= f(t_n + \alpha_2 h, y_n + h\beta_{2,1}K_1(t_n, y_n)) \\ K_3(t_n, y_n) &= f(t_n + \alpha_3 h, y_n + h[\beta_{3,1}K_1(t_n, y_n) + \beta_{3,2}K_2(t_n, \ y_n)]) \end{aligned} \tag{18b}$$

and where [see equation (12c)] $0 < \alpha_2 \le 1, \ 0 < \alpha_3 \le 1, \ \beta_{2,1} = \alpha_2,$ and $\beta_{3,1} + \beta_{3,2} = \alpha_3$. It can be shown (see Lambert[5]) that this three-stage Runge-Kutta method has order 3 if

---

[5] John D. Lambert, *Numerical Methods for Ordinary Differential Systems* (Chichester, England: Wiley, 1991).

the following four equations are satisfied:

$$
\begin{aligned}
A_1 + A_2 + A_3 &= 1 \\
\alpha_2 A_2 + \alpha_3 A_3 &= \tfrac{1}{2} \\
\alpha_2^2 A_2 + \alpha_3^2 A_3 &= \tfrac{1}{3} \\
\alpha_2 \beta_{3,2} A_3 &= \tfrac{1}{6}.
\end{aligned}
\tag{19}
$$

One way to find a solution of this system of four nonlinear equations is first to select values for α_2 and α_3. [Note that α_2 and α_3 determine the t-coordinate of the sampling points defining $K_2(t_n, y_n)$ and $K_3(t_n, y_n)$, respectively.] Once α_2 and α_3 are chosen, the first three equations in (19) can be solved for A_1, A_2, and A_3. The parameters α_2 and α_3 are nonzero; if they are distinct, then there are unique values A_1, A_2, and A_3 that satisfy the first three equations. Having A_1, A_2, and A_3, you can determine $\beta_{3,2}$ from the fourth equation and $\beta_{3,1}$ from the condition $\beta_{3,1} + \beta_{3,2} = \alpha_3$.

31. Consider a three-stage Runge-Kutta method. Show that if the first equation in system (19) holds, then the method has order at least 1.

32. Consider a three-stage Runge-Kutta method. Show that if the first and second equations in system (19) hold, then the method has order at least 2. [Note: In order to obtain order 3, the last two equations in (19) must hold as well.]

33. Determine the values of α_2 and α_3 that give rise to the three-stage third order Runge-Kutta method (13). Then solve equations (19), and verify that Runge-Kutta method (13) results.

34. (a) Verify that the choice of $\alpha_2 = \frac{7}{10}$ and $\alpha_3 = \frac{2}{10}$ leads to another solution of (19) having the same weights A_1, A_2, and A_3 as (13).

(b) Use the values from part (a) to form another three-stage, third order Runge-Kutta method. Test this method on $y' = t/(y + 1), y(0) = 1$, using step size $h = 0.05$. Compute the error at $t = 1$ [the exact solution is $y(t) = -1 + \sqrt{t^2 + 4}$].

Appendix I   Convergence of One-Step Methods

In this appendix, we state a theorem that guarantees convergence of the one-step method,

$$
y_{n+1} = y_n + h\phi(t_n, y_n; h), \qquad n = 0, 1, 2, \ldots, N - 1.
\tag{1}
$$

The convergence theorem, Theorem 7.3, applies to an initial value problem

$$
y' = f(t, y), \qquad y(t_0) = y_0.
$$

Let $f(t, y)$ be a function defined on the rectangle R given by $a < t < b$, $\alpha < y < \beta$. The function f is said to satisfy a **Lipschitz condition in y** if there is a positive constant K such that

Whenever (t, y_1) and (t, y_2) are two points in R, then

$$
|f(t, y_1) - f(t, y_2)| \leq K|y_1 - y_2|.
\tag{2}
$$

The constant K in (2) is called a Lipschitz constant. Note that Lipschitz constants are not unique; if a particular constant K can be used in inequality (2), then so can $K + M$ for any positive constant M. A Lipschitz condition is not an overly restrictive assumption; if the partial derivative $f_y(t, y)$ exists on R, then (by the mean value theorem)

$$f(t, y_1) - f(t, y_2) = f_y(t, y^*)(y_1 - y_2),$$

where y^* is some value between y_1 and y_2. Thus, if we know that $|f_y(t, y)| \leq K$ for all (t, y) in R, then the Lipschitz condition (2) holds where the bound on $|f_y(t, y)|$ serves as a Lipschitz constant K.

Theorem 7.3

Consider the initial value problem

$$y' = f(t, y), \qquad y(t_0) = y_0, \qquad \text{(3)}$$

where $f(t, y)$ is analytic and Lipschitz continuous in the vertical infinite strip defined by $a < t < b$, $-\infty < y < \infty$. Assume that $a < t_0 < t_0 + T < b$.

Let $y_{n+1} = y_n + h\phi(t_n, y_n; h)$ be a pth order one-step method, and let $h = T/N$. Assume that for all step sizes less than some h_0 the increment function ϕ, when applied to the initial value problem (3), satisfies a Lipschitz condition in y with Lipschitz constant L. Then

$$\max_{0 \leq n \leq N} |y(t_n) - y_n| = O(h^p). \qquad \text{(4)}$$

In words, conclusion (4) says that the global error can be bounded by some constant multiple of h^p as long as $0 < h \leq h_0$. Also note that we are asking for a Lipschitz condition to hold on a vertical infinite strip. This rather restrictive condition simplifies the theorem, since it ensures that initial value problem (3) has a unique solution on $[t_0, t_0 + T]$ and that $\phi(t_n, y_n; h)$ is defined for all points t_n in $t_0 \leq t \leq t_0 + T$.

Although Theorem 7.3 was stated for the scalar problem, a similar result can be established for a system of differential equations.

Appendix 2   Stability of One-Step Methods

Theorem 7.3 shows that we can, in principle, achieve arbitrarily good accuracy by using a one-step method with a sufficiently small step size h. We now consider the opposite situation and show that results sometimes become disastrously bad if we inadvertently use a step size that is just a little too large.

In particular, numerical methods for initial value problems are subject to difficulties of "stability." When a numerical method is applied to a given differential equation, it can happen that there is a sharp division between a step size h that is too large and that produces terrible results and a step size h that is small enough to produce acceptable results. Such a stability boundary is illustrated in Figures 7.9 and 7.10. In each case, we used Euler's method to solve

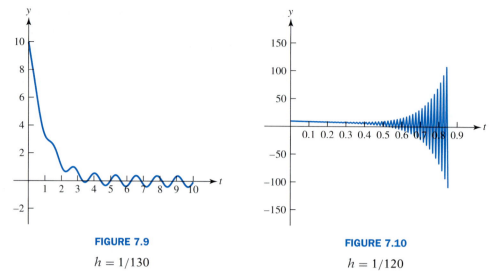

FIGURE 7.9

$h = 1/130$

FIGURE 7.10

$h = 1/120$

the initial value problem

$$y'' + 251y' + 250y = 500\cos 5t, \qquad y(0) = 10, \qquad y'(0) = 0. \tag{1}$$

This differential equation might model the forced vibrations of a spring-mass-dashpot system. (For the coefficients chosen, the spring constant and the damping coefficient per unit mass are relatively large.) As we know from Section 3.10, once the initial transients die out, the solution, $y(t)$, should tend toward a periodic steady-state solution. This expected behavior is exhibited by the results in Figure 7.9, but not by those in Figure 7.10. The only difference between the two computations is that the results in Figure 7.9 were obtained using a step size of $h = \frac{1}{130}$, whereas the results in Figure 7.10 were obtained using a slightly larger step size of $h = \frac{1}{120}$.

The sharp division between the accurate results of Figure 7.9 and the terrible results of Figure 7.10 can be explained by examining the behavior of Euler's method when it is applied to the homogeneous initial value problem $\mathbf{y}' = A\mathbf{y}, \mathbf{y}(0) = \mathbf{y}_0$. Assume that A is a (2×2) constant matrix, with distinct eigenvalues λ_1 and λ_2 and corresponding eigenvectors $\mathbf{u}_1$ and $\mathbf{u}_2$. Since the eigenvalues are distinct, the eigenvectors are linearly independent. Therefore, the initial condition can be represented as

$$\mathbf{y}_0 = \alpha_1 \mathbf{u}_1 + \alpha_2 \mathbf{u}_2 \tag{2}$$

for some constants α_1 and α_2. When applied to the initial value problem $\mathbf{y}' = A\mathbf{y}, \mathbf{y}(0) = \mathbf{y}_0$, Euler's method takes the form $\mathbf{y}_n = \mathbf{y}_{n-1} + hA\mathbf{y}_{n-1}$, or

$$\mathbf{y}_n = (I + hA)\mathbf{y}_{n-1}, \qquad n = 1, 2, \ldots. \tag{3}$$

It follows from (3) that $\mathbf{y}_n = (I + hA)^n \mathbf{y}_0, n = 1, 2, \ldots$. From (2), it follows that

$$\mathbf{y}_n = (I + hA)^n \mathbf{y}_0 = (I + hA)^n (\alpha_1 \mathbf{u}_1 + \alpha_2 \mathbf{u}_2)$$
$$= \alpha_1 (I + hA)^n \mathbf{u}_1 + \alpha_2 (I + hA)^n \mathbf{u}_2.$$

By Exercises 31 and 32 in Section 4.4, $(I + hA)^n \mathbf{u}_j = (1 + h\lambda_j)^n \mathbf{u}_j, j = 1, 2$. Thus,

$$\mathbf{y}_n = (I + hA)^n \mathbf{y}_0 = \alpha_1 (1 + h\lambda_1)^n \mathbf{u}_1 + \alpha_2 (1 + h\lambda_2)^n \mathbf{u}_2. \tag{4}$$

If the eigenvalues λ_1 and λ_2 are both negative, then the exact solution $\mathbf{y}(t)$ tends to $\mathbf{0}$ as t increases. Thus, the output from Euler's method [the sequence $\mathbf{y}_n$ in equation (4)] should also tend to $\mathbf{0}$ as t increases.

Assume that λ_1 and λ_2 are both negative and that $\lambda_1 < \lambda_2 < 0$. If α_1 and α_2 are both nonzero, having $\mathbf{y}_n \to \mathbf{0}$ as $n \to \infty$ requires $|1 + h\lambda_1| < 1$ and $|1 + h\lambda_2| < 1$. These two inequalities reduce to

$$-2 < h\lambda_1 < 0 \quad \text{and} \quad -2 < h\lambda_2 < 0.$$

Therefore, to obtain $\mathbf{y}_n \to \mathbf{0}$ as $n \to \infty$, we need to use a step size h that satisfies the inequality $h < -2/\lambda_1$.

When we write the homogeneous second order equation $y'' + 251y' + 250y = 0$ as a first order system $\mathbf{y}' = A\mathbf{y}$, $\mathbf{y}(0) = \mathbf{y}_0$, we find

$$A = \begin{bmatrix} 0 & 1 \\ -250 & -251 \end{bmatrix} \quad \text{and} \quad \lambda_1 = -250, \quad \lambda_2 = -1.$$

Therefore, for this problem, the critical step size is $\overline{h} = -2/(-250) = 1/125$. If we apply Euler's method to the homogeneous problem $y'' + 251y' + 250y = 0$, we expect to see results qualitatively similar to those shown in Figure 7.10 when we use a step size h, where $h > \overline{h} = 1/125$.

We now return to initial value problem (1), which we represent as $\mathbf{y}' = A\mathbf{y} + \mathbf{g}(t)$:

$$\mathbf{y}' = \begin{bmatrix} 0 & 1 \\ -250 & -251 \end{bmatrix} \mathbf{y} + (500 \cos 5t) \begin{bmatrix} 0 \\ 1 \end{bmatrix}, \quad \mathbf{y}(0) = \begin{bmatrix} 10 \\ 0 \end{bmatrix}.$$

Applying Euler's method to this problem, we obtain

$$\mathbf{y}_{n+1} = \mathbf{y}_n + h[A\mathbf{y}_n + \mathbf{g}(t_n)]$$
$$= [I + hA]^{n+1}\mathbf{y}_0 + h\sum_{j=0}^{n} [I + hA]^j \mathbf{g}(t_{n-j}).$$

If we represent the vectors $\mathbf{y}_0$ and $\mathbf{g}(t_{n-j})$ in terms of the eigenvectors of A,

$$\mathbf{y}_0 = \alpha_1 \mathbf{u}_1 + \alpha_2 \mathbf{u}_2 \quad \text{and} \quad \mathbf{g}(t_{n-j}) = (500 \cos 5t_{n-j})[\beta_1 \mathbf{u}_1 + \beta_2 \mathbf{u}_2],$$

Euler's method produces

$$\mathbf{y}_{n+1} = \alpha_1(1 + h\lambda_1)^{n+1}\mathbf{u}_1 + \alpha_2(1 + h\lambda_2)^{n+1}\mathbf{u}_2$$
$$+ 500h\beta_1 \sum_{j=0}^{n} \cos(t_{n-j})(1 + h\lambda_1)^j \mathbf{u}_1 + 500h\beta_2 \sum_{j=0}^{n} \cos(t_{n-j})(1 + h\lambda_2)^j \mathbf{u}_2.$$

Thus, as we saw with the homogeneous problem $\mathbf{y}' = A\mathbf{y}$, $\mathbf{y}(0) = \mathbf{y}_0$, if we do not choose a step size h such that $|1 + h\lambda_1| < 1$ and $|1 + h\lambda_2| < 1$, Euler's method will produce results qualitatively similar to those in Figure 7.10.

The ideas regarding stability that we have discussed in relation to Euler's method apply to one-step methods in general.

PROJECTS

Project 1: Projectile Motion

At some initial time, a projectile (such as a meteorite) is traveling above Earth; assume that its position and velocity at that instant are known. We consider a model in which the only force acting on the projectile is the gravitational force exerted by Earth. Given this assumption, the projectile's trajectory lies in the plane determined by the projectile's initial position vector and initial velocity vector. For simplicity, we assume this plane is the xy-plane. In our model, the projectile eventually strikes the surface of Earth. Our goal is to determine where and when the impact occurs.

The dynamics of the projectile can be described by the equations

$$x''(t) = \frac{-Gm_e x(t)}{\left[x^2(t) + y^2(t)\right]^{3/2}}, \qquad y''(t) = \frac{-Gm_e y(t)}{\left[x^2(t) + y^2(t)\right]^{3/2}}, \tag{1}$$

where G is the universal gravitational constant and m_e is the mass of Earth. The center of Earth is at the origin, and we let R_e denote the radius of Earth. The values of these constants are taken to be

$$G = 6.673 \times 10^{-11}\ \frac{\text{m}^3}{\text{kg} \cdot \text{s}^2}, \qquad m_e = 5.976 \times 10^{24}\ \text{kg}, \qquad R_e = 6.371 \times 10^6\ \text{m}.$$

According to equation (1), the projectile dynamics are governed by a pair of coupled nonlinear second order differential equations. We will solve the problem numerically.

1. The problem geometry and the nature of the force acting on the projectile suggest the use of polar coordinates. Let

$$x(t) = r(t) \cos[\theta(t)], \qquad y(t) = r(t) \sin[\theta(t)]. \tag{2}$$

   Show that equation (1) transforms into the following pair of equations for the polar variables:

$$r'' - (\theta')^2 r = -\frac{Gm_e}{r^2}, \qquad \theta'' + 2\frac{r'}{r}\theta' = 0. \tag{3}$$

2. Assume that the projectile is launched at time $t = 0$ at a point above Earth's surface, as shown in Figure 7.11. Thus, $r(0) = R_0 > R_e$, $\theta(0) = 0$. Show that

$$r'(0) = v_0 \cos \alpha \quad \text{and} \quad \theta'(0) = \frac{v_0}{R_0} \sin \alpha, \tag{4}$$

   where initial speed v_0 and angle α are as shown in the figure.

3. When performing the numerical calculations, we want to deal with variables whose magnitudes are comparable to unity. To achieve this, we will adopt Earth's radius, R_e, as the unit of length and the hour as the unit of time. For bookkeeping purposes, let

$$T = 3600\ \text{s/hr}.$$

   Define the scaled variables

$$\rho(t) = \frac{r(t)}{R_e} \quad \text{and} \quad \tau = \frac{t}{T}.$$

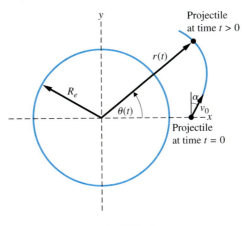

FIGURE 7.11

The initial conditions for the projectile whose trajectory is described by equation (3).

Thus, points on Earth's surface correspond to $\rho = 1$, while 3600 seconds corresponds to one unit of time on the τ-scale. Note, from the chain rule, that

$$\frac{d}{dt} = \frac{d\tau}{dt}\frac{d}{d\tau} = \frac{1}{T}\frac{d}{d\tau}.$$

Show that the initial value problem posed by (3) and (4) transforms into the following problem:

$$\frac{d^2\rho}{d\tau^2} - \left(\frac{d\theta}{d\tau}\right)^2\rho = -19.985\frac{1}{\rho^2}, \qquad \rho(0) = \frac{R_0}{R_e}, \qquad \frac{d\rho(0)}{d\tau} = \frac{v_0 T}{R_0}\cos\alpha$$

$$\frac{d^2\theta}{d\tau^2} + 2\left(\frac{\frac{d\rho}{d\tau}}{\rho}\right)\frac{d\theta}{d\tau} = 0, \qquad \theta(0) = 0, \qquad \frac{d\theta(0)}{d\tau} = \frac{v_0 T}{R_0}\sin\alpha. \tag{5}$$

The constant $Gm_e T^2/R_e^3 = 19.985$ has units of hr^{-2}.

4. Assume that the projectile is initially 9000 km above the surface of Earth with a speed $v_0 = 2000$ m/s and angle $\alpha = 10°$. Translate the assumed information into initial conditions for problem (5).

5. Recast initial value problem (5) as an initial value problem for a first order system, where

$$y_1(\tau) = \rho, \qquad y_2(\tau) = \frac{d\rho}{d\tau}, \qquad y_3(\tau) = \theta, \qquad y_4(\tau) = \frac{d\theta}{d\tau}.$$

6. Solve this problem using a fourth order Runge-Kutta method and a step size $h = 0.005$. The projectile will strike Earth when $\rho = 1$. Execute the program on a τ-interval sufficiently large to achieve this condition. [Hint: Gradually build up the size of the τ-interval. If too large an interval is used at the outset, the numerical solution will "blow up."]

7. Determine the polar coordinates of the impact point and the time of impact.

8. Suppose that the point $\rho = 1, \theta = 0$ corresponds to the point where the equator intersects the prime meridian, while the point $\rho = 1, \theta = \pi/2$ corresponds to the North Pole. Use a globe and determine the approximate location of impact.

Project 2: The Double Pendulum

The double pendulum shown in Figure 7.12 consists of one pendulum attached to another. Two bobs, having masses m_1 and m_2, are connected to frictionless pivots by rigid rods of lengths l_1 and l_2.

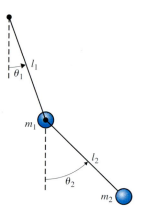

FIGURE 7.12

A double pendulum.

Assume that both pendulums can rotate freely about their pivots and that the masses of the two rigid rods are negligibly small. With respect to the coordinate system shown, the positions of the two bobs are

$$x_1 = l_1 \sin \theta_1, \qquad y_1 = -l_1 \cos \theta_1$$
$$x_2 = l_1 \sin \theta_1 + l_2 \sin \theta_2, \qquad y_2 = -l_1 \cos \theta_1 - l_2 \cos \theta_2. \tag{6}$$

Using g to denote gravitational acceleration, we can show that the angles θ_1 and θ_2 satisfy the following system of coupled nonlinear second order differential equations:

$$(m_1 + m_2)l_1^2 \frac{d^2\theta_1}{dt^2} + m_2 l_1 l_2 \frac{d^2\theta_2}{dt^2} \cos(\theta_1 - \theta_2) + m_2 l_1 l_2 \left(\frac{d\theta_2}{dt}\right)^2 \sin(\theta_1 - \theta_2) + l_1 g(m_1 + m_2) \sin\theta_1 = 0$$

$$m_2 l_2^2 \frac{d^2\theta_2}{dt^2} + m_2 l_1 l_2 \frac{d^2\theta_1}{dt^2} \cos(\theta_1 - \theta_2) - m_2 l_1 l_2 \left(\frac{d\theta_1}{dt}\right)^2 \sin(\theta_1 - \theta_2) + l_2 m_2 g \sin\theta_2 = 0.$$

Prescribing the initial angular position and velocity of each pendulum will complete the specification of the initial value problem.

1. As a check on the differential equations, determine what happens to these equations when

    (a) $m_2 = 0$ (b) $l_2 = 0$ (c) $l_1 = 0$

    In each of these cases, are the equations consistent with what you would expect on purely physical grounds? Note that, in the case of interest, l_1 and l_2 are both positive. Therefore, we can remove an l_1 factor from the first equation and an l_2 factor from the second.

2. Transform the differential equations into an equivalent pair of equations of the form

$$\theta_1'' = f_1(\theta_1, \theta_2, \theta_1', \theta_2')$$
$$\theta_2'' = f_2(\theta_1, \theta_2, \theta_1', \theta_2'). \tag{7}$$

Hint: The original differential equations can be written as

$$
\begin{bmatrix} (m_1 + m_2)l_1 & m_2 l_2 \cos(\theta_1 - \theta_2) \\ m_2 l_1 \cos(\theta_1 - \theta_2) & m_2 l_2 \end{bmatrix} \begin{bmatrix} \theta_1'' \\ \theta_2'' \end{bmatrix}
$$
$$
= -\begin{bmatrix} m_2 l_2 (\theta_2')^2 \sin(\theta_1 - \theta_2) + g(m_1 + m_2) \sin\theta_1 \\ -m_2 l_1 (\theta_1')^2 \sin(\theta_1 - \theta_2) + m_2 g \sin\theta_2 \end{bmatrix}.
$$

Is the determinant of the (2×2) matrix ever zero?

3. Rewrite system (7) as an equivalent four-dimensional first order system by defining

$$ y_1 = \theta_1, \qquad y_2 = \theta_1', \qquad y_3 = \theta_2, \qquad y_4 = \theta_2'. $$

4. Let $m_1 = m_2 = 2$ kg, $l_1 = l_2 = 0.5$ m, and let $g = 9.8$ m/s^2. Assume initial conditions

$$ \theta_1(0) = \frac{\pi}{2}, \qquad \theta_1'(0) = 0, \qquad \theta_2(0) = \frac{5\pi}{6}, \qquad \theta_2'(0) = 0. $$

Solve the initial value problem for **y** on the time interval $0 \le t \le 10$ using the fourth order Runge-Kutta method. Plot $\theta_1(t)$ and $\theta_2(t)$ versus t on separate graphs.

5. To obtain a better insight into how the double pendulum actually moves, use the numerical solutions obtained in part 4 and equations (6) to determine the bob coordinates $(x_i(t), y_i(t)), i = 1, 2$. Create parametric plots of the two bob trajectories on the same graph over the ten-second interval. On this graph, sketch the double pendulum configurations at initial and final times, $t = 0$ and $t = 10$.

Series Solutions of Linear Differential Equations

8.1 Introduction

In this chapter, attention is focused mainly on problems involving second order linear differential equations with variable coefficients,

$$y'' + p(t)y' + q(t)y = 0, \qquad a < t < b$$
$$y(t_0) = y_0, \qquad y'(t_0) = y_0',$$

where $a < t_0 < b$. In most of the cases considered, the coefficient functions $p(t)$ and $q(t)$ are rational functions (that is, ratios of polynomials). For the moment, however, we make no assumptions about $p(t)$ and $q(t)$.

Chapter 7 discussed techniques for generating numerical approximations to the solution of initial value problems. In this chapter, we look for solutions that have the form of a **power series**,

$$y(t) = \sum_{n=0}^{\infty} a_n(t - t_0)^n = a_0 + a_1(t - t_0) + a_2(t - t_0)^2 + a_3(t - t_0)^3 + \cdots.$$

To find such a power series solution of an initial value problem, we must answer several questions:

1. What properties must the coefficient functions $p(t)$ and $q(t)$ possess to guarantee that a power series solution does, in fact, exist?
2. If a power series solution of the initial value problem exists, how do we compute the coefficients $\{a_n\}_{n=0}^{\infty}$?
3. For what values of t does the resulting power series converge?

To answer these questions, we begin with a review of power series and the properties of functions defined by a power series.

For any fixed value of t, the power series

$$\sum_{n=0}^{\infty} a_n(t - t_0)^n \tag{1}$$

is an infinite series of constants. If, for a fixed value of t, the sequence of partial sums

$$S_N(t) = \sum_{n=0}^{N} a_n(t - t_0)^n \tag{2}$$

approaches a limit as $N \to \infty$, we say that the series **converges**. If the series does not converge, we say it **diverges**. Note that the power series (1) always converges for $t = t_0$, since $S_N(t_0) = a_0$ for all N. In general, we want to know those values of t for which the power series converges and those values for which it diverges.

Convergence Possibilities

There are three distinct possibilities for power series (1):

(a) The series $\sum_{n=0}^{\infty} a_n(t - t_0)^n$ might converge only at $t = t_0$ and diverge for all $t \neq t_0$.
(b) The series might converge for all t in an interval of the form $|t - t_0| < R$ for some $0 < R < \infty$ and diverge for all t satisfying $|t - t_0| > R$. The number R is called the **radius of convergence**.
(c) The series might converge for all t, $-\infty < t < \infty$.

It can be shown that every power series of the form (1) falls into exactly one of these three categories. [It is customary to say that the radius of convergence is $R = 0$ in case (a) and $R = \infty$ in case (c).]

The power series (1) is said to be **absolutely convergent** at a value t if the infinite series

$$\sum_{n=0}^{\infty} |a_n||t - t_0|^n$$

converges. As the terminology suggests, absolute convergence implies convergence. The converse is not true, however; convergence need not imply absolute convergence. In case (b), it can be shown that the power series is absolutely convergent in the interval $|t - t_0| < R$. In case (c), the power series is absolutely convergent for $-\infty < t < \infty$.

The **ratio test** is frequently used to test for absolute convergence of an infinite series. By way of review, consider the infinite series

$$\sum_{n=0}^{\infty} c_n \tag{3}$$

and suppose that

$$\lim_{n \to \infty} \left| \frac{c_{n+1}}{c_n} \right| = L, \tag{4}$$

where L is either a nonnegative finite constant or $+\infty$. If $L < 1$, the infinite series (3) is absolutely convergent. If $L > 1$, the infinite series (3) is divergent. If $L = 1$, the ratio rest is inconclusive.

The ratio test can be used to determine the radius of convergence of a power series, as we see in Example 1.

EXAMPLE 1

Use the ratio test to determine the radius of convergence of the power series

$$\sum_{n=1}^{\infty} nt^n = t + 2t^2 + 3t^3 + \cdots.$$

Solution: The power series clearly converges at $t = 0$. Applying the ratio test at an arbitrary value t, $t \neq 0$, we find

$$\lim_{n \to \infty} \left| \frac{(n+1)t^{n+1}}{nt^n} \right| = \lim_{n \to \infty} \left(1 + \frac{1}{n}\right) |t| = |t|.$$

Therefore, by the ratio test, the power series converges if $|t| < 1$ and diverges if $|t| > 1$. The radius of convergence is $R = 1$. ❖

Suppose a power series has a finite and positive radius of convergence R. The preceding discussion says nothing about whether or not the power series converges at the points $t = t_0 \pm R$. In fact, no general statements can be made about convergence or divergence at these points, which separate the open interval of absolute convergence from the semi-infinite intervals of divergence. For instance, the power series

$$\sum_{n=0}^{\infty} nt^n$$

considered in Example 1 diverges at $t = \pm 1$. In general, a power series might converge absolutely, converge **conditionally** (that is, converge but not converge absolutely), or diverge at the point $t = t_0 + R$. The same statement can be made with regard to the point $t = t_0 - R$.

Operations with Power Series

Every power series defines a function $f(t)$. The domain of $f(t)$ is the set of t-values for which the series converges. Consider the function $f(t)$ defined by

$$f(t) = \sum_{n=0}^{\infty} a_n (t - t_0)^n$$
$$= a_0 + a_1 (t - t_0) + a_2 (t - t_0)^2 + a_3 (t - t_0)^3 + \cdots.$$

Assume that the power series defining $f(t)$ has radius of convergence R, where $R > 0$. The following results are established in calculus and say, roughly, that power series can be treated like polynomials with respect to the operations of addition, subtraction, multiplication, and division.

Power Series Can Be Added and Subtracted

If $f(t)$ and $g(t)$ are given by

$$f(t) = \sum_{n=0}^{\infty} a_n (t - t_0)^n \quad \text{and} \quad g(t) = \sum_{n=0}^{\infty} b_n (t - t_0)^n,$$

with both series converging in $|t - t_0| < R$, then the sum and difference functions are given by

$$(f \pm g)(t) = \sum_{n=0}^{\infty} (a_n \pm b_n)(t - t_0)^n,$$

where the sum and difference both converge absolutely in $|t - t_0| < R$.

Power Series Can Be Multiplied

If $f(t)$ and $g(t)$ are given by

$$f(t) = \sum_{n=0}^{\infty} a_n (t - t_0)^n \quad \text{and} \quad g(t) = \sum_{n=0}^{\infty} b_n (t - t_0)^n,$$

with both series converging in $|t - t_0| < R$, then the product function, $(fg)(t)$, has a power series representation

$$(fg)(t) = \sum_{n=0}^{\infty} c_n (t - t_0)^n,$$

which likewise converges in $|t - t_0| < R$. Moreover, the coefficients c_n can be obtained by formally multiplying the power series for $f(t)$ and $g(t)$ as if they were polynomials and grouping terms. In other words,

$$c_0 + c_1 (t - t_0) + c_2 (t - t_0)^2 + \cdots$$
$$= [a_0 + a_1 (t - t_0) + a_2 (t - t_0)^2 + \cdots][b_0 + b_1 (t - t_0) + b_2 (t - t_0)^2 + \cdots]$$
$$= a_0 b_0 + (a_0 b_1 + a_1 b_0)(t - t_0) + (a_0 b_2 + a_1 b_1 + a_2 b_0)(t - t_0)^2 + \cdots.$$

Therefore,

$$c_0 = a_0 b_0, \qquad c_1 = a_0 b_1 + a_1 b_0, \qquad c_2 = a_0 b_2 + a_1 b_1 + a_2 b_0,$$

and, in general,

$$c_n = \sum_{i=0}^{n} a_i b_{n-i}, \qquad n = 0, 1, 2, \ldots.$$

The product power series $(fg)(t)$ is called the Cauchy[1] product.

In Some Cases Power Series Can Be Divided

If $f(t)$ and $g(t)$ are given by

$$f(t) = \sum_{n=0}^{\infty} a_n (t - t_0)^n \quad \text{and} \quad g(t) = \sum_{n=0}^{\infty} b_n (t - t_0)^n,$$

with both converging in $|t - t_0| < R$, and if $g(t_0) = b_0 \neq 0$, then the quotient function $(f/g)(t)$ has a power series representation

$$(f/g)(t) = \sum_{n=0}^{\infty} d_n (t - t_0)^n,$$

which converges in some neighborhood of t_0. Again, we can determine the coefficients d_n by formally manipulating the power series as if they were polynomials. We have

$$\frac{a_0 + a_1(t - t_0) + a_2(t - t_0)^2 + \cdots}{b_0 + b_1(t - t_0) + b_2(t - t_0)^2 + \cdots} = d_0 + d_1(t - t_0) + d_2(t - t_0)^2 + \cdots$$

or, after multiplying by the denominator series,

$$a_0 + a_1(t - t_0) + a_2(t - t_0)^2 + \cdots$$
$$= [b_0 + b_1(t - t_0) + b_2(t - t_0)^2 + \cdots][d_0 + d_1(t - t_0) + d_2(t - t_0)^2 + \cdots].$$

The coefficients d_n can be recursively determined by forming the Cauchy product of the two series on the right and solving the resulting hierarchy of linear equations. We obtain

$$a_0 = b_0 d_0 \quad \text{and hence} \quad d_0 = \frac{a_0}{b_0},$$

$$a_1 = b_0 d_1 + b_1 d_0 \quad \text{and hence} \quad d_1 = \frac{a_1 - b_1 d_0}{b_0},$$

$$a_2 = b_0 d_2 + b_1 d_1 + b_2 d_0 \quad \text{and hence} \quad d_2 = \frac{a_2 - b_1 d_1 - b_2 d_0}{b_0},$$

$$\vdots$$

Notice how the statement made in the case of division differs from that made in the previous cases. In particular, even though the series for $f(t)$ and $g(t)$

---

[1] Augustin Louis Cauchy (1789–1857) was a scientific giant whose life was enmeshed in the political turmoil of early nineteenth-century France. He contributed to many areas of mathematics and science and is considered to be the founder of the theory of functions of a complex variable. Numerous terms in mathematics bear his name, such as the Cauchy integral theorem, the Cauchy-Riemann equations, and Cauchy sequences. His collected works, when published, filled 27 volumes.

converge in $|t - t_0| < R$, it does *not* necessarily follow that the quotient series also converges in $|t - t_0| < R$. All we can say in general is that the quotient series converges in some neighborhood of t_0.

A Function Defined by a Power Series Can Be Differentiated Termwise

Let $f(t)$ be given by the power series

$$f(t) = \sum_{n=0}^{\infty} a_n(t - t_0)^n, \tag{5}$$

which converges in $|t - t_0| < R$. The function $f(t)$ has derivatives of all orders on the interval $t_0 - R < t < t_0 + R$. We can obtain these derivatives by termwise differentiation of the original power series. That is,

$$f'(t) = \sum_{n=1}^{\infty} na_n(t - t_0)^{n-1} = a_1 + 2a_2(t - t_0) + 3a_3(t - t_0)^2 + \cdots,$$

$$f''(t) = \sum_{n=2}^{\infty} n(n - 1)a_n(t - t_0)^{n-2} = 2a_2 + 6a_3(t - t_0) + \cdots,$$

and so forth. Each of these derived series also converges absolutely in the interval $t_0 - R < t < t_0 + R$. The derived series can be used to express the coefficient a_n in terms of the nth derivative of $f(t)$ evaluated at $t = t_0$. In particular, by evaluating the derived series at $t = t_0$, we see that

$$f(t_0) = a_0, \quad f'(t_0) = a_1, \quad f''(t_0) = (2 \cdot 1)a_2, \quad f'''(t_0) = (3 \cdot 2 \cdot 1)a_3, \quad \ldots.$$

In general, for $f(t)$ given by (5),

$$f^{(n)}(t_0) = n!a_n, \qquad n = 0, 1, 2, \ldots.$$

Some Functions Are Defined by a Taylor Series

If $f(t)$ is a function defined by the power series (5), then $f^{(n)}(t_0) = n!a_n$, $n = 0, 1, 2, \ldots$. Conversely, if we are given a function $f(t)$ that is defined and infinitely differentiable on an interval $t_0 - R < t < t_0 + R$, then we can associate $f(t)$ with its formal **Taylor series**

$$\sum_{n=0}^{\infty} \frac{f^{(n)}(t_0)}{n!}(t - t_0)^n. \tag{6}$$

Recall from calculus that the Taylor series for $f(t)$ need not necessarily converge to $f(t)$. However, for most of the functions considered in this chapter, the Taylor series converges to $f(t)$, so

$$f(t) = \sum_{n=0}^{\infty} \frac{f^{(n)}(t_0)}{n!}(t - t_0)^n, \qquad t_0 - R < t < t_0 + R.$$

If $t_0 = 0$, the Taylor series is usually referred to as a Maclaurin series. For later reference, we list the Maclaurin series for several functions. We also give the interval of absolute convergence for the series.

$$e^t = \sum_{n=0}^{\infty} \frac{t^n}{n!} = 1 + t + \frac{t^2}{2!} + \frac{t^3}{3!} + \frac{t^4}{4!} + \cdots, \quad -\infty < t < \infty \tag{7a}$$

$$\sin t = \sum_{n=0}^{\infty} (-1)^n \frac{t^{2n+1}}{(2n+1)!} = t - \frac{t^3}{3!} + \frac{t^5}{5!} - \frac{t^7}{7!} + \cdots, \quad -\infty < t < \infty \tag{7b}$$

$$\cos t = \sum_{n=0}^{\infty} (-1)^n \frac{t^{2n}}{(2n)!} = 1 - \frac{t^2}{2!} + \frac{t^4}{4!} - \frac{t^6}{6!} + \cdots, \quad -\infty < t < \infty \tag{7c}$$

$$\frac{1}{1-t} = \sum_{n=0}^{\infty} t^n = 1 + t + t^2 + t^3 + t^4 + \cdots, \quad -1 < t < 1 \tag{7d}$$

$$\ln(1+t) = \sum_{n=1}^{\infty} (-1)^{n-1} \frac{t^n}{n} = t - \frac{t^2}{2} + \frac{t^3}{3} - \frac{t^4}{4} + \cdots, \quad -1 < t < 1 \tag{7e}$$

$$\tan^{-1} t = \sum_{n=0}^{\infty} (-1)^n \frac{t^{2n+1}}{2n+1} = t - \frac{t^3}{3} + \frac{t^5}{5} - \frac{t^7}{7} + \cdots, \quad -1 < t < 1 \tag{7f}$$

Note that these basic series can be used to find the Taylor series of certain simple compositions. For example, by (7d),

$$\frac{1}{3-t} = \frac{1}{1-(t-2)} = \sum_{n=0}^{\infty} (t-2)^n, \quad |t-2| < 1$$

and

$$\frac{1}{1+4t^2} = \sum_{n=0}^{\infty} (-1)^n 4^n t^{2n}, \quad |t| < \frac{1}{2}.$$

A Function Defined by a Power Series Can Be Integrated Termwise

Let $f(t)$ be given by the power series

$$f(t) = \sum_{n=0}^{\infty} a_n (t - t_0)^n,$$

which converges in $|t - t_0| < R$. The function $f(t)$ has antiderivatives defined on the interval $t_0 - R < t < t_0 + R$. We can obtain these antiderivatives by termwise integration of the original power series. For example,

$$\int_{t_0}^{t} f(s)\,ds = \int_{t_0}^{t} \sum_{n=0}^{\infty} a_n (s - t_0)^n \, ds = \sum_{n=0}^{\infty} a_n \int_{t_0}^{t} (s - t_0)^n \, ds = \sum_{n=0}^{\infty} a_n \frac{(t - t_0)^{n+1}}{n+1}.$$

The integrated series can be shown to also converge absolutely on the interval $t_0 - R < t < t_0 + R$.

Uniqueness of the Power Series Representation of a Function

A power series representation of a function is unique. Let

$$\sum_{n=0}^{\infty} a_n(t - t_0)^n \quad \text{and} \quad \sum_{n=0}^{\infty} b_n(t - t_0)^n$$

be two power series having the same radius of convergence R. If

$$\sum_{n=0}^{\infty} a_n(t - t_0)^n = \sum_{n=0}^{\infty} b_n(t - t_0)^n \quad \text{for all } t \text{ such that} \quad |t - t_0| < R,$$

then it follows that the coefficients must be equal; that is,

$$a_n = b_n, \qquad n = 0, 1, 2, \ldots.$$

As an important special case, if

$$\sum_{n=0}^{\infty} a_n(t - t_0)^n = 0 \quad \text{for all } t \text{ such that} \quad |t - t_0| < R,$$

then $a_n = 0, n = 0, 1, 2, \ldots$.

Power Series Solutions of Linear Differential Equations

The next example introduces the ideas associated with finding a power series solution of a linear differential equation. In the sections that follow, we will elaborate on the theoretical foundations of the method and point out some of the potential difficulties.

EXAMPLE 2

Consider the equation

$$y'' + \omega^2 y = 0,$$

where ω is a positive constant. Assuming this equation has a solution of the form

$$y(t) = \sum_{n=0}^{\infty} a_n t^n,$$

determine the coefficients, $a_0, a_1, a_2, \ldots$. Can you also determine the *general solution*?

Solution: We look for a solution of the form $y(t) = \sum_{n=0}^{\infty} a_n t^n$, assuming that the power series has a positive radius of convergence. The actual radius of convergence will be determined once we find the coefficients, $a_0, a_1, a_2, \ldots$.
Differentiating termwise, we obtain

$$y' = \sum_{n=1}^{\infty} a_n n t^{n-1} \quad \text{and} \quad y'' = \sum_{n=2}^{\infty} a_n n(n - 1)t^{n-2}.$$

In these series, we have adjusted the lower limit in the summation to correspond to the first nonzero term in the series.

Substituting these expressions into the differential equation $y'' + \omega^2 y = 0$, we find

$$\sum_{n=2}^{\infty} a_n n(n-1)t^{n-2} + \omega^2 \sum_{n=0}^{\infty} a_n t^n = 0. \tag{8}$$

We want to combine the two summations in (8) and eventually use the consequences of uniqueness. In order to combine the two series, we adjust the summation index of the first series so that powers of t are in agreement.

In particular, to match the powers of t, we can make a change of index, $k = n - 2$, in the first series. Doing so, we see from (8) that the lower limit of $n = 2$ transforms to a new lower limit of $k = 0$ (the upper limits remain at ∞). Thus, the first series in (8) can be rewritten as

$$\sum_{n=2}^{\infty} a_n n(n-1)t^{n-2} = \sum_{k=0}^{\infty} a_{k+2}(k+2)(k+1)t^k = \sum_{n=0}^{\infty} a_{n+2}(n+2)(n+1)t^n. \tag{9}$$

In the last step, we have used the fact that the summation index is a dummy index and can be called n instead of k. Using (9) in (8) leads to

$$\sum_{n=0}^{\infty} a_{n+2}(n+2)(n+1)t^n + \omega^2 \sum_{n=0}^{\infty} a_n t^n = 0,$$

or

$$\sum_{n=0}^{\infty} [a_{n+2}(n+2)(n+1) + \omega^2 a_n]t^n = 0. \tag{10}$$

Since equality (10) is assumed to hold in some interval containing the origin, each coefficient of the series must vanish. We obtain the infinite set of equalities

$$(n+1)(n+2)a_{n+2} + \omega^2 a_n = 0, \qquad n = 0, 1, 2, \ldots,$$

or

$$a_{n+2} = -\frac{\omega^2}{(n+1)(n+2)} a_n, \qquad n = 0, 1, 2, 3, \ldots \tag{11}$$

The set of equations (11) is referred to as a **recurrence relation**. Solving for the unknown coefficients recursively allows us to find all the coefficients $\{a_n\}_{n=0}^{\infty}$ in terms of the coefficients a_0 and a_1. In particular, from (11) we find

$$n = 0: \quad a_2 = -\frac{\omega^2}{2 \cdot 1} a_0 \qquad\qquad n = 1: \quad a_3 = -\frac{\omega^2}{3 \cdot 2} a_1$$

$$n = 2: \quad a_4 = -\frac{\omega^2}{4 \cdot 3} a_2 = \frac{\omega^4}{4!} a_0 \qquad\qquad n = 3: \quad a_5 = -\frac{\omega^2}{5 \cdot 4} a_3 = \frac{\omega^4}{5!} a_1$$

$$n = 4: \quad a_6 = -\frac{\omega^2}{6 \cdot 5} a_4 = -\frac{\omega^6}{6!} a_0 \qquad\qquad n = 5: \quad a_7 = -\frac{\omega^2}{7 \cdot 6} a_5 = -\frac{\omega^6}{7!} a_1$$

$$\vdots \qquad\qquad\qquad\qquad\qquad \vdots$$

(continued)

(continued)

The emerging pattern is clear. For the even index $n = 2k$, we have

$$a_{2k} = (-1)^k \frac{\omega^{2k}}{(2k)!} a_0.$$

For the odd index $n = 2k + 1$, we have

$$a_{2k+1} = (-1)^k \frac{\omega^{2k}}{(2k+1)!} a_1 = (-1)^k \frac{\omega^{2k+1}}{(2k+1)!} \left(\frac{a_1}{\omega}\right).$$

Therefore, we can write the solution of the differential equation $y'' + \omega^2 y = 0$ as

$$y(t) = a_0 \left[1 - \frac{(\omega t)^2}{2!} + \frac{(\omega t)^4}{4!} - \frac{(\omega t)^6}{6!} + \cdots \right]$$

$$+ \left(\frac{a_1}{\omega}\right) \left[\omega t - \frac{(\omega t)^3}{3!} + \frac{(\omega t)^5}{5!} - \frac{(\omega t)^7}{7!} + \cdots \right]. \tag{12}$$

The first series is the Maclaurin series expansion of $\cos \omega t$ [see equation (7c)]. The second series is that of $\sin \omega t$ [see equation (7b)]. Therefore, identifying arbitrary constants c_1 with a_0 and c_2 with a_1/ω, we obtain the general solution familiar from Chapter 3,

$$y(t) = c_1 \cos \omega t + c_2 \sin \omega t.$$

Applying the ratio test, you can verify that the radius of convergence is $R = \infty$ for both of these power series. ❖

In the sections that follow, the same basic manipulations will be used to obtain series solutions of differential equations having variable coefficients.

Shifting the Index of Summation

We frequently find it convenient, as in Example 2, to shift the index of summation so that the general term in a series is a constant multiple of t^n. For example, consider the function $f(t) = t^3(e^t - 1)$. Using equation (7a), we see the series for $f(t) = t^3(e^t - 1)$ has the form

$$f(t) = t^3(e^t - 1) = t^3 \sum_{n=1}^{\infty} \frac{t^n}{n!} = \sum_{n=1}^{\infty} \frac{t^{n+3}}{n!}.$$

In order to rewrite the Maclaurin series so that the general term involves t^n, we make the change of index $k = n + 3$. With this shift of index, $n = k - 3$. Therefore, the terms in the summation are all of the form $t^k/(k-3)!$. We also must transform the limits of the summation. At the lower limit, $n = 1$ implies that $k = 4$. At the upper limit, $n = \infty$ implies that $k = \infty$ also. Thus, we can rewrite the series for $f(t) = t^3(e^t - 1)$ as

$$f(t) = \sum_{n=1}^{\infty} \frac{t^{n+3}}{n!} = \sum_{k=4}^{\infty} \frac{t^k}{(k-3)!} = \sum_{n=4}^{\infty} \frac{t^n}{(n-3)!}. \tag{13}$$

The summation index is a dummy index. In the last step of (13), therefore, we can replace k by n.

EXERCISES

Exercises 1–12:

As in Example 1, use the ratio test to find the radius of convergence R for the given power series.

1. $\displaystyle\sum_{n=0}^{\infty} \frac{t^n}{2^n}$

2. $\displaystyle\sum_{n=1}^{\infty} \frac{t^n}{n^2}$

3. $\displaystyle\sum_{n=0}^{\infty} (t-2)^n$

4. $\displaystyle\sum_{n=0}^{\infty} (3t-1)^n$

5. $\displaystyle\sum_{n=0}^{\infty} \frac{(t-1)^n}{n!}$

6. $\displaystyle\sum_{n=0}^{\infty} n!(t-1)^n$

7. $\displaystyle\sum_{n=1}^{\infty} \frac{(-1)^n t^n}{n}$

8. $\displaystyle\sum_{n=0}^{\infty} \frac{(-1)^n (t-3)^n}{4^n}$

9. $\displaystyle\sum_{n=1}^{\infty} (\ln n)(t+2)^n$

10. $\displaystyle\sum_{n=0}^{\infty} n^3(t-1)^n$

11. $\displaystyle\sum_{n=0}^{\infty} \frac{\sqrt{n}}{2^n}(t-4)^n$

12. $\displaystyle\sum_{n=1}^{\infty} \frac{(t-2)^n}{\arctan n}$

Exercises 13–16:

In each exercise, functions $f(t)$ and $g(t)$ are given. The functions $f(t)$ and $g(t)$ are defined by a power series that converges in $-R < t - t_0 < R$, where R is a positive constant. In each exercise, determine the largest value R such that $f(t)$ and $g(t)$ both converge in $-R < t - t_0 < R$. In addition,

(a) Write out the first four terms of the power series for $f(t)$ and $g(t)$.

(b) Write out the first four terms of the power series for $f(t) + g(t)$.

(c) Write out the first four terms of the power series for $f(t) - g(t)$.

(d) Write out the first four terms of the power series for $f'(t)$.

(e) Write out the first four terms of the power series for $f''(t)$.

13. $f(t) = \displaystyle\sum_{n=0}^{\infty} t^n, \quad g(t) = \displaystyle\sum_{n=0}^{\infty} n^2 t^n$

14. $f(t) = \displaystyle\sum_{n=0}^{\infty} n t^n, \quad g(t) = \displaystyle\sum_{n=0}^{\infty} (-1)^n n t^n$

15. $f(t) = \displaystyle\sum_{n=0}^{\infty} (-1)^n 2^n (t-1)^n, \quad g(t) = \displaystyle\sum_{n=0}^{\infty} (t-1)^n$

16. $f(t) = \displaystyle\sum_{n=0}^{\infty} 2^n (t+1)^n, \quad g(t) = \displaystyle\sum_{n=0}^{\infty} n(t+1)^n$

Exercises 17–23:

By shifting the index of summation as in equation (9) or (13), rewrite the given power series so that the general term involves t^n.

17. $\displaystyle\sum_{n=0}^{\infty} 2^n t^{n+2}$

18. $\displaystyle\sum_{n=0}^{\infty} (n+1)(n+2) t^{n+3}$

19. $\displaystyle\sum_{n=0}^{\infty} a_n t^{n+2}$

20. $\displaystyle\sum_{n=1}^{\infty} n a_n t^{n-1}$

21. $\displaystyle\sum_{n=2}^{\infty} n(n-1) a_n t^{n-2}$

22. $\displaystyle\sum_{n=0}^{\infty} (-1)^n a_n t^{n+3}$

23. $\displaystyle\sum_{n=0}^{\infty} (-1)^{n+1}(n+1) a_n t^{n+2}$

Exercises 24–27:

Using the information given in (7), write a Maclaurin series for the given function $f(t)$. Determine the radius of convergence of the series.

24. $f(t) = t^2(t - \sin t)$ **25.** $f(t) = 1 - \cos(3t)$

26. $f(t) = \dfrac{1}{1 + 2t}$ **27.** $f(t) = \dfrac{1}{1 - t^2}$

28. Use series (7a) to determine the first four nonvanishing terms of the Maclaurin series for

     (a) $\sinh t = \dfrac{e^t - e^{-t}}{2}$ (b) $\cosh t = \dfrac{e^t + e^{-t}}{2}$

29. Consider the differential equation $y'' - \omega^2 y = 0$, where ω is a positive constant. As in Example 2, assume this differential equation has a solution of the form $y(t) = \sum_{n=0}^{\infty} a_n t^n$.

     (a) Determine a recurrence relation for the coefficients $a_0, a_1, a_2, \ldots$.

     (b) As in equation (12), express the general solution in the form

$$y(t) = a_0 y_1(t) + \left(\frac{a_1}{\omega}\right) y_2(t).$$

     What are the functions $y_1(t)$ and $y_2(t)$? [Hint: Recall the series in Exercise 28.]

Exercises 30–35:

In each exercise,

     (a) Use the given information to determine a power series representation of the function $y(t)$.

     (b) Determine the radius of convergence of the series found in part (a).

     (c) Where possible, use (7) to identify the function $y(t)$.

30. $y'(t) = \displaystyle\sum_{n=1}^{\infty} nt^{n-1} = 1 + 2t + 3t^2 + \cdots, \quad y(0) = 1$

31. $y'(t) = \displaystyle\sum_{n=0}^{\infty} \frac{(t-1)^n}{n!} = 1 + (t-1) + \frac{(t-1)^2}{2!} + \cdots, \quad y(1) = 1$

32. $y''(t) = \displaystyle\sum_{n=0}^{\infty} (-1)^n \frac{t^n}{n!} = 1 - t + \frac{t^2}{2!} - \frac{t^3}{3!} + \cdots, \quad y(0) = 1, \ y'(0) = -1$

33. $y'(t) = \displaystyle\sum_{n=2}^{\infty} (-1)^n \frac{(t-1)^n}{n!} = \frac{(t-1)^2}{2!} - \frac{(t-1)^3}{3!} + \frac{(t-1)^4}{4!} - \frac{(t-1)^5}{5!} + \cdots, \quad y(1) = 0$

34. $y(t) = \displaystyle\int_0^t f(s)\,ds$, where $f(s) = \displaystyle\sum_{n=0}^{\infty} (-1)^n s^{2n} = 1 - s^2 + s^4 - s^6 + \cdots$

35. $\displaystyle\int_0^t y(s)\,ds = \displaystyle\sum_{n=1}^{\infty} \frac{t^n}{n} = t + \frac{t^2}{2} + \frac{t^3}{3} + \cdots$

Exercises 36–41:

In each exercise, an initial value problem is given. Assume that the initial value problem has a solution of the form $y(t) = \sum_{n=0}^{\infty} a_n t^n$, where the series has a positive radius of convergence. Determine the first six coefficients, $a_0, a_1, a_2, a_3, a_4, a_5$. Note that $y(0) = a_0$ and that $y'(0) = a_1$. Thus, the initial conditions determine the arbitrary constants. In Exercises 40 and 41, the exact solution is given in terms of exponential functions. Check your answer by comparing it with the Maclaurin series expansion of the exact solution.

36. $y'' - ty' - y = 0, \quad y(0) = 1, \ y'(0) = -1$

37. $y'' + ty' - 2y = 0, \quad y(0) = 0, \ y'(0) = 1$

38. $y'' + ty = 0$, $y(0) = 1$, $y'(0) = 2$

39. $y'' + (1 + t)y' + y = 0$, $y(0) = -1$, $y'(0) = 1$

40. $y'' - 5y' + 6y = 0$, $y(0) = 1$, $y'(0) = 2$, $y(t) = e^{2t}$

41. $y'' - 2y' + y = 0$, $y(0) = 0$, $y'(0) = 2$, $y(t) = 2te^{t}$

8.2   Series Solutions Near an Ordinary Point

Consider the linear differential equation

$$y'' + p(t)y' + q(t)y = 0 \tag{1}$$

in an open interval containing the point t_0. Frequently, t_0 is the point where the initial conditions are imposed. For our present discussion, however, t_0 is an arbitrary but fixed point. We are interested in answering the question "When is it possible to represent the general solution of (1) in terms of power series that converge in some neighborhood of the point t_0?"

Ordinary Points and Singular Points

Recall (see Section 7.3) that a function $f(t)$ is called **analytic at t_0** if $f(t)$ has a Taylor series expansion

$$f(t) = \sum_{n=0}^{\infty} \frac{f^{(n)}(t_0)}{n!}(t - t_0)^n,$$

with radius of convergence R, where $R > 0$. For later use, we also recall that

> If $f(t)$ and $g(t)$ are analytic at t_0, then the functions $f(t) \pm g(t)$ and $f(t)g(t)$ are also analytic at t_0. Furthermore, the quotient $f(t)/g(t)$ is analytic at t_0 if $g(t_0) \neq 0$. Polynomial functions are analytic at all points. Rational functions are analytic at all points where the denominator polynomial is nonzero. (When discussing rational functions, we assume the denominator and numerator have no factors in common.) If the denominator is nonzero at t_0, the radius of convergence is equal to the distance from t_0 to the nearest zero (either real or complex) of the denominator.

As we will show, the general solution of $y'' + p(t)y' + q(t)y = 0$ can be expressed in terms of power series that converge in a neighborhood of t_0 whenever both $p(t)$ and $q(t)$ are analytic at t_0. The point t_0 is called an **ordinary point** when both $p(t)$ and $q(t)$ are analytic at t_0. If $p(t)$ and/or $q(t)$ is not analytic at t_0, t_0 is called a **singular point**.

As noted in Section 8.1, if a function $f(t)$ is defined by a power series, $f(t) = \sum_{n=0}^{\infty} a_n(t - t_0)^n$, and if this power series has radius of convergence $R > 0$, then $f(t)$ has derivatives of all orders at $t = t_0$. Therefore, if some derivative of $f(t)$ fails to exist at a point t_0, then $f(t)$ cannot be analytic at t_0. To help identify ordinary points, we can use some facts noted earlier: Sums, differences, and products of functions analytic at t_0 are again analytic at t_0. Quotients of functions analytic at t_0 are also analytic at t_0 if the denominator function is nonzero at t_0.

E X A M P L E

1

Consider the differential equation

$$(1 - t^2)y'' + (\tan t)y' + t^{5/3}y = 0$$

in the open interval $-2 < t < 2$. Classify each point in this interval as an ordinary point or a singular point.

Solution: We first rewrite the equation in the form (1),

$$y'' + \frac{\tan t}{1 - t^2}y' + \frac{t^{5/3}}{1 - t^2}y = 0.$$

Therefore, the coefficient functions $p(t)$ and $q(t)$ are

$$p(t) = \frac{\tan t}{1 - t^2} \quad \text{and} \quad q(t) = \frac{t^{5/3}}{1 - t^2}.$$

The function $p(t)$ fails to be analytic at the points $t = \pm 1$ (where the denominator vanishes) and $t = \pm\pi/2$ (where the graph of $y = \tan t$ has vertical asymptotes). The function $q(t)$ is not analytic at $t = \pm 1$ or at $t = 0$. (The numerator function $t^{5/3}$ is not analytic at $t = 0$; it is continuous and has a continuous first derivative at $t = 0$, but its second derivative does not exist at $t = 0$.) Thus, in the interval $-2 < t < 2$, the five points $t = 0, \pm 1, \pm\pi/2$ are singular points and all other points are ordinary points. ❖

Series Solutions Near an Ordinary Point

Theorem 8.1 shows that, in a neighborhood of an ordinary point, we can represent the general solution of equation (1) in terms of convergent power series. The proof of Theorem 8.1 is given in more advanced texts, such as Birkhoff and Rota.[2]

Theorem 8.1

Let $p(t)$ and $q(t)$ be analytic at t_0, and let R denote the smaller of the two radii of convergence of their respective Taylor series representations. Then the initial value problem

$$y'' + p(t)y' + q(t)y = 0, \qquad y(t_0) = y_0, \qquad y'(t_0) = y'_0 \qquad \textbf{(2)}$$

has a unique solution that is analytic in the interval $|t - t_0| < R$.

According to Theorem 8.1, if t_0 is an ordinary point, then initial value problem (2) has a power series solution of the form

$$y(t) = \sum_{n=0}^{\infty} a_n(t - t_0)^n = a_0 + a_1(t - t_0) + a_2(t - t_0)^2 + \cdots.$$

---

[2]Garrett Birkhoff and Gian-Carlo Rota, *Ordinary Differential Equations*, 4th ed. (New York: Wiley, 1989).

Note that the first two coefficients, a_0 and a_1, are determined by the initial conditions in (2), since $y(t_0) = a_0$ and $y'(t_0) = a_1$. Theorem 8.1 assures us that the power series we obtain by solving the recurrence relation for the remaining coefficients $a_2, a_3, \ldots$ converges in the interval $|t - t_0| < R$, where R is the smaller of the radii of convergence of the coefficient functions $p(t)$ and $q(t)$. Note that Theorem 8.1 does not rule out the possibility that the power series for $y(t)$ may converge on a larger interval. This happens in some cases. The following corollary is a consequence of Theorem 8.1.

Corollary

Let $p(t)$ and $q(t)$ be analytic at t_0, and let R denote the smaller of the two radii of convergence of their respective Taylor series representations. The general solution of the differential equation

$$y'' + p(t)y' + q(t)y = 0 \tag{3}$$

can be expressed as

$$y(t) = \sum_{n=0}^{\infty} a_n(t - t_0)^n = a_0 y_1(t) + a_1 y_2(t),$$

where the constants a_0 and a_1 are arbitrary. The functions $y_1(t)$ and $y_2(t)$ form a fundamental set of solutions, analytic in the interval $|t - t_0| < R$.

The solutions $y_1(t)$ and $y_2(t)$ forming the fundamental set can be obtained by adopting the particular initial conditions $y_1(t_0) = 1$, $y_1'(t_0) = 0$ and $y_2(t_0) = 0$, $y_2'(t_0) = 1$.

EXAMPLE

2

Consider the initial value problem

$$y'' + \frac{t+1}{t^3 + t}y' + \frac{1}{t^2 - 4t + 5}y = 0, \qquad y(2) = y_0, \qquad y'(2) = y_0'.$$

If $y(t) = \sum_{n=0}^{\infty} a_n(t - 2)^n$ is the solution, determine a lower bound for the radius of convergence R of this series.

Solution: Since $t^3 + t = t(t^2 + 1)$, the coefficient function $p(t)$ has denominator zeros at $t = 0$, $t = -i$, and $t = i$. Likewise, the coefficient function $q(t)$ has denominator zeros at $t = 2 \pm i$. The radius of convergence R_p of the Taylor series expansion for $p(t)$ is equal to the distance from $t_0 = 2$ to the nearest denominator zero; that is, R_p is the smaller of $|2 \pm 0| = 2$ and $|2 - i| = \sqrt{5}$. Thus, $R_p = 2$. (See Figure 8.1.) Similarly, the radius of convergence of the Taylor series for $q(t)$ is $R_q = |2 - (2 \pm i)| = 1$. Thus, by Theorem 8.1, the radius of convergence of the Taylor series for $y(t)$ is guaranteed to be no smaller than $R = 1$.

(continued)

(continued)

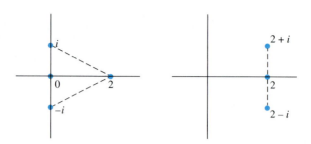

FIGURE 8.1

The radius of convergence of the expansion of $p(t)$ is equal to the distance from $t_0 = 2$ to the nearest of the denominator zeros, $0, i$, and $-i$. For $q(t)$, the two nearest denominator zeros, $2 + i$ and $2 - i$, are equidistant from $t_0 = 2$. The radius of convergence for the series in Example 2 is $R = 1$. ❖

When the coefficient functions of $y'' + p(t)y' + q(t)y = 0$ possess certain symmetries, some useful observations can be made; see Theorem 8.2.

Theorem 8.2

Consider the differential equation $y'' + p(t)y' + q(t)y = 0$.

(a) Let $p(t)$ be a continuous odd function defined on the domain $(-b, -a) \cup (a, b)$, where $a \geq 0$. Let $q(t)$ be a continuous even function defined on the same domain. If $f(t)$ is a solution of the differential equation on the interval $a < t < b$, then $f(-t)$ is a solution on the interval $(-b, -a)$.

(b) Let the coefficient functions $p(t)$ and $q(t)$ be analytic at $t = 0$ with a common radius of convergence $R > 0$. Let $p(t)$ be an odd function and $q(t)$ an even function. Then the differential equation has even and odd solutions that are analytic at $t = 0$ with radius of convergence R.

Recall the definitions of even and odd functions. We are assuming that $p(-t) = -p(t)$ and $q(-t) = q(t)$ for all t in $(-b, -a) \cup (a, b)$. The proof of Theorem 8.2 is outlined in Exercises 31–32.

In Example 2 of Section 8.1, we obtained a power series solution of $y'' + \omega^2 y = 0$ and observed that the ratio test could be used to show that each of the two series forming the general solution has an infinite radius of convergence. This fact is also an easy consequence of Theorem 8.1, since the differential equation $y'' + \omega^2 y = 0$ has coefficient functions $p(t) = 0$ and $q(t) = \omega^2$ that are analytic on $-\infty < t < \infty$. Moreover, since $p(t)$ is an odd function and $q(t)$ is an even function, it follows from Theorem 8.2 that even and odd solutions of this differential equation exist; they are $\cos \omega t$ and $\sin \omega t$, respectively.

Polynomial Solutions

Some of the second order linear differential equations that arise in mathematical and scientific applications (such as the Legendre equation, the Hermite equation, and the Chebyshev equation) have polynomial solutions. The next example treats the Chebyshev equation. Other equations are considered in the Exercises.

EXAMPLE 3

The Chebyshev[3] differential equation is

$$(1 - t^2)y'' - ty' + \mu^2 y = 0, \tag{4}$$

where μ is a constant. Find a Maclaurin series solution of (4). Show that if μ is an integer, the Chebyshev differential equation (4) has a polynomial solution.

Solution: Rewriting the equation as

$$y'' - \frac{t}{1 - t^2}y' + \frac{\mu^2}{1 - t^2}y = 0,$$

we see that $t = \pm 1$ are singular points. All other points are ordinary points. In addition, we deduce from the structure of the differential equation itself that it possesses solutions with even symmetry and solutions with odd symmetry (see Theorem 8.2).

By Theorem 8.1, the general solution of the Chebyshev equation can be represented in terms of Maclaurin series that we know will converge in $(-1, 1)$. Let $y(t) = \sum_{n=0}^{\infty} a_n t^n$. Substitution into differential equation (4) leads to

$$(1 - t^2) \sum_{n=2}^{\infty} a_n n(n-1) t^{n-2} - t \sum_{n=1}^{\infty} a_n n t^{n-1} + \mu^2 \sum_{n=0}^{\infty} a_n t^n = 0. \tag{5}$$

Equation (5) can be rewritten as

$$\sum_{n=2}^{\infty} a_n n(n-1) t^{n-2} - \sum_{n=0}^{\infty} [a_n n(n-1) + n a_n - \mu^2 a_n] t^n = 0,$$

or, after adjusting the index in the first summation and collecting terms,

$$\sum_{n=0}^{\infty} [a_{n+2}(n+2)(n+1) - a_n(n^2 - \mu^2)] t^n = 0.$$

The recurrence relation is therefore

$$a_{n+2} = \frac{n^2 - \mu^2}{(n+1)(n+2)} a_n, \qquad n = 0, 1, 2, \ldots. \tag{6}$$

The recurrence relation determines all the even-indexed coefficients to be

(continued)

---

[3] Pafnuty Lvovich Chebyshev (1821–1894) was appointed to the University of St. Petersburg in 1847. He contributed to many areas of mathematics and science, including number theory, mechanics, probability theory, special functions, and the calculation of geometric volumes.

(continued)

multiples of a_0 and all the odd-indexed coefficients to be multiples of a_1. The general solution is

$$y(t) = a_0 \left(1 + \frac{-\mu^2}{2}t^2 + \frac{-(4 - \mu^2)\mu^2}{24}t^4 + \cdots \right)$$

$$+ a_1 \left(t + \frac{1 - \mu^2}{6}t^3 + \frac{(1 - \mu^2)(9 - \mu^2)}{120}t^5 + \cdots \right). \tag{7}$$

The ratio test, in conjunction with recurrence relation (6), can be used to show that the power series in (7) have radius of convergence $R = 1$ (except in the case when they terminate after a finite number of terms).

If μ is an even integer, we see from (6) that all the even coefficients having index greater than μ will vanish. For example, if $\mu = 4$, then $a_6 = 0, a_8 = 0, \ldots$. Thus, when $\mu = 4$, recurrence relation (6) leads to

$$a_2 = -8a_0, \qquad a_4 = -a_2 = 8a_0, \qquad a_6 = a_8 = a_{10} = \cdots = 0.$$

From this, we see that the fourth degree polynomial $P(t) = a_0(1 - 8t^2 + 8t^4)$ is a solution of the Chebyshev equation. Similarly, if μ is an odd positive integer, then we obtain an odd polynomial solution of degree μ. These polynomial solutions, generated as μ ranges over the nonnegative integers, are known as **Chebyshev polynomials of the first kind**. The Nth degree Chebyshev polynomial of the first kind is usually denoted $T_N(t)$. The first few Chebyshev polynomials are

$$T_0(t) = 1, \qquad T_1(t) = t, \qquad T_2(t) = 2t^2 - 1, \qquad T_3(t) = 4t^3 - 3t,$$
$$T_4(t) = 8t^4 - 8t^2 + 1.$$

The Chebyshev polynomials are normalized; that is, the arbitrary constant is selected so that $T_N(1) = 1$. ❖

REMARKS:

1. Even though the differential equation has singular points at $t = \pm 1$, the Chebyshev polynomial solutions are well behaved at these points. The polynomial solutions are analytic with infinite radius of convergence. It is important to remember that solutions need not necessarily behave badly at singular points.

2. Chebyshev polynomials find important application in the design of antenna arrays and electrical filters. Consider, for example, the low-pass filtering problem illustrated in Figure 8.2. We want to build an electrical network having the power transfer function shown in Figure 8.2(a). Energy at all frequencies less than the cutoff frequency f_c should pass through the network unscathed, while the passage of energy at all frequencies above f_c should be completely blocked. The problem, however, is that the network elements we have available to build the network only allow us to realize power transfer functions that are rational functions of frequency. The design problem is to find a rational function that closely approximates the ideal behavior in Figure 8.2(a).

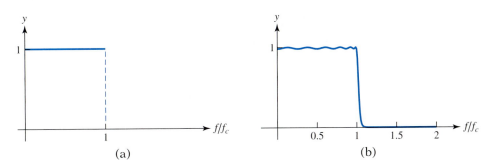

FIGURE 8.2

(a) The graph of an ideal power transfer function. (b) The graph of an approximation of the form (8) to the function graphed in (a).

Chebyshev polynomials are particularly suited for such problems because they possess an "equal ripple" property (see Exercise 27). The polynomial $T_N(t)$ oscillates between ± 1 for t in the range $-1 \le t \le 1$ and grows monotonically in magnitude when $|t| \ge 1$. Therefore, a power transfer function of the form

$$\frac{1}{1 + \varepsilon\, T_N^2\left(\dfrac{f}{f_c}\right)}, \tag{8}$$

with positive constant ε sufficiently small and integer N sufficiently large, can serve as a good rational approximation to the ideal power transfer function. Figure 8.2(b) illustrates the particular choice $\varepsilon = 0.02, N = 12$.

EXERCISES

Exercises 1–6:

Identify all the singular points of $y'' + p(t)y' + q(t)y = 0$ in the interval $-10 < t < 10$.

1. $y'' + (\sec t)y' + \dfrac{t}{t^2 - 4}y = 0$

2. $y'' + t^{2/3}y' + (\sin t)y = 0$

3. $(1 - t^2)y'' + ty' + (\csc t)y = 0$

4. $(\sin 2t)y'' + e^t y' + \dfrac{t}{25 - t^2}y = 0$

5. $(1 + \ln|t|)y'' + y' + (1 + t^2)y = 0$

6. $y'' + \dfrac{t}{1 + |t|}y' + (\tan t)y = 0$

Exercises 7–12:

In each exercise, $t = t_0$ is an ordinary point of $y'' + p(t)y' + q(t)y = 0$. Apply Theorem 8.1 to determine a value $R > 0$ such that an initial value problem, with the initial conditions prescribed at t_0, is guaranteed to have a unique solution that is analytic in the interval $t_0 - R < t < t_0 + R$.

7. $y'' + \dfrac{1}{1 + 2t}y' + \dfrac{t}{1 - t^2}y = 0, \quad t_0 = 0$

8. $(1 - 9t^2)y'' + 4y' + ty = 0, \quad t_0 = 1$

9. $y'' + \dfrac{1}{4 - 3t}y' + \dfrac{3t}{5 + 30t}y = 0, \quad t_0 = -1$

10. $y'' + \dfrac{1}{1 + 4t^2}y' + \dfrac{t}{4 + t}y = 0, \quad t_0 = 0$

11. $y'' + \dfrac{1}{1 + 3(t - 2)}y' + (\sin t)y = 0, \quad t_0 = 2$

12. $y'' + \dfrac{t + 3}{1 + t^2}y' + t^2 y = 0, \quad t_0 = 1$

Exercises 13–21:

In each exercise, $t = 0$ is an ordinary point of $y'' + p(t)y' + q(t)y = 0$.

(a) Find the recurrence relation that defines the coefficients of the power series solution $y(t) = \sum_{n=0}^{\infty} a_n t^n$.

(b) As in equation (7), find the first three nonzero terms in each of two linearly independent solutions.

(c) State the interval $-R < t < R$ on which Theorem 8.1 guarantees convergence.

(d) Does Theorem 8.2 indicate that the differential equation has solutions that are even and odd?

13. $y'' + ty' + y = 0$ **14.** $y'' + 2ty' + 3y = 0$ **15.** $(1 + t^2)y'' + ty' + 2y = 0$

16. $y'' - 5y' + 6y = 0$ **17.** $y'' - 4y' + 4y = 0$ **18.** $(1 + t)y'' + y = 0$

19. $(3 + t)y'' + 3ty' + y = 0$ **20.** $(2 + t^2)y'' + 4y = 0$ **21.** $y'' + t^2 y = 0$

Exercises 22–25:

In each exercise, $t = 1$ is an ordinary point of $y'' + p(t)y' + q(t)y = 0$.

(a) Find the recurrence relation that defines the coefficients of the power series solution $y(t) = \sum_{n=0}^{\infty} a_n (t - 1)^n$.

(b) Find the first three nonzero terms in each of two linearly independent solutions.

(c) State the interval $-R < t - 1 < R$ on which Theorem 8.1 guarantees convergence.

22. $y'' + (t - 1)y' + y = 0$ **23.** $y'' + y = 0$

24. $(t - 2)y'' + y' + y = 0$ **25.** $y'' + y' + (t - 2)y = 0$

26. Recall Chebyshev's equation from Example 3, $(1 - t^2)y'' - ty' + \mu^2 y = 0$. As you saw in Example 3, this equation has a polynomial solution, $T_N(t)$, when $\mu = N$ is a nonnegative integer. Using recurrence relation (6), find $T_5(t)$ and $T_6(t)$.

27. The Equal Ripple Property of Chebyshev Polynomials   Consider the Chebyshev differential equation $(1 - t^2)y'' - ty' + N^2 y = 0$, where N is a nonnegative integer.

(a) Show by substitution that the function $y(t) = \cos(N \arccos t)$ is a solution for $-1 < t < 1$.

(b) Show, for $N = 0, 1, 2$, that the function $\cos(N \arccos t)$ is a polynomial in t and that $T_N(t) = \cos(N \arccos t)$. This result holds, in fact, for all nonnegative integers N. It can be shown that

$$T_N(t) = \begin{cases} \cos(N \arccos t), & -1 \le t \le 1 \\ \cosh(N \operatorname{arccosh} t), & 1 < |t|. \end{cases}$$

(c) Use a computer graphics package to plot $T_N(t)$ for $N = 2, 5$, and 8 and for $-1.2 \le t \le 1.2$.

(d) What serves as a bound for $|T_N(t)|$ when $-1 \le t \le 1$? What is the behavior of $|T_N(t)|$ when $1 < |t|$?

28. Legendre's Equation   Legendre's equation is $(1 - t^2)y'' - 2ty' + \mu(\mu + 1)y = 0$. By Theorem 8.1, this equation has a power series solution of the form $y(t) = \sum_{n=0}^{\infty} a_n t^n$ that is guaranteed to be absolutely convergent in the interval $-1 < t < 1$. As in Example 3,

(a) Find the recurrence relation for the coefficients of the power series.

(b) Argue, when $\mu = N$ is a nonnegative integer, that Legendre's equation has a polynomial solution, $P_N(t)$.

(c) Show, by direct substitution, that the Legendre polynomials $P_0(t) = 1$ and $P_1(t) = t$ satisfy Legendre's equation for $\mu = 0$ and $\mu = 1$, respectively.

(d) Use the recurrence relation and the requirement that $P_n(1) = 1$ to determine the next four Legendre polynomials, $P_2(t), P_3(t), P_4(t), P_5(t)$.

29. Hermite's Equation   Hermite's equation is $y'' - 2ty' + 2\mu y = 0$. By Theorem 8.1, this equation has a power series solution of the form $y(t) = \sum_{n=0}^{\infty} a_n t^n$ that is guaranteed to be absolutely convergent in the interval $-\infty < t < \infty$. As in Example 3,

(a) Find the recurrence relation for the coefficients of the power series.

(b) Argue that, when $\mu = N$ is a nonnegative integer, Hermite's equation has a polynomial solution, $H_N(t)$.

(c) Show, by direct substitution, that the Hermite polynomials $H_0(t) = 1$ and $H_1(t) = 2t$ satisfy Hermite's equation for $\mu = 0$ and $\mu = 1$, respectively.

(d) Use the recurrence relation and the requirement that $H_n(t) = 2^n t^n + \cdots$ to determine the next four Hermite polynomials, $H_2(t), H_3(t), H_4(t), H_5(t)$.

30. Consider the differential equation $y'' + p(t)y' + q(t)y = 0$. In some cases, we may be able to find a power series solution of the form $y(t) = \sum_{n=0}^{\infty} a_n(t - t_0)^n$ even when t_0 is not an ordinary point. In other cases, there is no power series solution.

(a) The point $t = 0$ is a singular point of $ty'' + y' - y = 0$. Nevertheless, find a nontrivial power series solution, $y(t) = \sum_{n=0}^{\infty} a_n t^n$, of this equation.

(b) The point $t = 0$ is a singular point of $t^2 y'' + y = 0$. Show that the only solution of this equation having the form $y(t) = \sum_{n=0}^{\infty} a_n t^n$ is the trivial solution.

Exercises 31 and 32 outline the proof of parts (a) and (b) of Theorem 8.2, respectively. In each exercise, consider the differential equation $y'' + p(t)y' + q(t)y = 0$, where p and q are continuous on the domain $(-b, -a) \cup (a, b), a \geq 0$.

31. Let $f(t)$ be a solution on the interval (a, b).

(a) Let t lie in the interval $(-b, -a)$ and set $\tau = -t$, so that $a < \tau < b$. Show that if

$$\frac{d^2 f(\tau)}{d\tau^2} + p(\tau)\frac{df(\tau)}{d\tau} + q(\tau)f(\tau) = 0, \qquad a < \tau < b,$$

then

$$\frac{d^2 f(-t)}{dt^2} - p(-t)\frac{df(-t)}{dt} + q(-t)f(-t) = 0, \qquad -b < t < -a.$$

(b) Use the fact that p and q are odd and even functions, respectively, to show that $f(-t)$ is a solution of the given differential equation on the interval $-b < t < -a$.

32. Now let p and q be analytic at $t = 0$ with a common radius of convergence $R > 0$, where p is an odd function and q is an even function.

(a) Let $f_1(t)$ and $f_2(t)$ be solutions of the given differential equation, satisfying initial conditions $f_1(0) = 1, f_1'(0) = 0, f_2(0) = 0, f_2'(0) = 1$. What does Theorem 8.1 say about the solutions $f_1(t)$ and $f_2(t)$?

(b) Use the results of Exercise 31 to show that $f_1(-t)$ and $f_2(-t)$ are also solutions on the interval $-R < t < R$.

(c) Form the functions $f_e(t) = [f_1(t) + f_1(-t)]/2$ and $f_o(t) = [f_2(t) - f_2(-t)]/2$. Show that $f_e(t)$ and $f_o(t)$ are even and odd analytic solutions, respectively, on the interval $-R < t < R$.

(d) Show that $f_e(t)$ and $f_o(t)$ are nontrivial solutions by showing that $f_e(t) = f_1(t)$ and $f_o(t) = f_2(t)$. [Hint: Use the fact that solutions of initial value problems are unique.]

(e) Show that the solutions $f_e(t)$ and $f_o(t)$ form a fundamental set of solutions.

Exercises 33–38:

Suppose a linear differential equation $y'' + p(t)y' + q(t)y = 0$ satisfies the hypotheses of Theorem 8.2(b), on the interval $-\infty < t < \infty$. Then, by Exercise 32, we can assume the general solution of $y'' + p(t)y' + q(t)y = 0$ has the form

$$y(t) = c_1 y_e(t) + c_2 y_o(t), \tag{9}$$

where $y_e(t)$ is an even solution of $y'' + p(t)y' + q(t)y = 0$ and $y_o(t)$ is an odd solution. In each of the following exercises, determine whether Theorem 8.2(b) can be used to guarantee that the given differential equation has a general solution of the form (9). If your answer is no, explain why the equation fails to satisfy the hypotheses of Theorem 8.2(b).

33. $y'' + (\sin t)y' + t^2 y = 0$　**34.** $y'' + (\cos t)y' + ty = 0$　**35.** $y'' + t^2 y = 0$

36. $y'' + y' + t^2 y = 0$　**37.** $y'' + ty = 0$　**38.** $y'' + e^t y' + y = 0$

39. Consider the differential equation $y'' + ay' + by = 0$, where a and b are constants. For what values of a and b will the differential equation have nontrivial solutions that are odd and even?

40. Consider the initial value problem $(1 + t^2)y'' + y = 0, y(0) = 1, y'(0) = 0$.

(a) Show, by Theorem 8.1, that this initial value problem is guaranteed to have a unique solution of the form $y(t) = \sum_{n=0}^{\infty} a_n t^n$, where the series converges absolutely in $-1 < t < 1$.

(b) The recurrence relation has the form $r(n)a_{n+2} = s(n)a_n$, where $r(n)$ and $s(n)$ are quadratic polynomials. Therefore,

$$\left| \frac{a_{n+2}}{a_n} \right| = \left| \frac{s(n)}{r(n)} \right|.$$

Show that this ratio tends to 1 as n tends to ∞, and conclude, therefore, that the series solution of the initial value problem diverges for $1 < |t|$.

(c) Note, however, that the coefficient functions $p(t) = 0$ and $q(t) = (1 + t^2)^{-1}$ are continuous for $-\infty < t < \infty$, and hence, by Theorem 3.1, a unique solution exists for all values t. Do the conclusions of Theorem 8.1 and Theorem 3.1 contradict each other? Explain.

8.3 The Euler Equation

The second order linear homogeneous equation

$$t^2 y'' + \alpha t y' + \beta y = 0, \qquad t \neq 0 \tag{1}$$

is known as the **Euler equation** (it is also known as the **Cauchy-Euler equation** or the **equidimensional equation**). In equation (1), α and β are real constants. Note the special structure of the differential equation; the power of each monomial coefficient matches the order of the derivative that it multiplies—for example, t^2 multiplies y''. This special structure enables us to obtain an explicit representation for the general solution. In this section, we present two related approaches to deriving the general solution.

If we rewrite the Euler equation in the form $y'' + p(t)y' + q(t)y = 0$, then

$$p(t) = \frac{\alpha}{t} \quad \text{and} \quad q(t) = \frac{\beta}{t^2}.$$

The coefficients $p(t)$ and $q(t)$ are analytic at every point except $t = 0$. Ignoring the trivial case where α and β are both zero, we see that $t = 0$ is a singular point and all other values of t are ordinary points. Note as well that $p(t)$ and $q(t)$ are not continuous at $t = 0$. Therefore, the basic existence-uniqueness result, Theorem 3.1, alerts us to possible problems—solutions may or may not exist at $t = 0$.

The Euler equation arises in a variety of applications. An example is the problem of determining the time-independent (or steady-state) temperature distribution within a circular geometry (such as the interior of a circular pipe) from a knowledge of the temperature on the boundary; see the Projects at the end of this chapter. The Euler equation is also of interest because it serves as a prototype or model equation. In Section 8.4, we will define a special type of singular point, called a regular singular point; our understanding of the Euler equation will serve as the basis for studying regular singular points.

The General Solution of the Euler Equation

If $f(t)$ is a solution of the Euler equation (1), then so is $f(-t)$ (see Theorem 8.2 in Section 8.2). Therefore, we assume our interval of interest is $t > 0$. Once the general solution is obtained for the interval $t > 0$, we can obtain the general solution in $t < 0$ by replacing t with $-t$.

We present two approaches to deriving the general solution of (1). In retrospect, you will see that the two approaches are closely related. Nevertheless, both points of view are worthy of consideration.

Solutions of the Form $y(t) = t^\lambda$

The special structure of the Euler equation makes it possible to find solutions of the form $y(t) = t^\lambda$, where λ is a constant to be determined and where $t > 0$ is assumed. Substitution of this trial form into equation (1) leads to

$$t^2\lambda(\lambda - 1)t^{\lambda - 2} + \alpha t\lambda t^{\lambda - 1} + \beta t^\lambda = [\lambda(\lambda - 1) + \alpha\lambda + \beta]t^\lambda = 0, \qquad t > 0. \quad \textbf{(2)}$$

Since t^λ is not identically zero on $0 < t < \infty$ for any real or complex value λ, we must have

$$\lambda^2 + (\alpha - 1)\lambda + \beta = 0, \tag{3}$$

or

$$\lambda_{1,2} = \frac{-(\alpha - 1) \pm \sqrt{(\alpha - 1)^2 - 4\beta}}{2}. \tag{4}$$

In other words, $y(t) = t^\lambda$ is a solution of the Euler equation whenever λ is a root of equation (3). We refer to equation (3) as the **characteristic equation** or the **indicial equation**. There are three possibilities for the roots of the characteristic equation:

1. There are **two real distinct roots** if the discriminant, $(\alpha - 1)^2 - 4\beta$, is positive. In this case, the Wronskian of the solution set is

$$W(t) = \begin{vmatrix} t^{\lambda_1} & t^{\lambda_2} \\ \lambda_1 t^{\lambda_1 - 1} & \lambda_2 t^{\lambda_2 - 1} \end{vmatrix} = (\lambda_2 - \lambda_1) t^{\lambda_1 + \lambda_2 - 1}.$$

Since $W(t)$ is nonzero for $t > 0$, the functions $y_1(t) = t^{\lambda_1}$ and $y_2(t) = t^{\lambda_2}$ form a fundamental set. The general solution of equation (1) is $y(t) = c_1 t^{\lambda_1} + c_2 t^{\lambda_2}, t > 0$, where c_1 and c_2 are arbitrary constants. As previously mentioned, we can obtain the general solution for $t < 0$ by replacing t by $-t$; the general solution for $t < 0$ is $y(t) = c_1(-t)^{\lambda_1} + c_2(-t)^{\lambda_2}$, where c_1 and c_2 again denote arbitrary constants. Since $-t > 0$ when $t < 0$, the general solution can be expressed in the form

$$y(t) = c_1 |t|^{\lambda_1} + c_2 |t|^{\lambda_2}, \qquad t \neq 0. \tag{5}$$

2. There is **one real repeated root** if $(\alpha - 1)^2 - 4\beta = 0$. In this case, one solution is $y_1(t) = t^\lambda, t > 0$, where $\lambda = -(\alpha - 1)/2$. We can use reduction of order (see Section 3.4) to find a second linearly independent solution, $y_2(t) = t^\lambda \ln t, t > 0$ (see Exercise 1 of this section). The general solution in the repeated root case is therefore

$$y(t) = c_1 |t|^\lambda + c_2 |t|^\lambda \ln|t|, \qquad t \neq 0. \tag{6}$$

3. There are **complex conjugate roots** if $(\alpha - 1)^2 - 4\beta < 0$. For brevity, let

$$\lambda_{1,2} = \frac{-(\alpha - 1) \pm i\sqrt{4\beta - (\alpha - 1)^2}}{2} = \gamma \pm i\delta.$$

In this case,

$$t^{\gamma + i\delta} = e^{(\gamma + i\delta)\ln t} = e^{\gamma \ln t} e^{i\delta \ln t} = t^\gamma e^{i\delta \ln t}, \qquad t > 0.$$

Hence, by Euler's formula (see Section 3.5),

$$t^{\gamma + i\delta} = t^\gamma [\cos(\delta \ln t) + i \sin(\delta \ln t)], \qquad t > 0. \tag{7}$$

From equation (7), we obtain the two real-valued solutions $y_1(t) = t^\gamma \cos(\delta \ln t)$ and $y_2(t) = t^\gamma \sin(\delta \ln t)$, which can be shown to form a fundamental set on $0 < t < \infty$ (see Exercise 2). Therefore, the general solution is

$$y(t) = c_1 |t|^\gamma \cos(\delta \ln|t|) + c_2 |t|^\gamma \sin(\delta \ln|t|), \qquad t \neq 0. \tag{8}$$

EXAMPLE

1

Find the general solution of each of the Euler equations

(a) $t^2 y'' - 2ty' + 2y = 0$ (b) $t^2 y'' + 5ty' + 4y = 0$ (c) $t^2 y'' + 3ty' + 5y = 0$

Solution:

(a) For this equation, $\alpha = -2$ and $\beta = 2$. Looking for solutions of the form $t^\lambda, t > 0$ leads to $\lambda^2 - 3\lambda + 2 = (\lambda - 1)(\lambda - 2) = 0$. The general solution is therefore

$$y(t) = c_1 t + c_2 t^2, \qquad -\infty < t < \infty. \tag{9}$$

Since the general solution involves integer powers of t, we can dispense with the absolute value signs. Note further that solution (9) is defined for all values of t, including the singular point $t = 0$. For this equation, every solution is well behaved at $t = 0$.

(b) For this differential equation, $\alpha = 5$ and $\beta = 4$. The characteristic equation $\lambda^2 + 4\lambda + 4 = 0$ has real repeated roots $\lambda_1 = \lambda_2 = -2$. The general solution is

$$y(t) = c_1 t^{-2} + c_2 t^{-2} \ln|t|, \qquad t \neq 0.$$

(c) For this equation, $\alpha = 3$ and $\beta = 5$. The characteristic equation $\lambda^2 + 2\lambda + 5 = 0$ has complex conjugate roots $\lambda_{1,2} = -1 \pm 2i$. The general solution is

$$y(t) = c_1 t^{-1} \cos(2 \ln|t|) + c_2 t^{-1} \sin(2 \ln|t|), \qquad t \neq 0.$$

Neither (b) nor (c) has solutions that exist at $t = 0$. ❖

Change of Independent Variable

After a change of independent variable, the Euler equation can be transformed into a new homogeneous constant coefficient equation. We begin with the Euler equation $t^2 y'' + \alpha t y' + \beta y = 0$ for $t > 0$ and introduce a new independent variable $z = \ln t$, or (equivalently) $t = e^z$. The t-interval $0 < t < \infty$ transforms into the z-interval $-\infty < z < \infty$.

Let $y(t) = y(e^z) = Y(z)$. Then, using the chain rule, we have

$$\frac{dy(t)}{dt} = \frac{dy(e^z)}{dz} \frac{dz}{dt} = \frac{dY(z)}{dz} \left(\frac{dt}{dz} \right)^{-1} = e^{-z} \frac{dY(z)}{dz}.$$

Therefore, $t\,(dy(t)/dt)$ transforms into

$$e^z e^{-z} \frac{dY(z)}{dz} = \frac{dY(z)}{dz}, \tag{10}$$

while $t^2\,(d^2 y(t)/dt^2)$ transforms into

$$\frac{d^2 Y(z)}{dz^2} - \frac{dY(z)}{dz}. \tag{11}$$

Under this change of independent variable, the Euler equation (1) transforms into the constant coefficient equation

$$Y'' + (\alpha - 1)Y' + \beta Y = 0. \tag{12}$$

Therefore, we can solve the Euler equation (1) for $t > 0$ by

1. making the change of independent variable $t = e^z$,
2. solving the constant coefficient equation (12) using solution procedures developed in Chapter 3, and then
3. using the inverse map to obtain the desired solution, $y(t) = Y(\ln t)$.

Looking for solutions of $t^2 y'' + \alpha t y' + \beta y = 0$ having the form $y(t) = t^\lambda$ is equivalent to looking for solutions of $Y'' + (\alpha - 1)Y' + \beta Y = 0$ having the form $Y(z) = e^{\lambda z}$. The characteristic equation (3) results in either case.

Each of the two solution approaches has its utility. The first approach, looking for solutions of the form $|t|^\lambda$, serves as a guide in the next section when we study the behavior of solutions of differential equations in a neighborhood of certain types of singular points. The change of independent variable approach is useful because it permits us to use the techniques developed in Chapter 3.

As the next example shows, the change of variables approach can be applied to nonhomogeneous Euler equations.

EXAMPLE 2

Obtain the general solution of

$$4t^2 y'' + 8t y' + y = t + \ln t, \qquad t > 0$$

by making the change of independent variable $t = e^z$.

Solution: Under the change of variable $t = e^z$, the nonhomogeneous term transforms into $e^z + z$. Thus, using (10) and (11) yields for the given differential equation $4(Y'' - Y') + 8Y' + Y = e^z + z$, or

$$4Y'' + 4Y' + Y = e^z + z, \tag{13}$$

where $Y' = dY/dz$. As seen in Chapter 3, the solution of (13) is the sum of a complementary solution $Y_C(z)$ and a particular solution $Y_P(z)$. The complementary solution is

$$Y_C(z) = c_1 e^{-z/2} + c_2 z e^{-z/2}. \tag{14}$$

A particular solution $Y_P(z)$ can be obtained using the method of undetermined coefficients. We find

$$Y_P(z) = \tfrac{1}{9} e^z + z - 4.$$

The general solution is therefore

$$Y(z) = c_1 e^{-z/2} + c_2 z e^{-z/2} + \tfrac{1}{9} e^z + z - 4.$$

Using $z = \ln t$ to convert to the original independent variable t, we have

$$y(t) = c_1 t^{-1/2} + c_2 t^{-1/2} \ln t + \tfrac{1}{9} t + \ln t - 4, \qquad t > 0. \; ❖$$

Generalizations of the Euler Equation

There are two natural ways to generalize the Euler equation. One such generalization is given by

$$(t - t_0)^2 y'' + (t - t_0)\alpha y' + \beta y = 0. \tag{15}$$

In equation (15), t_0 is a point in the interval $-\infty < t < \infty$. To solve equation (15) for $t > t_0$, we assume a solution of the form $y(t) = (t - t_0)^\lambda$. (We could also adopt the change of independent variable $t - t_0 = e^z$.)

Another natural generalization is a higher order version of the Euler equation, such as

$$t^3 y''' + \alpha t^2 y'' + \beta t y' + \gamma y = 0. \tag{16}$$

Examples of each are given in the exercises.

EXERCISES

1. Consider the Euler equation $t^2y'' - (2\alpha - 1)ty' + \alpha^2 y = 0$.

    (a) Show that the characteristic equation has a repeated root $\lambda_1 = \lambda_2 = \alpha$. One solution is therefore $y(t) = t^\alpha, t > 0$.

    (b) Use the method of reduction of order (Section 3.4) to obtain a second linearly independent solution for the interval $0 < t < \infty$.

    (c) Compute the Wronskian of the set of solutions $\{t^\alpha, t^\alpha \ln t\}$, and show that it is a fundamental set of solutions on $0 < t < \infty$.

2. Let $y_1(t) = t^\gamma \cos(\delta \ln t)$ and $y_2(t) = t^\gamma \sin(\delta \ln t), t > 0$, where δ and γ are real constants with $\delta \neq 0$. Solutions of this form arise when the characteristic equation has complex roots. Compute the Wronskian of this pair of solutions, and show that they form a fundamental set of solutions on $0 < t < \infty$.

Exercises 3–18:

Identify the singular point. Find the general solution that is valid for values of t on either side of the singular point.

3. $t^2y'' - 4ty' + 6y = 0$ 4. $t^2y'' - 6y = 0$ 5. $t^2y'' - 3ty' + 4y = 0$

6. $t^2y'' - ty' + 5y = 0$ 7. $t^2y'' - 3ty' + 29y = 0$ 8. $t^2y'' - 5ty' + 9y = 0$

9. $t^2y'' + ty' + 9y = 0$ 10. $t^2y'' + 3ty' + y = 0$ 11. $t^2y'' + 3ty' + 17y = 0$

12. $y'' + \dfrac{11}{t}y' + \dfrac{25}{t^2}y = 0$ 13. $y'' + \dfrac{5}{t}y' + \dfrac{40}{t^2}y = 0$ 14. $t^2y'' - 2ty' = 0$

15. $(t - 1)^2y'' - (t - 1)y' - 3y = 0$ 16. $(t - 1)^2y'' + 3(t - 1)y' + 17y = 0$

17. $(t + 2)^2y'' + 6(t + 2)y' + 6y = 0$ 18. $(t - 2)^2y'' + (t - 2)y' + 4y = 0$

Exercises 19–21:

A Euler equation $(t - t_0)^2y'' + \alpha(t - t_0)y' + \beta y = 0$ is known to have the given general solution. What are the constants t_0, α, and β?

19. $y(t) = c_1(t + 2) + c_2\dfrac{1}{(t + 2)^2}, \quad t \neq -2$

20. $y(t) = c_1 + c_2 \ln|t - 1|, \quad t \neq 1$

21. $y(t) = c_1t^2 \cos(\ln|t|) + c_2t^2 \sin(\ln|t|), \quad t \neq 0$

Exercises 22–23:

A nonhomogeneous Euler equation $t^2y'' + \alpha ty' + \beta y = g(t)$ is known to have the given general solution. Determine the constants α and β and the function $g(t)$.

22. $y(t) = c_1t^2 + c_2t^{-1} + 2t + 1, \quad t > 0$ 23. $y(t) = c_1t^2 + c_2t^3 + \ln t, \quad t > 0$

Exercises 24–29:

Find the general solution of the given equation for $0 < t < \infty$. [Hint: You can, as in Example 2, use the change of variable $t = e^z$.]

24. $t^2y'' - 2y = 2$ 25. $t^2y'' - ty' + y = t^{-1}$

26. $t^2y'' + ty' + 9y = 10t$ 27. $t^2y'' - 6y = 10t^{-2} - 6$

28. $t^2y'' - 4ty' + 6y = 3\ln t$ 29. $t^2y'' + 8ty' + 10y = 36(t + t^{-1})$

Exercises 30–33:

Solve the given initial value problem. What is the interval of existence of the solution?

30. $t^2y'' - ty' - 3y = 8t + 6, \quad y(1) = 1, \quad y'(1) = 3$

31. $t^2y'' - 5ty' + 5y = 10$, $y(1) = 4$, $y'(1) = 6$

32. $t^2y'' + 3ty' + y = 8t + 9$, $y(-1) = 1$, $y'(-1) = 0$

33. $t^2y'' + 3ty' + y = 2t^{-1}$, $y(1) = -2$, $y'(1) = 1$

34. Consider the third order equation $t^3y''' + \alpha t^2 y'' + \beta t y' + \gamma y = 0$, $t > 0$. Make the change of independent variable $t = e^z$ and let $Y(z) = y(e^z)$. Derive the third order constant coefficient equation satisfied by Y.

Exercises 35–38:

Obtain the general solution of the given differential equation for $0 < t < \infty$.

35. $t^3y''' + 3t^2y'' - 3ty' = 0$

36. $t^3y''' + ty' - y = 0$

37. $t^3y''' + 3t^2y'' + ty' = 8t^2 + 12$

38. $t^3y''' + 6t^2y'' + 7ty' + y = 2 + \ln t$

8.4   Solutions Near a Regular Singular Point and the Method of Frobenius

We introduced the term *singular point* in Section 8.2 to denote a point t where at least one of the coefficient functions of $y'' + p(t)y' + q(t)y = 0$ fails to be analytic. Near a singular point, the possible types of solution behavior are diverse and complicated.

In this section, we restrict our attention to a particular type of singular point, one known as a *regular singular point*. Definitive statements can be made about the behavior of solutions near regular singular points. Many of the important equations of mathematical physics, such as the Euler equation, Bessel's equation, and Legendre's equation, possess regular singular points.

Regular Singular Points

We begin with the Euler equation

$$t^2y'' + \alpha t y' + \beta y = 0,$$

where α and β are constants, not both zero. Since the coefficient functions are

$$p(t) = \frac{\alpha}{t} \quad \text{and} \quad q(t) = \frac{\beta}{t^2},$$

$t = 0$ is a singular point and all other points are ordinary points. The Euler equation serves as a model for a differential equation having a regular singular point.

In general, let t_0 be a singular point of the differential equation

$$y'' + p(t)y' + q(t)y = 0.$$

The point t_0 is a **regular singular point** if both of the functions

$$(t - t_0)p(t) \quad \text{and} \quad (t - t_0)^2 q(t)$$

are analytic at t_0. A singular point that is not a regular singular point is called an **irregular singular point**.

If $(t - t_0)p(t)$ and/or $(t - t_0)^2 q(t)$ is indeterminate at $t = t_0$ but the limits $\lim_{t \to t_0}(t - t_0)p(t)$ and $\lim_{t \to t_0}(t - t_0)^2 q(t)$ exist, then we are tacitly assuming that the functions being considered are defined to equal these limits at $t = t_0$. In many important cases, the functions $p(t)$ and $q(t)$ are rational functions (that is, ratios of polynomials). As previously noted, a rational function is analytic at every point where the denominator polynomial is nonzero. In this case, to show that a singular point t_0 is a regular singular point, it suffices to show that the limits $\lim_{t \to t_0}(t - t_0)p(t)$ and $\lim_{t \to t_0}(t - t_0)^2 q(t)$ both exist.

E X A M P L E

1

For each differential equation, identify the singular points and classify them as regular or irregular singular points. In these equations, v and μ are constants.

(a) Bessel's equation[4]: $t^2 y'' + ty' + (t^2 - v^2)y = 0$

(b) Legendre's equation[5]: $(1 - t^2)y'' - 2ty' + [v(v + 1) - \mu^2(1 - t^2)^{-1}]y = 0$

(c) $(t - 1)y'' + \left(\tan\frac{\pi}{2}t\right)y' + t^{5/3}y = 0,$ $-2 < t < 2$

Solution:

(a) For Bessel's equation,

$$p(t) = \frac{1}{t} \quad \text{and} \quad q(t) = \frac{t^2 - v^2}{t^2}.$$

Therefore, $t = 0$ is a singular point. All other points are ordinary points. Because $p(t)$ and $q(t)$ are rational functions, we need only determine whether $\lim_{t \to 0} tp(t)$ and $\lim_{t \to 0} t^2 q(t)$ exist in order to establish analyticity. Since

$$\lim_{t \to 0} tp(t) = 1 \quad \text{and} \quad \lim_{t \to 0} t^2 q(t) = -v^2,$$

it follows that $t = 0$ is a regular singular point.

(b) For Legendre's equation,

$$p(t) = \frac{-2t}{(1 - t^2)} \quad \text{and} \quad q(t) = \frac{v(v + 1)}{(1 - t^2)} - \frac{\mu^2}{(1 - t^2)^2}.$$

Therefore, the points $t = \pm 1$ are singular points. All other points are ordinary points. We first consider the singular point $t = 1$. Since $p(t)$ and $q(t)$ are rational functions, we check the two limits

$$\lim_{t \to 1}(t - 1)p(t) = \lim_{t \to 1}\frac{2t}{1 + t} = 1$$

and

$$\lim_{t \to 1}(t - 1)^2 q(t) = \lim_{t \to 1}\frac{v(v + 1)(1 - t^2) - \mu^2}{(1 + t)^2} = \frac{-\mu^2}{4}.$$

(continued)

---

[4]Friedrich Wilhelm Bessel (1784–1846) was a German scientist noted for important contributions to the fields of astronomy, celestial mechanics, and mathematics. His mathematical analysis of what is now known as the Bessel function arose during his studies of planetary motion. Bessel's achievements seem even more remarkable when one realizes that his formal education ended at age 14.

[5]Adrien-Marie Legendre (1752–1833) was a French scientist whose research interests included projectile dynamics, celestial mechanics, number theory, and analysis. What are today called Legendre polynomials appeared in a 1784 paper on celestial mechanics. Legendre authored influential textbooks on Euclidean geometry and number theory.

(continued)

Since both limits exist, $t = 1$ is a regular singular point. The analysis of the other singular point, $t = -1$, is very similar, and we find that $t = -1$ is also a regular singular point.

(c) We have

$$p(t) = \frac{\tan \frac{\pi}{2} t}{t - 1} = \frac{\sin \frac{\pi}{2} t}{(t - 1) \cos \frac{\pi}{2} t} \quad \text{and} \quad q(t) = \frac{t^{5/3}}{t - 1}.$$

Note that $p(t)$, viewed as $p(t) = \sin(\pi t/2)/[(t - 1) \cos(\pi t/2)]$, is a quotient of functions analytic in $-2 < t < 2$. Therefore, $p(t)$ is analytic at all points in the interval $-2 < t < 2$ except $t = 1$ (where the denominator is zero). The function $q(t)$ fails to be analytic at $t = 0$ (where $t^{5/3}$ is not analytic) and $t = 1$ (where the denominator is zero). Consider first the singular point $t = 0$; $t^2 q(t) = (t - 1)^{-1} t^{11/3}$ is not analytic at $t = 0$, and therefore $t = 0$ is an irregular singular point.

Next, consider the singular point $t = 1$. In this case, $p(t)$ and $q(t)$ are both quotients of functions analytic at $t = 1$. Therefore, $t = 1$ is a regular singular point if both of the limits exist:

$$\lim_{t \to 1} (t - 1) p(t) \quad \text{and} \quad \lim_{t \to 1} (t - 1)^2 q(t).$$

The first limit,

$$\lim_{t \to 1} (t - 1) p(t) = \lim_{t \to 1} \tan \frac{\pi}{2} t,$$

does not exist. Therefore, $t = 1$ is also an irregular singular point. ❖

The Method of Frobenius

Just as the Euler equation served as a model for defining regular singular points, the general solution of the Euler equation will serve to introduce the method of Frobenius.[6] This method prescribes the type of solution to look for near a regular singular point.

For simplicity, consider solution behavior near $t = 0$. Suppose we know that the differential equation

$$y'' + p(t) y' + q(t) y = 0 \tag{1a}$$

has a regular singular point at $t = 0$. Then $tp(t)$ and $t^2 q(t)$ are analytic at $t = 0$. Let the Maclaurin series for $tp(t)$ and $t^2 q(t)$ be

$$tp(t) = \alpha_0 + \alpha_1 t + \alpha_2 t^2 + \cdots = \sum_{n=0}^{\infty} \alpha_n t^n$$

and

$$t^2 q(t) = \beta_0 + \beta_1 t + \beta_2 t^2 + \cdots = \sum_{n=0}^{\infty} \beta_n t^n,$$

---

[6] Ferdinand Georg Frobenius (1849–1917), a German mathematician, served on the faculties at the University of Zurich and the University of Berlin. He is remembered for his contributions to group theory and differential equations.

where both series converge in some interval $-R < t < R$. Multiply differential equation (1a) by t^2 to obtain

$$t^2 y'' + t[tp(t)]y' + [t^2 q(t)]y = 0. \tag{1b}$$

Inserting the Maclaurin series for $tp(t)$ and $t^2 q(t)$, we arrive at

$$t^2 y'' + t[\alpha_0 + \alpha_1 t + \alpha_2 t^2 + \cdots]y' + [\beta_0 + \beta_1 t + \beta_2 t^2 + \cdots]y = 0. \tag{2}$$

The method of Frobenius can be motivated by a heuristic argument. Very close to the singular point $t = 0$, we expect that

$$\alpha_0 + \alpha_1 t + \alpha_2 t^2 + \cdots \approx \alpha_0 \quad \text{and} \quad \beta_0 + \beta_1 t + \beta_2 t^2 + \cdots \approx \beta_0. \tag{3}$$

If we use approximations (3) in equation (1b), we recover the Euler equation. Thus, we reason: If the two differential equations are nearly the same near $t = 0$, shouldn't the solutions themselves have similar behavior near $t = 0$?

Recall that solutions of the Euler equation were obtained by looking for solutions of the form $y(t) = |t|^\lambda$, $t \neq 0$. For definiteness, we consider $t > 0$. The method of Frobenius consists in looking for solutions in which the factor t^λ is multiplied by an infinite series. In other words, near the regular singular point $t = 0$, we look for solutions of $y'' + p(t)y' + q(t)y = 0$ that have the form

$$y(t) = t^\lambda[a_0 + a_1 t + a_2 t^2 + \cdots] = \sum_{n=0}^{\infty} a_n t^{\lambda+n}. \tag{4}$$

In representation (4), λ is a constant (possibly complex-valued) that is to be determined, along with the constants $a_0, a_1, \ldots$. Also, since λ has not been specified, we can assume without any loss of generality that $a_0 \neq 0$. [That is, there must be a "first nonzero term" in series (4), and we simply take that term to be $a_0 t^\lambda$.]

Implementing the Method of Frobenius

Substituting (4) into differential equation (1b) creates the following three terms:

$$t^2 y'' = t^2 \sum_{n=0}^{\infty} a_n(\lambda + n)(\lambda + n - 1)t^{\lambda+n-2} = \sum_{n=0}^{\infty} a_n(\lambda + n)(\lambda + n - 1)t^{\lambda+n} \tag{5a}$$

$$t[tp(t)]y' = \left[\sum_{m=0}^{\infty} \alpha_m t^m\right]\left[\sum_{n=0}^{\infty} a_n(\lambda + n)t^{\lambda+n}\right] \tag{5b}$$

$$[t^2 q(t)]y = \left[\sum_{m=0}^{\infty} \beta_m t^m\right]\left[\sum_{n=0}^{\infty} a_n t^{\lambda+n}\right]. \tag{5c}$$

The three terms in (5) must be added, and the sum equated to zero. The series products are computed using the Cauchy product defined in Section 8.1, and, as before, coefficients of like powers of t are grouped together. We obtain

$$t^2 y'' + t[tp(t)]y' + [t^2 q(t)]y = [\lambda(\lambda - 1) + \alpha_0\lambda + \beta_0]a_0 t^\lambda + [\lambda(\lambda + 1)a_1$$
$$+ \alpha_0 a_1(\lambda + 1) + \alpha_1 a_0\lambda + \beta_0 a_1 + \beta_1 a_0]t^{\lambda+1}$$
$$+ [(\lambda + 2)(\lambda + 1)a_2 + \alpha_0 a_2(\lambda + 2) + \alpha_1 a_1(\lambda + 1)$$
$$+ \alpha_2 a_0\lambda + \beta_0 a_2 + \beta_1 a_1 + \beta_2 a_0]t^{\lambda+2} + \cdots, \quad t > 0. \tag{6}$$

When we equate the right-hand side of (6) to zero and invoke the uniqueness property of power series representations, it follows that the coefficient of each power of t must vanish. Setting the first coefficient equal to zero leads to

$$[\lambda^2 + (\alpha_0 - 1)\lambda + \beta_0]a_0 = 0. \tag{7}$$

Since $a_0 \neq 0$, it follows that representation (4) is a solution of the differential equation only if the exponent λ is a root of the equation $F(\lambda) = 0$, where

$$F(\lambda) = \lambda^2 + (\alpha_0 - 1)\lambda + \beta_0. \tag{8}$$

The equation $F(\lambda) = 0$ is called the **characteristic equation** or the **indicial equation**. (The latter term is used more frequently in this context, and we will use it as well.) Note that this equation (8) is consistent with the heuristic argument used to motivate representation (4). The equation $F(\lambda) = 0$ is precisely the indicial equation for the Euler equation that results from using the two approximations $tp(t) \approx \alpha_0$ and $t^2q(t) \approx \beta_0$.

Once we choose a root λ of the indicial equation, we set the coefficients of the higher powers of t equal to zero in (6). This gives us a recurrence relation for finding the coefficients $a_1, a_2, a_3, \ldots$ in terms of a_0. For instance, setting the coefficient of $t^{\lambda+1}$ equal to zero, we have

$$[(\lambda + 1)^2 + (\alpha_0 - 1)(\lambda + 1) + \beta_0]a_1 + [\alpha_1\lambda + \beta_1]a_0 = 0,$$

or

$$a_1 = -\frac{[\alpha_1\lambda + \beta_1]a_0}{F(\lambda + 1)}. \tag{9a}$$

Similarly, knowing a_1, we determine a_2 from

$$[(\lambda + 2)^2 + (\alpha_0 - 1)(\lambda + 2) + \beta_0]a_2 + [\alpha_1(\lambda + 1) + \beta_1]a_1 + [\alpha_2\lambda + \beta_2]a_0 = 0,$$

obtaining

$$a_2 = -\frac{[\alpha_1(\lambda + 1) + \beta_1]a_1 + [\alpha_2\lambda + \beta_2]a_0}{F(\lambda + 2)}. \tag{9b}$$

Note the difference between the procedure for constructing the general solution near an ordinary point and the method of Frobenius for constructing the general solution near a regular singular point. For example, if $t = 0$ is an ordinary point, then we look for a solution of the form

$$y(t) = \sum_{n=0}^{\infty} a_n t^n.$$

The recurrence relation obtained determines the coefficients a_n in terms of two of these coefficients, typically a_0 and a_1. Since a_0 and a_1 are arbitrary, both members of a fundamental set of solutions are obtained concurrently.

If $t = 0$ is a regular singular point, however, the method of Frobenius leads to the indicial equation, $F(\lambda) = 0$. Let the roots of this equation be denoted by λ_1 and λ_2. The recurrence relation found using (6) is used twice, first with $\lambda = \lambda_1$ and then with $\lambda = \lambda_2$. In this way, we seek two linearly independent solutions,

$$y_1(t) = \sum_{n=0}^{\infty} a_n^{(1)} t^{\lambda_1 + n} \quad \text{and} \quad y_2(t) = \sum_{n=0}^{\infty} a_n^{(2)} t^{\lambda_2 + n},$$

where the coefficients $a_1^{(1)}, a_2^{(1)}, a_3^{(1)}, \ldots$ and $a_1^{(2)}, a_2^{(2)}, a_3^{(2)}, \ldots$ are generated, respectively, in terms of arbitrary constants $a_0^{(1)}$ and $a_0^{(2)}$.

There are cases where the method of Frobenius, as outlined in the foregoing text, does not produce the general solution. An obvious case occurs when the two roots of the indicial equation are equal. In this repeated-root case, the foregoing procedure yields only one member of the fundamental set. A second (less obvious) case occurs when the indicial equation possesses two (real) roots that differ by an integer. These special cases will be discussed (and the entire method summarized) in the next section. We conclude with an example in which the procedure previously outlined can be used to obtain the general solution.

E X A M P L E

2

Use the method of Frobenius to obtain the general solution of

$$2t^2 y'' - t y' + (1 + t^2) y = 0 \quad \text{for} \quad t > 0.$$

Solution: The point $t = 0$ is a singular point. Since $p(t)$ and $q(t)$ are rational functions and since $\lim_{t \to 0} t p(t) = -\frac{1}{2}$ and $\lim_{t \to 0} t^2 q(t) = \frac{1}{2}$, it follows that $t = 0$ is a regular singular point.

Using the method of Frobenius, we look for solutions of the form

$$y(t) = \sum_{n=0}^{\infty} a_n t^{\lambda+n}, \qquad t > 0.$$

Substituting this series into the differential equation leads to

$$\sum_{n=0}^{\infty} a_n (\lambda+n)(\lambda+n-1) t^{\lambda+n} - \frac{1}{2} \sum_{n=0}^{\infty} a_n (\lambda+n) t^{\lambda+n} + \frac{1}{2} \sum_{n=0}^{\infty} a_n t^{\lambda+n}$$
$$+ \frac{1}{2} \sum_{n=0}^{\infty} a_n t^{\lambda+n+2} = 0.$$

Rewriting the last series as $\sum_{n=0}^{\infty} a_n t^{\lambda+n+2} = \sum_{n=2}^{\infty} a_{n-2} t^{\lambda+n}$ and combining terms where possible, we obtain

$$\sum_{n=0}^{\infty} \left[(\lambda+n)(\lambda+n-1) - \frac{1}{2}(\lambda+n) + \frac{1}{2} \right] a_n t^{\lambda+n} + \sum_{n=2}^{\infty} \frac{1}{2} a_{n-2} t^{\lambda+n} = 0, \quad \textbf{(10)}$$

or

$$\left[\lambda^2 - \frac{3}{2}\lambda + \frac{1}{2} \right] a_0 t^\lambda + \left[(\lambda+1)^2 - \frac{3}{2}(\lambda+1) + \frac{1}{2} \right] a_1 t^{\lambda+1}$$
$$+ \sum_{n=2}^{\infty} \left[\left\{ (\lambda+n)^2 - \frac{3}{2}(\lambda+n) + \frac{1}{2} \right\} a_n + \frac{1}{2} a_{n-2} \right] t^{\lambda+n} = 0.$$

The indicial equation, $F(\lambda) = \lambda^2 - \frac{3}{2}\lambda + \frac{1}{2} = 0$, has roots $\lambda_1 = \frac{1}{2}$ and $\lambda_2 = 1$. We now examine the recurrence relations associated with each of these roots.

(continued)

(continued)

Case 1 Let $\lambda = \frac{1}{2}$ in (10). We set the coefficient of $t^{\lambda+n} = t^{(1/2)+n}$ equal to zero for each value of n and use $a_n^{(1)}$ to denote the coefficients associated with this value of λ. The coefficient $a_0^{(1)}$ is arbitrary. The coefficient multiplying $t^{3/2}$ must be zero, and thus

$$\left[\left(\tfrac{1}{2}+1\right)^2 - \tfrac{3}{2}\left(\tfrac{1}{2}+1\right) + \tfrac{1}{2}\right] a_1^{(1)} = F\left(\tfrac{3}{2}\right) a_1^{(1)} = 0.$$

Since $F\left(\tfrac{3}{2}\right) \neq 0$, it follows that $a_1^{(1)} = 0$. As for the remaining coefficients, we have

$$\left[\left(\tfrac{1}{2}+n\right)^2 - \tfrac{3}{2}\left(\tfrac{1}{2}+n\right) + \tfrac{1}{2}\right] a_n^{(1)} + \tfrac{1}{2}a_{n-2}^{(1)} = 0, \qquad n = 2, 3, 4, \ldots,$$

and so

$$a_n^{(1)} = -\frac{a_{n-2}^{(1)}}{2F\left(\tfrac{1}{2}+n\right)}, \qquad n = 2, 3, 4, \ldots. \tag{11}$$

Since $F(\tfrac{1}{2}+n) \neq 0$ for all $n \geq 2$, recurrence relation (11) is well defined. The coefficients $a_n^{(1)}$ for n even can ultimately be expressed in terms of $a_0^{(1)}$. The coefficients $a_n^{(1)}$ for n odd are all zero because $a_1^{(1)} = 0$. Thus, we find a solution $y_1(t)$:

$$y_1(t) = a_0^{(1)}\left[t^{1/2} - \tfrac{1}{6}t^{5/2} + \tfrac{1}{168}t^{9/2} - \cdots\right], \qquad t > 0.$$

Case 2 Let $\lambda = 1$ in (10). We repeat the sequence of computations just completed with this new value of λ and with the coefficients now denoted by $a_n^{(2)}$. As in case 1, it follows that $a_1^{(2)} = 0$ since $F(2) \neq 0$. In general,

$$\left[(1+n)^2 - \tfrac{3}{2}(1+n) + \tfrac{1}{2}\right] a_n^{(2)} + \tfrac{1}{2}a_{n-2}^{(2)} = 0, \qquad n = 2, 3, 4, \ldots,$$

and thus

$$a_n^{(2)} = -\frac{a_{n-2}^{(2)}}{2F(1+n)}, \qquad n = 2, 3, 4, \ldots. \tag{12}$$

The solution obtained is

$$y_2(t) = a_0^{(2)}\left[t - \tfrac{1}{10}t^3 + \tfrac{1}{360}t^5 - \cdots\right]. \tag{13}$$

It is clear that the two solutions obtained are linearly independent on $t > 0$ since one solution is not a constant multiple of the other. Therefore, the two solutions form a fundamental set, and the general solution is

$$y(t) = a_0^{(1)}\left[t^{1/2} - \tfrac{1}{6}t^{5/2} + \tfrac{1}{168}t^{9/2} - \cdots\right]$$

$$+ a_0^{(2)}\left[t - \tfrac{1}{10}t^3 + \tfrac{1}{360}t^5 - \cdots\right], \qquad t > 0. \tag{14}$$

Note that the differential equation possesses the symmetries discussed in Theorem 8.2; that is, $p(t) = -2/t$ is an odd function, and $q(t) = (1+t^2)/t^2$ is an even function. Therefore, to find a solution for $t < 0$, we need only replace t by $-t$ in (14). As a final observation, note that recurrence relations (11) and (12), together with the ratio test, can be used to show that the two series in general solution (14) converge absolutely in $0 < t < \infty$. ❖

EXERCISES

Exercises 1–10:

In each exercise, find the singular points (if any) and classify them as regular or irregular.

1. $ty'' + (\cos t)y' + y = 0$

2. $t^2 y'' + (\sin t)y' + y = 0$

3. $(t^2 - 1)y'' + (t - 1)y' + y = 0$

4. $(t^2 - 1)^2 y'' + (t + 1)y' + y = 0$

5. $t^2 y'' + (1 - \cos t)y' + y = 0$

6. $|t|y'' + y' + y = 0$

7. $(1 - e^t)y'' + y' + y = 0$

8. $(4 - t^2)y'' + (t + 2)y' + (4 - t^2)^{-1}y = 0$

9. $(1 - t^2)^{1/3}y'' + y' + ty = 0$

10. $y'' + y' + t^{1/3}y = 0$

Exercises 11–13:

In each exercise, determine the polynomial $P(t)$ of smallest degree that causes the given differential equation to have the stated properties.

11. $y'' + \dfrac{\sin 2t}{P(t)}y' + y = 0$ — There is a regular singular point at $t = 0$ and irregular singular points at $t = \pm 1$. All other points are ordinary points.

12. $y'' + \dfrac{1}{t}y' + \dfrac{1}{P(t)}y = 0$ — There is a regular singular point at $t = 0$. All other points are ordinary points.

13. $y'' + \dfrac{1}{tP(t)}y' + \dfrac{1}{t^3}y = 0$ — There are irregular singular points at $t = 0$ and $t = \pm 1$. All other points are ordinary points.

Exercises 14–15:

In each exercise, the exponent n in the given differential equation is a nonnegative integer. Determine the possible values of n (if any) for which

(a) $t = 0$ is a regular singular point.

(b) $t = 0$ is an irregular singular point.

14. $y'' + \dfrac{1}{t^n}y' + \dfrac{1}{1 + t^2}y = 0$

15. $y'' + \dfrac{1}{\sin t}y' + \dfrac{1}{t^n}y = 0$

Exercises 16–23:

In each exercise,

(a) Verify that $t = 0$ is a regular singular point.

(b) Find the indicial equation.

(c) Find the recurrence relation.

(d) Find the first three nonzero terms of the series solution, for $t > 0$, corresponding to the larger root of the indicial equation. If there are fewer than three nonzero terms, give the corresponding exact solution.

16. $2t^2 y'' - ty' + (t + 1)y = 0$

17. $4t^2 y'' + 4ty' + (t - 1)y = 0$

18. $16t^2 y'' + t^2 y' + 3y = 0$

19. $t^2 y'' + ty' + (t - 9)y = 0$

20. $ty'' + (t + 2)y' - y = 0$

21. $t^2 y'' + 3ty' + (2t + 1)y = 0$

22. $t^2 y'' + t(t - 1)y' - 3y = 0$

23. $ty'' + (t - 2)y' + y = 0$

Exercises 24–27:

In each exercise,

(a) Verify that $t = 0$ is a regular singular point.

(b) Find the indicial equation.

(c) Find the first three terms of the series solution, for $t > 0$, corresponding to the larger root of the indicial equation.

24. $t^2 y'' - (2 \sin t)y' + (2 + t)y = 0$ **25.** $ty'' - 4y' + e^t y = 0$

26. $(\sin t)y'' - y' + y = 0$ **27.** $(1 - e^t)y'' + \frac{1}{2}y' + y = 0$

8.5   The Method of Frobenius Continued: Special Cases and a Summary

There are two important special cases where the method of Frobenius (as described in the previous section) may not yield the general solution of $y'' + p(t)y' + q(t)y = 0$ near a regular singular point. These special cases arise when the indicial equation has two real roots that are equal or that differ by an integer. We will use Bessel's equation as a vehicle to illustrate these cases. We then conclude this section by summarizing, for all cases, the structure of the general solution near a regular singular point.

The General Solution of Bessel's Equation

Consider Bessel's equation,

$$t^2 y'' + ty' + (t^2 - v^2)y = 0, \qquad t > 0,$$

where v is a real nonnegative constant. As we saw in Section 8.4, $t = 0$ is a regular singular point. In addition, note that the coefficient functions

$$p(t) = \frac{1}{t} \quad \text{and} \quad q(t) = \frac{t^2 - v^2}{t^2}$$

are odd and even functions, respectively. Therefore, by Theorem 8.2, we can obtain solutions for $t < 0$ by finding solutions for $t > 0$ and replacing t with $-t$.

We apply the method of Frobenius to Bessel's equation. Substituting $y(t) = \sum_{n=0}^{\infty} a_n t^{\lambda+n}$ leads to

$$[\lambda^2 - v^2]a_0 t^{\lambda} + [(\lambda + 1)^2 - v^2]a_1 t^{\lambda+1}$$

$$+ \sum_{n=2}^{\infty} [\{(\lambda + n)^2 - v^2\}a_n + a_{n-2}]t^{\lambda+n} = 0, \qquad t > 0. \tag{1}$$

Without loss of generality, we assume $a_0 \neq 0$. The indicial equation,

$$F(\lambda) = \lambda^2 - v^2 = 0, \tag{2}$$

has roots $\lambda_1 = v$ and $\lambda_2 = -v$. From (1), we also obtain the equations

$$F(\lambda + 1)a_1 = 0 \tag{3a}$$

and

$$F(\lambda + n)a_n + a_{n-2} = 0. \tag{3b}$$

Equation (3b) leads to the recurrence relation

$$a_n = \frac{-a_{n-2}}{F(\lambda + n)}, \qquad n = 2, 3, 4, \ldots. \tag{3c}$$

The special cases that arise can be illustrated by selecting particular values for the constant v.

Equal Roots ($\lambda_1 = \lambda_2$)

Consider Bessel's equation with $\nu = 0$. In this case, $\lambda_1 = \lambda_2 = 0$. Since the root is repeated, the method of Frobenius gives us only one member of the fundamental set of solutions; the second linearly independent solution has a structure different from $\sum_{n=0}^{\infty} a_n t^{\lambda+n}$.

Since $\nu = 0$, the indicial polynomial reduces to $F(\lambda) = \lambda^2$. Therefore, $F(1)$ is nonzero, and we see from (3a) that $a_1 = 0$. From (3c), we have

$$a_n = \frac{-a_{n-2}}{n^2}, \qquad n = 2, 3, 4, \ldots. \tag{4}$$

Since $a_1 = 0$, it follows from (4) that all odd-indexed coefficients are zero. The even-indexed coefficients can be expressed as multiples of a_0:

$$a_2 = \frac{-1}{2^2} a_0, \qquad a_4 = \frac{-1}{4^2} a_2 = \frac{(-1)^2}{2^4 (2!)^2} a_0, \qquad a_6 = \frac{-1}{6^2} a_4 = \frac{(-1)^3}{2^6 (3!)^2} a_0,$$

and, in general,

$$a_{2n} = \frac{(-1)^n}{2^{2n} (n!)^2} a_0, \qquad n = 0, 1, 2, \ldots.$$

The solution we thus obtain is

$$y_1(t) = a_0 \left[1 - \frac{t^2}{4} + \frac{t^4}{64} - \frac{t^6}{2304} + \cdots \right] = a_0 \left[1 + \sum_{n=1}^{\infty} \frac{(-1)^n t^{2n}}{2^{2n} (n!)^2} \right] = a_0 J_0(t), \tag{5a}$$

where

$$J_0(t) = 1 + \sum_{n=1}^{\infty} \frac{(-1)^n t^{2n}}{2^{2n} (n!)^2}. \tag{5b}$$

The function $J_0(t)$ is called the **Bessel function of the first kind of order zero**.

Having one solution, we could, in principle, use the method of reduction of order (see Section 3.4) to construct a second solution. We shall, however, simply state a form of the second solution that is commonly used in applications—the **Bessel function of the second kind of order zero** (also called Weber's function). The Bessel function of the second kind of order zero is given by

$$Y_0(t) = \frac{2}{\pi} \left[\gamma + \ln \frac{t}{2} \right] J_0(t) + \frac{2}{\pi} \sum_{n=1}^{\infty} (-1)^{n+1} k_n \frac{\left(\frac{t^2}{4} \right)^n}{(n!)^2}, \qquad t > 0. \tag{6}$$

In (6), $k_1 = 1$ and, in general,

$$k_n = 1 + \frac{1}{2} + \frac{1}{3} + \frac{1}{4} + \cdots + \frac{1}{n}, \qquad n = 2, 3, \ldots.$$

The constant γ in equation (6) is known as the **Euler-Mascheroni constant**[7] and is defined by the limit

$$\gamma = \lim_{n \to \infty} \left[\sum_{j=1}^{n} \frac{1}{j} - \ln n \right] \approx 0.5772 \ldots.$$

---

[7]Lorenzo Mascheroni (1750–1800) was an ordained priest, poet, and teacher of mathematics and physics. He became professor of algebra and geometry at the University of Pavia in 1786 and later became its rector. In 1790, Mascheroni correctly calculated the first 19 decimal places of Euler's constant. This accomplishment has caused his name to be linked with the constant.

Note that $Y_0(t)$ can ultimately be expressed as a constant multiple of $J_0(t)\ln t = y_1(t)\ln t$ added to a series of the form $\sum_{n=0}^{\infty} b_n t^n$. Since $J_0(0) = 1$, $Y_0(t)$ has a logarithmic singularity at $t = 0$. If we recall the heuristic argument used to motivate the method of Frobenius, the presence of the function $\ln t$ in (6) and the corresponding logarithmic singularity at $t = 0$ is not surprising since the general solution of the Euler equation having repeated roots $\lambda_1 = \lambda_2 = 0$ is $y(t) = c_1 + c_2 \ln t$, $t > 0$. Figure 8.3 shows graphs of $J_0(t)$ and $Y_0(t)$.

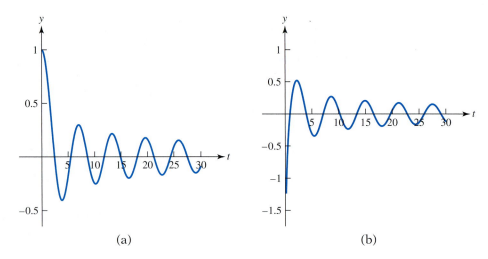

(a) (b)

FIGURE 8.3

(a) The graph of $J_0(t)$, the Bessel function of the first kind of order zero.
(b) The graph of $Y_0(t)$, the Bessel function of the second kind of order zero.
Note that $J_0(t)$ is defined for all t, whereas $Y_0(t)$ has a logarithmic singularity at the regular singular point $t = 0$.

Roots Differing by Unity ($\lambda_1 - \lambda_2 = 1$)

As an illustration of the case where roots differ by unity, let $\nu = \frac{1}{2}$ in Bessel's equation. The indicial equation,

$$F(\lambda) = \lambda^2 - \tfrac{1}{4},$$

has roots $\lambda_1 = \frac{1}{2}$ and $\lambda_2 = -\frac{1}{2}$, and therefore $\lambda_1 - \lambda_2 = 1$. Consider first the larger root, $\lambda_1 = \frac{1}{2}$, which corresponds to a solution of the form

$$y_1(t) = \sum_{n=0}^{\infty} a_n^{(1)} t^{(1/2)+n}, \qquad t > 0.$$

We can assume $a_0^{(1)} \neq 0$. Since $F(\lambda_1 + 1) = F\left(\frac{3}{2}\right) \neq 0$, it follows from (3a) that $a_1^{(1)} = 0$. From (3c), the recurrence relation is

$$a_n^{(1)} = \frac{-a_{n-2}^{(1)}}{\left(n + \frac{1}{2}\right)^2 - \frac{1}{4}} = \frac{-a_{n-2}^{(1)}}{n(n+1)}, \qquad n = 2, 3, 4, \ldots. \tag{7}$$

Equation (7) allows us to determine all the even-indexed coefficients as multiples of $a_0^{(1)}$ and implies that all the odd-indexed coefficients are zero. Solving recurrence relation (7) leads to the solution

$$y_1(t) = a_0^{(1)} \left[t^{1/2} - \frac{t^{5/2}}{3!} + \frac{t^{9/2}}{5!} - \frac{t^{13/2}}{7!} + \cdots \right]$$

$$= a_0^{(1)} t^{-1/2} \left[t - \frac{t^3}{3!} + \frac{t^5}{5!} - \frac{t^7}{7!} + \cdots \right] \qquad \text{(8)}$$

$$= a_0^{(1)} t^{-1/2} \sin t, \qquad t > 0.$$

Consider now the smaller root, $\lambda_2 = -\frac{1}{2}$, where we look for a solution of the form

$$y_2(t) = \sum_{n=0}^{\infty} a_n^{(2)} t^{-(1/2)+n}, \qquad t > 0.$$

As before, $F(\lambda) = \lambda^2 - \frac{1}{4}$. This time, however,

$$F(\lambda_2 + n) = \left(n - \tfrac{1}{2} \right)^2 - \tfrac{1}{4} = n(n-1),$$

and we see that $F(\lambda_2 + 1) = F(\lambda_1) = F\left(\tfrac{1}{2} \right) = 0$. Therefore [see equation (3a)], $a_1^{(2)}$ need not be zero. From (3c), the recurrence relation

$$a_n^{(2)} = \frac{-a_{n-2}^{(2)}}{\left(n - \tfrac{1}{2} \right)^2 - \tfrac{1}{4}} = \frac{-a_{n-2}^{(2)}}{n(n-1)}, \qquad n = 2, 3, 4, \ldots \qquad \text{(9)}$$

allows us to determine all even-indexed coefficients as multiples of $a_0^{(2)}$ and all odd-indexed coefficients as multiples of $a_1^{(2)}$. We find

$$a_{2k}^{(2)} = \frac{(-1)^k}{(2k)!} a_0^{(2)} \quad \text{and} \quad a_{2k+1}^{(2)} = \frac{(-1)^k}{(2k+1)!} a_1^{(2)}, \qquad k = 1, 2, 3, \ldots. \qquad \text{(10)}$$

The resulting solution $y_2(t)$ has the form

$$y_2(t) = a_0^{(2)} t^{-1/2} \left[1 - \frac{t^2}{2!} + \frac{t^4}{4!} - \frac{t^6}{6!} + \cdots \right]$$

$$+ a_1^{(2)} t^{-1/2} \left[t - \frac{t^3}{3!} + \frac{t^5}{5!} - \frac{t^7}{7!} + \cdots \right] \qquad \text{(11)}$$

$$= a_0^{(2)} t^{-1/2} \cos t + a_1^{(2)} t^{-1/2} \sin t, \qquad t > 0.$$

We therefore obtain a second linearly independent solution, $t^{-1/2} \cos t$, added to a multiple of the solution previously obtained, $t^{-1/2} \sin t$. In this example, the method of Frobenius has produced both members of a fundamental set of solutions.

The **Bessel functions of order one-half** are defined to be

$$J_{1/2}(t) = \sqrt{\frac{2}{\pi t}} \sin t \quad \text{and} \quad J_{-1/2}(t) = \sqrt{\frac{2}{\pi t}} \cos t, \qquad t > 0.$$

The general solution of Bessel's equation in this case is usually expressed as

$$y(t) = c_1 J_{1/2}(t) + c_2 J_{-1/2}(t). \qquad \text{(12)}$$

Note that $J_{1/2}(t)$ is well behaved near $t = 0$ since

$$\lim_{t \to 0^+} \frac{\sin t}{\sqrt{t}} = 0.$$

By contrast, $\lim_{t \to 0^+} J_{-1/2}(t) = +\infty$. Figure 8.4 shows the behavior of these two functions.

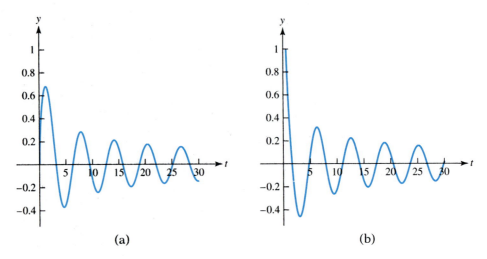

FIGURE 8.4

(a) The graph of the Bessel function $J_{1/2}(t)$.
(b) The graph of the Bessel function $J_{-1/2}(t)$.

Roots Differing by an Integer Greater than 1 ($\lambda_1 - \lambda_2 = N \geq 2$)

Let $\nu = M \geq 1$ in Bessel's equation. The two roots of the indicial equation, $\lambda^2 - M^2 = 0$, are $\lambda_1 = M$ and $\lambda_2 = -M$. The roots differ by an integer greater than 1, since $\lambda_1 - \lambda_2 = 2M \geq 2$. In this case, the method of Frobenius provides us with a solution corresponding to the larger root, $\lambda_1 = M$. It will fail, however, when we try to apply the method to the smaller root, $\lambda_2 = -M$.

Consider first the larger root, $\lambda_1 = M$. We look for a solution of the form

$$y_1(t) = \sum_{n=0}^{\infty} a_n^{(1)} t^{M+n}, \qquad t > 0.$$

We assume $a_0^{(1)} \neq 0$. Since $F(\lambda_1 + 1) = F(M + 1) \neq 0$, we know from equation (3a) that $a_1^{(1)} = 0$. Using equation (3c), we obtain the recurrence relation

$$a_n^{(1)} = \frac{-a_{n-2}^{(1)}}{[(n+M)^2 - M^2]} = \frac{-a_{n-2}^{(1)}}{[n(n+2M)]}, \qquad n = 2, 3, 4, \ldots. \tag{13}$$

Equation (13) allows us to determine all the even-indexed coefficients as multiples of $a_0^{(1)}$ and tells us that the odd-indexed coefficients are zero. Therefore, we obtain the solution

$$y_1(t) = a_0^{(1)} t^M \left\{ 1 - \frac{t^2}{2^2(M+1)} + \frac{t^4}{2^4 2!(M+2)(M+1)} \right.$$

$$\left. - \frac{t^6}{2^6 3!(M+3)(M+2)(M+1)} + \cdots + \frac{(-t^2)^k M!}{2^{2k} k!(M+k)!} + \cdots \right\}. \tag{14}$$

Choosing $a_0^{(1)} = (1/2^M)M!$ in (14) leads to the **Bessel function of the first kind of order M**:

$$J_M(t) = \left(\frac{t}{2} \right)^M \sum_{k=0}^{\infty} \frac{\left(\frac{-t^2}{4} \right)^k}{k!(M+k)!}. \tag{15}$$

Note that $J_M(t)$ is analytic at $t = 0$ and the series (15) has an infinite radius of convergence. The function $J_M(t)$ vanishes at $t = 0$ when $M \geq 1$; $J_M(t)$ is an even function when M is an even integer and an odd function when M is an odd integer.

Suppose we now consider the smaller root, $\lambda_2 = -M$, and look for a solution of the form

$$y_2(t) = \sum_{n=0}^{\infty} a_n^{(2)} t^{-M+n}.$$

Assume, without loss of generality, that $a_0^{(2)} \neq 0$. Since $M \geq 1, F(\lambda_2 + 1) = F(-M+1) \neq 0$ and $a_1^{(2)} = 0$. The difficulty arises when we try to use equation (3c) to evaluate $a_2^{(2)}, a_4^{(2)}, a_6^{(2)}, \ldots$ in terms of $a_0^{(2)}$. Using $\lambda = -M$ and setting $n = 2k$, we have for the recurrence relation (3c)

$$a_{2k}^{(2)} = \frac{-a_{2k-2}^{(2)}}{4k(k-M)}, \qquad k = 1, 2, 3, \ldots. \tag{16}$$

The trouble occurs when we try to evaluate (16) for $k = M$. The breakdown of recurrence relation (16) at $k = M$ tells us that the assumed form of the solution is incorrect and that the second linearly independent solution does not have the structure assumed by the method of Frobenius.

The second linearly independent solution of Bessel's equation customarily used is called the **Bessel function of the second kind of order M** and is denoted by $Y_M(t)$. It is defined as

$$Y_M(t) = \frac{2}{\pi} \left[\gamma + \ln \frac{t}{2} \right] J_M(t) - \frac{\left(\frac{t}{2} \right)^{-M}}{\pi} \sum_{k=0}^{M-1} \frac{(M-k-1)!}{k!} \left(\frac{t^2}{4} \right)^k$$

$$- \frac{\left(\frac{t}{2} \right)^M}{\pi} \sum_{n=0}^{\infty} (-1)^n \left(k_{M+n} + k_n \right) \frac{\left(-\frac{t^2}{4} \right)^n}{n!(M+n)!}, \qquad t > 0, \tag{17}$$

where the constants k_n are defined as $k_0 = 0, k_n = 1 + 2^{-1} + 3^{-1} + \cdots + n^{-1}$, $n \geq 1$ and where γ is the Euler-Mascheroni constant.

Figure 8.5 displays graphs of $J_1(t)$ and $Y_1(t)$.

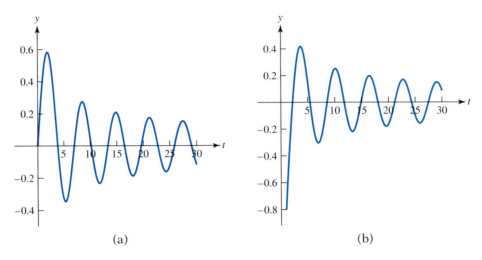

(a)   (b)

FIGURE 8.5

(a) The graph of $J_1(t)$.
(b) The graph of $Y_1(t)$.

REMARK: For reasonably large values of t (say $t > 5$), all of the Bessel functions plotted in Figures 8.3–8.5 seem to behave like damped sinusoids—that is, sine or cosine waves having amplitudes that decrease with increasing t (see Exercises 25 and 26).

Summary of the General Solution Near a Regular Singular Point

Our study of Bessel's equation for different values of ν provided some examples of the behavior of solutions near a regular singular point. Although these examples are representative, they are neither exhaustive nor completely general since Bessel's equation possesses a specific structure that is not necessarily present in the general case. The following summary, which we present without proof, describes the general behavior of solutions of $y'' + p(t)y' + q(t)y = 0$.

Consider the differential equation $y'' + p(t)y' + q(t)y = 0$, where we assume $t = 0$ is a regular singular point. Let $tp(t)$ and $t^2 q(t)$ be real-valued analytic functions with Maclaurin series

$$tp(t) = \sum_{n=0}^{\infty} \alpha_n t^n \quad \text{and} \quad t^2 q(t) = \sum_{n=0}^{\infty} \beta_n t^n$$

that converge in $-R < t < R$. The corresponding indicial equation is

$$\lambda^2 + (\alpha_0 - 1)\lambda + \beta_0 = (\lambda - \lambda_1)(\lambda - \lambda_2) = 0.$$

The roots λ_1 and λ_2 either are real or form a complex conjugate pair. In the event that λ_1 and λ_2 are real, we assume $\lambda_1 \geq \lambda_2$.

Then, in either of the intervals $(-R, 0)$ or $(0, R)$,

(a) There exists a solution having the form

$$y_1(t) = |t|^{\lambda_1} \sum_{n=0}^{\infty} a_n t^n, \qquad a_0 \neq 0, \tag{18}$$

where the series converges at least in $|t| < R$.

(b) The form of the second linearly independent solution, $y_2(t)$, depends on the difference $\lambda_1 - \lambda_2$.

   (i) If $\lambda_1 - \lambda_2$ is not an integer, then

$$y_2(t) = |t|^{\lambda_2} \sum_{n=0}^{\infty} b_n t^n, \qquad b_0 \neq 0, \tag{19}$$

   where the series converges at least in $|t| < R$.

   (ii) If $\lambda_1 = \lambda_2$, then

$$y_2(t) = y_1(t) \ln|t| + |t|^{\lambda_2} \sum_{n=0}^{\infty} c_n t^n, \tag{20}$$

   where the series converges at least in $|t| < R$.

   (iii) If $\lambda_1 - \lambda_2$ equals a positive integer, then

$$y_2(t) = Cy_1(t) \ln|t| + |t|^{\lambda_2} \sum_{n=0}^{\infty} d_n t^n, \qquad d_0 \neq 0, \tag{21}$$

   where C is a constant, possibly zero (if $C = 0$, there is no logarithmic term). Moreover, the series converges at least in $|t| < R$.

REMARKS:

1. We have assumed, for convenience, that $t = 0$ is a regular singular point. In general, if $t = t_0$ is a regular singular point, then the results in the summary are valid when $t - t_0$ replaces t in the formulas.

2. When the roots λ_1 and λ_2 form a complex conjugate pair—for instance, $\lambda_1 = \gamma + i\delta, \lambda_2 = \gamma - i\delta$—the difference $\lambda_1 - \lambda_2 = 2\delta i$ is purely imaginary and thus not equal to zero or a positive integer. In that case, the two complex-valued solutions

$$y_1(t) = |t|^{\gamma+i\delta} \sum_{n=0}^{\infty} a_n t^n \quad \text{and} \quad y_2(t) = |t|^{\gamma-i\delta} \sum_{n=0}^{\infty} b_n t^n,$$

obtained using the recurrence relations, can be used to create an equivalent real-valued fundamental set. Using Euler's formula, we have

$$|t|^{\gamma+i\delta} = |t|^{\gamma}[\cos(\delta \ln|t|) + i \sin(\delta \ln|t|)].$$

As mentioned in Section 8.1, linear differential equations with variable coefficients play an important role in applications. Because of their importance, the functions that emerge as solutions of these equations, usually collectively referred to as special functions, have been studied exhaustively. There are books devoted to a particular special function,[8] as well as general handbooks for special functions.[9] Modern software packages have many of these functions available as built-in functions—as accessible as the familiar exponential, logarithmic, and trigonometric functions. In that sense, these so-called special functions are fortunately becoming less and less special.

Linear differential equations with variable coefficients, such as Bessel's equation and Legendre's equation, arise when the technique called separation of variables is applied to the partial differential equations that model certain physical problems. The Projects at the end of this chapter study steady-state heat conduction between concentric cylinders and provide a brief introduction to this circle of ideas.

EXERCISES

Exercises 1–12:

In each exercise,

(a) Verify that the given differential equation has a regular singular point at $t = 0$.

(b) Determine the indicial equation and its two roots. (These roots are often called the **exponents at the singularity**.)

(c) Determine the recurrence relation for the series coefficients.

(d) Consider the interval $t > 0$. If the two exponents obtained in (c) are unequal and do not differ by an integer, determine the first two nonzero terms in the series for each of the two linearly independent solutions. If the exponents are equal or differ by an integer, obtain the first two nonzero terms in the series for the solution having the larger exponent.

(e) When the given differential equation is put in the form $y'' + p(t)y' + q(t)y = 0$, note that $tp(t)$ and $t^2q(t)$ are polynomials. Do the series, whose initial terms were found in part (d), converge for all t, $0 < t < \infty$? Explain.

1. $2ty'' - (1 + t)y' + 2y = 0$ 2. $2ty'' + 5y' + 3ty = 0$

3. $3t^2y'' - ty' + (1 + t)y = 0$ 4. $6t^2y'' + ty' + (1 - t)y = 0$

5. $t^2y'' - 5ty' + (9 + t^2)y = 0$ 6. $4t^2y'' + 8ty' + (1 + 2t)y = 0$

7. $t^2y'' - 2ty' + (2 + t)y = 0$ 8. $ty'' + 4y' - 2ty = 0$

9. $t^2y'' + ty' - (1 + t^2)y = 0$ 10. $t^2y'' + 5ty' + (4 - t^2)y = 0$

11. $t^2y'' + ty' - (16 + t)y = 0$ 12. $8t^2y'' + 6ty' - (1 - t)y = 0$

Exercises 13–16:

In each exercise,

(a) Determine all singular points of the given differential equation and classify them as regular or irregular singular points.

[8] See, for example, George N. Watson, *A Treatise on the Theory of Bessel Functions*, 2nd ed. (Cambridge: Cambridge University Press, 1966).

[9] Milton Abramowitz and Irene A. Stegun, editors, *Handbook of Mathematical Functions with Formulas, Graphs, and Mathematical Tables* (New York: Dover Publications, 1970) and Wilhelm Magnus, Fritz Oberhettinger, and Raj Pal Soni, *Formulas and Theorems for the Special Functions of Mathematical Physics* (Berlin and New York: Springer-Verlag, 1966).

(b) At each regular singular point, determine the indicial equation and the exponents at the singularity.

13. $(t^3 + t)y'' - (1 + t)y' + y = 0$

14. $t^2 y'' + (\sin 3t)y' + (\cos t)y = 0$

15. $(t^2 - 4)^2 y'' - y' + y = 0$

16. $t^2(1 - t)^{1/3} y'' + ty' - y = 0$

17. The Legendre differential equation $(1 - t^2)y'' - 2ty' + \alpha(\alpha + 1)y = 0$ has regular singular points at $t = \pm 1$; all other points are ordinary points.

(a) Determine the indicial equation and the exponent at the singularity $t = 1$.

(b) Assume that $\alpha \neq 0, 1$. Find the first three nonzero terms of the series solution in powers of $t - 1$ for $t - 1 > 0$. [Hint: Rewrite the coefficient functions in powers of $t - 1$. For example, $1 - t^2 = -(t - 1)(t + 1) = -(t - 1)((t - 1) + 2)$.]

(c) What is an exact solution of the differential equation when $\alpha = 1$?

18. The Chebyshev differential equation $(1 - t^2)y'' - ty' + \alpha^2 y = 0$ has regular singular points at $t = \pm 1$; all other points are ordinary points.

(a) Determine the indicial equation and the exponent at the singularity $t = 1$.

(b) Assume α is nonzero and not an integer multiple of $\frac{1}{2}$. Find two linearly independent solutions for $t - 1 > 0$. (Use the hint in Exercise 17.)

(c) On what interval of the form $0 < t - 1 < R$ do the solutions found in part (b) converge?

(d) What is an exact solution of the differential equation when $\alpha = \frac{1}{2}$?

19. The Laguerre[10] differential equation $ty'' + (1 - t)y' + \alpha y = 0$ has a regular singular point at $t = 0$.

(a) Determine the indicial equation and show that the roots are $\lambda_1 = \lambda_2 = 0$.

(b) Find the recurrence relation. Show that if $\alpha = N$, where N is a nonnegative integer, then the series solution reduces to a polynomial. Obtain the polynomial solution when $N = 5$. The polynomial solutions of this differential equation, when properly normalized, are called **Laguerre polynomials**.

(c) Is the polynomial obtained in part (b) for $\alpha = N = 5$ an even function, an odd function, or neither? Would you expect even and odd solutions of the differential equation based on its structure and the conclusions of Theorem 8.2? Explain.

Exercises 20–23:

In each exercise, use the stated information to determine the unspecified coefficients in the given differential equation.

20. $t^2 y'' + t(\alpha + 2t)y' + (\beta + t^2)y = 0.$ $t = 0$ is a regular singular point. The roots of the indicial equation at $t = 0$ are $\lambda_1 = 1$ and $\lambda_2 = 2$.

21. $t^2 y'' + \alpha ty' + (\beta + t - t^3)y = 0.$ $t = 0$ is a regular singular point. The roots of the indicial equation at $t = 0$ are $\lambda_1 = 1 + 2i$ and $\lambda_2 = 1 - 2i$.

22. $t^2 y'' + \alpha ty' + (2 + \beta t)y = 0.$ $t = 0$ is a regular singular point. One root of the indicial equation at $t = 0$ is $\lambda = 2$. The recurrence relation for the series solution corresponding to this root is
$$(n^2 + n)a_n - 4a_{n-1} = 0, \quad n = 1, 2, \ldots.$$

[10]Edmond Laguerre (1834–1886) attended the Ecole Polytechnique in Paris and returned there after ten years of service as a French artillery officer. He worked in the areas of analysis and geometry and is best remembered for his study of the polynomials that bear his name.

23. $ty'' + (1 + \alpha t)y' + \beta ty = 0.$ The recurrence relation for a series solution is

$$n^2 a_n - (n - 1)a_{n-1} + 3a_{n-2} = 0, \quad n = 2, 3, \ldots.$$

24. Modified Bessel Equation The differential equation $t^2 y'' + ty' - (t^2 + v^2)y = 0$ is known as the modified Bessel equation. Its solutions, usually denoted by $I_v(t)$ and $K_v(t)$, are called **modified Bessel functions**. This equation arises in solving certain partial differential equations involving cylindrical coordinates.

(a) Do you anticipate that the modified Bessel equation will possess solutions that are even and odd functions of t? Explain.

(b) The point $t = 0$ is a regular singular point of the modified Bessel equation; all other points are ordinary points. Determine the indicial equation for the singularity at $t = 0$ and find the exponents at the singularity.

(c) Obtain the recurrence relation for the modified Bessel equation. How do the exponents and recurrence relation for this equation compare with their counterparts for Bessel's equation?

25. Consider Bessel's equation, $t^2 y'' + ty' + (t^2 - v^2)y = 0$ for $t > 0$.

(a) Define a new dependent variable $u(t)$ by the relation $y(t) = t^{-1/2}u(t)$. Show that $u(t)$ satisfies the differential equation

$$u'' + \left[1 - \frac{\left(v^2 - \frac{1}{4}\right)}{t^2}\right]u = 0.$$

(b) Solve the differential equation in part (a) when $v^2 = \frac{1}{4}$. What is the corresponding solution of Bessel's equation in this case?

(c) Suppose that t is large enough to justify neglecting the term $(v^2 - \frac{1}{4})/t^2$ in the differential equation obtained in part (a). Show that neglecting $(v^2 - \frac{1}{4})/t^2$ leads to the approximation $y(t) \approx t^{1/2}R\cos(t - \delta)$ when t is large.

26. This exercise asks you to use computational software to show that Bessel functions behave like $R\cos(t - \delta)/\sqrt{t}$ for appropriate choices of constants R and δ and for t large enough. We restrict our attention to $J_0(t)$.

(a) Locate the abscissa of the first maximum of $J_0(t)$ in $t > 0$; call this point t_m. Since $J_0'(t) = -J_1(t)$, this point can be found by applying a root-finding routine to $J_1(t)$.

(b) Evaluate the constants R and δ by setting $t_m - \delta = 2\pi$ and $R = \sqrt{t_m}J_0(t_m)$.

(c) Plot the two functions $J_0(t)$ and $R\cos(t - \delta)/\sqrt{t}$ on the same graph for $t_m \le t \le 50$. How do the two graphs compare?

27. For the special case $v = \frac{1}{2}$, consider the modified Bessel equation for $t > 0$, $t^2 y'' + ty' - (t^2 + \frac{1}{4})y = 0$.

(a) Define a new dependent variable $u(t)$ by the relation $y(t) = t^{-1/2}u(t)$. Show that $u(t)$ satisfies the differential equation $u'' - u = 0$.

(b) Show that the differential equation has a fundamental set of solutions

$$\frac{\sinh t}{\sqrt{t}}, \qquad \frac{\cosh t}{\sqrt{t}}, \qquad t > 0.$$

PROJECTS

Project 1: The Simple Centrifuge Revisited

We revisit a problem that was studied numerically in Project 2 of Chapter 3. At time $t = 0$, a horizontally mounted tube of length l begins to rotate with a constant (positive) angular acceleration α. At that time, a particle of mass m is located a radial distance r_0 from the pivot and moving with a radial velocity r'_0. The radial distance of the particle from the pivot, $r(t)$, is a solution of the following initial value problem:

$$r'' - \alpha^2 t^2 r = 0, \qquad r(0) = r_0, \qquad r'(0) = r'_0. \tag{1}$$

We now develop a series solution of this problem. The point $t = 0$ is an ordinary point. Assume a solution of the form

$$r(t) = \sum_{n=0}^{\infty} a_n t^n.$$

1. Develop the recurrence relation for the coefficients $\{a_n\}_{n=0}^{\infty}$. What is the radius of convergence of the series obtained?

2. The series solution will have the form $r(t) = a_0 r_1(t) + a_1 r_2(t)$, with $a_0 = r_0$ and $a_1 = r'_0$. Determine the first four nonvanishing terms for each of the series $r_1(t)$ and $r_2(t)$.

3. Do the functions defined by series $r_1(t)$ and $r_2(t)$ possess even or odd symmetry? Could Theorem 8.2 have been used to predict the existence of even and odd solutions of differential equation (1)? Explain.

4. Assume the following numerical values:

$$\alpha = \pi \text{ rad/s}^2, \qquad l = 2 \text{ m}, \qquad r_0 = 10 \text{ cm}, \qquad r'_0 = 20 \text{ cm/s}.$$

   Use computer software and the polynomial approximation developed in part 2 to estimate the time at which the particle will exit the tube and the exit velocity.

   Since we are using a truncated series for these computations, the exit time and exit velocity computed will be approximate values. Is it possible to use this computed information to get a rough estimate of the exit time error? Explain.

5. The results of parts 1 and 2 show that the solution $r(t)$ is actually a function of $\alpha^{1/2} t$. Show that this could have been anticipated by studying the structure of the differential equation itself. Specifically, make the change of independent variable $\tau = \alpha^p t$ (or $t = \alpha^{-p}\tau$). Show that with this change of variable, the choice $p = \frac{1}{2}$ transforms the differential equation into the equation

$$\frac{d^2 r}{d\tau^2} - \tau^2 r = 0.$$

What does the initial condition $r'(0) = r'_0$ transform into?

Project 2: Steady-State Heat Flow between Concentric Cylinders

This exercise gives you a brief glimpse into an application involving variable coefficient linear differential equations.

   Consider the two concentric cylinders shown in Figure 8.6. The inner cylinder has radius $a > 0$, while the outer cylinder has radius $b > a$. Assume that these cylinders represent the inner and outer surfaces of a pipe and that the pipe itself is designed to function as part of a simple cooling system. Heat, or thermal energy, is drawn from the region exterior to the pipe by the presence of coolant flowing within the pipe. We assume that the coolant is "well-stirred" and that the inner surface of the pipe is maintained at the coolant temperature. Suppose we know the temperature of the region outside the pipe as well as the temperature of the coolant. For fixed values of the radii a and b, we would like to know the rate at which heat is drawn from the exterior region. For a fixed coolant temperature, we also want to know how the heat transfer varies as a function of the two pipe radii.

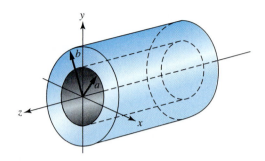

FIGURE 8.6

The concentric cylinders represent the inner and outer surfaces of a pipe. Heat is drawn from the exterior region by the presence of a coolant flowing through the pipe.

In general, a temperature reading depends on where and when the reading is taken. Thus, if T represents temperature (measured, perhaps, in degrees Centigrade), then $T = T(x, y, z, t)$.

We shall assume, however, that our cooling system is operating in a steady-state mode; that is, system operation has "settled down" to the point where the temperature everywhere remains constant in time. We further assume that temperature does not change in the axial or z-direction. With these assumptions, $T = T(x, y)$.

Because of the cylindrical geometry, we introduce polar coordinates. We set $x = r \cos \theta$ and $y = r \sin \theta$, as in Figure 8.7. We view temperature as a function of the polar variables, $T = T(r, \theta)$. The domain of interest is the annular region between the cylinders, described in polar coordinates by $a \le r \le b$, $0 \le \theta < 2\pi$. We assume that the temperature at the outer radius of the pipe is known; that is,

$$T(b, \theta) = T_b(\theta), \qquad 0 \le \theta < 2\pi,$$

where T_b is a known function of the angle θ. We also assume that the temperature at the inner pipe radius is a known constant; that is,

$$T(a, \theta) = T_a, \qquad 0 \le \theta < 2\pi.$$

Since the coolant is to draw heat from the exterior region, we assume that $T_b(\theta) > T_a$, $0 \le \theta < 2\pi$.

The problem is to determine the rate at which the coolant draws heat through the pipe from the exterior region. We first determine the temperature $T(r, \theta)$ within the annular cross-section of the pipe and then use this information to compute the required heat flow.

Within the annular region of the pipe, the steady-state temperature must be a solution of a partial differential equation known as Laplace's equation. In polar coordinates, Laplace's equation is

$$\frac{\partial^2 T}{\partial r^2} + \frac{1}{r} \frac{\partial T}{\partial r} + \frac{1}{r^2} \frac{\partial^2 T}{\partial \theta^2} = 0. \tag{2}$$

To determine the steady-state temperature within the annular pipe region, we need to

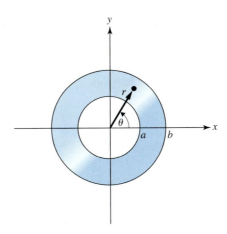

FIGURE 8.7

A cross-section of the pipe shown in Figure 8.6. We introduce polar coordinates into the problem because of the cylindrical geometry.

solve partial differential equation (2) subject to the boundary conditions

$$T(a, \theta) = T_a, \qquad T(b, \theta) = T_b(\theta), \qquad 0 \le \theta < 2\pi. \tag{3}$$

It should be emphasized that the problems we ultimately solve in this exercise are not initial value problems since constraints are imposed at two different values of the independent variable.

Once we find the temperature distribution within the annular region,

$$T(r, \theta), \qquad a \le r \le b, \qquad 0 \le \theta < 2\pi,$$

we will compute the heat transfer rate into the coolant at the inner radius a. Heat flows "downhill," from hotter to cooler regions. Moreover, the rate of heat flow is proportional to the temperature gradient—the steeper the gradient, the greater the rate of heat transfer. At a point on the inner pipe radius, the rate of heat transfer per unit surface area is given by

$$\kappa \frac{\partial T(a, \theta)}{\partial r}, \tag{4}$$

where κ is a positive constant (known as the thermal conductivity) that depends on the nature of the pipe material. It follows that the rate of heat transfer per unit axial length of the pipe can be found by integrating (4) around the inner pipe radius, obtaining

$$\int_0^{2\pi} \kappa \frac{\partial T(a, \theta)}{\partial r} a \, d\theta = \kappa a \int_0^{2\pi} \frac{\partial T(a, \theta)}{\partial r} \, d\theta. \tag{5}$$

1. **The Case of Constant Exterior Temperature** Assume that the temperature at the outer pipe radius is constant; that is,

$$T_b(\theta) = T_b, \qquad 0 \le \theta < 2\pi.$$

Since neither boundary condition varies with angle, we expect the temperature in the annular pipe region to likewise be independent of θ. Assume a solution of the form $T = T(r), a \le r \le b$.

(a) Substitute $T = T(r)$ into Laplace's equation (2), obtaining

$$\frac{d^2 T(r)}{dr^2} + \frac{1}{r} \frac{dT(r)}{dr} = 0, \qquad a < r < b. \tag{6}$$

Obtain the general solution of this Euler equation, and impose the boundary constraints

$$T(a) = T_a, \qquad T(b) = T_b. \tag{7}$$

The constraints (7) will determine the two arbitrary constants in the general solution.

(b) Compute dT/dr, and evaluate the integral (5).

(c) Let $T_a = 40°F$, $T_b = 120°F$, $a = 1$ in., and $b = 1.5$ in. For these parameter values, plot $T(r)$ versus r for $a \le r \le b$. Do the maximum and minimum temperatures occur where you expect them to occur?

2. **The Case of Varying Exterior Temperature**   Assume now that the temperature distribution at the outer pipe radius is not constant. As a specific case, assume that

$$T(b, \theta) = (1 + \alpha \sin \theta) T_b, \qquad 0 \le \theta < 2\pi,$$

where $T_b = 120°F$ and where $0 \le \alpha < 1$.

(a) Do you think the heat transfer rate will differ from that obtained in part (a) of the constant exterior temperature case?

(b) Assume a solution of Laplace's equation of the form

$$T(r, \theta) = T_0(r) + T_1(r) \sin \theta \tag{8}$$

within the annular pipe region. The unknown functions $T_0(r)$ and $T_1(r)$ must be determined. Substitute (8) into (2), obtaining

$$\left[\frac{d^2 T_0(r)}{dr^2} + \frac{1}{r} \frac{dT_0(r)}{dr} \right] + \left[\frac{d^2 T_1(r)}{dr^2} + \frac{1}{r} \frac{dT_1(r)}{dr} - \frac{1}{r^2} T_1(r) \right] \sin \theta = 0,$$

$$a < r < b, \qquad 0 \le \theta < 2\pi.$$

Assume for the moment that the radial variable has an arbitrary but fixed value. The set of functions $\{1, \sin \theta\}$ is linearly independent in $0 \le \theta < 2\pi$, and so this equation implies, for the particular fixed value of r, that

$$\frac{d^2 T_0(r)}{dr^2} + \frac{1}{r} \frac{dT_0(r)}{dr} = 0 \tag{9a}$$

and

$$\frac{d^2 T_1(r)}{dr^2} + \frac{1}{r} \frac{dT_1(r)}{dr} - \frac{1}{r^2} T_1(r) = 0. \tag{9b}$$

Since r is assumed to be arbitrary, these equations must hold for $a < r < b$. Find the general solution for each of the two Euler equations (9a) and (9b).

(c) Apply the boundary constraints (3). In particular, we have

$$T(a, \theta) = T_0(a) + T_1(a) \sin \theta = T_a$$
$$T(b, \theta) = T_0(b) + T_1(b) \sin \theta = T_b + T_b \alpha \sin \theta, \qquad 0 \le \theta < 2\pi.$$

Use the same linear independence argument employed in (b) to obtain boundary conditions for the functions T_0 and T_1. Impose these boundary conditions on the general solutions of the Euler equations obtained in (b), and determine $T(r, \theta)$.

(d) Determine the heat transfer rate at the inner pipe radius. Is the heat transfer rate the same as that obtained for the constant exterior temperature case?

Answers to Odd-Numbered Exercises

CHAPTER 1

Section 1.2, page 6

1. Order is 2.

3. Order is 1.

5. $k = -2$

7. $k = \frac{1}{2}$

9. (b) $C = 2e^{-1}$

11. (b) $C_1 = 3, \ C_2 = 1$

13. $c = 0$ and $c = 1$

15. $r = 1$ and $r = 2$

17. $y = e^{2t} + e^{-2t} = 2\cosh 2t$

19. $y = 3e^{-2t}$

21. $m = -2, y_0 = 1, y(t) = 2 - t$

23. $t_{\text{impact}} = \sqrt{2y_0/g}, \quad v_{\text{impact}} = -\sqrt{2gy_0}$

Section 1.3, page 12

1. (a) Autonomous
   (b) $y = 1$
   (c)

3. (a) Autonomous
   (b) $y = 0, \ y = \pm\pi, \ y = \pm 2\pi, \ldots$
   (c)

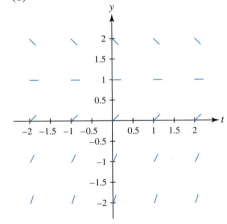

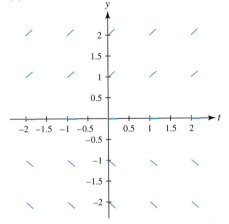

5. (a) Autonomous
   (b) There are none.

(c)

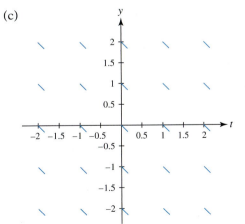

7. (a) The requested isoclines are
the lines $y = 2, y = 1$, and $y = 0$.
(b)

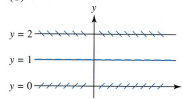

9. (a) The requested isoclines
are the hyperbolas
$y^2 - t^2 = -1, y^2 - t^2 = 0$, and $y^2 - t^2 = 1$.
(b)

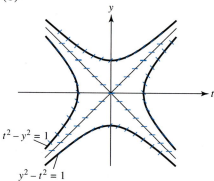

11. One possibility is $y' = -(y-1)^2$.   13. One possibility is $y' = \sin(2\pi y)$.

15. Direction Field F   17. Direction Field B   19. Direction Field E

CHAPTER 2

Section 2.1, page 17

1. Linear and nonhomogeneous   3. Nonlinear

5. Nonlinear   7. Nonlinear

9. Linear and nonhomogeneous

11. (a) $-\infty < t < \infty$ (b) $-\infty < t < \infty$ (c) $-\infty < t < \infty$

13. (a) $3 < t < \infty$ (b) $-2 < t < 2$ (c) $-2 < t < 2$ (d) $-\infty < t < -2$
(e) $-2 < t < 2$

15. $p(t) = -2t$ and $y_0 = 3$ 17. $y(t) = 0, a < t < b$

Section 2.2, page 26

1. (a) $y = Ce^{-3t}$ (b) $y = -3e^{-3t}$ 3. (a) $y = Ce^{t^2}$ (b) $y = 3e^{-1}e^{t^2} = 3e^{(t^2-1)}$

5. (a) $y = -2 + Ce^{3t}$ (b) $y = -2 + 3e^{3t}$

7. (a) $y = \frac{1}{5}e^t + Ce^{-3t/2}$ (b) $y = (e^t - e^{-3t/2})/5$

9. (a) $y = -3 + Ce^{-0.5\sin t}$ (b) $y = -3 - e^{-0.5\sin t}$

11. $y = Ct^{-4}$ 13. $y = Ce^{\sin 2t}$ 15. $y = Ce^{t^3 + 3t}$ 17. $y = 0.5 + Ce^{-2t}$

19. $y = te^{-2t} + Ce^{-2t}$ 21. $y = \frac{1}{4}t^2 + Ct^{-2}$ 23. $y = t - 1 + Ce^{-t}$

25. (a) 2   (b) 3   (c) 1   27. $\alpha = 2$ and $y_0 = \frac{1}{4}$

29. (a) $B' = -kB$, $B(0) = -A^*$ (b) $A(c) = A^*(1 - e^{-kc})$. $A(c)$ never exceeds A^*.
(c) $c = (1/k)\ln 20$

31. $p(t) = 2$, $g(t) = 2t + 3$ 33. $p(t) = t^{-1}$, $g(t) = t^{-1}$

35. $g(t) = 2e^t + \sin t + \cos t$, $y_0 = -1$ 37. $\lim_{t \to \infty} y(t) = -1$

39. A finite limit exists whenever $\lambda > 0$. In this case, the limit is equal to $1/\lambda$.

41. $y = \begin{cases} 1 + 2e^{-1+\cos t}, & 0 \le t \le \pi \\ -1 + 2e^{1+\cos t} + 2e^{-1+\cos t}, & \pi < t \le 2\pi \end{cases}$

43. $y = \begin{cases} 3e^{-t^2+t}, & 0 \le t \le 1 \\ 3, & 1 < t \le 3 \\ t, & 3 < t \le 4 \end{cases}$

45. $y = e^{t^2}\left[2 + \dfrac{\sqrt{\pi}}{2}\mathrm{erf}(t)\right]$

Section 2.3, page 37

1. (a) $Q(10) = 20(1 - e^{-0.3}) \approx 5.18$ lb

   (b) $\lim_{t\to\infty} Q(t) = 20$ and the limiting concentration is 0.2 lb/gal.

3. The required inflow rate is $r = (14{,}000/3)\ln(100) \approx 21{,}491$ m³/min. The fraction vented per minute is $r/v = (1/30)\ln(100) \approx 15.4\%$.

5. (a) $Q(t) = 500t^2 e^{-t/50}$ mg

   (b) The maximum value occurs at $t = 100$ min. The maximum concentration is about 135.3 mg/gal.

   (c) Yes, a graph of concentration versus time shows that $c(t) > 100$ for $60 \le t \le 160$.

7. (a) $t = 600$ min

   (b) $c(300) = Q(300)/V(300) = 197.5/400 \approx 0.494$ lb/gal

   (c) $0.5 - (40/700)(1/49) \approx 0.4988$ lb/gal

9. (a) $Q(0) = 0$ (b) $c_i(t) = 0.05$ lb/gal

11. (a) $Q' = (15/500)(\alpha - 1)Q$ (b) $\alpha = 1 - (1/5.4)\ln 100 \approx 0.1472$

13. (a) $Q_A' = -1000(Q_A/500{,}000)$, $Q_A(0) = 1000$
    $Q_B' = 1000(Q_A/500{,}000) - 1000(Q_B/200{,}000)$, $Q_B(0) = 0$

    (b) $Q_A(t) = 1000e^{-t/500}$ lb, $Q_B(t) = (2000/3)(e^{-t/500} - e^{-t/200})$ lb

    (c) The maximum value is attained at $t = (1000/3)\ln 2.5 \approx 305.4$ hr.

    (d) About 4056 hours, or approximately 169 days, is required.

15. (a) No, we do not expect the concentration to stabilize, since the inflow rate is varying.

    (b) $Q' = 0.6(1 + \sin t) - (3/200)Q$, $Q(0) = 10$

    (c) $Q(t) = 40 - 30e^{-(3/200)t} + (1/1.000225)[0.6(e^{-(3/200)t} - \cos t) + 0.009\sin t]$ lb

17. An oven temperature of $70 - 80/(\sqrt{15/23} - 1) \approx 485°$F

19. (a) $\theta(0) = 340°$F   (b) $\theta(t) \to S_0 = 70°$F as $t \to \infty$

21. (a) $\theta(0) = 40°$F   (b) $\theta(t) \to S_0 = 80°$F as $t \to \infty$

23. The times are the same.

Section 2.4, page 45

1. $P(30) = 10{,}000{,}000e^{6\ln(1.1)} = 17{,}715{,}610$

3. $t = (2\ln 3)/\ln 1.3 \approx 8.375$ weeks   5. It will take an additional 9.6 days.

7. $Q(0) = 20\sqrt{32} = 113.137\ldots$ g   9. After 45 days

11. (a) For Strategy I, $M_I = kP_0$. For Strategy II, $M_{II} = (e^k - 1)P_0$.

    (b) For Strategy I, the profit will be $500{,}000(0.3172)(0.75) = \$118{,}950$. For Strategy II, the profit will be $500{,}000(e^{0.3172} - 1)(0.6) \approx \$111{,}983$.

13. (a) $t = (5730/\ln 2)\ln(10/3) \approx 9953$ years   (b) $9901 \le t \le 10{,}005$ years
    (c) $Q(60{,}000)/Q(0) \approx 7.04 \times 10^{-4}$

15. Approximately 38.9 micrograms

Section 2.5, page 53

1. (a) $f(t, y) = (1 - 2t \cos y)/3$ (b) $f_y(t, y) = (2t \sin y)/3$ (c) The entire ty-plane

3. (a) $f(t, y) = -2t/(1 + y^2)$ (b) $f_y(t, y) = 4ty/(1 + y^2)^2$ (c) The entire ty-plane

5. (a) $f(t, y) = -ty^{1/3} + \tan t$ (b) $f_y(t, y) = -\frac{1}{3}ty^{-2/3}$
   (c) $-\pi/2 < t < \pi/2, \quad 0 < y < \infty$

7. (a) $f(t, y) = (2 + \tan t)/\cos y$ (b) $f_y(t, y) = (2 + \tan t) \sec y \tan y$
   (c) $-\pi/2 < t < \pi/2, \quad -\pi/2 < y < \pi/2$

9. (a) $0 < t < \infty, \ -\infty < y < \infty$

   (b) There is no contradiction. Just because the hypotheses are not satisfied on the entire t-axis does not mean that "bad things must happen."

11. $\bar{y}(t) = 2/\sqrt{1 - (t - 1)}$. Therefore, $\bar{y}(0) = \sqrt{2}$.

13. (a) Using $v = y^{-1}$, we obtain $v' + 2v = 1, v(0) = 1$. Solving for v and transforming back yields $y = 2/(1 + e^{-2t})$.

    (b) $-\infty < t < \infty$

15. (a) Using $v = y^{-1}$, we obtain $v' - v = -e^t, v(-1) = -1$. Solving for v and transforming back yields $y = -1/[(t + 1)e^t + e^{t+1}]$.

    (b) $-(1 + e) < t < \infty$

17. (a) Using $v = y^3$, we obtain $tv' + 3v = 3t^3, v(1) = 1$. Solving for v and transforming back yields $y = [0.5(t^3 + t^{-3})]^{1/3}$.
    (b) $0 < t < \infty$

19. (a) Following the hint, we obtain $z' = -z + tz^{-2}, z(0) = 2$. Using $v = z^3$, we obtain $v' = -3v + 3t, v(0) = 8$. Solving and transforming back to z and then to y, we have $y = \left(\frac{25}{3}e^{-3t} + t - \frac{1}{3}\right)^{1/3} - 1$.

    (b) $-\infty < t < \infty$

Section 2.6, page 60

1. (a) $y^2 = 4 - 2\cos t; \quad y = -\sqrt{4 - 2\cos t}$ (b) $-\infty < t < \infty$

3. (a) $y^2 + 2y + 2(t - 1) = 0; \quad y = -1 + \sqrt{3 - 2t}$ (b) $-\infty < t \le 1.5$

5. (a) $y^{-2} + t^2 = \frac{1}{4}; \quad y = 2/\sqrt{1 - 4t^2}$ (b) $-\frac{1}{2} < t < \frac{1}{2}$

7. (a) $\tan^{-1} y = t - \pi/2; \quad y = \tan(t - \pi/2)$ (b) $0 < t < \pi$

9. (a) $|(y + 1)/(y - 1)| = 3e^{t^2}; \quad y = (3e^{t^2} - 1)/(3e^{t^2} + 1)$ (b) $-\infty < t < \infty$

11. (a) $e^y = e^t + e - 1; \quad y = \ln(e^t + e - 1)$ (b) $-\infty < t < \infty$

13. (a) $\tan y = e^{-t}; \quad y = \tan^{-1}(e^{-t})$ (b) $-\infty < t < \infty$

15. (a) $ye^y = 2e^2 + [1 - (t - 2)^2]/2$; there is no explicit solution.
    (b) Approximately $-3.5 < t < 7.5$

17. (a) An implicit solution is $\ln(1 + e^y) = t - 2 + \ln 2$. This can be unraveled to yield $y = \ln(2e^{t-2} - 1)$.

    (b) $2 - \ln 2 < t < \infty$

19. $\alpha = \frac{2}{3}, n = 3, y_0 = 1$ 21. $(1 + y)e^y y' + 2t - \cos t = 0, \quad y(0) = 0$

23. (a) The equation has the form $[f(y)]^{-1}y' = 1$.

    (b) $y = 2e^{2t-4}/(1 + e^{2t-4})$

25. $y = -2 + \tan[(t^2/2) - \pi/4], \ -\sqrt{3\pi/2} < t < \sqrt{3\pi/2}$

27. The half life is $\tau = 3/(2kQ_0^2)$, and it does depend on Q_0.

29. (a) It is nonlinear and separable.
    (b) The two curves are the same and are
    (c) The two curves are different and are

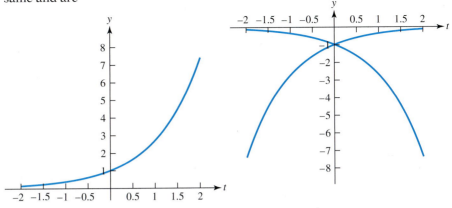

31. $S + K \ln S = -\alpha t + S_0 + K \ln S_0$

33. The equation has the form $y' = [(y/t) - 1]/[(y/t) + 1]$. Using the change of variables $z = y/t$, we obtain $z + tz' = (z - 1)/(z + 1), z(2) = 1$. Solving for z and transforming back to y yields the solution $\tan^{-1}(y/t) + 0.5 \ln [1 + (y/t)^2] + \ln t = (\pi/4) + 1.5 \ln 2$.

35. The equation has the form $y' = (y + t)^2 - 1$. Using the change of variables $z = y + t$, we obtain $z' - 1 = z^2 - 1,\ z(1) = 3$. Solving for z and transforming back to y yields the solution $y = (3 - 4t + 3t^2)/(4 - 3t)$.

37. The equation has the form $y' = (2t + y) + 1/(2t + y)$. Using the change of variables $z = y + 2t$, we obtain $z' - 2 = z + (1/z),\ z(1) = 3$. Solving for z and transforming back to y yields the solution $(1 + 2t + y)^{-1} + \ln |1 + 2t + y| = t - 0.75 + \ln 4$.

Section 2.7, page 68

1. $H_t = 2t - y$ and, therefore, $H = t^2 - yt + p(y)$. Since $H_y = -y + p'(y)$, it follows that $p(y) = y^2$. Thus, a family of solutions is given by $t^2 - yt + y^2 = C$. Imposing the initial condition leads to the implicit solution $t^2 - yt + y^2 = 1$. This solution can be "unraveled."

3. Rewrite the equation as $(y^2 + 1)^{-1}y' - (3t^2 + 1) = 0,\ y(0) = 1$. $H_t = -(3t^2 + 1)$ and, therefore, $H = -(t^3 + t) + p(y)$. Since $H_y = p'(y)$, it follows that $p(y) = \tan^{-1}y$. Thus, a family of solutions is given by $-(t^3 + t) + \tan^{-1}y = C$. Imposing the initial condition leads to the implicit solution $-(t^3 + t) + \tan^{-1}y = \pi/4$. This solution can be "unraveled."

5. $H_t = e^t e^y + 3t^2$ and, therefore, $H = e^t e^y + t^3 + p(y)$. Since $H_y = e^t e^y + p'(y)$, it follows that $p(y) = y^2$. Thus, a family of solutions is given by $e^t e^y + t^3 + y^2 = C$. Imposing the initial condition leads to the implicit solution $e^t e^y + t^3 + y^2 = 1$. This solution cannot be "unraveled."

7. $H_t = ty^2 + \cos t$ and, therefore, $H = \frac{1}{2}t^2y^2 + \sin t + p(y)$. Since $H_y = t^2y + p'(y)$, it follows that $p(y) = \frac{1}{2}e^{2y}$. Thus, a family of solutions is given by $\frac{1}{2}(t^2y^2 + e^{2y}) + \sin t = C$. Imposing the initial condition leads to the implicit solution $t^2y^2 + e^{2y} + 2\sin t = 3$. This solution cannot be "unraveled."

9. $H_t = y^2 - 1$ and, therefore, $H = (y^2 - 1)t + p(y)$. Since $H_y = 2yt + p'(y)$, it follows that $p(y) = \ln |y|$. Thus, a family of solutions is given by $(y^2 - 1)t + \ln |y| = C$. Imposing the initial condition leads to the implicit solution $(y^2 - 1)t + \ln |y| = 0$. This solution cannot be "unraveled."

11. $M(t, y) = 2y + q(t)$

13. $M(t, y) = e^y + y + q(t)$

15. $N(t, y) = -2y \cos t + p(y)$

17. $y_0 = 2$ or $y_0 = -2$; $M(t, y) = H_t(t, y) = 3t^2 y + e^t$, $N(t, y) = H_y(t, y) = t^3 + 2y$

19. $y_0 = 1$; $M(t, y) = 2(2t + y)^{-1} + 2t + ye^{yt}$, $N(t, y) = (2t + y)^{-1} + te^{yt}$

23. (c) $y = \sqrt{\frac{1}{3}(t + 2t^{-1/2})}$

25. (b) An integrating factor is $\mu(t, y) = y$.

   (c) $3y^2 t + y^3 = 27$. In principle, this implicit solution can be "unraveled." In practice, unless we needed to evaluate $y(t)$ at many times t, it would be more efficient to solve it numerically.

27. (b) An integrating factor is $\mu(t, y) = t$.   (c) $y = -\sqrt{t^{-2}(4 + 2e^t - 2te^t)}$

Section 2.8, page 74

1. It will take $55.645\ldots$ years.

3. The initial population was about 1.791×10^6 individuals.

5. (a) The equilibrium populations are $P = \frac{1}{4}$ and $P = \frac{3}{4}$.   (b) $P(t) \to \frac{3}{4}$ as $t \to \infty$.

7. (a) The equilibrium population is $P = \frac{1}{2}$.   (b) $P(t) \to \frac{1}{2}$ as $t \to \infty$.

9. (a) The equilibrium population is $P = 2$.   (b) $P(t) \to 2$ as $t \to \infty$.

11. $P_e = 3$ and $M = -\frac{2}{3}$.   13. $P_e = 1$ and $M = 2$.

17. $P = 1/\left(1 + 3e^{-[t + (1 - \cos 2\pi t)/(2\pi)]}\right)$. $P(t) \to 1$ as $t \to \infty$.

19. The infected individuals will number about 1.3763×10^6.

Section 2.9, page 85

1. $v = -(mg/k)(1 - e^{-kt/m})$; therefore, $t = (m/k) \ln 2$.

3. (a) $\kappa = 0.2469\ldots$ lb-sec^2/ft^2 (b) 562.4 ft

5. $y(t_m) = \int_0^{t_m} v(t)\, dt = -(mg/k)t_m + (m/k)[v_0 + (mg/k)][1 - e^{-(k/m)t_m}]$

7. The impact velocity is $-\sqrt{2y_0 g}$.

9. The transformed equation is $dv/dx = -(k/m)xv$. Thus, $v(x) = v_0 e^{-(kx^2)/(2m)}$. Since $v_0 > 0$, $x_f = \infty$.

11. The transformed equation is $dv/dx = -(k/m)(1 + x)^{-1}$; $v(x) = v_0 - (k/m) \ln(1 + x)$, leading to a stopping position of $x_f = e^{mv_0/k} - 1$.

13. (a) Let v_I denote the impact velocity. Then $\kappa v_I^2 = mg\left(1 - e^{-2\kappa y_0/m}\right)$.

   (b) $400.11\ldots$ ft

15. (a) $mv \dfrac{dv}{dx} + \kappa_0 xv^2 = 0$ (b) $\kappa_0 = (2m/d^2) \ln 100$

17. (a) $-187.26\ldots$ ft/sec or approximately 127.7 mph

   (b) $-128.18\ldots$ ft/sec or approximately 87.39 mph

19. Assuming the chute opens instantaneously, the dragster will slow to 50 mph after $5.556\ldots$ sec. In the other model, it requires $6.249\ldots$ sec.

21. Terminal velocity is $-\sqrt{mg/\kappa}$; therefore, we need $\kappa = 0.929\ldots$ lb-sec^2/ft^2.

23. 7.39 radians/sec

Section 2.10, page 98

1. (a) $y_{k+1} = y_k + h(2t_k - 1)$, $t_0 = 1$, $y_0 = 0$

   (b) $y_1 = 0.1000$, $y_2 = 0.2200$, $y_3 = 0.3600$

   (c) $y = t^2 - t$

3. (a) $y_{k+1} = y_k - ht_k y_k, t_0 = 0, y_0 = 1$

   (b) $y_1 = 1.0000, y_2 = 0.9900, y_3 = 0.9702$

   (c) $y = e^{-t^2/2}$

5. (a) $y_{k+1} = y_k + hy_k^2, t_0 = 0, y_0 = 1$

   (b) $y_1 = 1.1000, y_2 = 1.2210, y_3 = 1.3701\ldots$

   (c) $y = 1/(1-t)$

13. $Q(2) = 23.7556\ldots$ oz

19. (a) $y_1^E = 1.0000, y_1^{RK} = 0.9950$ (b) $y(t) = e^{-t^2/2}$

21. (a) $y_1^E = 1.1000, y_1^{RK} = 1.1111\ldots$ (b) $y(t) = 1/(1-t)$

23. $y(2) \approx y_{20} = 0.6399\ldots$ 25. $y(5) \approx y_{40} = 4.0000\ldots$ 27. $y(0.9) \approx y_9 = 9.9291\ldots$

Chapter 2 Review Exercises, page 100

1. $y = Ce^{-2t} + 3$ 3. $y^2 = 2t^3 + C$ 5. $y = Ce^{-t^2} + 1$

7. $ty + y^3 + t^2 = C$ 9. $y = 4e^{t^3 + 3t}$ 11. $\sqrt{y} = \frac{1}{3}t^{3/2} + C$

13. $t\cos y = C$ 15. $y = \begin{cases} e^{-2t}, & 0 \le t < 1 \\ e^{-(t+1)}, & 1 \le t \le 2 \end{cases}$

17. $y = \sqrt{1/(1-2t)}$ 19. $y = Ce^{-\sin t} + 1$

21. $y = (t^4 + 14t^2 + 65)/16$ 23. $y = -8 + Ce^{\sqrt{t}}$

25. $y = 12 - 7e^{-t}$ 27. $y^2 = 4t - t^3 + 1$

29. $y = 1 + 3e^{(\cos 2t)/2}$

CHAPTER 3

Section 3.1, page 112

1. $-\infty < t < \infty$ 3. $-\infty < t < -1$

5. (a) $0 < t < \infty$ (c) No

7. No, since Theorem 3.1 guarantees a solution on $-\infty < t < 3$.

9. $y = -2\cos 2t$

11. (a) Note that $y''(0) = 1$. Thus, the solution is decreasing and concave up at $t = 0$. Graph B is the appropriate one.

    (b) Graph D

    (c) Graph A

    (d) Graph C

13. (a) Drum 1 will bob more rapidly.

    (b) Drum 1 will bob more rapidly.

Section 3.2, page 120

1. (b) $W(t) = -8$. Therefore, the two functions form a fundamental set.

   (c) The general solution is $y = c_1 e^{2t} + c_2(2e^{-2t})$. The solution of the initial value problem is $y = e^{-2t}$.

3. (b) $W(t) = 0$. Therefore, the two functions do not form a fundamental set.

5. (b) $W(t) = e^{4t}$. Therefore, the two functions form a fundamental set.

   (c) The general solution is $y = c_1 e^{2t} + c_2 te^{2t}$. The solution of the initial value problem is $y = 2e^{2t} - 4te^{2t}$.

7. (b) $W(t) = 2e^{3t}$. Therefore, the two functions form a fundamental set.

   (c) The solution of the initial value problem is $y = 2e^{t+1} - e^{2(t+1)}$.

9. (b) $W(t) = -t^{-1} \ln 3$. Therefore, the two functions form a fundamental set.

   (c) The general solution is $y = c_1 \ln t + c_2 \ln 3t$. The solution of the initial value problem is $y = 18 \ln t - 9 \ln 3t$ or, equivalently, $y = 9 \ln (t/3)$.

11. (b) $W(t) = 4t$. Therefore, the two functions form a fundamental set.

   (c) The general solution is $y = c_1 t^3 + c_2(-t^{-1})$. The solution of the initial value problem is $y = -0.5t^3 + 0.5t^{-1}$.

13. (b) $W(t) = -3$. Therefore, the two functions form a fundamental set.

   (c) The general solution is $y = c_1(t + 1) + c_2(-t + 2)$. The solution of the initial value problem is $y = (t + 1) + 2(-t + 2)$, or $y = 5 - t$.

15. (b) $W(t) = e^{-t}$. Therefore, the two functions form a fundamental set.

   (c) The general solution is $y = c_1 e^{-t/2} + c_2 t e^{-t/2}$. The solution of the initial value problem is $y = 0.5e^{-(t-1)/2}(1 + t)$.

17. (b) $c_1 = 2 + \ln 3$, $c_2 = 1$ 19. $\alpha = 0$, $\beta = -9$

Section 3.3, page 125

1. (a) The general solution is $y = c_1 e^{-2t} + c_2 e^t$.

   (b) The solution of the initial value problem is $y = 2e^{-2t} + e^t$.

   (c) $\lim_{t \to -\infty} y(t) = \infty$ and $\lim_{t \to \infty} y(t) = \infty$

3. (a) The general solution is $y = c_1 e^t + c_2 e^{3t}$.

   (b) The solution of the initial value problem is $y = -2e^t + e^{3t}$.

   (c) $\lim_{t \to -\infty} y(t) = 0$ and $\lim_{t \to \infty} y(t) = \infty$

5. (a) The general solution is $y = c_1 e^{-t} + c_2 e^t$.

   (b) The solution of the initial value problem is $y = e^{-t}$.

   (c) $\lim_{t \to -\infty} y(t) = \infty$ and $\lim_{t \to \infty} y(t) = 0$

7. (a) The general solution is $y = c_1 e^{-3t} + c_2 e^{-2t}$.

   (b) The solution of the initial value problem is $y = -e^{-3t} + 2e^{-2t}$.

   (c) $\lim_{t \to -\infty} y(t) = -\infty$ and $\lim_{t \to \infty} y(t) = 0$

9. (a) The general solution is $y = c_1 e^{-2t} + c_2 e^{2t}$.

   (b) The solution of the initial value problem is $y = 0e^{-2t} + 0e^{2t}$, or $y(t) \equiv 0$.

   (c) $\lim_{t \to -\infty} y(t) = 0$ and $\lim_{t \to \infty} y(t) = 0$

11. (a) The general solution is $y = c_1 + c_2 e^{1.5t}$.

   (b) The solution of the initial value problem is $y = 3 + 0e^{1.5t}$, or $y(t) \equiv 3$.

   (c) $\lim_{t \to -\infty} y(t) = 3$ and $\lim_{t \to \infty} y(t) = 3$

13. (a) The general solution is $y = c_1 e^{(-2-\sqrt{2})t} + c_2 e^{(-2+\sqrt{2})t}$.

   (b) The solution of the initial value problem is $y = -\sqrt{2}\, e^{(-2-\sqrt{2})t} + \sqrt{2}\, e^{(-2+\sqrt{2})t}$.

   (c) $\lim_{t \to -\infty} y(t) = -\infty$ and $\lim_{t \to \infty} y(t) = 0$

15. (a) The general solution is $y = c_1 e^{-t/\sqrt{2}} + c_2 e^{t/\sqrt{2}}$.

   (b) The solution of the initial value problem is $y = -2e^{-t/\sqrt{2}}$.

   (c) $\lim_{t \to -\infty} y(t) = -\infty$ and $\lim_{t \to \infty} y(t) = 0$

17. (a) $y_2(t) = e^{3t}$ (b) $\alpha = -2,\quad \beta = -3$ (c) $y = e^{-t} + 2e^{3t}$

19. $y = c_1 e^{2t} + c_2 e^{3t} + c_3$

Section 3.4, page 131

1. (a) The general solution is $y = c_1 e^{-t} + c_2 t e^{-t}$.
   (b) The solution of the initial value problem is $y = te^{-(t-1)}$.
   (c) $\lim_{t\to-\infty} y(t) = -\infty$ and $\lim_{t\to\infty} y(t) = 0$

3. (a) The general solution is $y = c_1 e^{-3t} + c_2 t e^{-3t}$.
   (b) The solution of the initial value problem is $y = (2+4t)e^{-3t}$.
   (c) $\lim_{t\to-\infty} y(t) = -\infty$ and $\lim_{t\to\infty} y(t) = 0$

5. (a) The general solution is $y = c_1 e^{t/2} + c_2 t e^{t/2}$.
   (b) The solution of the initial value problem is $y = (-6+2t)e^{(t-1)/2}$.
   (c) $\lim_{t\to-\infty} y(t) = 0$ and $\lim_{t\to\infty} y(t) = \infty$

7. (a) The general solution is $y = c_1 e^{t/4} + c_2 t e^{t/4}$.
   (b) The solution of the initial value problem is $y = (-4+4t)e^{t/4}$.
   (c) $\lim_{t\to-\infty} y(t) = 0$ and $\lim_{t\to\infty} y(t) = \infty$

9. (a) The general solution is $y = c_1 e^{5t/2} + c_2 t e^{5t/2}$.
   (b) The solution of the initial value problem is $y = (2+t)e^{2.5(t+2)}$.
   (c) $\lim_{t\to-\infty} y(t) = 0$ and $\lim_{t\to\infty} y(t) = \infty$

11. $\alpha = -\frac{1}{2},\ y_0 = 0,\ y_0' = 4,\ y(t) = 4te^{-t/2}$

13. $y(t) = (2-t)e^{-t/2}$, and therefore $y(0) = 2, y'(0) = -2$.

15. (a) $y_2(t) = t\ln|t|,\ \ t \neq 0$ (b) $W(t) = t, t \neq 0$

17. (a) $y_2(t) = (t+1)^3$ (b) $W(t) = (t+1)^4$

19. (a) $y_2(t) = 1/(t-2)^2$ (b) $W(t) = -4/(t-2)$

Section 3.5, page 139

1. (a) $1 + \sqrt{3}\,i$ (b) $-2 + 2i$ (c) $-1 - 2i$ (d) $-\left(\sqrt{6} + \sqrt{2}\,i\right)/8$ (e) $-2 + 2\sqrt{3}\,i$

3. (a) $\lambda = \pm 2i$ (b) $y = c_1 \cos 2t + c_2 \sin 2t$ (c) $y = -0.5\cos 2t - 2\sin 2t$

5. (a) $\lambda = \pm i/3$ (b) $y = c_1 \cos(t/3) + c_2 \sin(t/3)$ (c) $y = 2\sqrt{3}\cos(t/3) + 2\sin(t/3)$

7. (a) $\lambda = (-1 \pm i\sqrt{3})/2$
   (b) $y = c_1 e^{-t/2}\cos\left(\sqrt{3}\,t/2\right) + c_2 e^{-t/2}\sin\left(\sqrt{3}\,t/2\right)$
   (c) $y = -2e^{-t/2}\cos\left(\sqrt{3}\,t/2\right) - 2\sqrt{3}\,e^{-t/2}\sin\left(\sqrt{3}\,t/2\right)$

9. (a) $\lambda = (-1 \pm i)/3$ (b) $y = c_1 e^{-t/3}\cos(t/3) + c_2 e^{-t/3}\sin(t/3)$
   (c) $y = -e^{-(t-3\pi)/3}\sin(t/3)$

11. (a) $\lambda = \sqrt{2} \pm i$ (b) $y = c_1 e^{\sqrt{2}t}\cos t + c_2 e^{\sqrt{2}t}\sin t$
    (c) $y = -0.5e^{\sqrt{2}t}\cos t + 1.5\sqrt{2}\,e^{\sqrt{2}t}\sin t$

13. $a = 0,\ b = 1,\ y_0 = (\sqrt{2}-2)/2,\ y_0' = (\sqrt{2}+2)/2$

15. $a = 4,\ b = 5,\ y_0 = 1,\ y_0' = -3$

17. $a = 0,\ b = \pi^2,\ y_0 = -1,\ y_0' = -\sqrt{3}\,\pi$

19. $a = -2,\ \ b = 1 + \pi^2,\ \ y_0 = 2e,\ \ y_0' = 2e$

21. $a = 4,\ \ b = 5,\ \ y_0 = 0,\ \ y_0' = 3$

23. $y = \sqrt{2}\,\cos(\pi t - 7\pi/4)$ 25. $y = 2e^{-t}\cos(t - 2\pi/3)$

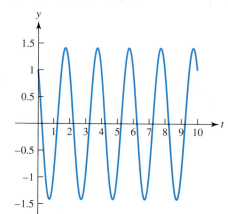

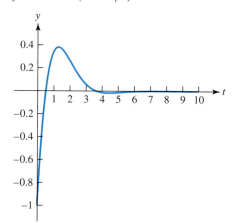

27. $a = 0$, $b = \pi^2/4$, $y_0 = 2$, $y_0' = 0$

29. $a = 0$, $b = 4$, $y_0 = 0.5\cos(5\pi/6)$, $y_0' = \sin(5\pi/6)$

31. $y = c_1\cos 3t + c_2\sin 3t$ 33. $y = c_1 e^{it} + c_2 e^{-5it}$

Section 3.6, page 151

1. (a)

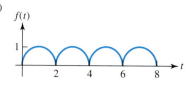

(b) $f(t)$

(c) $f(t)$

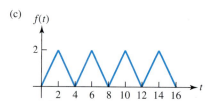

(d) $f(t)$

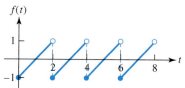

(e) $f(t)$

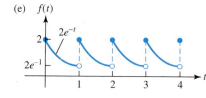

(f)

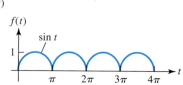

(g) $f(t)$

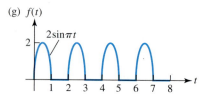

3. (a) $y = (\cos 10t + \sin 10t)/10$ (b) $t = \pi/40$ (c) $\sqrt{2}/10$

5. (a) $y_0 = -\frac{1}{9}$ ft/sec   (b) period $= 6\pi$

7. $y_0 = \sqrt{3}/8$ cm, $y_0' = 1/4$ cm/s, $k = 16$ N/m, $T = \pi$ s

9. (a) $10y'' + 7y' + 100y = 0$, $y(0) = 0.5$, $y'(0) = 1$

   (b) The general solution is $y = c_1 e^{\alpha t} \cos \beta t + c_2 e^{\alpha t} \sin \beta t$, where $\alpha = -0.35$ and $\beta = \sqrt{40 - 0.49}/2 \approx 3.1428$. The solution of the initial value problem requires $c_1 = 0.5, c_2 \approx 0.3739$. $\lim_{t \to \infty} y(t) = 0$; this limit is to be expected since damping dissipates energy, causing the motion to decrease.

   (c) Zooming in on the graph shows that τ is about 5.25 seconds.

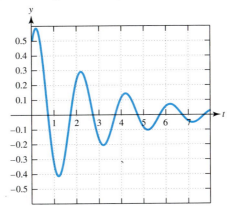

11. $\gamma = \sqrt{(1600)(0.44)/1.44}$ kg/s

13. (a) $\gamma_{\mathrm{crit}} = 2$

   (b) As γ increases, the solution tends to approach the constant solution $y(t) = 1$. This behavior is consistent with that predicted in Exercise 12.

Section 3.7, page 156

1. (b) $y_C = c_1 e^{-t} + c_2 e^{3t}$ (c) $y = 1.5 e^{-t} + 0.5 e^{3t} + 3t - 1$

3. (b) $y_C = c_1 e^{-t} + c_2 e^{2t}$ (c) $y = e^{-t} - 3e^{2t} + 2e^{4t}$

5. (b) $y_C = c_1 e^{-t} + c_2$ (c) $y = 2e^{-(t-1)} + t^2 - 2t$

7. (b) $y_C = c_1 \cos t + c_2 \sin t$ (c) $y = -\cos t - 2\sin t + 2t + \cos 2t$

9. (b) $y_C = c_1 e^t \cos t + c_2 e^t \sin t$ (c) $y = -5e^t \cos t - 5e^t \sin t + 5(t+1)^2$

11. (b) $y_C = c_1 e^t + c_2 t e^t$ (c) $y = -2e^t + 4te^t + t^2 e^t / 2$

15. $y_P = \frac{1}{2} u_2 + \frac{1}{3} u_3$ 17. $g(t) = 5e^{2t} + t^2 - 2t - 2$

19. $g(t) = 3e^t + 6t$ 21. $g(t) = t^2 + t + \sin t$

23. $\alpha = 1$, $\beta = 0$, $g(t) = 2 + 2t$ 25. $\alpha = -2$, $\beta = 2$, $g(t) = e^t - 2\cos t + \sin t$

Section 3.8, page 165

1. (a) $y_C = c_1 e^{-2t} + c_2 e^{2t}$ (b) $y_P = -t^2 - 0.5$ (c) $y = c_1 e^{-2t} + c_2 e^{2t} - t^2 - 0.5$

3. (a) $y_C = c_1 \cos t + c_2 \sin t$ (b) $y_P = 4e^t$ (c) $y = c_1 \cos t + c_2 \sin t + 4e^t$

5. (a) $y_C = c_1 e^{2t} + c_2 t e^{2t}$ (b) $y_P = \frac{1}{2} t^2 e^{2t}$

7. (a) $y_C = c_1 e^{-t} \cos t + c_2 e^{-t} \sin t$ (b) $y_P = (t^3 - 3t^2 + 3t)/2$

9. (a) $y_C = c_1 e^{-t} \cos t + c_2 e^{-t} \sin t$ (b) $y_P = e^{-t} + (\cos t + 2\sin t)/5$

11. (a) $y_C = c_1 e^{t/2} + c_2 e^{2t}$ (b) $y_P = 2te^{t/2}$

13. (a) $y_C = c_1 e^{t/3} + c_2 t e^{t/3}$ (b) $y_P = \frac{1}{6} t^3 e^{t/3}$

15. (a) $y_C = c_1 e^{-2t} \cos t + c_2 e^{-2t} \sin t$ (b) $y_P = 2e^{-2t} + (\cos t + \sin t)/8$

17. (a) $y_C = c_1 \cos 3t + c_2 \sin 3t$

(b) $y_P = t(A_2 t^2 + A_1 t + A_0) \cos 3t + t(B_2 t^2 + B_1 t + B_0) \sin 3t + C \cos t + D \sin t$

19. (a) $y_C = c_1 e^t \cos t + c_2 e^t \sin t$

(b) $y_P = Ae^{-t} \cos 2t + Be^{-t} \sin 2t + C_1 t + C_0 + e^{-t}(D_1 t + D_0) \cos t + e^{-t}(E_1 t + E_0) \sin t$

21. (a) $y_C = c_1 \cos 2t + c_2 \sin 2t$

(b) $y_P = At \cos 2t + Bt \sin 2t + C + D \cos 4t + E \sin 4t$

23. $\alpha = -1$, $\beta = -2$, $y = c_1 e^{-t} + c_2 e^{2t} - 2t + 1$

25. $\alpha = 4$, $\beta = 4$, $y = c_1 e^{-2t} + c_2 t e^{-2t} - (4 \cos t - 3 \sin t)/5$

27. $\alpha = 2$, $\beta = 5$, $y = c_1 e^{-t} \cos 2t + c_2 e^{-t} \sin 2t + 2e^{-t}$

29. $\alpha = 0$, $\beta = -4$ 31. $y = \frac{1}{5} e^{-t}$ 33. $y = \frac{3}{2}$

35. $\omega \geq 1/\sqrt{3}$

37. (a) Graph C   (b) Graph E   (c) Graph A   (d) Graph B   (e) Graph D

39. (a) $A = -\frac{1}{5}$ (b) $y = -\frac{1}{5}(\cos 2t + i \sin 2t)$

41. (a) $y_P = \frac{1}{3} e^{it}$ 43. (a) $y_P = \frac{1}{10}(-2 + i)e^{-2it}$

Section 3.9, page 173

1. (a) $y_C = c_1 \cos 2t + c_2 \sin 2t$ (b) $y_P = 1$

3. (a) $y_C = c_1 e^t + c_2 t^2 e^t$ (b) $y_P = t^3 e^t / 3$

5. (a) $y_C = c_1 e^{-t} + c_2 e^t$ (b) $y_P = -(1/4)e^t + (t/2)e^t$

7. (a) $y_C = c_1 e^t + c_2 t e^t$ (b) $y_P = (t^2/2)e^t$

9. (a) $y_C = c_1 \sin t + c_2 t \sin t$ (b) $y_P = (t^3/6) \sin t$

11. (a) $y_C = c_1 t + c_2 e^t$ (b) $y_P = [(t^2/2) - t]e^t$

13. (a) $y_C = c_1(t-1)^2 + c_2(t-1)^3$ (b) $y_P = (3t-2)/6$

17. $\alpha = 0$, $\beta = -1$, $y_0 = 1$, $y_0' = -1$

Section 3.10, page 184

3. (b) $y'' + 100y = 2e^{-t}$, $y(0) = 0$, $y'(0) = 0$, $y = \frac{2}{101}(-\cos 10t + 0.1 \sin 10t + e^{-t})$

(c) $|y|_{\max} \approx 0.035$ m

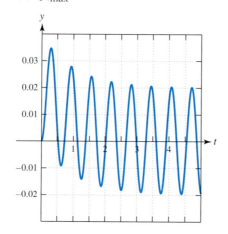

5. (b) $y'' + 100y = 2$, $y(0) = 0$, $y'(0) = 0$, $0 \le t \le \pi$
   $y = \frac{1}{50}(1 - \cos 10t)$, $0 \le t \le \pi$
   $y'' + 100y = 0$, $y(\pi) = 0$, $y'(\pi) = 0$, $\pi \le t$
   $y(t) = 0$, $\pi \le t$

(c) $|y|_{max} = 0.04$ m

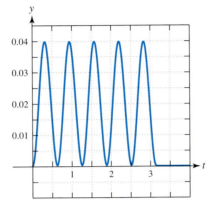

7. (a) $y = \frac{1}{60}e^{-2t}(9 \cos 6t - 13 \sin 6t) + \frac{1}{20}(-3 \cos 8t + 4 \sin 8t)$

(b) $\lim_{t \to \infty} y(t)$ does not exist. For large values of t, $y \approx \frac{1}{20}(-3 \cos 8t + 4 \sin 8t)$

9. (a) $y = \frac{1}{74}e^{-2t}(30 \cos 6t + 5 \sin 6t) + \frac{1}{74}(-30 \cos 6t + 5 \sin 6t)$

(b) $\lim_{t \to \infty} y(t)$ does not exist. For large values of t, $y \approx \frac{1}{74}(-30 \cos 6t + 5 \sin 6t)$

13. (b) Approximately 52.42 ft/sec   (c) Approximately 85.86 ft

15. $I = \frac{12}{7}[\cos(t/2) - \cos 3t]$ mA   17. $V = 2e^{-t}(1 - \cos t)$ V

Section 3.11, page 193

1. (a) $W(t) = 4$. Therefore, the functions form a fundamental set on $(-\infty, \infty)$.

   (b) $y = 4 + 2(t - 1)$

3. (a) $W(t) = 32$. Therefore, the functions form a fundamental set on $(-\infty, \infty)$.

   (b) $y = -1 + t + \cos 2t - \sin 2t$

5. (a) $W(t) = 2t^{-3}$. Therefore, the functions form a fundamental set on $(0, \infty)$.

   (b) $y = 2 - t + t^{-1}$

7. $W(1) = \begin{vmatrix} y_1(1) & y_2(1) \\ y_1'(1) & y_2'(1) \end{vmatrix} = \begin{vmatrix} 2 & -4 \\ -1 & 2 \end{vmatrix} = 0$. No, they do not form a fundamental set.

9. $W(0) = \begin{vmatrix} 1 & 0 & 2 \\ -1 & 0 & -2 \\ 0 & 2 & 1 \end{vmatrix} = 0$. No, they do not form a fundamental set.

11. By Abel's theorem [see equation (7)], $W(t) = e^{-t^2/4}$. Therefore, $W(4) = e^{-4}$.

13. By Abel's theorem, $W(t) = e^{-t}$. Therefore, $W(4) = e^{-4}$.

15. By Abel's theorem, $W(t) = t^2 + 1$. Therefore, $W(4) = 17$.

17. $\bar{y}_1(t) = \cos t + \sin t$, $\bar{y}_2(t) = \cos t - \sin t$

19. $\bar{y}_1(t) = -e^{-2t}$, $\bar{y}_2(t) = e^{-2t} + te^{-2t}$

21. (a) $A = \begin{bmatrix} 1 & 2 & 0 \\ -2 & 1 & 0 \\ 0 & 0 & e^{-2} \end{bmatrix}$

(b) Since the determinant of A is equal to $5e^{-2}$, $\{\bar{y}_1, \bar{y}_2, \bar{y}_3\}$ is a fundamental set.

23. $W(t) = e^{3(t-1)}$ 25. $W(t) = 3t^{-1}$

Section 3.12, page 199

1. (a) $y = c_1 + c_2 e^{2t} + c_3 e^{-2t}$ 3. (a) $y = c_1 e^{-t} + c_2 \cos 2t + c_3 \sin 2t$

5. (a) $y = c_1 \cos(t/2) + c_2 t \cos(t/2) + c_3 \sin(t/2) + c_4 t \sin(t/2)$

7. (a) $y = c_1 e^t + c_2 e^{-t} + c_3 e^{2t}$

9. (a) $y = c_1 e^{-2t} + c_2 e^t \cos \sqrt{3}\, t + c_3 e^t \sin \sqrt{3}\, t$

11. (a) $y = c_1 + c_2 \cos t + c_3 \sin t$

13. (a) $y = c_1 e^t + c_2 e^{-t} + e^{t/2}[c_3 \cos(\sqrt{3}t/2) + c_4 \sin(\sqrt{3}t/2)]$
$\qquad + e^{-t/2}[c_5 \cos(\sqrt{3}t/2) + c_6 \sin(\sqrt{3}t/2)]$

15. (a) $y = c_1 + c_2 e^{-t} + c_3 t e^{-t}$ (b) $y = 1 - e^{-t} - t e^{-t}$

17. (a) $y = c_1 e^{-t} + c_2 t e^{-t} + c_3 t^2 e^{-t}$ (b) $y = t e^{-t} + t^2 e^{-t}$

19. (a) $y = c_1 e^{\alpha t} + c_2 t e^{\alpha t} + c_3 t^2 e^{\alpha t}$ (b) $W(0) = 2$

21. $y^{(4)} + 9y'' = 0$ 23. $y^{(4)} - 2y'' + y = 0$

25. $y^{(4)} - 4y''' + 6y'' - 4y' + y = 0$ 27. (a) $n = 5$

29. (a) $n = 7$ 31. $a = -1, \quad n = 1$

33. $a = 0, \quad n = 4$ 35. $a = -1, \quad n = 3$

Section 3.13, page 204

1. (a) $y_C = c_1 + c_2 e^t + c_3 e^{-t}$ (b) $y_P = \frac{1}{6} e^{2t}$

3. (a) $y_C = c_1 + c_2 e^t + c_3 e^{-t}$ (b) $y_P = -2t^2$

5. (a) $y_C = c_1 + c_2 t + c_3 e^{-t}$ (b) $y_P = 6t e^{-t}$

7. (a) $y_C = c_1 + c_2 e^t + c_3 t e^t$ (b) $y_P = \frac{1}{2} t^2 + 2t + 2t^2 e^t$

9. (a) $y_C = c_1 e^t + e^{-t/2}[c_2 \cos(\sqrt{3}\, t/2) + c_3 \sin(\sqrt{3}\, t/2)]$ (b) $y_P = \frac{1}{3} t e^t$

11. (a) $y_C = c_1 e^t + c_2 e^{-t} + c_3 \cos t + c_4 \sin t$ (b) $y_P = -t - 1$

13. (a) $y_C = c_1 e^{-t} + e^{t/2}[c_2 \cos(\sqrt{3}\, t/2) + c_3 \sin(\sqrt{3}\, t/2)]$ (b) $y_P = t^3 - 6$

15. (a) $y_C = c_1 + c_2 e^{2t} + c_3 t e^{2t}$ (b) $y_P = t(At^3 + Bt^2 + Ct + D) + t^2(Et^2 + Ft + G)e^{2t}$

17. (a) $y_C = c_1 e^{2t} + c_2 e^{-2t} + c_3 \cos 2t + c_4 \sin 2t$

(b) $y_P = t(At + B)\sin 2t + t(Ct + D)\cos 2t$

19. (a) $y_C = c_1 e^t + c_2 e^{-t} + c_3 \cos t + c_4 \sin t$

(b) $y_P = t(At + B)e^{-t} + t(Ct + D)\cos t + t(Et + F)\sin t$

21. (a) $y_C = e^t(c_1 \cos t + c_2 \sin t) + e^{-t}(c_3 \cos t + c_4 \sin t)$

(b) $y_P = t e^t (A \cos t + B \sin t)$

23. $y''' - y'' + 4y' - 4y = -2 + 8t - 4t^2$ 25. $t^3 y''' - t^2 y'' = 12t^4$

27. (a) Calculating the Wronskian at $t = 1$, we obtain $W(1) = 6$.

(b) $y = c_1 t + c_2 t^2 + c_3 t^4 - \frac{16}{21} t^{1/2}$

29. $y = c_1 t + c_2 t^2 + c_3 t^4 - 3t^3$ 31. $y = 0.5 (e^{-t} + \cos t - \sin t)$

33. $y = e^{-t}(1 + t/3)$

Chapter 3 Review Exercises, page 206

1. $y = e^{-t}(c_1 \cos t + c_2 \sin t)$

3. $y = 3 \cos 2t + \sin 2t$

5. $y = c_1 e^{3t} + c_2 e^{2t}$

7. $y = c_1 e^{3t} + c_2 e^{-3t} + c_3 \cos 3t + c_4 \sin 3t$

9. $y = c_1 \cos 3t + c_2 \sin 3t$

11. $y = c_1 e^t + c_2 t e^t + t e^t \ln t$

13. $y = 2e^{-t} + 3te^{-t} + 8$

15. $y = c_1 + c_2 t + t^3 + 2t^2$

17. $y = c_1 e^t + c_2 e^{-t} + c_3 \cos t + c_4 \sin t - 4t$

19. $y = c_1 \cos t + c_2 \sin t + \ln(\cos t)(\cos t) + t \sin t + t$

21. $y = c_1 e^{3t} + c_2 t e^{3t}$

23. $y = c_1 e^t + c_2 e^{-t} + t^2 - 4$

25. $y = c_1 e^{10t} + c_2 t e^{10t}$

27. $y = c_1 + c_2 e^{4t}$

29. $y = c_1 + c_2 t + c_3 e^{2t} + c_4 t e^t$

CHAPTER 4

Section 4.1, page 220

1. $\begin{bmatrix} -3t^2 + 2t - 2 & 2t^2 + 3t \\ 4 & -3t^2 - 2t + 2 \end{bmatrix}$

3. $\begin{bmatrix} -1 \\ 1 \end{bmatrix}$

5. $\det[A(t)B(t)] = \det[A(t)]\det[B(t)] = -t(t+1)(t+2)$

7. $A^{-1}(t) = \dfrac{1}{(t-4)(t+1)} \begin{bmatrix} t-3 & -2 \\ -2 & t \end{bmatrix}, \quad t \neq 4, t \neq -1$

9. $A(t)$ cannot be inverted for any value of t.

11. $\begin{bmatrix} 0 & 0 \\ -2 & 1 \end{bmatrix}$

13. $A(t)$ is defined for $-\infty < t < 0, 0 < t \leq 1$. $A'(t) = \begin{bmatrix} 0 & 1/t \\ -\frac{1}{2}(1-t)^{-1/2} & 3e^{3t} \end{bmatrix}$,

$A''(t) = \begin{bmatrix} 0 & -1/t^2 \\ -\frac{1}{4}(1-t)^{-3/2} & 9e^{3t} \end{bmatrix}, \quad -\infty < t < 0, 0 < t < 1.$

15. $P(t) = \begin{bmatrix} t^{-1} & t^2 + 1 \\ 4 & t^{-1} \end{bmatrix}, \quad \mathbf{g}(t) = \begin{bmatrix} t \\ 8t \ln t \end{bmatrix}$

17. $A(t) = \left[t - 1, \frac{1}{2}t^2 + 1, e^t - 1 \right]$

19. $A(t) = \begin{bmatrix} 2 + \ln t & 2t^2 + 3 \\ 5t - 4 & t^3 - 3 \end{bmatrix}$

21. $A(t) = \begin{bmatrix} t^3/3 & t + 1 - \sin t \\ t & 0 \end{bmatrix}$

23. $A(t) = \begin{bmatrix} t^2 & \sin t & 2t \\ 5t & \ln|t+1| & t^3 \end{bmatrix}$

29. (a) For Tank 1, $V_1(t) = 100 + 5t$ gal; for Tank 2, $V_2(t) = 100 - 5t$ gal

(b) $0 \leq t \leq 20$ min

(c) $Q_1' = 2.5 - \dfrac{10Q_1}{100 + 5t} + \dfrac{10Q_2}{100 - 5t}, \quad Q_1(0) = 0$

$Q_2' = \dfrac{10Q_1}{100 + 5t} - \dfrac{15Q_2}{100 - 5t}, \quad Q_2(0) = 0$

31. (a) For Tank 1, $V_1(t) = 100 + 5t$ gal; for Tank 2, $V_2(t) = 100$ gal

(b) $0 \leq t \leq 80$ min

(c) $Q_1' = 2.5 - \dfrac{5Q_1}{100 + 5t} + \dfrac{5Q_2}{100}, \quad Q_1(0) = 0$

$Q_2' = \dfrac{5Q_1}{100 + 5t} - \dfrac{5Q_2}{100}, \quad Q_2(0) = 0$

Section 4.2, page 227

1. $\pi/2 < t < 3\pi/2$ 3. $0 < t < \pi/2$

7. (a) $A = \begin{bmatrix} 4 & 1 \\ 1 & 4 \end{bmatrix}$ (b) $\mathbf{y} = c_1 \begin{bmatrix} e^{5t} \\ e^{5t} \end{bmatrix} + c_2 \begin{bmatrix} e^{3t} \\ -e^{3t} \end{bmatrix}$

9. $c_1 = 1, c_2 = 2$ 11. $P(t) = \begin{bmatrix} 0 & 1 \\ -4 & -t^2 \end{bmatrix}, \quad \mathbf{g}(t) = \begin{bmatrix} 0 \\ \sin t \end{bmatrix}$

13. $P(t) = \begin{bmatrix} 0 & 1 & 0 \\ 0 & 0 & 1 \\ -e^{-t}\tan t & -t^{-1}e^{-t} & -5e^{-t} \end{bmatrix}, \quad \mathbf{g}(t) = \begin{bmatrix} 0 \\ 0 \\ e^{-t} \end{bmatrix}$

15. $y'' - 2y' + 3y = 2\cos 2t, \quad y(-1) = 1, \quad y'(-1) = 4$

17. $y^{(4)} - (y'')(y'' + \sin y) - y' = 0, \quad y(1) = 0, \quad y'(1) = 0, \quad y''(1) = -1, \quad y'''(1) = 2$

19. $P(t) = \begin{bmatrix} 0 & 1 & 0 & 0 \\ 4 & t^{-1} & -t & \sin t \\ 0 & 0 & 0 & 1 \\ 1 & 0 & 0 & -5 \end{bmatrix}, \quad \mathbf{g}(t) = \begin{bmatrix} 0 \\ e^{2t} \\ 0 \\ 0 \end{bmatrix}$

21. $P(t) = \begin{bmatrix} 0 & 1 & 0 & 0 \\ 4 & -3 & -5 & -2 \\ 0 & 0 & 0 & 1 \\ 5 & 6 & -2 & 1 \end{bmatrix}, \quad \mathbf{g}(t) = \begin{bmatrix} 0 \\ t^2 \\ 0 \\ -t \end{bmatrix}$

Section 4.3, page 235

1. (a) $\mathbf{y}' = \begin{bmatrix} 9 & -4 \\ 15 & -7 \end{bmatrix} \mathbf{y}$ 3. (a) $\mathbf{y}' = \begin{bmatrix} 1 & 4 \\ -1 & 1 \end{bmatrix} \mathbf{y}$ 5. (a) $\mathbf{y}' = \begin{bmatrix} 0 & 1 & 1 \\ -6 & -3 & 1 \\ -8 & -2 & 4 \end{bmatrix} \mathbf{y}$

7. Yes, they are solutions, and the Wronskian is $W(t) = 2$.

9. No, they are solutions, but the Wronskian is $W(t) = 0$.

11. No, they are not solutions.

13. Yes, they are solutions, and the Wronskian is $W(t) = 1$.

15. (b) $W(t) = 4e^{2t}$ (c) $\mathbf{y}(t) = \begin{bmatrix} 2e^{3t} & 2e^{-t} \\ 3e^{3t} & 5e^{-t} \end{bmatrix} \begin{bmatrix} c_1 \\ c_2 \end{bmatrix}$

(d) $\mathbf{c} = \begin{bmatrix} \frac{3}{4} \\ -\frac{1}{4} \end{bmatrix}, \quad \mathbf{y}(t) = \dfrac{3}{4} \begin{bmatrix} 2e^{3t} \\ 3e^{3t} \end{bmatrix} - \dfrac{1}{4} \begin{bmatrix} 2e^{-t} \\ 5e^{-t} \end{bmatrix} = \begin{bmatrix} (3e^{3t} - e^{-t})/2 \\ (9e^{3t} - 5e^{-t})/4 \end{bmatrix}$

17. (b) $W(t) \equiv 0$; therefore, these solutions do not form a fundamental set.

19. (b) $W(t) \equiv 1$ (c) $\mathbf{y}(t) = \begin{bmatrix} e^t & e^{-t} \\ -2e^t & -e^{-t} \end{bmatrix} \begin{bmatrix} c_1 \\ c_2 \end{bmatrix}$

(d) $\mathbf{c} = \begin{bmatrix} 2e^{-1} \\ -e \end{bmatrix}$, $\mathbf{y}(t) = 2e^{-1} \begin{bmatrix} e^t \\ -2e^t \end{bmatrix} - e \begin{bmatrix} e^{-t} \\ -e^{-t} \end{bmatrix} = \begin{bmatrix} 2e^{t-1} - e^{-(t-1)} \\ -4e^{t-1} + e^{-(t-1)} \end{bmatrix}$

21. (b) $W(t) = -t^2$ (c) $\mathbf{y}(t) = \begin{bmatrix} t^2 - 2t & t - 1 \\ 2t & 1 \end{bmatrix} \begin{bmatrix} c_1 \\ c_2 \end{bmatrix}$

(d) $\mathbf{c} = \begin{bmatrix} 1 \\ -2 \end{bmatrix}$, $\mathbf{y}(t) = \begin{bmatrix} t^2 - 2t \\ 2t \end{bmatrix} - 2 \begin{bmatrix} t - 1 \\ 1 \end{bmatrix} = \begin{bmatrix} t^2 - 4t + 2 \\ 2t - 2 \end{bmatrix}$

23. (b) $W(t) = -11e^t$ (c) $\mathbf{y}(t) = \begin{bmatrix} 5e^t & e^t & e^{-t} \\ -11e^t & 0 & -e^{-t} \\ 0 & 11e^t & 5e^{-t} \end{bmatrix} \begin{bmatrix} c_1 \\ c_2 \\ c_3 \end{bmatrix}$

(d) $\mathbf{c} = \begin{bmatrix} 1 \\ -1 \\ -1 \end{bmatrix}$, $\mathbf{y}(t) = \begin{bmatrix} 5e^t \\ -11e^t \\ 0 \end{bmatrix} - \begin{bmatrix} e^t \\ 0 \\ 11e^t \end{bmatrix} - \begin{bmatrix} e^{-t} \\ -e^{-t} \\ 5e^{-t} \end{bmatrix} = \begin{bmatrix} 4e^t - e^{-t} \\ -11e^t + e^{-t} \\ -11e^t - 5e^{-t} \end{bmatrix}$

25. (a) $W(t) = 2e^{6t}$ (b) 6   27. (a) $W(t) = -6e^{4t}$ (b) 4

29. (b) $C = \begin{bmatrix} \frac{1}{2} & \frac{1}{2} \\ -\frac{1}{2} & \frac{1}{2} \end{bmatrix}$ (c) $\hat{\Psi}(t)$ is also a fundamental matrix.

31. (b) $C = \begin{bmatrix} 0 & 0 \\ 2 & 0 \end{bmatrix}$ (c) $\hat{\Psi}(t)$ is not a fundamental matrix.

33. $C = \begin{bmatrix} \frac{1}{2} & \frac{1}{2} \\ \frac{1}{2} & -\frac{1}{2} \end{bmatrix}$

Section 4.4, page 245

1. $\lambda_1 = -1$, $\lambda_2 = 2$ 3. $\lambda_1 = 0$, $\lambda_2 = 3$ 5. $\lambda_1 = -2i$, $\lambda_2 = 2i$

7. $\lambda_1 = -1$, $\lambda_2 = 4$, $\lambda_3 = 5$ 9. $\lambda_1 = 0$, $\lambda_2 = 2$, $\lambda_3 = 3$

11. $\mathbf{x} = \begin{bmatrix} 1 \\ 2 \end{bmatrix}$ 13. $\mathbf{x} = \begin{bmatrix} 1 \\ 4 \end{bmatrix}$ 15. $\mathbf{x} = \begin{bmatrix} 1 \\ -2 \\ 1 \end{bmatrix}$ 17. $\mathbf{x} = \begin{bmatrix} -1 \\ -1 \\ 1 \end{bmatrix}$

19. (a) $\lambda_1 = 1$, $\mathbf{x}_1 = \begin{bmatrix} 1 \\ 2 \end{bmatrix}$; $\lambda_2 = -1$, $\mathbf{x}_2 = \begin{bmatrix} 3 \\ 8 \end{bmatrix}$

(b) $\mathbf{y}_1(t) = e^t \begin{bmatrix} 1 \\ 2 \end{bmatrix}$; $\mathbf{y}_2(t) = e^{-t} \begin{bmatrix} 3 \\ 8 \end{bmatrix}$

(c) Yes

21. (a) $\lambda_1 = -1$, $\mathbf{x}_1 = \begin{bmatrix} -1 \\ 1 \end{bmatrix}$; $\lambda_2 = 1$, $\mathbf{x}_2 = \begin{bmatrix} 1 \\ 1 \end{bmatrix}$

(b) $\mathbf{y}_1(t) = e^{-t} \begin{bmatrix} -1 \\ 1 \end{bmatrix}$; $\mathbf{y}_2(t) = e^t \begin{bmatrix} 1 \\ 1 \end{bmatrix}$

(c) Yes

23. (a) $\lambda_1 = -2, \quad \mathbf{x}_1 = \begin{bmatrix} -2 \\ 1 \end{bmatrix}; \quad \lambda_2 = 1, \quad \mathbf{x}_2 = \begin{bmatrix} -1 \\ 1 \end{bmatrix}$

(b) $\mathbf{y}_1(t) = e^{-2t} \begin{bmatrix} -2 \\ 1 \end{bmatrix}; \quad \mathbf{y}_2(t) = e^t \begin{bmatrix} -1 \\ 1 \end{bmatrix}$

(c) Yes

25. (a) $\lambda_1 = 1, \quad \mathbf{x}_1 = \begin{bmatrix} 0 \\ 0 \\ 1 \end{bmatrix}$

(b) $\lambda_2 = 3, \quad \lambda_3 = 5$

(c) $\mathbf{x}_2 = \begin{bmatrix} 1 \\ 1 \\ 0 \end{bmatrix}, \quad \mathbf{x}_3 = \begin{bmatrix} 1 \\ 2 \\ 0 \end{bmatrix}$

(d) Yes

27. (a) $\lambda_1 = 2, \quad \mathbf{x}_1 = \begin{bmatrix} -1 \\ 1 \\ 0 \end{bmatrix}$

(b) $\lambda_2 = 1, \quad \lambda_3 = -2$

(c) $\mathbf{x}_2 = \begin{bmatrix} -1 \\ 2 \\ 3 \end{bmatrix}, \quad \mathbf{x}_3 = \begin{bmatrix} -1 \\ 5 \\ 6 \end{bmatrix}$

(d) Yes

Section 4.5, page 252

1. The general solution is $\mathbf{y}(t) = \begin{bmatrix} -e^t & -e^{2t} \\ 2e^t & 3e^{2t} \end{bmatrix} \begin{bmatrix} c_1 \\ c_2 \end{bmatrix}$.

   The solution of the initial value problem is $\mathbf{y}(t) = \begin{bmatrix} -2e^{2t} \\ 6e^{2t} \end{bmatrix}$.

3. The general solution is $\mathbf{y}(t) = \begin{bmatrix} -e^{-t} & -e^t \\ 2e^{-t} & 3e^t \end{bmatrix} \begin{bmatrix} c_1 \\ c_2 \end{bmatrix}$.

   The solution of the initial value problem is $\mathbf{y}(t) = \begin{bmatrix} 3e^{-(t-1)} - 2e^{(t-1)} \\ -6e^{-(t-1)} + 6e^{(t-1)} \end{bmatrix}$.

5. The general solution is $\mathbf{y}(t) = \begin{bmatrix} -e^{-t} & e^{3t} \\ e^{-t} & e^{3t} \end{bmatrix} \begin{bmatrix} c_1 \\ c_2 \end{bmatrix}$.

   The solution of the initial value problem is $\mathbf{y}(t) = \begin{bmatrix} 2e^{3(t+1)} \\ 2e^{3(t+1)} \end{bmatrix}$.

7. The general solution is $\mathbf{y}(t) = \begin{bmatrix} e^{5t} & -e^{2t} & -e^{2t} \\ e^{5t} & e^{2t} & 0 \\ e^{5t} & 0 & e^{2t} \end{bmatrix} \begin{bmatrix} c_1 \\ c_2 \\ c_3 \end{bmatrix}$.

   The solution of the initial value problem is $\mathbf{y}(t) = \begin{bmatrix} 2e^{5t} - 3e^{2t} \\ 2e^{5t} - e^{2t} \\ 2e^{5t} + 4e^{2t} \end{bmatrix}$.

9. The general solution is $\mathbf{y}(t) = \begin{bmatrix} 0 & -2e^{-3t} & 2e^{2t} \\ 0 & 2e^{-3t} & -e^{2t} \\ e^t & e^{-3t} & 5e^{2t} \end{bmatrix} \begin{bmatrix} c_1 \\ c_2 \\ c_3 \end{bmatrix}$.

   The solution of the initial value problem is $\mathbf{y}(t) = \begin{bmatrix} -2e^{-3t} - 4e^{2t} \\ 2e^{-3t} + 2e^{2t} \\ e^t + e^{-3t} - 10e^{2t} \end{bmatrix}$.

13. (a) $\begin{bmatrix} Q_1' \\ Q_2' \end{bmatrix} = \dfrac{r}{V} \begin{bmatrix} -2 & 1 \\ 1 & -2 \end{bmatrix} \begin{bmatrix} Q_1 \\ Q_2 \end{bmatrix}$

 (b) $\mathbf{Q}(t) = \begin{bmatrix} -e^{(-3r/V)t} & e^{(-r/V)t} \\ e^{(-3r/V)t} & e^{(-r/V)t} \end{bmatrix} \begin{bmatrix} c_1 \\ c_2 \end{bmatrix}$

 (c) $t \approx 250.55$ sec, or about 4.18 min

15. (b) $\hat{\mathbf{y}}(2) = \begin{bmatrix} e^3 - 2e^{-3} \\ 3e^3 + e^{-3} \end{bmatrix}$

17. (a) Direction Field 4   (b) Direction Field 3   (c) Direction Field 2

 (d) Direction Field 1

19. The solution is $\mathbf{y}(t) = \begin{bmatrix} 0 \\ -2e^t \end{bmatrix}$. The phase plane solution trajectory is

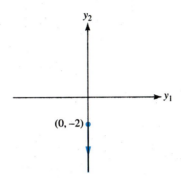

21. The solution is $\mathbf{y}(t) = \begin{bmatrix} 4 - e^{-t} \\ 2 + e^{-t} \end{bmatrix}$. The phase plane solution trajectory is

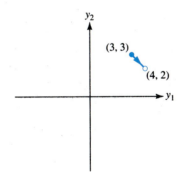

Section 4.6, page 263

1. $\lambda_1 = 2 + i$, $\mathbf{x}_1 = \begin{bmatrix} -i \\ 1 \end{bmatrix}$; $\quad \lambda_2 = 2 - i$, $\mathbf{x}_2 = \begin{bmatrix} i \\ 1 \end{bmatrix}$

3. $\lambda_1 = -1 + i$, $\mathbf{x}_1 = \begin{bmatrix} 1 - i \\ 2i \end{bmatrix}$; $\quad \lambda_2 = -1 - i$, $\mathbf{x}_2 = \begin{bmatrix} 1 + i \\ -2i \end{bmatrix}$

5. $\lambda_1 = -2 + i$, $\mathbf{x}_1 = \begin{bmatrix} -1 - 3i \\ 5i \end{bmatrix}$; $\quad \lambda_2 = -2 - i$, $\mathbf{x}_2 = \begin{bmatrix} -1 + 3i \\ -5i \end{bmatrix}$

7. $\lambda_1 = 2$, $\mathbf{x}_1 = \begin{bmatrix} 0 \\ 0 \\ 1 \end{bmatrix}$; $\lambda_2 = -1 + 0.5i$, $\mathbf{x}_2 = \begin{bmatrix} i \\ 1 \\ 0 \end{bmatrix}$; $\lambda_3 = -1 - 0.5i$, $\mathbf{x}_3 = \begin{bmatrix} -i \\ 1 \\ 0 \end{bmatrix}$

9. $\lambda_1 = -1$, $\mathbf{x}_1 = \begin{bmatrix} -3 \\ 0 \\ 1 \end{bmatrix}$; $\lambda_2 = -1 + i$, $\mathbf{x}_2 = \begin{bmatrix} -5 - i \\ -1 - i \\ 2 \end{bmatrix}$; $\lambda_3 = -1 - i$, $\mathbf{x}_3 = \begin{bmatrix} -5 + i \\ -1 + i \\ 2 \end{bmatrix}$

11. $\mathbf{y}_1(t) = e^{4t} \begin{bmatrix} 4\cos 2t \\ -\cos 2t - \sin 2t \end{bmatrix}$, $\mathbf{y}_2(t) = e^{4t} \begin{bmatrix} 4\sin 2t \\ -\sin 2t + \cos 2t \end{bmatrix}$

13. $\mathbf{y}_1(t) = \begin{bmatrix} -\cos 2t + \sin 2t \\ \cos 2t \end{bmatrix}$, $\mathbf{y}_2(t) = \begin{bmatrix} -\cos 2t - \sin 2t \\ \sin 2t \end{bmatrix}$

15. $\mathbf{y}_1(t) = e^{2t} \begin{bmatrix} 1 \\ 0 \\ -1 \end{bmatrix}$, $\mathbf{y}_2(t) = e^{2t} \begin{bmatrix} -5\cos 3t - 3\sin 3t \\ 3\cos 3t - 3\sin 3t \\ 2\cos 3t \end{bmatrix}$, $\mathbf{y}_3(t) = e^{2t} \begin{bmatrix} -5\sin 3t + 3\cos 3t \\ 3\sin 3t + 3\cos 3t \\ 2\sin 3t \end{bmatrix}$

17. $\mathbf{y}(t) = e^{2t} \begin{bmatrix} 7\sin t + 4\cos t \\ 7\cos t - 4\sin t \end{bmatrix}$ 19. $\mathbf{y}(t) = e^{-t} \begin{bmatrix} 2\cos t + 4\sin t \\ 2\cos t - 6\sin t \end{bmatrix}$

21. $\mathbf{y}(t) = e^{-2t} \begin{bmatrix} 4\sin t \\ -2\cos t - 6\sin t \end{bmatrix}$ 23. $\mathbf{y}(t) = \begin{bmatrix} e^{-t}[-3\sin(t/2) + 2\cos(t/2)] \\ e^{-t}[3\cos(t/2) + 2\sin(t/2)] \\ -e^{2t} \end{bmatrix}$

25. $\mathbf{y}(t) = e^{-t} \begin{bmatrix} 54 - 42\cos t + 76\sin t \\ 2\cos t + 24\sin t \\ -18 + 22\cos t - 26\sin t \end{bmatrix}$

29. (a) Real when $1 + 4\mu > 0$ and complex when $1 + 4\mu < 0$
  (b) For $\mu < 6$

31. (a) Real for all μ (b) For no values of μ

33. It moves around the origin on an elliptical orbit.

35. It moves around the origin on an elliptical orbit.

Section 4.7, page 274

1. (a) $\lambda_1 = \lambda_2 = 1$, $\mathbf{x}_1 = \begin{bmatrix} 1 \\ -1 \end{bmatrix}$ (c) $\mathbf{y}(t) = \begin{bmatrix} 3e^t + 2te^t \\ -e^t - 2te^t \end{bmatrix}$

3. (a) $\lambda_1 = \lambda_2 = -2$, $\mathbf{x}_1 = \begin{bmatrix} 1 \\ 0 \end{bmatrix}$ (c) $\mathbf{y}(t) = \begin{bmatrix} e^{-2t} - te^{-2t} \\ -e^{-2t} \end{bmatrix}$

5. (a) $\lambda_1 = \lambda_2 = 6$, $\mathbf{x}_1 = \begin{bmatrix} 0 \\ 1 \end{bmatrix}$ (c) $\mathbf{y}(t) = \begin{bmatrix} -2e^{6t} \\ -4te^{6t} \end{bmatrix}$

7. (a) $\lambda_1 = \lambda_2 = -3$, $\mathbf{x}_1 = \begin{bmatrix} 1 \\ 1 \end{bmatrix}$ (c) $\mathbf{y}(t) = e^{-(3+3t)} \begin{bmatrix} -t \\ -t - 1 \end{bmatrix}$

9. (a) $\lambda_1 = \lambda_2 = 2$, $\lambda_3 = 1$, $\mathbf{x}_1 = \begin{bmatrix} 1 \\ 0 \\ 0 \end{bmatrix}$, $\mathbf{x}_3 = \begin{bmatrix} 0 \\ 0 \\ 1 \end{bmatrix}$ (c) $\mathbf{y}(t) = \begin{bmatrix} e^{2t}(1 + 3t) \\ 3e^{2t} \\ -2e^t \end{bmatrix}$

11. (a) $\lambda_1 = \lambda_2 = 3$, $\lambda_3 = 0$, $\mathbf{x}_1 = \begin{bmatrix} 0 \\ 1 \\ 2 \end{bmatrix}$, $\mathbf{x}_3 = \begin{bmatrix} 1 \\ 0 \\ 0 \end{bmatrix}$ (c) $\mathbf{y}(t) = \begin{bmatrix} 4 \\ e^{3t}(1+t) \\ e^{3t}(1+2t) \end{bmatrix}$

13. (a) $\lambda_1 = \lambda_2 = \lambda_3 = 2$, $\mathbf{x} = \begin{bmatrix} 1 \\ 0 \\ 0 \end{bmatrix}$, geometric multiplicity is 1.

(b) $\lambda_1 = \lambda_2 = \lambda_3 = 2$, $\mathbf{x}_1 = \begin{bmatrix} 1 \\ 0 \\ 0 \end{bmatrix}$, $\mathbf{x}_2 = \begin{bmatrix} 0 \\ 0 \\ 1 \end{bmatrix}$, geometric multiplicity is 2.

15. (a) $y_3 = c_3 e^{2t}$, $y_2 = c_2 e^{2t}$, $y_1 = c_1 e^{2t} + c_2 t e^{2t}$

(b) $\mathbf{y}(t) = \begin{bmatrix} e^{2t} & te^{2t} & 0 \\ 0 & e^{2t} & 0 \\ 0 & 0 & e^{2t} \end{bmatrix} \begin{bmatrix} c_1 \\ c_2 \\ c_3 \end{bmatrix}$, $W(t) = e^{6t}$, and therefore $\Psi(t)$ is a fundamental matrix.

17. $\lambda_1 = \lambda_2 = \lambda_3 = 5$, $\mathbf{x}_1 = \begin{bmatrix} 0 \\ 0 \\ 1 \end{bmatrix}$. Thus, geometric multiplicity is 1, and the matrix does not have a full set of eigenvectors.

19. $\lambda_1 = \lambda_2 = \lambda_3 = 5$, $\mathbf{x}_1 = \begin{bmatrix} 1 \\ 0 \\ 0 \end{bmatrix}$, $\mathbf{x}_2 = \begin{bmatrix} 0 \\ 1 \\ 0 \end{bmatrix}$. Thus, geometric multiplicity is 2, and the matrix does not have a full set of eigenvectors.

21. The eigenvalues are $\lambda = 2$ (algebraic multiplicity 2, geometric multiplicity 1) and $\lambda = 3$ (algebraic multiplicity 2, geometric multiplicity 1). A does not have a full set of eigenvectors.

23. The eigenvalues are $\lambda = 2$ (algebraic multiplicity 3, geometric multiplicity 3) and $\lambda = 3$ (algebraic multiplicity 1, geometric multiplicity 1). A does have a full set of eigenvectors.

25. $A = \begin{bmatrix} 1 & 4 \\ -1 & -3 \end{bmatrix}$ 27. $A = \begin{bmatrix} 2 & -1 \\ 1 & 0 \end{bmatrix}$

29. (a) Direction Field 3 (b) Direction Field 4 (c) Direction Field 1
(d) Direction Field 2

31. $A = \begin{bmatrix} 1.5 & 0 \\ 0 & 1.5 \end{bmatrix}$ 33. $A = \begin{bmatrix} 0 & 0 \\ 0 & 0 \end{bmatrix}$

35. $\mathbf{y}_1(t) = \begin{bmatrix} 0 \\ 0 \\ e^{4t} \end{bmatrix}$, $\mathbf{y}_2(t) = \begin{bmatrix} 0 \\ e^{4t} \\ 3te^{4t} \end{bmatrix}$, $\mathbf{y}_3(t) = \begin{bmatrix} 3e^{4t} \\ (-1+6t)e^{4t} \\ 9t^2 e^{4t} \end{bmatrix}$

Section 4.8, page 285

1. (c) $\mathbf{y}(t) = \begin{bmatrix} -e^{-3t} & e^{-t} \\ e^{-3t} & e^{-t} \end{bmatrix} \begin{bmatrix} c_1 \\ c_2 \end{bmatrix} + \begin{bmatrix} 1 \\ 1 \end{bmatrix}$ (d) $\mathbf{y}(t) = \begin{bmatrix} 1 + e^{-t} + e^{-3t} \\ 1 + e^{-t} - e^{-3t} \end{bmatrix}$

3. (c) $\mathbf{y}(t) = \begin{bmatrix} e^{3t} & e^{t} \\ e^{3t} & -e^{t} \end{bmatrix} \begin{bmatrix} c_1 \\ c_2 \end{bmatrix} + \frac{e^{-t}}{8} \begin{bmatrix} -3 \\ 1 \end{bmatrix}$ (d) $\mathbf{y}(t) = \frac{1}{8} \begin{bmatrix} e^{3t} + 2e^{t} - 3e^{-t} \\ e^{3t} - 2e^{t} + e^{-t} \end{bmatrix}$

5. (c) $\mathbf{y}(t) = \begin{bmatrix} e^t & -e^{-t} \\ e^t & e^{-t} \end{bmatrix} \begin{bmatrix} c_1 \\ c_2 \end{bmatrix} + \begin{bmatrix} 0 \\ -t \end{bmatrix}$ (d) $\mathbf{y}(t) = \dfrac{1}{2} \begin{bmatrix} e^t + 3e^{-t} \\ e^t - 3e^{-t} - 2t \end{bmatrix}$

7. (c) $\mathbf{y}(t) = \begin{bmatrix} -e^t & -e^{-t} \\ 2e^t & e^{-t} \end{bmatrix} \begin{bmatrix} c_1 \\ c_2 \end{bmatrix} + \dfrac{1}{2} \begin{bmatrix} 3\sin t - \cos t \\ -4\sin t \end{bmatrix}$

(d) $\mathbf{y}(t) = \dfrac{1}{2} \begin{bmatrix} -e^t + 2e^{-t} + 3\sin t - \cos t \\ 2e^t - 2e^{-t} - 4\sin t \end{bmatrix}$

9. (c) $\mathbf{y}(t) = \begin{bmatrix} e^t & 0 & te^t \\ 0 & 0 & e^t \\ 0 & e^t & 0 \end{bmatrix} \begin{bmatrix} c_1 \\ c_2 \\ c_3 \end{bmatrix} + \begin{bmatrix} t+1 \\ -t-1 \\ 0 \end{bmatrix}$ (d) $\mathbf{y}(t) = \begin{bmatrix} te^t + t + 1 \\ e^t - t - 1 \\ 2e^t \end{bmatrix}$

11. $\mathbf{y}_0 = \begin{bmatrix} 1 \\ e^{\pi/2} - 1 \end{bmatrix}$, $\mathbf{g}(t) = \begin{bmatrix} -2e^t \\ e^t + 2 \end{bmatrix}$ 13. $P(t) = \begin{bmatrix} 1 & e^t \\ 0 & -1 \end{bmatrix}$

15. $\mathbf{y}(t) = \dfrac{1}{4} \begin{bmatrix} e^{2t} + 2te^{2t} - 1 \\ -e^{2t} + 2te^{2t} + 1 \end{bmatrix}$ 17. $\mathbf{y}(t) = \dfrac{1}{4} \begin{bmatrix} 3e^{2t} + 2t + 1 \\ -3e^{2t} + 2t - 1 \end{bmatrix}$

19. $\mathbf{y}(t) = \begin{bmatrix} 1 - \cos t + 3\sin t \\ -2 + 3\cos t + \sin t \end{bmatrix}$ 21. $\mathbf{y}(t) = \dfrac{1}{2} \begin{bmatrix} -\sin 2t \\ 1 + \cos 2t \end{bmatrix}$

23. $\mathbf{y}(t) \equiv \begin{bmatrix} 10 \\ -3 \end{bmatrix}$ 25. $\mathbf{y}(t) \equiv \begin{bmatrix} 0 \\ 2 \end{bmatrix} + c \begin{bmatrix} 1 \\ 1 \end{bmatrix}$

27. $\mathbf{y}(t) \equiv \begin{bmatrix} 2 \\ 0 \\ 0 \end{bmatrix} + c \begin{bmatrix} 0 \\ -1 \\ 1 \end{bmatrix}$ 29. $\Psi(t) = \begin{bmatrix} \sin 2t & -\cos 2t - \sin 2t \\ \cos 2t & -\cos 2t + \sin 2t \end{bmatrix}$

31. $\Psi(t) = \begin{bmatrix} -e^{-t} + 2e^t & 2e^{-t} - 2e^t \\ -e^{-t} + e^t & 2e^{-t} - e^t \end{bmatrix}$

33. (a) $\mathbf{Q}' = \dfrac{1}{100} \begin{bmatrix} -2 & 1 \\ 1 & -2 \end{bmatrix} \mathbf{Q} + \begin{bmatrix} 5 \\ 0 \end{bmatrix}$

(b) $\mathbf{Q}(t) = \dfrac{1}{3} \begin{bmatrix} -250e^{-3t/100} - 750e^{-t/100} + 1000 \\ 250e^{-3t/100} - 750e^{-t/100} + 500 \end{bmatrix}$, $\mathbf{Q}(t) \to \dfrac{1}{3} \begin{bmatrix} 1000 \\ 500 \end{bmatrix}$

35. (a) $\mathbf{Q}' = \dfrac{1}{100} \begin{bmatrix} -2 & 1 \\ 1 & -2 \end{bmatrix} \mathbf{Q} + 5e^{-2t/100} \begin{bmatrix} 1 \\ 0 \end{bmatrix}$

(b) $\mathbf{Q}(t) = \begin{bmatrix} -260e^{-3t/100} + 260e^{-t/100} \\ 260e^{-3t/100} + 260e^{-t/100} - 500e^{-2t/100} \end{bmatrix}$, $\mathbf{Q}(t) \to \begin{bmatrix} 0 \\ 0 \end{bmatrix}$

37. (a) $\mathbf{I}' = \begin{bmatrix} -6 & 4 \\ 4 & -6 \end{bmatrix} \mathbf{I} + \begin{bmatrix} 4e^{-2t} \\ 0 \end{bmatrix}$, $\mathbf{I}(0) = \mathbf{0}$ (b) $\Psi(t) = \begin{bmatrix} e^{-2t} & e^{-10t} \\ e^{-2t} & -e^{-10t} \end{bmatrix}$

(c) $\mathbf{I}(t) = \begin{bmatrix} 2te^{-2t} + (e^{-2t} - e^{-10t})/4 \\ 2te^{-2t} - (e^{-2t} - e^{-10t})/4 \end{bmatrix}$ mA

Section 4.9, page 297

1. (a) $\mathbf{y}_{k+1} = \mathbf{y}_k + h \left\{ \begin{bmatrix} 1 & 2 \\ 2 & 3 \end{bmatrix} \mathbf{y}_k + \begin{bmatrix} 1 \\ 1 \end{bmatrix} \right\}$, $\mathbf{y}_0 = \begin{bmatrix} -1 \\ 1 \end{bmatrix}$

(b) $t_k = k(0.01)$; $k = 0, \ldots, 100$

(c) $\mathbf{y}_0 = \begin{bmatrix} -1 \\ 1 \end{bmatrix}$, $\mathbf{y}_1 = \begin{bmatrix} -0.98 \\ 1.02 \end{bmatrix}$, $\mathbf{y}_2 = \begin{bmatrix} -0.9594 \\ 1.041 \end{bmatrix}$

3. (a) $\mathbf{y}_{k+1} = \mathbf{y}_k + h \left\{ \begin{bmatrix} -t_k^2 & t_k \\ 2 - t_k & 0 \end{bmatrix} \mathbf{y}_k + \begin{bmatrix} 1 \\ t_k \end{bmatrix} \right\}$, $\mathbf{y}_0 = \begin{bmatrix} 2 \\ 0 \end{bmatrix}$

   (b) $t_k = 1 + k(0.01)$; $k = 0, 1, \ldots, 300$

   (c) $\mathbf{y}_0 = \begin{bmatrix} 2 \\ 0 \end{bmatrix}$, $\mathbf{y}_1 = \begin{bmatrix} 1.99 \\ 0.03 \end{bmatrix}$, $\mathbf{y}_2 = \begin{bmatrix} 1.9800\ldots \\ 0.0598\ldots \end{bmatrix}$

5. (a) $\mathbf{y}_{k+1} = \mathbf{y}_k + h \left\{ \begin{bmatrix} -t_k^{-1} & \sin t_k \\ 1 - t_k & 1 \end{bmatrix} \mathbf{y}_k + \begin{bmatrix} 0 \\ t_k^2 \end{bmatrix} \right\}$, $\mathbf{y}_0 = \begin{bmatrix} 0 \\ 0 \end{bmatrix}$

   (b) $t_k = 1 + k(0.01)$; $k = 0, 1, \ldots, 500$

   (c) $\mathbf{y}_0 = \begin{bmatrix} 0 \\ 0 \end{bmatrix}$, $\mathbf{y}_1 = \begin{bmatrix} 0.00 \\ 0.01 \end{bmatrix}$, $\mathbf{y}_2 = \begin{bmatrix} 0.00008\ldots \\ 0.02030\ldots \end{bmatrix}$

7. (a) $\mathbf{y}' = \begin{bmatrix} 0 & 1 \\ -t^2 & -1 \end{bmatrix} \mathbf{y} + \begin{bmatrix} 0 \\ 2 \end{bmatrix}$, $\mathbf{y}(0) = \begin{bmatrix} 1 \\ 1 \end{bmatrix}$

   (c) $\mathbf{y}_0 = \begin{bmatrix} 1 \\ 1 \end{bmatrix}$, $\mathbf{y}_1 = \begin{bmatrix} 1.01 \\ 1.01 \end{bmatrix}$, $\mathbf{y}_2 = \begin{bmatrix} 1.0201 \\ 1.0199 \end{bmatrix}$

9. (a) $\mathbf{y}' = \begin{bmatrix} 0 & 1 \\ -e^{-t} & -1 \end{bmatrix} \mathbf{y} + \begin{bmatrix} 0 \\ 2 \end{bmatrix}$, $\mathbf{y}(0) = \begin{bmatrix} -1 \\ 1 \end{bmatrix}$

   (c) $\mathbf{y}_0 = \begin{bmatrix} -1 \\ 1 \end{bmatrix}$, $\mathbf{y}_1 = \begin{bmatrix} -0.99 \\ 1.02 \end{bmatrix}$, $\mathbf{y}_2 = \begin{bmatrix} -0.9798 \\ 1.0396\ldots \end{bmatrix}$

19. (c) $\mathbf{y}(1) - \bar{\mathbf{y}}_{2n} = \begin{bmatrix} 0.00768\ldots \\ 0.00584\ldots \end{bmatrix}$, $\bar{\mathbf{y}}_{2n} - \mathbf{y}_n = \begin{bmatrix} 0.00762\ldots \\ 0.00577\ldots \end{bmatrix}$

21. (c) $\mathbf{y}(1) - \bar{\mathbf{y}}_{2n} = \begin{bmatrix} -0.1718\ldots \\ -0.0624\ldots \end{bmatrix}$, $\bar{\mathbf{y}}_{2n} - \mathbf{y}_n = \begin{bmatrix} -0.1693\ldots \\ -0.0583\ldots \end{bmatrix}$

23. (a) $\mathbf{Q}' = \begin{bmatrix} -15/(200 - 10t) & 5/(500 - 20t) \\ 15/(200 - 10t) & -35/(500 - 20t) \end{bmatrix} \mathbf{Q}$, $\mathbf{Q}(0) = \begin{bmatrix} 40 \\ 40 \end{bmatrix}$

25. Solid curve is $y_1(t)$,
    dashed curve is $y_2(t)$.

27. Solid curve is $y_1(t)$,
    dashed curve is $y_2(t)$.

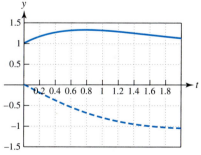

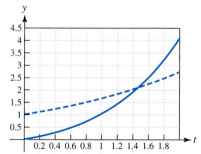

29. $\mathbf{y}(1) - \mathbf{y}_{100} = \begin{bmatrix} 0.1399\ldots \times 10^{-8} \\ 0.1401\ldots \times 10^{-8} \end{bmatrix}$, $\mathbf{y}(1) - \mathbf{y}_{200} = \begin{bmatrix} 0.8802\ldots \times 10^{-10} \\ 0.8812\ldots \times 10^{-10} \end{bmatrix}$. Dividing

these vectors componentwise, we obtain the vector $\begin{bmatrix} 0.062892\ldots \\ 0.062890\ldots \end{bmatrix}$. Note that

$\left(\frac{1}{2}\right)^4 = 0.0625$ and, therefore, the errors are reduced by a factor of about $\left(\frac{1}{2}\right)^4$ when the step size h is cut in half.

Section 4.10, page 308

1. (a) $T = \begin{bmatrix} 2 & 1 \\ 1 & 1 \end{bmatrix}$, $D = \begin{bmatrix} 2 & 0 \\ 0 & -1 \end{bmatrix}$ (b) $e^{At} = \begin{bmatrix} 2e^{2t} - e^{-t} & -2e^{2t} + 2e^{-t} \\ e^{2t} - e^{-t} & -e^{2t} + 2e^{-t} \end{bmatrix}$

3. (a) $T = \begin{bmatrix} -1 & 1 \\ 1 & 1 \end{bmatrix}$, $D = \begin{bmatrix} 0 & 0 \\ 0 & 2 \end{bmatrix}$ (b) $e^{At} = \frac{1}{2}\begin{bmatrix} 1 + e^{2t} & -1 + e^{2t} \\ -1 + e^{2t} & 1 + e^{2t} \end{bmatrix}$

5. (a) $T = \begin{bmatrix} 1 & -1 \\ 1 & 1 \end{bmatrix}$, $D = \begin{bmatrix} 5 & 0 \\ 0 & -1 \end{bmatrix}$ (b) $e^{At} = \frac{1}{2}\begin{bmatrix} e^{5t} + e^{-t} & e^{5t} - e^{-t} \\ e^{5t} - e^{-t} & e^{5t} + e^{-t} \end{bmatrix}$

7. (a) $T = \begin{bmatrix} 1 & 0 \\ 1 & 1 \end{bmatrix}$, $D = \begin{bmatrix} 2 & 0 \\ 0 & 1 \end{bmatrix}$ (b) $e^{At} = \begin{bmatrix} e^{2t} & 0 \\ e^{2t} - e^{t} & e^{t} \end{bmatrix}$

9. (a) $T = \begin{bmatrix} 1 & 1 \\ i & -i \end{bmatrix}$, $D = \begin{bmatrix} 2i & 0 \\ 0 & -2i \end{bmatrix}$ (b) $e^{At} = \begin{bmatrix} \cos 2t & \sin 2t \\ -\sin 2t & \cos 2t \end{bmatrix}$

11. (a) $T = \begin{bmatrix} 2 & 3 \\ 1 & 2 \end{bmatrix}$. The uncoupled system is $\mathbf{z}' = \begin{bmatrix} 3 & 0 \\ 0 & 2 \end{bmatrix}\mathbf{z}$, $\mathbf{z}(0) = \begin{bmatrix} 11 \\ -7 \end{bmatrix}$.

   (b) $\mathbf{y}(t) = T\mathbf{z}(t) = \begin{bmatrix} 22e^{3t} - 21e^{2t} \\ 11e^{3t} - 14e^{2t} \end{bmatrix}$

13. (a) $T = \begin{bmatrix} -1 & -2 \\ 1 & 1 \end{bmatrix}$. The uncoupled system is $\mathbf{z}' = \begin{bmatrix} 2 & 0 \\ 0 & -1 \end{bmatrix}\mathbf{z} + \begin{bmatrix} -e^{2t} \\ 0 \end{bmatrix}$, $\mathbf{z}(0) = \mathbf{0}$.

   (b) $\mathbf{y}(t) = T\mathbf{z}(t) = \begin{bmatrix} te^{2t} \\ -te^{2t} \end{bmatrix}$

15. (a) $T = \begin{bmatrix} -1 & -5 & 0 \\ 1 & 8 & 0 \\ 0 & 0 & 1 \end{bmatrix}$, giving $\mathbf{z}' = \begin{bmatrix} -4 & 0 & 0 \\ 0 & -1 & 0 \\ 0 & 0 & 1 \end{bmatrix}\mathbf{z} + \begin{bmatrix} -\frac{5}{3} \\ \frac{1}{3} \\ 2 \end{bmatrix}$, $\mathbf{z}(0) = \begin{bmatrix} -\frac{8}{3} \\ \frac{1}{3} \\ 0 \end{bmatrix}$.

   (b) $\mathbf{y}(t) = T\mathbf{z}(t) = \frac{1}{4}\begin{bmatrix} 9e^{-4t} - 5 \\ -9e^{-4t} + 9 \\ 8e^{t} - 8 \end{bmatrix}$

17. By (8), $\mathbf{y}(t + \Delta t) = e^{A\Delta t}\mathbf{y}(t)$. By Example 2, $e^{A\Delta t} = \begin{bmatrix} e^{2\Delta t} & \Delta te^{2\Delta t} \\ 0 & e^{2\Delta t} \end{bmatrix}$. Therefore, $\mathbf{y}(4) = \begin{bmatrix} e^{6} & 3e^{6} \\ 0 & e^{6} \end{bmatrix}\mathbf{y}(1) = e^{6}\begin{bmatrix} 7 \\ 2 \end{bmatrix}$. Similarly, $\mathbf{y}(-1) = e^{-4}\begin{bmatrix} -3 \\ 2 \end{bmatrix}$.

23. (a) $AB \neq BA$

    (b) $e^{At}e^{Bt} \neq e^{(A+B)t}$

25. (a) $T = \begin{bmatrix} 1 & 1 \\ 1 & -1 \end{bmatrix}$. The uncoupled system is $\mathbf{z}'' + \begin{bmatrix} 3 & 0 \\ 0 & 1 \end{bmatrix}\mathbf{z} = \mathbf{0}$.

    (b) $\mathbf{y}(t) = T\mathbf{z}(t) = \left(c_1 \cos \sqrt{3}\,t + c_2 \sin \sqrt{3}\,t\right)\begin{bmatrix} 1 \\ 1 \end{bmatrix} + (c_3 \cos t + c_4 \sin t)\begin{bmatrix} 1 \\ -1 \end{bmatrix}$

27. (a) $\mathbf{z}'' + \begin{bmatrix} 3 & 0 \\ 0 & 1 \end{bmatrix}\mathbf{z} = \frac{1}{2}\begin{bmatrix} 1 \\ 1 \end{bmatrix}$, $\mathbf{z}(0) = \frac{1}{2}\begin{bmatrix} 1 \\ 1 \end{bmatrix}$, $\mathbf{z}'(0) = \frac{1}{2}\begin{bmatrix} 1 \\ -1 \end{bmatrix}$

    (b) $\mathbf{y}(t) = \begin{bmatrix} (1/3)\cos\sqrt{3}\,t + (1/2\sqrt{3})\sin\sqrt{3}\,t - (1/2)\sin t + (2/3) \\ (1/3)\cos\sqrt{3}\,t + (1/2\sqrt{3})\sin\sqrt{3}\,t + (1/2)\sin t - (1/3) \end{bmatrix}$

29. (a) $\mathbf{z}'' + \begin{bmatrix} 6 & 0 \\ 0 & 2 \end{bmatrix} \mathbf{z}' = \frac{1}{2} \begin{bmatrix} 1 \\ -1 \end{bmatrix}$ (b) $\mathbf{y}(t) = \begin{bmatrix} c_1 + c_2 e^{-6t} + c_3 + c_4 e^{-2t} - t/6 \\ c_1 + c_2 e^{-6t} - c_3 - c_4 e^{-2t} + t/3 \end{bmatrix}$

Chapter 4 Review Exercises, page 310

1. $\mathbf{y}(t) = \begin{bmatrix} 2e^t & e^{-t} \\ e^t & e^{-t} \end{bmatrix} \begin{bmatrix} c_1 \\ c_2 \end{bmatrix}$

3. $\mathbf{y}(t) = \begin{bmatrix} \cos 4t & -\sin 4t \\ \sin 4t & \cos 4t \end{bmatrix} \begin{bmatrix} c_1 \\ c_2 \end{bmatrix}$

5. $\mathbf{y}(t) = \begin{bmatrix} e^{2t} & (t+1)e^{2t} \\ e^{2t} & te^{2t} \end{bmatrix} \begin{bmatrix} c_1 \\ c_2 \end{bmatrix}$

7. $\mathbf{y}(t) = \begin{bmatrix} 2e^t & e^{2t} \\ e^t & e^{2t} \end{bmatrix} \begin{bmatrix} c_1 \\ c_2 \end{bmatrix} - \begin{bmatrix} 2 \\ 3 \end{bmatrix}$

9. $\mathbf{y}(t) = \begin{bmatrix} e^{4t} & -e^t & -e^t \\ e^{4t} & e^t & 0 \\ e^{4t} & 0 & e^t \end{bmatrix} \begin{bmatrix} c_1 \\ c_2 \\ c_3 \end{bmatrix}$

11. The general solution is $\mathbf{y}(t) = \begin{bmatrix} 2e^{-t} & e^{2t} \\ e^{-t} & e^{2t} \end{bmatrix} \begin{bmatrix} c_1 \\ c_2 \end{bmatrix}$. The solution of the initial value problem is $\mathbf{y}(t) = \begin{bmatrix} 4e^{-t} + 3e^{2t} \\ 2e^{-t} + 3e^{2t} \end{bmatrix}$.

13. The general solution is $\mathbf{y}(t) = \begin{bmatrix} 5\cos 2t & 5\sin 2t \\ 3\cos 2t - \sin 2t & \cos 2t + 3\sin 2t \end{bmatrix} \begin{bmatrix} c_1 \\ c_2 \end{bmatrix}$. The solution of the initial value problem is $\mathbf{y}(t) = \begin{bmatrix} 5\cos 2t + 10\sin 2t \\ 5\cos 2t + 5\sin 2t \end{bmatrix}$.

15. $\mathbf{y}(t) = \begin{bmatrix} e^{3t} & e^{3t} & -1 \\ -e^{3t} & 0 & -1 \\ 0 & e^{3t} & 1 \end{bmatrix} \begin{bmatrix} c_1 \\ c_2 \\ c_3 \end{bmatrix}$

17. $\mathbf{y}(t) = \begin{bmatrix} 2e^{3t} & (2t-1)e^{3t} \\ e^{3t} & te^{3t} \end{bmatrix} \begin{bmatrix} c_1 \\ c_2 \end{bmatrix} + \begin{bmatrix} 1 \\ -2 \end{bmatrix}$

19. $\mathbf{y}(t) = \begin{bmatrix} 2e^{2t} & (2t-1)e^{2t} \\ e^{2t} & te^{2t} \end{bmatrix} \begin{bmatrix} c_1 \\ c_2 \end{bmatrix}$

21. $\mathbf{y}(t) = \begin{bmatrix} e^{-t} & (t-1)e^{-t} \\ e^{-t} & te^{-t} \end{bmatrix} \begin{bmatrix} c_1 \\ c_2 \end{bmatrix}$

CHAPTER 5

Section 5.1, page 327

1. $\mathcal{L}\{1\} = 1/s, \quad s > 0$

3. $\mathcal{L}\{te^{-t}\} = 1/(s+1)^2, \quad s > -1$

5. The Laplace transform does not exist.

7. $\mathcal{L}\{|t-1|\} = 2e^{-s}s^{-2} + s^{-1} - s^{-2}, \quad s > 0$

9. $\mathcal{L}\{f(t)\} = e^{-s}s^{-1}, \quad s > 0$

11. $\mathcal{L}\{f(t)\} = e^{-s}s^{-1} - e^{-2s}s^{-1}, \quad s \neq 0; \quad \mathcal{L}\{f(t)\} = 1, \quad s = 0$

15. (b) $\mathcal{L}\{t^2\} = 2/s^3, \quad \mathcal{L}\{t^3\} = 6/s^4, \quad \mathcal{L}\{t^4\} = 24/s^5, \quad \mathcal{L}\{t^5\} = 120/s^6, \quad s > 0$

   (c) $\mathcal{L}\{t^m\} = m!/s^{m+1}, \quad s > 0$

17. $\mathcal{L}\{\sin \omega t\} = \omega/(s^2 + \omega^2), \quad s > 0$

19. $\mathcal{L}\{\sin[\omega(t-2)]\} = (\omega \cos 2\omega - s \sin 2\omega)/(s^2 + \omega^2), \quad s > 0$

21. $\mathcal{L}\{e^{-2t} \cos 4t\} = (s+2)/[(s+2)^2 + 16], \quad s > -2$

23. $R(s) = 5\mathcal{L}\{e^{-7t}\} + \mathcal{L}\{t\} + 2\mathcal{L}\{e^{2t}\} = 5/(s+7) + (1/s^2) + 2/(s-2), \quad s > 2$

25. f is continuous and exponentially bounded. $M = 1, a = 1$

27. f is continuous and exponentially bounded. $M = 1, a = 2$ (since $\cosh 2t \leq e^{2t}$)

29. f is piecewise continuous and exponentially bounded. $M = 1$, $a = 2$

31. f is neither continuous nor exponentially bounded.

33. The integral diverges.   35. The integral converges to $\frac{1}{2}$.

37. $\mathcal{L}^{-1}\{F(s)\} = -2t + e^{-t}$, $t \geq 0$ 39. $\mathcal{L}^{-1}\{F(s)\} = e^{t} - e^{-t}$, $t \geq 0$

Section 5.2, page 341

1. $F(s) = (s^2 + 2s + 6)/s^3$, $s > 0$ 3. $F(s) = s^{-1} + 3/(9 + s^2)$, $s > 0$

5. $F(s) = e^{-s}(2/s^3)$, $s > 0$ 7. $F(s) = 2/(s + 2)^2$, $s > -2$

9. $F(s) = e^{-2s}(2s^{-2} + 4s^{-1})$, $s > 0$ 11. $F(s) = e^{3-s}/(s - 3)$, $s > 3$

13. $\mathcal{L}^{-1}\{F(s)\} = 3 + 4t^3$, $t \geq 0$ 15. $\mathcal{L}^{-1}\{F(s)\} = 2e^{2t}\cos 3t$, $t \geq 0$

17. $\mathcal{L}^{-1}\{F(s)\} = h(t - 2)\sin[3(t - 2)]$, $t \geq 0$

19. $\mathcal{L}^{-1}\{F(s)\} = \frac{2}{3}e^{t}[6\cos 3t - \sin 3t]$, $t \geq 0$

21. $\mathcal{L}^{-1}\{F(s)\} = 2(t - 3)^4 h(t - 3) + 4(t - 5)^4 h(t - 5)$, $t \geq 0$

23. f

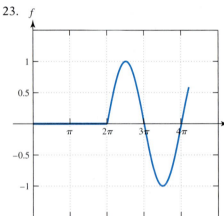

$$F(s) = e^{-2\pi s}/(s^2 + 1), \quad s > 0$$

25. f

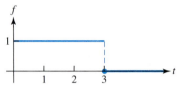

$$F(s) = (1 - e^{-3s})/s$$

27. f

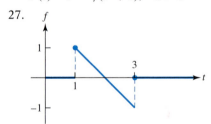

$$F(s) = [e^{-s}(s - 1) + e^{-3s}(s + 1)]/s^2$$

29. f

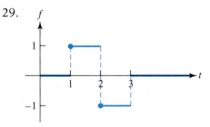

$$F(s) = (e^{-s} - 2e^{-2s} + e^{-3s})/s$$

31.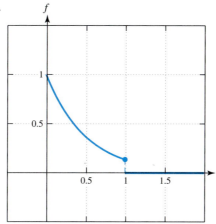

33.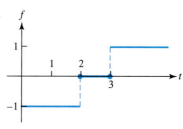

$F(s) = (e^{-2s} + e^{-3s} - 1)/s, \quad s > 0$

$F(s) = (1 - e^{-(s+2)})/(s+2)$

35. $f(t) = h(t-1) + h(t-2) - 2h(t-3), \quad F(s) = (e^{-s} + e^{-2s} - 2e^{-3s})/s$

37. $f(t) = (1-t)[h(t) - h(t-1)] + (2-t)[h(t-1) - h(t-2)]$, or, after expanding,
    $f(t) = 1 - t + h(t-1) + (t-2)h(t-2), \quad t \geq 0,$
    $F(s) = (1/s) - (1/s^2) + (e^{-s}/s) + (e^{-2s}/s^2)$

39. $F(s) = (2/s) - [2/(s+2)], \quad \mathcal{L}^{-1}\{F(s)\} = 2 - 2e^{-2t}, \quad t \geq 0$

41. $F(s) = [10e^{-s}/(s-3)] - [10e^{-s}/(s-2)],$
    $\mathcal{L}^{-1}\{F(s)\} = 10h(t-1)[e^{3(t-1)} - e^{2(t-1)}], \quad t \geq 0$

43. $Y(s) = [1/(s-1)] + e^{12-4s}/[(s-3)(s-1)],$
    $y(t) = \mathcal{L}^{-1}\{Y(s)\} = e^t + h(t-4)[e^{3t} - e^{t+8}]/2, \quad t \geq 0$

45. $Y(s) = s/[(s-1)(s^2 - 2s - 8)],$
    $y(t) = \mathcal{L}^{-1}\{Y(s)\} = (-e^{-2t} - e^t + 2e^{4t})/9, \quad t \geq 0$

47. $G(s) = F(s)/s^2, \quad s > \max\{a, 0\}$ where $|f(t)| \leq Me^{at}$

49. (a) No, they differ at $t = 3$.   (b) $F(s) = G(s) = (1 - e^{-3s})/s$

Section 5.3, page 349

1. $F(s) = \dfrac{A}{s-1} + \dfrac{B_2}{(s-2)^2} + \dfrac{B_1}{(s-2)}$

3. $F(s) = \dfrac{A_2}{s^2} + \dfrac{A_1}{s} + \dfrac{Bs+C}{(s+1)^2 + 9}$

5. $F(s) = \dfrac{A_2}{(s-3)^2} + \dfrac{A_1}{(s-3)} + \dfrac{B_2}{(s+3)^2} + \dfrac{B_1}{(s+3)}$

7. $F(s) = \dfrac{Bs+C}{(s+4)^2 + 1} + \dfrac{Ds+E}{(s+3)^2 + 4}$

9. $\mathcal{L}^{-1}\{F(s)\} = 2e^{3t}, \quad t \geq 0$ 11. $\mathcal{L}^{-1}\{F(s)\} = 4\cos 3t + \tfrac{5}{3}\sin 3t, \quad t \geq 0$

13. $\mathcal{L}^{-1}\{F(s)\} = e^{-3t} + 2e^{-t}, \quad t \geq 0$

15. $\mathcal{L}^{-1}\{F(s)\} = 2 + \cos 2t + \tfrac{1}{2}\sin 2t, \quad t \geq 0$

17. $\mathcal{L}^{-1}\{F(s)\} = te^t + \tfrac{1}{2}t^2 e^t, \quad t \geq 0$

19. $y(t) = 4e^{3t} - 3\cos 2t + 2\sin 2t$ 21. $y(t) = e^{3t} + te^{3t}$

23. $y(t) = 2t + 2\cos 2t + 2\sin 2t$

25. $y(t) = \cos 2t - \frac{1}{4}t\cos 2t + \frac{1}{8}\sin 2t$

27. $y(t) = te^{-t} + \frac{1}{2}t^2 e^{-t}$

29. $y(t) = t + \cos t - \sin t + h(t-2)[-t + 2\cos(t-2) + \sin(t-2)]$

31. $\alpha = 0, \beta = -4, y_0 = 0, y_0' = 3$

Section 5.4, page 355

1. $F(s) = \dfrac{3(1-e^{-2s})^2}{s(1-e^{-4s})} = \dfrac{3(1-e^{-2s})}{s(1+e^{-2s})}$

3. $F(s) = \dfrac{-2e^{-4s} + 5e^{-2s} - 3}{s(1-e^{-4s})}$

5. $F(s) = \dfrac{1-e^{-s}}{s^2(1+e^{-s})}$

7. $F(s) = \dfrac{1 + (2e^{-s}/s) + (2e^{-s}/s^2) - (2/s^2)}{s(1-e^{-2s})}$

9.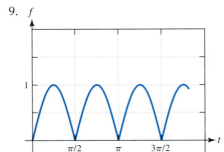

$F(s) = \dfrac{2(1+e^{-\pi s/2})}{(s^2+4)(1-e^{-\pi s/2})}$

11.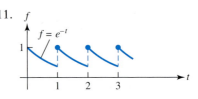

$F(s) = \dfrac{1-e^{-(s+1)}}{(s+1)(1-e^{-s})}$

13.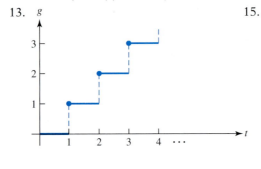

15.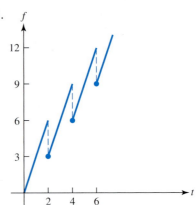

$\mathcal{L}^{-1}\{F(s)\} = 3t - 3[h(t-2) + h(t-4) + h(t-6) + \cdots]$

17. (a) $v(t) = e^{-t} + \displaystyle\sum_{n=0}^{\infty} (-1)^n [1 - e^{-(t-n)}]h(t-n)$

19. (a) $v(t) = 1.5e^{-t} - 0.5(1-t) + \displaystyle\sum_{n=1}^{\infty} (-1)^n [e^{-(t-2n)} - 1]h(t-2n)$

21. (a) $q' + 0.1q = 5c_i(t),\ \ q(0) = 0,$ where $c_i(t) = \begin{cases} 1, & 0 \le t < \frac{1}{2}, \\ 0, & \frac{1}{2} \le t < 1, \end{cases}$ $c_i(t+1) = c_i(t)$

(b) $Q(s) = \dfrac{5(1 - e^{-s/2})}{s(s+0.1)(1 - e^{-s})}$

(c) $q(t) = \begin{cases} 50[1 - e^{-t/10} + e^{-(t-1/2)/10} - e^{-(t-1)/10}], & 1 \le t < \frac{3}{2} \\ 50[-e^{-t/10} + e^{-(t-1/2)/10} - e^{-(t-1)/10} + e^{-(t-3/2)/10}], & \frac{3}{2} \le t < 2 \end{cases}$

25. $\Phi(s) = 1/(s+1)^2$

27. (a) $\Phi(s) = 1/(s^2+4)$ (b) $Y(s) = 2/[s^3(s^2+4)]$

29. (a) $\Phi(s) = 1/(s+2)^2$ (b) $Y(s) = [1 - (s+1)e^{-s}]/[s^2(1 - e^{-s})(s+2)^2]$

31. (a) $\Phi(s) = 1/[s(s^2+4)]$ (b) $Y(s) = 1/(s^2+4)^2$

33. $b = 0, c = 4, y_0 = 1, y_0' = 0$

Section 5.5, page 365

1. $\mathbf{Y}(s) = \begin{bmatrix} s/(s^2+1) \\ 1/s^2 \\ 1/(s-1)^2 \end{bmatrix}$

3. $\mathbf{Y}(s) = \begin{bmatrix} (2 - se^{-2s})/s^2 \\ 2e^{-2s}/s \end{bmatrix}$

5. $\mathbf{Y}(s) = \begin{bmatrix} e^{-s}/(1+s^2) \\ 1/[(s-1)e] - 2/s^2 \end{bmatrix}$

7. $\mathcal{L}^{-1}\{\mathbf{Y}(s)\} = h(t-1)\begin{bmatrix} 1 - \sin(t-1) \\ 2\sin(t-1) \end{bmatrix}$

9. $\mathbf{y}(t) = \begin{bmatrix} 4 + e^t \\ 5 + e^t \end{bmatrix}$

11. $\mathbf{y}(t) = \begin{bmatrix} t + e^t - e^{2t} \\ 1.5t - 0.25 + e^t - 0.75e^{2t} \end{bmatrix}$

13. $\mathbf{y}(t) = e^t \begin{bmatrix} 2\cos 2t \\ -\sin 2t \end{bmatrix}$

15. $\mathbf{y}(t) = \begin{bmatrix} 2e^{3(t-1)} + 3e^{-2(t-1)} \\ 2e^{3(t-1)} + 8e^{-2(t-1)} \end{bmatrix}$

17. $\mathbf{y}(t) = \dfrac{1}{120}\begin{bmatrix} t^5 - 5t^4 + 20t^3 \\ t^5 - 5t^4 + 60t^2 \end{bmatrix}$

19. $\mathbf{y}(t) = \begin{bmatrix} 5e^{-t} - 3e^t \\ -7e^{-t} + 3e^t \\ -e^{-2t} \end{bmatrix}$

21. (a) $\lambda_1 = 6,\ \lambda_2 = 3$ (b) $A = \begin{bmatrix} 7 & -1 \\ 4 & 2 \end{bmatrix}$

23. $A = \begin{bmatrix} 6 & -3 \\ 8 & -5 \end{bmatrix}$

25. (b) $\mathbf{i}(t) = 0.25e^{-t}\begin{bmatrix} t^2 \\ t^2 - 2t \end{bmatrix}$

Section 5.6, page 375

3. $f * g = t^4/12$

5. $f * g = t - \sin t$

7. $f * g = (t^2/2) - 0.5h(t-1)(t-1)^2$

9. $\begin{bmatrix} t^3/6 \\ 1 - \cos t \end{bmatrix}$

11.

$f * g = (t-2)h(t-2) - 2(t-3)h(t-3) + (t-4)h(t-4)$

13. $t^5/120$ 15. $-t + (e^t - e^{-t})/2$ 17. $n = 5, C = \frac{1}{24}, \alpha = -1$

19. $y(t) = 2 - 2e^{-t} - 2t^2e^{-t}$ 21. $y(t) = 6t$ or $y(t) = -6t$

23. $y(t) = 2 - 2e^{-t} - te^{-t}$ 25. $y(t) = \frac{1}{3}e^{-t} + \frac{2}{3}e^{t/2} \cos(\sqrt{3}\,t/2)$

27. $y(t) = 1 + (t^4/24)$

Section 5.7, page 383

1. (a) $1 + e^{-2}$ (b) 0 (c) $\begin{bmatrix} 1 \\ 0 \end{bmatrix}$ (d) $\begin{bmatrix} e^{-4} - 2 \\ e^2 + 1 \\ 0 \end{bmatrix}$

3. One possible choice is $t_0 = \frac{1}{6}$.

5.

$f(t) = 1 - h(1 - t)$

7. $k(t) = h(2 - t) - h(1 - t)$

9. (a) $y(t) = -1 + e^t$ (b) $\phi(t) = e^t$

11. (a) $y(t) = -t - 1 + e^t$ (b) $\phi(t) = e^t$

13.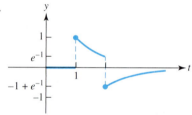

$y(t) = h(t - 1)e^{-(t-1)} - h(t - 2)e^{-(t-2)}$

15.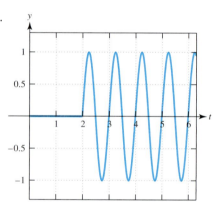

$y(t) = h(t - 2)\sin[2\pi(t - 2)]$

19. $\mathbf{y}(t) = \dfrac{h(t - 1)}{2} \begin{bmatrix} 1 + e^{2(t-1)} \\ -1 + e^{2(t-1)} \end{bmatrix}$

17.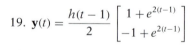

$y(t) = h(t - 1)e^{-(t-1)} \sin(t - 1)$

CHAPTER 6

Section 6.1, page 397

1. (a) $\mathbf{y}' = \begin{bmatrix} y_2 \\ -2y_1 - ty_2 \end{bmatrix}$, $\mathbf{y}(0) = \begin{bmatrix} 1 \\ 2 \end{bmatrix}$

   (b) $\partial f_1/\partial y_1 = 0$, $\partial f_1/\partial y_2 = 1$,
   $\partial f_2/\partial y_1 = -2$, $\partial f_2/\partial y_2 = -t$

   (c) The hypotheses of Theorem 6.1 are satisfied everywhere.

3. (a) $\mathbf{y}' = \begin{bmatrix} y_2 \\ -ty_1 + \sin y_2 \end{bmatrix}$, $\mathbf{y}(0) = \begin{bmatrix} 0 \\ 1 \end{bmatrix}$

   (b) $\partial f_1/\partial y_1 = 0$, $\partial f_1/\partial y_2 = 1$,
   $\partial f_2/\partial y_1 = -t$, $\partial f_2/\partial y_2 = \cos y_2$

   (c) The hypotheses of Theorem 6.1 are satisfied everywhere.

5. (a) $\mathbf{y}' = \begin{bmatrix} y_2 \\ (e^{-t}/t) - 1/[(1 + y_1 + 2y_2)t] \end{bmatrix}$, $\mathbf{y}(2) = \begin{bmatrix} 2 \\ 1 \end{bmatrix}$

   (b) $\partial f_1/\partial y_1 = 0$, $\partial f_1/\partial y_2 = 1$,
   $\partial f_2/\partial y_1 = (1/t)(1 + y_1 + 2y_2)^{-2}$, $\partial f_2/\partial y_2 = (2/t)(1 + y_1 + 2y_2)^{-2}$

   (c) The hypotheses of Theorem 6.1 are satisfied except when $t = 0$ or $1 + y_1 + 2y_2 = 0$.

7. (a) $\mathbf{y}' = \begin{bmatrix} y_2 \\ y_3 \\ -y_1^2 - y_2 \end{bmatrix}$, $\mathbf{y}(-1) = \begin{bmatrix} 0 \\ 1 \\ 0 \end{bmatrix}$

   (b) $\partial f_1/\partial y_1 = 0$, $\partial f_1/\partial y_2 = 1$, $\partial f_1/\partial y_3 = 0$,
   $\partial f_2/\partial y_1 = 0$, $\partial f_2/\partial y_2 = 0$, $\partial f_2/\partial y_3 = 1$,
   $\partial f_3/\partial y_1 = -2y_1$, $\partial f_3/\partial y_2 = -1$, $\partial f_3/\partial y_3 = 0$

   (c) The hypotheses of Theorem 6.1 are satisfied everywhere.

9. (a) $\mathbf{y}' = \begin{bmatrix} y_2 \\ y_3 \\ -2t^{1/3}/[(y_1 - 2)(y_3 + 2)] \end{bmatrix}$, $\mathbf{y}(2) = \begin{bmatrix} 0 \\ 2 \\ 2 \end{bmatrix}$

   (b) $\partial f_1/\partial y_1 = 0$, $\partial f_1/\partial y_2 = 1$, $\partial f_1/\partial y_3 = 0$,
   $\partial f_2/\partial y_1 = 0$, $\partial f_2/\partial y_2 = 0$, $\partial f_2/\partial y_3 = 1$,
   $\partial f_3/\partial y_1 = 2t^{1/3}/[(y_1 - 2)^2(y_3 + 2)]$, $\partial f_3/\partial y_2 = 0$,
   $\partial f_3/\partial y_3 = 2t^{1/3}/[(y_1 - 2)(y_3 + 2)^2]$

   (c) The hypotheses of Theorem 6.1 are satisfied except when $y_1 = 2$ or $y_3 = -2$.

11. $y'' = e^{y'} + y' \tan y$, $y(0) = 0$, $y'(0) = 1$

13. $y''' = (y'y'' + t^2)^{1/2}$, $y(1) = 1$, $y'(1) = \frac{1}{2}$, $y''(1) = 3$

15. No, because the differential equation is nonlinear.

19. (b) $a' = -k_1 e_0 a + k_1 ac + k_1' c$
    $c' = k_1 e_0 a - k_1 ac - (k_1' + k_2)c$

Section 6.2, page 410

1. $(0, 0)$, $(1, 1)$ 3. $(1, -1)$, $(3, -1)$ 5. $(0, 0)$

7. $(2, 2)$, $(2, -2)$, $(-2, 2)$, $(-2, -2)$ 9. $(0, 1, 2)$, $(-1, 1, 2)$

11. $x' = y$

   $y' = -x - x^3$

   The only equilibrium point is $(0, 0)$.

13. $x' = y$

   $y' = 1 - x^2 - 2y/(1 + x^4)$

   The only equilibrium points are $(1, 0)$ and $(-1, 0)$.

15. $x' = y$

   $y' = z$

   $z' = y^2 + (x^2 - 4)/(y^2 + 2)$

   The only equilibrium points are $(-2, 0, 0)$ and $(2, 0, 0)$.

17. $\alpha = -1$, $\beta = -1$, $\gamma = \frac{1}{2}$, $\delta = \frac{1}{2}$

19. $\alpha = 1$, $\beta = 1$, $\gamma = -2$ 21. $\alpha = 8$

23. (b) The velocity vector is a constant multiple of the position vector. Therefore, the velocity vector is oriented along the line.

25. Direction Field B   27. Direction Field A

29. (a) $x' = y$ 31. (a) $x' = y$

   $y' = -x - x^3$ $y' = 1 - 2\sin^2 x$

   (c)   (c)

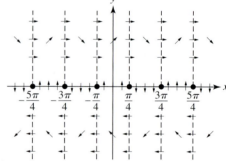

33. (a) Equilibrium point $(1, 1)$ 35. (a) Equilibrium points $(0, 0)$ and $(2, -2)$

   (b)   (b)

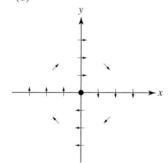

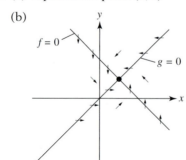

Section 6.3, page 421

1. (a) $0.5(x')^2 + 2x^2 = C$
   (b) $x' = y$
   $\quad y' = -4x$
   (c) $\mathbf{v}(1, 1) = \mathbf{i} - 4\mathbf{j}$

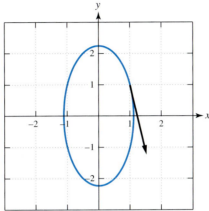

(d) The graph is an ellipse. The maximum value of $|x(t)|$ is $\sqrt{5}/2$.

3. (a) $0.5(x')^2 + 0.25x^4 = C$
   (b) $x' = y$
   $\quad y' = -x^3$
   (c) $\mathbf{v}(1, 1) = \mathbf{i} - \mathbf{j}$

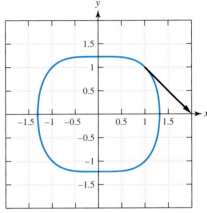

(d) The maximum value of $|x(t)|$ is $(3)^{1/4}$.

5. (a) $3(x')^2 + 2x^3 = C$
   (b) $x' = y$
   $\quad y' = -x^2$
   (c) $\mathbf{v}(1, 1) = \mathbf{i} - \mathbf{j}$

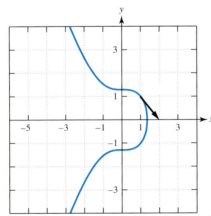

(d) $|x(t)|$ is unbounded.

7. $2x'' - x^2 \sin x + 2x \cos x = 0$

9. (a)

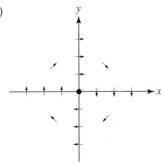

(c) $H(x, y) = 0.5(x^2 + y^2) + (0.25)x^4$

11. $0.5(y'')^2 + F(y') = C$, where $F(u)$ is an antiderivative of $f(u)$

13. (b) $H(x, y) = 2xy$ 15. (b) $H(x, y) = xy - x^2y + y - 2x^2$

17. (b) $H(x, y) = y^2 \cos x$ 19. (b) $H(x, y) = -x^3 - y^2$

21. $H(x, y) = x^3y - \cos(2x + 3y)$ 23. Not a Hamiltonian system

25. $H(x, y) = 0.5(y^2 - x^2) + x^3/3$ 27. $H(x, y) = F(y) - G(x)$

29. $H(x, y) = 3F(y) - xy^2 - G(x) - x$ 31. Yes, $K(x, y)$

Section 6.4, page 431

1. (a) $\delta = \varepsilon/2$ (b) Not asymptotically stable

3. (a) $x' = y$

$\qquad y' = -x - \gamma y$

   (b) Asymptotically stable if $\gamma > 0$, stable if $\gamma = 0$, unstable if $\gamma < 0$

5. Asymptotically stable   7. Unstable

9. Unstable   11. Unstable

13. Stable but not asymptotically stable   15. Asymptotically stable

17. Unstable   19. Unstable

21. Asymptotically stable   23. Asymptotically stable

25. $\mathbf{y}_e = \begin{bmatrix} -2 \\ 0 \end{bmatrix}$, asymptotically stable   27. $\mathbf{y}_e = \begin{bmatrix} 2 \\ -2 \end{bmatrix}$, unstable

29. $\mathbf{y}_e = \begin{bmatrix} 0 \\ 0 \\ 0 \end{bmatrix}$, unstable

31. $\mathbf{y}_e = \begin{bmatrix} 0 \\ 0 \\ 0 \\ 0 \end{bmatrix}$, stable but not asymptotically stable

35. $a_{12} = \frac{1}{2}$, $a_{22} = 1$, $a_{21} = 2$

Section 6.5, page 445

1. (a) $(4, 4)$, $(-4, -4)$

   (b) Using the order given in (a),

$$\mathbf{z}' = \begin{bmatrix} 8 & 8 \\ -1 & 1 \end{bmatrix} \mathbf{z}, \quad \mathbf{z}' = \begin{bmatrix} -8 & -8 \\ -1 & 1 \end{bmatrix} \mathbf{z}.$$

   (c) $(4, 4)$ and $(-4, -4)$ are each unstable equilibrium points.

3. (a) $(1, 1)$, $(-1, 1)$, $(-1, -1)$, $(1, -1)$

   (b) Using the order given in (a),

$$\mathbf{z}' = \begin{bmatrix} -2 & 0 \\ 2 & 2 \end{bmatrix} \mathbf{z}, \quad \mathbf{z}' = \begin{bmatrix} 2 & 0 \\ -2 & 2 \end{bmatrix} \mathbf{z}, \quad \mathbf{z}' = \begin{bmatrix} 2 & 0 \\ -2 & -2 \end{bmatrix} \mathbf{z}, \quad \mathbf{z}' = \begin{bmatrix} -2 & 0 \\ 2 & -2 \end{bmatrix} \mathbf{z}.$$

   (c) $(1, 1)$ is an unstable equilibrium point, $(-1, 1)$ is an unstable equilibrium point, $(-1, -1)$ is an unstable equilibrium point, $(1, -1)$ is an asymptotically stable equilibrium point.

5. (a) $(2, 1)$, $(2, -1)$, $(-6, 3)$

(b) Using the order given in (a),

$$\mathbf{z}' = \begin{bmatrix} -2 & 0 \\ 0 & 4 \end{bmatrix} \mathbf{z}, \quad \mathbf{z}' = \begin{bmatrix} -4 & 0 \\ -2 & -4 \end{bmatrix} \mathbf{z}, \quad \mathbf{z}' = \begin{bmatrix} 0 & -8 \\ 2 & 4 \end{bmatrix} \mathbf{z}.$$

(c) $(2, 1)$ is an unstable equilibrium point, $(2, -1)$ is an asymptotically stable equilibrium point, $(-6, 3)$ is an unstable equilibrium point.

7. (a) $(0, 0)$, $(-2, -4)$

(b) Using the order given in (a),

$$\mathbf{z}' = \begin{bmatrix} 4 & -8 \\ 2 & -1 \end{bmatrix} \mathbf{z}, \quad \mathbf{z}' = \begin{bmatrix} 0 & 6 \\ 2 & -1 \end{bmatrix} \mathbf{z}.$$

(c) $(0, 0)$ and $(-2, -4)$ are each unstable equilibrium points.

9. (a) $(0, 0)$, $(1, 1)$

(b) Using the order given in (a),

$$\mathbf{z}' = \begin{bmatrix} -1 & 0 \\ 0 & -1 \end{bmatrix} \mathbf{z}, \quad \mathbf{z}' = \begin{bmatrix} -1 & 2 \\ 2 & -1 \end{bmatrix} \mathbf{z}.$$

(c) $(0, 0)$ is an asymptotically stable equilibrium point, $(1, 1)$ is an unstable equilibrium point.

13. (a) $A = \begin{bmatrix} 5 & -14 \\ 3 & -8 \end{bmatrix}$, $\quad \mathbf{g}(\mathbf{z}) = \begin{bmatrix} z_1 z_2 \\ z_1^2 + z_2^2 \end{bmatrix}$

(c) The limit is 0. $\mathbf{z}' = A\mathbf{z} + \mathbf{g}(\mathbf{z})$ is an almost linear system.

(d) $\mathbf{z} = \mathbf{0}$ is an asymptotically stable equilibrium point.

15. (a) $A = \begin{bmatrix} -1 & 3 \\ -1 & -5 \end{bmatrix}$, $\quad \mathbf{g}(\mathbf{z}) = \begin{bmatrix} z_2 \cos \sqrt{z_1^2 + z_2^2} \\ z_1 \cos \sqrt{z_1^2 + z_2^2} \end{bmatrix}$

(c) The limit is 1. $\mathbf{z}' = A\mathbf{z} + \mathbf{g}(\mathbf{z})$ is not an almost linear system.

17. (a) $A = \begin{bmatrix} 0 & 2 \\ -2 & 0 \end{bmatrix}$, $\quad \mathbf{g}(\mathbf{z}) = \begin{bmatrix} z_2^2 \\ z_1 z_2 \end{bmatrix}$

(c) The limit is 0. $\mathbf{z}' = A\mathbf{z} + \mathbf{g}(\mathbf{z})$ is an almost linear system.

(d) No conclusion can be drawn using Theorem 6.4.

19. (a) $A = \begin{bmatrix} 9 & 5 \\ -7 & -3 \end{bmatrix}$, $\quad \mathbf{g}(\mathbf{z}) = \begin{bmatrix} z_1 z_2 \\ z_1^2 \end{bmatrix}$

(c) The limit is 0. $\mathbf{z}' = A\mathbf{z} + \mathbf{g}(\mathbf{z})$ is an almost linear system.

(d) $\mathbf{z} = \mathbf{0}$ is an unstable equilibrium point.

21. (b) $\mathbf{z}' = \begin{bmatrix} -1 & 1 \\ 1 & -2 \end{bmatrix} \mathbf{z}$, asymptotically stable at $\mathbf{z} = \begin{bmatrix} 0 \\ 0 \end{bmatrix}$.

(d) The given nonlinear system is asymptotically stable at $(0, 0)$.

23. (a) $0.5y^2 - x + \frac{2}{5}(1+x)^{5/2} = C$

(b) The graphs are consistent with bobbing motion. The origin appears to be a stable equilibrium point but not an asymptotically stable equilibrium point.

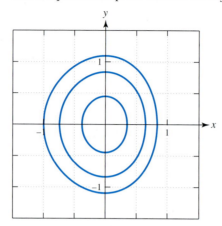

27. $x = [\cos(c_1 - t)]/\sqrt{2(c_2 - \alpha t)}$, $y = [\sin(c_1 - t)]/\sqrt{2(c_2 - \alpha t)}$; c_1 and c_2 are arbitrary constants.

29. $x(1) = \exp(e^{-2})\cos(\pi/4 - 1)$, $y(1) = \exp(e^{-2})\sin(\pi/4 - 1)$

Section 6.6, page 454

1. (a) Saddle point, unstable
   (c)

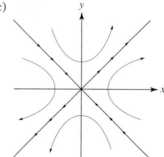

3. (a) Unstable improper node
   (c)

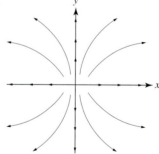

5. (a) Saddle point, unstable
   (c)

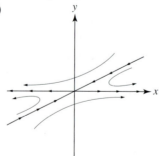

7. (a) $\lambda = 1$, $\lambda = 2$ (b) Improper node, unstable

9. (a) $\lambda = 3i$, $\lambda = -3i$ (b) Center, stable but not asymptotically stable

11. (a) $\lambda = -2$, $\lambda = -3$ (b) Improper node, asymptotically stable

13. (a) $\lambda = 4i$, $\lambda = -4i$ (b) Center, stable but not asymptotically stable

15. (a) $\lambda = 1$, $\lambda = 3$ (b) Improper node, unstable

17. (a) $\lambda = -2$, $\lambda = -2$ (b) Improper node, asymptotically stable

19. (a) $\lambda = 1 + 4i$, $\lambda = 1 - 4i$ (b) Spiral point, unstable

21. (a) Direction Field 2   (b) Direction Field 4   (c) Direction Field 1
    (d) Direction Field 3

23. $\alpha > \frac{9}{2}$ 25. $\alpha < 8$

27. $\mathbf{y}_e = \begin{bmatrix} 6 \\ -8 \end{bmatrix}$ is an unstable saddle point.

29. $\mathbf{y}_e = \begin{bmatrix} 2 \\ 2 \end{bmatrix}$ is an unstable saddle point.

31. (c) The solution point moves clockwise.

33. (a) $\lambda = \pm 2\sqrt{2}\,i$ (b) Center, stable but not asymptotically stable
    (c) $H(x, y) = (3x^2 + 2xy + 3y^2)/2$

Section 6.7, page 462

1. (a) x approaches an equilibrium population of 1 in the absence of y. If y is present, it is harmful to x. y approaches an equilibrium population of $\frac{1}{3}$ in the absence of x. If x is present, it is harmful to y.

    (b) $(0, 0)$, $(1, 0)$, $\left(0, \frac{1}{3}\right)$, $\left(\frac{4}{5}, \frac{1}{5}\right)$

    (c) The origin is an unstable proper node of the linearized system $\mathbf{z}' = \begin{bmatrix} 1 & 0 \\ 0 & 1 \end{bmatrix} \mathbf{z}$.

    Therefore, the given system is unstable at the origin.

3. (a) x approaches an equilibrium population of 1 in the absence of y. If y is present, it is harmful to x. y approaches an equilibrium population of 0 in the absence of x. If x is present, it is beneficial to y.

    (b) $(0, 0)$, $(1, 0)$

    (c) The origin is an unstable saddle point of the linearized system $\mathbf{z}' = \begin{bmatrix} 1 & 0 \\ 0 & -1 \end{bmatrix} \mathbf{z}$.

    Therefore, the given system is unstable at the origin.

5. (a) $\alpha_1 = 0$, $\alpha_2 = 0$, $r_1 = \frac{1}{2}$, $r_2 = -1$, $\beta_1 = -\frac{1}{3}$, $\beta_2 = -\frac{1}{2}$

    (b) x is beneficial to y, but y is harmful to x.

7. (a) $(0, 0)$, $(1/\alpha, 0)$, $(0, [r - \mu]/[\alpha r])$,
    $([\alpha r - \beta r + \beta \mu]/[\alpha^2 r - \beta^2 r], [\alpha r - \beta r - \alpha \mu]/[\alpha^2 r - \beta^2 r])$

    (c) $(0, 0)$ is an unstable saddle point, $(1/\alpha, 0)$ is an asymptotically stable improper node.

9. $x' = a_1 x - b_1 x^2 - c_1 xy - d_1 xz$

    $y' = a_2 y - b_2 y^2 - c_2 xy - d_2 yz$

    $z' = -a_3 z + c_3 xz + d_3 yz$

11. (a) $s' = -\alpha si - \gamma s - \gamma i + \gamma N$

    $i' = \alpha si - \beta i$

    (b) $(s, i) = (9, 0)$ and $(s, i) = (1, 4)$

    (c) For $(9, 0)$, $\mathbf{z}' = \begin{bmatrix} -1 & -10 \\ 0 & 8 \end{bmatrix} \mathbf{z}$, unstable saddle point

    For $(1, 4)$, $\mathbf{z}' = \begin{bmatrix} -5 & -2 \\ 4 & 0 \end{bmatrix} \mathbf{z}$, asymptotically stable spiral point

CHAPTER 7

Section 7.2, page 477

1. (a) $y = t^2 - t$
   (b) $y_{n+1} = y_n + (h/2)[(2t_n - 1) + (2t_{n+1} - 1)]$
   (c) $y_{n+1} = y_n + h[2(t_n + h/2) - 1]$
   (d) $y_1 = 0.11, \quad y_2 = 0.24, \quad y_3 = 0.39$
   (e) $y_1 = 0.11, \quad y_2 = 0.24, \quad y_3 = 0.39$
   (f) $y(t_1) = 0.11, \quad y(t_2) = 0.24, \quad y(t_3) = 0.39$

3. (a) $y = e^{-t^2/2}$
   (b) $y_{n+1} = y_n + (h/2)[-t_n y_n - t_{n+1}(y_n - ht_n y_n)]$
   (c) $y_{n+1} = y_n - h(t_n + 0.5h)(y_n - 0.5ht_n y_n)$
   (d) $y_1 = 0.9950, \quad y_2 = 0.9801\ldots, \quad y_3 = 0.9559\ldots$
   (e) $y_1 = 0.9950, \quad y_2 = 0.9801\ldots, \quad y_3 = 0.9558\ldots$
   (f) $y(t_1) = 0.9950\ldots, \quad y(t_2) = 0.9801\ldots, \quad y(t_3) = 0.9559\ldots$

5. (a) $y = (1 - 1.5t^2)^{1/3}$
   (b) $y_{n+1} = y_n + (h/2)\left[-t_n y_n^{-2} - t_{n+1}(y_n - ht_n y_n^{-2})^{-2}\right]$
   (c) $y_{n+1} = y_n - h(t_n + 0.5h)(y_n - 0.5ht_n y_n^{-2})^{-2}$
   (d) $y_1 = 0.9950, \quad y_2 = 0.9796\ldots, \quad y_3 = 0.9529\ldots$
   (e) $y_1 = 0.9950, \quad y_2 = 0.9796\ldots, \quad y_3 = 0.9530\ldots$
   (f) $y(t_1) = 0.9949\ldots, \quad y(t_2) = 0.9795\ldots, \quad y(t_3) = 0.9528\ldots$

7. (a) $y = \sqrt{9 - t^2}$
   (b) $\left|y(1) - y_{20}^{\text{Euler}}\right| = 0.009146\ldots,$

   $\left|y(1) - y_{20}^{\text{Heun}}\right| = 6.902 \times 10^{-7}, \quad \left|y(1) - y_{20}^{\text{modEul}}\right| = 1.375 \times 10^{-5}$

9. (a) $y = 2e^{-t^2}$
   (b) $\left|y(1) - y_{20}^{\text{Euler}}\right| = 0.01300\ldots,$

   $\left|y(1) - y_{20}^{\text{Heun}}\right| = 6.022 \times 10^{-4}, \quad \left|y(1) - y_{20}^{\text{modEul}}\right| = 3.329 \times 10^{-4}$

11. Heun's method, with $t_0 = 1, T = 5, f(t, y) = ty^2 + 1$

13. Euler's method, with $t_0 = 2, T = 1, f(t, y) = y/(t^2 + y^2)$

17. (a) The initial value problem is $Q' = 12 - 6\cos(\pi t) - Q/(90 + 5t), Q(0) = 0.$
    (c) About 23.75 lb

19. About 1.5003 million individuals

Section 7.3, page 491

1. (a) $y'(0) = 1, y''(0) = -1, y'''(0) = 1, y^{(4)}(0) = -1$
   $P_4(t) = 1 + t - \frac{1}{2}t^2 + \frac{1}{6}t^3 - \frac{1}{24}t^4$
   (c) $E(0.1) = 8.196 \times 10^{-8}$

3. (a) $y'(0) = 0, y''(0) = 1, y'''(0) = 0, y^{(4)}(0) = 6$
   $P_4(t) = 1 + \frac{1}{2}t^2 + \frac{1}{4}t^4$
   (c) $E(0.1) = 1.256 \times 10^{-7}$

5. (a) $y'(0) = 1, y''(0) = \frac{1}{2}, y'''(0) = 0, y^{(4)}(0) = 0$
   $P_4(t) = 1 + t + \frac{1}{4}t^2$

   (c) $E(0.1) = 0$

7. (a) $y'(0) = 1, y''(0) = 2, y'''(0) = 2, y^{(4)}(0) = 1$
   $P_4(t) = 1 + t + t^2 + \frac{1}{3}t^3 + \frac{1}{24}t^4$

   (c) $E(0.1) = 8.615 \times 10^{-8}$

9. (a) $y'(0) = 2, y''(0) = 4, y'''(0) = 16, y^{(4)}(0) = 80$
   $P_4(t) = 1 + 2t + 2t^2 + \frac{8}{3}t^3 + \frac{10}{3}t^4$

   (c) $E(0.1) = 4.888 \times 10^{-5}$

11. (a) $P_5(t) = 1 - t^2 - t^3 - \frac{7}{12}t^4 - \frac{1}{4}t^5$

    (b) $E(0.1) = -8.867 \times 10^{-8}$

13. (a) $P_5(t) = 1 + 2t + \frac{1}{3}t^3 + \frac{1}{60}t^5$

    (b) $E(0.1) = 3.968 \times 10^{-11}$

15. $r = 1$ 17. $r = 2$

21. (b) At $t = 1$, the errors are (order 1) $1.8054\ldots \times 10^{-3}$, (order 2) $-4.0475\ldots \times 10^{-4}$, (order 3) $6.8372\ldots \times 10^{-6}$.

23. (b) At $t = 1$, the errors are (order 1) $-7.2978\ldots \times 10^{-3}$, (order 2) $8.2708\ldots \times 10^{-4}$, (order 3) $-3.0262\ldots \times 10^{-5}$.

25. At $t = 1$, the error ratio is $E_2/E_1 = 8.4648\ldots \times 10^{-7}/6.8372\ldots \times 10^{-6} = 0.1238\ldots$

27. At $t = 1$, the error ratio is $E_2/E_1 = -3.6501\ldots \times 10^{-6}/-3.0262\ldots \times 10^{-5} = 0.1206\ldots$

Section 7.4, page 502

1. (a) $y_1 = 1.09516666\ldots$ (b) $y_1 = 1.09516250\ldots$ (c) No
   (d) $y(t_1) = 1.09516258\ldots$

3. (a) $y_1 = 1.00503350\ldots$ (b) $y_1 = 1.00502513\ldots$ (c) No
   (d) $y(t_1) = 1.00502512\ldots$

5. (a) $y_1 = 1.10249901\ldots$

   (b) $y_1 = 1.10249998\ldots$

   (c) Yes, since methods of order p or higher yield exact results if the solution is a polynomial of degree p or less.

   (d) $y(t_1) = 1.10250000\ldots$. Note that the numerical results do not agree with our answer in part (c). This discrepancy is due to the finite precision of computer arithmetic.

7. (a) $y_1 = 1.11032909\ldots$ (b) $y_1 = 1.11033743\ldots$ (c) No
   (d) $y(t_1) = 1.11033758\ldots$

9. (a) $y_1 = 1.22304273\ldots$ (b) $y_1 = 1.22304891\ldots$ (c) No
   (d) $y(t_1) = 1.22304888\ldots$

11. $\mathbf{y}_1 = \begin{bmatrix} 0.89534540\ldots \\ -1.08953454\ldots \end{bmatrix}$ 13. $\mathbf{y}_1 = \begin{bmatrix} 2.10571842\ldots \\ 1.22088665\ldots \end{bmatrix}$

15. $\mathbf{y}_1 = \begin{bmatrix} 0.99500416\ldots \\ -0.09983354\ldots \\ -0.99501250\ldots \end{bmatrix}$

19. (b) $y_{20} = 1.23606797\ldots, \ y(1) = 1.23606797\ldots$

21. (b) $y_{20} = 1.41421356\ldots, \ y(1) = 1.41421356\ldots$

23. (b) $y_{20} = -0.81020240\ldots$, $y(1) = -0.81019930\ldots$
    $y_{20}' = -0.42549625\ldots$, $y'(1) = -0.42549942\ldots$

25. (b) $y_{10} = 4.613705448\ldots$, $y(1) = 4.613705638\ldots$
    $y_{10}' = 3.30685272\ldots$, $y'(1) = 3.30685281\ldots$

27.

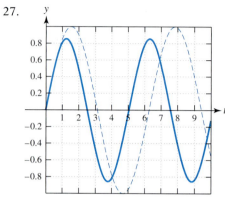

Solid is numerical solution, dotted is $\sin(t)$.

29.

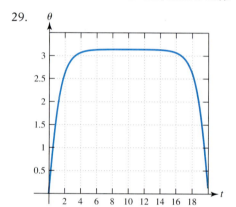

Graph of theta versus time

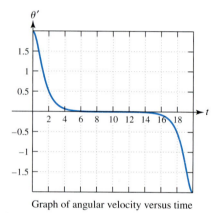

Graph of angular velocity versus time

CHAPTER 8

Section 8.1, page 525

1. $R = 2$ 3. $R = 1$ 5. $R = \infty$

7. $R = 1$ 9. $R = 1$ 11. $R = 2$

13. $R = 1$

  (a) $f(t) = 1 + t + t^2 + t^3 + \cdots$
  $g(t) = t + 4t^2 + 9t^3 + 16t^4 + \cdots$

  (b) $f(t) + g(t) = 1 + 2t + 5t^2 + 10t^3 + \cdots$

  (c) $f(t) - g(t) = 1 - 3t^2 - 8t^3 - 15t^4 - \cdots$

  (d) $f'(t) = 1 + 2t + 3t^2 + 4t^3 + \cdots$

  (e) $f''(t) = 2 + 6t + 12t^2 + 20t^3 + \cdots$

15. $R = 1/2$

    (a) $f(t) = 1 - 2(t - 1) + 4(t - 1)^2 - 8(t - 1)^3 + \cdots$
       $g(t) = 1 + (t - 1) + (t - 1)^2 + (t - 1)^3 + \cdots$

    (b) $f(t) + g(t) = 2 - (t - 1) + 5(t - 1)^2 - 7(t - 1)^3 + \cdots$

    (c) $f(t) - g(t) = -3(t - 1) + 3(t - 1)^2 - 9(t - 1)^3 + 15(t - 1)^4 - \cdots$

    (d) $f'(t) = -2 + 8(t - 1) - 24(t - 1)^2 + 64(t - 1)^3 - \cdots$

    (e) $f''(t) = 8 - 48(t - 1) + 192(t - 1)^2 - 640(t - 1)^3 + \cdots$

17. $\displaystyle\sum_{n=2}^{\infty} 2^{n-2} t^n$ 19. $\displaystyle\sum_{n=2}^{\infty} a_{n-2} t^n$

21. $\displaystyle\sum_{n=0}^{\infty} (n + 2)(n + 1)a_{n+2} t^n$ 23. $-\displaystyle\sum_{n=2}^{\infty} (-1)^n (n - 1)a_{n-2} t^n$

25. $f(t) = \displaystyle\sum_{n=1}^{\infty} (-1)^{n+1}(3t)^{2n}/(2n)!, \quad R = \infty$

27. $f(t) = \displaystyle\sum_{n=0}^{\infty} t^{2n}, \quad R = 1$

29. (a) $a_{n+2} = \dfrac{\omega^2}{(n + 2)(n + 1)} a_n$ (b) $y_1(t) = \cosh \omega t, \; y_2(t) = \sinh \omega t$

31. (a) $y(t) = \displaystyle\sum_{n=0}^{\infty} (t - 1)^n/n!$ (b) $R = \infty$ (c) $y(t) = e^{t-1}$

33. (a) $y(t) = -\displaystyle\sum_{n=3}^{\infty} (-1)^n (t - 1)^n/n!$ (b) $R = \infty$

    (c) $y(t) = -e^{-(t-1)} + 1 - (t - 1) + (t - 1)^2/2$

35. (a) $y(t) = \displaystyle\sum_{n=0}^{\infty} t^n$ (b) $R = 1$ (c) $y(t) = 1/(1 - t)$

37. $a_0 = 0, \; a_1 = 1, \; a_2 = 0, \; a_3 = \frac{1}{6}, \; a_4 = 0, \; a_5 = -\frac{1}{120}$

39. $a_0 = -1, \; a_1 = 1, \; a_2 = 0, \; a_3 = -\frac{1}{3}, \; a_4 = \frac{1}{12}, \; a_5 = \frac{1}{20}$

41. $a_0 = 0, \; a_1 = 2, \; a_2 = 2, \; a_3 = 1, \; a_4 = \frac{1}{3}, \; a_5 = \frac{1}{12}$

Section 8.2, page 533

1. $\pm\pi/2, \; \pm 2, \; \pm 3\pi/2, \; \pm 5\pi/2$ 3. $0, \; \pm 1, \; \pm\pi, \; \pm 2\pi, \; \pm 3\pi$

5. $0, \; \pm e^{-1}$ 7. $R = \frac{1}{2}$

9. $R = \frac{5}{6}$ 11. $R = \frac{1}{3}$

13. (a) $a_{n+2} = -a_n/(n + 2)$

    (b) $y_1(t) = a_0\left[1 - \frac{1}{2}t^2 + \frac{1}{8}t^4 + \cdots\right]$
       $y_2(t) = a_1\left[t - \frac{1}{3}t^3 + \frac{1}{15}t^5 + \cdots\right]$

    (c) $R = \infty$

    (d) Yes

15. (a) $a_{n+2} = -(n^2 + 2)a_n/[(n + 2)(n + 1)]$

    (b) $y_1(t) = a_0\left[1 - t^2 + \frac{1}{2}t^4 + \cdots\right]$
       $y_2(t) = a_1\left[t - \frac{1}{2}t^3 + \frac{11}{40}t^5 + \cdots\right]$

    (c) $R = 1$

    (d) Yes

17. (a) $(n+2)(n+1)a_{n+2} = 4(n+1)a_{n+1} - 4a_n$

   (b) $y_1(t) = a_0\left[1 - 2t^2 - \frac{8}{3}t^3 + \cdots\right]$

       $y_2(t) = a_1\left[t + 2t^2 + 2t^3 + \cdots\right]$

   (c) $R = \infty$

   (d) No

19. (a) $3(n+2)(n+1)a_{n+2} = -n(n+1)a_{n+1} - (3n+1)a_n$

   (b) $y_1(t) = a_0\left[1 - \frac{1}{6}t^2 + \frac{1}{54}t^3 + \cdots\right]$

       $y_2(t) = a_1\left[t - \frac{2}{9}t^3 + \frac{1}{27}t^4 + \cdots\right]$

   (c) $R = 3$

   (d) No

21. (a) $(n+2)(n+1)a_{n+2} = -a_{n-2}$

   (b) $y_1(t) = a_0\left[1 - \frac{1}{12}t^4 + \frac{1}{672}t^8 + \cdots\right]$

       $y_2(t) = a_1\left[t - \frac{1}{20}t^5 + \frac{1}{1440}t^9 + \cdots\right]$

   (c) $R = \infty$

   (d) Yes

23. (a) $(n+2)(n+1)a_{n+2} = -a_n$

   (b) $y_1(t) = a_0\left[1 - \frac{1}{2}(t-1)^2 + \frac{1}{24}(t-1)^4 + \cdots\right]$

       $y_2(t) = a_1\left[(t-1) - \frac{1}{6}(t-1)^3 + \frac{1}{120}(t-1)^5 + \cdots\right]$

   (c) $R = \infty$

25. (a) $(n+2)(n+1)a_{n+2} = -(n+1)a_{n+1} + a_n - a_{n-1}$

   (b) $y_1(t) = a_0\left[1 + \frac{1}{2}(t-1)^2 - \frac{1}{3}(t-1)^3 + \cdots\right]$

       $y_2(t) = a_1\left[(t-1) - \frac{1}{2}(t-1)^2 + \frac{1}{3}(t-1)^3 + \cdots\right]$

   (c) $R = \infty$

29. (a) $(n+2)(n+1)a_{n+2} = -2(\mu - n)a_n$

   (d) $H_2(t) = 4t^2 - 2$, $H_3(t) = 8t^3 - 12t$, $H_4(t) = 16t^4 - 48t^2 + 12$,
      $H_5(t) = 32t^5 - 160t^3 + 120t$

33. Yes   35. Yes

37. No   39. $a = 0$, b is arbitrary.

Section 8.3, page 541

3. $y = c_1 t^2 + c_2 t^3$ 5. $y = c_1 t^2 + c_2 t^2 \ln|t|$

7. $y = c_1 t^2 \cos(5 \ln|t|) + c_2 t^2 \sin(5 \ln|t|)$ 9. $y = c_1 \cos(3 \ln|t|) + c_2 \sin(3 \ln|t|)$

11. $y = c_1 t^{-1} \cos(4 \ln|t|) + c_2 t^{-1} \sin(4 \ln|t|)$

13. $y = c_1 t^{-2} \cos(6 \ln|t|) + c_2 t^{-2} \sin(6 \ln|t|)$

15. $y = c_1(t-1)^{-1} + c_2(t-1)^3$ 17. $y = c_1(t+2)^{-2} + c_2(t+2)^{-3}$

19. $t_0 = -2$, $\alpha = 2$, $\beta = -2$ 21. $t_0 = 0$, $\alpha = -3$, $\beta = 5$

23. $\alpha = -4$, $\beta = 6$, $g(t) = -5 + 6 \ln t$

25. The transformed equation is $Y'' - 2Y' + Y = e^{-z}$.
     $y = c_1 t + c_2 t \ln t + \frac{1}{4}t^{-1}$

27. The transformed equation is $Y'' - Y' - 6Y = 10e^{-2z} - 6$.
     $y = c_1 t^{-2} + c_2 t^3 + 1 - 2t^{-2} \ln t$

29. The transformed equation is $Y'' + 7Y' + 10Y = 36(e^z + e^{-z})$.
     $y = c_1 t^{-5} + c_2 t^{-2} + 2t + 9t^{-1}$

31. $y = t + t^5 + 2, \ -\infty < t < \infty$

33. $y = -2t^{-1} - t^{-1}\ln t + t^{-1}(\ln t)^2, \ 0 < t < \infty$

35. $y = c_1 + c_2 t^{-2} + c_3 t^2, \ t \neq 0$

37. $y = c_1 + c_2 \ln|t| + c_3(\ln|t|)^2 + t^2 + 2(\ln|t|)^3$

Section 8.4, page 549

1. $t = 0$ is a regular singular point.

3. $t = 1$ and $t = -1$ are regular singular points.

5. $t = 0$ is a regular singular point.

7. $t = 0$ is a regular singular point.

9. $t = 1$ and $t = -1$ are irregular singular points.

11. $P(t) = (t^2 - 1)^2 t^2$ 13. $P(t) = (t-1)^2(t+1)^2$

15. (a) $n = 0, 1, 2$ (b) $n > 2$

17. (b) $F(\lambda) = 4\lambda^2 - 1 = 0$

　　(c) $F(\lambda + n)a_n = -a_{n-1}$

　　(d) $y_1 = a_0\left[t^{1/2} - \frac{1}{8}t^{3/2} + \frac{1}{192}t^{5/2} + \cdots\right]$

19. (b) $F(\lambda) = \lambda^2 - 9 = 0$

　　(c) $F(\lambda + n)a_n = -a_{n-1}$

　　(d) $y_1 = a_0\left[t^3 - \frac{1}{7}t^4 + \frac{1}{112}t^5 + \cdots\right]$

21. (b) $F(\lambda) = \lambda^2 + 2\lambda + 1 = 0$

　　(c) $F(\lambda + n)a_n = -2a_{n-1}$

　　(d) $y_1 = a_0[t^{-1} - 2 + t + \cdots]$

23. (b) $F(\lambda) = \lambda^2 - 3\lambda = 0$

　　(c) $F(\lambda + n)a_{n+1} = -(\lambda + n + 1)a_n$

　　(d) $y_1 = a_0\left[t^3 - t^4 + \frac{1}{2}t^5 + \cdots\right]$

25. (b) $\lambda^2 - 5\lambda = 0$

　　(c) $y_1 = a_0\left[t^5 - \frac{1}{6}t^6 - \frac{5}{84}t^7 + \cdots\right]$

27. (b) $\lambda(\lambda - 1.5) = 0$

　　(c) $y_1 = a_0\left[t^{3/2} + \frac{1}{2}t^{5/2} - \frac{17}{96}t^{7/2} + \cdots\right]$

Section 8.5, page 558

1. (b) $\lambda(2\lambda - 3) = 0$

　　(c) $[(\lambda + n)(2\lambda + 2n - 3)]a_n = (\lambda + n - 3)a_{n-1}$

　　(d) $y_1 = a_0^{(1)}[1 + 2t - t^2]$

　　　　$y_2 = a_0^{(2)}t^{3/2}\left[1 - \frac{1}{10}t - \frac{1}{280}t^2 + \cdots\right]$

　　(e) Yes

3. (b) $(\lambda - 1)(3\lambda - 1) = 0$

　　(c) $[3(\lambda + n)^2 - 4(\lambda + n) + 1]a_n = -a_{n-1}$

　　(d) $y_1 = a_0^{(1)}\left[t - \frac{1}{5}t^2 + \frac{1}{80}t^3 + \cdots\right]$

　　　　$y_2 = a_0^{(2)}[t^{1/3} - t^{4/3} + \frac{1}{8}t^{7/3} + \cdots]$

　　(e) Yes

5. (b) $(\lambda - 3)(\lambda - 3) = 0$

   (c) $[(\lambda + n)^2 - 6(\lambda + n) + 9]a_n = -a_{n-2}$

   (d) $y_1 = a_0 t^3 \left[1 - \frac{1}{4}t^2 + \frac{1}{64}t^4 + \cdots\right]$

   (e) Yes

7. (b) $(\lambda - 2)(\lambda - 1) = 0$

   (c) $[(\lambda + n)^2 - 3(\lambda + n) + 2]a_n = -a_{n-1}$

   (d) $y_1 = a_0 \left[t^2 - \frac{1}{2}t^3 + \frac{1}{12}t^4 + \cdots\right]$

   (e) Yes

9. (b) $(\lambda + 1)(\lambda - 1) = 0$

   (c) $[(\lambda + n)^2 - 1]a_n = a_{n-2}$

   (d) $y_1 = a_0 \left[t + \frac{1}{8}t^3 + \frac{1}{192}t^4 + \cdots\right]$

   (e) Yes

11. (b) $(\lambda - 4)(\lambda + 4) = 0$

    (c) $[(\lambda + n)^2 - 16]a_n = a_{n-1}$

    (d) $y_1 = a_0 \left[t^4 + \frac{1}{9}t^5 + \frac{1}{180}t^6 + \cdots\right]$

    (e) Yes

13. (a) $t = 0$ is a regular singular point.

    (b) $\lambda(\lambda - 2) = 0$

15. (a) $t = 2$ and $t = -2$ are irregular singular points.

17. (a) $\lambda^2 = 0$

    (b) $y_1 = a_0 \left[1 + \frac{1}{2}(\alpha^2 + \alpha)(t - 1) + \cdots\right]$

    (c) $y_1 = a_0 t$

19. (a) $\lambda^2 = 0$ (b) $y_1 = a_0 \left[1 - 5t + 5t^2 - \frac{5}{3}t^3 + \frac{5}{24}t^4 - \frac{1}{120}t^5\right]$

21. $\alpha = -1, \beta = 5$ 23. $\alpha = -1, \beta = 3$

Index

Terms are defined on pages with boldfaced numbers.